Society of Earth Scientists Series

Series Editor

Satish C. Tripathi, Lucknow, India

The Society of Earth Scientists Series aims to publish selected conference proceedings, monographs, edited topical books/text books by leading scientists and experts in the field of geophysics, geology, atmospheric and environmental science, meteorology and oceanography as Special Publications of The Society of Earth Scientists. The objective is to highlight recent multidisciplinary scientific research and to strengthen the scientific literature related to Earth Sciences. Quality scientific contributions from all across the Globe are invited for publication under this series. Series Editor: Dr. Satish C. Tripathi.

More information about this series at http://www.springer.com/series/8785

A. K. Jain · D. M. Banerjee ·
Vivek S. Kale

Tectonics of the Indian Subcontinent

The Society of Earth Scientists
2008

Springer

A. K. Jain
CSIR-Central Building Research Institute
Roorkee, Uttarakhand, India

Vivek S. Kale
Advanced Center for Water Resources
Development and Management
Pune, Maharashtra, India

D. M. Banerjee
Department of Geology
University of Delhi
Delhi, India

ISSN 2194-9204 ISSN 2194-9212 (electronic)
Society of Earth Scientists Series
ISBN 978-3-030-42844-0 ISBN 978-3-030-42845-7 (eBook)
https://doi.org/10.1007/978-3-030-42845-7

This Springer imprint is published by the registered company Springer Nature Switzerland AG
The registered company address is: Gewerbestrasse 11, 6330 Cham, Switzerland

Series Editor Foreword

The Indian subcontinent has been a unique landmass in the history of the evolution of Earth carrying oldest to youngest rock records and imprints of various tectonic episodes. Unraveling the geological mystery has been the continuous endeavor of geoscientists. Active Himalayan chain, peninsular cratonic shield, sedimentary basins of almost all ages, Precambrian and Phanerozoic plate-tectonic movements and associated tectonism offered natural laboratory to carry out geological studies. As such extensive research has been attempted in the recent past. The compilation and interpretation of existing data and opening new vista have been the ultimate objective of any book. The volume on the tectonics of the Indian subcontinent, thus, is a praiseworthy attempt by a group of renowned geoscientists. I believe that the book will act as a classic compilation for both students and researchers equally. I express my personal gratitude to all the authors for their exemplary contribution to the field of geology and tectonics of India.

The mega-event of 36th International Geological Congress 2020 in India opened a new chapter on the geology of India. On such an occasion, the Society of Earth Scientists Series by Springer decided to bring out 36th IGC Commemorative Volumes on various recent geological and geophysical studies of India. As such, veteran geoscientists were requested to prepare comprehensive accounts as monographs or edited volumes. I am personally thankful to all the editors and authors for the timely submission of high-quality manuscripts for inviting the interest of the global community of geoscientists.

Lucknow, India

Satish C. Tripathi
Series Editor

Preface

It is now a common knowledge that the Archean Eon (4.0–2.5 Ga) witnessed the preservation of thick piles of rocks, following the early cooling and emergence of the Earth's crust during the Hadean times. Several workers have postulated that water released during mantle outgassing and its accumulation in liquid form within the shallow depressions on the early Archean mafic crust is one of the key differentiators in the unique trajectory of crustal evolution on the Earth and the other planets (Taylor and McLennan 2009). Records of Earth's primitive atmosphere possibly seem to have been successfully documented in the Eoarchean records, known only from Australia and India to date. Different forms of cyanobacteria are identified in the 3.75–3.5 Ga Paleoarchean rocks. Magmatism, sedimentation in the shallow marine and terrestrial basins, dominated the geological scenario throughout the Archean when magmatic greenstone belts developed along with the emergence of sizeable continental crust. As rightly pointed out by Eriksson and Mazumdar (2019) "—the cratonic nuclei (proto-cratons) in some parts of the globe had attained a measure of geodynamic stability in the Meso- and Neoarchean; early supercratons heralded their much larger successor supercontinents which emerged during the succeeding Proterozoic Eon."

The progressive growth of continental crust led to changes in the style of tectonics gradually across the Archean–Proterozoic transition, from vertical/lid tectonics to Phanerozoic plate-tectonic style (Condie, 2018). Many geodynamic and metallogenic events in the Earth's history have close linkage with the transition time between the Archean and Proterozoic. Widespread oxidation of the paleo-ocean and -atmosphere represents the Great Oxidation Event followed by worldwide (?) glaciation as evident from Archean–Proterozoic glacial diamictites in many parts of the world including some terrains of central India. Climatic changes in cratons led to the creation of a highly modified hydrosphere which allowed the proliferation of biota and precipitation of iron in large basins across the continents. Although the Archean–Proterozoic boundary is chronologically placed at 2.5 Ga, this time plane is not sacrosanct in terms of tectonic styles, which varies in different continents, depending upon the evolutionary traits shown by the rocks across this boundary. These changes in tectonic regime and style appear reflected in the Archean and Proterozoic rock records of Peninsular India. This ~2.45–2.3 Ga period is considered as a time of global magmatic shutdown or slowdown with concomitant influences on the supracrustal record. It is, however, necessary to realize that in spite of such vast differences in tectonics, climate and basin configurations, both the Archean and Proterozoic basins show significant similarities with the Phanerozoic basins, barring the prominent influence of biota in the latter. Hence, Phanerozoic analogs can be discerned all over the Precambrian terrains of the Indian subcontinent.

The geological evolution of the Indian Peninsula during the Phanerozoic is one long hiatus (punctuated by relatively localized events), indicating that this crustal block remained an emergent continental block throughout the last 600+ million years. The exceptions are the events resulting from the breakup of Gondwana, which left behind coal-rich intracontinental Gondwana basins, including the separation from the Australia–Antarctica blocks. This event is followed by rifting of this block from Madagascar and then other smaller microcontinents on the west and its passage over hotspots (Kerguelan and later Marion–Reunion), currently located in the Indian Ocean that led to the development of one of the largest flood basalt provinces in the world during the Cretaceous–Paleogene transition. Only the northern edge of the Indian Peninsular was tectonically active during the Phanerozoic with a series of basin forming, collisional events that remain preserved in the Himalayan domain. The northward flight of the Indian Plate and its underthrusting/subduction beneath Asia in the last 100 million years has resulted in dramatic changes, marked by the rise of the spectacular Himalayan mountains and the wide frontal alluvial belt. This has affected the climatic systems of the northern Indian Ocean during the Quaternary.

Significant advancements have been made in the field of paleobiology, sedimentology, stratigraphy, crust–mantle interactions, paleomagnetism, and geodynamics in recent years. These new data supported by precise radiometric dates have enriched geological literature on the Indian subcontinent in the past two decades. Several books have been published recently which focus on specific tectonics and/or stratigraphy of this region. Excellent compilation by Ramakrishnan and Vaidyanadhan (2008) and Valdiya (2016) dealt with all aspects of the geology of India, including physiography, stratigraphy, tectonics, mineral deposits, and evolution of life. Sharma (2009) and Roy and Purohit (2018) focused on the Archean cratons and Proterozoic mobile belts. Yet, much of the recent data has not yet found a place in a single compilation.

Our book has made an attempt to fill this gap for researchers and students. We have followed the philosophy which believes that the behavior of the parent affects its children all through their existence. Here, we consider the Archean Eon, which resides in the cratons as the geological parent, gave rise to the mobile belts in the subsequent era while other Proterozoic and younger fold belts formed in a well-defined sequence determined by the movement of the tectonic plates. The geodynamic processes led to the creation of specific terrains displaying regional as well as local characters. Final configuration and vulnerability of these terrains to the dynamic Earth processes like effusion/intrusion of magma, uplift, burial, weathering, erosion, transportation, and sedimentation remain solely controlled by the tectonics. Stratigraphy, the backbone of all geological studies, is also exclusively tectonic-dependent; hence, this book on Tectonics of the Indian subcontinent has substantial stratigraphy inputs.

We have divided the book into chapters beginning with the description of the oldest parent, the Archean cratonic terrains of the Aravalli, Bundelkhand, Meghalaya, Bastar, Singhbhum, and Dharwar. We have mentioned about a Marwar craton, suspected to exist west of the Aravalli craton. This is followed by the chapter on the immediate offspring of deformed, metasedimentary belt associated with the effusive and intrusive igneous rocks. These represent the mobile belts, exclusively confined to the Proterozoic Era and inferred to have deposited on the cratonic Archean basement. Mobile belts of the Aravalli–Delhi, Satpura–Sausar (Central Indian Tectonic Zone), Singhbhum, Eastern Ghats, and Pandyan are the major constituents which are spread all over the Peninsular India. Each one of these mobile belts has been treated as independent as well as interlinked entities with reference to tectonism and stratigraphy. These Proterozoic mobile belts as well as the cratons supported a large number of intracratonic basins comprising undisturbed to mildly disturbing sediments with occasional volcanic episodes preserved as dikes, sills, and ash beds. Some of the prominent basins are the Marwar, Vindhyan, Bijawar, Gwalior, Chhattisgarh, Cuddapah, etc., along with a number of smaller basins resting over the mobile belts or directly on the Archean basement. The rocks range from Stratherian to Ediacaran periods. In some of these basins, Fortunian age trace fossils have been recorded. The Proterozoic basin of Pranhita-Godavari valley is a rift-related basin and served as the depository of the Gondwana Paleo-Mesozoic coal, which was deposited in many other half-grabens of Central India.

The Himalayan Orogen with extensive mobile belt characters of folding, faulting, and thrusting has been treated as an exclusive segment and is primarily covered by the Proterozoic rocks with sporadic Phanerozoic outliers. In the continuity of the Himalayan Orogen, the Indus–Ganga–Brahmaputra foredeep sediments are affected by deep-seated tectonism which controlled the distribution of Neogene sediments within the linear E-W trending belt. The delta front of this basin stretches into the Bangladesh territory.

The Deccan Volcanic Province occupying vast plateaux has a long history and has been the focus of the attention of volcanologists, petrologists, and paleontologists given its temporal proximity to the terminal Cretaceous mass extinctions across the world. A description of this domain has been included highlighting areas that require attention in future studies to unravel the controversies associated with it. The closely-linked and greatly influenced by the Deccan volcanism, the Western Ghats (Sahyadri Ranges), gave the basic shape to the Indian Peninsula and therefore dealt in an independent Chapter.

Geology, stratigraphy, and subsurface tectonics of predominantly alluvial Bangladesh have been included as an independent chapter. In contrast to alluvial Bangladesh, Sri Lanka geology and tectonics are primarily related to the high-grade metamorphics of Archean–Proterozoic ages. Nepal and Bhutan have been included under the Himalaya chapter. The two other countries of the Indian subcontinent, the Myanmar and the Pakistan, could not be covered due to logistic constraints.

Our journey for writing this book on Tectonics of the Indian subcontinent began with a chance interaction of A. K. Jain with Satish C. Tripathi who reposed faith to take up the task of completing a book under the Society of Earth Scientists Series for the Springer. In turn, Jain invited D. M. Banerjee to join him in this endeavor. Midway through this journey, A. K. Jain and D. M. Banerjee decided to invite Vivek S. Kale to be the third author for this book. All three of us express our gratitude to Satish C. Tripathi and thank him profusely for this invitation. In the same go, we wish to thank M/s. Springer for publishing this book linking it with the 36th International Geological Congress in Delhi and Ms. Monica Janet Michael for carrying out all the necessary correction at various stages.

In the course of writing, we consulted a large number of books, papers, and reports and used valuable information and data contained in them. We are grateful to all the unnamed authors and publishers whose data sets were used in this book. We thank all our colleagues in the workplace and in other institutions for sincere advice and giving permission to use their published and unpublished data for improving the scientific content of the book, including the original drawings. M. Jayananda, Jayanta Pati, Naresh Pant, Maibam Bidyananda, J. P. Srivastava, and several anonymous knowledgeable went through different chapters of this book and offered valuable suggestions for improvement.

A. K. Jain would like to thank Sandeep Singh for his contributions to the Himalayan Tectonics in the field, and otherwise over the years, Rahul Dixit and Gargi Deshmukh who gave him technical assistance in formatting the text, figures, tables, and indexing as and when required. D. M. Banerjee would like to thank Anupam Chattopadhaya, Partha Chakravarty, Naresh Pant, and Fareduddin for scientific discussion on many aspects of Proterozoic geology. D. M. Banerjee wants to thank Aditi from DU for some CorelDRAW drawing, his granddaughter Kritika for assisting in desktop-related problems which he could not tackle and to Aniruddha and Roma for physical, logistic, and moral support during late-night sittings on the home Desktop. D. M. Banerjee remains indebted to his wife Late Anjali for her moral and physical support all through his professional life. A. K. Jain is grateful and highly obliged to his wife Kamlesh for encouraging him to achieve the target but patiently suffering through long lonely hours which was imposed on her during the last many months. He thanks his sons Manish and Ankur in the USA for a very academic house environment where a considerable part of this book was completed.

Vivek wishes to place on record his deep gratitude to his teacher Prof. V. V. Peshwa for his teaching and guidance and thanks to Kanchan Pande, Makarand Bodas, Abhay Mudholkar, Anand Kale, Himanshu Kulkarni, Raymond Duraiswami, Gauri Dole, Shilpa Patil Pillai, Shreyas Mangave, Ninad Bondre, and Vinit Phadnis for collaborating research on the Deccan Traps and related geology.

Both A. K. Jain and D. M. Banerjee thank the Indian National Science Academy for awarding the Honorary Scientist positions, during the tenure of which this book was written.

Roorkee, India A. K. Jain
Delhi, India D. M. Banerjee
Pune, India Vivek S. Kale

References

Condie KC (2018) A planet in transition: the onset of Plate Tectonics on Earth between 3 and 2 Ga? Geoscie Front 9:51−60

Eriksson PG, Mazumder R (2019) Preface Earth-Sci Rev. https://doi.org/10.1016/j.earscirev.2019.103058

Ramakrishnan M, Vaidyanadhan (2008) Geology of India, vols 1 and 2. Geological Society of India, Bangalore, pp 1−556, 557−994

Roy AB, Purohit R (2018) Indian shield: precambrian evolution and planetary reconstitution. Elsevier, ISBN 978-0-12-809889-4

Sharma R (2009) Cratons and fold belts of India. Lecture notes in earth sciences. Springer, Berlin, pp 1–304

Taylor SR, McLellan SM (2009) Planetary crusts: their composition. Origin and evolution. Cambridge University Press p 378

Valdiya KS (2016) The making of India geodynamic evolution. Society of earth scientists series, 2nd edn. Springer, Berlin, pp 1–924

Contents

About the Authors

Prof. A. K. Jain obtained his Ph.D. in Geology from the erstwhile University of Roorkee and now the IIT Roorkee. After a brief career as a Scientist at the Wadia Institute of Himalayan Geology, Delhi (now at Dehradun), his interest in teaching brought him to the University of Roorkee in 1974 where he taught structural geology. He retired from the IIT Roorkee in 2007 as Professor and remained at the Institute as Emeritus Fellow till 2011.

He was a Fellow, Alexander von Humboldt Foundation (1979–1981) at the University of Karlsruhe, Germany, where he carried out postdoctoral research on Rift Tectonics of the Upper Rhine Graben. He was a Visiting Young Scientist at Bedford College, London, in 1979 under UGC-British Council Exchange Program, the Visiting Scientist at the Lund University, Sweden (1991), and a Senior JSPS Fellow (2004) at the Hokkaido University, Japan. He was a member of many Selection Committees and Project Advisory Committees of Earth Science Division of the Department of Science and Technology (DST), Government of India, New Delhi, besides Chairman of some of them. He is a Fellow of the Indian National Science Academy (FNA), New Delhi, and is currently active as the INSA Honorary Scientist at the CSIR-Central Building Research Institute, Roorkee.

He has written and edited five books on the Himalayan Geology and authored more than 120 research publications. He has supervised 15 Ph.D. students besides many for their dissertations. Some of his students are currently occupying high positions in teaching and government. His main area of research interest remained the

Himalayan Tectonics since the beginning. The book on the *An Introduction to Structural Geology*, published by the Geological Society of India, Bangalore, is an outcome of his teaching and research in the subject.

Prof. D. M. Banerjee obtained his Ph.D. Degree in 1967 from the University of Lucknow on a thesis on Himalayan Geology. He taught for a year in Dehradun and then spent nearly 2 years in the Geological Survey of India working on geohydrology and Precambrian geology. He joined the University of Delhi in 1970 and retired as Professor in 2007. He taught sedimentology, sediment geochemistry, and Precambrian stratigraphy.

He was a Visiting Lecturer in 1976 under the UGC-British Council Young Scientist Scheme at the University of St. Andrews, Scotland, and spent part of this tenure at the Universities of Glasgow, Reading, Keel, and Institute of Geological Sciences (now British Geological Survey). He was Fellow of the Alexander von Humboldt Foundation between 1982 and 1984 at the Max Planck Institute für Chemie, Mainz, West Germany, and had several short stints as AvH Visiting Professor at the Universities of Munster, Munich, and TU Berlin. In 1997, he initiated an IGCP-156 program on Phosphorite, served as its National Convener, and guided the IGCP-386 as the Global Co-Leader. He conducted a DFG funded long-term research program on Phosphorite at the MPI, Germany, and British Council-UGC funded program on groundwater Arsenic at the University College London, UK. He was a Visiting Professor at the University of Shizuoka under the INSA-JSPS exchange program. He Chaired the Department at the Delhi University and mentored many Ph.D. and M.Sc. level students between 1970 and 2007. He served as an expert member in the Scientific Advisory Committees of scientific departments of the Government of India and the University Grants Commission. He co-chaired the UGC committee for the revision of the geology syllabus in Indian universities. As Chairman of the National Committee for IUGS, INQUA, and ILP, he led the Indian delegation to the 34th International Geological Congress where India won the bid to hold the Congress in 2020. He was elected Fellow of the Indian National Science Academy (INSA), New Delhi, in 1999, INSA Senior Scientist in 2007, and Honorary Scientist in 2012 and will join as Emeritus Scientist in 2020.

He edited and co-edited six books on subjects related to phosphorite, stromatolite, sedimentology, and Earth Sciences. He served as a scientific reviewer for many Indian and International journals and served in International selection panels. Presently, he concentrates on themes concerning geoethics and geoheritage conservation.

Prof. Vivek S. Kale is a former Reader of Geology at the University of Pune and a former Adjunct Professor of Earth Sciences at the Indian Institute of Technology, Mumbai, with more than 30 years of teaching experience. He, currently, is the Managing Trustee at the Advance Center for Water Resources Development and Management, Pune (ACWADAM) and serves as the Head of Geospatial at Kalyani Global Engineering Private Limited. He is a Member/Fellow of Geological Society of India, Geological Society of America, Indian Geophysical Union, and Indian Society of Remote Sensing.

He specializes in tectonics, geomorphology and has extensively worked on the Purana Basins and Deccan Traps of India. He has recently authored the book on *Processes, Products, and Cycles of Tectonic Geomorphology* focusing on the process of landscape evolution, linking modified Davisian landscape, and Wilson tectonic cycles, sustainable land-use planning, and hazard mitigation.

Chapter 1
Tectonics of the Indian Subcontinent: An Introduction

1.1 Introduction

Geology and Tectonics of the Indian Sub-continent constitute a significant component of the Indian Plate straddling the equator as a part of the Gondwanaland that broke into fragments at ~100 Ma and started moving northwards. Once fused with adjacent Australia to form a single Indo-Australian Plate, recent studies suggest that India and Australia have been separate plates for at least 3 million years and likely longer. The Indian Plate includes most of South Asia—the Indian Sub-continent—and a portion of the basin under the Indian Ocean, parts of Tibet (South China) and western Indonesia, and extending up to but not including Ladakh, Kohistan, and Baluchistan (Fig. 1.1). Indus-Tsangpo Suture Zone marks the boundary of this plate with the northern Asian Plate. The plate has a convergent margin along the Himalaya in the north, transform margins in the west and the east, and the oceanic ridge in the Indian Ocean.

In this book, we have defined the Indian Sub-continent as the southern part of Asia comprising India, Pakistan, Nepal, Bhutan, Bangladesh, Sri Lanka, and Myanmar. Tibet, once considered a part of this sub-continent, is linked to this landmass before Permian time. Although Myanmar, traditionally linked to the Indian Sub-continent, has many tectonic commonalities with the terrains of Thailand and Malaysia. From the geological point of view, all these countries share many standard geological features that transgress political boundaries. Countries surrounding the main Indian landmass invariably share the same tectonic and stratigraphic elements. In other words, the evolutionary history of this south Asian geographic segment is common, represented today by trans-country mountain ranges, rivers, coastline, and desert.

Contemporary workers have identified several physiographic/geological terrains, which display distinct lithological assemblages, specific orders of superposition, characteristic geomorphology, and tectonic history. We have classified these terrains confined by the Himalayan orogen, Trans-Himalaya, and the Karakoram in the north,

A. K. Jain et al., *Tectonics of the Indian Subcontinent*, Society of Earth Scientists Series, https://doi.org/10.1007/978-3-030-42845-7_1

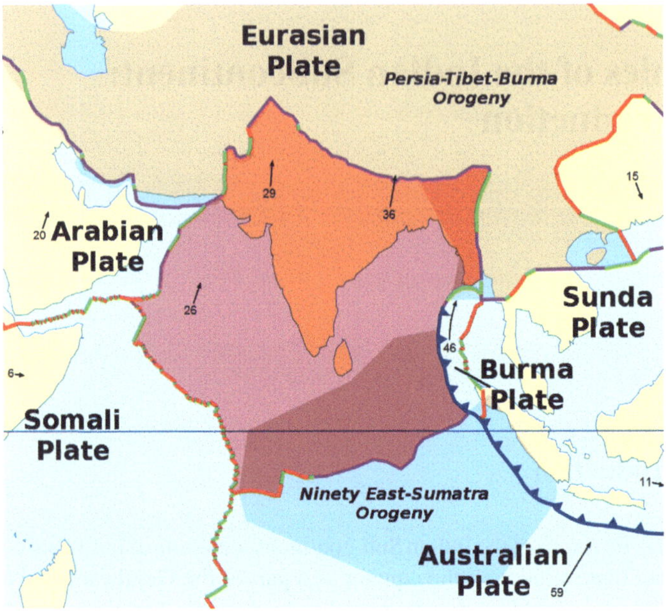

Fig. 1.1 The Indian Plate and its configuration. *Source* Tectonic plates boundaries detailed-fr.svg created by Sting under CC-BY-SA. File: Indian Plate map-fr.png

Kirthar and Sulaiman ranges in the northwest and west, Indo-Myanmar Ranges including Patkai, Naga, Arakan Yoma Mountains in the east extending into the Gulf of Thailand through the Bay of Bengal and the Arabian Sea in west and the Indian Ocean embracing it from the south.

Traditionally, the physiographic characteristics of the Indian landmass encompass the following (Fig. 1.2):

(i) Peninsular India with uplands and seacoasts,
(ii) The Himalayan Range including those in Pakistan and Myanmar, and
(iii) The Indus–Ganga–Brahmaputra Plain sandwiched between the two.

1.2 Peninsular India

1.2.1 Mountains and Plateaus

The NE–SW-trending Aravalli Range with Archean and Proterozoic rocks extends from Palanpur in Gujarat to Delhi and beyond subsurface up to Haridwar in the Gangetic alluvial plain. Granites extensively intrude these Precambrian rocks. The Aravalli range forms the water divide between the rivers of the Ganga and Indus

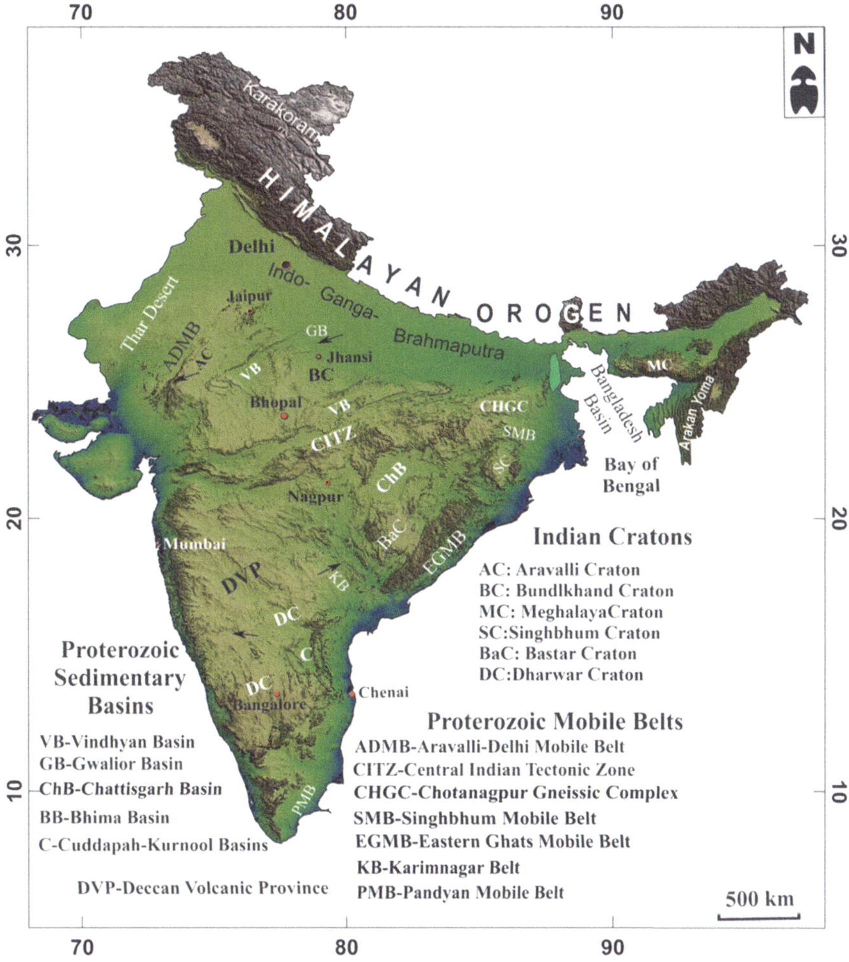

Fig. 1.2 DEM of the Indian sub-continent with major geological and tectonic units. Basic map courtesy Ajay Manglik (NGRI Hyderabad)

drainage systems. The Aravalli Range bends southeastwards to join the ENE–WSW-trending *Satpura Range*. Faulting along the Narmada valley disconnected the two mountain ranges. The Satpura Ranges extend ENE from southern Gujarat to central India and beyond. In the east, it embraces the *Rajmahal Hill* and *Chotanagpur terrain*. The *Shillong Plateau or* the *Meghalaya massif* borders Bangladesh and consists of the Garo Hills, the Khasi–Jaintia Hills and the Mikir Hills. The Shillong massif acted as a pivot for deflecting the Brahmaputra River westwards to skirt around the upland before joining the Ganga in Bangladesh.

In the central part of the Peninsula, the Satpura Range divides the landmass into the northern plateaus of Malwa, Bundelkhand, and Vindhyachal; the southern part comprises the Deccan, Mysore and Telangana plateaus. *The Deccan lavas also cover the Malwa Plateau. The Bundelkhand Upland with* Late Archean gneisses and granites occur in the deeply dissected ravines of Chambal River. The Kaimur and Bhainer Plateaus with steep escarpments occur in the *Vindhyan Range.* A large tract south of the Satpura Ranges and east of the Sahyadri is known as the *Deccan Plateau.* The upper Mesozoic lavas formed a flat terraced landscape with extensive bauxite and laterite cappings. The *Mysore Plateau, in the south,* made up of highly metamorphosed Archaean gneisses and granites form isolated hillocks of the Bababudan, the *Chitradurga Range* and the *Kolar Range* (Fig. 1.3).

From the Tapti Valley in the north to Kanyakumari in the south, the rugged N–S range is the *Sahyadri Range.* The upper Mesozoic Deccan basalt forms the *Northern Sahyadri*, the Archaean gneisses and high-grade metamorphic rocks that occur in the *Central Sahyadri* and late Proterozoic charnockites and khondalites form *Southern Sahyadri.* The NNW–SSE-trending fractures and faults make the Sahyadri Range a horst mountain. Its west-facing steep to near-vertical escarpments is known as the *Western Ghats.* East-flowing Godavari, Krishna, Tungabhadra and Kaveri rivers originate in the Western Ghats while some small rivers flow to the west. Between the towering Nilgiri massif and the Annamalai Range is the 25–30 km wide *Palghat Gap.*

Disconnected charnockitic *Shevaroy Hills, Eastern Hills,* and highly tectonized Proterozoic *Nallamalai Hills constitute the East Coast.*

Several ancient Archean cratons constitute Peninsular India, each of these cratons acting as the basement for the deposition of the Proterozoic sediments which either ended up as variably metamorphosed and deformed mobile belts or remained as undisturbed intracratonic basins (Fig. 1.2). The entire Peninsular terrain suffered multiple orogenic cycles followed by sedimentation, erosion and weathering cycles all through its geological history. Mountains like Nilgiri–Aravalli, Sahyadri range, hills like Pachmarhi, plateaus like Deccan and Rajmahal, intra-hills plains, river basins fill, delta plains like Bengal–Mahanadi–Krishna–Godavari–Kaveri deltas, Western and the Eastern Ghats, coastal low lands like Sundarbans and manifestations of ancient tectonic sutures reflect the physiography of this entire segment. This terrain also supports freshwater riverine and lacustrine deposits which formed in the rifted basins in the central part of the Peninsula. The Gondwana Basin formed in these basins created during the Paleozoic–Mesozoic tectonics. This region is covered partially by late Mesozoic volcanic lava which has carved out distinct physiographic features dotted by pop-up Archean and Proterozoic inliers. In the northern part of the Peninsula, some of the Proterozoic rocks occur at considerable depth under the thick alluvial fills of the Ganga–Jamuna Plains and possibly extended much to the north and got mixed up with the rocks which formed the present-day Himalaya.

The late Paleozoic–early Mesozoic tectonics, which caused the formation of horsts and grabens in the central part of the Indian shield, got accelerated due to further northward drift of this rigid mass. In the process, some remnant upper Mesozoic basins were pushed northward and got buried in the alluvial plains. In the southern end of east-central India, bordering the alluvium of the Ganga Basin, the Damodar River valleys acted as the repository of a large quantity of riverine sediments. Likewise,

Fig. 1.3 Physiography of the Indian sub-continent showing major mountains, plateaus and basins

Mahanadi, Son, Pranhita and Godavari rivers carved out many linear horsts and grabens, trending E–W to NNW–SSE. These basins allowed the deposition of an exceptional thickness of fluvial sediments and preserved significant organic matter which turned into coal. These events began in late Permian time. Glacial deposits along with occasional intermixed marine sediments occur in some sectors which mark the waning phase of sedimentation in these basins. At the beginning of Permian, the glaciated Gondwanaland started sagging that allowed the oceanic waters to enter the heart of the continental landmass. This upper Paleozoic marine incursion is distributed along the Umaria–Manedragarh belt in central India and even extend up to Meghalaya in the east.

During the Upper Cretaceous, as a result of large-scale volcanic eruptions and outpouring of mobile lava, mafic volcanic rocks covered a large part of the Indian shield. These flat-topped plateau basalts formed by the volcanic eruption which started in Cretaceous time, around 69 Ma continued erupting till 61 Ma. These rocks identified as the *Deccan Traps* created by pulsating volcanic activities and allowed preservation of fossil-bearing sedimentary rocks during the quiescent period. These are Intertrappean and Infratrappean beds. The Deccan Traps are distributed all over the Western Ghats including the Konkan Coast. Kachchh–Saurashtra, Malwa Plateau, Rajmahal Hills and Sylhet Plateau also expose similar volcanic rock assemblages. Associated features like dyke swarms seen along the Konkan coast, Narmada and Tapti River valleys acted as conduits for outpouring lavas to spread over an extensive area. Along numerous lakes and swamps which came up due to damming of flowing rivers, supported a variety of plants and animals including giant reptiles like dinosaurs.

1.2.2 Coastal Plains

The Indian Sub-continent has more than 6000 km long coastline between the Indus delta to the southern tip of the Peninsula, the Bengal delta, along the Arakan coast, to the Irrawaddy delta (Fig. 1.4). Sri Lanka has a broad fringe of coastal plains all around. The *Makran Coast*, near the Indus deltaic plain, form an arcuate zone of older tidal-flat deposits. East-southeast of the Indus deltaic plain is the *Rann of Kachchh*, a salty marshy tidal flat.

The *West Coast* is mainly a *coast of submergence represented by* cliff faces and near-shore islands, estuaries, and backwaters. The coastal plain of Gujarat is the prolongation of the alluvial plains of the Sabarmati and the Mahi rivers. The *Gujarat Coast with* strand lines 8–10 m above the present sea level mark the northern end of the West Coast. A narrow strip of the *Konkan Coast* is a rocky shore of cliffs, bays, coves with small mudflats and small beaches.

The East Coast is a *coast of emergence*, characterized by well-defined beaches, dunes and sand spits, and many lagoonal lakes associated with backwater swamps. Pulicat Lake and the Chilka Lake are large coastal lakes on the eastern coast. A product of the combined contributions of the Ganga and the Brahmaputra, the Ganga–Brahmaputra–Meghna Delta, is called the *Sundarban Delta* which forms part of the coastal plain, creating the head of the Bengal Basin.

Fluvial Upper Jurassic to Quaternary sedimentary rock successions dots the Indian coast from west to south. These are pericratonic basins that formed in response to faulting along the coastal belt accompanied by intermittent sagging of the basin floor. The Middle Jurassic sediments in Kachchh and the Upper Jurassic strata in the Coromandal Belt are well preserved. Marine transgression also affected parts

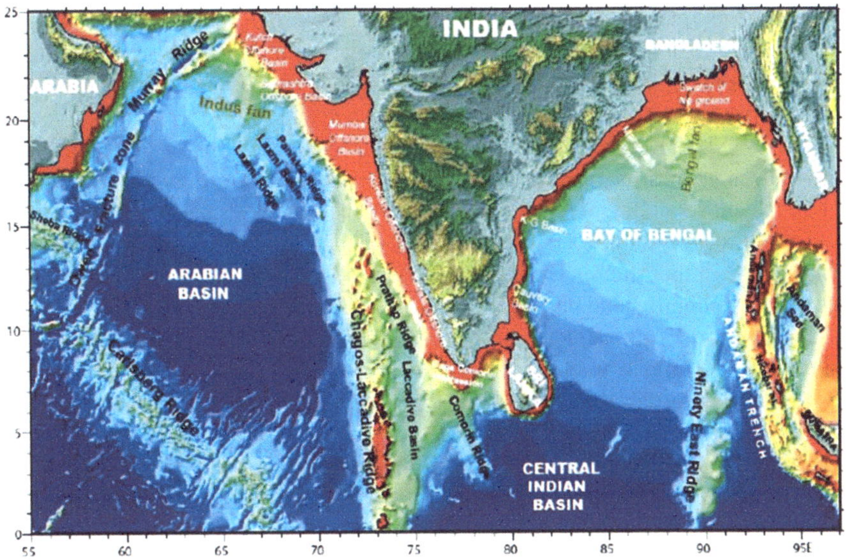

Fig. 1.4 Morphology off the littoral regions of India. Rifted passive margin controls east coast morphology whereas the west coast is marked by intense volcanism near the K–T boundary. After Fainstein et al. (2019)

of Bangladesh, southern Meghalaya and Western Rajasthan at different times. A narrow strip of marine bed in the Narmada river valley received sediments to form the Mesozoic Bagh Beds. The upper parts of these sedimentary piles are represented by Paleogene and Neogene sediments, which cover larger areas and form the continental shelves and continental slopes in eastern and western sea coasts.

1.2.3 Himalaya

The *Himalaya–Karakoram–Arakan* ranges formed as a result of the Indian and Asian Plate convergence and pushing up the mobile sedimentary fills of the Paleozoic Tethys Ocean with Peninsular rocks as the seafloor. The Himalayan mountains grew in height through multiple spasmodic episodes, and each spasm left its signature in the form of characteristic sedimentation, deformation, metamorphism patterns, and metallogeny. The Early Cenozoic was the period when docking and collision/subduction of the two major plates started leading to the initiation of the Himalayan tectonic domain. Extensive folding, large-scale thrusting, inversion of strata and subsequent erosion and degradation mark the history of this mountain chain. All these activities were repeatedly activated by seismic tremors that dominated the tectonic scenarios all through its history since birth from an oceanic cradle.

Beyond the Tethyan Himalayan Belt and across the *Indus* and *Tsangpo valleys*, a vast plateau region belongs to Tibet which is part of a different Asian landmass. This peneplaned uplifted plateau of 90–45 Ma old granites and granodiorites along our intermixed with 60–80 Ma old volcanic rocks.

The NW–SE to E–W trending Himalayan range show bending of strata on a very large-scale in western as well as eastern ends. On the northwestern end in the Kashmir region, the entire mountain chain shows abrupt bending of all the strata making an acute angle to the south. Even the structural features like faults and folds show a sharp change in their trend. This zone is called the *western syntaxis* around Nanga Parbat. The southerly bend of the Himalayan Range gradually splays out to the west as Sulaiman and Kirthar Ranges. In contrast, the *eastern syntaxis* developed at the east end of the Himalayan chain beyond the Arunachal Pradesh; both the syntaxes control the present-day U-shape bending and deep gorge of the Indus and Tsangpo Rivers, respectively.

In the east, the NW–SE trending Lohit–Mishmi Ranges in Arunachal Himalaya meets the NNE–SSW trending Patkai–Naga and Arakan–Yoma Ranges which undergo clockwise rotation, originated through an altogether different mechanism. These ranges extend into Myanmar, embracing the Indo–Myanmar Ranges along the coast with a westward bulge, a depressed Myanmar Central Belt region and the uplifted Shan Plateau in the east. With the opening of the Andaman Sea along the ridge, the system defines the Burmese microplate.

West of the Indus River in northern Pakistan lies the N–S trending ranges of Kohistan, Swat, and Dir with very high mountain peaks and steep slope faces. The Pir Panjal Range of Kashmir, after the syntaxial bend, is the *Hazara Ranges* trending NE–SW and comprising a series of hills. The Hazara belt of Pakistan is comparable to the Lesser Himalaya terrain in India. The Attock–Cherat Range and the Khyber Pass Range along with *Peshawar Plain* of Northwest Frontier Province are other mountainous terrains in Pakistan. South of the Hazara Ranges highly dissected *Kohat–Potwar Plateau* has the Salt Range on its southern margin. The N–S-oriented arcuate *Sulaiman Range* forms the eastern part of Balochistan. The Quetta Hills are west oriented. The Sulaiman Range and Kirthar Range are in the south. The *Kirthar Range* is an N–S-trending 400-km-long belt. The NNE–SSW-trending Las Bela Range associated with the Kalat Plateau in southern Balochistan gives way to the *Makran Range* that trends E–W parallel to the coast.

1.2.4 Indus–Ganga–Brahmputra Plain

A significant lithospheric flexure developed on the foothills or south of the rising Himalaya Mountains leading to the deposition of the Cenozoic Siwalik in a foreland basin, filled in subsequently by the Quaternary sediments of the great Himalayan rivers. The vast spread of undulating sedimentary terrain extending from the Indus in the west to the Brahmaputra in the east and Ganga–Jamuna plains in between the two, formed in the foredeep which was created by the rising Himalaya and complete draining of the Tethys in the early Cenozoic. The foredeep was filled up by sediments derived primarily from the tectonically active Himalaya in the north and partly from the stable Archean massifs and Proterozoic platforms of the northern front of the Indian shield.

The width of the IGBP varies from 150 to more than 300 km, being most comprehensive in Panjab and very narrow (90–100 km) in Assam. In the Indus basin, the alluvial plain gives way south-eastwards to the arid plain of western Rajasthan and adjoining Sindh—the *Thar Desert*. The monotonous flatness of the *Ganga Plain* is relieved, in limited areas, by bluffs, levees and abandoned channels, oxbow lakes, and ravines or badlands.

In the central Ganga domain, two distinct physiographical units the *Older Alluvium* with Bhangar and *Newer Alluvium* with *Khadar* and *Bhabar*, show a different chronology of events. These are sandy and gravelly deposits of Late Holocene period, riverine in origin which occasionally forms inland deltas and fan deposits. In the Bengal–Bangladesh region, the fluvio-deltaic plain with a fringe of 30-m-high Pleistocene terrace contains reddish-brown ferruginous concretions. These terraces are called the *Barind and Madhupur* in Assam and Bangladesh and are comparable to the Bhangar formation. The *Indus Plain*, 180–210 m above mean sea level in the north and 3–4 m at the delta end, is mostly the Khadar expanse which is known as *Chung* in the Panjab interfluve areas. An extensive *Piedmont Belt*, the analog of the Bhabhar in the Ganga domain, forms an apron of coalescing gravel fans along the foot of the Sulaiman and Kirthar Ranges.

The *Luni River* has developed a 650 × 300 km stretch of alluvial plain in the arid part of western Rajasthan along the foot of the Aravali mountain. The plain slopes southwestwards from the north to form very low-lying terrain at the lower end. Wind action coupled with neotectonics in the fluvial regime has given rise to anomalous drainage patterns and the formation of many saline lakes and playas, including the Sambhar, Didwana, Degana and Lunkaransar lakes. The playas, which turn into water bodies during the rainy season, are called *rann*. The biggest one of these is the *Rann of Kachchh* at the terminal point of the seasonal rivers Luni and Ghaghgar Nara. The western part of the arid land—*Thar Desert*—is covered with a thick mantle of sand dunes of a variety of shapes and sizes. They are longitudinal, transverse, parabolic and barchans.

Between the *Patkai–Naga–Chin–Arakan Yoma Ranges* in the west and the *Kachin–Shan–Tenasserim Ranges* in the east lie N–S-trending plains built by the Irrawaddy, its tributary Chindwin and the Sittang rivers. The Irrawaddy Basin comprises several plains—the Putao and the Hukawang plains with Indaw Lake in the upper reaches, the Irrawaddy, and the Chindwin.

Exposed in terrains with 10–100 m elevation, the sediments of the three river systems, the Indus, Ganga and Brahmaputra were deposited in the *Quaternary* period. The thickness of the alluvial sedimentary pile varies from less than 10 m to >2500 m in the northern sectors adjacent to the Himalayan foothills. Four major units of this alluvial domain have distinctive geomorphic features, characteristic sequencing of sedimentary strata, and well-defined physical composition. The *Pleistocene package* of iron oxide coated concretions, and coarse white sand was the first to form. In the south, this package show distribution along the margins of the uplifted Proterozoic Vindhyan mountain range. These sediments occur on the surface but more extensively in the subsurface and provided the floor for the deposition of the Holocene alluvium. The Holocene gravel and sand apron, known as *Bhabar,* occupy the northern tract up to the foothills of the Siwalik range. The region between the *Bhabar* and the Pleistocene package lies the river terraces formed within the flood plain, channel and lake deposits. The vast and more or less flat valley fills are the *Khadar flats*. The southern extension of the Indus Plains, enlarged by the deposits formed by ancient rivers, turned into a desert the *Thar Desert* in the Western Rajasthan, Gujarat and parts of Narmada valley. The *Bengal Delta Plain*, the coastal plain sediments, delta plains of the Peninsular rivers were all covered by Quaternary sediments.

1.3 Geology and Tectonics

Figure 1.5 presents salient geological and tectonic characters of the Indian Subcontinent.

1.3.1 Ancient Cratons

Indian Peninsula is commonly described as an ancient rigid shield which formed largely during the Archean time (Ramakrishnan and Vaidyanadhan 2008). For a long time, rocks of these shield area were considered unaffected by deformation processes, many tell-tale signatures of tectonism are now being identified. Rocks exposed in these shield areas unravel the story of the primitive earth. They are represented by a variety of magmatic rocks, both effusive and intrusive and often tectonized and metamorphosed. These are invariably interlayered with metamorphosed clastic and carbonate sediments. Receptacles of these rock assemblages are known as Cratons. Six distinct cratons have been identified in the Peninsular India. From north to south these are:

(i) Aravalli Craton, (ii) Bundelkhand Craton, (iii) Meghalaya Craton, (iv) Bastar Craton, (v) Singhbhum Craton, (vi) Dharwar Craton, (vii) (A suspected) Marwar Craton

Extensive oceanic/crustal components occur on the floors of the >3 Ga old (Paleoarchean) Cratons. The lithological types include granitic gneisses, basic and ultrabasic rock complexes and occasional metasediments trapped between the magmatic flows. Extensive petro-mineralogical and isotopic studied have helped in establishing the time of stabilization of these rigid blocks by late Paleoarchean to the early Mesoarchean era. New investigations have identified Eoarchean rocks in the Singhbhum craton in the north along with presence of the Hadean zircons. There is a consensus amongst scientists that this terrain got stabilized by about 3.0 billion years accompanied by large-scale emplacement of granites and related magma. Extensive

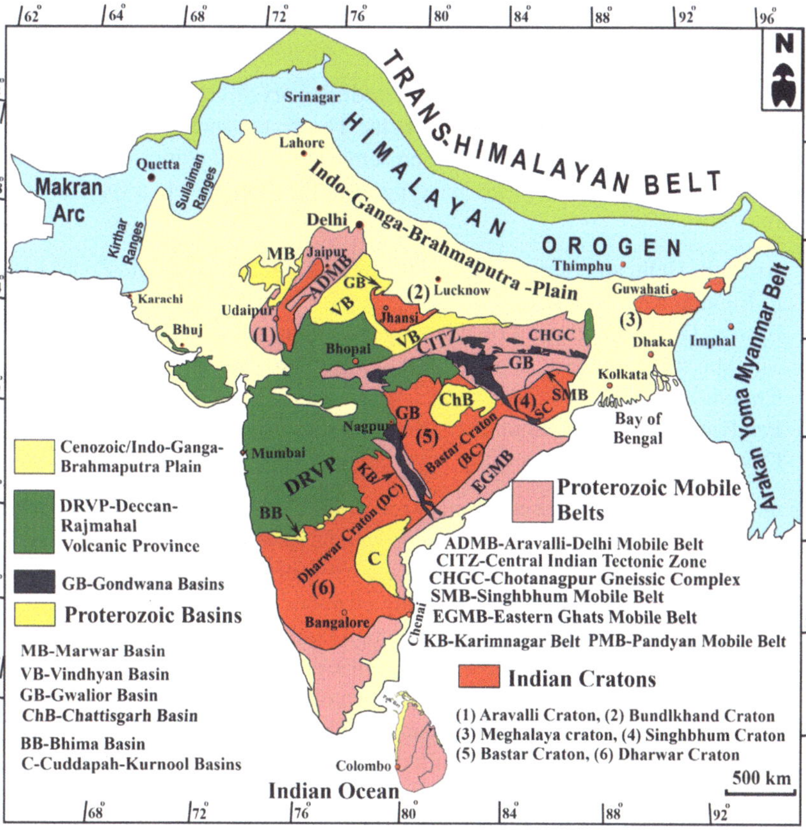

Fig. 1.5 Geology and tectonics of the Indian sub-continent, showing its main characters. Map is compiled from various published sources

basic volcanism characterize end phase of cratonization characterize leading to rifting and formation of linear depressions. The end phase of cratonization leading to rifting and formation of linear depressions. These negative physiographic features acted as repository of clastic sediments eroded and transported from the cratonic interior. As iron minerals occur in abundance in some of these cratons, the early formed sediments in the linear marginal depositional belts form ironstones-silica rich sedimentary rocks. Layered iron-rich lithology is the sole representative of this sedimentation regime. All the Cratons show the emplacement of plutonic bodies belonging to the Neoarchean era. Some sectors of these cratons display intense metamorphic effects.

1.3.2 Proterozoic Mobile Belts

Between the cratons, metasedimentary and sedimentary folded rocks that hold the terrain is known as Mobile belts or Fold belts. The craton-mobile belt boundary is invariably unconformable, sheared, or represents detachment planes. The following mobile belts have been delineated depending upon the proximity to a specific craton margin,

(i) Aravalli–Delhi and Satpura Mobile belt (Central Indian Tectonic Zone) linked to Bundelkhand Craton
(ii) Eastern Ghats Mobile belt on the margins of Dharwar and Bastar Cratons
(iii) Singhbhum Mobile belt (including Chotanagpur belt) related to the Singhbhum Craton
(iv) Pandyan Mobile belt (Southern Granulite Terrain) connected to the Dharwar Craton.

In most cases, the tectono-sedimentary regime between the cratonic basement is a Paleoproterozoic event. High-grade metamorphics, metasediments, rifting, the outpouring of basic and acidic volcanics, emplacements of ultrabasic and alkaline rocks intricate generations of folding and shearing characterize the rocks of these mobile belts. These Paleo-Mesoproterozoic belts contain sandstones of variable maturity and tectonic affinities, like pebbly conglomerate, meta-argillites, dolomite, and marble. Several Neoproterozoic granites and effusive rhyolite were emplaced all over these cratons. The Tonian, and early Cryogenian periods exhumed the whole of the mobile belt. Ediacaran sediments are not recorded in any mobile belt of the Indian Peninsula although the Pandyan Mobile Belt witnessed a significant orogeny in the Lower Cambrian. Very high-grade metamorphic rocks, including charnockite, khondalite and granulite, intimately associated with granite gneisses are found in the Eastern Ghats and Pandyan Mobile Belts. Widespread emplacement of granitic, alkaline and ultrabasic magmas within the proximity of shear zones reflect the Pan-African orogeny.

1.3.3 Intracratonic Sedimentary Basins

The Proterozoic Mobile Belts developed large linear and non-linear sedimentation troughs in front of the mobile belts and over the cratons. These are primarily formed with unmetamorphosed and undeformed sediments of the Neoproterozoic era although some of these basins belong to the late Paleoproterozoic and Mesoproterozoic. The orogenic belt-basin contact show localized deformation and at places, well-developed thrust separates the two.

The initiation of most of these basins started in and around 1700 Ma. A few of these intracratonic setups, like Bijawar and Gwalior basins, originated at a much earlier date and therefore traditionally excluded from being classified under this category of basins. In Indian stratigraphy, these rocks are called Puranas (meaning old or ancient or may have originated from the ancient Hindu scripture called "Puran") a term that should deemed to be redundant. Except for their being old, the term Purana does not indicate any geological characteristics of these rocks or the basin. In this book, we will persist with the term "Purana Basin" for all the intracratonic sedimentary basin that piggyback the mobile belt. We have discussed the following basins from North to South of Peninsular India:

(i) Marwar Basin, (ii) Vindhyan Basin, (iii) Gwalior Basin, (iv) Bijawar Basin, (v) Kolhan Basin, (vi) Chattisgarh Basin, (vii) Indravati Basin, (viii) Cuddapah Basin, (ix) Pranhita-Godavari Basin, (x) Kurnool Basin (xi) Kaladgi Basin (xii) Pakhal and Penganga Basins, (xiii) Bhima Basin, (xiv) Sullavay Basin, (xv) Khairar Basin, (xvi) Sausar Basin

These basins formed through the accumulation of clastic, carbonate and argillaceous sediments derived from the hinterland of fold belts and interlayered volcanics, tuffs, and agglomerates. These strata maintained a shallow marine depositional profile with intermittent fluvial and aeolian formative phases. The ages range from the youngest Stratherian to the base of the Tonian. However, some Precambrian paleontologists extend some of these successions to the uppermost Ediacaran, with one basin preserving the lower Cambrian (Stage 2) strata. U–Pb detrital zircon dating of several volcanic tuff in recent years suggested a virtual capping of the uppermost layers in these basins at $\pm$1100 Ma. Disconformities and unconformities of variable magnitude and duration separated internal sedimentary packages within these unmetamorphosed and mostly flat-lying successions. The carbonates are characterized by stromatolites at different stratigraphic levels that indicate their Mesoproterozoic age. The Neoproterozoic kimberlite pipe intruding the Upper Vindhyan sediments implies that the terrestrial sedimentation commenced well before 1100 Ma. The spectacularly developed stromatolites in the carbonate rocks in the upper horizons some of these assemblages suggest a time range of 680–570 Ma. While most sediments are of shallow marine in origin, some sedimentation units show marked fluvial influence. Across the Aravalli Ranges in western Rajasthan, wide distribution of several Neoproterozoic intrusive granites of Erinpura, Jalore, Mount Abu etc. and acid volcanics like Malani Rhyolite Granite indicate the increased but sporadic tectonic activity accompanied by igneous activity.

1.3.4 Himalaya Mobile Belt

It is almost well established that an arcuate Himalayan mountain range formed between 65 and 55 Ma, 53.3 $\pm$ 0.7 Ma to be more precise (Leech et al. 2005), when the Indian continental lithosphere subducted beneath the Asian Plate. Strongly deformed and highly crushed oceanic rocks and sediments mark the northern limit of this orogen, while the southern boundary is demarcated by many thrusts against the northern Indian plains. Between these two tectonic boundaries, the Himalaya

Mobile Belt can be divided into five lithotectonical terrains, each displaying distinctive geomorphological characters. Each terrain is limited in space by regional level thrust faults (Jain 2014, 2017; Valdiya 2016). These are:

(i) The Tethys Himalaya in the north, represented by mildly deformed to an undeformed thick pile of sedimentary rocks, range in age from lower Cambrian to end-Cretaceous. Paleogene rocks are also recorded in some parts of this belt. In recent times, many disconformities have been identified in this vast sedimentary pile. Widespread Permian volcanic eruptions in the Pir Panjal Range in Kashmir and the Abor Volcanics in the far east rocks area a significant but isolated volcanic event in the Tethys Himalaya. The **Tethyan sediments** were deposited on the highly metamorphosed crystalline rocks of the Higher Himalaya.

(ii) South of the Tethyan belt lies the ***Great Himalaya* or *Himadri*** which is conventionally identified as the *Higher Himalayan Crystalline* ***or Central Crystalline Axis***. These rocks are found in highly rugged, perennially snow capped high mountains. A NW to E–W trending belt define this entire zone. At places, the contact between the Tethyan sedimentary rocks and crystalline rocks is marked by normal faults and multiple shears. These rocks had provided the floor for the deposition of the Tethyan sedimentary pile. It includes the celebrated peaks of Nanga Parbat (8126 m), Nun-Kun (7135 m), Kedarnath (6900 m), Badrinath (7138 m), Nanda Devi (7817 m), Dhaulagiri (8172 m), Sagarmatha or Everest (8848 m), Kanchanjangha (8598 m) and Namcha Barwa (7756 m).

(iii) The terrain south of the deformed **Central Crystalline Axis** is included in the Lesser Himalaya and a complexly folded Main Central Thrust marks the boundary. **The Lesser Himalayan** terrain exposes the deformed, mildly metamorphosed Proterozoic sedimentary successions, capped by isolated belts of metamorphic rocks separated by thrust planes. Most of argillo-arenaceous and carbonate rocks of the Proterozoic Lesser Himalaya are of marine, shallow-water origin with sporadic small occurrences of deeper sea deposits. The Lesser Himalayan rocks occur in two distinct belts, a northern belt of Mesoproterozoic age with characteristic stromatolites and recently dated U–Pb detrital zircon. The southern belt is predominantly Neoproterozoic with glacial diamictite and capping of Lower Cambrian rocks exposed on isolated hill tops. The overthrust crystallines show mylonitized boundaries with the Lesser Himalayan metasediments. Sporadic mafic volcanism is manifested in some Lesser Himalayan domains as interbedded amphibolites or as basaltic lavas. Granites of Cambro-Ordovician age occur in many places.

(iv) South of the Lesser Himalaya, Cenozoic sedimentary rocks formed in the foreland basin which developed in front of the rising Himalayan Mountain chain. The Lesser and Greater Himalayan rocks provided the required detritus for filling up this trough. These sediments rose to form low lying hills, **the Siwalik Belt** and these were further affected by the south-directed thrust of the Lesser Himalaya. These thrust planes demarcate the southern boundary of the

Himalayan terrain. Then, there are long flat stretches of sedimentary fill parallel to the main Himalayan Range within the otherwise rugged Siwalik terrain. These gravelly and pebbly deposits are deposited in **the *duns***. These depositional lows in the synclinal valleys were filled up by the sediments of the vanished lakes and ponded water bodies.

(v) Last phase in evolution of the Himalaya witnessed the formation of another large foreland basin—the **Indo–Gangetic–Brahmaputra Plain**.

Based on the geographic location, these Himalayan ranges are also described region-wise. Hence, from west to east we have Kashmir/Himachal Himalaya, Garhwal Himalaya, Kumaon Himalaya, Nepal Himalaya, Sikkim Himalaya, Bhutan Himalaya, and Arunachal Himalaya. North to south linear belt has already been described above and identified as the Tethys Himalaya, Greater or Higher Himalaya, Lesser Himalaya and Outer Himalaya or Siwalik range.

1.3.5 Andaman Island Arc and Central Myanmar Plain

The Andaman Island Arc chain with Andaman-Nicobar Islands in the Bay of Bengal is recognized as the Andaman Island Arc chain extends up to Indonesia in the southern extremity. This arcuate belt, composed of two parallel N–S belts, is a part of the Indonesia–Myanmar Mobile Belt. Several hundred tiny island of the Nicobar and Andaman groups constitute the western belt and exposes Cretaceous and Cenozoic sedimentary rocks, intermixed with ocean floor fragments and are disposed along steeply dipping rocks. Eastward underthrusting of the ocean floor along the Java-Andaman Trench produced these islands. Rock of this chain reappears on land in the north along the Indo-Myanmar Border Range. The eastern arc is represented by seamounts and dormant and active volcanoes fed by the actively spreading floor of the Andaman Sea. The volcanic island arc extends northwards into the Central Myanmar Plain, characterized by Miocene to Quaternary volcanoes in the alluvial expanse of the Irrawaddy River system (Barber et al. 2017). The Central Myanmar Plain was formed exclusively by the sediments brought down by the Irrawaddy River.

References

Barber AJ, Zaw K, Crow MJ (eds) (2017) Myanmar: geology, resources and tectonics. Geol Soc Lond Mem 48:19–52. https://doi.org/10.1144/M48.2

Fainstein R, Richards M, Kalra R (2019) Seismic imaging of Deccan-related lava flows at the K-T boundary, deepwater west India. Lead Edge 38:286–290. https://doi.org/10.1190/tle38040286.1

Jain AK (2014) When did India-Asia collide and make the Himalaya? Curr Sci 106(2):254–266

Jain AK (2017) Continental subduction in the NW-Himalaya and trans-Himalaya. Ital J Geosci 136(1):89–102. https://doi.org/10.3301/IJG.2015.43

Leech ML, Singh S, Jain AK, Klemperer SL, Manickavasagam RM (2005) The onset of India-Asia continental collision: early, steep subduction required by the timing of UHP metamorphism in W Himalaya. Earth Planet Sci Lett 234:83–97
Ramakrishnan M, Vaidyanadhan (2008) Geology of India, vol 1 & 2. Geol Soc India, Bangalore, pp 1–556, 557–994
Tectonic plates boundaries detailed-fr.svg (https://commons.wikimedia.org/wiki/File:Tectonic_plates_boundaries_detailed-fr.svg) created by Sting (https://commons.wikimedia.org/wiki/User:Sting) under CC-BY-SA. File: Indian Plate map-fr.png
Valdiya KS (2016) The making of India: geodynamic evolution. Society of earth scientists series, 2nd ed. Springer, pp 1–924

Chapter 2
Indian Cratons

2.1 Introduction

The Precambrian Peninsular India is comprised of a few ancient cratonic nuclei that were formed during prolonged geological history during Archean to Paleoproterozoic and are classified into two blocks: The North Indian Block (NIB) and the South Indian block (SIB) (Naqvi and Rogers 1987). The former comprises of the Bundelkhand and Aravalli Cratons and the latter is made up of the Dharwar, Bastar and Singhbhum cratons; these are all surrounded by younger Proterozoic Fold Belts. A prominent ENE–WSW trending Central India Tectonic Zone (CITZ) separates these blocks whose fabric extends eastward through Chhotanagpur Plateau, while the isolated Meghalaya makes the sixth craton (Sharma 2009).

In the reconstructed geological map, it is conceived that the southern cratons (Dharwar, Bastar, Singhbhum, Meghalaya) constitute a single cratonic landmass, while the northern mass is made up of Aravalli and Bundelkhand Cratons. Younger rifts like the Pranhita-Godavari and Mahanadi dissect the southern block and separate it into individual cratons. The northern landmass is now covered by the Proterozoic Vindhyan sedimentary sequence, the Cretaceous Deccan Volcanics and the Indo-Gangetic Plain to make the precise configuration difficult.

This chapter provides an account of geology, structure, geochemistry and geochronology of each of these six cratons, viz. the Aravalli (including Marwar) and Bundelkhand of the Northern Block, the Dharwar, Bastar, Singhbhum and Meghalaya of the Southern Block.

A. K. Jain et al., *Tectonics of the Indian Subcontinent*, Society of Earth Scientists Series, https://doi.org/10.1007/978-3-030-42845-7_2

2.2 Aravalli Craton

The Aravalli Craton is often described as the western component of the greater Bundelkhand Craton, but it is sensu stricto not directly comparable with it. Like any other Archean province, the Aravalli Craton has lithologies ranging in age from 3.5 to 2.5 Ga, along with its metamorphosed and tectonically modified lithologies. Because of its highly complicated geological relationships, Heron (1953) termed it as the 'the Banded Gneissic Complex (BGC)' to describe this cratonic mass, which consists of granulite, granite gneiss, aplo-granite and amphibolite.

2.2.1 Geological Framework

The southeastern boundary of the Aravalli Craton hugs around the unmetamorphosed Vindhyan sedimentary succession and the overlying Deccan Volcanics (Fig. 2.1) across the northwest-directed Great Boundary Fault (GBF). The craton, in the strict sense, does not possess a distinct boundary and is surrounded mostly by the Proterozoic Aravalli Supergroup belt, while its northwestern component around Sandmata is juxtaposed against the Delhi Supergroup along the Aravalli–Delhi Tectonic Contact.

The Aravalli Craton is comprised of the following units (Sinha-Roy 1985; Gopalan et al. 1990; Wiedenbeck and Goswami 1994; Wiedenbeck et al. 1996; Roy and Kröner 1996; Mondal and Raza 2013; Sharma and Mondal 2019): (i) Banded Gneissic Complex (BGC-I and BGC-II) of heterogeneous TTG gneissic complex of 3.3–3.2 Ga, (ii) a dismembered volcano-sedimentary greenstone belt of 2.8 Ga, and (iii) Berach granitoids of calc-alkaline to high-K character.

2.2.2 Banded Gneissic Complex (BGC)

The Aravalli Craton has been subjected to structural, petrological, geochemical and geochronological investigations and is represented by a stratigraphic unit known as the Banded Gneissic Complex (BGC—Heron 1953). It is made up of granite gneisses, migmatites, metasedimentary enclave and metabasics and referred to be the oldest stratigraphic unit of the craton. Subsequent investigations revealed much more complex nature of this basement due to complex depositional history of the inter-layered meta-sediments, timing and style of structural deformation, variations in metamorphic imprints, and conflicting and often contradicting radiometric ages.

Raja Rao et al. (1971) and Gupta et al. (1997) classified pre-Aravalli gneissic granitoids under the Mangalwar Complex of the Bhilwara Supergroup. The peneplaned terrain of Mewar near Mangalwar on the Udaipur–Chittaurgarh road is placed under the BGC. Three subgroups have been identified with seven formations: migmatite dominated, migmatite and meta-sediment dominated and metasediments

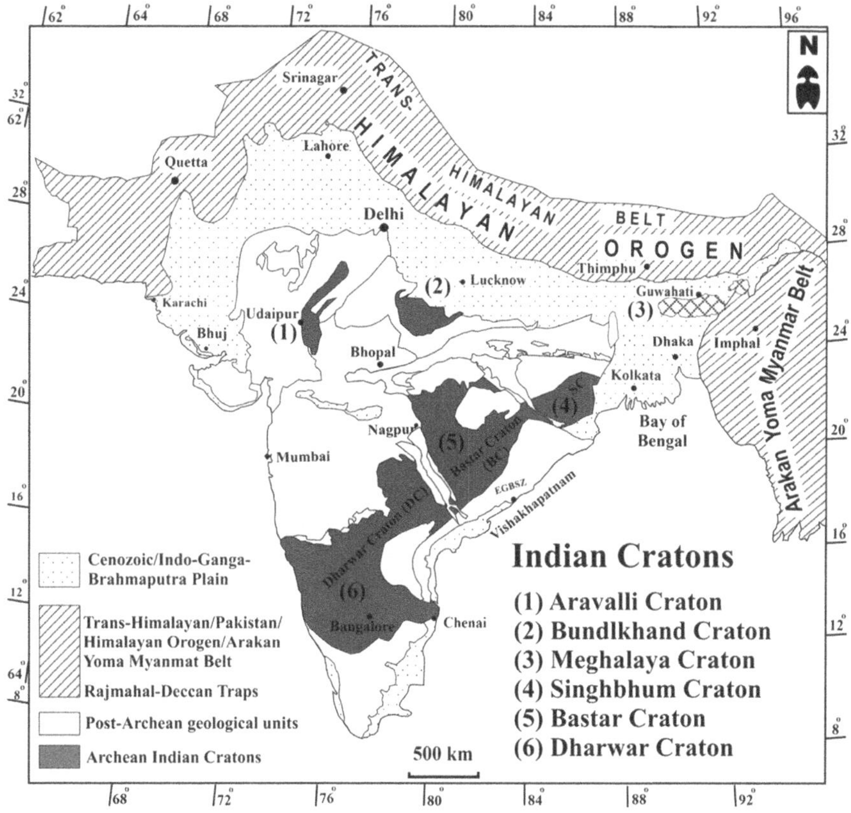

Fig. 2.1 Distribution of the Indian Archean Cratons. Redrawn from published sources

dominated with minor migmatites in sequential order. Roy (1988) named the pre-Aravalli granite gneiss–amphibolite bodies as the Mewar Gneiss (known as the BGC) and differentiated these from those developed by syntectonic migmatization of the Aravallis.

Validity of the term 'Banded Gneissic Complex' (BGC) has recently been critically evaluated in view various provisions in the stratigraphic codes (Fareeduddin and Banerjee 2020). It was realized that BGC cannot be treated as lithostratigraphic unit but only as lithodemic unit (not conforming to the Law of Superposition), as emphasized by (cf. Easton 2009; Kumpulainen 2017). The term 'Banded Gneissic Complex' (Heron 1953), with its vast expanse covering 1000s of km^2, could be assigned to a '*super suite*'. Nomenclatures like Mangalwar Group and Mangalwar Complex with their constituent units are equally untenable since these are not based on the Law of Order of Superposition. Their subdivisions show a curious mixture of lithostratigraphic and lithodemic classifications. Similarly, the term "Mewar Gneiss' does not define a formal rank, hence neither qualify to be a lithostratigraphic unit nor a lithodemic entity. Contemporary workers opined that the term BGC should be retained as it partially satisfies the stratigraphic code, enjoys the provision of priority enshrined in the code and is well-entrenched in the geological literature (Sharma and Mondal 2019).

In this part of Rajasthan, the BGC has been interpreted as follows: (i) Proterozoic in age as migmatized Aravalli and Raialo sediment (Crookshank 1948; Naha and Halyburton 1974; Roy et al. 1981; Buick et al. 2006), (ii) Archean in age showing widespread remobilization during the Paleoproterozoic (Guha and Bhattacharya 1995; Sharma 1988; Roy et al. 2005). Bhowmik and Dasgupta (2012) viewed these conflicting interpretations of the BGC due to different stages of structural and metamorphic transformations that modified and largely obliterated the original protolith history of magmatic and sedimentary components of the BGC and their stratigraphic relationships.

Gupta (1934) and Heron (1953) considered the BGC occurring to the north and south of the Nathdwara in Girwa Valley as two different entities. In these two situations, relationships between the basement and the overlying supracrustals, grade of metamorphism and their ages show marked differences (cf. Gopalan et al. 1990; Wiedenbeck and Goswami 1994; Wiedenbeck et al. 1996; Buick et al. 2006; Dharma Rao et al. 2011; Bhowmik and Dasgupta 2012). Hence, these were identified as the (1) BGC-I (exposed east and south of Nathdwara) and (2) BGC-II (exposed the north of Nathdwara), though these two basement domains of the Aravalli Craton have been a subject of debate (Roy et al. 2005; Buick et al. 2006; Dharma Rao et al. 2011; Ahmad and Mondal 2016). The age of formation, geological history and relationship between the Present-day configuration of the Aravalli Craton is an outcome of terrane accretion between the two distinct basement complexes.

The BGC-I (Gupta 1934; Bhilwara Gneiss of Sinha-Roy et al. 1995) displays tonalite–trondhjemite–granite (TTG), gneiss–migmatite–granitoid–amphibolite and intimately inter-layered supracrustal components (Heron 1953; Crawford 1970). Gopalan et al. (1990), Wiedenbeck and Goswami (1994) and Roy and Kröner (1996) identified an Archean 3.3–3.2 Ga nuclei in the BGC-1 and BGC-II, representing a single crustal block by the Late Archean time following extensive granite magmatism and contemporary metamorphism at 2.54–2.45 Ga. This cratonic block became a basement for the overlying Paleoproterozoic Aravalli succession.

2.2.3 Tectonothermal Events and Metamorphism in Sandmata and Mangalwar Complexes

The Aravalli Craton experienced tectonothermal events during the Proterozoic orogeny (ca. 1.7 Ga) (Buick et al. 2006). During this event, the BGC-I domain remained unaffected, while the BGC-II was largely reworked (Roy et al. 2005). As noted above, the BGC-I rocks are of Archean ages (~3.5 to 2.5 Ga: MacDougal et al. 1983; Gopalan et al. 1990; Wiedenbeck and Goswami 1994; Roy and Kröner 1996; Wiedenbeck et al. 1996; Roy et al. 2001) with oldest zircon of 3491 $\pm$ 7 Ma, derived from the Delwara volcanics (McKenzie et al. 2013). Gopalan et al. (1990) reported a Sm–Nd isochron age of 3.31 ± 0.07 Ga and an initial Nd isotopic ratio that is close to chondritic values for tonalitic to granodioritic basement gneisses from the Ahar River-type area of the Aravalli Supergroup. They also reported a younger isochron age of 2.83 ± 0.05 Ga and small positive initial ϵ_{Nd} for lenticular mafic amphibolites within the gneisses. This value is close to 2.95 Ga Untala granitoid. Similarities in ages of gneisses and associated amphibolites suggest that mafic volcanism is associated with the formation of the granitic protoliths. Roy et al. (2012) reported single zircon 'evaporation' age of 2905.4 ± 0.3 Ma for granite gneiss from Nathdwara. This age compares well with the age of ~2887 Ma for a granite gneiss from the Archaean basement outcrop southeast of Udaipur (Roy and Kröner 1996). U–Pb–Hf isotope studies by Kaur et al. (2019) and Wei et al. (2017) reveal that the TTG precursors of gneissic complex intruded during the Paleoarchean (3310 Ma) and Neoarchean (2563–2548 Ma). Wang et al. (2018) reported average U–Pb age of 2985 Ma for several zircon grains from basement granite underlying the Raialo Group in northeast Rajasthan. The 2985 Ma granite is the first report of an Archaean basement from the northern part of the Aravalli range. The BGC thus evolved under polycyclic regime between Paleo- to Mesoarchean time (~3.3 Ga). Precise ion microprobe $^{207}Pb/^{206}Pb$ zircon crystallization ages between 2562 ± 6 and 2440 ± 8 Ma for the south Aravalli Mountain granitoids were determined by Wiedenbeck et al. (1996). This study, together with other geochronological studies for the Berach Granite (Crawford 1970, 1975; Roy and Kröner 1996) indicated age bracket of 2.44–2.58 Ga for the end-Neoarchean craton stabilization event.

On the other hand, the BGC-II yields largely Proterozoic ages (~1.7 Ga) (Buick et al. 2006; Dharma Rao et al. 2011) with sporadic Archean remnants. Proterozoic tectono-thermal events did not leave any imprint on the BGC-I cratonic component. On the other hand, in central Rajasthan, the BGC-II reveals two contrasting metamorphic histories: (i) granulite facies metamorphosed pelites, charnockite, calc-silicates, minor mafic dykes (Sharma et al. 1987) and intrusive granitoids of the Sandmata Complex, and (ii) amphibolite facies assemblage within TTG gneisses, calc-silicates and meta-sediments of the Mangalwar Complex (Gupta 1934; Gupta et al. 1980, 1997). Subsequently, Dasgupta et al. (1997) and Roy et al. (2005) observed that these complexes is wrapped around by a regional level shear zone of variable thickness. The Mangalwar Complex with complex metamorphosed pelitic lithology cannot be placed under a single litho-tectonic unit. Consensus now exists about the Sandmata Complex with its granulite and related rock to represent a distinct litho-tectonic domain within the BGC-II. Based on lithological characteristics, Bhowmik and Dasgupta (2012, 2015) divided the Sandmata Complex into (i) Sandmata Gneissic Complex and (ii) Sandmata Metasedimentary Complex. The Mangalwar Complex (MC) is further divided into: (i) Asan Group comprising bimodal gneisses, (ii) older greenstone Sawadri Group, and (iii) younger greenstone Tanwan Group (Ahmad et al. 2016). The Banas Dislocation Zone (BDZ) delimits the MC to its north while the BGC-I and the Aravalli Supergroup lies to its south (Sinha-Roy et al. 1993; Mohanty and Guha 1995). Lithologies of the latter two groups are intruded by several generations of Ran Igneous Complex of the Sandmata Complex (Mohanty and Guha 1995). In some sectors in the Aravalli domain, the contact between the Mangalwar and Sandmata Complex has been interpreted as thrust (Sharma 1988). In contrast, Gupta and Choudhuri (2002) interpreted the shear zones wrapping the Sandmata Complex as conjugate system of oblique strike-slip faults whose orientations suggest bulk shortening in a NW–SE direction.

In recent years, integrated textural–mineralogical–geochemical and geochronological studies have helped in deciphering well-constrained thermobarometry (Bhowmik et al. 2010) and SHRIMP dates on monazite (Buick et al. 2006; Bhowmik et al. 2010; Dharma Rao et al. 2011) helped in proper quantification of metamorphism in these rocks. These studies show that Sandmata granulites are polymetamorphic, with an early medium-pressure (MP) granulite facies metamorphism in the stability field of sillimanite. On the other hand, Mangalwar Complex rocks are monocyclic having suffered only one high-pressure metamorphism (Bhowmik and Dasgupta 2012).

In order to propose a new tectonic model, Bhowmik and Dasgupta (2012) tried to establish relationship between tectono-thermal events at ~1.85 to 1.8 Ga, ~1.72 Ga and 0.95–0.88 Ga and the development of three sedimentary basins before 1.85 Ga (Aravalli Basin), between 1.8 and1.74 Ga (Sandmata Basin) and between 1.72 and 0.95 Ga (Delhi and supposed equivalent of Mangalwar basin rocks).

2.2.4 Sedimentation Over the Rifted Aravalli Basement

The predominantly gneissic outcrops in this part of Rajasthan terrain represent the most important component of the BGC (≡Mewar Gneiss/Mawli Gneiss). Gopalan et al. (1990) accurately dated some of these gneisses as 3310 ± 0.007 Ma. These basement gneisses on culmination of the granitic activity show telltale evidence of exhumation, rifting, volcanism and erosion, followed by weathering. Extensive development of paleosol on top of the gneisses is represented by pyrophyllite, talc schist and clays, and marked the major geological event towards the end of Archean (Banerjee 1996; Pandit et al. 2008). This erosional hiatus was preceded by intermittent emplacement of basic volcanic lava indicating the culmination of rifting phase. The paleosol units at the base of the Aravalli sequences vary in thickness from a meter to tens of meter when clastic sediments of this mobile belt were deposited over this paleosol surface.

2.2.5 Nature of Tectonic Contacts

A thrust contact was identified as the Banas Lineament Zone between Hindoli-Jahajpur succession and the Mangalwar Complex (Sinha-Roy and Malhotra 1989), hence it became the youngest litho-succession. Similarly, the Sandmata Complex is juxtaposed against the Aravalli mobile belt due to the Dislocation Kaliguman (Shear)

Zone and the Delwara Dislocation Zone. Sharma (1988) and Sinha-Roy et al. (1993) interpreted the exhumation of the Sandmata Complex terrain through active faulting. Tectonic activity along the Kaliguman Shear Zone may have facilitated exhumation process and outcropping of the Mangalwar and Sandmata Complexes. The Banas Dislocation Zone restricted the domain of Delwara Dislocation Zone and Rakhabdeo Suture in the north. In south Rakhabdeo Lineament abuts against north Gujarat Suture (Sant and Karanth 1993) or extends up to western end of the Son-Narmada Fault, north of Gujarat

2.2.6 Volcano-Sedimentary Greenstone Belts

Like the Dharwar Craton, large linear belts overlying the BGC are reported to possess greenstone-like features. Upadhyay et al. (1992) and Roy and Jakhar (2002) identified these dismembered remnants as the Archean greenstones exposed in the southeast and northeast terrain of Udaipur. Supposedly 'large disjointed bodies' shown in the maps do not show spread more than a couple of hundred meters, whereas the 'greenstone belts' are much larger linear belts. It is worth noting that mafic and ultramafic enclaves of this belt are tholeiitic and komatiitic in composition (Ahmad et al. 2008), hence the above-mentioned rocks represent the ancient supracrustals like that of Sargur supracrustals in the Dhawar Craton. Sinha-Roy et al. (1998) also used the term greenstone for some rock sequences in the BGC from Sawdri and Tanwan areas. Earlier, Sinha-Roy et al. (1993) interpolated a large-scale Archaean granite-greenstone set up within the basement rocks of the Aravalli Craton. According to him, the basement rocks display a two-tier greenstone development with widespread migmatization. The Hindoli Group is also interpreted as a secondary greenstone belt juxtaposed with Mangalwar terrane with a prominent DSZ. Amphibolites and metasedimentary rocks occur as linear belt of volcano-sedimentary sequence on the western margin of the BGC-I (Ahmad et al. 2016).

The metasedimentaries in contact with the Berach Granite in the west, show low grade metamorphism and mild deformation. Such low grade metasediments, known in the contemporary literature as the Jahajpur Formation and the Hindoli Group, pass into amphibolite grade migmatized rock of the Mangalwar Complex in the W and NW. Many recent workers, including Heron (1953), considered these low-grade rocks of the Hindoli-Jahajpur area as equivalent of the Aravalli succession, exposed in the type area of Girwa Valley. Possibility of the Hindoli rocks being less metamorphosed facies of the Mangalwar Complex cannot be ruled out. Those advocating this idea identify gradual increase in metamorphism from east to the west.

2.2.7 *The Berach Granite—Base of the Aravalli Metasediments*

The southeastern boundary of the Aravalli Craton hugs around the unmetamorphosed Vindhyan sedimentary succession and the overlying Deccan Volcanics (Fig. 2.2) across the northwest-directed Great Boundary Fault (GBF) where the Berach Granite crops out close to the GBT near Chittaurgarh and has either been interpreted as the basement (Pascoe 1950) for the overlying Aravalli metasediments or last phase of cratonization event (Sinha-Roy 1985).

This granitic body is largely made up of non-foliated massive, pink leucogranite whose age varies between 2506 and 2585 Ma to make it almost contemporaneous with granites of the Bundelkhand Craton. Hence, it was previously designated as the Bundelkhand Granite and, therefore, considered basement for the Aravalli metasediments as well as time equivalent of the BGC by Heron (1953). Subsequently, Mondal and Raza (2013) opined that the Berach granitoids are intrusive into the Banded Gneissic

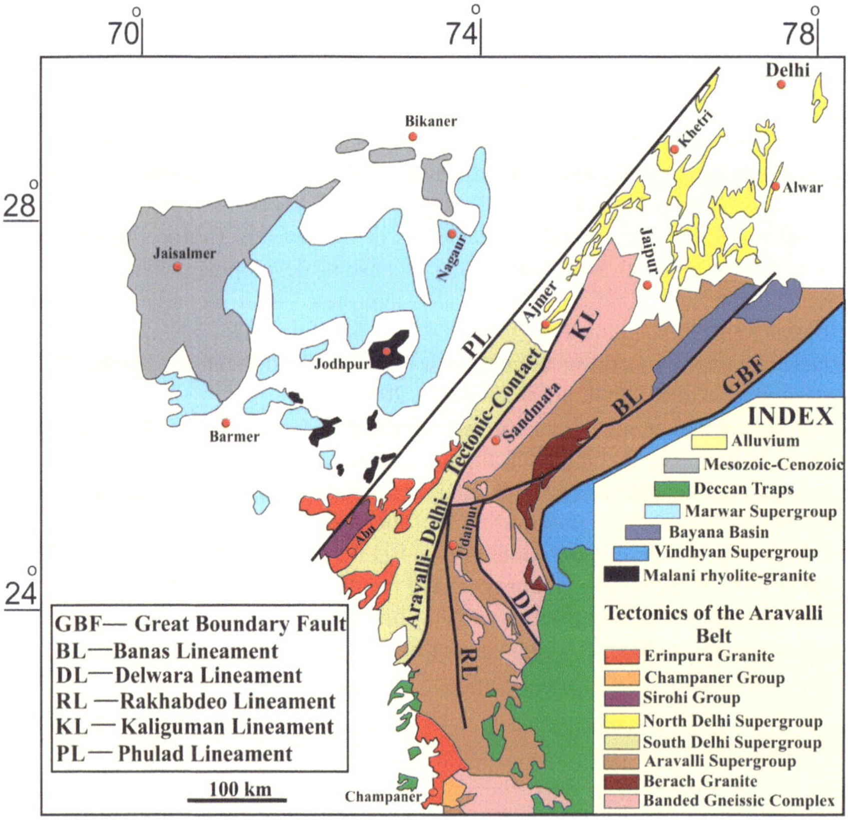

Fig. 2.2 Tectonic elements in the Aravalli Craton in western Rajasthan

Complex of the Aravalli Craton, and incorporates calc-alkaline sanukitoid series, enriched in both compatible as well as incompatible elements. Geochemically, the Berach granitoids are strikingly like other Archaean sanukitoid series of the world and appear to have formed due to partial melting of metasomatized subcontinental lithospheric mantle, enriched in large ion lithophile elements. Gupta et al. (1997) reported a distinct intrusive relationship of this granite with enveloping low-grade mildly deformed meta-sediments of the Hindoli Group. From this granite, Kaur et al. (2016, 2019) obtained $^{207}Pb/^{206}Pb$ ages between 2545 ± 18 and 2182 ± 24 Ma which are significantly older than the zircon SIMS age of 2440 ± 8 Ma (cf., Wiedenbeck et al. 1996). The Berach Granite appears to have crystallized during a single magmatic event from a homogeneous melt and represents a pervasive terminal phase of cratonization activity in this region (also Roy and Kröner 1996). On the contrary, the Untala Granitoid appears to be derived from an already evolved Paleoarchean crust, intruded by granitoids at ca. 3100 Ma, and was intensely reworked during the Neoarchean at 2544 Ma (Kaur et al. 2019).

2.2.8 *The Mangalwar Complex: The Sequence Above*

To the west, the Mangalwar Complex shows increased high-pressure metamorphism with granulite and amphibolite and is designated as the Sandmata Complex (Gupta et al. 1980). Heron's Banded Gneissic Complex was renamed as the Bhilwara Supergroup, comprising a variety of metasediments, calc silicates, high grade metamorphics, metavolcanics, granites and gneisses (Raja Rao 1967; Gupta et al. 1980). They correlated the base-metal and organic matter-rich carbonates and pelitic chemogenic meta-sediments of Rajpura-Dariba, Pur-Banera and Sawar with this supergroup, which appeared to be incorrect based on geochemical and geochronological criteria. The Bhilwara Supergroup was divided into numerous groups, sub-groups and formations based on the locality where they were studied or described.

2.2.9 *Place in the Global Paleogeography*

The age of a continent is deduced from the oldest rocks exposed in that continent. Ur is the earliest known supercontinent, which was formed at ~3.0 Ga ago. With newly published dates, is it possible to consider Bundelkhand–Aravalli nucleus as part of the Ur supercontinent (Mondal 2003, 2009). Conventionally, nuclei of the Indian Plate that are grouped together in the Ur assembly include the cratons of Western Dharwar, Eastern Dharwar, Bhandara–Bastar and Singhbhum. Mondal (2009) questioned as to why the Bundelkhand–Aravalli nucleus is not considered in the Ur assembly, although it has all the characteristics to be grouped in the earliest continental reconstructions. It is now widely accepted that by 3.3 Ga, the gneisses of the Bundelkhand–Aravalli nucleus attained stability so that it could act as a stable crust to sustain the continent and provide a foundation for the evolution of Aravalli metasediments. This is a valid argument to include the Bundelkhand–Aravalli nucleus as constituents of the blocks in the Ur assembly.

2.3 Bundelkhand Craton

2.3.1 Spatial Distribution and Geological Framework

Northwestern parts of the Indian Shield is comprised of two main cratons: (i) the Aravalli Craton (≡Rajasthan Craton) and (ii) the Bundelkhand Craton (Sharma 2009; Valdiya 2016). Eastern triangular-shaped Bundelkhand Craton covers an area of about 26,000 km^2. Some workers attribute this separation to the effects of the Great Boundary Fault (GBF), which delimits the Vindhyan rocks against Aravalli metasediments. This premise is also based on the correlation of the Berach Granite in Rajasthan with the Bundelkhand granite. However, contemporary workers prefer to deal with Bundelkhand and Aravalli Cratons as two separate entities because of uncertainty of connectivity under the intervening arcuate belt of Mesoproterozoic-Early Neoproterozoic Vindhyan sedimentary succession on its north, west and south. Very-low grade Paleoproterozoic metasedimentaries of the Bijawar (=Sonrai) and Gwalior Basins occur on the margin of this craton, while unmetamorphosed sedimentary rocks of the Vindhyan basin wrap the craton in the north, southwest and west.

Folded, faulted and variably metamorphosed NE–SW trending Aravalli–Delhi Fold Belt (ADFB) and the ENE–WSW trending Central Indian Tectonic Zone (CITZ) of Paleoproterozoic age directly overlie the Archean gneissic and volcanic rock assemblage of this craton. The Aravalli Craton, Bastar Craton and Chotanagpur Gneissic Belt occur in the near vicinity of the Bundelkhand Craton. The southern margin of the craton shows a steep gravity gradient which reflects the nearby Son-Narmada Fault (Ramakrishnan and Vaidyanadhan 2008). Numerous boreholes drilled by the ONGC and other agencies in the Ganga Plains indicated extension of rocks of this craton underneath the alluvial plains and possible extension into the base of the Himalaya.

Following the review by Basu (1986, 2007) there had been keen interest in geology and tectonics (Roday et al. 1995), geochemical investigations of granitoids (Rahman and Zainuddin 1993; Sarkar et al. 1993, 2007), and geochronology indicating the presence of a Paleoarchean component (Sarkar et al. 1984, 1996), besides presence of the Archean ophiolite (Malviya et al. 2006), and world's seventh oldest meteoritic impact Dhala structure (Pati et al. 2008).

The oldest component of Bundelkhand craton is manifested by the TTG magmatism at ca 3.27 Ga and granitoid emplacement at ca 2.55–2.52 Ga indicating the stabilization of the craton (Mondal et al. 2002, 2008). Joshi et al. (2017) classified TTGs into two categories by their REE patterns: (i) low-HREE TTG and (ii) enriched TTG. Additional geochronology indicate that these TTG gneisses vary in age between 3.59 and 2.66 Ga (Kaur et al. 2016; Joshi et al. 2017; Saha et al. 2016; Verma et al. 2016) and are grouped into four distinct episodes of TTG magmatism (Kaur et al. 2016).

2.3.2 *Stratigraphic and Tectonic Units*

The Bundelkhand Craton consists of three distinct litho-tectonic units: (i) highly deformed Paleoarchean to Neoarchean Bundelkhand greenstone belt and a gneissic complex (Bundelkhand Greenstone Belt), surrounded by (ii) an abundant undeformed multiphase late Neoarchean Bundelkhand Igneous Complex, and (iii) intrusive Proterozoic mafic dykes and quartz reefs (Fig. 2.3; Sharma and Rahman 2000; Meert et al. 2010; Meert and Pandit 2015).

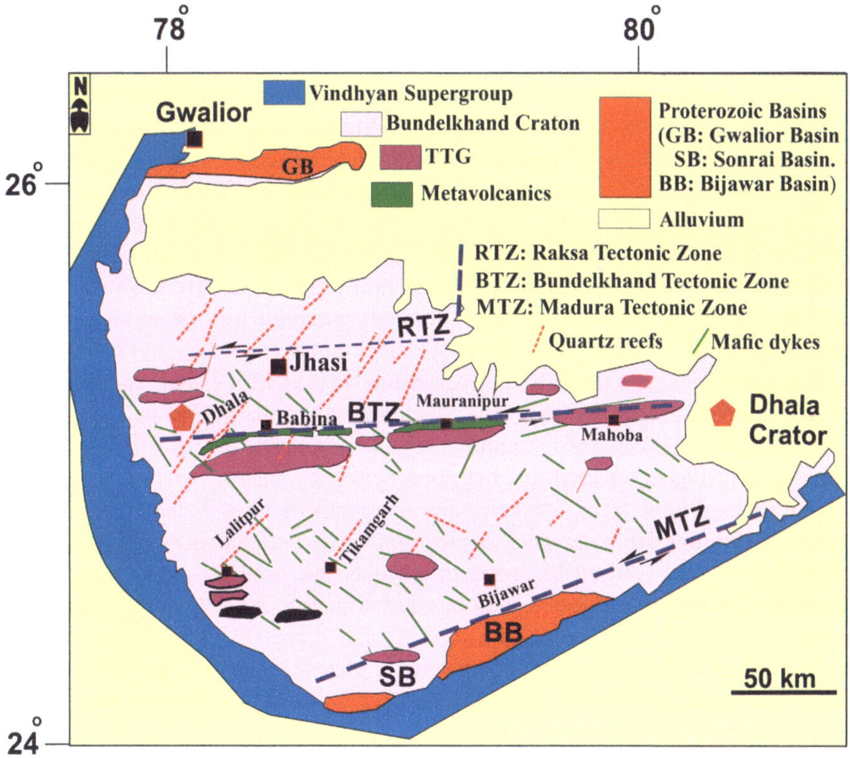

Fig. 2.3 Simplified geological map of the Bundelkhand Craton, showing major tectonic boundaries, shear zones, and location of Dhala Crater. Modified after Pati et al. (2008)

2.3.2.1 Bundelkhand Greenstone Belt (Enclave Suite)

Three greenstone ~E–W-trending complexes of metasedimentary and metavolcanic rocks have been recognized from the Bundelkhand craton: (i) Karera Greenstone Belt in northern parts, (ii) Central Bundelkhand Greenstone Complex (CBGC) in the middle, also known as the Babina Greenstone Belt (Singh and Slabunov 2015, 2016), and (iii) Southern Bundelkhand Greenstone Complex (SBGC). One of the most significant association of this suite is observed within the E–W trending Bundelkhand Tectonic Zone, the Babina Greenstone Belt, for nearly 200 km across the central BC and South Bundelkhand (Fig. 2.3; Girar–Madaura Greenstone Belt). Malviya et al. (2004, 2006) and Singh and Slabunov (2015) were the first to discuss the greenstone belts in the Bundelkhand craton and classified them into two complexes viz. Central Bundelkhand (Babina and Mauranipur belts) and South Bundelkhand (Girar and Madaura belts) complex.

The Babina–Mauranipur–Mahoba greenstone belt comprises disrupted sequences of amphibolites, banded iron formation, pillow basalts, komatiitic basalts, calc-silicate rocks, white schists, quartzites, anatectic granites and metapelites and ~3.55 to 3.20 Ga TTG intrusives (Absar et al. 2009).

Similar trondhejemite-tonalite gneisses from Mahoba and Kuraicha yielded 3270 ± 3 Ma and 3297 ± 8 Ma ages for zircons (Mondal et al. 1998, 2002). Around Mauranipur, deformed and metamorphosed basaltic andesitic pillow and massive lavas, serpentinized and basaltic komatiite, volcaniclastics, tuffs and BIF are located like ophiolite belts from other Archean cratons (Malviya et al. 2006; Singh and Slabunov 2015; Slabunov and Singh 2018; Singh et al. 2019b).

Normally disposed elongated flow lobes in pillows are about 50 cm with cm-thick rim with intercalated metasediments. Ultramafics exhibit well-preserved original mineralogy of olivine and chrome spinel. Distinct mm to cm-scale interlayered ironstone and quartzite mark the BIF sequence Alfimova et al. (2019) and is occasionally interbedded with pillows. Stratigraphic sequence within the Babina belt appears to be normal. Though greenschist–amphibolite metamorphism (and local high-pressure metamorphism) have affected this belt, geochemical signatures are still preserved in these rocks. Pillow lavas and volcanics are subalkalic, low-K tholeiitic basaltic andesite, exhibiting HFS elemental depletion, relative enrichment of LIL elements, and nearly flat REE pattern like modern intra-oceanic arcs. Strong compositional similarities of high-Mg andesite and basaltic komatiite with modern equivalents strongly suggest that the Mauranipur supracrustals represent an Archean ophiolite sequence in an island arc setting of converging plates (Malviya et al. 2004, 2006). Additional geochemical data suggest that differentiation of tholeiitic to calc-alkaline mafic-felsic volcanics and basalt-andesite-dacite-rhyolite took place in a mature arc in a convergent tectonic setting where the melts were derived from partial melting of thick basaltic crust metamorphosed to amphibolite-eclogite facies (Singh et al. 2018). These deformed and metamorphosed felsic volcanics yielded poorly fractionated ($La_N/Lu_N = 11–16$) pattern with small negative Eu anomaly ($Eu/Eu^* = 0.68–0.85$), and chondrite-normalized rare earth element distribution, which is characteristic of subduction volcanics (Singh and Slabunov 2015). U–Pb SHRIMP dating of zircon indicates that these Babina greenstone calc-alkaline felsic dacites erupted at 2542 ± 17 Ma, while their geochemical characters point their eruptions in subduction zone settings (Singh and Slabunov 2015).

Three generations of gneisses are thought to have formed at 3.2–3.3, 2.7 and 2.5 Ga respectively, as indicated by $^{207}Pb/^{206}Pb$ isotopic data (Gopalan et al. 1990; Mondal and Zainuddin 1997; Mohan et al. 2012). Combined U–Pb zircon ages and Lu–Hf isotope characters of trondhjemite gneiss from the BC yielded even older felsic crust of 3551 ± 6 Ma (Kaur et al. 2014). Along with highly fractionated REE pattern [$(La/Yb)_N = 34.4$], low HREE contents, positive Eu ($Eu/Eu^* = 1.52$) and Sr anomalies, they opined that the melt was extracted from a garnet-bearing and plagioclase-free source by reworking of the Eoarchaean mafic crust. This is evident from slightly subchondritic Hf compositions ($\varepsilon Hf_{3.55\ Ga} = -0.8 \pm 0.3$) and Hf model ages between 3.80 and 3.95 Ga.

Additional geochronology of the TTG rocks revealed at least 4 Paleoarchean magmatic episodes within the BC at ca. 3.55, 3.44–3.40, 3.30 and 3.20 Ga at regular intervals of 100–150 Myr (Kaur et al. 2016). Hf-isotope (εHf_t of about +5.0 at 3.0 Ga) and whole-rock geochemical data of the TTGs record reworking of older mafic and felsic crust between 3.55 and 3.20 Ga, crust formation and stabilization. An oceanic crust was possibly partially melted and derived from chondritic mantle sources, shortly before formation of the TTG. Thus, the TTGs of this craton are remnants of Paleo- to Neoarchean TTG-greenstone microcrontinents that were formed by melting of island arc or oceanic basalts and amalgamated between 3.3 and 2.7 Ga (Joshi et al. 2017). Intense melting of the heterogeneous TTG crust resulted in coeval formation of granodioritic gneisses and voluminous granites during the Neoarchean.

Verma et al. (2016) observed that one TTG sample dated ~2.7 Ga from northern parts of this craton, and happens to be the youngest age, like the Precambrian basement TTG gneisses of the Aravalli Craton, thereby indicating widespread crust formation in this timeframe.

Pb–Pb zircons from the Bansi Rhyolite yielded an age of 2517 ± 7 Ma (Mondal et al. 2002), which is almost the same age as that of the Bundelkhand Granite. The 2.5 Ga age is also considered as the stabilization age of the Bundelkhand Craton, as it overlaps with the ages of undeformed granitoid plutons in the Bundelkhand (see Sect. 2.3.2.2) and the BGC–Aravalli cratons.

Detached isolated bodies and disrupted volcano-sedimentary sequences are also recorded elsewhere and are considered as part of greenstone belts within the BC. Along the Mauranipur-Babina greenstone belt, detailed mineralogy and geochemical compositions of metasedimentaries reveal arkose, greywacke and mafic-rich metapelites (greywacke) (Raza and Mondal 2018). It is likely that this basin received immature detritus by eroding the young TTG and granitic batholithic sources along with syn-depositional volcanic centers in an active tectonic environment such as an island arc.

In Lalitpur region, volcanogenic Mehroni Group appears to have been deposited on the cratonic basement, and is represented by lower meta-sedimentaries, gneisses and meta-volcanics of the Kuraicha Formation (=Rajaula) (Misra and Sharma 1974), having a thick upper unit of conglomerate, marble, BHQ and intercalated basaltic rocks. Metamorphosed ultramafic rocks, often altering into serpentinite, occur in the southern margin of the craton. East–west trending and north-dipping supracrustal enclaves are found as northern (Karera to Kabrai) and southern belts (Madura to Baraitha = Madaura Formation), and also near Panna and Kalinjer.

Peridotite, dunite and pyroxenite and gabbro represent the ultramafic suites, and are intimately inseparable from metabasics (Mohanty et al. 2018), which are prominently exposed both in the southern and northern belts. Banded quartz magnetite rocks, representing the BIF, in the enclaves occur along with meta-quartz arenites, metapelites, minor lenses of marble and calc-silicates. Gneisses are either grey or pink are often streaky in appearance and show banding. These gneisses closely match with the Archean TTG suite. There is a great variation in the mineralogical and chemical composition of these gneisses.

Though detailed metamorphic conditions in the Bundelkhand Craton are poorly known, high-pressure metamorphism affected the Babina belt, with corundum-bearing phlogopite-chlorite schists recording 18–20 kbar and ca. 630 °C from outer collar and lower metamorphic P–T conditions of 11 ± 3 kbar, ca. 630 °C from inner collar (Saha et al. 2011). Metamorphic monazite and re-equilibrated zircon date the HP–corona formation ca. 2.78 Ga and the Neoarchean high-P metamorphism, indicating evidences for the Archean plate tectonics.

2.3.2.2 Bundelkhand Igneous Complex

After the emplacement of the TTGs of island arc or oceanic origin between 3.5 and 3.3 and 2.7 Ga, and after a time gap of about 130 Ma (Joshi et al. 2017), the BC witnessed widespread intrusion of the Bundelkhand Igneous Complex. It makes up the bulk of exposed granite of a batholithic dimension within BC (Goodwin 1991; Joshi and Slabunov 2019). Coarse to fine grained and grey to pink granites are associated with interlayered contemporaneous rhyolite as dykes and sills. Based on characteristic composition and REE patterns several granitoid types are now recognized, e.g. sanukitoid, Closepet-type granitoid, low-HREE monzogranite, low-Eu monzogranite and monzogranite with U–Pb-zircon age of emplacement ca. 2.5 Ga (Joshi et al. 2017; Verma et al. 2016). Small sanukitoids plutons of dioritic to graniodioritic compositions intrude the Bundelkhand granite-greenstone belts at ~2.95 to 2.55 Ga and, in turn, by pink granite, pegmatite and quartz veins (Joshi et al. 2017; Joshi and Slabunov 2019; Singh et al. 2019a). Based on bulk-rock geochemistry, Sm–Nd isotopic data and U–Pb-zicon ages, Singh et al. (2019b) concluded that sanukitoids originated and emplaced in subduction-related tectonic setting between 2.58 and 2.50 Ga. Possibly, mixing of metasomatized melts from the mantle with anatectic granitic melts and their homogenization at shallow crustal level, resulted in the formation of sanukitoids.

Considerable U–Pb zircon ages are now available from the Bundelkhand Igneous Province to indicate episodic granite emplacement between 2.57 and 2.45 Ga (Mondal et al. 2002; Kaur et al. 2016; Joshi et al. 2017; Verma et al. 2016; Chauhan et al.

2018), and possibly its division into the northern and southern blocks by intervening Bundelkhand Tectonic Zone. Verma et al. (2016) obtained high precision zircon U–Pb geochronological data for the pinkish porphyritic granites around 2.5 Ga.

2.3.2.3 Mafic Dykes and Quartz Reefs

(i) **Mafic dolerite dykes**: Mafic dykes intrude all the lithologies of the Bundelkhand Craton as the youngest magmatic activity (Basu 1986). These dykes display chilled margins with an aphanitic groundmass of plagioclase microlite laths, granular Fe–Ti oxides and fine clinopyroxene needles, which are aligned parallel to the wall-rock contact (Mondal and Ahmad 2001; also Rao et al. 2005). These mostly trend NW–SE, but also trend WSW–ENE and NE–SW (Basu 1986; Mondal and Zainuddin 1996; Mondal et al. 2002; Rao 2004; Pati et al. 2008; Pradhan et al. 2012; Ernst 2014). Earlier K–Ar data loosely constrain the age of most of these dykes ~2.1 and 1.5 Ga (Crawford and Compston 1970; Sarkar et al. 1996, 1997; Rao 2004). An earlier study by Sarkar (1993, 1997) reported two distinct K–Ar age clusters (ca. 1.8 and 1.56 Ga) on mafic dykes intruding this craton. Based on laser ablation $^{40}Ar/^{39}Ar$ isotopic analyses, Rao et al. (2005) suggested that most of mafic dykes were emplaced in two phases at 2.15 Ga and 2.09 Ga, while one NE–SW trending dyke shows a mean age of ~2.0 Ga. U–Pb geochronology on well-faceted zircon grains from the NW–SE trending dykes (the Jhansi Swarm–Ernst 2014) yielded a U–Pb concordia age of 1979 $\pm$ 8 Ma and a mean $^{207}Pb/^{206}Pb$ age of 1113 $\pm$ 7 Ma for ENE–WSW trending dykes (the Mahoba Swarm), confirming at least two distinct dyke emplacement events within the BC (Pradhan et al. 2012). A U–Pb rutile age of 2100 $\pm$ 11 Ma was interpreted as a minimum age of emplacement for a metamorphosed dyke (French 2007), while high precision U–Pb baddeleyite and zircon ages indicated emplacement of two unmetamorphosed dolerite dykes at 1891.1 $\pm$ 0.9 and 1888 $\pm$ 1.4 Ma (Meert et al. 2011). These dykes apparently represent a post-cratonization event.

(ii) **Quartz Reefs**: In the Bundelkhand Craton, giant quartz veins (GQV), also known as the quartz reefs, constitute spectacular landforms due to their resistance to erosion, spatial homogeneous distribution and NE–SW trending preferred orientation over nearly 29,000 km^2 (Basu 1986; Pati et al. 2007). There are more than 1500 mappable and exposed quartz veins. Outcrop width varies from $\leq$1 to 70 m and pervasively extend over tens of kilometers along the strike over the entire craton. Numerous younger thin quartz veins with somewhat similar orientation cut across the GQV, and exhibit evidences of strong brittle to ductile–brittle deformation. Base metal, gold incidences and pyrophyllite-diaspore mineralization are associated with these veins.

Roday et al. (1995) observed that quartz reefs are controlled by brittle and brittle–ductile shear zones as basically co-genetic melts from microgranite. Bhattacharya and Singh (2013) have opined that the GQV represent sinistral strike-slip dominated vertical to subvertical shear zones, while openings of these reefs represent an extensional phase. Sharma and Rahman (1996) and Pati et al. (2007) opined that silica precipitation of hydrothermal fluids within GQV was late-stage tectonically controlled polyphase activity throughout the BC, though source(s) of such fluids remained ambiguous.

Based on the K–Ar geochronology, GQV and associated hydrothermal activity took place in three phases: (a) 1930 ± 40 to 2010 ± 80 Ma, (b) 1790 ± 40 to 1850 ± 35 Ma and (c) 1480 ± 35 to 1660 ± 40 Ma (Pati et al. 1997).

2.3.3 *Deformation*

2.3.3.1 Regional Pattern

This craton underwent polyphase deformation (Roday et al. 1995), and at least three phases of folding are observed in the supracrustal rocks, though details of various events are poorly known. The ~2.5 Ga granitoids exhibit submagmatic fabric not affected by later deformation events (Sarkar et al. 2017). Scattered exposures of the TTG gneisses around Babina contain centimeter- to meter-scale foliated mafic magmatic enclaves (MMEs) (Ramiz and Mondal 2017), and reveal polyphase deformation events (Fig. 2.4a and b; Nasipuri et al. 2019). Gneissic foliation (STG1) defines the earliest recognizable fabric in the TTG as alternating leucocratic and mesocratic layers and is axial planar to an earlier compositional layering (Fig. 2.4a). The STG1 is folded by FTG2 folds with N–S trending axial planar foliation STG2 and is marked by biotite and hornblende. In high-strain domains, STG1 gneissic fabric becomes parallel with STG2 foliation. Hook-shaped folds indicate coaxial character of two deformation events (Fig. 2.4a). East- or west-verging open folds represent last phase of deformation (FTG3) and sinistral shearing, which produces steep-dipping axial planes. E–W trending mylonitic foliation planes (STG3M) is coeval with the STG3 in the TTG; all the structural elements are schematically shown in Fig. 2.4c (Nasipuri et al. 2019).

2.3.3.2 Ductile Shear Zones

Though the Bundelkhand Craton is dissected by three major E–W trending ductile shear zones (Raksa Tectonic Zone—RTZ, Bundelkhand Tectonic Zone—BTZ, and Madura Tectonic Zone—MTZ), details have not yet worked out about their structures, geometry and kinematics. Singh and Bhattacharya (2017) opined three major shear events in the craton, out of which oldest E–W trending vertical shear system is marked by various types of mylonites, while younger NE–SW trending shear system is quartz reef-controlled and NW–SE system is marked by mafic dykes. Granitoid mylonites (Pati et al. 1997) from the BTZ have alternating ultramylonite

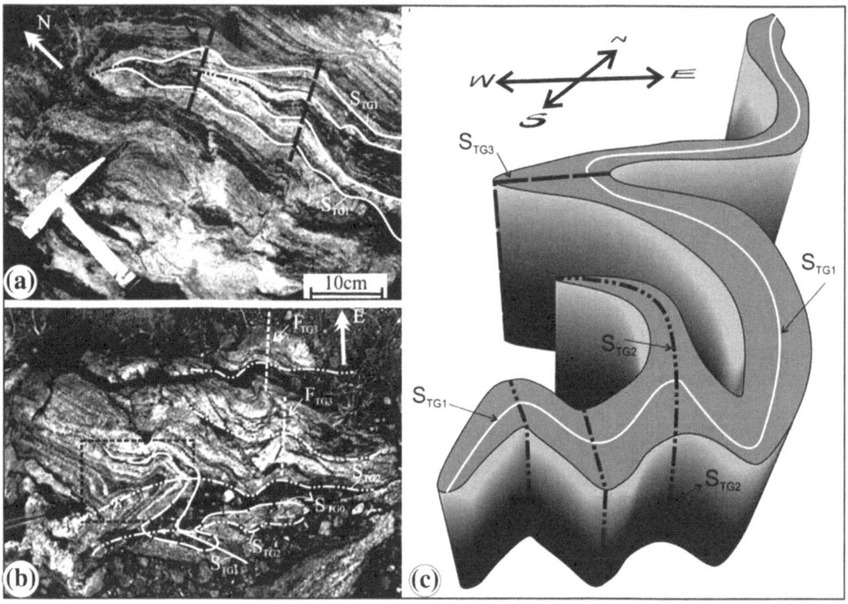

Fig. 2.4 Field photographs of the TTG. **a** Gneissic banding (STG1) in the TTG consists of alternating mesocratic and leucocratic layers. **b** Development of isoclinal folds on STG0 (FTG2), hook shaped FTG2 folds on $S_{TG0||1}$ and open folds on TTG. F_{TG3} and F_{TG2} axes are oriented nearly perpendicular. **c** Cartoon showing summary of the three stages of folding in the TTGs. After Nasipuri et al. (2019)

(dark) and mylonite foliated layers, having relict feldspar megacrysts with sinistral shear sense inferred from asymmetrical feldspars and S-C shear fabric in the central parts (Fig. 2.5a), or ultramylonite-mylonite alternating bands in a strongly-foliated mylonite, showing typical book-shelf gliding in opposite direction in a ductile sinistral shear zone (Fig. 2.5b). In addition, both δ- and σ-shaped megacrysts are also observed with ductile sinistral shear sense in fine-grained mylonite (Fig. 2.5c and d).

2.3.3.3 Deep Structure Beneath Bundelkhand Craton

Gokarn et al. (2013) carried out magnetotelluric profiling across the Bundelkhand Craton for its sub-surface configuration and observed a high resistivity sub-structure, like other Archean-Proterozoic regions (Fig. 2.6). Northern parts of this craton are marked by high resistivity layer to a depth of about 60 km and three-layered resistivity structure overlying a conductive bottom in its southern part; junction between two blocks is delineated by an E–W trending ultramafic and mafic belt of the Bundelkhand Tectonic Zone (BTZ) or the Babina shear zone. Bouguer gravity and seismic soundings studies in combination with geoelectric structure reveal that the Vindhyan

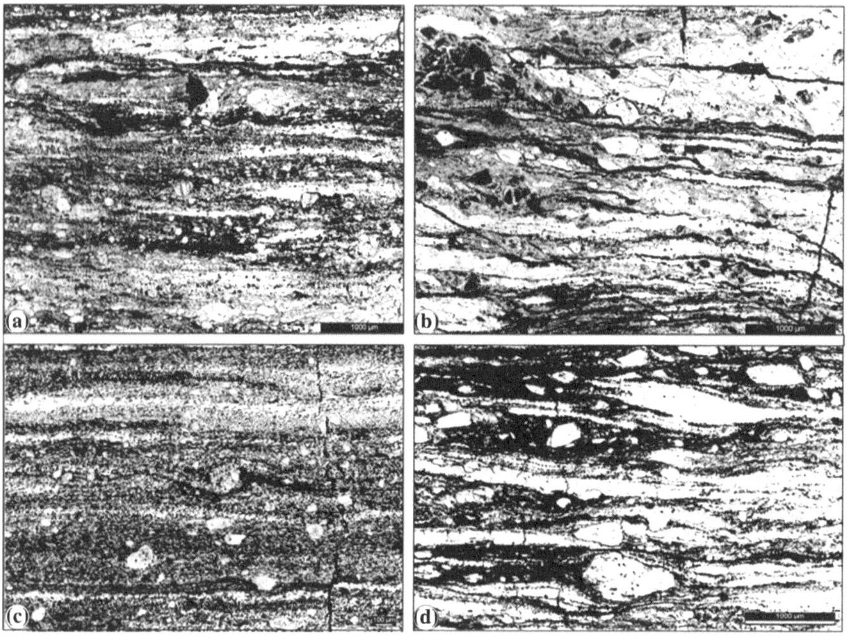

Fig. 2.5 Granitoid mylonites from Bundelkhand Tectonic Zone (BTZ). All sections are XZ thin sections. **a** Typical mylonite with alternating ultramylonite (dark) and mylonite foliated layers, having relict feldspar megacrysts. Ductile sinistral shear sense inferred from asymmetrical feldspars and S-C shear fabric in the central parts. Loc. Kabrai, Mahoba, U.P. **b** Ultramylonite-mylonite alternating bands in a strongly foliated mylonite, showing typical book-shelf gliding in opposite direction in a ductile sinistral shear zone. Loc. Kabrai, Mahoba, U.P. **c** δ-shaped megacryst showing ductile sinistral shear sense in fine-grained mylonite. Loc. Babina, Jhansi, U.P. **d** σ-shape feldspar megacryst in mylonite. Loc. SE of Mauranipur, U.P. *Courtesy* Photomicrographs by J. K. Pati, University of Allahabad, Allahabad, India

Supergroup limit the resistive Bundelkhand craton in the south and does not extend beneath it.

Geologically, we are of the opinion that the BTZ is a major structure of the Bundelkhand Craton where northward-dipping large-scale thrusts dominate and subsequently reactivated as the sinistral strike-slip ductile shear zone. Northern and southern BC blocks are marked by the BTZ as a suture zone, which is characterized by the oldest ~3.5 Ga crustal rocks and mafic-ultramafic rocks.

2.3.4 *Dhala Impact Crater*

Deeply-eroded Dhala Crator is a near-circular structure of about 11 km diameter in western parts of the Bundelkhand Craton, and has been interpreted either as a cryptovolcanic explosion structure (cauldron structure) (Jain et al. 2001; Markandeyulu

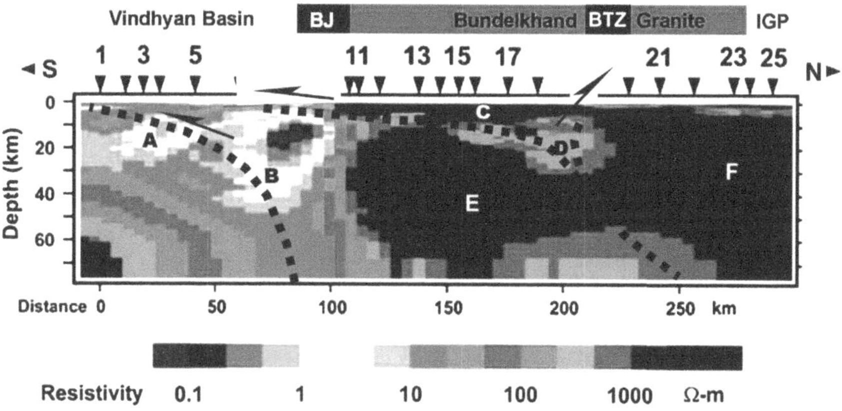

Fig. 2.6 Magnetotelluric cross-section of the Bundelkhand Craton, showing low resistive zones A, B and D beneath the Vindhyan Basin and BTZ, while craton exhibits high resistive zones at E and F (Gokarn et al. 2013). Possible tectonic boundaries and thrusts are shown by thick dashed lines. Section runs NW–SE through Mau-Tikamgarh

et al. 2013) or world's seventh oldest meteoritic impact structure (Pati 2005; Pati et al. 2008, 2017). The crater has a distinct (i) central elevated area (CEA) (Fig. 2.7a and b) of the Paleoproterozoic Vindhyan Supergroup, (ii) pre-Vindhyan flat-lying alternating argillaceous-arenaceous sediments and lateritized conglomerate, followed by (iii) rings of areas of impact melt veins and large breccia on the impacted Archean granitoid (Pati 2005; Pati et al. 2008).

Pseudotachylitic breccia (PTB) veins from the Dhala structure (Fig. 2.8a), also

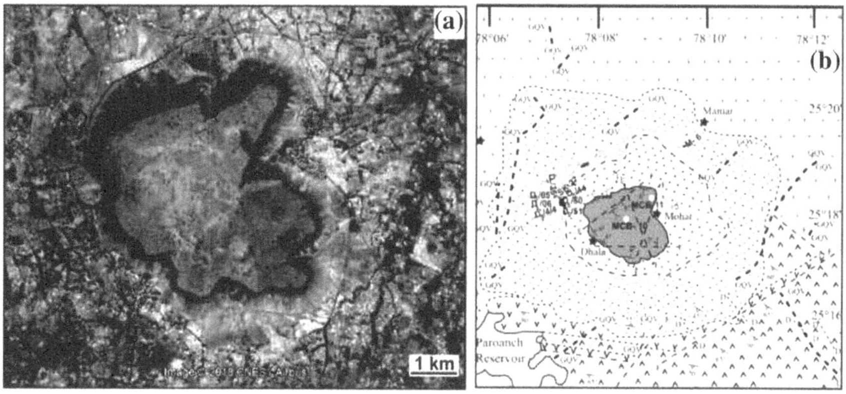

Fig. 2.7 Dhala Impact Crator on **a** Google Earth showing central elevated area (CEA) with an impacted rim (light coloured). **b** Geological map of the Dhala Impact Crator with uplifted region (grey shades) in the middle, followed by rimmed areas of siltsone (dots), monomict granite breccia (grey patches in dot-plus region), and undeformed granite (both fine- and coarse-grained). GQV-Giant quartz veins. For more details, please see Fig. 2 of Pati et al. (2010)

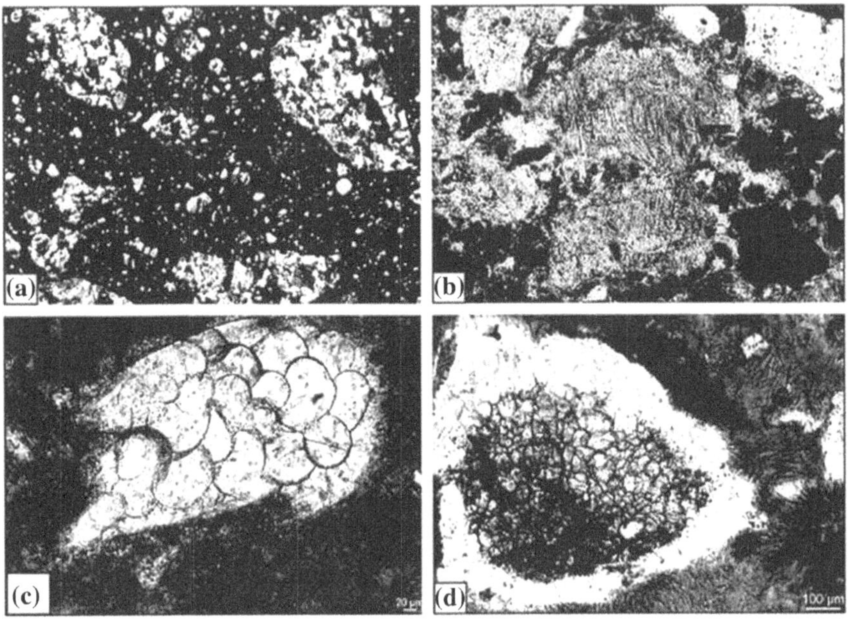

Fig. 2.8 Textures of impacted granite from Dhala Crator. **a** Impact melt breccia having angular to subrounded lithic clasts of granitic composition embedded in dark gray aphanitic matrix (Pati et al. 2015). **b** Shocked quartz (toasted) grains with multiple sets of PDF and clouded feldspar grains (dark colour) in impact melt breccia. **c** Melted quartz clast having characteristic ballen texture and a few some clear domains. **d** Quartz clast with ballen-texture having an inner toasted and an outer clear domain, set in aphanitic matrix with spherulites. Photomicrographs **b–d** are from Pati et al. (2008)

known from other large impact structures, occur as veins, pods and anastomozing veinlets (Pati et al. 2015). Multiple cross-cutting sharp PTB veins rip off clastic debris from the walls. These veins have contrasting major element chemistry and identical REE pattern like host granitoids and were locally generated due to frictional melting, or as shock veins.

Presence of diagnostic features of shock metamorphism in many lithic and mineral clasts within the melt breccia, such as multiple sets of planar deformation features (PDFs) in quartz and feldspar (Fig. 2.8b), melt breccia, ballen-textured quartz (Fig. 2.8c and d), checkerboard feldspar, coesite, granular zircon, and shock-melted lithic clasts confirm the Dhala structure as of impact origin (Pati et al. 2008, 2019).

Li et al. (2018a) identified impactites in the Dhala structure as brecciated granite, melt breccia and dacite to rhyolite (melt-rich), with reidite (high pressure zircon polymorph), and pervasive K_2O metasomatism at high temperatures. U–Pb zircon ages reveal magmatic basement at ~2.50 to 2.47 Ga and Pb loss characters due to impact between ca. 2.44 and 2.24 Ga, and a thermal event after the impact between

1826 ± 97 and 1767 ± 100 Ma. Though previous studies have revealed crust formation of the Bundelkhand craton ~2.50 Ga, Pati et al. (2018) questioned chronological evidences regarding the above timing of the Dhala impact at ~2.44 Ga.

2.3.5 *Crustal Evolution*

U–Pb zircon dating from the Bundelkhand Craton indicate the presence of the ~3.55 Ga TTG (Kaur et al. 2016), though presence of zircon xenocrysts of 3.59 Ga and zircon Hf model age (two-stage; Wei et al. 2017) up to 3.95 Ga suggest an older crust formation history going back to the Eoarchean (Saha et al. 2011; Joshi et al. 2017; Kaur et al. 2016). Occurrence of both low-, high-HREE and near chondritic zircon εHf(T) value in the Paleoarchean TTGs indicate variable depth of melting of an oceanic crust. The TTG originated from a depleted-mantle source at ~3.95 Ga; younger (3.45–3.2 Ga) TTGs are explained as the product of melting of the enriched mantle at variable depths followed by minor assimilation of older felsic crustal material (Kaur et al. 2016). A mantle plume-dominated tectonic setting is suggested to elucidate the source of heat for melting. However, the presence of enriched (transitional) TTGs (Joshi et al. 2017) suggests a reworking of older crust also.

In the Bundelkhand Craton the Neoarchaean is marked by appearance of a variety of mantle- and crust-derived granitoids including TTGs, sanukitoids, granites, anatectic K-rich leucogranites and A-type granites (Joshi et al. 2017; Kaur et al. 2016). The chemical nature of granitoids suggests widespread crust-mantle interactions, pervasive fluid flow and reworking of older continental crust (Joshi et al. 2017) during Neoarchean. The Neoarchaean greenstone belts also host igneous rocks displaying geochemical signatures of diverse tectonic settings like plume, arc and mid-oceanic ridge (Malviya et al. 2006; Ramiz et al. 2018; Raza and Mondal 2019). The lithological association and geochemistry indicate operation of Neoarchaean subduction at the margin of a Paleoarchean to Mesoarchean TTG continent. The 2.51 Ga K-rich anatectic leucogranites mark the reworking and cratonization of Archean crust.

As there is ample evidence for crust formational events on the Bundelkhand craton around 2.49–2.55 Ga (Mondal et al. 2002; Pati et al. 2010; Kumar et al. 2011; Kaur et al. 2016), it stands to reason that all chronological evidences, generated by Li et al. (2018a and Chauhan et al. 2018), refer to granitoid formation and subsequent secondary overprint—and not to the timing of the Dhala impact.

It is unclear if the Aravalli and Bundelkhand Cratons shared a complete common history. The ages of gneissic (TTG) rocks and late Archaean granitoids in both the BGC and Bundelkhand cratons span the same broad age range of 3.3–2.5 Ga. Perhaps the most important link between the two cratonic nuclei is that both are overlain by the c. 1.85 Ga Hindoli Group (Mondal 2003). On that basis, we posit that the BGC–Bundelkhand sectors were a single block by at least c. 1.9 Ga and perhaps earlier (Meert and Pandit 2015).

2.4 Meghalaya Craton

Though unrecognized as a distinct craton so far, possibly due to its remoteness, Sharma (2009) named the Precambrian succession of the Shillong-Mikir Hills (Meghalaya State) as a distinct Meghalaya Craton (MC). Covering an exposed area of approximately 33,000 km^2, the ENE-trending and rectangular-shaped plateau is extensively covered by the Holocene alluvium of the Brahmaputra River in the north and east-northeast, which separates it from the Himalayan Orogenic Belt (Fig. 2.9). The southern margin is partly covered by the Cretaceous Sylhet Trap (Evans 1964) (133–100 Ma), Cenozoic petroliferous sedimentary succession of the Arakan-Yoma Belt, and enormously thick sediments of the Bangladesh Basin.

2.4.1 Limits of the Craton

Southern margin of the craton is demarcated by E–W trending Dauki fault system, which extends further NE as the Haflong Fault and merges into the Naga Thrust system of the Naga-Patkai Ranges (Fig. 2.10). On the western margin, the Jamuna Fault controls flow of the Brahmaputra River where this river swings southwards. NW–SE trending Kopili Fault separates the Shillong Plateau from the Mikir Hills in the east. Northern margin of the craton is dissected by the E–W trending Brahmaputra

Fig. 2.9 Meghalaya Craton as the MODIS image by NASA's Terra satellite. West-flowing Brahmaputra River swings around the craton to flow southwards into the Bay of Bengal, while it's Holocene alluviam covers it from three sides. Cenozoic sedimentary sequence of Bangladesh and Tripura are tightly folded into N–S trending folds in south and separated by the E–W trending Dauki fault along southern margin of the craton which is, itself, dissected by deep-seated faults and lineaments. *NASA image courtesy* Jeff Schmaltz

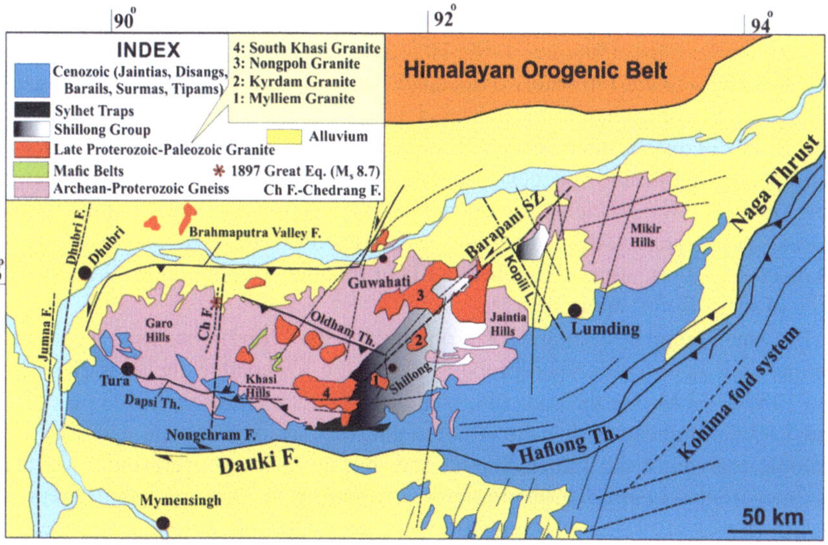

Fig. 2.10 Regional geological and tectonic map of the Meghalaya Craton. Tectonic boundaries: Dauki Fault and Haflong Thrust in south; Oldham Thrust and Brahmaputra Fault in north. Craton is sandwiched between the E–W trending and south-moving Himalayan Orogenic Belt and north-moving Haflong-Naga Thrust systems. Also shown is location of the 1897 great earthquake (M_S 8.7). Redrawn after Mazumder (1976), Ghosh et al. (1991), and other published sources

Fault System (Nandy 1981) or the Oldham Fault (Bilham and England 2001). The Shillong Plateau is also traversed by numerous fractures and fault systems such as the Nongchram Fault in the west, the Barapani Shear Zone in the center, and the Um-Ngot Lineament in the eastern parts (Kayal 1998).

2.4.2 Geological Framework

Geological investigations by Oldham (1856), Medlicott (1869), Desikachar (1974), Mazumder (1976), Dasgupta and Biswas (2000) and Nandy (2001) have established the following main units in the Meghalaya Craton: (i) Archean Gneissic Complex, (ii) Proterozoic Shillong Group, (iii) Mafic belts, (iv) late Neoproterozoic-Paleozoic granitoids, (v) Carbonatite complexes, and (vi) Sylhet Traps.

2.4.2.1 Archean Gneissic Complex

The Archean Gneissic Complex, also called as the Older Metamorphic Group (Mazumder 1976, 1986), consists of granite gneiss, augen gneisses and upper amphibolite to granulite facies metamorphics. Cordierite–sillimanite ± corumdum gneiss and schist, quartz–feldspar gneiss (orthogneiss), amphibolite, granulite and biotite ± hornblende schists (Rahman 1999; Hussain et al. 2019) are dominant lithologies of the craton and are intimately associated with granite gneiss in Khasi Hills. It is migmatitic biotite gneiss, with subordinate augen gneiss and banded gneiss. It consists mainly of quartz, microperthite, sodic plagioclase and biotite with accessory zircon, sphene, apatite and opaques. Gneisses also contain enclaves of Older Metamorphic Group (OMG) (ancient supracrustal rocks), including cordierite–biotite ± sillimanite gneiss, quartz–sillimanite schist, Mg–Al pelites, etc., and minor bands of BIF (Banded Iron Formation), though the TTG (Tonalite–Trondhjemite–Granodiorite) suite of rocks is not reported so far.

Bidyananda and Deomurari (2007) obtained $^{207}Pb/^{206}Pb$ zircon Neoarchean–Paleoproterozoic ages by ion microprobe from quartzo-feldspathic gneiss as 2637 ± 55 and 2230 ± 13 Ma for the core-rim, while other zircon grains ranged between 1.98 and 1.67 Ga, indicating that the cratonic gneisses are Archaean in age, like other cratonic regions of the Indian Shield, and that the thermal overprinting on these basement rocks occurred during Proterozoic event(s). ~1.7 Ga age was also reported by Ghosh et al. (1994) by Rb–Sr method for the granitoid emplacement in the craton. One gneissic sample from Rhesu area of Garo Hills yielded $^{207}Pb/^{206}Pb$ zircon ages between 1.84 and 1.54 Ga and are within the age range of the Umling gneiss.

Mesoproterozoic monazite age of 1596 ± 15 Ma with matrix monazite rims having younger ages of 1032–1273 Ma were recorded from granulite facies metapelites of Garo-Goalpara Hills (Chatterjee et al. 2007). These rocks have counterclockwise pressure–temperature path and near-peak conditions of 7–8 kbar and 850 °C. In addition, P–T conditions for Sonapahar granulite facies corundum–spinel–sapphirine metapelites, occurring as enclaves within the granite–granodiorite suite, were somewhat at lower 5 kbar and 750 °C (Lal et al. 1978). In this region, homogenous monazite grains in granulite facies metapelites yielded tightly clustered date at 500 ± 14 Ma, which is much younger than the Garo region, and nearly coincides with ~480 Ma Rb–Sr dates of the porphyritic granite that intruded the Meghalaya gneiss complex (Chatterjee et al. 2007). A charnockite from Nongstoin–Riangdo road gave zircon ages between 1.28 and 1.08 Ga, like the Neoproterozoic imprints from the Eastern Ghats (Bidyananda and Deomurari 2007).

2.4.2.2 Proterozoic Shillong Group

The Archean Gneissic Complex is unconformably overlain by NE–SW trending Proterozoic Shillong Group, containing siliciclastic association of mica schist, phyllite, quartzite and slate. Out of four phases of folding, earliest very tight to isoclinal F1 folds on bedding plane So develop penetrative axial plane foliation S1 (Mitra 2005). These F1 folds are coaxially deformed by open to tight upright F2 folds, having NNE-striking crenulation foliation S2 as the axial planes. In more foliated rocks, NE-trending 3rd generation recumbent folds F3 develop on both S1 and S2 foliations, followed by last-stage F4 upright conjugate folds and kink bands. The maximum time limit of the Shillong Group is given by Rb–Sr whole-rock isochron age of 1150 ± 26 Ma granite gneiss that occurs at the base of the Shillong Group (see Ghosh et al. 1991).

The overgrowth on detrital zircons from the cover rocks of the Shillong Group gave 1.5–1.7 Ga which is considered to be the age of metamorphism related to the Mesoproterozoic orogeny that affected both basement and cover rocks of the Indian shield (cf. Ghosh et al. 1994). Maximum depositional age of this group is indicated by the youngest zircon of 1.48 Ga from a suite of very few grains ranging in age from 1.98 and 1.48 Ga (Bidyananda and Deomurari 2007), which may not be enough for determining the age of sedimentation. In contrast, Yin et al. (2010) noted Pb–Pb zircons ages from 3.3 to 1.1 Ga, with two sub-groups of 1.1–1.25 and 1.5–1.75 Ga, and worked out deposition post-dating 1.1 Ga in one sample, while the other sample gave much younger age of deposition around 0.56 Ga.

2.4.2.3 Mafic Belts

The Proterozoic mafic rocks of the Meghalaya Craton, also known as the Khasi Greenstone (Mazumder 1986), are best exposed at Laitlyngkot in East Khasi Hills as lensoidal intrusives within the low-grade Shillong Group. These hornblende gabbros occur both as massive and foliated varieties, whose geochemical and REE characters reveal tholeiitic lineage within a subduction-related arc signatures (Ray et al. 2013). These indicate the generation of the parent melt by slab dehydration and wedge melting processes. Tectonically, their generation and emplacement took place in a continental margin arc setting, involving subduction of oceanic plate beneath continental lithosphere, slab dehydration and mantle wedge melting associated with assimilation of crustal components and fractional crystallization.

2.4.2.4 Late Neoproterozoic–Paleozoic Granitoids

Numerous granitic plutons of different sizes intrude the Archean Gneissic Complex and the overlying Shillong Group; the largest being the South Khasi granitoid body (Medlicott 1869; Mazumder 1976; Chimote et al. 1988; Ghosh et al. 1991). Typical characters of these plutons are exemplified by the Kyrdem Granite—a grey and pink, very coarse grained, porphyritic granitoid with abundant K-feldspar megacryst and minor amphibole, quartz, biotite and plagioclase. Fine grained granite, aplite and pegmatite veins constitute the late felsic pulses. These granitoids are conformably aligned E–W to ENE–WSW with major lineaments of the craton (GSI 1993). Composition of these bodies vary from quartz monzonite, monzogranite and granodiorite. Mazumder (1976, 1986) opined that felsic Mylliem, Nongpoh, Kyrdem and South Khasi plutons as late- to post-tectonic, fracture-controlled diapirs due to episodic thermal events because of mantle upwelling. Rahman (1999) reported a narrow ~400 m wide thermal aureole, suggesting shallow depth of the granite intrusion.

The Meghalaya granite bodies were earlier dated by Rb–Sr whole-rock isochron method, which gave an age of 607 ± 13.14 Ma for the Mylliem granite—an intrusive into the Shillong Group (Crawford 1969; Chimote et al. 1988). The pink granite from Songsak of East Garo hills, known for uraninite occurrence, has been dated at 500 ± 40 Ma (Kumar et al. 2017a). Van Breemen et al. (1989) reported Rb–Sr Proterozoic-Early Paleozoic plutonic granitic activity in the region and inferred possible petrogenetic processes that led to their evolution in space and lime. Ghosh et al. (1991, 1994, 2005) provided Rb–Sr whole-rock isotopic ages for these plutons (Kyrdem 479 ± 26 Ma; Nongpoh 550 ± 15 Ma; Mylliem 607 ± 13 Ma; South Khasi 690 ± 19 Ma), and suggested a protracted thermal event of ca 200 Ma duration (Proterozoic-Early Palaeozoic; ca 700–500 Ma) in Meghalaya, and was probably related to Pan African-Caledonian orogeny. These felsic plutons appear younging in age from southwest to northeast and contain abundant microgranular enclaves (Kumar and Rino 2006).

U–Pb SHRIMP zircon geochronology and geochemistry of the Rongjeng, Guwahati and Longavalli granite gneisses, and Sonsak, Kaziranga, South Khasi, Kyrdem and Nongpoh granitoids provided two distinct age groups: (i) 1778 ± 37 Ma and (ii) 535 ± 11 to 516 ± 9.0 Ma, with inherited zircon cores of 2566.4 ± 26.9, 1630.8 ± 16 and 1430.4 ± 9.6 Ma that were incorporated either from older continental crust or magmatic bodies (Kumar et al. 2017a). Medium to coarse grained, equigranular to porphyritic Cambrian granite plutons intruded the basement granite gneisses and Shillong Group. These plutons have peraluminous to metaluminous compositions, syn- to post-collisional affinities, and record crustal growth of the Columbia and Gondwana supercontinents in the Meghalaya craton. Further, within the South Khasi granitoid (519.5 ± 9.7 Ma) occur rounded to elongate, fine- to medium-grained, mafic to porphyritic microgranular enclaves of almost the same age, i.e. 515 ± 13 Ma (Kumar et al. 2017b). Identical trace element patterns of these enclaves and the granitoids reveal chemical re-equilibration through diffusion during synchronous mixing–fractionation and mingling of co-existed Cambrian mafic and felsic magmas. Chemical (U–Th–Pb) monazite age of 501 ± 5 Ma mafic magmatic enclaves ascertained the age of the magma hybridization event in the Nongpoh granitoids due to igneous activity at 500 Ma during amalgamation of Eastern Gondwana (Sadiq et al. 2017).

Further east in the Mikir (Karbi) Hills, late phase A-type bimodal granitoids profusely intrude the Shillong Group along with dolerite and amphibolites (metabasalts) of 515.1 ± 3.3 Ma age (Mazumder and Dutta 2016). These massive and non-porphyritic granitoids, exhibit flow banding, degassing pots and amphibolite breccias with chilled margins, and possess metaluminous to peraluminous, shoshonitic to high potassium, and calc-alkaline composition. Binary plots of Y + Nb–Rb and Y–Nb confirm their within-plate granite (WPG) character, with enriched total REE, Y, and moderately negative Eu anomalies. Zircons have an initial $^{176}Hf/^{177}Hf$ ratio from 0.2819–0.2821 and εHf(t) values between −12.20 and −18.90, indicating melting of a 2.25–2.67 Ga magmatic crust during post-collisional late Pan-African Gondwana continental restructuring event.

In western parts of the Meghalaya Craton, the West Garo pluton is A-Type post-tectonic biotite-monzogranite and biotite-syenogranite with low Ca, alkaline and peraluminous character (Choudhury et al. 2012), with Rb–Sr age of 616 ± 86 Ma corresponding to widespread Middle to Upper Pan African activity. Table 2.3 summarizes the petrographic and geochemical characters of some of the Pan-African granitoids of the Meghalaya Craton.

2.4.2.5 Carbonatite Complexes

The Meghalaya Craton witnessed Early Cretaceous (105–107 Ma) igneous activity along N–S and E–W trending fractures, which controlled the emplacement of ultramafic-alkaline-carbonatite complexes (UACC) (Chattopadhyay and Hashimi 1984; Mazumder 1986; Gupta and Sen 1988; Dasgupta and Biswas 2000; Srivastava and Sinha 2004; Sadiq et al. 2014). Among the four such bodies, the Sung Valley carbonatite was emplaced into the Proterozoic Shillong Group and consists of ultramafics (serpentinized peridotite, pyroxenite, and melilitolite), alkaline rocks (ijolite and nepheline syenite), and carbonatites. The latter are the youngest intrusive phase in the complex, and occur as oval-shaped bodies, small dykes and veins. These are classified as calcite carbonatite with minor apatite, phlogopite, pyrochlore and ilmenite. Carbonatites are, Petrological, geochemical isotopic data suggest that these have been derived from a primary carbonate magma generated by the low-degree melting of a metasomatized mantle peridotite (Srivastava and Sinha 2004). Presence of REE carbonates, phosphates and associated REE-Nb bearing pyrochlore enhances the economic potential of the Sung Valley carbonatites (Sadiq et al. 2014).

Srivastava et al. (2019) used in situ U–Pb ages and Sr–Nd–Hf isotopic data on mineral phases of the Sung Valley and Jasra UACC intrusions from the Meghalaya Craton to determine petrogenesis of the Cretaceous volcanism where perovskite of dunite, ijolite and uncompahgrite yielded U–Pb ages of 109.1 ± 1.6, 104.0 ± 1.3 and 101.7 ± 3.6 Ma, respectively. The Sung Valley nepheline syenite and perovskite in the Jasra clinopyroxenite have U–Pb zircon ages of 106.8 ± 1.5 and 101.6 ± 1.2 Ma; the latter is different from the U–Pb age of 106.8 ± 0.8 Ma on zircon of Jasra syenites. Bulk-rock isotopic compositions of these rocks are initial $^{87}Sr/^{86}Sr =$ 0.70472 to 0.71080, $\varepsilon_{Ndi} = -10.85$ to $+0.86$ and $\varepsilon_{Hfi} = -7.43$ to $+1.52$ and appear to be derived from the lithospheric mantle.

Gupta and Sen (1988) favored strong tectonic control on emplacement of the UACC and postulated a genetic relation to the Ninety-East Ridge in the Indian Ocean, and association with the Kerguelen plume igneous activity (Srivastava and Sinha 2004).

2.4.2.6 Sylhet Trap

Outcrops of Cretaceous flood basalt volcanics belonging to the Sylhet Traps are restricted within a narrow east–west tract, 60–80 km long and 4 km wide along the southern edge of the Meghalaya Craton. These lavas occur above the eroded Precambrian basement and are overlain by the Cretaceous-Eocene sediments. The Rajmahal Traps, basaltic rocks in subsurface Bengal Basin and the Sylhet Traps

(RBST) bear strong geochemical similarity and belong to a Large Igneous Province (LIP), which is related to the Kerguelen hotspot activity during the Early Cretaceous (Bakshi 1995; Frey et al. 2000; Kent et al. 2002) and affected the Archean East Indian cratonic margin. Over one million sq. km area around the Bengal Basin is covered by the RBST as alkalic–ultrabasic intrusives in the west and Sikkim in the north, and Sylhet basalts and alkalic–carbonatitic–ultramafic complexes in the Meghalaya Craton farther east of the Rajmahal–Bengal Traps.

The oldest volcanism attributed to the Kerguelen mantle plume are the 132 and 123 Ma old Bunbury Basalts (Coffin et al. 2002), the ~118 Ma old Rajmahal-Bengal Traps (Kent et al. 2002) and 115.8 ± 3.2 Ma for the Sylhet Traps by $^{40}Ar/^{39}Ar$ (Ray et al. 2011). Together with the Ninety-East Ridge, these have been associated with break-up of the Eastern Gondwanaland and opening of the Indian Ocean (Fig. 2.11).

New geochemical and isotopic data from these bodies show similarity with the previously-published data of these bodies and indicate a relatively primitive Kerguelen Plume (KP) source (Ghatak and Basu 2013). Lack of an asthenospheric MORB component in the RBST province is indicated by various trace element and Nd–Sr isotopic ratios. The KP, thus, reveals heterogeneities in lower mantle-derived plume; the carbonated components yielding ultrabasic melts at greater depths and subsequent rise of plume at shallower depths with tholeiitic flood basalt volcanism. Thus, the KP head influenced ~1000 km around the Bengal Basin, as represented by the widely scattered and diverse rock types of the RBST (Ghatak and Basu 2013).

2.4.3 *Tectonics*

2.4.3.1 Tectonics of the Meghalaya Craton

Seismically active Shillong Plateau (Meghalaya Craton) witnessed the 1897 great earthquake (M_S 8.7), sourced along its northwestern edge, and was considered to have occurred due to movements of a north-dipping fault along its southern margin (Oldham 1899). However, detailed analysis of geodetic measurements led Bilham and England (2001) to postulate a ‘pop-up’ of the Shillong Plateau by 11 m by this earthquake between steep southerly-dipping Oldham Fault and the northerly-dipping Dauki-Dapsi Thrust (Fig. 2.12). This fault underwent 16 m of reverse slip. On the basis of gravity modeling and broadband seismological data, Nayak et al. (2008) examined crustal structure of the Shillong Plateau and suggested that depth of the Moho is shallower (~35 km) beneath the plateau in contrast to its depth of ~42 km beneath the Bengal basin and the Assam valley. This thinning was due to consequences of the ‘pop-up’ mechanism during the great earthquake 1897 due to movements along steeply-dipping Oldham Fault and Dauki-Dapsi Thrust, which have re-activated and exhumed the pop-up structure rapidly since 15–9 Ma (Biswas et al. 2007; Clark and Bilham 2008; Adlakha et al. 2011) and even captured the course of Brahmaputra River. Due to the uplift of the Shillong Plateau, the Brahmaputra reached Dungsam Chu (DC) section in the foothills of eastern Bhutan between 4.9 and

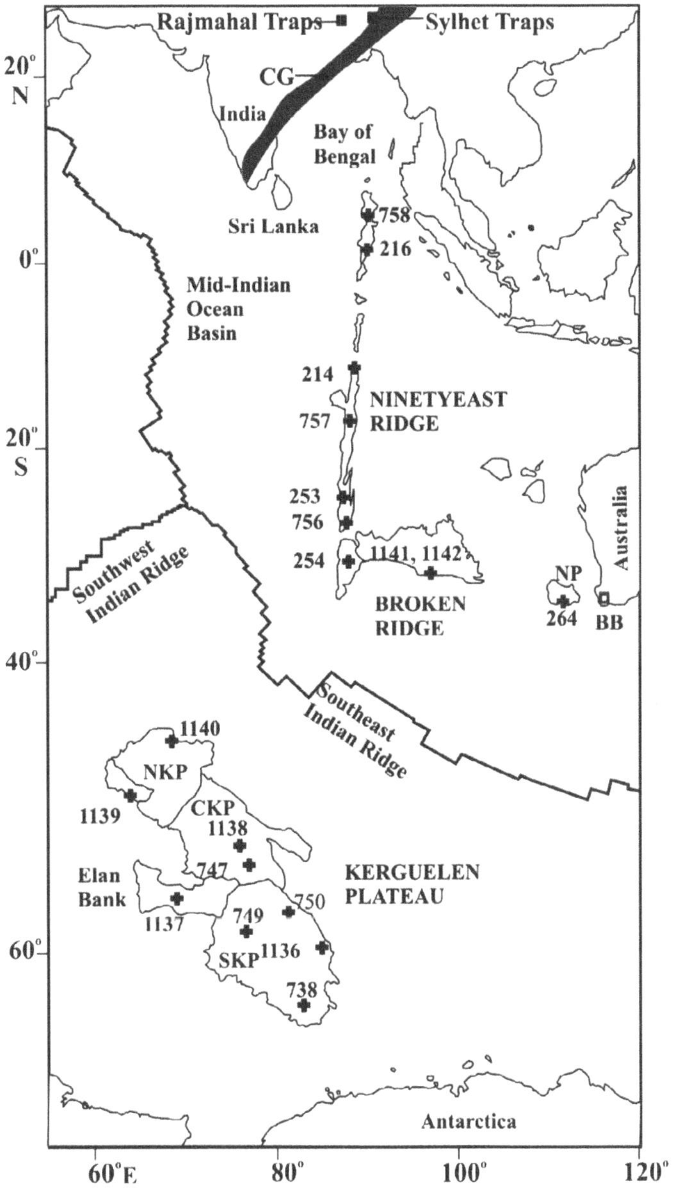

Fig. 2.11 Basalt provinces of the Indian Ocean and surrounding continents due to the Kerguelen plume (Frey et al. 2002). These include the Kerguelen Plateau (NKP—North Kerguelen Plateau; CKP—Central Kerguelen Plateau; SKP—South Kerguelen Plateau), Broken Ridge, Ninetyeast Ridge, Bunbury basalts (BB), Rajmahal and Sylhet Traps. Black crosses: ODP sites. After Ghatak and Basu (2013)

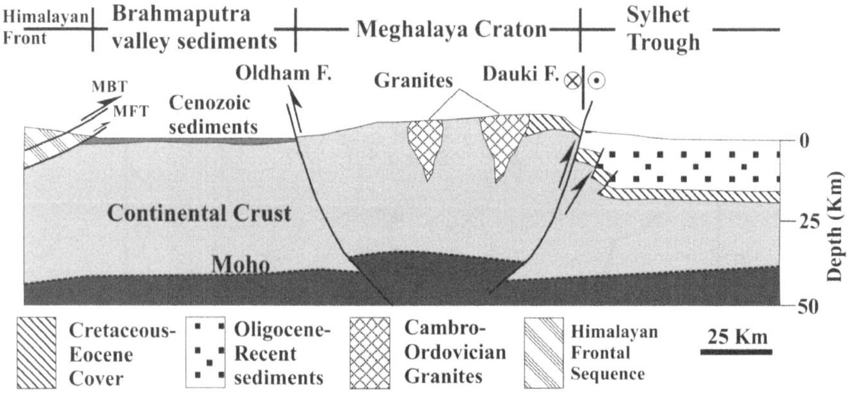

Fig. 2.12 'Pop-up' structure of the Shillong plateau, showing opposite dipping Dauki–Dapsi Fault and Oldham Fault along with distribution of earthquakes. Brahmaputra Fault is located vide Nandy (2001). After Nayak et al. (2008), based on Bilham and England (2001)

5.2 Ma, deflecting the river to the northward and westward (Govin et al. 2018). This scenario is in contrast to the magnetotelluric imaging where the Meghalaya Craton is considered to have thrust over the Sylhet basin sediments and its underlying resistant mafic crust along N-dipping tectonic boundary in the south—the Dauki Thrust/Fault; craton might have been emplaced about 50 km length of the oceanic crust (Gokarn et al. 2008). Northerly and gentle-dipping resistivity zones indicating the presence of Oldham and Brahmaputra Faults are delineated along northern margin of the craton.

2.4.3.2 Dauki Fault Zone (DFZ)

E–W trending Dauki Fault Zone (DFZ) is located along southern margin of the Meghalaya Craton, and separates the plateau of about 1.2–1.5 km elevation from the sea-level exposed sedimentary sequences of the Sylhet/Bangladesh Plains. It merges with the NE-trending Haflong thrust belt further northeast. Very sharp linear E–W trending features demarcate the two domains with a highly-dissected plateau. The fault zone also controls the sedimentation within Bangladesh Plains since many kilometers thick sequences are deposited in this basin, whose northern margin was controlled by continuous subsidence, besides controlling the emplacement of the Sylhet Traps (Das et al. 1995; Dasgupta and Biswas 2000; Hossain et al. 2018).

That the Dauki FZ has a dextral strike-slip component also is indicated by eastward rotation of the N–S trending fold hinges of the Bangladesh and Tripura, as is observed in the field and the Google Earth imageries (Das et al. 1995).

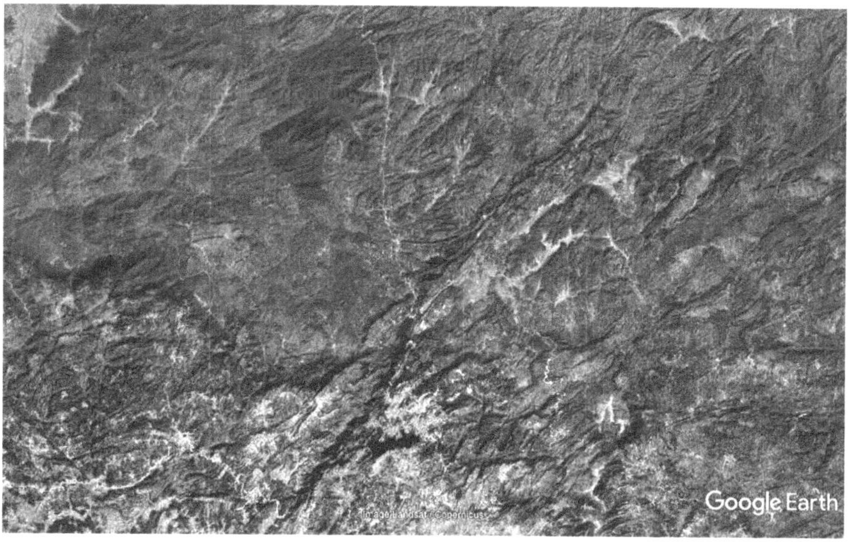

Fig. 2.13 Barapani Shear Zone (BSZ) on Google imagery showing deeply dissected terrain in the center along with curved ridges in upper central parts

2.4.3.3 Barapani Shear Zone (BSZ)

NE–SW trending Barapani Shear Zone (BSZ) is developed within the Shillong Group metamorphics and characterized by distinct curviplanar landform features indicating its sinistral strike-slip character (Das et al. 1995). Nearly subvertical and highly fractured phyllites show polyphase deformation and shearing, exhibiting narrow long ridges north of Barapani reservoir. These ridges meet the shear zone tangentially indicating sinistral strike-slip movements along the shear zone (Fig. 2.13).

2.4.3.4 Urn Ngot Fault (UGF)

N–S trending slip linear dextral strike-slip Urn Ngot fault controls the flow of river for more than 30 km before taking a westward turn and entering Bangladesh. The fault affects the Precambrian basement and appears to be pre-Eocene in age. Krishnamurthy (1985) and Gupta and Sen (1988) opined that this deep-seated fault controls the emplacement of alkaline-ultramafic-carbonatite complex, which is genetically related to the 90° E Ridge. The UGF intersects the Barapani Shear Zone in the north.

2.4.3.5 Other Fault and Lineaments Systems

In addition to the above, the Meghalaya Craton is marked by numerous N–S oriented fractures (faults/lineaments); many of these trend almost throughout the plateau and dissect it into different N–S narrow blocks moving northwards. Some of the most important zones are as follows (Das et al. 1995).

(i) **Jamuna Fault**: Almost N–S trending fault along the western margin of the Shillong plateau controls straight course of the Brahmaputra River (known as the Jumna River) for nearly 200 km, which changes its E–W course on its northern margin. The fault extends northwards into the Bhutan Himalaya (Nandy 1980).

(ii) **Dudhnai Fault**: N–S trending dextral fault controls river course for ~40 km and displaces the Dauki fault dextrally in the south. In the north, this fault terminates into several NE–SW trending fracture traces.

(iii) **Kopili Fault**: Separating the Shillong plateau from the Mikir Hills and covered under alluvial deposits, nearly 300-km long NW–SE trending zone controls the course of the Kopili river flows and coincides with the well-known Kopili graben. The fault has been zone of intense seismicity with large earthquakes in 1869 and 1943 (Bhattacharya et al. 2008; Kayal 1998). An "Assam seismic gap" has been identified along this fault, having potential to experience earthquake in future (Khattri and Wyss 1978). Kumar et al. (2016) measured three time intervals for causative seismic events around 250 ± 25 year BP, between 400 and 770 year BP and 900 ± 50 year. BP, besides the above historical earthquakes for better understanding of paleoseismic history of this region.

(iv) **Guwahati Fault**: A prominent NW–SE trending sinistral fault traverses the Shillong plateau around Guwahati and controls course of the Brahmaputra River beyond this city.

2.5 Bastar Craton (BC)

2.5.1 Tectonic Boundaries

The trapezoidal-shaped Bastar Craton (BC) occupies vast area of ~130,000 km^2 and is located to the northeast and west of the Dharwar Craton and the Singhbhum Craton, respectively. Eastern and southeastern boundaries of the craton are marked by the prominent Proterozoic Eastern Ghats Mobile Belt (EGMB), which appears to have tectonic contact with the craton as a thrust (Biswal et al. 2000; Chetty 2017). In the north, ENE-trending Central Indian Tectonic Zone–CITZ (Roy and Hanuma Prasad 2003) or the Central Indian Shear Zone (Jain et al. 1991) tectonizes the Proterozoic Satpura Mobile Belt. Two NW-trending Pranhita–Godavari and Mahanadi rift zones, containing the Gondwana sediments, bound the craton on its southwestern and northeastern margins respectively, while the Deccan Traps cover part of its westernmost parts around Nagpur (Fig. 2.14).

Details of its geology and tectonics were sketchy till recently, possibly due to its remoteness and difficult working conditions.

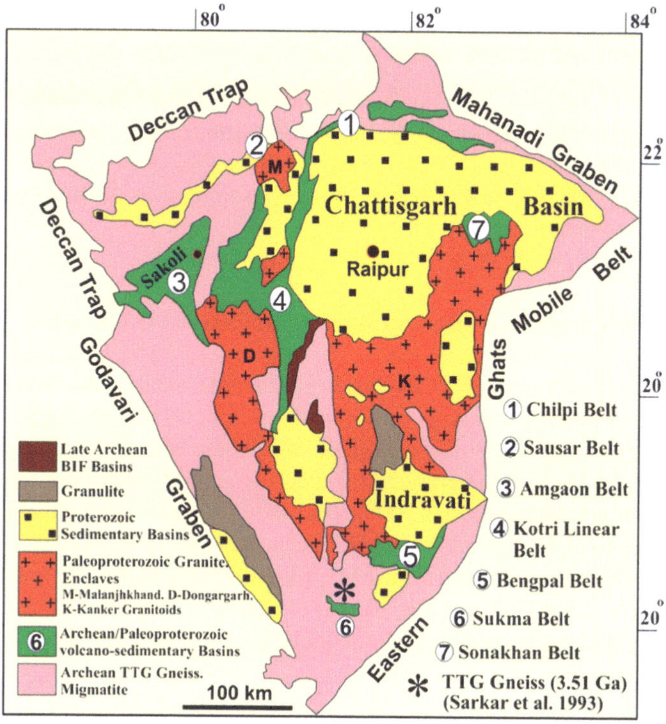

Fig. 2.14 Generalized geological map of the Bastar Craton (also known as the Bastar–Bhandara Craton, showing its important surroundings and distribution of Archean gneissic rocks, Paleoproterozoic volcano-sedimentary belts and Proterozoic sedimentary basins. Redrawn after Ramakrishnan (1990), Ramachandra et al. (2001) and Roy and Hanuma Prasad (2003)

2.5.2 *Major Litho-tectonic Components*

This craton contains dominantly granite gneisses, basic and acidic volcanics, associated metasedimentary rocks; together these are exposed into greenstone belts of different ages. Several generations of granites, gneisses and mafic dyke swarms categorized the craton and can be identified basis of geology, geochemistry and geochronological data. The BC essentially is comprised of the following important litho-tectonic components (Fig. 2.14):

(i) Oldest gneissic complex like Sukma and Amgaon with tonalite–trondhjemite–granodiorite (TTG) suite,
(ii) Proterozoic supracrustal volcano-sedimentary greenstone mobile belts: the Bengpal–Sukma, the Kotri–Dongargarh, the Amgaon, the Sausar–Chilpi and the Sonakhan belts,
(iii) Paleoproterozoic younger granitoid intrusives (Dongarhgarh, Malanjkhand etc.),
(iv) Granulite belts,

(v) Mafic dike swarms, and
(vi) Proterozoic sedimentary basins.

References to the geology of the Bastar Craton are found in the works by V. Ball (1877), W. King (1881–86) and P. N. Bose (1899–1900), H. Crookshand and P. K. Ghosh (1932–38), and Ghosh (1941), besides others. Please see Ramakrishnan and Vaidyanadhan (2008) for further details. Table 2.1 summarizes the litho-tectonics of the Bastar Craton (Mohanty 2015).

2.5.2.1 Oldest Gneissic Complex (Sukma–Amgaon TTG Complex)

The oldest 'Gneissic Complex' of the Bastar craton, referred to as the Sukma Gneiss, is dominated by tonalite–trondhjemite–granodiorite (TTG) suite (Sukma Granite-I), a K-rich granite (Sukma Granite-II), granulite, gabbro and gabbroic anorthosites (Mohanty 2013). These are interpreted to reflect a major interval of crustal accretion (Ramakrishnan and Vaidyanadhan 2008). The Sukma gneisses contain mainly quartz, plagioclase, K-feldspar, hornblende and biotite, and are high-Al, silicic ($SiO_2 = 69$–73 wt%) trondhjemites (Hussain et al. 2003) with moderate Mg# (~30) and moderate to highly fractionated REE patterns (Sarkar et al. 1993; Mondal et al. 2006). They were possibly derived from high-pressure melting of a mafic crustal source without mantle involvement.

Sarkar et al. (1993) reported high-Al_2O_3 trondhjemitic gneisses as enclaves within granites at Markampara with moderate to highly fractionated REE pattern [$(La/Yb)_N = 60$–83], with 3509^{+14}_{-7} Ma as their primary crystallization U–Pb zircon age possibly due to partial melting of an amphibolitic protolith. Tonalite gneiss from this suite in central parts yielded a U–Pb upper intercept age of 3561 ± 11 Ma (Ghosh 2004), which is thought to reflect the oldest age of the gneissic protolith. A whole-rock 3018 ± 61 Ma Pb–Pb age by Sarkar et al. (1990) for gneisses from Sukma area as crystallization age possibly reveal two generations of gneisses at 3.6–3.5 and 3.0 Ga. Further, these TTG's and associate supracrustal enclaves are intruded by true undeformed silica-rich, mildly peraluminous calc-alkaline granite of 3582.6 ± 4 Ma (zircon U–Pb age) at Dalli-Rajhara, which has been linked to the Neoarchean collisional tectonics involving crustal thickening (Rajesh et al. 2009). Thus, the ~3.5 Ga TTG, the calc-alkaline granitoids and the associate enclaves are the oldest elements of the Bastar craton recorded so far.

Engulfed within the vast Archaean TTG sequence are ancient supracrustals of the Sukma Group and Amgaon Group in the south and north, respectively containing quartzite–carbonate–pelite (QCP), BIF and minor mafic-ultramafic rocks. These are exposed as scattered enclaves and narrow belts (Ramakrishnan and Vaidyanadhan 2008). Near Nagaras, it contains quartzite, granulite and amphibolite, and reveals 3 deformation phases on primary stratification S0 as axial plane foliation S1 to the F1 folds, while large-scale geometry is controlled by F3 folds producing Type 1 dome and basin interference patterns (Thakur et al. 2006).

Table 2.1 Archean-Proterozoic sequence of the Bastar Craton. After Mohanty (2015)

Stratigraphic units	Rock types	Age (in Ma)	Depositional/tectonic environment
Chhattisgarh Supergroup	Limestone, shale, sandstone and tuff	1011 ± 19	Peritidal carbonate platform
	Conglomerate, sandstone and shale	c. 1500	Fan delta, shore face shelf
	Unconformity		
	Conglomerate, porcellanite, sandstone, shale, limestone		Braid plain, alluvial fan, shallow-marine shelf
Unconformity			
Abujhmar Group	Fe-tholeiite dykes (BD-2B) Mafic dykes (BD-2A), quartz veins Paliam–Darba granite	1883 ± 2 2100 ± 11 2275 ± 80	Orogenic collapse (extension) Orogenic collapse (extension) Collisional orogen Lithospheric plume Collisional orogen BIF, high sea-level, marine transgression–shelf sedimentation
	High-Mg mafic dykes (BD-1B)		
	Sukma granite IV (Markampara granite) Coarse clastics, volcanics, and BIF	2480 ± 3 2490 ± 48	
Unconformity			
Bailadila Group	Katekalyan migmatite; granulites BIF Chlorite, sericite phyllites Conglomerate, mafic-ultramafic rocks Feldspathic quartzite	2672 ± 54	Collisional orogen BIF High sea-level Marine shelf transgression
Unconformity			
Bengpal Group	Subalkaline mafic dykes (BD-1A) Sukma granite III (ambhibole/bio granite) Andalusite schist, andalusite–grt quartzite Chloritoid–chlorite schist Metabasalts (amygdular, pillow, spinifex) Quartz–sericite schist/quartzite	3018 ± 61	Orogenic collapse (extension) Collisional orogen Stable continental rift Regressive continental deposition High freeboard

(continued)

Table 2.1 (continued)

Stratigraphic units	Rock types	Age (in Ma)	Depositional/tectonic environment
Unconformity			
Sukma Group	Ultramafics, mafics BIF Amphibolite, ultramafics Cordierite–anthophyllite rock Cordierite–andalusite–grt–bio schist/gneiss Calc-silicate rocks and amphibolites Sillimanite quartzite		Marine flooding Transgressive systems tract Low-stands systems tract
Unconformity			
Sukma Gneiss	Sukma granite II (K-rich)	3509 ± 14	
	Sukma granite I (TTG) gneisses and high-grade gneisses	3561 ± 11	

[a]See Table 11.2 for details regarding events, ages and references (Mohanty 2015)

2.5.2.2 Proterozoic Supracrustal Volcano-Sedimentary Greenstone Belts

The Bastar Craton incorporates many low- to high-grade metamorphosed Proterozoic supracrustal greenstone mobile belts such as the Bengpal–Sukma belt in south, Kotri–Dongargarh belt in north and center, the Sonakhan belt in the east, the Amgaon belt in west, and the Chilpi belt in north along with numerous scattered enclaves (Ramakrishnan and Vaidyanadhan 2008).

Large expansive of ancient scattered supracrustal enclaves and narrow belts of quartzite-carbonate-pelite (QCP) with BIF and minor mafic-ultramafic rocks, collectively called Sukma Group in the south and Amgaon Group in the north, occur within the Archean gneiss and granites. The Bengpal and the Sukma Groups are the oldest of these supracrustal rocks (Ramakrishnan 1990).

The Bengpal Group, unconformably overlying the Sukma Group, contains amygdaloidal meta-basalts, metagabbroic dikes/sills with intercalated andalusite-, chloritoid-, and nodular fibrolite schists, interlayered immature arkose, quartz- and lithic wackes, metapelites and BIF. The Sukma Group consists of sillimanite quartzite, cordierite–sillimanite, cordierite anthophyllite rocks, calc-gneisses and widespread banded iron formations (BIF). The sequence is intruded by granites. Unconformably overlain on the Bengpal Group schists are the BIF (Bailaddila Group).

These supracrustals are scattered within the gneisses, with the earliest deformation phase marked by isoclinal folds and later tight to open folds. Low-pressure metamorphism ranges from middle to upper amphibolite facies and occasionally grading into granulite facies. No geochronological data are available for these enclaves but these may be around 3.0 Ga or even older. This volcano-sedimentary belt represents the older greenstone complex of the craton. The basement gneisses and greenstone belts were involved in deformation and granulite metamorphism. Amphibole-bearing granite and biotite granite of 3018 ± 61 Ma age, intruding these gneissic complexes, provide the age limits of a possible orogenic event (Amgaon/Bengpal orogeny) (Mohanty 2015).

Largest Kotri-Dongargarh belt of the Bastar Craton unconformably overlies the Archean gneisses (Sukma Gneiss/Amgaon Gneiss), whose oldest unit, the Nandgaon Group, is associated with bimodal volcanics. The lower felsic unit constitutes the Bijli Formation and the upper mafic unit forms the Pitepani Formation. WR Rb–Sr isochron ages of the Bijli rhyolite are 2180 ± 25 and 2503 ± 35 Ma, while the Dongarhgarh volcanics date 2465 ± 22 and 2270 ± 90 Ma (Sarkar et al. 1981; Krishnamurthy et al. 1988) The Nandgaon Group is intruded by the Dongargarh Granite (2465 ± 22 Ma) in the south and the Malanjkhand Granite (2490 ± 8 Ma) in the north. Intrusion of high-Mg mafic dykes of ~2450 Ma are found in both these granites.

The younger greenstone belts (Bailadila Group, Sonakhan Group, and Abujhmar Group) comprise mafic and felsic volcanic rocks and metasedimentary sequence containing Banded Iron Formations (BIF) and resemble the Archaean greenstone sequences. The craton was involved in another orogenic event marked by granulite facies metamorphism at the end Archaean time (2672 ± 54 Ma), followed by intrusion of alkali feldspar megacrystic granites and pink granites (Sukma Granite, Keskal, Darbha, Sitagaon, Dongargarh and Malanjkhand granitoids) of 2450–2650 Ma age (Mohanty 2015).

The Neoarchean Sonakhan Greenstone Belt (SGB) is dominated by basalt, andesite, dacite, rhyolite and volcano-sedimentary rocks. In lower parts, the Baghmara Formation contains an association of entirely basaltic volcanics in lower unit, while upper unit is comprised of basalt-andesite-dacite-rhyolite (BADR) series (Mondal and Raza 2009). Geochemically, lower tholeiitic basalts are depleted in highly incompatible elements, Nb-maxima (Nb/Nb* = 0.93–1.48, relative to Th and La) and near flat REE patterns; these are distinguished as plume-related oceanic plateau basalt. The upper volcanics from mafic to felsic characters are enriched with highly incompatible elements, large negative HFSE anomalies, depletion of Nb (Nb/Nb* = 0.10–0.68) relative to Th and La and light REE enrichment patterns with island arc geochemical signatures. Plume-generated thickened, hot, buoyant oceanic plateau basalts coexisted with mafic and felsic melts, which were generated in an island-arc setting in the Sonakhan greenstone belt (Mondal and Raza 2009; Deshmukh et al. 2018).

2.5.2.3 Paleoproterozoic Younger Granitoid Intrusives

The Paleoproterozoic felsic magmatism in the Bastar Craton is represented mainly by the Malanjhkhand, Dongargarh, and Kanker granitoids of which the former is investigated thoroughly due its well-known Cu–Mo–Au mineralization.

(i) **Malanjkhand Granitoids (MG)**: N–S trending and about 1500 km^2 pluton in Balaghat district (Madhya Pradesh) intrudes the older Amgaon schist/gneiss, and is exposed in the northwest and west of the pluton, whereas the Nandgaon Group bimodal volcano-sedimentary lithounits are exposed in the southern and eastern margins of the pluton. The MG are overlain by the Chilpi metasediments in the southeastern and western parts of the pluton. It is composed mainly of granodiorite to granite and is linked with the Central Indian Tectonic Zone–CITZ (Yedekar et al. 1990; Jain et al. 1995). The MG is intruded by quartz veins and basic dikes.

The MG is coarse-grained equigranular, porphyritic to fine-grained granitoid with magmatic fabric, xenoliths and enclaves, and appear to have crystallization depths of ~11.7 km, using hornblende–plagioclase geothermobarometry (Pandit and Panigrahi 2012). The MG is metaluminous to peraluminous, calc-alkaline, porphyry copper-bearing felsic igneous plutons. It originated in a subduction zone with mixing of mafic and felsic magma in various proportions in a dynamic magma chamber (Kumar and Rino 2006). Geochemically, these are classified into diorite, granodiorite to granite and a minor component of sanukitoid, which exhibit calc-alkaline, high-K to shoshonitic characters (Chaki et al. 2008). These are enriched in LILE, Mg, Ni and Cr, and depleted in Nb and Ti. Tectonic discrimination diagrams indicate volcanic arc granite field whose parental magma was derived from an enriched metasomatized mantle, overlying a northward-dipping subducting oceanic slab along the southern margin of the craton.

Panigrahi et al. (2004) obtained new SHRIMP RG $^{207}Pb/^{206}Pb$ zircon data from the MG and suggested an age of granitic activity at ca 2.48 Ga (2476 ± 7 Ma, 2477 ± 8 Ma), while whole rock Rb–Sr ages are 2362 ± 58, 2467 ± 38 and 2243 ± 217 Ma (ca 2.4 Ga), and can be attributed to hydrothermal overprinting (Ghosh et al. 1986; Panigrahi et al. 1993). Chaki et al. (2008) obtained WR Rb–Sr isochron age of 2347 ± 16 Ma from 35 samples of these granitoids ($^{87}Sr/^{86}Sr$ initial ratio = 0.70482 ± 0.00038). Mineralization in the MG appears to be along quartz reef, veins and stringers, more akin to lode-type, and is almost contemporaneous at 2490 ± 8 Ma, 2494 ± 8 to 2449 ± 8 Ma from the Re–Os molybdenite ages (Zimmerman et al. 2002; Stein et al. 2004).

(ii) **Dongargarh Granite**: Lying to the southwest of the Nandgaon Group, this granite intrudes both the Nandgaon and the Amgaon gneiss and is nonconformably overlain by the Khairagarh Group. The pluton is essentially granite, containing enclaves of numerous lithologies like the rhyolites. Geochemically, and appear to have crystallization depths of ~20 km, using hornblende–plagioclase geothermobarometry (Pandit and Panigrahi 2012).

(iii) **Kanker Granite**: This granite consists of quartz, K-feldspar, plagioclase and biotite with accessory titanite, apatite and zircon (Ramachandra and Roy 1998). Sarkar et al. (1993) obtained a U–Pb zircon age of 2480 ± 3 Ma from southernmost part of this body. It is a peraluminous, silicic (SiO_2 = 66–76 wt%) granite with moderate K_2O and Mg# with fractionated LREE, negative Eu anomaly and flat HREE patterns, suggesting origin through shallow crustal melting of continental crust (Mondal et al. 2006).

2.5.2.4 Granulite Belts

The Bastar Craton exposes two granulite belts: (i) the Bhopalpatnam granulite belt (BGB) (~300 km long, 20–40 km wide) along the southwestern border of the craton adjoining the Godavari Graben, and (ii) the Kondagaon granulite belt (KGB) (~70 km long, 10 km wide); the former belt along with the Karimnagar granulites to the south of the Godavari Graben conceal mutual relations with the Dharwar Craton (Ramakrishnan and Vaidyanadhan 2008). Grt-Opx- and Grt-Cpx-bearing enderbitic charnockites, Grt-Bt-gneiss and Grt-Sil-Rt-gneiss with corundum characterize the BGB, and yielded peak metamorphism of 5–6 kbar, 700–750 °C for the BGB rocks, and 4–6 kbar, 700 °C for the KGB (Santosh et al. 2004). From the BGB, the EPMA chemical dates on zircons have 1.9 Ga cores with new growths at ~1.7 Ga around rims, while its monazite defines a clear isochron of 1.59 ± 0.03 Ga. From the KGB, monazite and uraninite show clear peaks with well-defined ages in narrow range between 2.42 ± 0.08 Ga and 2.47 ± 0.03 Ga; these are much older than the BGB (Santosh et al. 2004).

2.5.2.5 Mafic Dike Swarms

A number of NW–SE trending dike swarms traverse the supracrustals and granitoids of the Bastar Craton. Two sets are sub-alkaline mafic dikes with medium to high-grade metamorphism of ~2.9 Ga (BD1) and 1.88–1.89 Ga (BD2) and the third set is ~2.4 to 2.5 Ga (BN) boninite–norite dikes (Srivastava and Singh 2004; French et al. 2008; Gautam and Srivastava 2011; Srivastava and Gautam 2015, 2016); others are tholeiitic with variable Mg# (41–60), TiO_2 (0.83–1.74 wt%) and low CaO (<10.9 wt%) indicating that chemical variability is primarily due to crystal fractionation and crustal contamination (Shellnutt et al. 2018). Similarity in geochemistry, northwesterly trends and age make the dikes and sills from the southern Bastar Craton (1883.0 ± 1.4; 1883.5 ± 4.4; 1885.4 ± 3.1 Ma) and the Yilgarn Craton (1888 $\pm$ 9) of Western Australia into one group, and support the concept that Southern India and Western Australia were once connected during Paleoproterozoic and made a Large Igneous Province (Shellnutt et al. 2018). These dikes possibly belong to an extensional rifting event in the Bastar Craton with the melt derived from garnet spinel-bearing transition zone, with crustal contamination (Fig. 2.15).

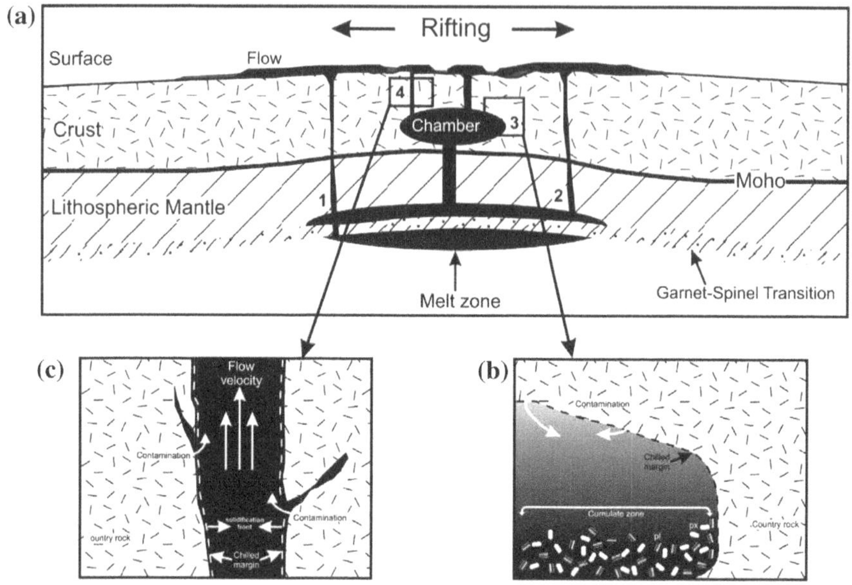

Fig. 2.15 Emplacement model for 1.88 Ga dikes and rifting of the Bastar Craton. **a** Melt derived from garnet (1) or spinel-bearing (2) transition zone and injected into crust. **b** Additional crystal fractanation of pyroxene and plagioclase. **c** Crustal contamination either during residence (3) in magma chamber or during emplacement (4). After Shellnutt et al. (2018)

2.5.2.6 Proterozoic Sedimentary Basins

The Bastar Craton is dotted with numerous Proterozoic sedimentary basins, including the largest Chattisgarh, Khariar, Ampani, Keskar, Singanpur, Abujhmar, Indravati and Sabari basins (Ramakrishnan and Vaidyanadhan 2008), though opinions differ whether these were once connected and part of a single basin (Ramakrishnan 1987a, b) or evolved individually (Kale 1991). Details of these basins will be dealt separately in Chap. 5.

2.6 Singhbhum Craton

2.6.1 Location

The Singhbhum Craton (SC), also known as the Singhbhum–Orissa Craton, encompasses N–S elongated and pear-shaped area of approximately 50,000 km^2 in Jharkhand, Orrisa and near-by regions in Eastern India. Sandwiched between two Proterozoic mobile belts—the Eastern Ghats Mobile Belt (EGMB) in the south and the Singhbhum Mobile Belt (SMB) or Northern Singhbhum Orogen (NSO), this Archean craton has a core 8000 km^2 region of the Singhbhum granite, which is encircled by supracrustal rocks. The craton is covered by the Indo-Gangetic alluvium along the eastern margin, while Gondwana Mahanadi graben separates it from another important Bastar Craton along its western margin (Fig. 2.16; Saha 1994; Ramakrishnan and Vaidyanadhan 2008).

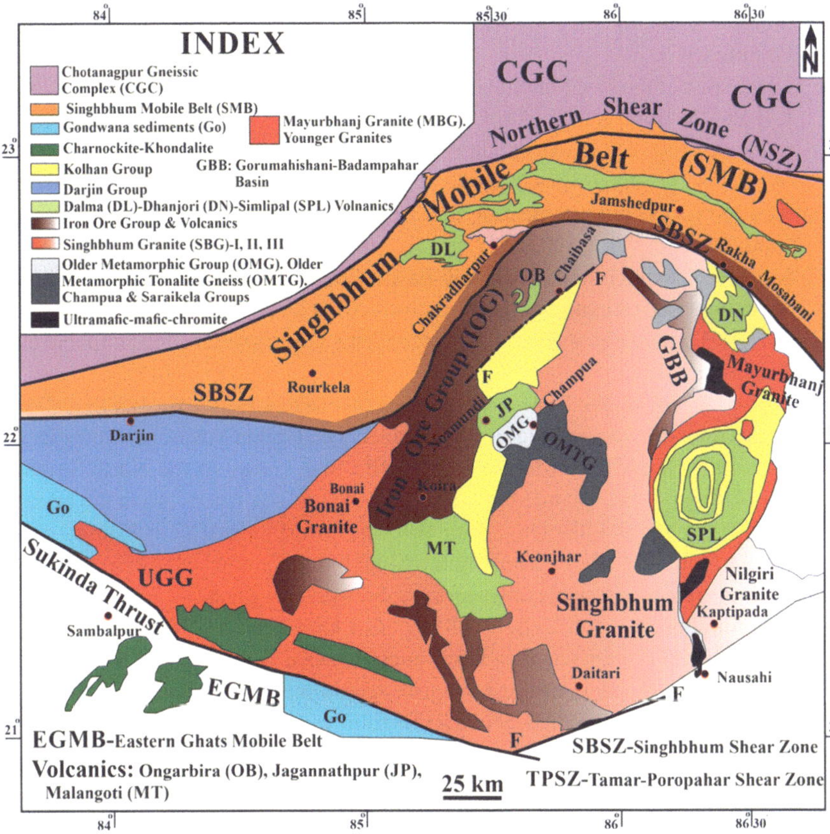

Fig. 2.16 Simplified geological map of the Singhbhum Craton, NE India, showing its main litho-units and tectonic margins. Abbreviations: OMG—Older Metamorphic Group. OMTG—Older Metamorphic Tonalite Group. Singhbhum Granite Complex—SBG with SBG-I, SBG-II SBG-III. IOG—Iron Ore Group. UGG—Undifferentiated Granite Gneiss. NSO—Northern Singhbhum Orogen. CGGC—Chotanagpur Granite Gneiss Complex. Map compiled after Saha (1994), Ramakrishnan and Vaidyanadhan (2008) and other published literature

2.6.2 Geological and Tectonic Framework

The Singhbhum Craton (SC) is a complex of numerous lithounits (Dunn 1940) of different ages and evolved during the Singhbhum Orogeny (Saha 1994). It can be divided into the following associations (Table 2.2).

(i) Older Metamorphic Group (OMG) of sediments—volcanics and Older Metamorphic Tonalite Gneiss (OMTG), occurring as enclaves into the Singhbhum Granite (SG).
(ii) Large span of the Singhbhum Granite (SG) pluton, intruded into three phases—I, II and III: Type A (SG–A) of Phases-I and II (~3.33 Ga) and Type-B (SG-B) of Phase-III (~3.1 Ga). Granodiorite-adamellite-granite suite characterized the SG. It also occurs as the Chakradhar granite in the north, Bonai granite in the west, Nilgiri granite in the east, and large unclassified Pala Lahara gneiss the southwest.
(iii) Iron Ore Group (IOG) of thee sub-basins, located in the west, south and east and comprised of banded iron formations (BIF) and banded hematite quartzite (BHQ). These are located at Noamundi-Koira, Gorumahisani-Badampahar and Tonka-Daitari.
(iv) Dalma, Dhanjori and Simlipal Volcanics of variable ages, exposed in isolated basins and also known as greenstone belts. The Dalma Volcanics are exposed within the Singhbhum Mobile Belt (SMB), also known as the North Singhbhum Orogen (NSO), while the latter two volcanic sequences are in the eastern parts above the SG; the Ongarbira Volcanics overlie the IOG at Koira.
(v) Newer Dolerite Dykes traversing the Singhbhum Craton.
(vi) Kolhan Group sediments, unconformably overlying the SG.
(vii) The Singhbhum Mobile Belt (SMB), also known as the North Singhbhum Orogen (NSO), starts with a basal conglomerate over the SG with younger low-grade metasediments further north. A prominent ductile Singhbhum Shear Zone (SSZ) is exposed in between and is followed by the Singhbhum Group flyschoids of the Chaibasa and Dhalbhum Formations. Further north, the Dalma

Table 2.2 Simplified chronostratigraphic succession for the Singhbhum Craton, eastern India. After Saha (1994)

Kolhan Group	c. 2.1–2.2 Ga
Jagannathpur Lava–Dhanjori Lava	
Singhbhum Group	2.3–2.4 Ga (?)
Unconformity	
Singhbhum Granite (phase III)	c. 3.1 Ga
Iron Ore Group	
Singhbhum Granite (phase I and II)	c. 3.3 Ga
Older Metamorphic Group and Older Metamorphic Tonalite Gneiss	c. 3.4–3.5 Ga

Volcanics are overlain by younger clastics, cherts, felsic volcanics and mafic-ultramafic schists. The Tamat–Poropahar Shear Zone (TPSZ) is located near the norther margin of the NSO, where the Chotanagpur Granite Gneiss of the Satpura Mobile Belt delimits this orogen in the north (Fig. 2.16). Details of the SMB are given in Chap. 4 on the Proterozoic Fold Belt.

Tectonically, the Singhbhum Craton is limited by two major thrust zones: (i) most prominent ductile Singhbhum Shear Zone (SBSZ) and also known as the Copper Belt Thrust along its northern margin with the Singhbhum Mobile Belt (SMB), and (ii) the Sukinda Thrust (ST) along its southern margin against the Eastern Ghats Mobile Belt.

Extensive work is now available on the Singhbhum Craton, and the readers are advised to refer to the contributions made by Naqvi and Rogers (1987), Saha (1994), Mishra et al. (1999), Mukhopadhyay (2001), Mukhopadhyay et al. (2008), Mondal et al. (2007), Chatterjee et al. (2008), Ramakrishnan and Vaidyanadhan (2008), Sharma (2009), Acharyya et al. (2010), Saha and Mazumder (2012), Mazumder et al. (2012), Hoffmann and Mazumdar (2015), Meert and Pandit (2015), Chaudhury et al. (2018), Miller et al. (2018), Sharma and Mondal (2019 and references therein), Sreenivas et al. (2019) and many others.

In view of distinct tectonic boundaries of the Singhbhum Craton, this part of the chapter restricts descriptions and discussions to the main cratonic region, while the Singhbhum Mobile Belt (SMB)/North Singhbhum Orogen (NSO) and other units will be taken up elsewhere.

2.6.2.1 Older Metamorphic Group (OMG—~3.5 Ga)

The OMG crops out as widely and irregularly scattered supracrustal enclaves within the SG pluton, with the largest enclave at Champua in about 200 km^2 area in the western parts. These enclaves of quartzite (fuchsite-bearing), mica schist, amphibolite (both ortho- and para-amphibolite) and calc-silicates have undergone intense deformation, development of foliation and lineation, and steeply plunging to vertical sheath folds (Roy and Bhattacharya 2012). Granite bodies profusely intrude the OMG extensively and completely engulf amphibolite/calc-silicates as agmite and intrusives (Fig. 2.17a and b).

An association of quartz-kyanite, sillimanite-muscovite, quartz-feldspar aggregates and fuchsite-bearing quartzite indicate metamorphosed clay and regolith palaeosols having ‘granite protoliths’ (cf., Misra 2006), and indicate an erosion unconformity.

Low-K tholeiitic (LKT) amphibolite of the OMG exhibits slightly fractionated light rare earth element (LREE) patterns and almost flat heavy rare earth element (HREE) with minor negative Eu-anomaly (Saha and Ray 1984). These are enriched in large ion lithophile elements (LILE) with Rb/Sr (0.08–0.10), Ba/Sr (1.04–2.77) and K/Ba (11.56–39.75). The amphibolites were produced from smaller degrees of

Fig. 2.17 **a** Agmatite formed by anastoming intrusive veins of granitoid around angular (breccia-like) pieces of calc-silicate rock. **b** Massive, porphyritic granitoid engulfing large fragments of folded calc silicate rock. **c** Non-conformity between the Singbhum Granite Complex and the overlying IOG. **d** Exposure of a BIF preserving its pristine depositional character. All photographs after Roy and Bhattacharya (2012) and converted to black and white

partial melting of mantle peridotite with higher content of LILE or higher degrees of partial melting of LKT.

Detrital zircons from 2 samples of the OMG supracrustal metasediments, muscovite–biotite schist and meta-quartzite, from the Champua region range between 3627 ± 39 Ma and ca. 3550 Ma, and two younger clusters of 3.4 and 3.2 Ga (Goswami et al. 1995; Mishra et al. 1999; Mondal et al. 2007), thus establishing the oldest maximum depositional age of the OMG sediments around 3.5 Ga on an unknown granitic basement. The younger ages were taken as metamorphic events, which were confirmed by rutile U–Pb age of 3317 ± 8 Ma from mica schist of the OMG as the post-metamorphic cooling subsequent to the high-grade amphibolite-facies overprint. This age is broadly synchronous with the emplacement of expansive granitoids of the OMTG and SG (Upadhyay et al. 2014).

2.6.2.2 Older Metamorphic Tonalite Gneiss (OMTG—3.45–3.30 Ga)

Highly scattered tonalite gneiss enclaves occur extensively within the SGC as the Older Metamorphic Tonalite Gneiss (OMTG) (Saha 1994). These contain thinly layered medium grained, granodiorite and tonalite bodies near Champua, with development of quartzo-feldspathic patches along small-scale ductile shear zones and in necks of foliation boudins, thus indicating partial melting during deformation. Amphibolite bodies interleave tectonically with the OMTG. Strong penetrative planar fabric and associated mineral lineation in gneisses also parallel the OMG rocks, and plunge steeply on highly dips to the NE (Saha 1994).

Close to the detrital zircon dates from the OMC is 3664 ± 79 Ma Pb–Pb isochron age from the OMTG from by Ghosh et al. (1996). Apart from these, all other ages, determined from the gneissic granitoids, indicate much younger ages. These include: (i) whole-rock Rb–Sr and Pb/Pb isochron ages of 3280 ± 130 and 3378 ± 98 Ma, respectively for the OMTG (Moorbath et al. 1986), (ii) 3448 ± 2.2 Ma U–Pb TIMS zircon age, and ca. 3380 Ma Pb–Pb whole-rock age from a 'tonalite-xenolith' hosted within Bonai granite (Sengupta et al. 1996), (iii) 3442 ± 26 Ma Pb–Pb age of granitoid (Ghosh et al. 1996), (iv) 3457 ± 35 Ma and 3437 ± 9 Ma Pb–Pb zircon ages (Mishra et al. 1999) and (v) U–Pb SHRIMP zircon age at 3380 ± 11 Ma from the tonalitic gneiss (Nelson et al. 2007), (vi) 3448 ± 19 and 3527 ± 17 Ma U–Pb zircon ages of tonalitic gneiss from the northern most part (Acharyya et al. 2010), (vii) $^{207}Pb/^{206}Pb$ (SHRIMP) zircon age of 3.45 Ga from tonalite gneiss of the OMTG near as age of emplacement of the gneiss precursor (Saha and Mazumder 2012).

$^{207}Pb/^{206}Pb$ ion probe zircon ages of the OMTG also cluster around 3.2 Ga, which is considered as growth during metamorphism (Mishra et al. 1999). In agreement with this age, are monazite ages of 3306 ± 22 Ma from the OMTG as crystallized during metamorphism (Prabhakar and Bhattacharya 2013).

Tonalite gneiss of the OMTG sequence from the Champua also revealed 4.24–4.03 Ga xenocrystic zircons, indicating records of oldest precursor of Hadean age from Singhbhum (Chaudhury et al. 2018). In addition, one single Hadean zircon is recorded with a $^{207}Pb/^{206}Pb$ age of 4015 ± 9 Ma in modern sediments of the Baitarani River, draining the Singhbhum craton (Miller et al. 2018). They further measured initial εHf of −5.30 in this grain indicating an episode of Hadean felsic crust formation in the Singhbhum craton like the Jack Hills of the Yilgarn craton (Australia). Chaudhury et al. (2018) observed εHf[t] = −2.5 to −5.2 and unradiogenic $^{176}Hf/^{177}Hf$ initial compositions (0.27995 ± 0.0009 to 0.28001 ± 0.0007 and interpreted existence of an enriched reservoir during Hadean eon in the Singhbhum cratonic mantle. In addition, Sreenivas et al. (2019) discovered detrital zircon grains between 3.95 Ga and 2.91 Ga from quartzite and dacite from the Iron Ore Group. In combination with the above reported Hadean, Eoarchean xenocrystic (up to 4.24 Ga) and modern detritus zircon grains from the Singhbhum craton, it is likely that the Eoarchean crust was generated by recycling of the Hadean felsic crust formed at ~4.3 to 4.2 and ~3.95 Ga. Thus, the Singhbhum Craton so far becomes the oldest Hadean–Archean craton within the Indian Sub-continent. It further preserves ca. 3.7 Ga sialic crust (Goswami et al. 1995; Ghosh et al. 1996; Mishra et al. 1999) and

c. 3.1 Ga cratonization history. A near-continuous Archaean rock record is, thus, preserved within the SC and can, therefore, be used for any future correlation and supercontinent reconstructions by researchers like the Kapvaal and Pilbara cratons (Ghosh et al. 2015).

Analyzed zircons from the Baitarani River sediments by LA-ICP-MS for their U–Pb systematics yielded 3.62–3.55 Ga (5%), 3.50–3.22 Ga (87%), and 3.10–3.06 Ga (6%) populations with bulk of ~3.50 Ga zircons sourced from the 3.47 Ga tonalite gneiss enclaves within the 3.35–3.30 Ga Singhbhum Granite near Keonjhar (Miller et al. 2018).

2.6.2.3 Singhbhum Granite (SG) Pluton

Central parts of the Singhbhum Craton occupy nearly 8000 km^2 area of the north–south elongated outcrop of the Singhbhum Granite (SBG) complex of nearly 12 separate magmatic bodies (Saha 1994). On the basis of petrography, geochemistry and age relationships, at least three multi-stage emplacement of granitic phases have been distinguished within the SG pluton: SG I (3.35 Ga), SG II (3.1 Ga), and SG III (2.9 Ga), spanning a long period of about 500 Ma (Saha and Ray 1984; Saha 1994; Misra and Johnson 2005; Mondal et al. 2007; Acharyya et al. 2010; Upadhyay et al. 2014; Dey et al. 2017).

The rocks belonging to Phase-I SG intrusions are relatively K-poor granodiorite-tronjodiorite, while the Phase-II and Phase-III plutons are granodiorite grading to adamellite and leucogranite. Two distinct types of granitoids (Type A and Type B) have been identified geochemically within the SG complex. Type A granitic phases (SG I and SG II) exhibit rare earth element (REE) patterns, which are characterized by gentle slopes and slight heavy rare earth element (HREE) depletion and very minor negative or no Eu-anomaly, like the tonalite belonging to OMG (Saha and Ray 1984). Dey et al. (2017) further noted that the ~3.47 Ga Na-rich, gneissic tonalite of the Archaean TTG suite represents first phase of granitoid magmatism with high-HREE, negative Eu anomaly, low Sr/Y ratio and positive zircon Hf isotope signature (εHf_t = +2.1 to +4.8) in central part of the craton. This magma originated by shallow melting of a juvenile mafic source. Second phase at 3.35 Ga exhibits variable K/Na, low Y and high Sr/Y, moderately fractionated HREE and positive zircon εHf_t (+1.8 to +4.0), due to high temperature melting of a heterogeneous, juvenile source consisting of tonalities and mafic rocks at lower crustal depth (Saha and Ray 1984; Dey et al. 2017). Lastly, ~3.30 Ga non-porphyritic ferroan, silica-rich biotite granite, having moderately high K and fractionated HREE, low Ca and Sr/Y, and zircon εHf_t (+0.8 to +3.7) formed by high-pressure melting of a tonalite-dominated source with short crustal residence. Episodic plume-related mafic-ultramafic magma underplating and intraplating in an oceanic plateau setting is suggested as the possible mechanism for formation of the granitoids (Dey et al. 2017).

Erosion of the Precambrian Singhbhum Craton from Keonjhar produced pyrophyllite-bearing paleosol between 3.29 and 3.02 Ga, prior to unconformable deposition of an overlying sequence of shallow marine siliciclastic sediments (Mukhopadhyay et al. 2014).

The SG is rimmed by the Iron Ore Group (IOG) and volcanics as mantle melt extruded in the coexisting ocean where oxidation formed Banded Iron Formations (BIF).

2.6.2.4 Iron Ore Group

Surrounding the central Singhbhum gneissic Craton on its western, southern and eastern margins and non-conformably overlying it (Fig. 2.16c), three extensive belts constitute the Iron Ore Group (IOG) around Jamda-Koira, Tomka-Daitari and Gorumahisani–Badampahar (Ghosh and Mukhopadhyay 2007; Mukhopadhyay et al. 2012; references therein; Alfimova et al. 2019) (Fig. 2.16). This group contains massive pillow basalt, dacite, pyroclastics and ultramafics in the lower parts, and is overlain by predominantly BIF sequence (Fig. 2.16d); the whole belt was deposited under deep marine conditions (Mukhopadhyay et al. 2012). Very little geochronological data exists for the IOG. Opinions vary regarding the relative age of the IOG belts, which were coeval and deposited in three different basins (Acharyya 1993; Saha 1994). However, Iyengar and Murthy (1982) opined an earlier Gorumahisani-Badampahar and Tomka-Daitari sequences of supracrustal deposition before the Jamda-Koira belt. The latter belt is an NNE-trending horse-shoe syncline-anticline pair, which is subsequently folded along E–W trends to produce type 1 and type 2 patterns (Ghosh and Mukhopadhyay 2007). The Gorumahisani belt represents either a pinched-in synclinal cusp within the basement or a faulted graben (Mukhopadhyay 1976; Banerjee 1982). In the southern Tomla-Daiteri belt, regional folds reveal E–W striking axial surfaces, thus regional large-scale folds within different IOG belts curve round the Singhbhum Granite, indicating curvature of these supracrustal envelope around the rigid basement block.

Mukhopadhyay et al. (2008) obtained zircon U–Pb age of 3507 ± 3 Ma from of the Tomka-Daitari belt, indicating parts of the IOG sequence may even be older than the OMG. Other geochronological data on the IOG are the zircon U–Pb age of 3392 ± 25 Ma from a tuff of the Jamda-Koira belt (Basu et al. 2008) and ~3.3 to 3.17 ± 0.1 Ga granite intrusions in the Gorumahisani-Badampahar belt (Saha 1994; Nelson et al. 2014); thus age of the IOG brackets between 3.5 and 3.1 Ga.

Tectonic discrimination diagrams using Nb, Y, Zr and Ti suggest an oceanic arc–forearc geodynamic setting, like the Phanerozoic supra-subduction zone ophiolites where ophiolite development is controlled by slab roll-back. It is likely that the BIF was deposited near spreading centers over a rifted arc. Bimodal volcanics-BIF-ultramafic succession of the IOG in a convergent margin is an important example for Paleoarchean plate tectonics in the Indian Sub-continent.

2.6.2.5 Volcanic Sequences

Subsequent to the culmination of the Iron Ore Orogeny (~3.0 Ga) and unconformably overlying the periphery of the Singhbhum Craton, many intercalated metabasalts and fluvial–marginal marine sedimentary successions are exposed as the Simlipal, Dhanjori, Dalma, Ongarbira, Jagannathpur, Malangtoli Volcanics. These constitute possible Palaeoproterozoic intracratonic development within the Singhbhum Craton. As linear Dalma Volcanic belt is associated with the North Sinnghbhu Belt, it will be dealt separately. Since there is lack of consensus regarding their mutual relationships, these volcanics are briefly summarized below.

(i) **Simlipal Volcanics**: Southeastern parts the Singhbhum Craton contains a spectacular oval-shaped outcrop of alternating mafic lava–tuff–volcanic breccia with arenaceous intertrappean beds, having a quartz-pebble conglomerate and quartzite at the base (Iyengar et al. 1981; Mazumder 1996). Tholeiite and andesitic basalts along with spilitic lavas and tuffs make the bulk of the volcanic succession, while the top of the sequence is marked by ~800 m thick differentiated sill of dunite to quartz diorite in composition (Iyenger and Banerjee 1964). These undeformed and unmetamorphosed belt remains unaffected by any granitoid intrusion.

The Simlipal basin represents explosive volcanism in deep marine conditions on a continental platform (Saha 1994) and has further been described as a 'crypto-explosion structure' (Ramakrishnan and Vaidyanadhan 2008) or a plume-outburst related volcano-sedimentary basin (cf., Banerjee and Ghosh 1994).

(ii) **Dhanjori Volcanics**: Another large volcanic exposure is observed along the northeastern margin of the Singhbhum Craton, where it is thrust over by the Singhbhum Mobile Belt (SMB) along the Singhbhum Shear Zone (SSZ). It consists predominantly of volcano-sedimentary depository (Gupta et al. 1985; Alvi and Raza 1992; Majumdar and Sarkar 2004), which unconformably overlies the Singhbhum Granite Complex and the associated older supracrustals (Sarkar and Saha 1962). The Dhanjori Volcanics contain metapelites, psammites, ultramafics and mafics (gabbro/dolerite) in the lower part, while a volcanic sequence of mafic/ultramafic tuffs, intrusives, metabasalts and tuffaceous sediments make its upper parts. Distinct komatiitic ultramafics having spinifex textures are observed in its lower parts (Mazumder 1996). The upper Dhanjoris have basaltic suite of alkali olivine basalts, changing into K-poor oceanic tholeiite. Copper deposits occur within its low grade metavolcanic member.

An isolated mafic body occurring within the Dhanjori sediments yielded a whole-rock Sm–Nd age of 2072 $\pm$ 106 Ma (Roy et al. 2002), which may even be a younger intrusive body because Misra and Johnson (2005) obtained Pb–Pb and Sm–Nd whole-rock isochron ages of 2794 $\pm$ 210 and 2787 $\pm$ 270 Ma, respectively from the Dhanjori volcanics, like similar age bracket of 2837 $\pm$ 200 Ma upper intercept age of zircon from acid tuff (Acharyya et al. 2010). ~2800 Ma age is, therefore, more appropriate for the Dhanjori Volcanics.

(iii) **Ongarbira Volcanics**: Unconformably overlying the Singhbhum Craton, hook-shaped volcanic-sedimentary Ongarbira sequence is exposed within a km-sized and E–W trending synform. It is comprised of ultramafic lavas at the base, followed upwards by pillowed tholeiitic basalts. These are intercalated with thin bands of quartzite and slate; and have suffered greenschist metamorphism (Blackburn and Srivastava 1994). Geochemically, these volcanics probably belong to either marginal basin or arc tholeittes. The Ongarbira suite may represent outpouring of oceanic basalt in marginal rift basin coeval with the Dhanjori volcanics (Blackburn and Srivastava 1994).

(iv) **Jagannathpur Volcanics**: The Jagannathpur Volcanics cover an area of about 250 km^2 on southwestern flank of the craton and are overlain by the Kolhan Group sequence. The volcanics were emplaced in the Noamundi basin. Weakly-metamorphosed volcanics have variable composition from komatiite to tholeiite with minor andesite and basanite. Misra and Johnson (2005) obtained a whole-rock Pb–Pb four-point isochron age of 2250 ± 81 Ma as the formation age for these volcanics, which are stratigraphically equivalent to the Malangtoli volcanics.

(v) **Malangtoli Volcanics**: This volcanic field spans about 800 km^2 of the craton in its southwestern parts to the west of Keonjhar. These volcanics range from basalt to andesite of the tholeiite to calc-alkaline series. While the Malangtoli lavas are tholeiitic basalts with uniform composition, the Jagannathpur lavas vary in composition from tholeiite to andesite. These lavas are interpreted to have erupted under a continental flood basalt-like setting.

2.6.2.6 Newer Dolerite Dyke Swarm (NDDS)

The Singhbhum Granite pluton exhibits the Newer Dolerite Dyke Swarm (NDDS) which intruded in NNE–SSW, N–S, WNW–ESE and NNW–SSE; out of which NNE- and WNW-trending dykes are most prominent (Dunn and Dey 1942; Bose 2008), possibly resulting from NNW-trending σ_1 stresses (Fig. 2.18). Most of the dykes are of quartz dolerite with small occurrences of norite, granophyre, microgranite and syenodiorite. These are a few meters to 700 m in thickness and can extend for several kilometers. Two generations of dykes are decipherable from field and geochemical data, while Verma and Prasad (1974) inferred three different episodes of dyke intrusion from palaeomagnetic data. K–Ar and Rb–Sr whole-rock ages constrained these dykes poorly as 2613 ± 177 Ma on Keshargaria swarm (Roy et al. 2004), while K–Ar ages ranged between ~2100 ± 100, ~1500 ± 100 Ma, and 1100 ± 200 Ma (see Mallik and Sarkar 1994). However, recent precise $^{207}Pb/^{206}Pb$ baddeleyite ages on these dykes revealed the following: (i) 2800.2 ± 0.7 Ma NNE–SSW trending dyke and dykes between 2762.4 ± 2 and 2750 ± 0.9 Ma (Kumar et al. 2017a), (ii) $^{207}Pb/^{206}Pb$ 2.25 Ga baddeleyite age of NE–SW to ENE–WSW trending Kaptipada swarm (Srivastava et al. 2016), and (iii) younger WNW–ENE trending 'newer dolerite' Pipilia dyke swarm of ~1765 Ma (Shankar et al. 2014).

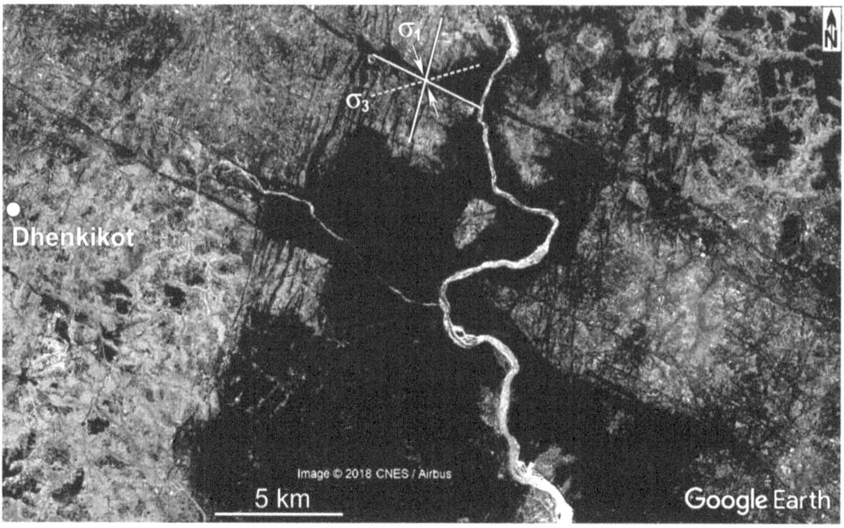

Fig. 2.18 Conjugate set of the Newer Dolerite Dyke Swarm (NDDS) around Dhenkikot (21° 54′ 52″:85° 50′ 06.24″) showing two sets of long linear dyke system, possibly originated due to NNW-trending principal stress σ_1; WNW-trending dykes orient in σ3 direction

Since 1.77–1.76 Ga dyke swarm is also observed from other North China craton, Rio de Plata Craton (Uraguay), Sarmatia (Baltica), Kimberley (Australia) and Roraima (Brazil), it might represent erosional remnants of a Large Igneous Province (LIP); the continents might belong to the Columbia Supercontinent. It may as well represent the cessation of tectonics, cooling and crustal stabilization of the Singhbhum Craton.

2.6.2.7 Post-Singhbhum Granite Intrusions

A few small granite bodies intrude the IOG within the Singhbhum Craton, of which well-known bodies are the Bonai Granite (BG), the Nilgiri Granite (NG), the Mayurbhanj Granite (MG), and Soda Granite, to name a few, while the Chakradharpur Granite Gneiss (CGG) is developed within the SMB.

(i) **Bonai Granite**: It is separated from the main Singhbhum Granite batholith by the IOG supracrustal belt of Jamda-Koira horse-shoe synclinorium (Sengupta et al. 1991; Saha 1994). Two phases of the Bonai Granite have been reported: Phase-I migmatitic Bonai granite, having 3369 ± 57 Ma Pb–Pb WR age. The Phase-II Bonai granite is porphyritic and biotite-rich with a 3163 ± 126 Ma Pb–Pb WR age (Sengupta et al. 1991, 1996); the latter intrudes the IOG supracrustals and contains large xenolithic blocks of these quartzites and metalavas.

(ii) **Nilgiri Granite**: Separated from the main SG by a narrow strip of 3–8 km wide IOG phyllite and quartzite, this body of granite is composed of TTG to granite in southeastern parts of the craton (Saha 1994). The Nilgiri pluton is intruded by the Mayurbhanj Granite along its margin. For the second and third phases of the Mayurbhanj granite, in situ Pb–Pb zircon ages are 3080 ± 8 Ma and 3092 ± 5 Ma, respectively (Misra and Johnson 2005).

(iii) **Soda Granite**: Along the SSZ, the Soda granite dates 2220 ± 54 Ma (Pb–Pb WR isochron age) and has been interpreted as the age of emplacement/crystallization of this body. A younger 1.68–1.63 Ga age is considered as the reset age of the Soda granite due to shearing along Singhbhum shear zone and associated metamorphism (Sarkar et al. 1985).

2.7 Dharwar Craton

2.7.1 Introduction

The Dharwar Craton is the largest preserved Archean cratonic block and one of the best-investigated cratons of the Indian Sub-continent covering an area of about 0.45 million sq. km^2 and preserves the crustal record from ca. 3.6 to 2.5 Ga. In the north, it is overlain by narrow strips of the Proterozoic sedimentary sequences of the Kaladgi and Bhima Basins and the Cretaceous Deccan Traps, while the Archean Karimnagar granulite belt makes its northeastern periphery. The eastern margin of the craton is again covered by the Proterozoic Cuddapah basin, while a vast Pan-African Pandyan Mobile Belt is juxtaposed against the craton along its southern margin. The region to the south of the orthopyroxene isograd, covering parts of the Moyar and Bhavani shear zones, Palghat-Cauvery shear zone and the Pandyan Mobile Belt is also called the South Granulite Terrain (SGT) (Fig. 2.19; Ramakrishnan and Vaidyanadhan 2008).

The Dharwar Craton is divided into two major blocks: The Western Block and the Eastern Block (Swami Nath et al. 1976), and are also termed as the Western Dharwar Craton (WDC) and the Eastern Dharwar Craton (EDC) (Swami Nath et al. 1976; Drury 1984; Naqvi and Rogers 1987; Chardon et al. 2008, 2011); two are separated by the Chitradurga Shear Zone along the eastern margin of the Chitradurga Schist Belt (Radhakrishna and Naqvi 1986). Based on new thermal, accretionary and geochronological data (Dey 2013; Jayananda et al. 2013a; Peucat et al. 2013), the EDC is further sub-divided in two sub-provinces: the Central Dharwar Province (CDP) and the Eastern Dharwar Province (EDP), called herein as the CDC and the EDC, respectively. The CDP incorporates the area between the WDC and the EDP and has a significant component of old crust (>3.0 Ga) which has undergone crustal remobilization at 2.56–2.51 Ga, like the EDP. The latter is comprised of the remaining Dharwar crust. These divisions are shown in Fig. 2.19. This chapter follows the earlier two-fold classification of the Dharwar Craton into the WDC and the EDC, with the granulite along their southern margin, making the SGT.

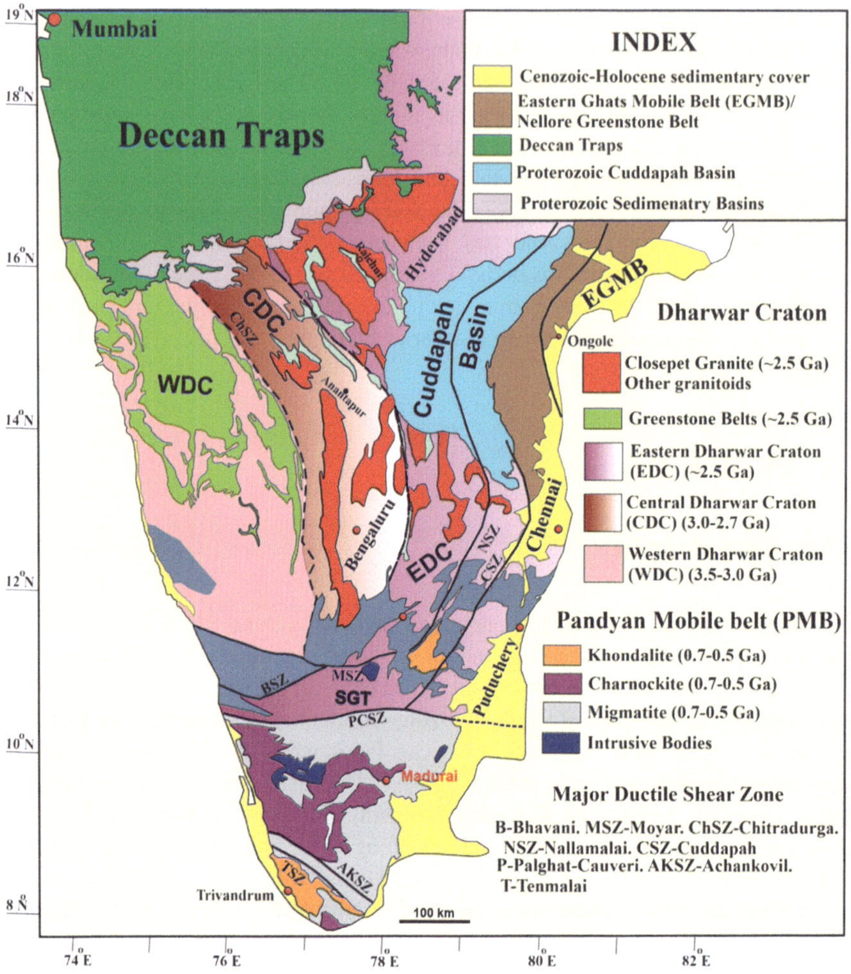

Fig. 2.19 Simplified geological map of Southern Indian shield showing Dharwar craton and its schist belts, Southern Granulite Terrain (SGT), shear zones and Pandyan Mobile belt. Redrawn after GSI and ISRO 1994

The Dharwar Craton exhibits the following main components: (i) polyphase TTG-type gneisses of the WDC (Jayananda et al. 2015; Guitreau et al. 2017), minor TTGs, abundant transitional TTGs and composite granitoids in the EDC (Jayananda et al. 2018, 2020), (ii) Southern Granulite Terrain to the south of the orthopyroxene isograd (Peucat et al. 2013; Ratheesh Kumar et al. 2016), (iii) Volcano-sedimentary greenstone belts (Swami Nath and Ramakrishnan 1981), (iv) Major Shear Zones, (v) ~2.5 Ga granitoid belts, (vi) Proterozoic mafic dikes, (vii) deformation patterns, (viii) metamorphism, and (ix) crustal evolution.

Vast literature is available on the geology of the Dharwar Craton, hence this chapter cannot justify full review of its geology and tectonics. Bruce Foote (1886) made a breakthrough by identifying the 'Fundamental granitoid gneiss' as the basement for unconformably overlying auriferous schistose belts of the 'Dharwar System', in contrast to Smeeth's viewpoints (1915–16) that the former intruded into the schistose units. These two contrasting thoughts prevailed upon in the Dharwar geology for very long period till 1976 when conceptual stratigraphic model (Swami Nath et al. 1976) and new geological map (Swami Nath and Ramakrishnan 1981) were presented and became backbone of further inputs in structural geology, igneous petrology, geochemistry, metamorphic petrology, remote sensing, crustal evolution, geochronology, geophysics etc. The readers are referred to the recent books and memoirs by Naqvi and Rogers (1983), Radhakrishna and Ramakrishnan (1990), Radhakrishna et al. (1990), Mahadevan (1994), Ramakrishnan and Vaidyanadhan (2008), Sharma (2009), Mahadevan (2003), Valdiya (2016) and others, besides some excellent papers by Jayananda et al. (2018, 2019, 2020a, b), and Bhaskar Rao et al. (2020).

2.7.2 Geological and Tectonic Framework

As has been elaborate above, the DC is composed of two sub-cratons: The Western Dharwar and the Eastern Dharwar Cratons which possess distinct geological histories. They expose an uninterrupted section through the Archean continental crust due to northward tilt; consequently, deepest granulite facies parts are exposed in the south. Gross lithologies of these cratons include (i) deformed Tonalitic-trondhjemitic-granodioritic (TTG) type Peninsular gneisses, (ii) two generations of volcanic-sedimentary greenstone sequences including older-Sargur Group and younger Dharwar Supergroup, and (iii) voluminous composite granitoids including calc-alkaline sanukitoids (high-Mg diorite, tonalite, granodiorite) and high-potassic granites.

Table 2.3 provides a simplified stratigraphic scheme of highly deformed and metamorphosed lithounits and comparison of the both the cratons. Salient characters of both the cratons are summarized below.

Like in most of the other Archean cratons the Dharwar Craton contains three major lithological associations (Rogers 1986): (i) Tonalite–trondhjemite–granodiorite TTG-type Peninsular gneisses, (ii) greenstone belts of volcano-sedimentary sequences and (iii) calc-alkaline to potassic granites.

2.7.2.1 TTG-Type Gneisses

The TTG-type gneisses are highly heterogeneous association of felsic rocks corresponds to most voluminous lithologies in the Dharwar Craton. Presence of older

Table 2.3 Simplified Precambrian stratigraphy of the Dharwar craton. After Ramakrishnan and Vaidyanadhan (2008) and Dey (2013)

Western Dharwar Craton (WDC)		Eastern Dharwar Craton (EDC)
Badami Group (0.6 Ga?)	Kaladgi Supergroup	Kurnool Group, Bhima Group (0.6 Ga?) Unconformity
Unconformity	(~Cuddapah Supergroup)	
Bagalkot Group Unconformity		Cuddapah Supergroup (1.9–1.4 Ga?) Unconformity
Younger granitoids (~2.6 Ga)		
Chitradurga Group (conglomerate, quartzite, carbonate-greywacke, argillite, BIF, basalt)		Transitional TTGs granites (2.7–2.6 Ga) Kolar Group (~2.7 and 2.545 Ga) (basalt, komatiite, tuffs, chert, BIF, greywacke-argillite, felsic volcanics)
Bababudan Group (conglomerate, quartzite, basalt, phyllite, BIF) Unconformity	Dharwar Supergroup (2.9–2.6 Ga)	–
Sargur Group (3.4–3.2? Ga) (quartzite, pelite, calcsilicate, basalt, komatiite, BIF, layered mafic-ultramafics) Gneisses (Gorur—3.6–3.0 Ga)		Vestiges of older Gneisses (~3.38 to 3.23 Ga) and supracrustals (~3.3 Ga?)

greenstone belts (Sargur-type dominated by ultramafic-mafic greenstone sequences, with interlayered quartzite-carbonate-pelite sequences interlayered TTG gneisses) distinguish an older Gorur Gneiss from the younger Gneiss. A suite of migmatitic TTG-type gneiss with numerous enclaves appear to be the oldest sequence in the WDC and is best exposed in stone quarries along the Gorur-Hassan-Holenarsipur region (Bhaskar Rao et al. 1991; Figs. 2.20 and 2.21). More recent U–Pb zircon dating by SHRIMP and LA-ICPMS show ages ranging 3607 ± 16 to 3280 ± 5 Ma (Jayananda et al. 2015; Guitreau et al. 2017). The diapiric trondhjemites intrude into the TTG gneisses and greenstones indicate U–Pb zircon ages of ca. 3223–3178 Ma (Jayananda et al. 2015; Guitreau et al. 2017). The migmatitic gneisses comprise two different compositional types including low-Al and high-Al gneisses. The low-Al gneisses with high SiO_2, MgO, Fe_2O_3, HREE but low Rb, Ba Sr, total REE, Sr/Y, $(La/Yb)_N$; their origin is attributed to shallow level melting of arc crust (Jayananda et al. 2015). On the contrary, the lower SiO_2, Fe_2O_3, MgO but higher total REE, Ba, Sr, higher Sr/Y $(La/Yb)_N$ of the high-Al gneisses have explained by melting of oceanic arc crust at deeper levels (Jayananda et al. 2015). The origin of 3600–3300 Ma gneisses is also interpreted by low degree of partial melting of basalts in lower crust. Guitreau et al. (2017) obtained magmatic zircons of 3410.8 ± 3.6 Ma

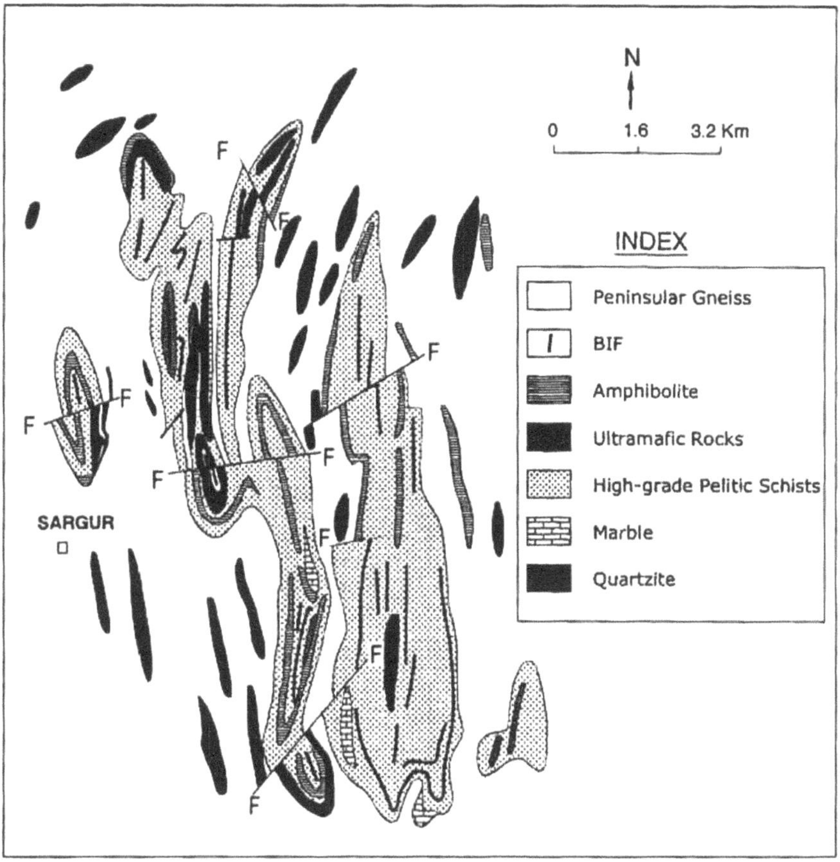

Fig. 2.20 Geological map of the Sargur schist belt around Sargur. After Vishwanatha and Ramakrishnan (1981)

age from granitic gneiss of the Holenarsipur Schist Belt (HSB) with inherited zircons with ages ranging from 3295 ± 18 to 3607 ± 16 Ma within a 3178 ± 10 Ma trondhjemitic gneiss and biotite-rich enclave.

2.7.2.2 Dharwar Batholith

To the east of the Chitradurga greenstone belt, greenstones and intervening gneisses and granitoids are called as the Dharwar Batholith of the Eastern Dharwar Craton (EDC) (Chadwick et al. 2000). The gneiss is comprised of mainly banded grey to dark grey gneisses and migmatite TTGs with whitish grey trondhjemite–grey granodiorite. However, to the east of Chitradurga shear zone in the EDC typical TTG progressively decreases with an increase of transitional TTGs with tonalitic

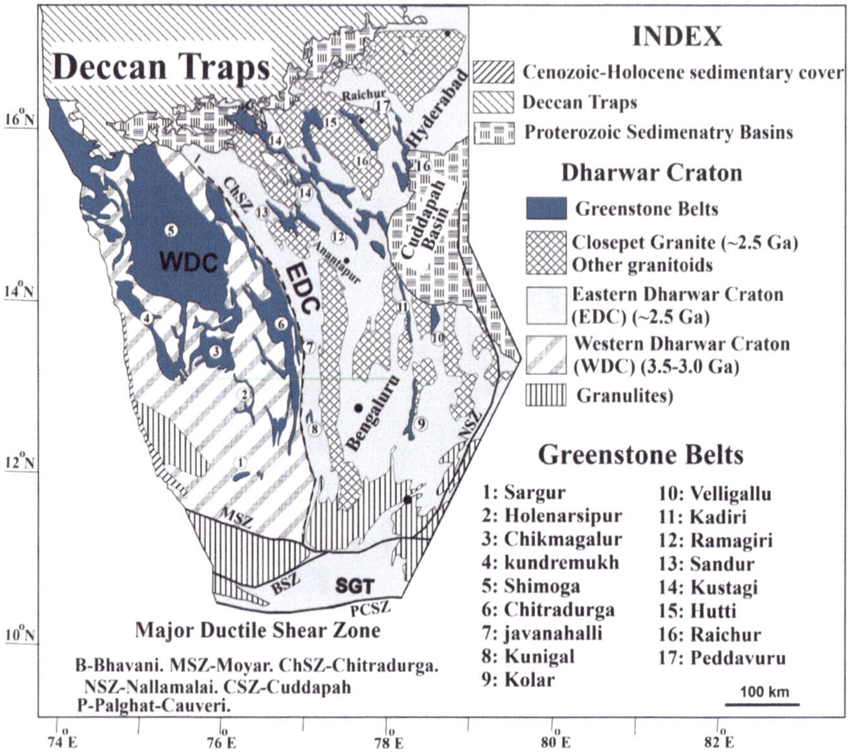

Fig. 2.21 Simplified geological map of the Western Dharwar (WDC) and Eastern Dharwar (EDC) Cratons showing distribution of main greenstone belts. Redrawn after Ramakrishnan and Vaidyanadhan (2008)

to granodioritic phases, having dioritic or mafic components. Regionally, the TTG gneisses are exposed as circular to elliptical domes with widely spaced shear zones in the WDC, while shallow to steep planar fabric are observed in the east with more closely spaced shear zones (Jayananda et al. 2018).

On the regional scale in the Dharwar craton, the TTGs show banded structure with an alternate light and dark layers. Trondhjemite is interlayered with tonalite and granodiorite that show cross-cutting relationship and form folded veins in tonalite-granodiorite. Geochronologically, from west to the east the following characters in the TTG are noteworthy: (i) in the WDC these were accreted widely during ca. 3.45–3.23 Ga (Peucat et al. 1993; Ishwar-Kumar et al. 2013; Jayananda et al. 2015; Maibam et al. 2016; Guitreau et al. 2017) with remnants of ca. 3.60 Ga due to melting of arc crust at different depths (Jayananda et al. 2015), (ii) in the central parts, these accreted during ca. 3.30–3.00 Ga, while transitional TTGs were emplaced at ca. 2700–2560 Ma (Balakrishnan et al. 1999; Peucat et al. 2013; Maibam et al. 2016; Ratheesh Kumar et al. 2016), (iii) in the EDC the gneisses accreted during 2.70–2.55 Ga (Krogstad et al. 1991; Peucat et al. 2013; Dey et al. 2016).

2.7.3 *Dharwar Greenstone Belts*

The Dharwar Craton contains numerous well-preserved greenstone belts of ages from 3.4–3.1, 2.9–2.7 and 2.7–2.5 Ga (Jayananda et al. 2013a and references therein), which are intruded by Neoarchean granitoid plutons (Fig. 2.19). Based on their lithological assemblages, grade of regional metamorphism and geochemistry the greenstone belts have been divided into Western Dharwar Craton (WDC) greenstone and Eastern Dharwar Craton (EDC) greenstone belts, separated by the Closepet granitoids (Jayananda et al. 2000; Naqvi 2005; Manikyamba et al. 2012 and references therein; Manikyamba and Kerrich 2012). The following greenstone belts characterize the Dharwar Craton:

(i) Mesoarchean mafic-ultramafic magmatism Sargur Group greenstone belt 3.4–3.1 Ga (WDC).
(ii) Neoarchaean Dharwar Supergroup bimodal mafic-felsic greenstone belts (2.9–2.7 Ga) (WDC) greenstone belts.
(iii) Eastern Dharwar greenstone belts 2.7–2.5 Ga (EDC).

2.7.3.1 WDC Dharwar Craton: Paleoarchean (WDC) Sargur Group 3.38–3.15 Ga Greenstone Belts

In the WDC Craton, the oldest greenstone bodies are associated with the Sargur Group, containing linear ultramafic-mafic volcanics with quartzite-carbonate-pelite and BIF in Sargur, Nuggihalli, Holenarsipur, Nagamangala, Krishnarajpet, J.C. Pura, Banasandra, Mercara and other smaller belts (Fig. 2.21). In these belts, komatiite and komatiite basalt dominate with minor spinifex textured flows and pillows/pillow breccia association (Jayananda et al. 2016) indicating submarine environmental eruptions as oceanic plateaus beneath or just above the sea levels. These have erupted over time range from 3.38 to 2.56 Ga (Taylor et al. 1984; Nutman et al. 1996; Jayananda et al. 2013a; Khanna et al. 2016; Jayananda et al. 2020).

Jayananda et al. (2016), Tushipokla and Jayananda (2013) recognized both Al-depleted Barberton-type and Al-undepleted Munro-type komatiites, based on major and trace element ratios such as CaO/Al_2O_3, Al_2O_3/TiO_2 and $(Gd/Yb)_N$. Elemental ratios like Nb/Th versus Zr/Nb, Nb/U versus Nb/Th Zr/Y versus Nb/Y along with their variable REE contents (slightly enriched, chondritic to depleted), flat to slightly fractionated REE patterns $[(Gd/Yb)_N = 0.7–2.3]$ suggest their derivation from heterogeneous sources of primitive to depleted to highly depleted mantle reservoirs (Jayananda et al. 2016; Tushipokla and Jayananda 2013). Al-depleted komatiites probably derived from greater depths of 350–400 km with residual garnet whilst Al-undepleted komatiites originated at shallow depth between 300 and 250 km) without any residual garnet. Nd isotope data $[(\varepsilon Nd(T) = -1.5 \text{ to } +5]$ suggest their derivation from primitive to depleted mantle sources with traces of crustal contamination.

2.7.3.2 WDC Dharwar Craton: 3.0–2.56 Ga Dharwar Supergroup Greenstone Belts

The younger Dharwar Supergroup comprises lower Bababudan Group and upper Chitradurga Group corresponding to two age groups of volcanics ca. 3.0–2.9 Ga and 2.74–2.67 Ga respectively (Kumar et al. 1996; Nutman et al. 1996; Jayananda et al. 2013a). The Dharwar–Shimoga basin corresponds to youngest volcano-sedimentary sequences which stratigraphically corresponds uppermost part of Chitradurga Group as the volcanics dated close to 2.61 Ga (Nutman et al. 1996). The 3.0–2.85 Ga Bababudan Group mafic volcanics are preserved in Kudremukh, Bababudan basin, Sigegudda and Kibbanahalli arm (Fig. 2.22). The 3.0 Ga Kudremukh greenstone belt is dominated by high-Mg basaltic rocks with minor komatiite and boninite whose elemental data such as LREE enrichment, significant negative Nb–Ta and Zr anomalies on primitive mantle normalized multi-element diagrams indicate arc signatures (Chandan Kumar and Ugarkar 2017). On the other hand, boninites show much lower incompatible element contents coupled with slight enrichment in LREE and HREE which could be attributed to their derivation from refractory mantle source. The spatial association of komatiites, high-Mg-basalts, tholeiites and boninites contrasting

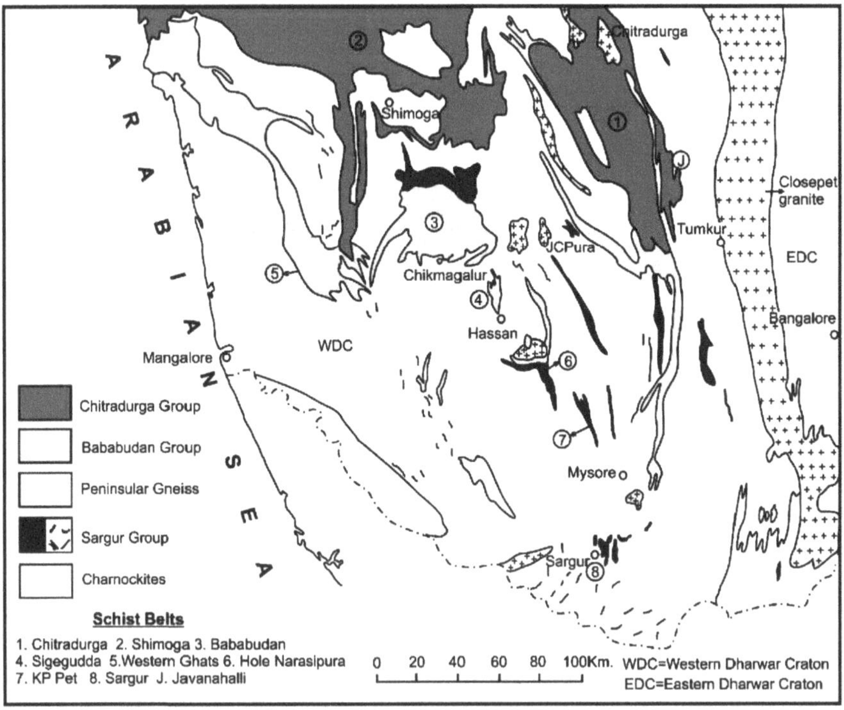

Fig. 2.22 Simplified geological map of the Western Dharwar Craton showing its Archean greenstone belts and ~2.5 Ga N–S trending Closepet Granite. After Ramakrishnan (2009)

characteristics have been attributed to plume–arc accretion during 3.0–2.8 Ga. The Bababudan basin comprises dominantly different types of vesicular to amygdaloidal basalts with minor ultramafic rocks in lower levels while felsic volcanics are developed at higher stratigraphic levels. Nd isotope data [εNd(T) = 2.90–3.30] suggest their derivation from depleted mantle source (Kumar et al. 1996).

Two different types of ultramafic rocks and alkaline basalts characterize the Sigegudda belt where the lower unit has lower MgO (12–29 wt%) and Ni (526–1150 ppm) with elevated TiO_2 and $(Gd/Yb)_N > 1$, which are similar to Ti-enriched komatiites, whilst upper unit is magnesian (22–26 wt%) with Ni (610–1000 ppm) and similar to Al-undepleted komatiites (Manikyamba et al. 2013). These komatiite flows were derived from different depths where the lower flow is formed by a lower degree of melting less than 90 km depth, while Al-undepleted komatiites originated greater than 90 km depth.

(i) **Chitradurga Group volcanics**: The Chitradurga–Gadag greenstone belt is about 400 km long and 40 km wide (Fig. 2.23) and covers an area of about 6000 km^2. The Chitradurga–Gadag greenstone belt preserves the upper 2.74–2.67 Ga Chitradurga Group volcanics while slightly younger 2.61 Ga volcanic-sedimentary sequences are developed in the Shimoga–Dharwar basin. These two belts reveal mafic volcanics of tholeiitic composition (Gupta et al. 2014). The Chitradurga belt reveal involvement of juvenile source with minor crustal contamination (Jayananda et al. 2013a, 2018). In the Gadag segment further north, metabasalts and intermediate to felsic volcanics show bimodal composition, and were derived from different juvenile sources in the arc settings (Ugarkar et al. 2000). Directly west of Chitradurga belt, elemental characteristics of 2.61 Ga felsic volcanics of Shimoga basin such as high LIL, low HFSE, negative Nb–Ta–Zr anomalies on primitive mantle normalized spider diagrams have been attributed to melting of thick basaltic arc crust at mantle depth (Manikyamba et al. 2014).

(ii) **Greenstone Belts from central Dharwar**: In this part, major greenstone belts include Ramagiri, Sandur, Kustagi, Penakacherla and Hungund belts; the former ~2.7 Ga belt extends northwards into the Penakacherla belt up to the Tungbhadra River. Mafic-ultramafics, layered basic intrusions, quartzite, calc-silicates siliceous schist and BIF make the lower unit in southwestern parts, pillowed metabasalts, greywacke and felsic volcanics make the bulk of these belts. Manikyamba et al. (2004) suggested that metabasalt of the Ramagiri belt formed from depleted sources like the N-MORB or depleted mantle wedge due to their depleted LREE, flat HREE, absence of any Eu anomalies, coupled with high Cr, Ni, V and negative Nb–Ti anomalies. Large variations in Nd and Pb isotopic compositions [εNd(T) = + 0.1 to +0.2 and −8 to −12] were caused possibly due to their alteration by external fluid source from granitoid intrusions of 2.54 Ga juvenile Closepet Batholith (Zachariah et al. 1995).

The Sandur Greenstone Belt (ca. 2.70 Ga) contains abundant basalt and felsic volcanics at higher stratigraphic levels with minor komatiite to komatiitic basalt which have been attributed to a fossil plume (Naqvi et al. 2002). Subsequently,

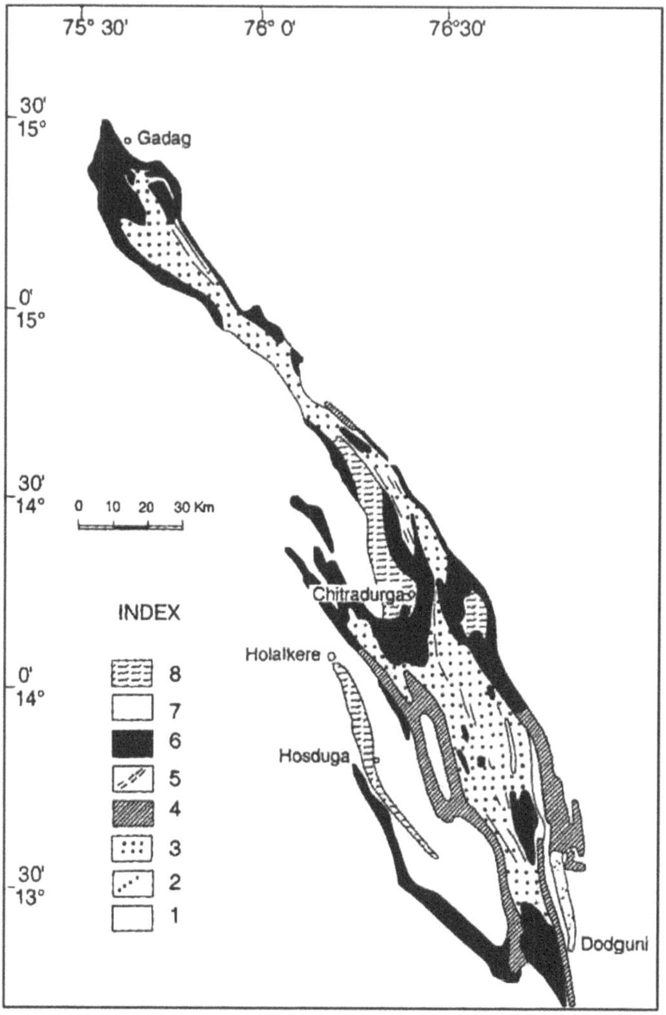

Fig. 2.23 Geological map of Chitradurga–Gadag greenstone belt. 1—Gneiss. 2, 3—Quartzite. 4—Phyllite with iron manganese marker horizon. 5—Banded Iron Formation. 6—Mafic flows. 7—Greywacke. 8—Intrusive granite. After Radhakrishna and Vaidyanadhan (1997)

Manikyamba et al. (2008) obtained distinct elemental signatures of plume–arc accretion in ultramafic-mafic and felsic volcanics. Geochemical data from other ca. 2.7 Ga mafic to felsic volcanics of greenstone belts in the central province suggest their derivation from heterogeneous mantle sources (enriched to depleted) of plume–arc settings (Manikyamba et al. 2012), whilst 2.55 Ga felsic volcanics formed by melting of thickened arc crust in continental margin settings (Dey et al. 2017; Jayananda et al. 2020).

(iii) **Greenstone belts from the Eastern Dharwar Craton (EDC)**: Numerous high-Mg basaltic greenstone belts dominate the eastern Dharawar with small amounts of komatiite and intermediate to felsic volcanics/pyroclastics around Kolar, Kadiri, Veligallu, Jonnagiri, Gadwal and Hutti (Fig. 2.21). These range in age from ~2.7 to 2.56 Ga (Manikyamba et al. 2017; Jayananda et al. 2013a, 2018 and references therein). Volcanics of these belts are interlayered with minor greywacke-argillite, carbonate and BIFs. Some of the important greenstone belts from the eastern Dharwar are described below.

Kolar Greenstone Belt (KGB): N–S trending narrow linear KGB (80 × 2 km) is one of the easternmost typical greenstone belts of the EDC, intruded by diapiric Patna–Bisanattam granitoids and granodioritic gneiss. It consists of amphibolite, pyroclastic felsic flows, polymictic conglomerate, greywacke-argillite, BIF, cordierite–sillimanite–fuchsite quartzite–calc-silicate and chert (Fig. 2.24a and b). This sequence is

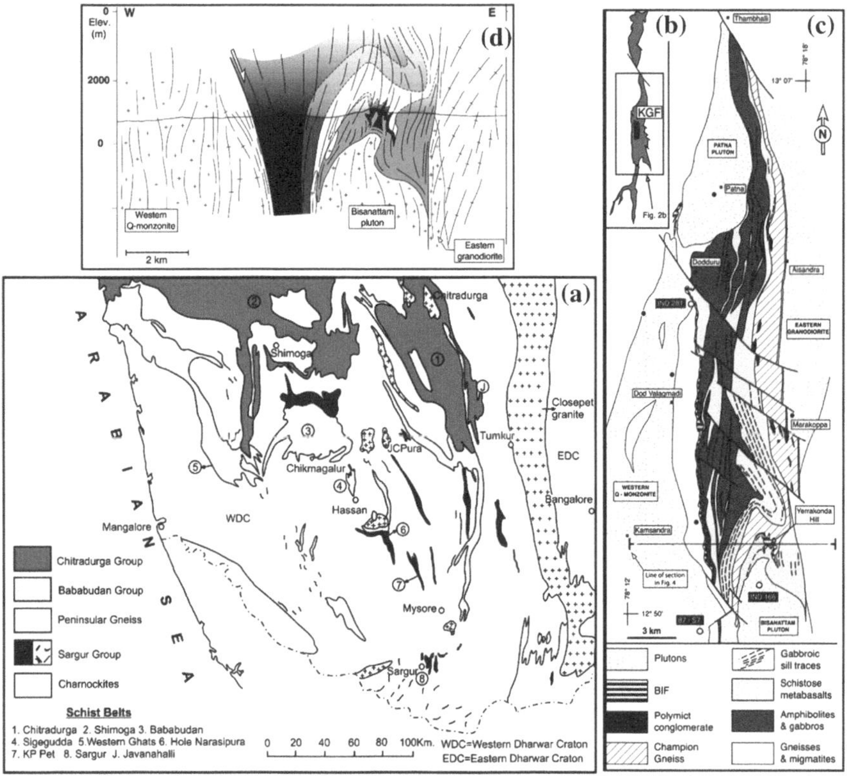

Fig. 2.24 Kolar Greenstone Belt. **a** Sketch map of the Kolar belt. **b** Geological map of the KGB and its neighbouring granite-gneiss terrains. **c** E–W section across the KGB with foliation traces. Pitch of stretching lineation changes from gentle to steep within the belt. **d** 3-D crustal-scale strain pattern and kinematics of the KGB, Krishnagiri (Kr), and Karimangalam (Ka) areas. All diagrams from Chardon et al. (2002) with geological map by Viswanatha (1978)

overlain by pillowed amphibolite, and intercalated high-Mg basalts of komatiitic and picritic affinity. Gold mineralization is associated with volcanics, quartz-carbonate veins and stratiform sulphide lodes in BIF along with shear zones.

The KGB is deformed by early two-stage tight-isoclinal folding and shearing during major magmatic accretion event at middle and lower crustal within the EDC due to E–W bulk inhomogeneous shortening causing N–S horizontal stretching and syntectonic juvenile pluton emplacement (Fig. 2.24c) between 2550 and 2530 Ma (Chardon et al. 2002). This has caused pervasive synchronous N–S trending vertical foliation and shallow stretching lineation coupled with conjugate strike-slip shear zone pattern (Fig. 2.24d). This was coeval and compatible with deformation during juvenile magmatic accretion, melting, and granulite metamorphism in the lower crust.

Rajamani et al. (1985) interpreted komatiitic to tholeiitic amphibolite from the KGB, originating by 10–25% melting of the heterogeneous mantle over >80 km depth and 1575 °C, whilst tholeiitic basalts originated from similar sources and melting conditions <80 km depth. Amphibolite and gneiss terranes across the KGB were juxtaposed by horizontal E–W compression between 2530 and 2420 Ma along this belt, which sutured discrete crustal terranes on its either sides (Krogstad et al. 1989).

Kadiri greenstone belt: North of the KGB, the 2.56 Ga Kadiri greenstone belt (75 × 2.5 km) disappears beneath the Cuddapah Basin and consists mainly of acid volcanics, polymictic conglomerate, sericite schist, tuffs, greywacke, BIF and basic volcanics (Jayananda et al. 2013a; Dey et al. 2015; Manikyamba et al. 2015). Type I basalts show flat to slightly depleted LREE patterns whilst Type II basalts exhibit enriched LREE patterns (Manikyamba et al. 2015). Their Ba/Th, Nb/Th, Th/Nb, La/Nb, Ce/Yb ratios have also been attributed to magma generation at spinel to garnet stability field in mantle for back-arc–rift system proximal to intra-oceanic arc (Fig. 2.25; Manikyamba et al. 2015). On the other hand, Dey et al. (2015) have

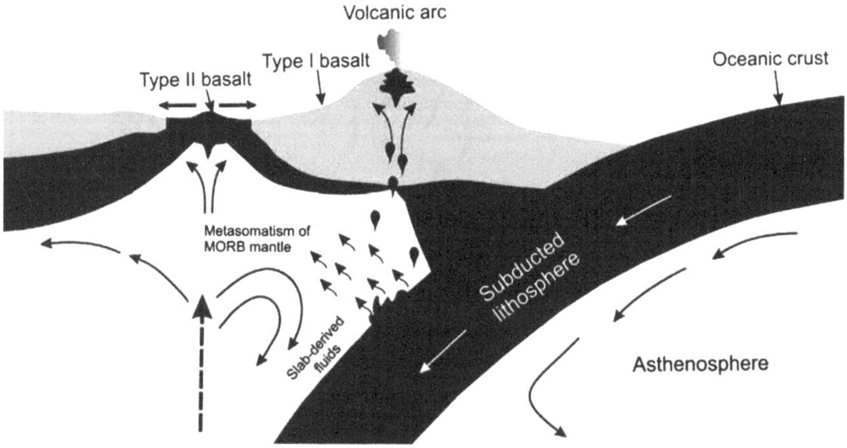

Fig. 2.25 Cartoon illustration showing the generation of type I and type II basalts of KGB in an arc–back arc setting. After Manikyamba et al. (2015)

identified tholeiitic basalts, calc-alkaline high-Mg# andesites and dominant dacites–rhyolites. The flat REE patterns and slight negative Nb–Ta anomalies of basaltic rocks have been attributed to plume fed ocean plateaus whilst high Mg# coupled with Cr–Ni, depleted HFSE and enriched LREE of andesites were explained by melting of mantle wedge over subducting slab. Further, the elemental characteristics of dacite-rhyolite suite (SiO_2 = 61–72 wt%, low Cr, Ni, highly fractionated REE patterns) are consistent with their derivation by melting of thickened arc crust in continental margin setting.

Veligallu greenstone belt: The Veligallu greenstone belt of approximately 480 km^2 area is exposed up to the Cuddapah Basin and is split into two branches due to a granite intrusion comprises ultramafic-mafic-felsic volcanic sequences (2.7–2.67 Ga) (Jayananda et al. 2013a; Khanna et al. 2016; Dey et al. 2018). Elemental data of Mafic-ultramafic rocks contain high MgO (24–34 wt%), Cr (2290–3855 ppm), Ni (604–966 ppm) coupled with moderate to low SiO_2 (52–49 wt%), Al_2O_3 (9.2–4.7 wt%), TiO_2 (0.30–0.14 wt%) and depleted middle REE due to derivation from mantle sources in intra-oceanic arc setting lacking critical geochemical evidence for plume origin (Khanna et al. 2016). On the other hand, Dey et al. (2018) suggested minor contamination of deep mantle-derived ultramafic lavas with sediments, while associated tholeiitic basalts in absence of crustal contamination are attributed to shallow melting of slightly depleted mantle.

Hutti greenstone belt: The Hutti greenstone belt (670 km^2) is the northernmost extension of the Kolar belt, and has significant gold mineralization. Pb–Pb and Sm–Nd whole rock isochrons and U–Pb zircon ages indicate ages ranging 2.67–2.57 Ga (Anand and Balakrishnan 2010; Jayananda et al. 2013a). It preserves bimodal tholeiitic basalts and calc-alkaline intermediate to felsic volcanics with tuffaceous chlorite schist and BIF (Rogers et al. 2007; Manikyamba et al. 2009). A younger sequence is made up of felsic volcanics, greywacke, and conglomerate. Two different types of basalts are identified and include (i) enriched basalts in LREE (20–50 times chondrite) and LREE depleted (12–20 times chondrite), and are fractionated (La/Sm = 12–17) with negative Nb, P, Ti anomalies, and (ii) LREE depleted basalts are poorly fractionated and show feeble Nb, P and Ti anomalies with $(Gd/Yb)_N$ ratios ranging 1.1–1.6. Depleted basalts show N-MORB signatures whilst enriched basalts depict less depleted arc mantle reservoirs which is consistent with Nd [$\varepsilon Nd(T)$ = 1.4–3.0] isotope data. Intermediate to felsic volcanics with low Y, high $(La/Yb)_N$ and (Zr/Sm) but Nb/Ta or without or with positive Eu anomalies could be probably related to their derivation by melting of thickened arc crust (Manikyamba et al. 2009).

Peddavura greenstone belt: NW–SE trending belt is 25 km in length and 0.5–2 km in width which is partly covered by Mesoproterozoic Cuddapah basin. The major lithologies in this belt are pillowed basalt, basaltic andesite, andesite and rhyolite. Imprecise U–Pb zircon age indicates 2.4 Ga age corresponding to the minimum age for their eruption (Jayananda et al. 2013b) whilst Rajamanickam et al. (2014) obtained a Rb–Sr WR isochron age of 2551 ± 19 Ma (MSWD = 1.16), which tallies with U–Pb zircon ages of rhyolite from other greenstone belts in the EDC (Jayananda et al. 2013b). Elemental data, REE abundances and HREE fractionation of basaltic to rhyolite sequences together with Nd–Sr isotope data are consistent with their derivation from heterogeneous sources (mantle and thickened arc crust) in the oceanic island arc setting (Rajamanickam et al. 2014).

In summary, ca. 2.7 Ga greenstone volcanics comprise different kinds of basaltic rocks, intermediate to felsic volcanics, minor boninite and komatiite, and is spatially associated with accretion of the surrounding ca. 2.7 Ga gneissic granitoids. Whole-rock geochemical and tracer isotope (Nd–Pb–Sr) data suggest derivation of this volcanic suite from the heterogeneous mantle and thickened arc crust. Their magmas generated either deep mantle in hotspot environments associated with plume or at different depths in shallow arc mantle and thickened island arc crust. On the other hand, 2.58–2.54 Ga felsic volcanics are spatially related to the emplacement of sanukitoids and generated by melting of arc crust at different depths with minor involvement of ancient crust in the western block and central block (Jayananda et al. 2013b).

Pb–Pb, Rb–Sr, Lu–Hf and Sm–Nd whole-rock isochron ages for mafic volcanics in all greenstone belts show a range of 2700–2640 Ma (Anand and Balakrishnan 2010; Khanna et al. 2016). U–Pb ages of detrital zircon grains from the BIFs are 2719 $\pm$ 19 to 2698 $\pm$ 50 Ma, indicating timing of deposition of pelagic sediments with ocean floor basalts (Yang and Santosh 2015). On the contrary, U–Pb zircon ages for felsic volcanics/pyroclastics have bimodal distribution with rhyolites from Sandur, Ramagiri, Veligallu greenstone belts showing eruption time as ca. 2700–2680 Ma and those from Kolar–Kadiri, Gadwal, Hutti-Muski, Peddavura belts from 2587 to 2545 Ma (Jayananda et al. 2013a; Dey et al. 2015; Yang and Santosh 2015).

2.7.4 Closepet Granite (Calc-alkaline to Potassic Plutons)

The Late Archaean Closepet Granite is a 400 km long and approximately 20–30 km wide nearly N–S to NNW–SSE trending arcuate batholith along the eastern sheared margin with the Western Dharwar Craton (Fig. 2.26a; Moyen et al. 2003). The exposed body is divided into northern and southern parts, separated near Kalyandurga by the Sandur Schist belt; however, both the parts appear to be lithologically similar (Naqvi and Rogers 1987). Since the main body crops out along crustal cross-section of about 10–13 km in depth, from granulitic lower crust in the south to greenschist-facies upper crust in the north, it provides an opportunity to study the emplacement of granitic magmas at different crustal levels.

Moyen et al. (2003) subdivided the Closepet Granite into three depth zones: (i) Root Zone, (ii) Transfer Zone, and (iii) Intrusion Zone (Fig. 2.26b).

2.7.4.1 Root Zone

This zone extends from the Cauvery River in the south to Magadi and consists mainly of coarse porphyritic monzogranite and minor amount of cpx-bearing monzonite in large rounded to elongate bodies. Marginally, pink and grey anatectic granite grades

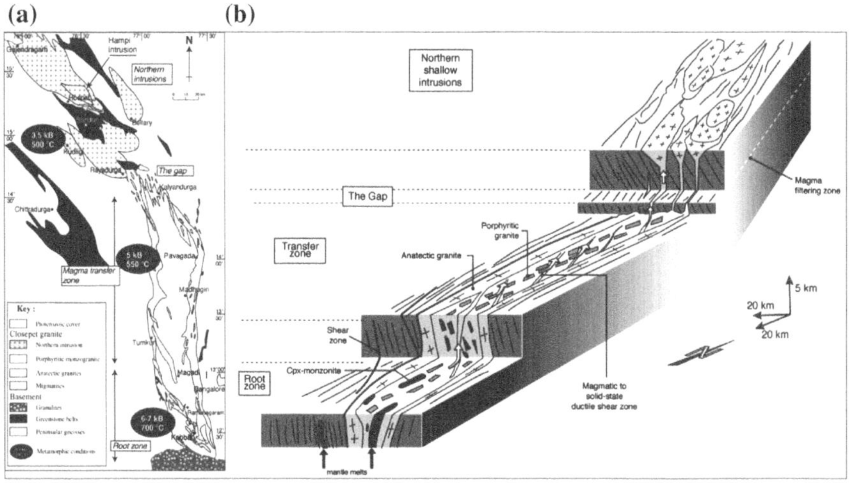

Fig. 2.26 Closepet Granite. **a** Simplified geological map of the Closepet Granite, showing its variation from north to south. **b** 3-dimensional cross-section showing variations from deep-seated root zone in south to northern shallow intrusions. After Moyen et al. (2003)

into basement gneiss through a thick zone of intense migmatization (Fig. 2.26a). The Root Zone is highly heterogenous with concentrated feldspar, microgranular enclaves, basement xenoliths, cumulate enclaves etc. This zone represents deepest crustal levels, where granite interacted with mantle derived magma and the lower crust on large-scale (Fig. 2.26b; Jayananda et al. 1995; Moyen et al. 2003). Moyen et al. (2001) deduced that mantle-derived mafic magma intruded the gneissic crust and induced its partial melting, followed by 5–10% of fractional crystallization, and mixing/mingling of mantle derived and crustal magmas to account for diverse chemical and petrological features of the Closepet Batholith.

2.7.4.2 Transfer Zone

This zone extends between Kalyandurga and Magadi to the magmatic gap area. Main rock type is again the porphyritic monzogranite, which is associated with anatectic granite along the periphery. Mafic enclaves, xenoliths and feldspar accumulations are restricted into narrow channels paralleling the main foliation in the Closepet Granite. These channels are likely zones of magma ascent. Despite local petrographic heterogeneities, physical continuity of the porphyritic monzogranite can be observed all over from root to transfer zones. Consequently, the Closepet Granite appears as a single magmatic body rather than a series of plutons.

2.7.4.3 The Gap

The geological map of the 'magmatic gap', between Kalyandurg and Rayadurga (Fig. 2.26a) shows (i) southern part of narrow apophyses of porphyritic monzogranite and abundant 10–50 m-wide pink and grey equigranular granite sheets from the main body into the Peninsular Gneiss, (ii) central part of only a network of grey or pink granite sheets within the Peninsular Gneiss, and (iii) predominant greyish granite near Rayadurga as an enclave- and schlieren-rich granite changing into homogeneous granite.

Thus, a link gets established between the main mass and the northern intrusions through 'magmatic gap', consisting of a network of granitic sheets connecting heterogeneous southern main mass of the batholith to the homogeneous northern intrusions.

2.7.4.4 Northern Granite Intrusions

Small 10–50 km long ellipsoidal medium-grained, equigranular or weakly porphyritic granite intrusions characterize the northern parts of the Closepet Granite, having sharp contacts with the Peninsular Gneisses and lacking any deformation and migmatization. Since northern group occur along strike of main body, they are classified as part of the Closepet batholith due to their nearly coeval ages, identical mineralogical, chemical composition and differentiation trends for both major and trace elements (Jayananda et al. 2019).

The Closepet Granite has been dated as 2513 ± 5 Ma (Friend and Nutman 1991; Jayananda et al. 1995) and is a part of widespread Neoarchaean phase of plutonism (Mojzsis et al. 2003; Maibam et al. 2011; Jayananda et al. 2000, 2013a, 2020) in the Eastern Craton and mark the stabilization of the craton. Numerous other N–S trending composite granitoid plutons intrude both the central and eastern basement of the Dharwar (Jayananda et al. 2020a). These magmatic bodies are composite plutons and comprised of most abundant juvenile intrusive facies with numerous Mafic Magmatic Enclaves (MME) and minor anatectic granites derived by reworking of pre-existing crust (Jayananda et al. 1992; Moyen et al. 2003). The intrusive facies including clinopyroxene-bearing quartz-monzonite and porphyritic monzogranite and are confined to the central part whilst anatectic pink and grey granites are confined to periphery of the plutons (Jayananda et al. 2019; and references therein). These plutons contain numerous disrupted/boudinaged synplutonic mafic to intermediate dykes and Mafic Magmatic Enclaves (MME) which show interaction, mixing and mingling with the host granitoids (Prabhakar et al. 2009; Jayananda et al. 2009, 2014). Like the main Closepet granite, U–Pb zircon age data of granitoids in the EDC reveal widespread late Neoarchaean juvenile plutonism during 2580–2510 Ma (Balakrishnan et al. 1999; Jayananda et al. 2000, 2020a; Krogstad et al. 1991, 1995) representing 2.68–2.60 Ga transitional TTGs, 2.58–2.53 Ga sanukitoids, 2.54–2.52 Ga Closepet-type mantle and crustal melting, and finally 2.53–2.52 Ga

K-rich granitoids due to crustal anatexis (Dey et al. 2016; Jayananda et al. 2020a, references therein).

2.7.5 *Major Shear Zones*

High-grade granulite and gneiss along with linear supracrustal belts of the Dharwar craton are dissected by some major shear zones and play an important role in the reconstructions of the continental assembly. Using satellite images and integrated fieldwork, Drury (1980), Drury and Holt (1980) and Drury et al. (1984) produced some of the early geological maps of shear zones and identified major structures like Nallamalai-Velikonda, Moyar-Bhavani, Attur, Palghat-Cauvery, and Achankovil. Detailed structural analysis led to the visualization of ductile shear senses along many of these shear zones (Figs. 2.27 and 2.28; Jain et al. 2003; Ramakrishnan 2003). Ramakrishnan (2003) recognized the following shear zones: Balehonnur Shear Zone (Bh), Babubudan Shear Zone (Bb), Chitradurga Boundary Shear Zone (CT) Chitradurga Shear Zone (CSZ), Mettur Shear Zone (MT)/Cuddapah Eastern Margin Shear Zone (C)/Nallamalai Shear Zone (N), Eastern Ghats Shear Zone (E), Bhavani Shear Zone (B), Moyar Shear Zone (M), Noyil-Cauveri (N-Ca)/Palghat-Cauveri Shear Zone (PC), and Achankovil Shear Zone (AK). A comprehensive structural mapping based on LANDSAT imagery, existing documentation and field data led Ramakrishnan (2003) and Chardon et al. (2008) to delineate anastomosing array of about 20 important shear zones (Fig. 2.27); some of these are briefly described below.

2.7.5.1 Balehonnur Shear Zone (BhSZ)

The BhSZ is the westernmost first-order shear zone in the Dharwar Craton with shallow plunging lineation and left-lateral kinematic indicators; fabric trajectories indicate shortening across the shear zone (Chardon et al. 2008). Southern strand of this zone possibly is deflected into the Moyar Shear Zone and merges with the Ranibennur shear zone NW of Honnali.

2.7.5.2 Chitradurga Shear Zone

This arcuate shaped of anastomosing shear zone separates the Eastern Dharwar Craton (EDC) from the Western Dharwar Craton (WDC) for about 400 km with a width between 1 and 4 km and is considered as the boundary between them (Swami Nath et al. 1976; Chadwick et al. 2000). It extends from Markonahalli near Kunigal to the western margin of the BR Hills massif into the junction of Moyar and Bhavani shear zones. Chadwick et al. (2000) observed that a narrow mylonite belt sharply borders the Chitradurga schist belt.

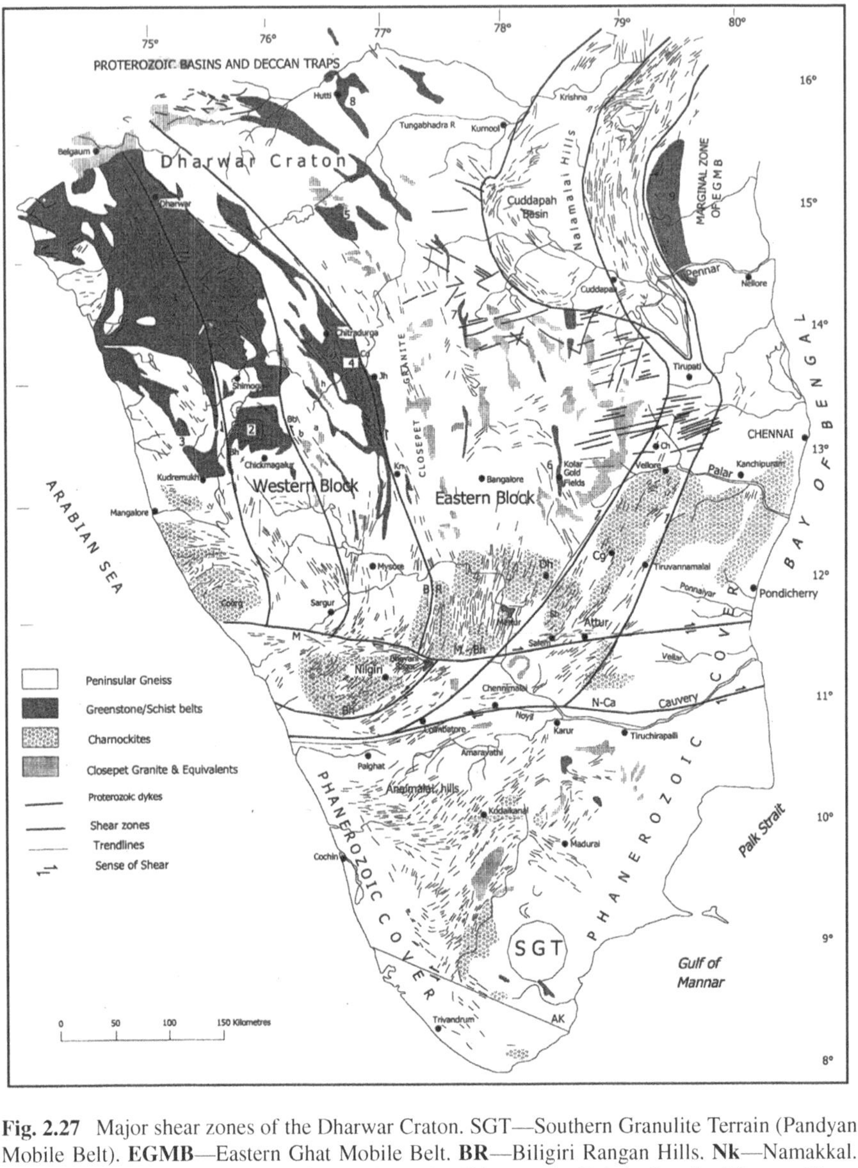

Fig. 2.27 Major shear zones of the Dharwar Craton. SGT—Southern Granulite Terrain (Pandyan Mobile Belt). **EGMB**—Eastern Ghat Mobile Belt. **BR**—Biligiri Rangan Hills. **Nk**—Namakkal. **Sh**—Shevaroy hills. **Greenstnone/Schist Belts**: 1—Shimoga. 2—Bababudan. 3—Western Ghats. 4—Chitradurga. 5—Sandur. 6—Kolar. 7—Ramagiri, 8—Hutti. 9—Nellore. 10—Veligallu. **Shear Zones**: N-Ca—Noyil-Cauvery. M—Moyar. B—Bhavani. CB—Cuddapah boundary shear zone. Mt—Mettur shear zone linking up with Nallamalai. Ak—Achankovil. Bh—Baiehonnur. Bb—Bababudan. Cd—Chitradurga. After Ramakrishnan (2003)

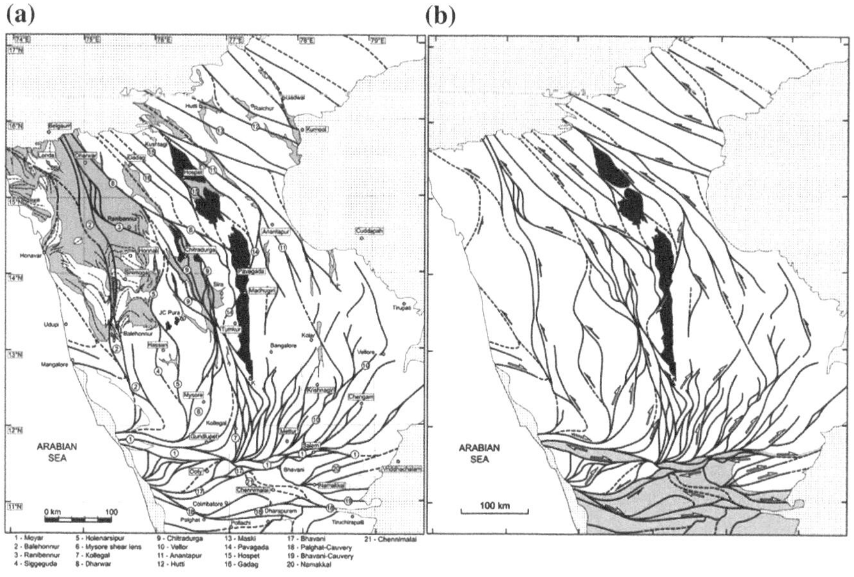

Fig. 2.28 **a** Major shear zones (thick lines) of the Dharwar Craton with axial traces of folds/domes shown by thin dashed lines. Many shear zone boundaries are marked by mylonite and lineation. **b** Shear zone pattern of the Dharwar Craton and its kinematics. Major shear zones are shown with their sense of movements. After Chardon et al. (2008)

Roy et al. (2008) located an intense zone of almost subvertical and N10° W trending mylonitic foliation and strong sub-horizontal stretching lineation on these foliations. Persistent sub-horizontal stretching lineation on subvertical mylonitic foliation, asymmetrical folds, orientation of shear bands, and asymmetry of porphyroclast tails indicate a dominant sinistral sense of strike-slip movement along the shear zone. Within the shear zone, mylonitic gneiss shows a gradual variation from protomylonite to mylonite and ultramylonite from margins to the central part of the shear zone with thin pseudotachylyte veins.

According to Chardon et al. (2008), this zone consists of northward-converging dextral and sinistral shear zones (dextral Mysore shear lens, Kollegal shear zone, and anastomosing shear system between Kollegal and Kabbaldurga) that merges around 13°30′ N and diverge northward to form a duplex that ends NE of Ranibennur. In the north, this system is confined to a single trace as the Dharwar shear zone around the eastern boundary of the Shimoga greenstone belt (Fig. 2.28a).

2.7.5.3 Mettur Shear Zone (MT)

Delimiting the deformation to the west of the Cuddapah basin, this shear zone is linked to the fault in the west of Nallamalai fold belt through Chittoor and Shevaroy Hills, and was designated as the Nallamalai shear zone (Drury et al. 1984;

Project Vasundhara map GSI–ISRO 1994). Also, it coincides with NE-trending Main Fault (cf. Grady 1971), shear zone near Vellore (cf. Drury and Holt 1980), Mettur-Palakkodu fault (Gopalakrishnan 1996) and Mettur-Dharmapuri shear zone (Chetty et al. 2003). This shear zone marks the boundary of basement deformation related to the Cuddapah event and hosts numerous younger intrusions (750 Ma), hence it was also called as the Dharmapuri suture-rift zone (Gopalakrishnan 1996). Seismic reflection profile between Krishnagiri and Bhavani regions reflects the evidence for late Archean collision along the "Mettur shear zone" where the belt of NE trending foliations is well developed (Drury et al. 1984; Vijaya Rao et al. 2006).

2.7.5.4 Moyar Shear Zone (MSZ)

Almost E–W trending MSZ is characterized by deflecting the NNE-trending fabric of the Northern massifs clockwise and possesses very steep east–west foliations with steeply to moderately plunging lineation (Bhadra 2000; Meißner et al. 2002; Jain et al. 2003). Gradual rotation of the regional trends on either side of the MSZ (Figs. 2.28 and 2.29) reveals its prominent dextral shear character in contrast to the postulated large-scale upthrust displacements (Naha and Srinivasan 1996). The shear zone skirts around the northern margin of Sankaridurg and extends uninterruptedly eastwards toward Salem and Attur (Fig. 2.29).

The MSZ is characterized by strong penetrative almost E–W trending mylonitic shear foliation and dips very steeply. A subordinate trend of foliation having N30° E orientation is also noteworthy and may correspond to the relict N–S foliation of the Northern massifs within the shear zone. Numerous shear criteria like asymmetric mafic boudins, S-C shear fabric, and rotation of axial surfaces of tight to isoclinal folds within the MSZ reveal its distinct ductile dextral shear sense of sub-horizontal movement, indicated by sub-horizontal to gently plunging mineral/stretching lineation. However, mylonitic foliation also contains variable sub-vertical to steeply down-dip plunging lineation towards NW or SE and its gradual rotation of due to dextral shearing within the MSZ. Further eastwards in the Satyamangalam region, orientation of dextral shear zones have slightly changed to S80° E with steep northerly dips, while sinistral shears remain oriented at N30° E with steep E/W dips. In the easternmost region, this shear zone is termed as the Salem-Attur Shear Zone (Drury et al. 1984), and trend S70° E with very steep to vertical dips and sinistral ductile shear zones are oriented N20° E. All along the MSZ, mineral/ stretching lineation is well developed on mylonite foliation and plunges either steeply down-dip (Naha and Srinivasan 1996; Bhadra 2000) or undergoes rotation to become moderate to gentle (Chetty and Bhaskar Rao 1998); in such cases, transpressive kinematics is postulated (Plavsa et al. 2015).

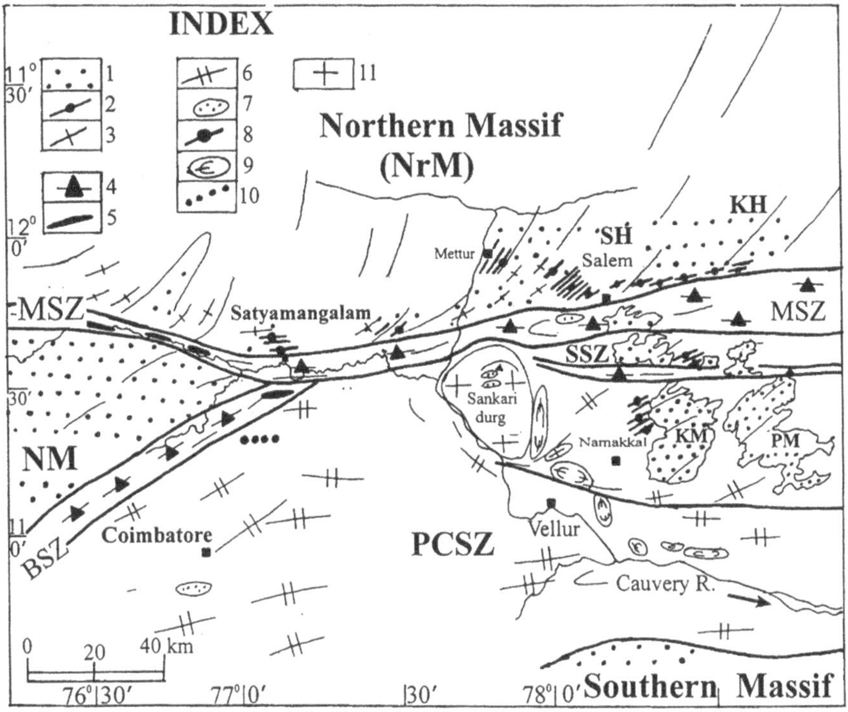

Fig. 2.29 Simplified geological map of major shear zones in the Nilgiri-Coimbatore-Namakkal area. Northern massif (NrM): Nilgiri massif (NM). Kollimalai (KM). Pachaimalai (PM). 1—Massive charnockite. 2—Basic granulite. 3—Quartzo-feldspathic gneiss. Moyar (MSZ)/Bhavani (BSZ) Shear Zones: 4—Mylonite, Phyllonite. 5—Pyroxenite/amphibolite. Palghat-Cauvery Shear Zone (PCSZ): 6—Hornblende-biotite gneiss, quartzo-feldspathic gneiss, mafic bands, amphibolite, sillimanite-biotite gneiss/schist, quartzite. 7—Calc-silicates, marble. 8—Mafic banded granulite, quartzo feldspathic gneiss; 9 Quartzo feldspathic gneiss and migmatite dome. 10—Mafic-ultramafic intrusives. 11—Granite. SH—Shevaroy Hills. KH—Kalrayan Hills. After Jain et al. (2003)

2.7.5.5 Bhavani Shear Zone (BSZ)

The Nilgiri Massif (NM) is marked by another prominent almost ENE-trending Bhavani Shear Zone (BSZ) along its southern margin and possibly truncated by the MSZ around Bhavani Sagar and Satyamangalam (Fig. 2.29). Numerous shear criteria within the shear zone reveal its distinct sinistral ductile character with steep mylonitic foliation dipping towards northwest and bear strong gently plunging lineation towards SW. Most prominent character of the MSZ, i.e. bending of regional trends of the Northern massifs, calls for the presence of a single shear zone system all through, as has been shown by Ramakrishnan (1993), unlike the opinion expressed by Drury et al. (1984) that the BSZ extends beyond its confluence with the MSZ. The BSZ cuts across granulite isograd of the NM and contains lenses of high-pressure mafic granulites (Raith et al. 1999).

2.7.5.6 Palghat-Cauvery Shear Zone (PCSZ)

Extensive low-lying expanse around Coimbatore-Namakkal-Tiruchirappalli and from east to west coast is characterized by E–W trends within the Palghat-Cauvery Shear Zone (Fig. 2.30; Drury et al. 1984; Chardon et al. 2008). In this domain, eastward draining Cauvery River flows along distinct E–W trending major shear zone boundary from south of Namakkal to Tiruchirapalli. Sigmoidal bending of NE-trends of Kollimalai and Pachaimalai hills along their southern margins indicates the presence of a distinct large-scale dextral shear zone, where numerous elliptical quartzo-feldspathic gneissic domes are located south of Namakkal (Fig. 2.30). Further, NNE-trends within these massifs and adjoining areas show undisputed evidences of the continuation and presence of the Northern massifs within the PCSZ like the one shown in the geological map of Project Vasundhara (GSI–ISRO 1994).

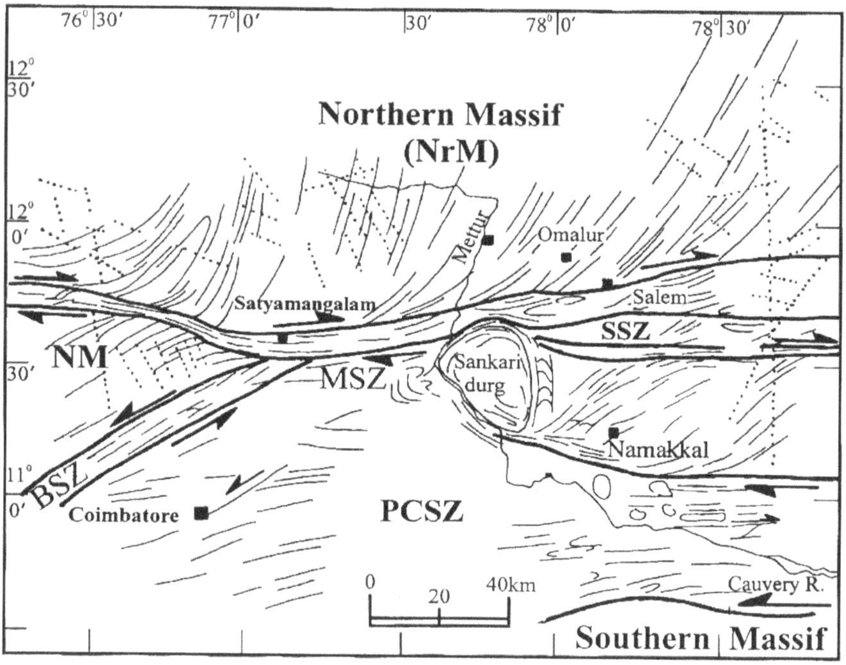

Fig. 2.30 Lineament map of major shear zones of the Southern Granulite Terrain. **MSZ**—Moyar Shear Zone. **BSZ**—Bhavani Shear Zone. **PCSZ**—Palghat-Cauvery Shear Zone. **SSZ**—Sweta Shear Zone. **NM**—Nilgiri massif. **KM**—Kollimalai massif. PM—Pachaimalai massif. **Na**—Namakkal. After Jain et al. (2003)

Within the PCSZ, the main foliation trends almost E–W with subordinate orientation towards NE and N and is marked by E or W gently plunging mineral lineation (D'Cruz et al. 2000; Jain et al. 2003). Dextral shear zones within the PCSZ are oriented almost E–W to ESE–WNW with steep to vertical dips; N30° E trending sinistral shear zones also characterize the PCSZ. These shear zone record UHT and HP metamorphism (Shimpo et al. 2006). Other characters of the PCSZ are as follows:

(i) Sheared asymmetrical pyroxenite and retrogressed amphibolite lenses within quartz-biotite gneiss indicate distinct dextral shearing. Such mafic and ultramafic bodies are either concordantly banded with biotite gneiss or sheared and folded along with the country rocks indicating their very early emplacement in the geological history of the PCSZ. In contrast, another group of long rectilinear mafic-ultramafic bodies appear to be dykes and subsequently intruded the high-grade terrain.

(ii) Widespread banded to streaky hornblende-biotite gneiss and quartzo-feldspathic gneiss within the PCSZ change into migmatite with leucosome either within the foliation boudins, irregular pods or ductile shear zones. In the undeformed segments of the PCSZ, isolated metadiorite and metagranodiorite still retain their igneous textures along with large undeformed charnockitic massifs. This pre-date the main ductile shearing and are only marginally sheared. Structureless patchy charnockite superpose many ductile shear zones, where both gneissic and sheared fabric are destroyed, thus indicating the possibility of youngest granulite metamorphism, post-dating the main deformational episodes.

(c) Within this terrain, numerous charnockite bodies range from a few cm to many km. One finds undisputed field evidence for charnockitization post-dating the intense deformation and ductile shearing and at least one phase of granulite metamorphism pre-dating this deformation.

(d) Some of the largest N–S and NNE-trending lineaments extend for about 150 km where dolerite dykes or pseudotachylite veins mark these features, such as Gangavalli Shear Zone near Salem (Behera et al. 2017). These post-date all the regional trends within the area and appear to be the youngest.

(e) Elliptical high strain zones around granite bodies are so distinct on the imageries that one can pick them up without much difficulty. The Sankaridurg granitoid complex and other domes along the Cauveri River are bounded by strong elliptical fabric within the country rocks, which appear to be highly strained due to magmatic emplacement (Fig. 2.30).

(f) U–Pb/Th zircon/monazite dating of the shear zones yielded Neoproterozoic to earliest Paleozoic ages provide ages between 800 and 525 Ma (early D2 deformation dated between 700 and 800 Ma and a later phase D3 between 550 and 600 Ma) (Ghosh et al. 2004), while Collins et al. (2007) identified the PCSZ as the boundary between crustal domains of contrasting origin at ~535 Ma. Cooling below biotite Rb–Sr closing temperature took place until 450 Ma (Meißner et al. 2002).

2.7.6 Proterozoic Mafic Dikes

Since the publication of dyke map in eastern Dharwar (Halls 1982), several sets of Proterozoic dykes are now mapped almost orthogonal to the regional trend of the Dharwar Craton (Halls et al. 2007; French and Heaman 2010; Kumar et al. 2012; Srivastava et al. 2014; Söderlund et al. 2019). These dykes cover an area of about 140,000 km^2 of the Proterozoic Large Igneous Province (LIP) (Srivastava et al. 2020) and are concentrated more in the southern and northeastern parts (Kumar et al. 2012). These have a wide range of ages between 2.43 and 2.10 Ga (Radhakrishna and Joseph 1996; Pandey et al. 1997; French and Heaman 2010; Kumar et al. 2012), with some ages clustering around 1.85 Ga (French et al. 2008).

Kumar et al. (2012) and Söderlund et al. (2019) have provided many U–Pb baddeleyite ages (ID-TIMS) on the mafic dykes from the Dharwar craton; the latter identified 8 different mafic dyke swarms between 2.37 and 1.79 Ga across the eastern Dharwar Craton (also, Nagaraju et al. 2018a, b; Samal et al. 2019): (i) NE–SW to ESE–WNW trending ca. 2.37 Ga Bangalore–Karimnagar swarm, (ii) N–S to NNE–SSW trending ca. 2.25 Ga Ippaguda-Dhiburahalli swarm, (iii) N–S to NNW–SSE trending ca. 2.22 Ga Kandlamadugu swarm, (iv) NW–SE to WNW–ESE trending ca. 2.21 Ga Anantapur-Kunigal swarm, (v) NW–SE to WNW–ESE trending ca. 2.18 Ga Mahbubnagar-Dandeli swarm, (vi) N–S, NW–SE, and ENE–WSW trending ca. 2.08 Ga Devarabanda swarm, (vii) E–W trending 1.88–1.89 Ga Hampi swarm, and (viii) NW–SE ca. 1.79 Ga Pebbair swarm (Fig. 2.31). In general, high precision baddeleyite ages range from 2.370 to 2.365 Ga around the Bangalore swarm, which was emplaced within ~5 Myr or less. An overall regional bending and clockwise rotation of 30° in their trends is noticed along with the rotation in regional strike of the Closepet Granite and greenstone belts, which has been explained by Söderlund et al. (2019) possibly either due to orthogonal extension of its eastern convex side or a zone of pervasive dextral shearing affecting its southern portion. Once their orientations are corrected, many of these 'radiating' dykes become approximately linear swarms, hence these do not represent the plume central location (Söderlund et al. 2019).

2.7.7 Deformation

Structural patterns of TTG-greenstone associations are extensively used to constrain the Archean tectonic regimes as well as thermal and mechanical behaviour of lithosphere (Chardon et al. 2008). Two contrasting large-scale deformation patterns characterize the Dharwar Craton: (i) extensive elongate domes and basins in the WDC Paleo- to Mesoarchean gneiss and greenstone belts preserving the 'dome and keel' structural pattern related to vertical (sagduction) tectonics, and (ii) superposed N–S trending strike-slip ductile shear zones both in the WDC and the EDC (Drury and Holt 1980; Naha et al. 1995; Bouhallier et al. 1993, 1995; Choukroune et al. 1997;

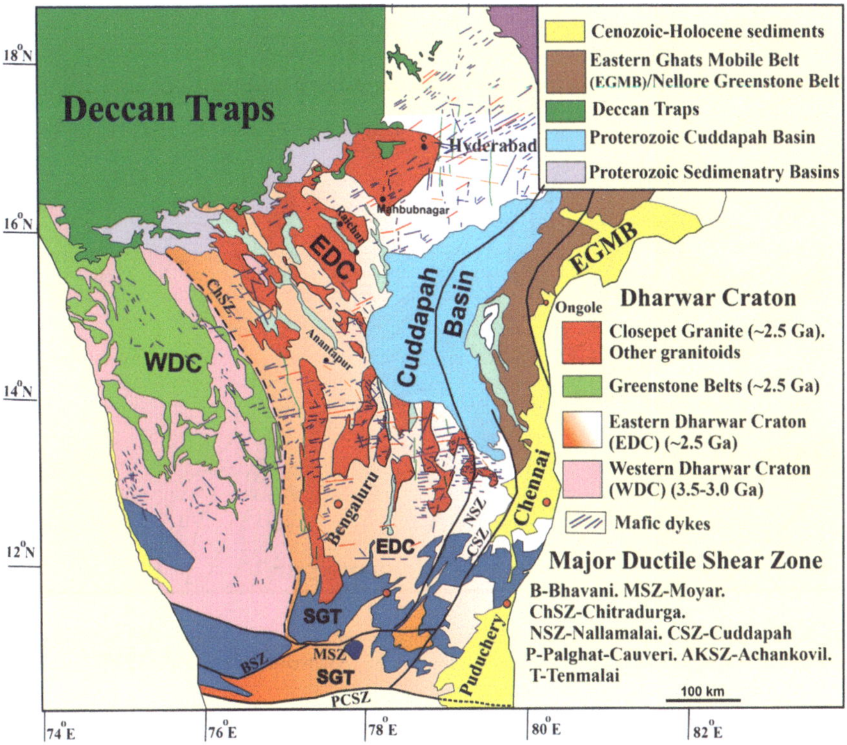

Fig. 2.31 Simplified geological map of the Dharwar Craton showing emplaced Paleoproterozoic mafic dyke swarms. WDC: Western Dharwar Craton; EDC: Eastern Dharwar Craton; SGT: South Granulite Terrain. Redrawn after French and Heaman (2010), Nagaraju et al. (2018b), Samal et al. (2019) and references therein

Chardon et al. 1996, 1998, 2002, 2008, 2011; Chadwick et al. 2000; Jayananda et al. 2006, 2018, 2020), preserving evidences of the Neoarchean convergent (plate) tectonics.

Some of the very early investigations, using LANDSAT-TM data and field investigations, indicated presence of dome and basin patterns, regional scale strike-slip shear zones and spectacular flat to shallow fabrics (Drury and Holt 1980). Detailed strain fabrics mapping and kinematic analysis in the western Dharwar craton reveal crustal scale dome and basin patterns where low-density intracrustal gneisses form domes whilst high-density greenstone form keels or sagducting ridges. Fabric analysis in Holenarsipur greenstone belt and surrounding TTG basement in the western Dharwar craton reveal world-class dome and basin patterns where low-density gneisses form circular to sub-circular domes whilst high-density greenstone volcanic-sedimentary sequences form keel structures (Fig. 2.32; Bouhallier et al. 1993). Directly south in the Gundlupet area, the gneisses form elliptical domes and the greenstone sequences trapped between them (Bouhallier et al. 1995). The dome and basin patterns have been

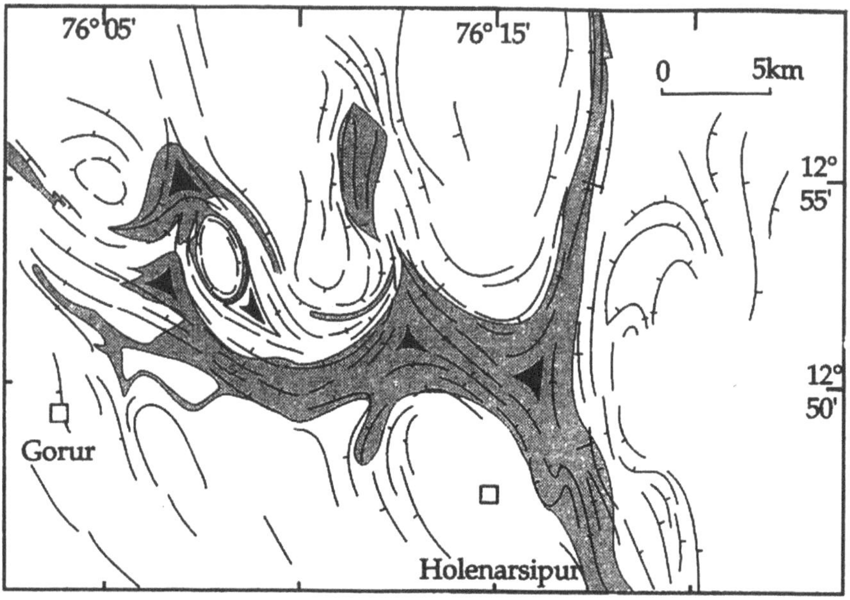

Fig. 2.32 Foliation trajectory map of the Holenarsipur belt, Karnataka (WDC), showing supracrustal belts (dark) and foliation triple areas (black). After Bouhallier et al. (1995)

attributed to gravitational collapse of high-density supracrustal material into low-density gneiss basement associated with reheating of crust during major tectonothermal events (Bouhallier et al. 1993, 1995; Choukroune 1994). Foliation trajectory maps (assumed λ1 λ2 trajectories) in the WDC indicate extensive dome-and-basin structures, having flat-lying foliations in supracrustal rocks, and vertical foliation triple points.

East of Chitradurga boundary shear zone in the area around Sira, Chardon et al. (2011) observed variable structural trends with (i) N–S steeply- to vertically-dipping mylonitic orthogneiss in the Chitradurga Shear Zone with shallow plunging stretching lineation, (ii) shallow to moderately easterly dipping foliations in the middle domains, pervasive westwards overturned folds having sub-horizontal NNE trending hinges with coaxial with L2 lineations and sheared backlimbs of the folds, and (iii) shallow-dipping foliation along western margin of the Closepet batholith, becoming steeper with eastern dips inside the batholith with southeast-plunging mineral lineation and melt-bearing extensional shear zones (Fig. 2.33). Such a deformation pattern is explained by lateral constrictional flow (LCF) as a combination of orogen-normal shortening (E–W), lateral constrictional stretching (N–S), and transtension, achieving gravity-driven flow, lateral escape, and 3-D mass redistribution in a viscous lower crust submitted to convergence.

Structural studies of plutons and surrounding basement in eastern parts of the WDC reveal that 2.62 Ga Chitradurga granite pre-date the development of 2.5 Ga regional shear zones (Jayananda et al. 2006). Plutons in the CDC and EDC are

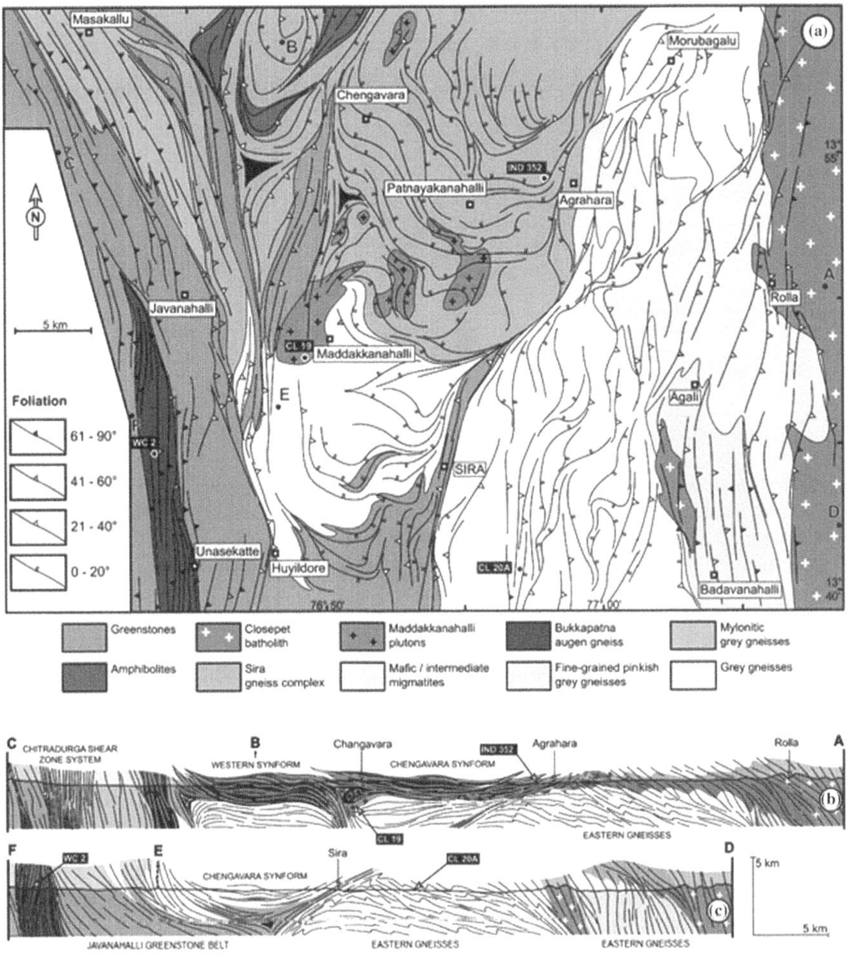

Fig. 2.33 **a** Geology and foliation trajectory map of the Sira area showing triple junction points. **b, c** Cross sections of the Sira area showing variations in foliation steepness. After Chardon et al. (2011)

contemporaneous with development of regional shear zones or slightly pre-date the shear deformation since their early facies were affected by shearing while late facies occupy the shear zones (Jayananda and Mahabaleswar 1991; Harish Kumar et al. 2003; Gireesh et al. 2012; Jayananda et al. 2014). Detailed structural studies of southern part of the Closepet Granite and surrounding basement reveal that horizontal E–W contractional deformation affected the mid crust and lower crust during major episode of juvenile magmatic accretion and regional partial melting of the basement (Chardon and Jayananda 2008). Development of constructional fabrics is explained

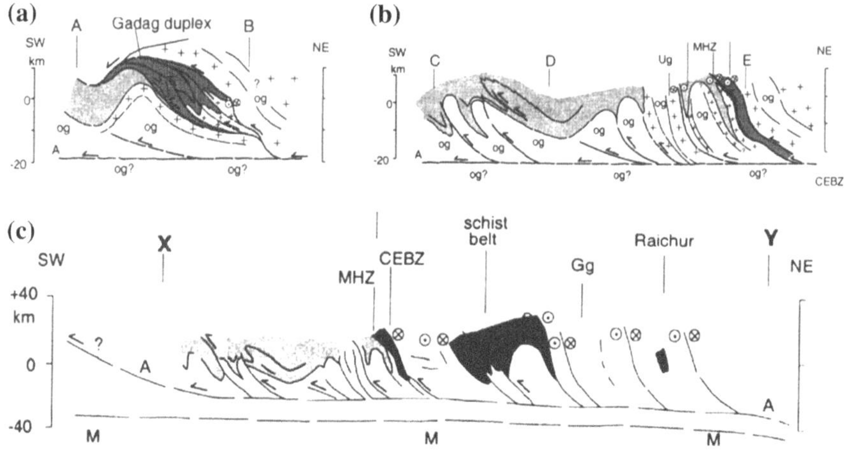

Fig. 2.34 Neoarchean foreland structural setup (WDC) and accretionary wedge (EDC). **a** Structure of the Gadak Duplex. **b** Imbricates and NE-verging thrust geometry. **c** Generalized thrust geometry of the WDC and EDC with a decollement at the Moho level. og—Orthogneiss. Ug—Uchchangidurga granite. Small circles with crosses and dots represent sinistral shear (dextral shear adjacent to the Uchchangidurga granite). After Chadwick et al. (2007)

by a combination of coeval shallow and steep planar fabrics sharing common horizontal elongation direction, two sets of conjugate shear zones and extensional shear zones (Chardon and Jayananda 2008).

Based on detailed structural analysis and mapping, Chadwick et al. (2007) modelled mutual relationships between granites, their emplacement and Ranibennur and Chitradurga schist belts in the WDC as a mid-crustal part of a craton-wide imbricate fold-thrust belt, exhibiting duplex, imbrication and eastward thickening (Fig. 2.34a and b). Steep high strain zones in the foreland are interpreted as listric structures that flatten and root into an attachment at a depth of ca. 18–20 km (Fig. 2.34b and c). New structural, zircon dating and isotopic age data show that inner margin (foreland) in the WDC and outer margin (accretionary complex) in the EDC are linked in orogen-parallel diffuse, steeply dipping boundary zone of least 200 km wide (Chadwick et al. 2007).

For the EDC, Krogstad et al. (1989), based on distinct geochemistry and isotopic characteristics, opined that two adjoining gneiss and amphibolite terranes with distinct histories were converged and sutured along the Archean Kolar Schist Belt by horizontal crustal shortening and shearing between 2.53 and 2.42 Ga, and that the plate tectonics operated on Earth by 2.5 Ga. On the contrary, Chardon et al. (2008) considered that shear zone patterns only reflect symmetry of the regional finite strain field of a wide hot orogen and did not record large horizontal displacements but only transverse shortening and vertical extrusion.

2.7.8 Metamorphism

Progressive regional metamorphism (~2.50 Ga) is remarkable in the Dharwar Craton, and reveals a continuous increase in P–T conditions from north to the south from greenschist facies to amphibolite and granulite facies (Pichamuthu 1965; Rollinson et al. 1981; Janardhan et al. 1982; Raith et al. 1982; Hansen et al. 1984; Raase et al. 1986; Harlov et al. 1997; Jayananda et al. 2013b, 2018). These remarkable E–W trending metamorphic isograds are superposed on the N–S structural and tectonic fabrics of the Dharwar Craton, though such an orthogonal pattern has remained unexplained, satisfactorily up till now.

Along a N–S traverse through Chitradurga–Nagamangala belts, Sargur and Nilgiri Hills, Raase et al. (1986) recorded an increase in metamorphism from greenschist (chl + act + ab + ep + carb, qtz), amphibolite (hbl + plag + qtz, gar, cumm) to mafic granulite (pyx + plag + gar ± hbl, qtz) with changes in mineral composition and geochemistry. Actinolite in the greenschist facies changes to pargasitic-hastingsitic hornblende in the granulite facies with increasing Ti content; plagioclase becomes more calcic with garnet having more pyrope and grossularite components along with the development of hypersthene in granulite zone; all these changes reveal higher pressures and temperatures southwards. P–T conditions increase continuously from 4 to 5 kbar/~500 °C in the greenschist facies to about 5–7 kbar/600 °C in amphibolite zone, increase to about 8 kbar/700–750 °C in hornblende-granulite facies transition zone, and to granulite facies at 8–9 kbar/800–850 °C. Increasing P–T conditions of metamorphism also changed fluid regime from hydrous-rich greenschist–amphibolite zones to water-deficient granulite (Raase et al. 1986). In the Sandur schist belt, co-existing mineral pairs, such as cordierite-gedrite-garnet-biotite, staurolite-biotite-garnet-muscovite-chlorite in metapelites yielded metamorphic P-T conditions of 4–5.2 kbar/ 550–600 °C (Hanuma Prasad et al. 1996).

Contrary to these simplistic views regarding metamorphism, Rollinson et al. (1981) documented moderate to high pressure–low temperature assemblages from the WDC and low pressure–high temperature assemblages in the EDC. X-ray diffraction study on structural state of feldspars from gneiss and granite reveal that the whole cratonal assemblages were affected by metamorphism (Kano et al. 2004).

Subsequent studies focused on the timeframe of metamorphism and crustal history of magmatic protoliths (e.g. Peucat et al. 2013). More recently, Jayananda et al. (2013a, 2018) have summarized metamorphic record of the craton including P–T conditions, timeframe of metamorphism through cooling. These studies reveal three cratonic blocks (western, central and eastern) in the Dharwar Craton with contrasting metamorphic record and thermal history. The WDC preserves garnet-kyanite assemblage indicating moderate to high-pressure (~7 kbar) and two thermal events ca. 3100 and 2500 Ma. The 3100 Ma event could be spatially related to emplacement of diapiric trondhjemite and development of dome-basin patterns (Jayananda et al. 2015).

Southern parts of the Chitradurga Shear Zone, dividing the WDC and CDC, records minimum peak metamorphic conditions of ~10 kbar/~820–875 °C, indicating equilibration at the base of thickened continental crust (Li et al. 2018a, b). U–Pb zircon and Pb–Pb monazite geochronology indicates crystallization of parent mafic magmas at c. 2.61–2.51 Ga and subsequent regional metamorphism of these intrusives to garnet-amphibolite and garnet-granulite facies at c. 2.48–2.44 Ga; these ages bracket the timing of micro-block accretion to the Archean–Proterozoic boundary in a supra-subduction zone environment.

On the contrary, the CDC reveals progressive increase of metamorphism from greenschist facies in the Sandur greenstone belt in the north to granulite facies in Kabbaldurga to the south (Pichamuthu 1965). Granulite contains orthopyroxene-sillimanite and spinel-quartz assemblages indicating moderate to low pressure–high temperature assemblages (Jayananda et al. 2011). The central block is affected by an upper amphibolite to granulite facies metamorphism close to 3211–3160 and 2640–2610 Ma high to ultra-high temperature event (Jayananda et al. 2013a). Further south in Kabbaldurga–BR Hills region, mineral assemblages in charnockite and metapelite have yielded amphibolite to granulite facies P-T conditions reaching 5–9 kbar/600–775 °C (Janardhan et al. 1982; Hansen et al. 1984). Texturally controlled in situ monazite dating together with Sm–Nd garnet-whole-rock isochron ages indicate three thermal events at ca. 3211–3160, 2640–2610 and 2520–2450 Ma (Jayananda et al. 2013a; Peucat et al. 2013).

The Archean Hutti-Maski greenstone belt of the EDC had undergone mid-amphibolite facies metamorphism at ca. 2564 ± 12 Ma with contemporaneous PT increase of up to ~6 kbar and ~620 °C followed by post-peak near-isothermal decompression (Hazarika et al. 2015). Northernmost parts of this belt reveal plagioclase–K-feldspar–hornblende–clinopyroxene bearing syn-plutonic dykes under amphibolite facies (Prabhakar et al. 2009). In comparison, the Kolar greenstone belt underwent a lower amphibolite facies metamorphism at ca. 2546 ± 12 Ma and synchronous PT increase of up to ~4.6 kbar/~600 °C, followed by decompressional cooling. The ages of felsic volcanism are constrained at ca. 2669 ± 22 and 2661 ± 32 Ma in the Hutti-Maski and Kolar belts, respectively (Hazarika et al. 2015). This block has cordierite-andalusite-biotite assemblage in pelites indicating low pressure–low to moderate temperature in western margin of Kolar Greenstone Belt and garnet-ortho-clinopyroxene mafic granulite and sapphirine-bearing pelitic granulite to the south in Krishnagiri-Dharmapuri region/Shevaroy Hills corresponding to amphibolite to granulite facies conditions with progressive increase from 650 to 800 °C (Harlov et al. 1997; Jayananda et al. 2003, 2013a). U–Pb monazite and zircon ages from the EDC indicate that it was involved only in one thermal event around 2500–2530 Ma (Peucat et al. 2013).

2.7.9 Crustal Evolution

Archean crustal evolution of the Dharwar Craton has the following end member process-based models (Jayananda et al. 2018), besides others:

(i) E–W trending greenstone belts as back-arc marginal basins in the Dharwar Craton adjacent to the E–W trending and north-verging subduction zones as crustal scale shear zones, and tectonically modified by N–S trending shear systems that produced current configuration of schist belts (Newton 1990; Naha et al. 1991).

(ii) Greenstone belts as signatures of mantle plumes in shallow subduction zones (Kolar and Ramagiri) in the EDC (Fig. 2.35; Dey et al. 2018; Jayananda et al. 2018, 2020a; Manikyamba and Kerrich 2012; Manikyamba et al. 2017; Naqvi et al. 2002). Such greenstone belts of the eastern Dharwar Craton contain komatiite and Mg-tholeiitic basalt volcanic association with deep water siliciclastic and BIF and are interpreted as composite tectonostratigraphic terranes of accreted plume-derived and convergent margin-derived magmatic sequences. Plumes melted at <90 km under thin rifted continental lithosphere to preserve intra-oceanic and continental margins. Associated alkaline basalts

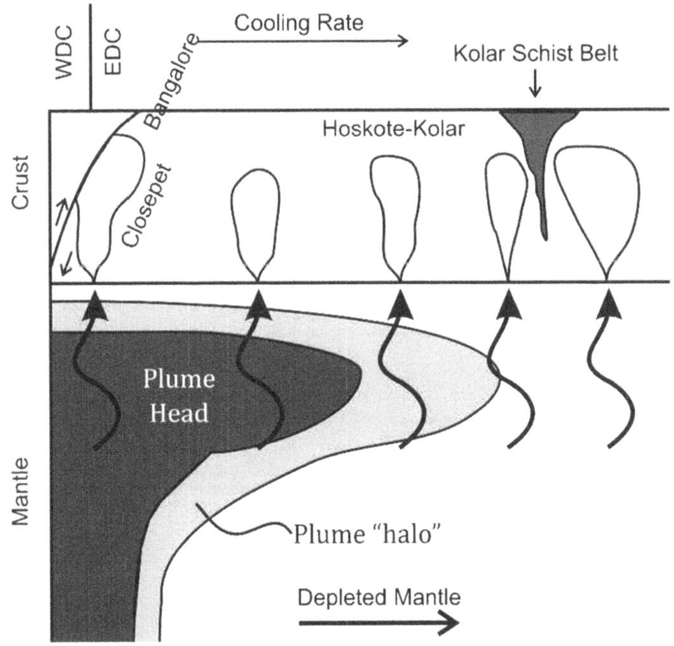

Fig. 2.35 Superplume model for generation of the Closepet-type granitic gneisses of the EDC. Mantle transitions from an enriched area near the plume to a depleted zone beneath the Kolar schist belt. Partial melting of lower lithosphere creates plutons that form the mainland of the EDC. From Meert et al. (2010), based on Jayananda et al. (2000)

record subduction-recycling of Mesoarchean oceanic crust, incubated in the asthenosphere, and erupted coevally with Mg-basalts from a heterogeneous

The CDC grew along the margin of already existing ca. 3.4–3.0 Ga WDC microcontinent through hotspot and active margin setting. In the oceanic environments, hotspot magmatism generated komatiite to basaltic magmatism (Naqvi et al. 2002) possibly as oceanic islands, while westward subduction of intervening oceanic crust along the margin of microcontinent generated arc crust. Continued subduction and slab breakoff caused melting of the base of the arc crust as well as surrounding sub-arc mantle which resulted in magmatic protoliths of transitional TTGs (Jayananda et al. 2018).

The EDC contains TTG–greenstone 2.7–2.54 Ga associations whose lithology, elemental and isotope signatures reveal affinity to hotspot magmatism to oceanic arcs. The 2.7 Ga komatiite to basaltic rocks probably formed in hot spot environment (Dey et al. 2018), though arc setting has also been proposed (Khanna et al. 2016). Subduction of the intervening oceanic crust and melting of its refractory harzburgite mantle could have probably generated oceanic arc crust with intermediate to felsic volcanics such as boninites and dacites (Manikyamba et al. 2005). Continued subduction and melting of arc crust and sub-arc mantle produced transitional TTGs (Jayananda et al. 2018). Assembly of the EDC, CDC and the WDC during 2.58–2.54 Ga triggered mantle upwelling its melting and generation of sanukitoid magma which was emplaced into the CDC as well as newly formed greenstone–transitional TTG crust (Jayananda et al. 2018). This upwelling caused crustal reworking and high-T and low-P metamorphism and cratonization of Archean crust.

Jayananda et al. (2020a) elaborated the geodynamic evolution of the CDC and EDC by hotspot magmatism, which led to the formation of oceanic plateaus (Fig. 2.36). This was followed by westward convergence of intervening slabs producing the arc crust. Melting along the mantle hot spot generated komatiite and alkali basalt magmas between ca. 2.75–2.70 Ga beneath ancient 3.3–3.0 Ga crust of the CDC and produced the Sandur and Ramagiri greenstone basins. During this period, komatiite to high-Mg basalt magmas erupted in oceanic environments and produced

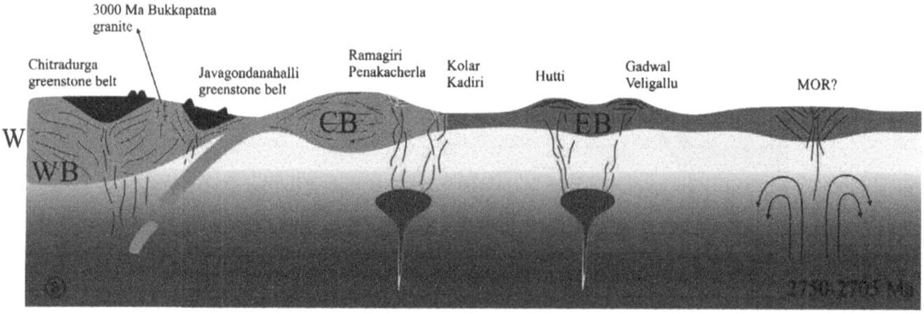

Fig. 2.36 Conceptual model showing westward convergence of hot oceanic lithosphere during Neoarchean assembly of micro-blocks, widespread mantle melting resulting in juvenile magmatic additions and evolution of greenstone belts. After Jayananda et al. (2020a)

volcanic plateaus such as Kolar, Kadiri, Veligallu, Gadwal greenstone belts in the EDC. Simultaneous westward convergence of intervening oceanic crust beneath these cratons led to the development of ca. 2.74–2.65 Ga Chitradurga back-arc and 2.74 Ga Javagondanahalli basins along eastern margin of the WDC and the Sandur–Ramagiri–Penakacherla–Hungund belt in an active continental margin setting (Jayananda et al. 2020a and references therein).

(iii) Greenstone belts containing komatiite, tholeiite and intrusive calc-alkaline gneiss as suture zones ~2.5 Ga between two continental blocks as Archean examples of plate tectonics and arc accretion, which resembles present-day plate tectonics (Krogstad et al. 1989). Based on distinct geochemistry and isotopic characteristics, they opined that Archean Kolar Schist Belt recorded distinct histories in two adjoining gneiss and amphibolite terranes which converged and sutured along this belt by horizontal crustal shortening and shearing between 2.53 and 2.42 Ga. More significantly, plate tectonics operated by 2.5 Ga on Earth. Different subduction polarity explains Neoarchean crustal accretion and northward-trending fabrics, which involves either westward oblique convergence (Chadwick et al. 2000, 2007) or eastward subduction (Maibam et al. 2016; Santosh and Li 2018).

(iv) Gravitationally 'sagduction' greenstone belts into gneissic basement and tectonic slides along margins in the WDC (Chardon et al. 1996). Greenstone belts of >3 Ga Sargur Group (WDC) and the 3.0–2.5 Ga Dharwar Supergroup within the DC exhibit distinct tectonic patterns. Dome-and-basin structures in the older belts are associated with diapiric-type gravitational instabilities due to incipient development of a 'sagducting' trough in the gneissic basement (Bouhallier et al. 1995; Chardon et al. 1996, 1998). Such subsiding Such subsiding basins exhibit vertical flattening and zone of horizontal constriction, while their margins reveal opposite sense of ductile shearing due to sinking denser greenstone material into gneissic basement (Fig. 2.37).

(v) Convergent tectonic model with the WDC as the foreland marginal basins (schist belt) and the EDC with greenstone belts as intra-arc basins, accreted on to the former ~2.75 to 2.51 Ga (Chadwick et al. 2000).

(vi) Continental rift model for the evolution of EDC schist belts as rifted small ocean basins that closed by NE–SW compression (Ramakrishnan 1987a, b). Although above models provide large-scale scenario of the DC, absence of fore arc accretionary prism, dismembered ophiolites and melange characterizing the suture zone led Ramakrishnan (1987a, b) to postulate that the greenstone belts were deposited in rifted small oceans.

(vii) An episode of crustal extension, fracturing and dyke emplacement between 2.37 and 1.79 Ga representing the large Igneous Province (Srivastava et al. 2019). Several sets of dykes are now mapped almost orthogonal to the regional trend of both the WDC and EDC and represent Large Igneous Province (LIP) as consequence of ~2.0 Ga crustal extension and fracturing of the lithospheric continental crust (Söderlund et al. 2019).

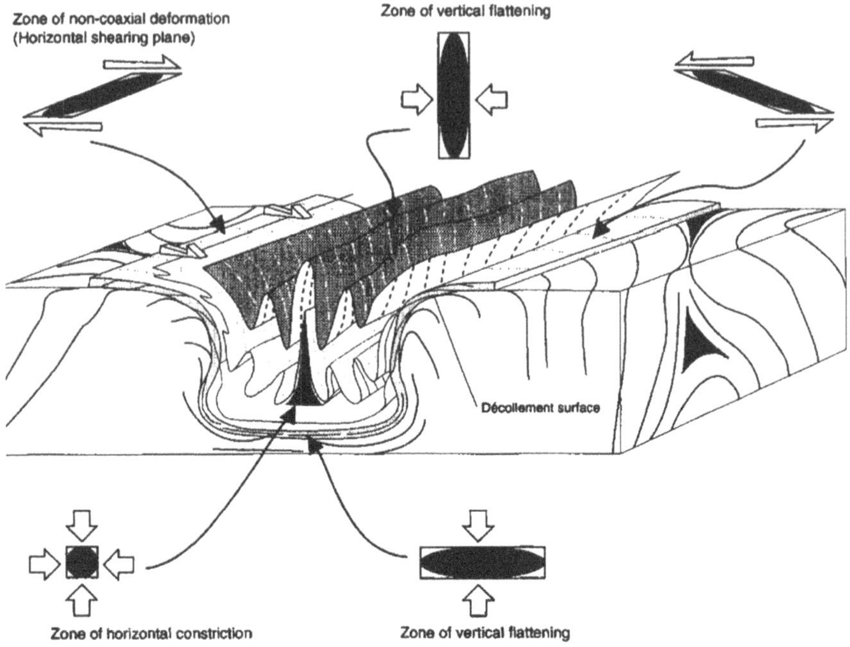

Fig. 2.37 Schematic block diagram illustrating spatial variations of structures, strain ellipsoid type and strain regime within a subsiding trough. This model is supported by the results of the subsurface structural analysis of the Kibbanahalli Arm. After Chardon et al. (1996)

Based on age zonation patterns, crustal thickness, metamorphic records, cratonization and assembly of micro-blocks into cratonization, most tectonic models invoked westward convergence of oceanic lithosphere, however a few studies also proposed an eastward subduction (Ratheesh Kumar et al. 2016). Further, it is likely that both the WDC and EDC accreted simultaneously (Maibam et al. 2016), although thermal history, U–Pb zircon data and Nd–Hf isotopes indicate their independent geological evolution (Dey et al. 2018; Jayananda et al. 2018).

To summarize, the Dharwar Craton was formed as two distinct terranes: The Western Dharwar Craton (WDC) and the Eastern Dharwar Craton (EDC) with transient signatures in the Central Dharwar Craton (CDC) having diffused geological boundaries. The most prevalent viewpoints are their evolution in westward converging plume and arc settings. Plumes dominated vertical accretion contributed to the growth of 3.45–3.2 Ga crust in the WDC, while arc dominated processes was an important process in the CDC and EDC with significant hot spot magmatism. These three pieces probably assemble into the Dharwar Cratonic framework during 2.56–2.5 Ga.

References

Absar N, Raza M, Roy M, Naqvi SM, Roy AK (2009) Composition and weathering conditions of Paleoproterozoic upper crust of Bundelkhand craton, Central India: records from geochemistry of clastic sediments of 1.9 Ga Gwalior Group. Precambrian Res 168:313–329

Acharyya SK (1993) Greenstones from Singhbhum Craton, their Archaean character, oceanic crustal affinity and tectonics. Proc Natl Acad Sci India 63A:211–222

Acharyya SK, Gupta A, Orihashi Y (2010) New U-Pb zircon ages from Paleo-Mesoarchean TTG gneisses of the Singhbhum Craton, eastern India. Geochem J 44:81–88

Adlakha V, Patel RC, Lal N (2011) Exhumation and thermal history of the Shillong Plateau: constraints from apatite fission track analysis. Mem Geol Soc India 77:57–73

Ahmad I, Mondal MEA (2016) Do the BGC-I and BGC-II domains of the Aravalli craton, northwestern India represent accreted terranes? Earth Sci India 9(IV):167–175

Ahmad T, Dragusanu C, Tanaka T (2008) Provenance of Proterozoic Basal Aravalli mafic volcanic rocks from Rajasthan, Northwestern India: Nd isotopes evidence for enriched mantle reservoirs. Precambrian Res 162:150–159

Ahmad I, Mondal MEA, Satyanarayanan M (2016) Geochemistry of Archean metasedimentary rocks of the Aravalli craton, NW India: implications for provenance, paleoweathering and supercontinent reconstruction. J Asian Earth Sci 126:58–73

Alfimova N, Raza MB, Felitsyn S, Matrenichev V, Bogomolov E, Nasipuri P, Saha L, Pati JK, Kumar V (2019) Isotopic Sm-Nd signatures of Precambrian Banded Iron Formation from the Fennoscandian shield, East-European Platform, and Bundelkhand craton, India. Precambrian Res. https://doi.org/10.1016/j.precamres.2019.04.008

Alvi SH, Raza M (1992) Geochemical evidence for the volcanic arc setting of the Dhanjori volcanics, Singhbhum craton, eastern India. Geol Mag 129:337–348

Anand R, Balakrishnan S (2010) Pb, Sr and Nd isotope systematics of metavolcanic rocks of the Hutti greenstone belt, Eastern Dharwar Craton: constraints on age, duration of volcanism and evolution of mantle sources during Late Archean. J Asian Earth Sci 39:1–11

Bakshi AK (1995) Petrogenesis and timing of volcanism in the Rajmahal flood basalt Province, northeastern India. Chem Geol 121:73–90

Balakrishnan S, Rajamani V, Hanson GN (1999) U-Pb ages for zircon and Titanite from the Ramagiri area, southern India: evidence for accretionary origin of the eastern Dharwar Craton during the late Archean. J Geol 107:69–86

Banerjee PK (1982) Stratigraphy, petrology and geochemistry of some Precambrian basic volcanic and associated rocks of Singhbhum district, Bihar, and Mayurbhanj and Kheonjhar districts, Orissa. Geol Surv India Mem 111:1–54

Banerjee DM (1996) A Lower Proterozoic paleosol at BGC-Aravalli boundary in South-central Rajasthan. J Geol Soc India 48:277–288

Banerjee PK, Ghosh SK (1994) Is the Simlipal Complex a product of shallow plume tectonics? J Geol Soc India 43:353–359

Basu AK (1986) Geology of parts of the Bundelkhand massif, central India. Rec Geol Surv India 117(2):61–124

Basu A (2007) Role of the Bundelkhand Granite Massif and the Son-Narmada megafault in Precambrian crustal evolution and tectonism in Central and Western India. J Geol Soc India 70:745–770

Basu AR, Bandyopadhyay PK, Chakri R, Zou H (2008) Large 3.4 Ga Algoma type BIF in the Eastern Indian Craton. Geochim Cosmochim Acta 72:A59

Behera BM, Thirukumaran V, Soni A, Mishra PK, Biswal TK (2017) Size distribution and roundness of clasts within pseudotachylites of the Gangavalli Shear Zone, Salem, Tamil Nadu: an insight into its origin and tectonic significance. J Earth Syst Sci 126(4):46

Bhadra BK (2000) Ductile shearing in Attur shear zone and its relation with Moyar shear zone, South India. Gondwana Res 3:361–369

Bhaskar Rao YJ, Naha K, Srinivasan R, Gopalan K (1991) Geology, geochemistry and geochronology of Peninsular Gneiss around Gorur, Hassan district, Karnataka. Proc Indian Acad Sci (Earth Planet Sci) 100:399–412

Bhaskar Rao YJ, Vijaya Kumar T, Sreenivas B, Babu EVSSK (2020) A review of Paleo- to Neoarchean crustal evolution in the Dharwar craton, Southern India and the transition towards a Plate Tectonic regime. Episodes 43(1):51–68, https://doi.org/10.18814/epiiugs/2020/020003

Bhattacharya AR, Singh SP (2013) Proterozoic crustal scale shearing in the Bundelkhand massif with special reference to quartz reefs. J Geol Soc India 82:474–484

Bhattacharya PM, Mukhopadhya S, Mazumdar RK, Kayal JR (2008) 3-D seismic structure of the northeast India region and its implications for local and regional tectonics. J Asian Earth Sci 33:25–41

Bhowmik SK, Dasgupta S (2012) Tectonothermal evolution of the BGC in central Rajasthan, NW India: present status and correlation. J Asian Earth Sci 49:339–348

Bhowmik SK, Bernhardt H-J, Dasgupta S (2010) Grenvillian age high pressure upper amphibolite-granulite metamorphism in the Aravalli-Delhi Mobile Belt, Northwestern India: new evidence from monazite chemical age and its implication. Precambrian Res 178:168–184

Bidyananda M, Deomurari MP (2007) Geochronological constraints on the evolution of Megahalaya massif, northeastern India: an ion microprobe study. Curr Sci 93(11):1620–1623

Bilham R, England P (2001) Plateau 'pop-up' in the great 1897 Assam earthquake. Nature 410:806–809

Biswal TK, Jena SK, Datta S, Das R, Khan K (2000) Deformation of the terrane boundary shear zone (Lakhna Shear Zone) between the Eastern Ghats Mobile belt and the Bastar craton, in Balangir and Kalahandi districts of Orissa. J Geol Soc India 55:367–380

Biswas S, Coutand I, Grujic D, Hager C, Stockli D, Grasemann B (2007) Exhumation and uplift of the Shillong Plateau and its influence on the eastern Himalayas: new constraints from apatite and zircon (U-Th-[Sm])/He and apatite fission track analyses. Tectonics 26:TC6013. https://doi.org/10.1029/2007tc002125

Blackburn WH, Srivastava DC (1994) Geochemistry and tectonic significance of the Ongarbira metavolcanic rocks, Singhbhum District, India. Precambrian Res 67:181–206

Bose MK (2008) Proterozoic dykes from Singhbhum granite pluton. In: Srivastava RK, Sivaji Ch, Rao C (eds) Indian dykes. Narosa Publishing House, New Delhi, pp 413–445

Bouhallier H, Choukroune P, Ballèvre M (1993) Diapirism, bulk homogeneous shortening and transcurrent shearing in the Archaean Dharwar craton: the Holenarsipura area, southern India. Precambrian Res 63:43–58

Bouhallier H, Chardon D, Choukroune P (1995) Strain patterns in Archaean dome-and-basin structures: the Dharwar craton (Karnataka, South India). Earth Planet Sci Lett 135:57–75

Bruce Foote R (1888, 1989) The Dharwar System, the chief auriferous series in south India. Rec Geol Surv India 21(2):34–56, 139–164

Buick IS, Allen C, Pandit M, Rubatto D, Herman J (2006) The Proterozoic magmatic and metamorphic history of the BGC, central Rajasthan, India: LA-ICP-MS U-Pb zircon constraints. Precambrian Res 151:119–142

Chadwick B, Vasudev VN, Hedge GV (2000) The Dharwar craton, southern India, interpreted as the result of Late Archaean oblique convergence. Precambrian Res 99:91–101

Chadwick B, Vasudev VN, Hedge GV, Nutman AP (2007) Structure and SHRIMP U/Pb zircon ages of granites adjacent to the Chitradurga schist belt: implications for Neoarchean convergence in the Dharwar craton, Southern India. J Geol Soc India 69:5–24

Chaki A, Shivkumar K, Dey S (2008) Geochemistry and Rb-Sr geochronology of granitoids from Northern Parts of Bastar Craton, Central India: implications to petrogenesis and geodynamics. Mem Geol Soc India 73:101–116

Chandan Kumar B, Ugarkar G (2017) Geochemistry of mafic-ultramafic magmatism in the Western Ghats Belt (Kuderemukh greenstone belt), western Dharwar Craton, India: implications for mantle sources and geodynamic setting. Int Geol Rev 59:1507–1531. https://doi.org/10.1080/00206814.2017.1278623

Chardon D, Jayananda M (2008) Three-dimensional field perspective on deformation, flow, and growth of the lower continental crust (Dharwar craton, India). Tectonics 27:TC1014. http://dx.doi.org/10.1029/2007TC002120

Chardon D, Choukroune P, Jayananda M (1996) Strain patterns, decollement and incipient sagducted greenstone terrains in the Archaean Dharwar craton (south India). J Struct Geol 18:991–1004

Chardon D, Choukroune P, Jayananda M (1998) Sinking of the Dharwar basin (South India): implications for Archaean tectonics. Precambrian Res 91:15–39

Chardon D, Peucat J-J, Jayananda M, Choukroune P, Fanning CM (2002) Archaean granite-greenstone tectonics at Kolar (South India): interplay of diapirism and bulk homogenous contraction during juvenile magmatic accretion. Tectonics 21(3):1016. http://dx.doi.org/10.1029/2001TC901032

Chardon D, Jayananda M, Chetty TRK, Peucat J-J (2008) Precambrian continental strain and shear zone patterns: South Indian case. J Geophys Res 113:B08402. https://doi.org/10.1029/2007JB005299

Chardon D, Jayananda M, Peucat J-J (2011) Lateral constrictional flow of hot orogenic crust: insights from the Neoarchean of south India, geological and geophysical implications for orogenic plateaux. Geochem Geophys Geosys 12:Q02005. https://doi.org/10.1029/2010GC003398

Chatterjee N, Mazumadar AC, Bhattacharya A, Saikia RR (2007) Mesoproterozoic granulites of the Shillong-Meghalaya plateau: evidence of westward continuation of the Prydz Bay Pan-African suture into northeastern India. Precambrian Res 152:1–26

Chatterjee N, Crowley J, Ghosh NC (2008) Geochronology of the 1.55 Ga Bengal anorthosite and Grenvillian metamorphism in the Chhotanagpur gneissic complex, eastern India. Precambrian Res 161:303–316

Chattopadhyay N, Hashimi S (1984) The Sung Valley alkaline-ultramafic-carbonatite complex, East Khasi and Jaintia Hills districts, Meghalaya. Rec Geol Surv India 113:24–33

Chaudhury T, Wan Y, Mazumder R, Ma M, Liu D (2018) Evidence of enriched, Hadean Mantle Reservoir from 4.2–4.0 Ga zircon xenocrysts from Paleoarchean TTGs of the Singhbhum Craton, Eastern India. Sci Rep 8:7069. https://doi.org/10.1038/s41598-018-25494-6

Chauhan H, Saikia A, Ahmad T (2018) Episodic crustal growth in the Bundelkhand craton of central India Shield: constraints from petrogensis of the tonalite-trondhjemite-granodiorite gneisses and K-rich granites of Bundelkhand Tectonic Zone. J Earth Syst Sci 127:1–34

Chetty TRK (2017) Proterozoic orogens of India. A critical window to Gondwana. Elsevier Inc., Berkeley, CA, p 395

Chetty TRK, Bhaskar Rao YJ (1998) Behaviour of stretching lineation in the Salem–Attur shear belt, Southern Granulite Terrane, south India. J Geol Soc India 52:443–448

Chetty TRK, Bhaskar Rao YJ, Narayana BL (2003) A structural cross-section along Krishnagiri-Palami corridor, Southern Granulite Terrain of India. Mem Geol Soc India 50:255–277

Chimote JS, Pandey BK, Bagchi AK, Basu AN, Gupta JN, Saraswat AC (1988) Rb-Sr whole-rock isochron age for the Mylliem Granite, E.K. Hills, Meghalaya. In: Proceedings of 4th national symposium on mass spectrometry, IISC, Bangalore

Choudhury DK, Pradhan AK, Zakaulla S, Umamaheswar K (2012) Geochemistry and petrogenesis of anorogenic (?) granitoids of west Garo Hills, Meghalaya. J Geol Soc India 80:276–286

Choukroune P (1994) Deformations et delplacements dans la croute terrestre. Masson, Paris, 226p

Choukroune P, Ludden JN, Chardon D, Calvert AJ, Bouhallier H (1997) Archaean crustal growth and processes: a comparison of the Superior province and the Dharwar craton India. In: Burg JP, Ford M (eds) Orogeny through time. Geol Soc London Spec Publ 121, pp 63–98

Clark MK, Bilham R (2008) Miocene rise of the Shillong Plateau and the beginning of the end for the Eastern Himalaya. Earth Planet Sci Lett 269:337–351

Coffin MF, Pringle MS, Duncan RA, Gladczenko TP, Storey M, Muller RD, Gahagan LA (2002) Kerguelen hotspot magma output since 130 Ma. J Petrol 43:1121–1139

Collins AS, Clark C, Sajeev K, Santosh M, Kelsey DE, Hand M (2007) Passage through India: the Mozambique Ocean suture, high-pressure granulites and the Palghat-Cauvery shear zone system. Terra Nova 19(2):141–147

Crawford AR (1969) Reconnaissance Rb-Sr dating of Precambrian rocks of southern peninsular India. J Geol Soc India 10:117–166
Crawford A (1970) The Precambrian geochronology of Rajasthan and Bundelkhand, northern India. Can J Earth Sci 7:91–110
Crawford AR, Compston W (1970) The age of Vindhyan System of Peninsular India. Quart J Geol Soc London 125:251–371
Crookshank H (1948) Minerals of the Rajputana pegmatites. Trans Min Geol Met Inst India 42:105–189
D'Cruz ED, Nair PKR, Kumar P (2000) Palghat Gap—a dextral shear zone from the South Indian granulite terrain. Gondwana Res 3:21–31
Das JD, Saraf AK, Jain AK (1995) Fault tectonics of the Shillong plateau and adjoining regions, north-east India using remote sensing data. Int J Remote Sens 16(9):1633–1646
Dasgupta AB, Biswas AK (2000) Geology of Assam. Geol Soc India, Bangalore, p 170
Dasgupta S, Guha D, Sengupta P, Miura H, Ehl J (1997) Pressure-temperature-fluid evolutionary history of the polymetamorphic Sandmata granulite complex, Northwestern India. Precambrian Res 83:267–290
Deshmukh SD, Hari KR, Diwan P, Manu Prasanth MP (2018) Geochemical constraints on the tectonic setting of the Sonakhan Greenstone Belt, Bastar Craton, Central India. Acta Geochim 37(3):489–499. https://doi.org/10.1007/s11631-017-0213-z
Desikachar SV (1974) A review of the tectonic and geologic history of eastern India in terms of plate tectonic theory. J Geol Soc India 15:137–149
Dey S (2013) Evolution of Archean crust in the Dharwar craton: the Nd isotope record. Precambrian Res 227:227–246
Dey S, Nandy J, Choudhary AK, Liu Y, Zong K (2015) Neoarchean crustal growth by combined arc–plume action: evidence from the Kadiri Greenstone Belt, EDC, India. In: Roberts NMW, Van Kranendonk M, Parman S, Shirey S, Clift PD (eds) Continent formation through time. Geol Soc London Spec Publ 389, pp 135–163
Dey S, Halla J, Kurhila M, Nandy J, Heilimo E, Pal S (2016) Geochronology of Neoarchean granitoids of the NW eastern Dharwar craton: implications for crust formation. In: Halla J, Whitehouse MJ, Ahmad T, Bagai Z (eds) Crust–mantle interactions and granitoid diversification: insights from Archaean cratons. Geol Soc London Spec Publ 449. http://doi.org/10.1144/SP449.9
Dey S, Topno A, Liu Y, Zong K (2017) Generation and evolution of Palaeoarchaean continental crust in the central part of the Singhbhum craton, eastern India. Precambrian Res 298:268–291. https://doi.org/10.1016/j.precamres.2017.06.009
Dey S, Pal S, Balakrishnan S, Halla J, Kurhila M, Heilimo E (2018) Both plume and arc: origin of Neoarchaean crust as recorded in Veligallu greenstone belt, Dharwar craton, India. Precambrian Res. https://doi.org/10.1016/j.precamres.2018.04.019
Dharma Rao CV, Santosh M, Purohit R, Wang J, Jiang X, Kusky T (2011) LA-ICP-MS U-Pb zircon age constraints on the Paleoproterozoic and Neoarchean history of the Sandmata Complex in Rajasthan within the NW Indian Plate. J Asian Earth Sci 42:286–305
Drury SA (1980) A regional tectonic study of the Archean Chitradurga Greenstone Belt, Karnataka, based on LANDSAT interpretation. J Geol Soc India 24:167–184
Drury SA, Holt RW (1980) Tectonic framework of south Indian craton: a reconnaissance involving LANDSAT imagery. Tectonophysics 65:T1–T15
Drury SA, Harris NB, Holt RW, Reeves-Smith GJ, Wightman RT (1984) Precambrian tectonics and crustal evolution in South India. J Geol 92:3–20
Dunn JA (1940) The stratigraphy of south Singhbhum. Geol Surv India Mem 63:303–369
Dunn JA, Dey AK (1942) The geology and petrology of eastern Singhbhum and surrounding areas. Mem Geol Surv India 69(2):281–450
Easton RM (2009) A guide to the application of lithostratigraphic terminology in Precambrian terrains. Stratigraphy 6(2):117–134
Ernst RE (2014) Large igneous provinces. Cambridge University Press, Cambridge
Evans P (1964) The tectonic framework of Assam. J Geol Soc India 5:80–96

Fareeduddin, Banerjee DM (2020) Aravalli Craton and Mobile Belt: an update. In: Faredduddin et al (eds) Geodynamic evolution of Indian subcontinent. Episodes 43(1), http://doi.org/10.18814/epiiugs/2020/20005

French JE (2007) U–Pb dating of Paleoproterozoic mafic dyke swarms of the South Indian shield: implications for paleocontinental reconstructions and identifying ancient mantle plume events. Ph.D. thesis, University of Alberta, Edmonton, Alberta, p 373

French JE, Heaman LM (2010) Precise U-Pb dating of Paleoproterozoic mafic dyke swarms of the Dharwar craton, India: implications for the existence of the Neoarchean supercraton Sclavia. Precambrian Res 183:416–441

French JE, Heaman LM, Chacko T, Srivastava RK (2008) 1891–1883 Ma Southern Bastar-Cuddapah mafic igneous events, India: a newly recognized large igneous province. Precambrian Res 160:308–322

Frey FA, Coffin MF, Wallace PJ, Weis D, Zhao X (2000) Origin and evolution of a submarine large igneous province: the Kerguelen Plateau and Broken Ridge, southern Indian Ocean. Earth Planet Sci Lett 176:73–89

Friend CRL, Nutman AP (1991) SHRIMP U-Pb geochronology of the Closepet granite and Peninsular gneisses, Karnataka, South of India. J Geol Soc India 38:357–368

Gautam GC, Srivastava RK (2011) Petrology, geochemistry and petrogenesis of Early Precambrian Mafic Dyke Swarm from Dondi-Bhanupratappur-Keshkal Area, Central Bastar Craton, India. In: Srivastava RK (ed) Dyke swarms: keys for geodynamic interpretation, pp 203–218. https://doi.org/10.1007/978-3-642-12496-9-12

Geological Survey of India (1973) Geological and mineral map of Arunachal Pradesh, Assam, Manipur, Meghalaya, Mizoram, Nagaland and Tripura, 1st ed. (1:2,000,000). Publ. G.S.I., Calcutta

Ghatak A, Basu AR (2013) Isotopic and trace element geochemistry of alkalic–mafic–ultramafic–carbonatitic complexes and flood basalts in NE India: origin in a heterogeneous Kerguelen plume. Geochim Cosmochim Acta 115:46–72

Ghosh JG (2004) 3.56 Ga tonalite in the central part of the Bastar Craton, India: oldest Indian date. J Asian Earth Sci 23:359–364

Ghosh G, Mukhopadhyay J (2007) Reappraisal of the structure of the Western Iron Ore Group, Singhbhum Craton, and eastern India: implications for the exploration of BIF-hosted iron ore deposits. Gondwana Res 12:525–532

Ghosh PK, Chandy KC, Bishui PK, Prasad R (1986) Rb-Sr age of granitoid gneiss in Malanjkhand area, Balaghat District, Madhya Pradesh, India. Indian Miner 40:1–8

Ghosh S, Chakraborty S, Bhalla JK, Paul DK, Sarkar A, Bishui PK, Gupta SN (1991) Geochronology and geochemistry of granite plutons from East Khasi Hills, Meghalaya. J Geol Soc India 73:331–342

Ghosh S, Chakraborty S, Paul DK, Bhalla JK, Bishui PK, Gupta SN (1994) New Rb-Sr isotopic ages and geochemistry of granitoid from Meghalaya and their significance in middle to late Proterozoic crustal evolution. Indian Miner 48:33–44

Ghosh DK, Sarkar SN, Saha AK, Ray SL (1996) New insights on the early Archaean crustal evolution in eastern India: re-evaluation of Pb-Pb, Sm-Nd and Rb-Sr geochronology. Indian Miner 50:175–188

Ghosh JG, De Wit MJ, Zartman RE (2004) Age and tectonic evolution of Neoproterozoic ductile shear zones in the Southern Granulite Terrain of India, with implications for Gondwana studies. Tectonics 23:TC3006

Ghosh S, Fallick AE, Paul DK, Potts PJ (2005) Geochemistry and origin of Neoproterozoic granitoids of Meghalaya, Northeast India: implication for linkage with amalgamation of Gondwana Supercontinent. Gondwana Res 8:421–432

Ghosh G, Ghosh B, Mukhopadhyay J (2015) Palaeoarchaean–Mesoproterozoic sedimentation and tectonics along the west-northwestern margin of the Singhbhum Granite body, eastern India: a synthesis. In: Mazumder R, Eriksson PG (eds) Precambrian basins of India: stratigraphic and tectonic context. Geol Soc London Mem 43, pp 121–138. https://doi.org/10.1144/m43.9

Gireesh RV, Kowete-u S, Jayananda M (2012) Anatomy of 2.57–2.52 Ga granitoids plutons in the eastern Dharwar craton, southern India: implications for magma chamber processes. Episodes 35–3:398–413
Gokarn SG, Gokarn Gupta G, Walia D, Sanabam SS, Hazarika N (2008) Deep geoelectric structure over the Lower Brahmaputra valley and Shillong Plateau, NE India using magnetotellurics. Geophys J Int 173:92–104. https://doi.org/10.1111/j.1365-246X.2007.03711.x
Gokarn SG, Rao CK, Selvaraj C, Gupta G, Singh BP (2013) Crustal evolution and tectonics of the Archean Bundelkhand Craton, Central India. J Geol Soc India 82:455–460
Goodwin AM (1991) Precambrian geology: the dynamic evolution of the continental crust. Academic Press, London
Gopalakrishnan K (1996) An overview of Southern Granulite Terrain, India—constraints in reconstruction of Precambrian assembly of Gondwanaland. Gondwana nine, vol 2. Oxford & IBH, New Delhi, pp 1003–1026
Gopalan K, Macdougall J, Roy A, Murali A (1990) Sm-Nd evidence for 3.3 Ga old rocks in Rajasthan, northwestern India. Precambrian Res 48:287–297
Goswami JN, Misra S, Wiedenbeck M, Ray SL, Saha AK (1995) 3.55 Ga old zircon from Singhbhum-Orissa Iron-Ore craton, eastern India. Curr Sci 69:1008–1011
Govin G, Najman Y, Copley A, Millar I, van der Beek P, Huyghe P, Grujic D, Davenport J (2018) Timing and mechanism of the rise of the Shillong Plateau in the Himalayan foreland. Geology 46(3):279–282
Grady JC (1971) Deep main faults in South India. J Geol Soc India 12(1):56–62
GSI (1993) Geological map of India, 7th ed., scale 1:2000,000. Geol Surv India, Hyderabad and Calcutta
GSI, ISRO (1994) Project Yasundhara: generalised geological map. Geol Surv India and Indian Space Res Organ Publ
Guha DB, Bhattacharya AK (1995) Metamorphic evolution and high-grade reworking of the Sandmata Complex granulites. In: Gupta KR, Sinha-Roy S (eds) Continental crust of northwestern and central India. Mem Geol Soc India 31, pp 163–198
Guitreau M, Mukusa SB, Loudin L, Krishnan S (2017) New constraints on early formation of western Dharwar craton (India) from igneous zircon U-Pb and Lu-Hf isotopes. Precambrian Res 302:33–49. https://doi.org/10.1016/j.precamres.2017.09.016
Gupta BC (1934) The geology of central Mewar. Geol Surv India Mem 65:107–169
Gupta P, Choudhuri AR (2002) Tectono-metamorphic evolution of the amphibolite- and granulite-facies rocks of the Sandmata Complex, Central Rajasthan. Indian Miner 56:137–160
Gupta RP, Sen AK (1988) Imprints of the Ninety-East ridge in the Shillong plateau, Indian shield. Tectonophysics 154:335–341
Gupta SN, Arora YK, Mathur RK, Iqbaluddin, Prasad B, Sahai TN, Sharma SB (1980) Lithostratigraphic map of Aravalli region, southern Rajasthan and NE Gujarat. Geol Surv India Publ, Hyderabad
Gupta A, Basu A, Singh SK (1985) Stratigraphy and petrochemistry of Dhanjori greenstone belt, Eastern India. Quart J Geol Min Met Soc India 57:248–263
Gupta SN, Arora YK, Mathur, RK, Iqbaluddin, Prasad B, Sahai TN, Sharma SB (1997) Precambrian geology of the Aravalli region, southern Rajasthan and northeastern Gujarat. Mem Geol Survey India, 262p
Gupta S, Jayananda M, Fareeduddin (2014) Tourmaline from the Archean G.R.Halli gold deposit, Chitradurga greenstone belt, Dharwar craton (India): implications for the gold metallogeny. Geosci Front 5:877–892
Halls HC (1982) The importance and potential of mafic dyke swarms in studies of geodynamic process. Geosci Can 9:145–154
Halls HC, Kumar A, Srinivasan R, Hamilton MA (2007) Paleomagnetism and U-Pb geochronology of easterly trending dykes in the Dharwar craton, India: feldspar clouding, radiating dyke swarms and the position of India at 2.37 Ga. Precambrian Res 155:47–68

Hansen EC, Newton RC, Janardhan AS (1984) Fluid inclusions in rocks from the amphibolite-facies gneiss to charnockite progression in southern Karnataka, India: direct evidence concerning the fluids of granulite metamorphism. J Metamorph Geol 2:249–264. https://doi.org/10.1111/j.1525-1314.1984.tb00299.x

Hanuma Prasad M, Krishna Rao B, Vasudev VN, Basavalingu B (1996) P-T conditions of metamorphism of supracrustal rocks in the Sandur Schist Belt, Dharwar Craton, Southern India. J Geol Soc India 48:617–628

Harish Kumar SB, Jayananda M, Kano T, Shadakshara Swamy N, Mahabaleshwar B (2003) Late Archean juvenile accretion process in the Eastern Dharwar Craton: Kuppam–Karimangala area. Mem Geol Soc India 50:375–408

Harlov DE, Newton RC, Hansen EC, Janardhan AS (1997) Oxide and sulfide minerals in highly oxidized, Rb-depleted, Archean granulites of the Shevaroy hills massif, South India: oxidation states and the role of metamorphicfluids. J Metamorph Geol 95:701–717

Hazarika P, Pruseth KL, Mishra B (2015) Neoarchean Greenstone metamorphism in the Eastern Dharwar Craton, India: constraints from monazite U-Th-Pb_{total} ages and PT pseudosection calculations. J Geol 123:429–461

Heron AM (1953) The geology of central Rajputana. Mem Geol Surv India 79, 389p

Hoffmann A, Mazumdar R (2015) A review of the current status of the Older Metamorphic Group and Older Metamorphic Tonalite Gneiss: insights into the Palaeoarchaean history of the Singhbhum craton, India. In: Mazumder R, Eriksson PG (eds) Precambrian basins of India: stratigraphic and tectonic context. Geol Soc London Mem 43, pp 103–107. https://doi.org/10.1144/m43.7

Hossain MdS, Sharif MdHK, Chowdhury KR, Abdullah R (2018) Synthesis of the tectonic and structural elements of the Bengal Basin and its surroundings. In: Mukherjee S (ed) Tectonics and structural geology: Indian context. Springer geology. Springer, Switzerland. https://doi.org/10.1007/978-3-319-99341-6-6

Hussain MF, Mondal MEA, Ahmad T (2003) Petrological and geochemical characteristics of Archean gneisses and granitoids from Bastar craton, Central India—implication for Subduction related magmatism. Gondwana Res 7:531–537

Hussain MF, Vanthangliana V, Mondal MEA (2019) Geochemical constraints on the petrogenesis of the metasedimentary rocks forming the basement of the Shillong Plateau, Northeast India. In: Mondal MEA (ed) Geological evolution of the Precambrian Indian Shield. Society of earth scientists series. Springer, Cham. https://doi.org/10.1007/978-3-319-89698-4-16

Ishwar-Kumar Windley BF, Horie K, Kato T, Hokada T, Itaya T, Yagi K (2013) A Rodinia suture in western India: new insights on India-Madagascar correlations. Precambrian Res 236:227–251

Iyengar SVP, Murthy YGK (1982) The evolution of the Archaean-Proterozoic crust in parts of Bihar and Orissa, eastern India. Rec Geol Surv India 112(3):1–5

Iyengar SYP, Grandy KC, Narayanaswamy R (1981) Geochronology and Rb-Sr systematics of the igneous rocks of the Simlipal complex, Orissa. Indian J Earth Sci 8(1):61–65

Iyenger SVP, Banerjee S (1964) Magmatic phases associated with the Precambrian tectonics of Mayurbhanj district, Orissa, India. In: 22nd international geological congress, vol 10, pp 515–538

Jain SC, Yedekar DB, Nair KKK (1991) Central Indian Shear Zone: a major Precambrian crustal boundary. J Geol Soc India 37:521–548

Jain SC, Nair KK, Yedekar DB (1995) Geology of Son-Narmada-Tapti lineament zone of Central India. Geol Surv India Spec Publ 10:1–154

Jain SC, Gaur VP, Srivastava SK, Nambiar KV, Saxena HP (2001) Recent find of a cauldron structure in Bundelkhand craton. Geol Surv India Spec Publ 64:289–297

Jain AK, Singh S, Manickavasagam RM (2003) Intracontinental shear zones in the Southern Granulite Terrain: their kinematics and evolution. Mem Geol Soc India 50:225–253

Janardhan AS, Newton RC, Hansen EC (1982) The transformation of amphibolite facies gneiss to charnockite in southern Karnataka and northern Tamil Nadu, India. Contrib Mineral Petrol 79:130–149

Jayananda M, Mahabaleswar B (1991) Relationship between shear zones and igneous activity: the Closepet granite of Southern India. Proc Indian Acad Sci (Earth Planet Sci) 100:31–36

Jayananda M, Martin H, Mahabaleswar B (1992) The mechanisms of recycling of Archaean continental crust: example of the Closepet granite, southern India. In: Ho SE, Glover JE (eds) Archaean: terrains, processes and metallogeny. Univ West Australia Publ 22, pp 213–222

Jayananda M, Martin H, Peucat J-J, Mahabaleswar B (1995) Late Archaean crust mantle interactions: geochemistry of LREE-enriched mantle derived magmas. Example of the Closepet batholith, southern India. Contrib Mineral Petrol 119:314–329

Jayananda M, Moyen JF, Martin H, Peucat JJ, Auvray B, Mahabaleswar B (2000) Late Archean (2550–2520) juvenile magmatism in the Eastern Dharwar Craton, southern India: constraints from geochronology, Nd–Sr isotopes and whole rock geochemistry. Precambrian Res 99:225–254

Jayananda M, Kano T, Kumar SBH, Mohan A, Swamy NS, Mahabaleswar B (2003) Thermal history of the 2.5 Ga juvenile continental crust in the Kuppam-Karimangalam area, Eastern Dharwar craton, southern India. Mem Geol Soc India 52:255–287

Jayananda M, Chardon D, Peucat J-J, Capdevila R (2006) 2.61 Ga Pottassic granites and Crustal reworking in the western Dharwar craton, southern India: tectonic, geochronologic and geochemical constraints. Precambrian Res 150:1–26

Jayananda M, Banerjee M, Pant NC, Dasgupta S, Kano T, Mahesha N, Mahabaleswar B (2011) 2.62 Ga high-temperature metamorphism in the central part of the Eastern Dharwar Craton: implications for late Archaean tectonothermal history. Geol J 47(2–3):213–236

Jayananda M, Peucat J-J, Chardon D, Krishna Rao B, Corfu F (2013a) Neoarchean greenstone volcanism, Dharwar craton, Southern India: constraints from SIMS zircon geochronology and Nd isotopes. Precambrian Res 227:55–76. https://doi.org/10.1016/j.precamres.2012.05.002

Jayananda M, Tsutsumi Y, Miyazaki T, Gireesh RV, Kapfo K-u, Hidaka H, Tushipokla, Kano T (2013b) Geochronological constraints on Meso-Neoarchean regional metamorphism and magmatism in the Dharwar craton, southern India. J Asian Earth Sci 78:18–38. https://doi.org/10.1016/j.jseaes.2013.04.033

Jayananda M, Gireesh RV, Sekhamo K-u, Miyazaki T (2014) Coeval felsic and mafic magmas in Neoarchean calc-alkaline magmatic arcs, Dharwar Craton, Southern India: field and petrographic evidence from mafic to hybrid magmatic enclaves and synplutonic mafic dykes. J Geol Soc India 84:5–28

Jayananda M, Chardon D, Peucat J-J, Fanning CM (2015) Paleo- to Mesoarchean TTG accretion and continental growth, western Dharwar craton, southern India: SHRIMP U-Pb zircon geochronology, whole-rock geochemistry and Nd-Sr isotopes. Precambrian Res 268:295–322. https://doi.org/10.1016/j.precamres.2015.07.015

Jayananda M, Duraiswami RA, Aadhiseshan KR, Gireesh RV, Prabhakar BC, Kafo K-u, Tushipokla, Namratha R (2016) Physical volcanology and geochemistry of Palaeoarchaean komatiite lava flows from the western Dharwar craton, southern India: implications for Archaean mantle evolution and crustal growth. Int Geol Rev 58(13):1569–1595

Jayananda M, Santosh M, Aadhiseshan KR (2018) Formation of Archean continental crust in the Dharwar craton, Southern India. Earth Sci Rev 181:12–42

Jayananda M, Guitreau M, Thomas TT, Martin H, Aadhiseshan KR, Gireesh RV, Peucat J-J, Satyanarayanan M (2019) Geochronology and geochemistry of Meso- to Neoarchean magmatic epidote-bearing potassic granites, Western Dharwar Craton (Bellur-Nagamangala-Pandavpura corridor), Southern India: implications for the successive stages of crustal reworking and cratonization. Geol Soc London Spec Publ. https://doi.org/10.1144/sp489-2018-125

Jayananda M, Aadhiseshan KR, Kusiak MA, Wilde SA, Sekhamo K-u, Guitreau M, Santosh M, Gireesh RV (2020a) Multi-stage crustal growth and Neoarchean geodynamics in the Eastern Dharwar Craton, southern India. Gondwana Res 78:228–260, https://doi.org/10.1016/j.gr.2019.09.005

Jayananda M, Dey S, Aadhiseshan KR (2020b) Evolving early Earth: Insights from Peninsular India. In: Tandon SK, Gupta N (eds) Geodynamics of the Indian plate: evolutionary perspectives. Springer, Geology 5–103, https://doi.org/10.1007/978-3-030-15989-4_2

Jayananda M, Miyazaki T, Gireesh RV, Mahesha N, Kano T (2009) Synplutonic mafic dykes from late Archaean granitoids in the Eastern Dharwar Craton, southern India. J Geol Soc India 73:117–130

Joshi KB, Slabunov AI (2019) Neoarchean Sanukitoids from the Karelian and Bundelkhand Cratons: comparison of composition, regional distribution and geodynamic setting. Trans KarRC RAS 2:5–25

Joshi KB, Bhattacharjee J, Rai G, Halla J, Ahmad T, Kurhila MI, Heilimo E, Choudhary AK (2017) The diversification of granitoids and plate tectonic implications at the Archaean-Proterozoic boundary in the Bundelkhand craton, Central India. In: Halla J, Whitehouse MJ, Ahmad T, Bagai Z (eds) Crust-mantle interactions and granitoid diversification: insights from Archaean Cratons. Geol Soc London Spec Publ 449, pp 123–157. http://doi.org/10.1144/SP449.8

Kale VS (1991) Constraints on the evolution of the Purana basins of peninsular India. J Geol Soc India 38:231–252

Kano T, Jayananda M, Nishimura K, Kesamaru T (2004) Modal and chemical compositions of peninsular gneisses and Closepet granite in the Dharwar Craton, in comparison with the Phanerozoic Island arc granitoids of Hida region central Japan: a preliminary report. In: Proceeding of the International Symposium on Gondwana Assembly and Dispersal. Gondwana Res, Suppl Issue, pp 1322–1323

Kaur P, Zeh A, Chaudhri N (2014) Characterisation and U-Pb–Hf record of the 3.55 Ga felsic crust from the Bundelkhand Craton, northern India. Precambrian Res 255:236–244

Kaur P, Zhe A, Choudhuri N (2016) Unraveling the record of Archaean crustal evolution of the Bundelkhand Craton, northern India using U-Pb zircon-monazite ages, Lu-Hf isotope systematics, and whole-rock geochemistry of granitoids. Precambrian Res 281:384–413. https://doi.org/10.1016/j.precamres.2016.06.005

Kaur P, Zeh A, Choudhuri N (2019) Archean crustal evolution of the Aravalli Banded Gneissic Complex, NW India: constraints from zircon U-Pb ages, Lu-Hf isotope systematics, and whole-rock geochemistry of granitoids. Precambrian Res 327:81–102

Kayal JR (1998) Seismicity of northeast India and surroundings—development over the past 100 years. J Geophy 19:9–34

Kent RW, Storey M, Saunders AD (2002) Large igneous provinces: sites of plume impact or plume incubation? Geology 20:891–894

Khanna TC, Sesha Sai VV, Bizimis M, Krishna AK (2016) Petrogenesis of ultramafics in the Neoarchean Veligallu greenstone terrane, eastern Dharwar craton, India: constraints from bulk-rock geochemistry and Lu-Hf isotopes. Precambrian Res 285:186–201

Khattri K, Wyss M (1978) Precursory variations of seismicity in Assam area, India. Geology 6:685–688

Krishnamurthy P (1985) Petrology of the carbonatites and associated rocks of Sung valley, Jaintia Hills District, Meghalaya, India. J Geol Soc India 26:361–379

Krishnamurthy P, Chaki A, Pandey BK, Chimote JS, Singh SN (1988) Geochronology of the granite–rhyolite suites of the Dongargarh supergroup central India. In: Proceedings of the 4th national symposium on mass spectrometry, pp EPS-2/1–EPS-2/3

Krogstad EJ, Balakrishnan B, Mukhopadhyay DK, Rajamani V, Hanson GN (1989) Plate tectonics 2.5 billion years ago: evidence at Kolar, South India. Science 243:1337–1340

Krogstad EJ, Hanson GN, Rajamani V (1991) U-Pb ages of zircon and sphene for two gneiss terrains adjacent to the Kolar Schist Belt, south India: evidence for separate crustal evolution histories. J Geol 99:801–816

Krogstad EJ, Hanson GN, Rajamani V (1995) Sources of continental magmatism adjacent to the late Archaean Kolar suture zone, South India: distinct isotopic and elemental signature of two late Archaean magmatic series. Contrib Mineral Petrol 122:159–173

Kumar S, Rino V (2006) Mineralogy and geochemistry of microgranular enclaves in Palaeoproterozoic Malanjkhand granitoids, central India: evidence of magma mixing, mingling, and chemical equilibration. Contrib Mineral Petrol 152:591–609

Kumar A, Bhaskar Rao YJ, Sivaraman TV, Gopalan K (1996) Sm-Nd ages of Archaean metavolcanic of the Dharwar craton, South India. Precambrian Res 80:206–215
Kumar S, Yi K, Raju K, Pathak M, Kim N, Lee TH (2011) SHRIMP U–Pb geochronology of felsic magmatic lithounits in the central part of Bundelkhand Craton, Central India. In: Molina JFS, Scarrow JH, Bea F, Montero P (eds) 7th Hutton symposium on granites and related rocks, Avila, Spain, p 83
Kumar A, Nagaraju E, Besse J, Bhaskar Rao YJ (2012) New age, geochemical and paleomagnetic data on a 2.21 Ga dyke swarm from south India: constraints on Paleoproterozoic reconstruction. Precambrian Res 220–221:123–138
Kumar D, Reddy DV, Pandey AK (2016) Paleoseismic investigations in the Kopili Fault Zone of North East India: evidences from liquefaction chronology. Tectonophysics 674:65–75
Kumar A, Parashuramulu V, Shankar R, Besse J (2017) Evidence for a Neoarchean LIP in the Singhbhum craton, eastern India: implications to Vaalbara supercontinent. Precambrian Res 292:163–174
Kumar S, Rino V, Hayasaka Y, Kimura K, Raju S, Terada K, Pathak M (2017a) Contribution of Columbia and Gondwana Supercontinent assembly and growth-related magmatism in the evolution of the Meghalaya Plateau and the Mikir Hills, Northeast India: constraints from U-Pb SHRIMP zircon geochronology and geochemistry. Lithos 277:356–377. https://doi.org/10.1016/j.lithos.2016.10.020
Kumar S, Pieru Th, Rino V, Hayasaka Y (2017b) Geochemistry and U–Pb SHRIMP zircon geochronology of microgranular enclaves and host granitoids from the South Khasi Hills of the Meghalaya Plateau, NE India: evidence of synchronous mafic–felsic magma mixing–fractionation and diffusion in a post-collision tectonic environment during the Pan-African orogenic cycle. In: Pant NC, Dasgupta S (eds) Crustal evolution of India and Antarctica: the supercontinent connection. Geol Soc London Spec Publ 457. https://doi.org/10.1144/SP457.10
Kumpulainen (ed) (2017) Guide for geological nomenclature in Sweden. GFF, vol 139, pp 1, 3–20. https://doi.org/10.1080/11035897.2016.1178666
Lal RK, Ackermand D, Seifert F, Haldar SK (1978) Chemographic relationships in sapphirine-bearing rocks from Sonapahar, Assam, India. Contrib Mineral Petrol 67:169–187
Li S-S, Keerthy S, Santosh M, Singh SP, Deering CD, Satyanarayanan M, Praveen MN, Aneeshkumar V, Indu GK, Anilkumar Y, Sajinkumar KS (2018a) Anatomy of impactites and shocked zircon grains from Dhala reveals Paleoproterozoic meteorite impact in the Archean basement rocks of Central India. Gondwana Res 54:81–101
Li S-S, Santosh M, Palin RM (2018b) Metamorphism during the Archean-Paleoproterozoic transition associated with microblock amalgamation in the Dharwar Craton. J Petrol, India. https://doi.org/10.1093/petrology/egy102
MacDougal JD, Gopalan K, Lugmair GW, Roy AB (1983) The Banded Gneissic Complex of India: early crust from the depleted mantle at 3.5 AE. Trans Amer Geophys Union 64:381
Mahadevan TM (1994) Deep continental structure of India: a review. Mem Geol Soc India 28:569
Mahadevan TM (2003) Kuppam-Palani transect programme and new insights into continental evolution. In: Mahadevan TM, Arora BR, Gupta KR (eds) Indian continental lithosphere: emerging research trends. Mem Geol Soc India 53, pp 99–114
Maibam B, Goswami JN, Srinivasan R (2011) Pb–Pb zircon ages of Archaean metasediments and gneisses from the Dharwar craton, southern India: implications for the antiquity of the eastern Dharwar craton. J Earth Syst Sci 120(4):643–661
Maibam B, Gerdes A, Goswami JN (2016) U–Pb and Hf isotope records in detrital and magmatic zircon from eastern and western Dharwar craton, southern India: evidence for coeval Archaean crustal evolution. Precambrian Res 275:496–512. https://doi.org/10.1016/j.precamres.2016.01.009
Majumdar R, Sarkar S (2004) Sedimentation history of the Palaeoproterozoic Dhanjori Formation, Singhbhum, India and its implications. Precambrian Res 130:267–287
Mallik AK, Sarkar A (1994) Geochronology and geochemistry of mafic dykes from the Precambrians of Keonjhar, Orissa. Indian Miner 48(1):13–24

Malviya VP, Arima M, Pati JK, Kaneko Y (2004) First report of metamorphosed basaltic pillow lava from central part of Bundelkhand craton, India: an Island arc setting of possible Late Archean age. Gondwana Res 7:1338–1340

Malviya VP, Arima M, Pati JK, Kaneko Y (2006) Petrology and geochemistry of metamorphosed basaltic pillow lava and basaltic komatiite in the Mauranipur area: subduction related volcanism in the Archean Bundelkhand craton, Central India. J Mineral Petrol Sci 101:199–217. https://doi.org/10.2465/jmps.101.199

Manikyamba C, Kerrich R (2012) Eastern Dharwar Craton, India: continental lithosphere growth by accretion of diverse plume and arc terranes. Geosci Front 3:225–240

Manikyamba C, Kerrich R, Naqvi SM, Ram Mohan M (2004) Geochemical systematics tholeiitic basalts from the 2.7 Ga Ramagiri-Hungund composite greenstone belt, Dharwar Craton. Precambrian Res 134:21–39

Manikyamba C, Naqvi SM, Subba Rao DV, Ram Mohan M, Khanna TC, Rao TG, Reddy GLN (2005) Neoarchaean Boninites: implications for Archaean subduction processes. Earth Planet Sci Lett 230:65–83

Manikyamba C, Kerrich R, Khanna TC, Krishna AK, Satyanarayanan M (2008) Geochemical systematics of komatiite-tholeiite and adakitic-arc basalt associations: the role of a mantle plume and convergent margin in formation of the Sandur Superterrane, Dharwar craton, India. Lithos 106:155–172

Manikyamba C, Kerrich R, Khanna TC, Satyanarayanan M, Krishna AK (2009) Enriched and depleted arc basalts, with high-Mg andesites and adakites: a potential paired arc-backarc of the 2.7 Ga Hutti greenstone terrane, India. Geochim Cosmochim Acta 73:1711–1736

Manikyamba C, Kerrich R, Polat A, Raju K, Satyanarayanan M, Krishna AK (2012) Arc picrite-potassic adakitic-shoshonitic volcanic association of the Neoarchean Sigegudda Greenstone Terrane, western Dharwar Craton: transition from arc wedge to lithosphere melting. Precambrian Res 212–213:207–224

Manikyamba C, Kerrich R, Polat A, Saha A (2013) Geochemistry of two stratigraphically-related ultramafic (komatiite) layers from the Neoarchean Sigegudda greenstone terrane, Western Dharwar Craton, India: evidence for compositional diversity in Archean mantle plumes. Lithos 177:120–135

Manikyamba C, Ganuly S, Santosh M, Rajanikanth Singh R, Saha A (2014) Arc-nascent back-arc signature in metabasalts from the Neoarchaean Jonnagiri greenstone terrane, Eastern Dharwar Craton, India. Geol J 50(5):651–669. https://doi.org/10.1002/gj.2581

Manikyamba C, Ganguly S, Santosh M, Saha A, Chatterjee A, Khelen AC (2015) Neoarchean arc-juvenile back-arc magmatism in eastern Dharwar Craton, India: geochemical fingerprints from the basalts of Kadiri greenstone belt. Precambrian Res 258:1–23

Manikyamba C, Ganguly S, Santosh M, Subramanyam KSV (2017) Volcano-sedimentary and metallogenic records of the Dharwar greenstone terranes, India: window to Archean plate tectonics, continent growth, and mineral endowment. Gondwana Res 50:38–66. https://doi.org/10.1016/j.gr.2017.06.005

Markandeyulu A, Chaturvedi AK, Raju BVSN, Parihar PS, Miller R, Gooch G (2013) Application of high resolution airborne geophysical data in geological modelling of Mohar Cauldron Complex, Bundelkhand Massif, central India: implications for uranium exploration. Explor Geophys 45:134–146

Mazumder SK (1976) A summary of the Precambrian geology of the Khasi Hills, Meghalaya. Geol Surv India Misc Publ 23:311–324

Mazumder SK (1986) The Precambrian framework of part of the Khasi Hills, Meghalaya. Rec Geol Surv India 117(2):1–59

Mazumder SK (1996) Precambrian geology of peninsular eastern India. Indian Miner 50:139–174

Mazumder D, Dutta P (2016) Geodynamic evolution of a Pan-African granitoid of extended Dizo Valley in Karbi Hills, NE India: evidence from Geochemistry and Isotope Geology. J Asian Earth Sci 117:256–268

Mazumder R, Van Loon AJ, Mallik L, Reddy SM, Arima M, Altermann W, Eriksson PG, De S (2012) Mesoarchaean–Palaeoproterozoic stratigraphic record of the Singhbhum crustal province, eastern India: a synthesis. In: Mazumder R, Saha D (eds) Palaeoproterozoic of India. Geol Soc London Spec Publ 365, pp 31–49. https://doi.org/10.1144/sp365.3
McKenzie NR, Hughes NC, Myrow PM, Banerjee DM, Deb M, Planavasky NJ (2013) New age constraints for the Proterozoic Aravalli-Delhi successions of India and their implications. Precambrian Res 238:12–28
Medlicott HB (1869) Geological sketch of the Shillong plateau in N.E. Bengal. Mem Geol Surv India 7(1):151–207
Meert JG, Pandit (2015) The Archaean and Proterozoic history of Peninsular India: tectonic framework for Precambrian sedimentary basins in India. In: Mazumder R, Eriksson PG (eds) Precambrian basins of India: stratigraphic and tectonic context. Geol Soc London Mem 43, pp 29–54. https://doi.org/10.1144/m43.3
Meert JG, Pandit MK, Pradhan VR, Banks J, Sirianni R, Stroud M, Newstead B, Gifford J (2010) Precambrian crustal evolution of Peninsular India: a 3.0 billion year odyssey. J Asian Earth Sci 39:483–515
Meert JG, Pandit MK, Pradhan VR, Kamenov G (2011) Preliminary report on the paleomagnetism of 1.88 Ga dykes from the Bastar and Dharwar cratons, Peninsular India. Gondwana Res 20:335–343
Meißner B, Deters P, Srikantappa C, Köhler H (2002) Geochronological evolution of the Moyar, Bhavani and Palghat Shear Zones of southern India: implications for East Gondwana correlations. Precambrian Res 114:149–175
Miller SR, Mueller PA, Meert JG, Kamenov GD, Pivarunas AF, Sinha AK, Pandit MK (2018) Detrital zircons reveal evidence of Hadean Crust in the Singhbhum Craton, India. J Geol 126:541–552
Mishra S, Deomurari MP, Wiedenbeck M, Goswami JN, Ray S, Saha AK (1999) $^{207}Pb/^{206}Pb$ zircon ages and the evolution of the Singhbhum craton, eastern India: an ion microprobe study. Precambrian Res 93:139–151
Misra S (2006) Precambrian chronostratigraphic growth of Singhbhum-Orissa craton, eastern Indian shield. J Geol Soc India 67:356–378
Misra S, Johnson PT (2005) Geochronological constraints on evolution of Singhbhum Mobile Belt and associated basic volcanics of eastern Indian shield. Gondwana Res 8:129–142
Misra RC, Sharma RP (1974) Petrochemistry of Bundelkhand granites and associated rocks of central India. Indian Miner 15:43–50
Mitra SK (2005) Tectonic setting of the Meghalaya Plateau and its sulphide mineralizations. J Geol Soc India 56:117–118
Mohan MR, Singh SP, Santosh M, Siddiqui MA, Balaram V (2012) TTG suite from the Bundelkhand Craton, Central India: geochemistry, petrogenesis and implications for Archaean crustal evolution. J Asian Earth Sci 58:38–50
Mohanty SP (2013) Spatio-temporal evolution of the Central Indian Shield and its correlation with South African and Western Australian cratons. In: 3rd Precambrian Continental Growth Tectonism 23–26 November 2013, Jhansi, India. Abstract Volume. IAGR Conf Ser 16:109–112
Mohanty SP (2015) Palaeoproterozoic supracrustals of the Bastar Craton: Dongargarh Supergroup and Sausar Group. In Mazumder R, Eriksson PG (eds) Precambrian basins of India: stratigraphic and tectonic context. Geol Soc, London, Mem 43, pp 151–164. http://dx.doi.org/10.1144/M43.11
Mohanty M, Guha DB (1995) Lithotectonic stratigraphy of the dismembered greenstone sequence of the Mangalwar Complex around Lawa-Sardargarh and Parasali areas, Rajsamand District, Rajasthan. Mem Geol Soc India 31:141–162
Mohanty N, Singh SP, Satyanarayanan M, Jayananda M, Korakoppa MM, Hiloidari S (2018) Chromian spinel compositions from Madawara ultramafics, Bundelkhand craton: implications on petrogenesis and tectonic evolution of the southern part of Bundelkhand craton, Central India. Geol J 53:1–25
Mojzsis SJ, Devaraju TC, Newton RC (2003) Ion microprobe U-Pb age determinations on zircon from the late Archean granulite facies transition zone of Southern India. J Geol 111(4):407–425

Mondal MEA (2003) Are the Bundelkhand craton and the Banded Gneissic Complex parts of one large Archaean protocontinent? Evidence from ion microprobe $^{207}Pb/^{206}Pb$ zircon ages. Newslett Deep Cont Stud India 13(2):6–10

Mondal MEA (2009) Was Bundelkhand-Aravalli nucleus par of Ur supercontinent? Curr Sci 96(1):33–35

Mondal MEA, Ahmad T (2001) Bundelkhand mafic dykes, Central Indian Shield: implications for the role of sediment subduction in Proterozoic crustal evolution. Island Arc 10:51–67

Mondal MEA, Raza M (2009) Tectonomagmatic evolution of the Bastar craton of Indian shield through plume-arc interaction: evidence from geochemistry of the mafic and felsic volcanic rocks of Sonakhan greenstone belt. In Ahmad T, Hirsch F, Charusiri P (eds) J Virtual Expl 32, Paper 7. https://doi.org/10.3809/jvirtex.2009.00245

Mondal MEA, Raza A (2013) Geochemistry of sanukitoid series granitoids from the Neoarchaen Berach granitoid batholiths, Aravalli craton, northwestern Indian shield. Curr Sci 105:102–108

Mondal MEA, Zainuddin SM (1997) Geochemical characteristics of the granites of Bundelkhand massif, central India. J Geol Soc India 50:69–74

Mondal MEA, Sharma KK, Rahman A, Goswami JN (1998) Ion microprobe $^{207}Pb/^{206}Pb$ zircon ages for gneiss–granitoid rocks from Bundelkhand massif: evidence for Archaean components. Curr Sci 74:70–75

Mondal MEA, Goswami JN, Deomurari MP, Sharma KK (2002) Ion Microprobe $^{207}Pb/^{206}Pb$ ages of zircons from Bundelkhand massif, northern India: implications for crustal evolution of the Bundelkhand-Aravalli protocontinent. Precamb Res 117:85–100

Mondal MEA, Hussain MF, Ahmad T (2006) Continental growth of Bastar craton, central Indian shield during Precambrian via multiphase subduction and lithospheric extension/rifting: evidence from geochemistry of gneisses, granitoids and mafic dykes. J Geosci Osaka City Univ 49:137–151

Mondal SK, Frei R, Ripley EM (2007) Os isotope systematics of mesoarchean chromitite-PGE deposits in the Singhbhum Craton (India): implications for the evolution of lithospheric mantle. Chem Geol 244:391–408

Mondal MEA, Chandra R, Ahmad T (2008) Precambrian mafic magmatism in Bundelkhand Craton. J Geol Soc India 72:113–122

Moorbath S, Taylor PN, Jones NW (1986) Dating the oldest terrestrial rocks—facts and fiction. Chem Geol 57:63–80

Moyen J-F, Martin H, Jayananda M (2001) Multi-element geochemical modelling of crust-mantle interactions during late-Archaean crustal growth: the Closepet granite (South India). Precambrian Res 112:87–105

Moyen J-F, Nedelec A, Martin H, Jayananda M (2003) Syntectonic granite emplacement at different structural levels: the Closepet granite, South India. J Struct Geol 25:611–631

Mukhopadhyay D (1976) Precambrian stratigraphy of Singhbhum—the problem and a prospect. Indian J Earth Sci 3:208–219

Mukhopadhyay D (2001) The Archaean nucleus of Singhbhum: the present state of knowledge. Gondwana Res 4:307–318

Mukhopadhyay J, Beukes NJ, Armstong RA, Zimmermann U, Ghosh G, Medda RA (2008) Dating of the oldest greenstone belt in India: a 3.51 Ga precise U-Pb SHRIMP zircon age for Dacitic Lava of the Southern Iron Ore Group, Singhbhum craton. J Geol 116:449–461

Mukhopadhyay J, Ghosh G, Zimmermann U, Guha S, Mukherje T (2012) A 3.51 Ga bimodal volcanic-BIF-ultramafic succession from Singhbhum Craton: implications for Palaeoarchaean geodynamic processes from the oldest greenstone succession of the Indian subcontinent. Geol J 47:284–311

Mukhopadhyay J, Crowley QG, Ghosh S, Ghosh G, Chakrabarti K, Misra B, Heron K, Bose S (2014) Oxygenation of the Archean atmosphere: new paleosol constraints from eastern India. Geology 42(10):923–926. https://doi.org/10.1130/G36091.1

Nagaraju E, Parashuramulu V, Kumar A, Srinivas Sarma D (2018a) Paleomagnetism and geochronological studies on a 450 km long 2216 Ma dyke from the Dharwar craton, southern India. Phys Earth Planet Inter 274:222–231

Nagaraju E, Parashuramulu V, Ramesh Babu N, Narayana AC (2018b) A 2207 Ma radiating mafic dyke swarm from eastern Dharwar craton. Southern India: drift history through Paleoproterozoic Precambrian Res 317:89–100

Naha K, Halyburton RV (1974) Early Precambrian stratigraphy of central and southern Rajasthan, India. Precambrian Res 1:55–73

Naha K, Srinivasan R (1996) Nature of Moyar and Bhavani Shear Zones with a note on its implication on the tectonics of the southern Indian Precambrian shield. Proc Indian Acad Sci (Earth Planet Sci) 105:173–189

Naha K, Srinivasan R, Jayaram S (1991) Sedimentological, structural and migmatitic history of the Archaean Dharwar tectonic province, southern India. Proc Indian Acad Sci (Earth Planet Sci) 100:413–433

Naha K, Mukhopadhyay D, Dastidar S, Mukhopadhyay RP (1995) Basement–cover relations between granite gneiss body and its metasedimentary envelope: a structural study from early Precambrian Dharwar tectonic province, southern India. Precambrian Res 72:238–299

Nandy DR (1980) Tectonic pattern of northeast India and adjoining region. Indian J Earth Sci 7:103–107

Nandy DR (2001) Geodynamics of Northeast India and the adjoining region. ABC Publications, Calcutta, pp 1–209

Naqvi SM (2005) Geology and evolution of the Indian Plate (Hadean to Holocene—4 Ga to 4 Ka). Capital Publ Co, New Delhi

Naqvi SM, Rogers JJW (eds) (1983) Precambrian of South India. Mem Geol Soc India 4, p 575

Naqvi SM, Rogers JJW (1987) Precambrian geology of India. Oxford University Press, New York

Naqvi SM, Manikyamba C, Rao G, Subba Rao DV, Ram Mohan M, Sarma S (2002) Geochemical and isotopic constraints on Neoarchean fossil plumes for the formation of volcanic rocks of Sandur Greenstone Belt, India. J Geol Soc India 60:27–56

Nasipuri P, Saha L, Xie H, Pati JK, Satyanarayanan M, Sarkar S, Bhandari A, Gaur Y (2019) Paleoarchean crustal evolution of the Bundelkhand craton, north-central India. In: van Kranendonk M, Bennett V, Hoffmann J (eds) Earth's oldest rock, 2nd ed. Elsevier, Amsterdam, pp 793–817

Nayak GK, Rao VK, Rambabu HV, Kayal JR (2008) Pop-up tectonics of the Shillong Plateau in the great 1897 earthquake (Ms 8.7): insights from the gravity in conjunction with the recent seismological results. Tectonics 27:TC1018. https://doi.org/10.1029/2006tc002027

Nelson D, Bhattacharya HN, Misra S, Dasgupta N, Alterman W (2007) New SHRIMP U-Pb dates from Singhbhum Craton, Jharkhand-Orissa region, India. In: International conference on Precambrian sedimentation and tectonics (abst) 2nd GPSS meeting, IIT-Bombay, p 47

Nelson D, Bhattacharya HN, Thern ER, Altermann W (2014) Geochemical and ion-microprobe U-Pb zircon constraints on the Archaean evolution of Singhbhum Craton, eastern India. Precambrian Res 255:412–432

Newton RC (1990) The late Archean high-grade terrain of south India and the deep structure of Dharwar craton. In: Salisbury MH, Fountain DM (eds) Exposed cross sections of the continental crust. Kluwer Acad Publ, Dordrecht, pp 305–326

Nutman AP, Chadwick B, Krishna Rao B, Vasudev VN (1996) SHRIMP U/Pb zircon ages of acid volcanic rocks in the Chitradurga and Sandur groups, and granites adjacent to the Sandur Schist belt, Karnataka. J Geol Soc India 47:153–164

Oldham TH (1856) Remarks on the classification of the rocks of Central India resulting from the investigation of the Geological Survey. J Asiat Soc Bengal 25:224–256

Oldham RD (1899) Report on the great earthquake of the 12th June 1897. Geol Surv India Mem 29. Geol Surv India, Calcutta, pp 1–379 (Reprinted 1981)

Pandey BK, Gupta JN, Sarma KJ, Sastry CA (1997) Sm–Nd, Pb–Pb and Rb–Sr geochronology and petrogenesis of the mafic dyke swarm of Mahbubnagar, South India: implications for Paleoproterozoic crustal evolution of the Eastern Dharwar Craton. Precambrian Res 84:181–196

Pandit D, Panigrahi MK (2012) Comparative petrogenesis and tectonics of Paleoproterozoic Malanjkhand and Dongargarh granitoids, Central India. J Asian Earth Sci 50:14–26

Pandit MK, de Wall H, Chauhan NK (2008) Paleosol at the Archean-Proterozoic contact in NW India revisited: evidence for oxidizing conditions during paleoweathering? J Earth System Sci 117:201–209

Panigrahi MK, Mookherjee A, Pantulu GVC, Gopalan K (1993) Granitoids around Malanjkhand copper deposit: types and age relationship. Proc Indian Acad Sci Earth Planet Sci Lett 102:399–413

Panigrahi MK, Bream BR, Misra KC, Naik RK (2004) Age of granitic activity associated with copper-molybdenum mineralization at Malanjkhand, Central India. Miner Depos 39:670–677

Pascoe EH (1950) A manual of geology of India and Burma. Geol Surv India Misc Publ 1:245–250

Pati JK (2005) The Dhala Structure, Bundelkhand craton, Central India—a new large Paleoproterozoic impact structure (abstract). Meteorit Planet Sci 40:A121

Pati JK, Raju S, Pruseth KL, Magngain VD, Shankar R (1997) Gold mineralization in parts of Bundelkhand Granitoid Complex (BGC). J Geol Soc India 50:601–606

Pati JK, Patel SC, Pruseth KL, Malviya VP, Arima M, Raju S, Pati P, Prakash K (2007) Geology and geochemistry of giant quartz veins from the Bundelkhand Craton, central India and their implications. J Earth Syst Sci 116(6):497–510

Pati JK, Reimold WU, Koeberl C, Pati P (2008) The Dhala structure, Bundelkhand craton, Central India—eroded remnant of a large Paleoproterozoic impact structure. Meteorit Planet Sci 43(8):1383–1398

Pati JK, Jourdan F, Armstrong RA, Reimold WU, Prakash K (2010) First SHRIMP U-Pb and $^{40}Ar/^{39}Ar$ chronological results from impact melt breccia from the Paleoproterozoic Dhala impact structure, India. In: Gibson RL, Reimold, WU (eds) Large meteorite impacts and planetary evolution IV. Geol Soc Amer Spec Paper 465, pp 571–591. https://doi.org/10.1130/2010.2465(27)

Pati JK, Reimold WU, Greshake A, Schmitt RT, Koeberl C, Pati P, Prakash K (2015) Pseudotachylitic breccia from the Dhala impact structure, north-central India: texture, mineralogy and geochemical characterization. Tectonophysics 649:18–32

Pati JK, Qu WJ, Koeberl C, Reimold WU, Chakarvorty M, Schmitt RT (2017) Geochemical evidence of an extraterrestrial component in impact melt breccia from the Paleoproterozoic Dhala impact structure, India. Meteorit Planet Sci 52:722–736. https://doi.org/10.1111/maps.12826

Pati JK, Reimold WU, Hauser N (2018) Comment on "Anatomy of impactites and shocked zircon grains from Dhala reveals Paleoproterozoic meteorite impact in the Archean basement rocks of Central India" by Li et al. (2018). Gondwana Res 54:81–101

Pati JK, Poelchau MH, Reimold WU, Nakamura N, Kuriyama Y, Singh AK (2019) Documentation of shock features in impactites from the Dhala impact structure, India. Meteorit Planet Sci 54(10):2312–2333. https://doi.org/10.1111/maps.13369

Peucat J-J, Mahabaleswar M, Jayananda M (1993) Age of younger tonalitic magmatism and granulite metamorphism in the amphibolite–granulite transition zone of southern India (Krishnagiri area): comparison with older peninsular gneisses of Gorur-Hassan area. J Metamorph Geol 11:879–888

Peucat J-J, Jayananda M, Chardon D, Capdevila R, Fanning MC, Paquette J-L (2013) The lower crust of Dharwar craton, south India: patchwork of Archean granulitic domains. Precambrian Res 227:4–29. https://doi.org/10.1016/j.precamres.2012.06.009

Pichamuthu CS (1965) Regional metamorphism and charnockitisation in Mysore state India. Indian Miner 6:116–126

Plavsa D, Collins AS, Foden JD, Clark C (2015) The evolution of a Gondwanan collisional orogen: a structural and geochronological appraisal from the Southern Granulite Terrane, South India. Tectonics 34(5):820–857

Prabhakar N, Bhattacharya A (2013) Paleoarchean partial convective overturn in the Singhbhum Craton, Eastern India. Precambrian Res 231:106–121

Prabhakar BC, Jayananda M, Shareef M, Kano T (2009) Synplutonic mafic injections into crystallizing granite pluton in the northern part of the eastern Dharwar craton: implications for the magma chamber processes. J Geol Soc India 74:171–188

Pradhan VR, Meert JG, Pandit MK, Kamenov G, Mondal MEA (2012) Paleomagnetic and geochronological studies of the mafic dyke swarms of Bundelkhand craton, central India: implications for the tectonic evolution and paleogeographic reconstructions. Precambrian Res 198–199:51–76

Raase P, Raith M, Ackermand D, Lal RK (1986) Progressive metamorphism of mafic rocks from greenschist to granulite facies in the Dharwar craton of South India. J Geol 94:261–282

Radhakrishna T, Joseph M (1996) Proterozoic paleomagnetism of the mafic dyke swarms in the high grade region of southern India. Precambrian Res 76:31–46. https://doi.org/10.1016/0301-9268(95)00022-4

Radhakrishna BP, Naqvi SM (1986) Precambrian continental crust of India and its evolution. J Geol 94:145–166

Radhakrishna BP, Ramakrishnan M (eds) (1990) Archean greenstone belts of South India (Bellur Rama Rao vol). Mem Goel Soc India, 19, p 497

Radhakrishna BP, Ramakrishnan M, Mahabaleswar B (eds) (1990) Granulites of South India (Pichamuthu vol). Geol Soc India, New Delhi, p 502

Radhakrishna BP, Vaidyanadhan R (1997) Geology of Karnataka. Geol Soc India (2):1–353

Rahman S (1999) The Precambrian rocks of the Khasi Hills, Megahalaya, Shillong Plateau. In: Verma PK (ed) Geological studies in the Eastern Himalayas. Pilgrims Book Pvt Ltd, Delhi, pp 59–65

Rahman A, Zainuddin SM (1993) Bundelkhand granites: an example of collision-related Precambrian magmatism and its relevance to the evolution of the central Indian shield. J Geol 101:413–419

Raith M, Raase P, Ackermand D, Lal RK (1982) The Archean Dharwar craton of Southern India: metamorphic evolution and P-T conditions. Geol Rundsch 71:280–290

Raith MM, Srikantappa C, Buhl D, Kaehler H (1999) The Nilgiri enderbites, south India: nature and age constraints on protolith formation, high grade metamorphism and cooling history. Precambrian Res 98:129–150

Raja Rao CS (1967) On the age of Precambrian groups of Rajasthan. J Mines Met Fuels 15(9):306–309

Raja Rao CS, Poddar BC, Basu KK, Dutta A (1971) Precambrian stratigraphy of Rajasthan—a review. Rec Geol Surv India 101:60–79

Rajamani V, Shivkumar K, Hanson GN, Shirey SB (1985) Geochemistry and petrogenesis of amphibolites, Kolar Schist Belt, South India: evidence for komatiitic magma derived by low percentage of melting of the mantle. J Petrol 96:92–123

Rajamanickam M, Balakrishnan S, Bhutani R (2014) Rb–Sr and Sm–Nd isotope systematics and geochemical studies on metavolcanic rocks from Peddavura greenstone belt: evidence for presence of Mesoarchean continental crust in easternmost part of Dharwar Craton, India. J Earth Syst Sci 123:989–1011

Rajesh HM, Mukhopadhyay J, Beukes NJ, Gutzmer J, Belyanin GA, Armstrong RA (2009) Evidence for an early Archaean granite from Bastar craton, India. J Geol Soc London 166:193–196. https://doi.org/10.1144/0016-76492008-089

Ramachandra HM, Roy A (1998) Geology of intrusive granitoids with particular reference to Dongargarh granite and their impact on the tectonic evolution of the Precambrian of central India. India Min 52:15–32

Ramachandra HM, Roy A, Misra VP, Dutta NK (2001) A critical review of the tectono-thermal evolution of the Bastar Craton. Geol Surv India Spec Publ 52(2):161–180

Ramakrishnan M (1987a) Stratigraphy, sedimentary environment and evolution of the Late Proterozoic Indravati basin. Geol Soc India Mem 6:139–160

Ramakrishnan M (1987b) Tectonic evolution of Proterozoic-type platform-fold belts of Archean age in Western Dharwar Craton. Indian Miner 27:1–9

Ramakrishnan M (1990) Crustal development in southern Bastar, central Indian craton. Geol Surv India Spec Publ 28:44–66

Ramakrishnan M (1993) Tectonic evolution of granulite terrains of southern India. Mem Goel Soc India 25:35–44

Ramakrishna M (2003) Craton-mobile belt relations in Southern Granulite Terrain. Geol Soc India Mem 50:1–24

Ramakrishnan M (2009) Precambrian mafic magmatism in the Western Dharwar Craton, Southern India. J Goel Soc India 73:101–116

Ramakrishnan M, Vaidyanadhan R (2008) Geology of India, vol 1. Geol Soc India, Bangalore, p 556

Ramiz MM, Mondal MEA (2017) Petrogenesis of mafic magmatic enclaves of the Bundelkhand granitoids near Orchha, Central Indian shield: evidence for rapid crystallization. In: Halla J, Whitehouse MJ, Ahmad T, Bagai Z (eds) Crust-mantle interactions and granitoid diversification: insights from Archaean Cratons. Geol Soc London Spec Publ 449, pp 159–173. https://doi.org/10.1144/SP449.6

Ramiz MM, Mondal MEA, Farooq SH (2018) Geochemistry of ultramafic-mafic rocks of the Madawara Ultramafic Complex in the southern part of the Bundelkhand Craton, Central Indian Shield: implications for mantle sources and geodynamic setting. Geol J 53:1–23

Rao JM (2004) The wide spread 2 Ga dyke activity in the Indian Shield—evidences from Bundelkhand Mafic Dyke Swarm, Central India and their tectonic implications. Gondwana Res 7(4):1219–1228

Rao JM, Pooranchandra Rao GVS, Widdowson M, Kelley SP (2005) Evolution of Proterozoic mafic dyke swarms of the Bundelkhand granite massif, central India. Curr Sci 88(3):502–506

Ratheesh Kumar RT, Santosh M, Yang Q-Y, Iswar Kumar C, Chen NS, Sajeev K (2016) Archean tectonics and crustal evolution of the Biligiri Rangan Block, Southern India. Precambrian Res 275:406–428

Ray J, Saha A, Koeberl C, Thoni M, Ganguly S, Hazra S (2013) Geochemistry and petrogenesis of Proterozoic mafic rocks from east Khasi Hills, Shillong Plateau, northeastern India. Precambrian Res 230:119–137

Raza A, Mondal MEA (2018) Geochemistry of the Archaean metasedimentary rocks of the Bundelkhand Mauranipur-Babina greenstone belt, central India: implications for provenance characteristics. J Indian Assoc Sedimentol 35:57–76

Raza A, Mondal MEA (2019) Geochemistry of the mafic metavolcanic rocks of Mauranipur-Babina Greenstone Belt, Bundelkhand Craton, Central India: implication for tectonic settings during the Archaean. In: Mondal MEA (ed) Geological evolution of the Precambrian Indian Shield. Society of earth scientists series. Springer, Cham. https://doi.org/10.1007/978-3-319-89698-4-22

Roday PP, Diwan P, Singh S (1995) A synkinematic model of emplacement of quartz reefs and subsequent deformation patterns in the central Indian Bundelkhand batholith. Proc Indian Acad Sci (Earth Planet Sci) 104:465–468

Rogers AJ, Kolb J, Meyer FM, Armstrong RA (2007) Tectono-magmatic evolution of the Hutti-Maski Greenstone Belt, India: constrained using geochemical and geochronological data. J Asian Earth Sci 31:55–70

Rollinson HR, Windley BF, Ramakrishnan M (1981) Contrasting high and intermediate pressures of metamorphism in the Archaean Sargur Schists of southern India. Contrib Mineral Petrol 76:420–429

Roy AB (1988) Stratigraphic and tectonic framework of the Aravalli mountain range. Mem Geol Soc India 7:3–32

Roy AB, Bhattacharya HN (2012) Tectonostratigraphic and geochronologic reappraisal constraining the growth and evolution of Singhbhum Archaean Craton, Eastern India. J Geol Soc India 80:455–469

Roy AB, Jakhar SR (2002) Geology of Rajasthan: Precambrian to recent. Scientific Publishers, Jodhpur, India, 421p

Roy AB, Kröner A (1996) Single zircon evaporation ages constraining the growth of the Archaean Aravalli craton, northwestern Indian shield. Geol Mag 133:333–342

Roy A, Hanuma Prasad M (2003) Tectonothermal events in Central Indian Tectonic Zone (CITZ) and its implications in Rodinian crustal assembly. J Asian Earth Sci 22:115–129
Roy AB, Somani MK, Sharma BL (1981) Aravalli and Pre-Aravalli relationship: a study from the Bhindar region, Southern Rajasthan. Indian J Earth Sci 8(2):119–130
Roy A, Sarkar A, Jeyakumar S, Aggrawal SK, Ebihara M (2002) Sm-Nd age and mantle source characteristics of the Dhanjori volcanic rocks, Eastern India. Geochem J 36:503–518
Roy A, Sarkar A, Jeyakumar S, Aggarwal SK, Ebihara M, Satoh H (2004) Late Archaean mantle metasomatism below eastern Indian craton: evidence from trace elements, REE geochemistry and Sr-Nd-O isotope systematics of ultramafic dykes. Proc Indian Acad Sci (Earth Planet Sci) 113:649–665
Roy AB, Kröner A, Bhattacharya PK, Rathore S (2005) Metamorphic evolution and zircon geochronology of early Proterozoic granulites in the Aravalli Mountains of northwestern India. Geol Mag 142:287–302
Roy A, Sengupta S, Mandal A (2008) Synchronous development of mylonite and pseudotachylyte: an example from the Chitradurga Eastern Margin Shear Zone, Karnataka. J Geol Soc India 72:447–457
Roy AB, Kroner A, Rathor S, Laul V, Purohit R (2012) Tectonometamorphic studies from Sandmata Complex northwest Indian shield: implications on exhumation of late Proterozoic granulites in an late-Archean-Early Paleoprotrozoic granite-gneiss terrane. J Geol Soc India 79:323–334
Sadiq M, Ranjith A, Umrao RK (2014) REE mineralization in the carbonatites of the Sung Valley ultramafic-alkaline-carbonatite complex, Meghalaya, India. Cent Eur J Geosci 6(4):457–475. https://doi.org/10.2478/s13533-012-0191-y
Sadiq M, Umrao RK, Sharma BB, Chakraborti S, Bhattacharyya S, Kundu A (2017) Mineralogy, geochemistry and geochronology of mafic magmatic enclaves and their significance in evolution of Nongpoh granitoids, Meghalaya, NE India. In: Sensarma S, Storey BC (eds) Large igneous provinces from Gondwana and adjacent regions. Geol Soc London Spec Publ 463. https://doi.org/10.1144/sp463.2
Saha AK (1994) Crustal evolution of Singhbhum-North Orissa, Eastern India. Geol Soc India Mem 27:339
Saha D, Mazumder R (2012) An overview of the Palaeoproterozoic geology of Peninsular India, and key stratigraphic and tectonic issues. In: Mazumder, R, Saha D (eds) Palaeoproterozoic of India. Geol Soc London Spec Publ 365, pp 5–29. https://doi.org/10.1144/sp365.2
Saha AK, Ray SL (1984) The structural and geochemical evolution of the Singhbhum granite batholithic complex, India. Tectonophysics 105:163–176
Saha L, Pant NC, Pati JK, Upadhyaya D, Berndt J, Bhattacharya A, Satyanarayan M (2011) Neoarchean high pressure margarite-phengitic muscovite-chlorite corona mantled corundum in quartz-free high-Mg, Al phlogopite-chlorite schists from the Bundelkhand craton, north-central India. Contrib Mineral Petrol 161:511–530
Saha L, Frei D, Gerdes A, Pati JK, Sarkar S, Patoke V, Bhandari A, Nasipur P (2016) Crustal geodynamics from the Archaean Bundelkhand Craton, India: constraints from zircon U-Pb–Hf isotope studies. Geol Mag 153:179–192
Samal AK, Srivastava RK, Ernst RE, Söderlund U (2019) Neoarchean-Mesoproterozoic mafic dyke swarms of the Indian Shield mapped using Google Earth™ images and ArcGIS™, and links with large igneous provinces. In: Srivastava RK, Ernst RE, Peng P (eds) Dyke swarms of the world: a modern perspective. Springer, Singapore, pp 335–390. https://doi.org/10.1007/978-981-13-1666-1-9
Sant DA, Karanth RV (1993) Drainage evolution of the Lower Narmada Valley, Western India. Geomorphology 8:221–244
Santosh M, Li SS (2018) Anorthosites from an Archean continental arc in the Dharwar Craton, southern India: implications for terrane assembly and cratonization. Precambrian Res. https://doi.org/10.1016/j.precamres.2018.02.011
Santosh M, Yokoyama K, Acharyya SK (2004) Geochronology and tectonic evolution of Karimnagar and Bhopalpatnam Granulite Belts, Central India. Gondwana Res 7(2):501–518

Sarkar SN, Saha AK (1962) A revision of the Precambrian stratigraphy and tectonics of Singhbhum and adjoining region. Quart J Geol Min Met Soc India 34:97–136
Sarkar SN, Gopalan K, Trivedi JR (1981) New data on the geochronology of Precambrian of Bhandara-Durg, Central India. Indian J Earth Sci 8:131–151
Sarkar SN, Ghosh DK, Lambert RJS (1985) Rubidium-Strontium and Lead isotopic studies of the Soda Granites from Mosabani, Singhbhum Copper Belt, India. Indian J Earth Sci 13:101–116
Sarkar A, Sarkar G, Paul DK, Mitra ND (1990) Precambrian geochronology of central Indian shield: a review. Geol Surv India Hyderabad Spec Publ 28:453–482
Sarkar A, Paul DK, Potts PJ (1995) Geochronology and geochemistry of the Mid-Archaean trondhjemitic gneisses from the Bundelkhand craton, central India. In: Saha AK (ed) Recent Res Geology, Hindustan Publ. Co., pp 76–92
Sarkar G, Corfu F, Paul DK, McNaughton NJ, Gupta SN, Bishui PK (1993) Early Archaean crust in Bastar craton, Central India: a geochemical and isotopic study. Precambr Res 62:127–137
Sarkar G, Matin A and Sensarma S (2017) Submagmatic fabric in the 2.6 Ga Bundelkhand granitoid, India: evidence from microstructure. Curr Sci 112:348–354
Sarkar SC, Kabiraj S, Bhattacharya S, Pal AB (1996) Nature, origin and evolution of the granitoid-hosted early Proterozoic copper-molybdenum mineralization at Malanjkhand, Central India. Mineral Deposita 31:419–425
Sengupta S, Paul DK, de Laeter JR, McNaughton NJ, Bandyopadhyay BK, de Smeth JB (1991) Mid-Archaean evolution of the eastern Indian craton: geochemical and isotopic evidence from Bonai pluton. Precambrian Res 49:23–37
Sengupta S, Corfu F, McNutt RH, Paul DK (1996) Mesoarchaean crustal history of the eastern Indian craton: Sm-Nd and U-Pb isotopic evidence. Precambrian Res 77:17–22
Shankar R, Vijayagopal B, Kumar A (2014) Precise Pb–Pb baddeleyite ages of 1765 Ma for a Singhbhum 'newer dolerite' dyke swarm. Curr Sci 106:1306–1310
Sharma RS (1988) Patterns of metamorphism in the Precambrian rocks of the Aravalli Mountain Belt. Mem Geol Soc India 7:33–75
Sharma RS (2009) Cratons and fold belts of India. Lecture notes in earth sciences, vol 127, Springer, Heidelberg, pp 1–304. https://doi.org/10.1007/978-3-642-01459-8-2
Sharma RS, Mondal MEA (2019) Evolution of the Indian Shield: A New Approach. In: Mondal MEA (ed) Geological evolution of the Precambrian Indian Shield. Society of earth scientists series. Springer, Cham. https://doi.org/10.1007/978-3-319-89698-4-2
Sharma KK, Rahman A (1996) Mafic dykes in Bundelkhand granitoids and mafic volcanics in Supra-batholithic volcano-sedimentaries (Bijawar). DST Newslett 15:17–19
Sharma KK, Rahman A (2000) The Early Archaean-Paleoproterozoic crustal growth of the Bundelkhand craton, northern Indian shield. In: Deb M (ed) Crustal evolution and metallogeny in the Northwestern Indian Shield. Narosa, New Delhi, pp 51–72
Sharma RS, Sills JD, Joshi M (1987) Mineralogy and metamorphic history of norite dykes within granulite facies gneisses from Sandmata, Rajasthan, NW India. Miner Mag 51:207–215
Shellnutt JG, Hari KR, Liao AC-Y, Denyszyn SW, Vishwakarma N (2018) A 1.88 Ga giant radiating mafic dyke swarm across southern India and Western Australia. Precambrian Res 308:58–74. https://doi.org/10.1016/j.precamres.2018.01.021
Shimpo M, Tsunogae T, Santosh M (2006) First report of garnet–corundum rocks from southern India: implications for prograde high-pressure (eclogite-facies?) metamorphism. Earth Planet Sci Lett 242(1):111–129
Singh SP, Bhattacharya AR (2017) N-S crustal shear system in the Bundelkhand massif: a unique crustal evolution signature in the northern Indian peninsula. J Earth Syst Sci 126:121. https://doi.org/10.1007/s12040-017-0900-5
Singh VK, Slabunov A (2015) The Central Bundelkhand Archean greenstone complex, Bundelkhand Craton, central India: geology, composition, and geochronology of supracrustal rocks. Int Geol Rev 57:1349–1364

Singh VK, Slabunov A (2016) Two types of Archean supracrustal Belts in the Bundelkhand Craton, India: geology, geochemistry, age and implication for Craton crustal evolution. J Geol Soc India 88:533–672
Singh SP, Subramanyam KSV, Manikyamba C, Santosh M, Rajanikanta Singh M, Chandan Kumar B (2018) Geochemical systematics of the Mauranipur-Babina greenstone belt, Bundelkhand Craton, Central India: insights on Neoarchean mantle plume-arc accretion and crustal evolution. Geosci Front 9:769–788
Singh PK, Verma SK, Moreno JA, Singh VK, Malviya VP, Oliveira EP, Mishra S, Arima M (2019a) Geochemistry and Sm-Nd isotope systematics of mafic-ultramafic rocks from the Babina and Mauranipur greenstone belts, Bundelkhand Craton, India: implications for tectonic setting and Paleoarchean mantle evolution. Lithos 330–331:90–107
Singh PK, Verma SK, Singh VK, Moreno JA, Oliveira EP, Mehta P (2019b) Geochemistry and petrogenesis of sanukitoids and high-K anatectic granites from the Bundelkhand Craton, India: implications for late-Archean crustal evolution. J Asian Earth Sci 174:263–282
Sinha-Roy S (1985) Granite-greenstone sequence and geotectonic development of SE Rajasthan. Bull Geol Min Met Soc India 53:115–123
Sinha-Roy S, Malhotra G (1989) Structural relations of the cover and its basement: an example from Jahajpur belt, Rajasthan. J Geol Soc India 34:233–244
Sinha-Roy S, Mohanty M, Guha DB (1993) Banas dislocation zone in Nathdwara-Khamor Area, Udaipur district, Rajasthan, and its significance on the basement cover relations in the Aravalli Fold Belt. Curr Sci 65(1):68–72
Sinha-Roy S, Malhotra G, Guha DB (1995) A transect across Rajasthan Precambrian terrain in relation to geology, tectonics and crustal evolution of south-central Rajasthan. In: Sinha-Roy S, Gupta KR (Eds) Continental Crust of Northwestern and Central India. Mem Geol Soc India 31:63–90
Sinha-Roy S, Malhotra G, Mohanty M (1998) Geology of Rajasthan. Geol Soc India, 278p
Slabunov A, Singh VK (2018) Meso-Neoarchaean crustal evolution of the Bundelkhand Craton, Indian Shield: new data from greenstone belts. Int Geol Rev 60:1–20
Söderlund U, Bleeker W, Demirer K, Srivastava RK, Hamilton MA, Nilsson M, Pesonen L, Samal AK, Jayananda M, Ernst RE, Srinivas M (2019) Emplacement ages of Paleoproterozoic mafic dyke swarms in eastern Dharwar craton, India: implications for paleoreconstructions and evidence for a ~30° internal block rotation. Precambrian Res 329:26–43. https://doi.org/10.1016/j.precamres.2018.12.017
Sreenivas B, Dey S, Bhaskar Rao YJ, Vijaya Kumar T, Babu EVSSK, Williams IS (2019) A new cache of Eoarchaean detrital zircons from the Singhbhum craton, eastern India and constraints on early Earth geodynamics. Geosci Front 10:1359–1370. https://doi.org/10.1016/j.gsf.2019.02.001
Srivastava RK, Gautam GC (2015) Geochemistry and petrogenesis of Paleo-Mesoproterozoic mafic dyke swarms from northern Bastar craton, central India: Geodynamic implications in reference to Columbia supercontinent. Gondwana Res 28:1061–1078
Srivastava RK, Gautam GC (2016) Petrogenesis and tectonic significance of an Early Paleoproterozoic high-Mg boninite-norite-diorite suite of rocks from the Bastar Craton, Central India. Acta Geol Sinica 90:116–116
Srivastava RK, Singh RK (2004) Trace element geochemistry and genesis of Precambrrian subalkaline mafic dikes from the central Indian craton: evidence for mantle metasomatism. J Asian Earth Sci 23:373–389. https://doi.org/10.1016/S1367-9120(03)00150-0
Srivastava RK, Sinha AK (2004) Early Cretaceous Sung Valley ultramafic-alkaline carbonatite complex, Shillong Plateau, northeastern India: petrological and genetic significance. Miner Petrol 80:241–263
Srivastava RK, Jayananda M, Gautam GC, Gireesh V, Samal AK (2014) Geochemistry of an ENE–WSW to NE–SW trending ~2.37 Ga mafic dyke swarm of the eastern Dharwar craton, India: does it represent a single magmatic event? Chem Erde-Geochem 74:251–265

Srivastava RK, Söderlund U, Ernst RE, Mondal SK, Samal AK (2016) Neoarchaean-Palaeoproterozoic mafic dyke swarms from the Singhbhum Granite Complex, Singhbhum Craton, Eastern India: implications for identification of Large igneous provinces and their possible continuation on other formerly adjacent crustal blocks. Acta Geol Sinica 90(1):17–18

Srivastava RK, Guarino V, Wu F-Y, Melluso L, Sinha AK (2019) Evidence of sub-continental lithospheric mantle sources and open-system crystallization processes from in-situ U-Pb ages and Nd–Sr–Hf isotope geochemistry of the Cretaceous ultramafic-alkaline-(carbonatite) intrusions from the Shillong Plateau, north-eastern India. Lithos 330–331:108–119

Srivastava RK, Samal AK, Ernst RE, Söderlund U, Mondal SK, Shankar R (2020) Spatial and temporal distribution of large igneous provinces in the Indian shield—highlights of recent investigations. Proc Indian Sci Acad (In Press)

Stein HJ, Hannah JL, Zimmerman A, Markey RJ, Sarkar SC, Pal AB (2004) A 2.5 Ga porphyry Cu-Mo-Au deposit at Malanjkhand, central India: implications for Late Archean continental assembly. Precambr Res 134:189–226

Swami Nath J, Ramakrishnan M, Viswanatha MN (1976) Dharwar stratigraphic model and Karnataka craton evolution. Rec Geol Surv India 107:149–175

Swaminath J, Ramakrishna M (eds) (1981) Early Precambrian Supracrustals of Southern Karnataka. Mem Geol Surv India 112:1–350

Taylor PN, Chadwick B, Moorbath S, Ramakrishnan M, Viswanatha MN (1984) Petrography, chemistry and isotopic ages of Peninsular Gneiss, Dharwar acid volcanic rocks and the Chitradurga granite with special reference to the late Archean evolution of the Karnataka craton. Precambrian Res 23:349–375

Thakur SS, Patel SC, Dasgupta S (2006) Structural pattern in Sukma supracrustal rocks near Nagaras, southern Bastar Craton. J Geol Soc India 68:769–773

Tushipokla, Jayananda M (2013) Geochemical constraints on komatiite volcanism from Sargur Group Nagamangala greenstone belt, western Dharwar craton, southern India: implications for Mesoarchean mantle evolution and continental growth. Geosci Front 4:321–340

Ugarkar AG, Paniskar DB, Ranganatha Gowda G (2000) Geochemistry, petrogenesis and tectonic setting of metavolcanics and their implications for gold mineralization in Gadag Gold Field, Southern India. Gondwana Res 3:371–384

Upadhyay R, Sharma BL Jr, Sharma BL, Roy AB (1992) Remnants of greenstone sequence from the Archean rocks of Rajasthan. Curr Sci 63(2):87–92

Upadhyay D, Chattopadhyay S, Kooijman E, Mezger K, Berndt J (2014) Magmatic and metamorphic history of Paleoarchean tonalite-trondhjemite-granodiorite (TTG) suite from the Singhbhum craton, eastern India. Precambrian Res 252:180–190

Valdiya KS (2016) The making of India. Geodynamic evolution, 2nd ed. Society of earth scientist series. Springer, Switzerland, pp 1–924. https://doi.org/10.1007/978-3-319-25029-8

Van Breemen O, Bowes DR, Bhattacharjee CC, Chowdhary PK (1989) Late Proterozoic-Early Paleozoic Rb–Sr whole rock and mineral ages for granite and pegmatite, Goalpara, Assam, India. J Geol Soc India 34:89–92

Verma RK, Prasad SN (1974) Palaeomagnetic study and chemistry of Newer Dolerites from Singhbhum, Bihar, India. Can J Earth Sci 11:1043–1054

Verma SK, Verma SP, Oliveira EP, Singh VK, Moreno JA (2016) LA-SF-ICP-MS zircon U-Pb geochronology of granitic rocks from the central Bundelkhand greenstone complex, Bundelkhand craton, India. J Asian Earth Sci 118:125–137

Vijaya Rao VK, Reddy SPR, Mooney WD (2006) Crustal structure and tectonics of the northern part of the Southern Granulite Terrain, India. Earth Planet Sci Lett 251:90–103

Vishwanatha MN, Ramakrishnan M (1981) Sargur and allied belts. Mem Geol Surv India 112:115–142

Viswanatha MN (1978) Geological map of parts of the Kolar Greenstone belt, Karnataka, scale 1:50,000. Geol Surv India, Bangalore

Wang W, Cawood PA, Pandit MK, Zhou MF, Zhao JH (2018) Evolving passive- and active-margin tectonics of the Paleoproterozoic Aravalli Basin, NW India. Geol Soc Amer Bull 313 (3-4):426–443

Wei W, Cawood PA, Pandit MK, Zhou M-F, Chen W-T (2017) Zircon U-Pb age and Hf isotope evidence for an Eoarchaean crustal remnant and episodic crustal reworking in response to supercontinent cycles in NW India. Geol Soc London. https://doi.org/10.1144/jgs2016-080

Wiedenbeck M, Goswami JN (1994) An ion-probe single zircon $^{207}Pb/^{206}Pb$ age from the Mewar Gneiss at Jhamarkotra, Rajasthan. Geochim Cosmochim Acta 58:2135–2141

Wiedenbeck M, Goswami JN, Roy AB (1996) Stabilisation of the Aravalli craton of the northwestern India at 2.5 Ga: an ion-microprobe zircon study. Chem Geol 129:325–340

Yang QY, Santosh M (2015) Zircon U-Pb geochronology and Lu-Hf isotopes from the Kolar greenstone belt, Dharwar craton, India: implications for crustal evolution in an ocean-trench-continental transect. J Asian Earth Sci 113(2):797–811

Yedekar DB, Jain SC, Nair KKK, Dutta KK (1990) The central Indian collision suture. In: Precambrrian of central India. Geol Surv India Spec Publ 28, pp 1–43

Yin A, Dubey CS, Webb AAG, Kelty TK, Grove M, Gehrels GE, Burgess WP (2010) Geologic correlation of the Himalayan orogen and Indian craton: Part 1. Structural geology, U-Pb zircon geochronology, and tectonic evolution of the Shillong Plateau and its neighboring regions in NE India. Geol Soc Amer Bull 122:336–359. https://doi.org/10.1130/B26460.1

Zachariah JK, Hanson GN, Rajamani V (1995) Post-crystallization disturbance in the neodymium and lead isotope systems of metabasalts from Ramagiri schist belt in southern India. Geochim Cosmochim Acta 59:3189–3203

Zimmerman A, Stein HJ, Markey RJ, Hannah JL, Sarkar SC, Pal AB (2002) Re–Os dating of the Malanjkhand Cu (Mo) deposit in central India: tracking formation and deformation during the Late Archean-Early Proterozoic. Geol Soc Amer Abst Program 34(6):187

Chapter 3
Tectonics of Sri Lanka

3.1 Introduction

Sri Lanka, an island nation in the Indian Ocean, is located on southern tip of the Indian subcontinent, and is made up many offshore islands, with Mannar Island as the largest island. It's terrain is mostly low, flat to rolling plains, with mountains in the south-central parts, with highest point at Pidurutalagala (2524.13 m). Three distinct topographic zones characterize Sri Lanka: the Central Highlands, the Coastal Plains, and the Coastal Belt (Fig. 3.1).

Rugged Central Highlands of highest mountains possess a north-south trending high plateau and a series of minor highlands, which descend into escarpments and ledges at 400–500 m before sloping down toward the coastal plains.

The Coastal Plains (30 and 200 m) are most of the island's surface where ridges and valleys gradually rise to merge with the Central Highlands and produce dissected plains. Extensive erosion in this area has worn down the ridges and deposited rich soil for agriculture downstream. In the southeast, a red, lateritic soil covers relatively level ground that is studded with bare, monolithic hills. In southeast parts, change between the two domains is abrupt, and Highlands appear to rise up like a wall.

A Coastal Belt, ~30 m above sea level surrounding the island, consists of scenic sandy beaches and indented coastal lagoons, including the Jaffna Peninsula. Elsewhere, extensive eroded metamorphic rocks produced dissected coast rocky cliffs, bays, and offshore islands.

Most of Sri Lanka's rivers are short, rising in the Central Highlands and flowing radially toward the sea. Only 16 principal rivers control about 75% discharge. Upper reaches of these rivers are wild and rugged. In the Coastal Plains, the rivers slow down and meander across flood plains and deltas.

A. K. Jain et al., *Tectonics of the Indian Subcontinent*, Society of Earth Scientists Series, https://doi.org/10.1007/978-3-030-42845-7_3

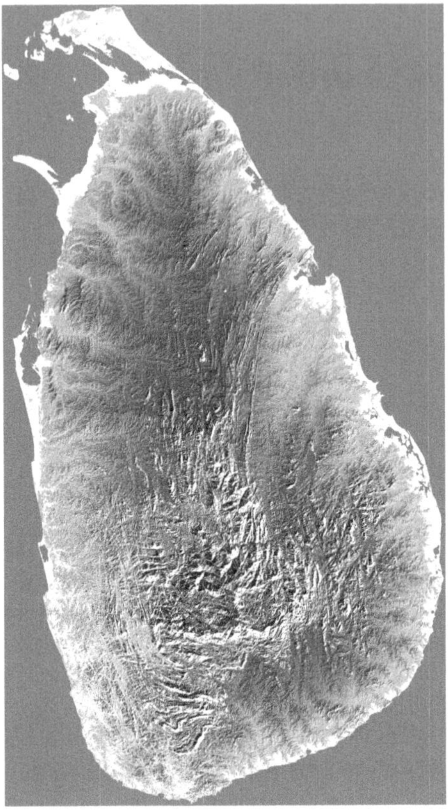

Fig. 3.1 Topography of Sri Lanka (Negative Print), obtained from the SRTM aboard the Space Shuttle Endeavour. *Source* http://photojournal.jpl.nasa.gov/catalog/PIA06670. NASA

3.2 Geological Framework

Sri Lanka is comprised of more than 90% Precambian high grade metamorphic and igneous rocks, which are exposed into almost N–S trending belts (Fig. 3.2) whose following nomenclature is now generally accepted (Cooray 1994).

3.2.1 Highland Complex (HC)

(Formerly known as the "Highland Series" and "Southwestern Group"). A sequence of supracrustal association of garnet-sillimanite gneiss, metaquartzite, marble and calc-silicates is intimately associated and interbanded with granitoid orthogneiss and charnockitic gneiss. These are metamorphosed under upper amphibolite to granulite

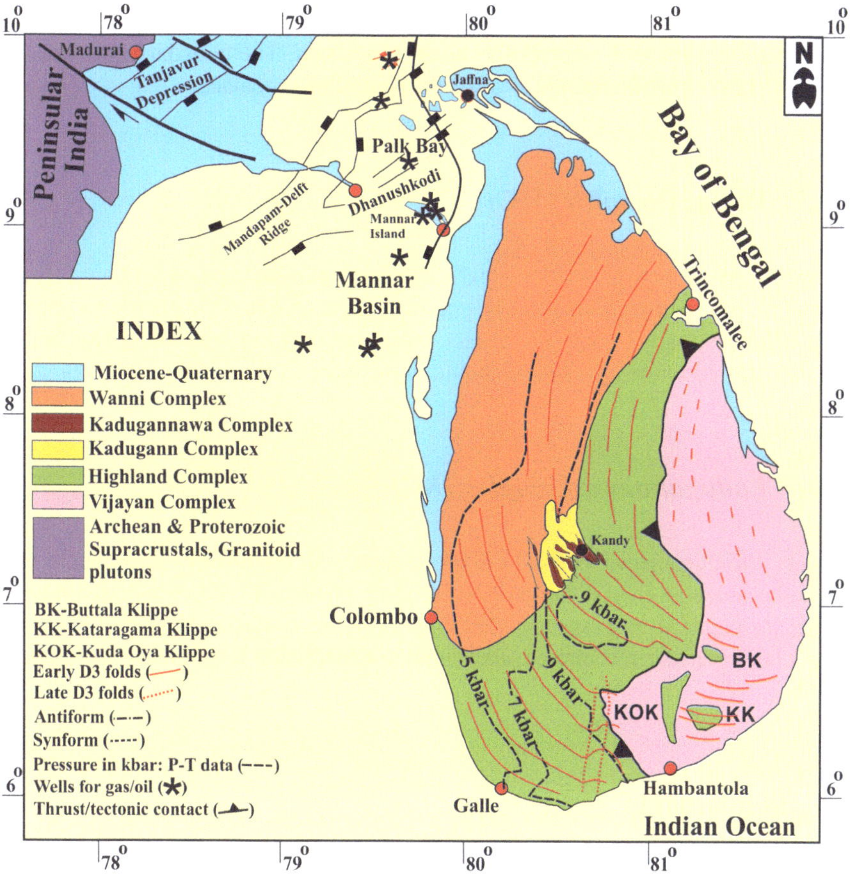

Fig. 3.2 Simplified geological map of Sri Lanka, showing main units and their generalized trends along with distribution of metamorphic pressures. Compiled after Geological map of Sri Lanka (1982), Kröner et al. (1991), Cooray (1994), NASA—http://photojournal.jpl.nasa.gov/catalog/PIA06670 and Berger and Jayasinghe (1976)

facies conditions. This belt extends from northeast to southwest through the central highlands. Some klippen of the HC are exposed within the Vijayan Complex at Buttala, Kataragama and Kuta Oya.

3.2.2 Vijayan Complex (VC)

(Formerly "Eastern Vijayan Complex"). Located in the eastern and southeastern parts of the island, the Vijayan Complex (VC) contains granitoids, migmatite and granitic gneiss, with scattered metaquartzite, amphibolite and calc-silicates. These are metamorphosed under upper amphibolite facies conditions. The boundary between the HC and VC is often interpreted as a west-dipping thrust contact.

3.2.3 *Wanni Complex (WC)*

(Formerly "Western Vijayan Complex"). The northwestern parts of Sri Lanka, lying west of the HC, consist of migmatite, granitic gneiss, charnockitic gneiss, and minor metasediments (garnet-cordierite gneiss, metaquartzite) and granitoids. This belt is metamorphosed under upper amphibolite to granulite facies conditions.

3.2.4 *Kadugannawa Complex (KC)*

(Former "Kadugannawa Gneiss"). This sequence is dominantly hornblende and biotite-hornblende gneiss, hypersthene-bearing gneiss and minor metasediments. These are exposed as elongated doubling-plunging synforms ("Arenas") around Kandy, and are metamorphosed under upper amphibolite to granulite facies conditions.

3.3 Deformation

Sri Lanka mainly contains three high-grade metamorphic terranes, namely the Highland Complex (HC), the Wanni Complex (WC) and the Vijayan Complex (VC), which have undergone different deformation histories, though some of the elements remain common.

3.3.1 *Structures of the Highland Complex (HC)*

Berger and Jayasinghe (1976) demonstrated, for the first time, that the HC has undergone polyphase deformation around Kandy, having dominant NNW–SSE trending L-S fabric, S-surfaces, stretching lineation, boudinage and rootless isoclinal folds; all of these were produced during D1-D2 deformations (Fig. 3.3a, b). Subsequent D3 deformation produced open to tight NNW-trending km-scale synforms and antiforms, which refolded these earlier structures and gave rise the morphology of the Central Highlands (Fig. 3.3c). Granulite facies metamorphism was opined as coeval with the D1-D2 deformations. Some other evidences of the D1 deformation incorporated inclusion trails in garnet porphyroblasts, relict minor folds and fold hinges

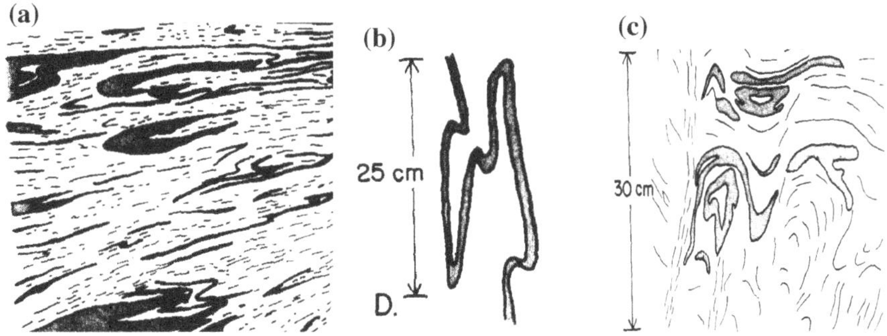

Fig. 3.3 Significant deformation structures from the HC. **a** Rootless isoclinal F1 folds in mafic rocks. **b** Type-3 fold interference pattern between co-axial F1-F2 and F3 folds near Gadaladeniya Temple. **c** Type-2 fold interference pattern in biotite gneiss at Dambulla. Field sketches after Berger and Jayasinghe (1976)

(Mathavan et al. 1999). This deformation is a pre-granulite phase (Kröner et al. 1994; Kehelpannala 1997), and was superposed by ductile to brittle D4 structures (Yoshida et al. 1990). The granulites have undergone very strong flattening of the order 1/10–1/25 and stretching of the order 10–15 times the original length (Voll and Kleinschrodt 1991; Kleinschrodt and Voll 1994), while shallow-dipping high-grade compositional layering in Sri Lanka were interpreted due to a combination of flattening and noncoaxial deformation during thrusting (Kröner et al. 1994).

Whether original stratigraphy could still be recognized in strongly deformation HC, is one of the most controversial aspects of this terrane. Kröner et al. (1994) considered that strong deformation and interleaved metasediments and granitoids preclude recognition of the original stratigraphy, while Kleinschrodt and Voll (1994) have demonstrated preservation of original stratigraphy.

3.3.2 Structures of the Vijayan Complex (VC)

The major structural trends in the VC are discontinuous and rather complexly oriented, forming several circular or dome like structures (see Yoshida et al. 1990). Though there is a distinct break in the structural pattern, some prominent trends of the HC appear to be continuous into the VC. Kriegsman (1991) has recognized two phases of deformation in the VC. The early phase produced N–S stretching isoclinal folds and a later phase produced upright folds with the same trend.

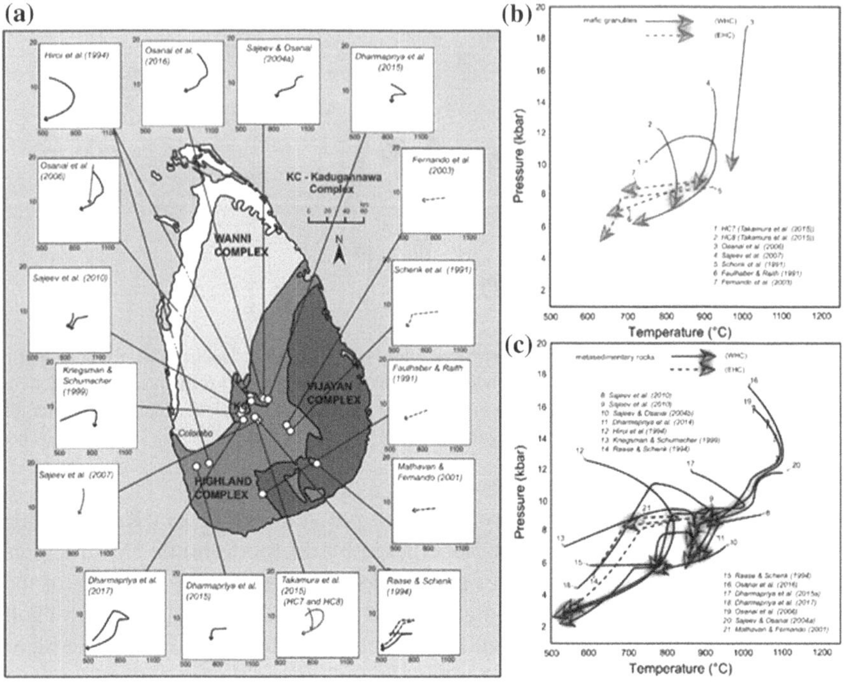

Fig. 3.4 P–T conditions of the Highland Complex (HC) granulites in Sri Lanka. **a** Compiled P–T trajectories of high-T to UHT granulites from the HC. **b** Existing P–T trajectories from mafic granulite. **c** Metasedimentary rocks. See references and compilation by He et al. (2018)

3.3.3 Structures of the Wanni Complex (WC)

The structural geology of the WC is not properly understood, but the works of many imply a similar history between the WC and HC, partly because a large part of the WC possibly represents former Highland Group rocks. Prominent and major upright folds in the HC are continuous into the eastern WC (see Fig. 3.4 of Kehelpannala 1991). Even the early structures such as tight to isoclinal folds and strong stretching lineation, are similar and have identical orientation in two complexes (see Kriegsman 1991; Voll and Kleinschrodt 1991).

3.4 Metamorphism

High-grade metamorphism during the Pan-African event at ca. 610–520 Ma has produced widespread granulite-facies assemblages that are now largely retrogressed and were affected by extensive late metamorphic K-metasomatism.

As has been discussed in Sect. 3.2, widely accepted classification of Sri Lanka terrane (cf. Cooray 1994), based on isotopic mapping (Milisenda et al. 1988) and geochronology (Kröner et al. 1991), is comprised of (i) Vijayan Complex (VC) in the east, (ii) the Highland Complex (HC) in the center, (iii) the Wanni Complex (WC) in the west, and (iv) Kaduganawa Complex (KC) (Fig. 3.2). The VC experienced amphibolite-facies metamorphism, whereas the other complexes underwent granulite-facies metamorphism with a retrograde amphibolite-facies overprint (Kehelpannala 1997; Sajeev et al. 2003 and references therein). From central parts of the HC, Sajeev et al. (2003) recorded textural and compositional evidences of high-pressure–ultrahigh-temperature (HP–UHT) crustal metamorphism and multi-stage retrogression in grn-cpx-qz granulite from peak P–T conditions of 12.5 kbar at 925 °C to 844–982 °C, thus indicating a multistage decompression after the HP–UHT metamorphism. Peak metamorphism was dated ca. 580 Ma (U–Pb zircon ages), followed by isothermal decompression (retrogression) at 534 ± 12 Ma (Sm–Nd isochron), thus providing important constraints to the Neoproterozoic orogeny associated with the Gondwana assembly.

Compiled P–T conditions from the UHT-controled metasedimentary and metabasic rocks in western HC range from <8 to >15 kbar and <800 to >1200 °C, though P–T conditions for the eastern HC remained poorly constrained (Fig. 3.4; He et al. 2018). Two contrasting P–T–t paths for metasedimentary and meta-igneous rocks indicated a clockwise path for the former with a near-isobaric prograde heating segment, followed by a near-isothermal down-pressure retrograde segment. And, meta-igneous rocks are marked by a near-isobaric cooling path, followed by a P–T decrease after peak metamorphism (see He et al. 2018 for details). In the eastern parts of the HC, long-lived HT metamorphism lasted possibly longer than 100 Ma between ca. 660–520 Ma.

A single ~610 to 456 Ma common metamorphic event has affected all these terranes as the consequence of collision of Sri Lankan terranes during the assembly of Gondwana. A double-sided subduction system was proposed by Santosh et al. (2014) with the Wanni and Vijayan magmatic arcs juxtaposing against each other along sutures with metamorphosed Highland accretionary sediments in the center.

The WC had undergone peak metamorphism around 700–830 °C and 5–7 kbar (Schenk et al. 1991; Raase and Schenk 1994). In central WC, incipient charnockite represents granulite formation within host amphibolite facies foliated gneiss along veins, patches, shear zones and foliation planes, thus indicating pervasive fluid influx in the Sri Lankan lower crust (Santosh et al. 1991; Kehelpannala 1999).

The Vijayan Complex (VC), exposed in eastern Sri Lanka, consists predominantly of meta-igneous rocks including upper amphibolite facies calc-alkaline suite of granitoid gneisses, augen-gneisses and minor amphibolites. The VC has a ~580 Ma Pan-African regional metamorphism (Kröner et al. 2013) over the 1100–1000 Ma protolith magmatic bodies. This granitoid suite was interpreted as a product of subduction-related magmatism. However, the VC gneisses show evidence for granulite-facies metamorphism, as well (Kröner et al. 2013).

Minor sedimentary enclaves such as metaquartzite, calc-silicate rocks and marble also occur within the VC, and granitoid gneisses show TTG (tonalite-trondhjemite-granodiorite)-dominated signature. Igneous charnockite bodies around SE margin of the VC are interpreted as tectonically infolded or inter-sliced fragments of the HC during the thrusting of the HC over the VC (Kröner et al. 1991).

3.5 Geochronology

Extensive LA-ICP-MS U–Pb dating of remnant zircon cores from the HC and WC high-grade metamorphics in Sri Lanka reveals that these domains have distinct characteristics: the HC is dominated by detrital zircon (DZ) ages of ca. 3.50–1.50 Ga from various gneisses, while its igneous bodies intruded ca. 2.00–1.80 Ga (Fig. 3.5; Kitano et al. 2018). In one of the earliest investigations in Sri Lanka, Kröner et al. (1987) identified DZ from metaquartzite of the HC, having U–Pb ages between 3.17 and 2.4 Ga, possibly derived from an unknown Archean terrane. Two pelitic gneisses yielded DZ up to 2.04 Ga with a 1.10 Ga metamorphic event around old cores and new zircon growths. A granite intrusive into the HC granulite records an emplacement age of 1.10–1.00 Ga. Zircons from a metaquartzite xenolith within the VC granitoid are not older than ~1.10 Ga; therefore it is neither Archean in age nor acted as basement to the HC. Kröner et al. (1987) suggested that the VC formed significantly later than the HC and that the two units were brought into contact through post–1.1 Ga thrusting.

In contrast, the WC terrane appears to be much younger with dominant DZ ages of ca. 1.10–0.70 Ga from paragneiss, indicating their different origins from the HC. Sedimentary sources for the WC were mainly eroded from local late Mesoproterozoic to Neoproterozoic igneous rocks with very minor components from an older 2.50–1.50 Ga craton.

The Vijayan Complex (VC) in eastern Sri Lanka is tectonically juxtaposed against the HC and has undergone intense ductile deformation of diorite to leucogranite bodies with a distinct calc-alkaline geochemical signature and are interpreted as a magmatic arc. Kröner et al. (2013) have undertaken extensive analyses of large section of the Vijayan gneisses, using zircon ages, whole-rock Nd and Hf-in-zircon isotopic systematics and geochemical data in order to characterize this complex. U–Pb zircon ages fall predominantly between 1.10 and 1.00 Ga with a few early Neoproterozoic intrusions, and indicate the VC as a Grenville-age magmatic arc. Many zircons experienced minor to significant lead-loss at ~580 Ma. Nd and Hf isotopic data indicate primitive origin for most Vijayan gneisses, but significant variations indicate source heterogeneities and involvement of minor amounts of older

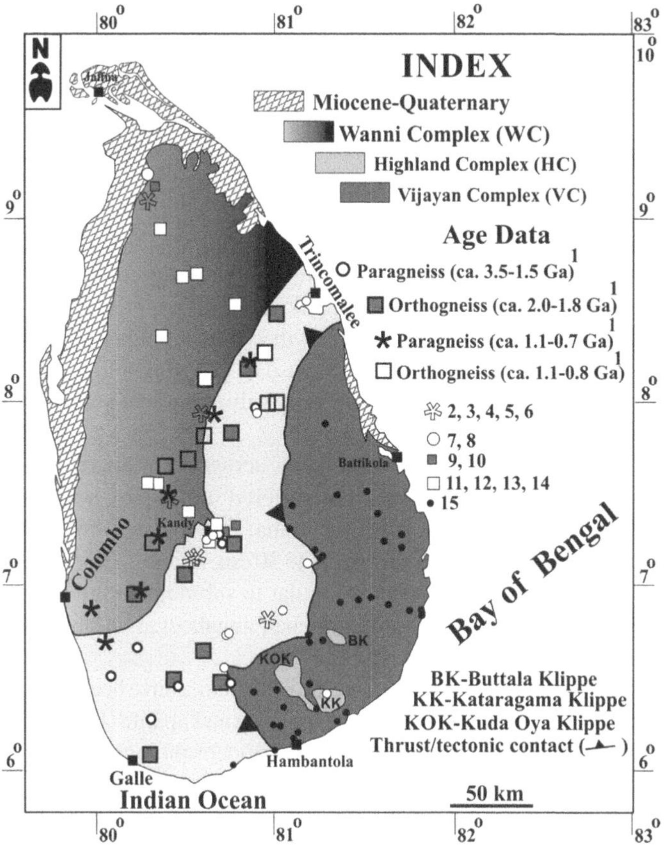

Fig. 3.5 Simplified geological map of Sri Lanka with distribution of U–Pb zircon ages from the Wanni Complex (WC), Highland Complex (HC) and Vijayan Complex (VC) terranes of Sri Lanka. Redrawn after Kitano et al. (2018). **Age Data**: 1—Kitano et al. (2018). 2—Osanai et al. (2016). 3, 7—Takamura et al. (2016). 4, 12—Dharmapriya et al. (2015). 5—Sajeev et al. (2010). 6, 8—Kröner et al. (1987). 9, 13—Santosh et al. (2014). 10—Baur et al. (1991). 11—He et al. (2016). 14—Kröner et al. (1994). 15—Kröner et al. (2013)

continental material (Kröner et al. 2013). After peak metamorphism, P–T trajectory from prograde to near isobaric cooling seems to occur within a time span from Late Cryogenian to Early Cambrian (ca. 665–500 Ma), coeval with the assembly of the Gondwana Supercontinent (Dharampriya et al. 2017).

3.6 Sri Lankan Thrusts and Faults

Kleinschrodt (1994) have opined that granulite-facies HC was thrust onto the Vijayan Complex (amphibolite-facies) along its eastern margin through deep crustal and subhorizontal thrust surface with N–S to NNW–SSE oriented folds and stretching lineation, which is most prominent element through its entire deformational history. Open to flexure-slip folds followed thrusting and generated subhorizontal fold envelop, thus thrust must underlie large parts of the HC. Remnants of such thrust are the occurrences of HC klippen such as the Kataragama klippe, indicating that thrust surface was subhorizontal and can be traced for more than 200 km parallel to the transport direction till the coastal regions. This thrusting appeared to have occurred during retrograde metamorphism and younger than formation of 577 ± 14 Ma old migmatite. Several serpentinite bodies are located near this boundary. The longest Mahaweli Ganga in Sri Lanka flows along this boundary for considerable distance.

All the high-grade tectonic units in Sri Lanka have undergone brittle deformation including a phase of cataclasis. Several sets of these faults range from less than 1 km to more than 100 km, controlling the present-day geomorphology (Kehelpannala et al. 2006). Many of these are characterized by sub-horizontal striations indicating a component of strike-slip faulting.

E–W trending vertical Randenigala Fault cuts across central parts of the granulite facies Highland Complex (HC), which has suffered ultra-high temperature (UHT) assemblage of the lower crust (Sanjeev and Osanai 2004). Black-colored cataclastic rocks characterize this fault and vary from 0.1 to 30 cm. These seams typically exhibit incohesive cataclastic texture with sharp angular to sub-angular quartz and feldspar clasts which are embedded in fine matrix (Kehelpannala et al. 2006), and appear to have evolved within 1–4 km crustal depth.

Brittle faults from southern and southwestern Sri Lanka have been recently investigated, using apatite and zircon fission track (FT) data and apatite (U–Th–Sm)/He dating methods to determine their kinematic and thermochronological patterns (Emmel et al. 2012). These data on faults reveal at least 5 thermal overprinting events at 159 ± 18, 144 ± 14, 120 ± 10, 94 ± 8, and 70 ± 10 Ma due to major N–S trending extension subsequent to the Gondwana breakup. The resulting brittle structures exert primary control on the island's geomorphology, especially on the Southern Escarpment of the Sri Lankan Highlands. Structural data also indicate variations in stress regime during the Late Jurassic and Late Cretaceous, as Sri Lanka separated from East Antarctica and progressively moved away from Madagascar and the Seychelles microcontinent.

3.7 Crustal Evolution

In the KC terrane, magmatic episodes between ca. 1.10 and 0.89 Ga were linked to the breakup of the Rodinia supercontinent as a consequent of large-scale rifting (Kröner et al. 2003). However, Santosh et al. (2014) demonstrated that these Neoproterozoic magmatic pulses in the Sri Lankan crustal complexes are distinctly linked to convergent margin setting.

Not considering a very few exceptions, Sri Lanka does not record widespread extension-related mafic or bimodal magmatism during mid-Neoproterozoic, hence a Rodinia break-up model becomes untenable (cf., Santosh et al. 2014). In support of their data, they propounded accreted oceanic components and arc magmatism in a convergent margin setting with the Wanni and Vijayan complexes as continental arcs underlain by Neoarchean–Paleoproterozoic basement. These terranes represent a double-sided subduction of an oceanic lithosphere with arc-related geochemical features during early- to mid-Neoproterozoic. Negative εNd(t) and εHf(t) values of these magmatic suites and Neoarchean–Paleoproterozoic two stage Hf crustal model ages are also in accordance with melting of underlying older basement in both of these blocks. This study also identifies juvenile magmatic additions during mid-Neoproterozoic and latest Neoproterozoic–Cambrian in post-collisional asthenospheric upwelling, which triggered widespread UHT metamorphism within the HC.

Oceanic lithosphere subduction generated protracted arc magmatism in a convergent margin setting since early Neoproterozoic, continued through mid-Neoproterozoic and culminated in Cambrian regional UHT collisional metamorphism in all the blocks with imprint along the collisional suture (HC). This might be related to the prolonged Pacific-type orogeny associated with the assembly of Gondwana supercontinent. Thus, convergent margin processes explains better the Neoproterozoic arc-related magmatic suites in contrast to the existing concepts of linking Sri Lanka with Rodinia rifting, associated with the assembly of Gondwana (Santosh et al. 2014).

Among various Gondwanland assemblies now available in literature including those based on distribution of graphite deposits and gemstones (Dissanayake and Chandrajith 1999), Kitano et al. (2018) compared available zircon ages from the VC in Sri Lanka with neighbouring Gondwana terranes for identification of premetamorphic configuration. Of significance are adjacent terranes to Sri Lanka such as the Southern Granulite Terrane (SGT) of southern India and Lützow-Holm Complex (LHC) of East Antarctica in the Gondwana reconstructions (Fig. 3.6). These fragments have been correlated previously, based on constituent rock types, geological structures, P–T conditions and paths, and various types of isotopic data including Nd model ages. However, new U–Pb zircon data and Th/U ratios of detrital zircon cores in paragneisses from the HC and WC by Kitano et al. (2018) indicate that protoliths from these terranes correlate closely with those of the Trivandrum Block, the Achankovil Shear Zone and Southern Madurai Block, respectively. Comparisons of the HC and WC regions with East Antarctica (Lützow-Holm Complex—LHC) broadly support correlation of their protoliths with the Skallen Group, the Ongul and Okuiwa Groups of the LHC.

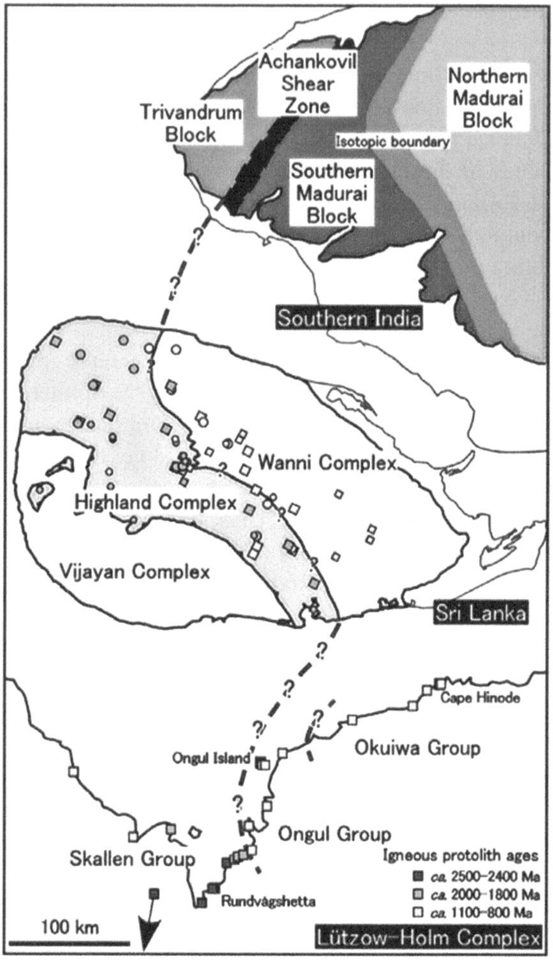

Fig. 3.6 Spatial relationships between Sri Lanka, southern India and the Lützow-Holm Complex (LHC) during Gondwana assembly. See text for details. After Kitano et al. (2018)

References

Baur N, Kröner A, Liew TC, Todt W, Williams IS, Hofmann AW (1991) U–Pb isotopic systematics of zircons from prograde and retrograde transition zones in high grade orthogneisses, Sri Lanka. J Geol 99:527–545

Cooray PG (1994) The Precambian of Sri Lanka: a historical review. Precamb Res 66:3–18

Dharmapriya PL, Malaviarachchi SPK, Santosh M, Tang L, Sajeev K (2015) Late-neoproterozoic ultrahigh-temperature metamorphism in the Highland Complex, Sri Lanka. Precamb Res 271:311–333

Dharmapriya PL, Malaviarachchi SPK, Kriegsman LM, Galli A, Sajeev K, Zhang C (2017) New constraints on the P–T path of HT/UHT metapelites from the Highland Complex of Sri Lanka. Geoscien Front 8:1405–1430

Dissanayake CB, Chandrajith R (1999) Sri Lanka-Madagascar Gondwana linkage: evidence for a Pan-African mineral belt. J Geology 107:223–235

Emmel B, Lisker F, Hewawasam T (2012) Thermochronological dating of brittle structures in basement rocks: a case study from the onshore passive margin of SW Sri Lanka. J Geophys Res 117:B10407. https://doi.org/10.1029/2012JB009136
Geological map of Sri Lanka (1982) Redrawn after Geological Survey Department of Sri Lanka
He X-F, Santosh M, Tsunogae T, Malaviarachchi SPK, Dharmapriya PL (2016) Neoproterozoic arc accretion along the 'eastern suture' in Sri Lanka during Gondwana assembly. Precamb Res 279:57–80
Kehelpannala KVW (1997) Deformation of a high-grade Gondwana fragment, Sri Lanka. Gondwana Res 1:47–68
Kehelpannala KVW (1999) Shear-zone controlled arrested charnockitization, retrogression and metasomatism of high-grade rocks. Gondwana Res 2:573–577
Kehelpannala KVW, Wada H, Ranaweera L, Hamana N (2006) Cataclastic rocks from the granulite terrain of Sri Lanka: evidence for younger brittle deformation of the exhumed lower crust. Geosci Repts Shizuoko Univ 33:9–19
Kitano I, Osanai Y, Nakano N, Adachi T, Fitzsimons ICW (2018) Detrital zircon and igneous protolith ages of high-grade metamorphic rocks in the Highland and Wanni Complexes, Sri Lanka: their geochronological correlation with southern India and East Antarctica. J Asian Earth Sci 156:122–144
Kleinschrodt R (1994) Large-scale thrusting in the lower crustal basement of Sri Lanka. Precamb Res 66:39–57
Kriegsman L (1991) Structural geology of the Sri Lankan basement—a preliminary review. In Kröner A (ed) The crystalline crust of Sri Lanka, part I, summary of research of the German-Sri Lankan consortium. Geological Survey Department, Sri Lanka, Professional Paper, vol 5, pp 52–68
Kröner A, Cooray PG, Vitanage PW (1991) Lithotectonic subdivision of the Precambrian basement in Sri Lanka. In: A. Kröner (ed) The Crystalline Crust of Sri Lanka, part I, summary of research of the German-Sri Lankan consortium. Geological Survey Department, Sri Lanka, Professional Paper, vol 5: pp 5–21
Kröner A, Jaeckel P, Williams IS (1994) Pb-loss patterns in zircons from a high-grade metamorphic terrain as revealed by different dating methods: U–Pb and Pb–Pb ages for igneous and metamorphic zircons from northern Sri Lanka. Precamb Res 66:299–337
Kröner A, Rojas-Agramonte Y, Kehelpannala KVW, Zack T, Hegner E, Geng HY, Wong J, Barth M (2013) Age, Nd–Hf isotopes, and geochemistry of the Vijayan Complex of eastern and southern Sri Lanka: a Grenville-age magmatic arc of unknown derivation. Precamb Res 234:288–321
Kröner A, Williams IS, Compston W, Baur N, Vitanage PW, Perera LRK (1987) Zircon ion microprobe dating of Precambrian high-grade rocks in Sri Lanka. J Geol 95:775–791
Mathavan V, Prame WKBN, Cooray PG (1999) Geology of the high grade proterozoic terrains of Sri Lanka and the assembly of Gondwana: an update on recent developments. Gondwana Res 2(2):237–250
Milisenda CC, Liew TC, Hoffman AW, Kröner A (1988) Isotopic mapping of age provinces in Precambrian high-grade terrains: Sri Lanka. J Geol 96:608–615
Osanai Y, Sajeev K, Nakano N, Kitano I, Kehelpannala KVW, Kato R, Adachi T, Malaviarachchi SPK (2016) UHT granulites from highland complex, Sri Lanka II: geochronological constraints and implications on Gondwana correlation. J Miner Petrol Sci 111:157–169
Raase P, Schenk V (1994) Petrology of granulite facies metapelites of the Highland Complex, Sri Lanka: implication for the metamorphic zonation and the P–T path. Precamb Res 66:265–294
Sajeev K, Osanai Y (2004) Ultrahigh-temperature metamorphism (1150 °C, 12 kbar) and multistage evolution of Mg-, Al-rich granulites from the Central Highland Complex Sri Lanka. J Petrol 45:1821–1844. https://doi.org/10.1093/petrology/egh035
Sajeev K, Osanai Y, Suzuki S, Kagami H (2003) Geochronological evidence for multistage-metamorphic events in ultrahigh-temperature granulites from central Highland Complex, Sri Lanka. Polar Geosci 16:137–148

Sajeev K, Williams IS, Osanai Y (2010) Sensitive high-resolution ion microprobe U–Pb dating of prograde and retrograde ultrahigh-temperature metamorphism as exemplified by Sri Lankan granulites. Geology 38:971–974

Santosh M, Tsunogae T, Malaviarachchi SPK, Zhang Z, Ding H, Tang L, Dharmapriya PL (2014) Neoproterozoic crustal evolution in Sri Lanka: insights from petrologic, geochemical and zircon U–Pb and Lu–Hf isotopic data and implications for Gondwana assembly. Precamb Res 255:1–29

Santosh M, Yoshida M, Nanda-Kumar V (1991) Fluid characteristics across a gneiss-charnockite reaction front in Sri Lanka: implication for granulite formation in Gondwana deep crust. J Miner Petrol Econ Geol 86:27–44

Schenk V, Raase P, Schumacher R (1991) Metamorphic zonation and P–T history of the Highland Complex in Sri Lanka. Geological Survey Department, Sri Lanka, Professional Paper, vol 5, pp 150–163

Takamura Y, Tsunogae T, Santosh M, Malaviarachchi SPK, Tsutumi Y (2016) U–Pb geochronology of detrital zircon in metasediments from Sri Lanka: Implications for the regional correlation of Gondwana fragments. Precamb Res 281:434–452

Yoshida M, Wilbert KV, Kehelpannala, Hiroi Y, Vitanage PW (1990) Sequence of Deformation and Metamorphism of Granulites of Sri Lanka. J Geosci Osaka City Univ 33 (3) 69–107

Voll G, Kleinschrodt R (1991) Sri Lanka: structural, magmatic and metamorphic development of a Gondwana fragment. Geol Surv Depart Sri Lanka Prof paper, pp 22–52

Chapter 4
Proterozoic Mobile Belts

4.1 Introduction

The Indian Subcontinent is comprised of the Archean-Proterozoic cratons, the Aravalli Craton, the Bundelkhand Craton, the Mehghalaya Craton, the Bastar Craton, the Singhbhum Craton and the Dharwar Craton. Many of these cratons are fringed by the Proterozoic Mobile Belts, some of these prominent belts are the Aravalli–Delhi Mobile Belt (ADMB), the Central Indian Tectonic Zone (CITZ)/or the Satpura Mobile Belt (SMB), the Singhbhum Mobile Belt (SMB), the Eastern Ghats Mobile Belt (EGMB), Nellore Mobile Belt, Pranhita-Godavari Belt and the Pandyan Mobile Belt (PMB) (Fig. 4.1). Though the Himalaya contains widespread Proterozoic sequences, the Himalaya is considered a Cenozoic mobile belt, hence it is excluded from this section.

This chapter briefly deals with salient characters of the Proterozoic Mobile Belts belts including their lithology, deformation, metamorphism and various tectonic models of their evolution.

4.2 Aravalli–Delhi Mobile Belt (ADMB)

4.2.1 Location

In parts of Gujarat and Rajasthan of western India, NE–SW trending ~800 km long Aravalli Mountain Range represent an ancient mountain chain with well preserved records of ~3 Ga years of the Precambrian evolutionary history. Three major chrono-stratigraphic terranes with distinct development histories and metallogenic events are represented by the Banded Gneissic Complex (BGC) as the basement, the Aravalli and Delhi Supergroups as the mobile/fold belt components (Fig. 4.2). The metamorphosed and deformed Archean BGC, differentiated into a low grade rock unit

A. K. Jain et al., *Tectonics of the Indian Subcontinent*, Society of Earth Scientists Series, https://doi.org/10.1007/978-3-030-42845-7_4

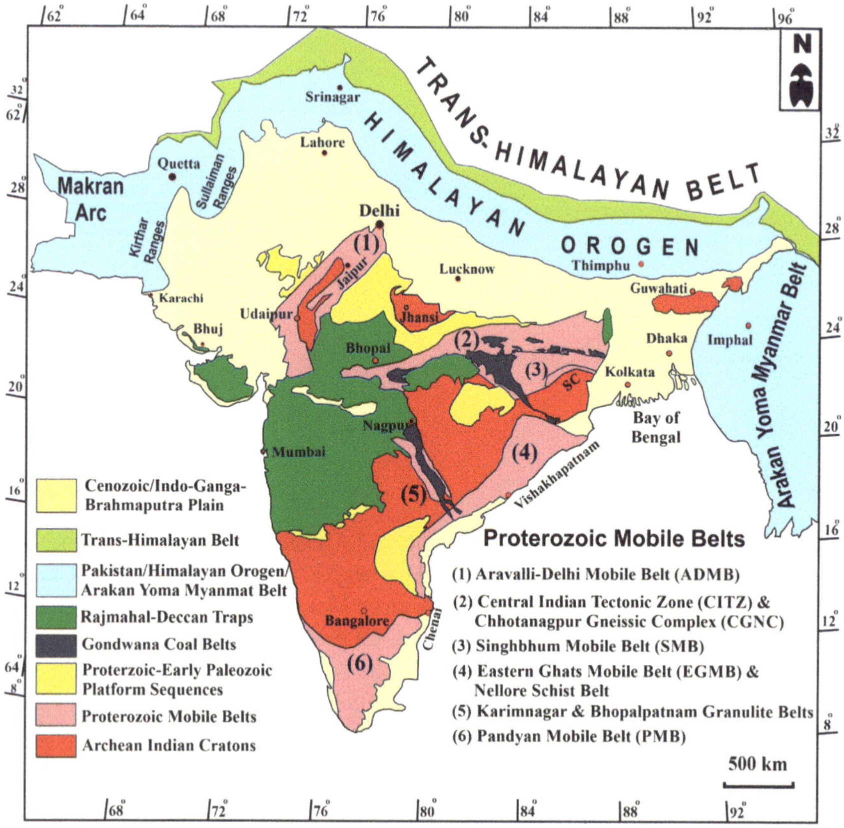

Fig. 4.1 Distribution of Proterozoic mobile belts in the Indian Sub-continent and their relationships with the cratons

adjacent to the 2.58 Ga old Berach Granite, was followed to the west by medium grade Mangalwar Complex (Ahmad et al. 2018b) and granulite facies (Joshi et al. 1993) Sandmata Complex in the south-central and southeastern Rajasthan, consisting of migmatites, composite gneisses, biotite schist, garnet sillimanite–staurolite schist, dolomite, marble, granite, charnockite, norite and pyroxenite (Gupta et al. 1980a, b). Boundary between the Sandmata and Mangalwar Complexes is marked by an isograd along the Delwara Lineament (Sinha-Roy et al. 1998). Occurring within the BGC terrain are two prominent granite masses at Untala and Gingla; the former is intruded by the Newania carbonatite. These granites initially correlated with post–Delhi Erinpura granitic suite are now placed in the Archean basement complex suite.

The Aravalli–Delhi Mobile Belt (ADMB) or simply the Aravalli Mobile Belt includes two distinct stratigraphic subunits designated as the older Aravalli Supergroup and the younger Delhi Supergroup, linked to the Aravalli Craton (Fig. 4.2). While sensu stricto Aravalli Supergroup is limited to the south-central and east-central parts of Rajasthan, the Delhi Supergroup has two components, northern and

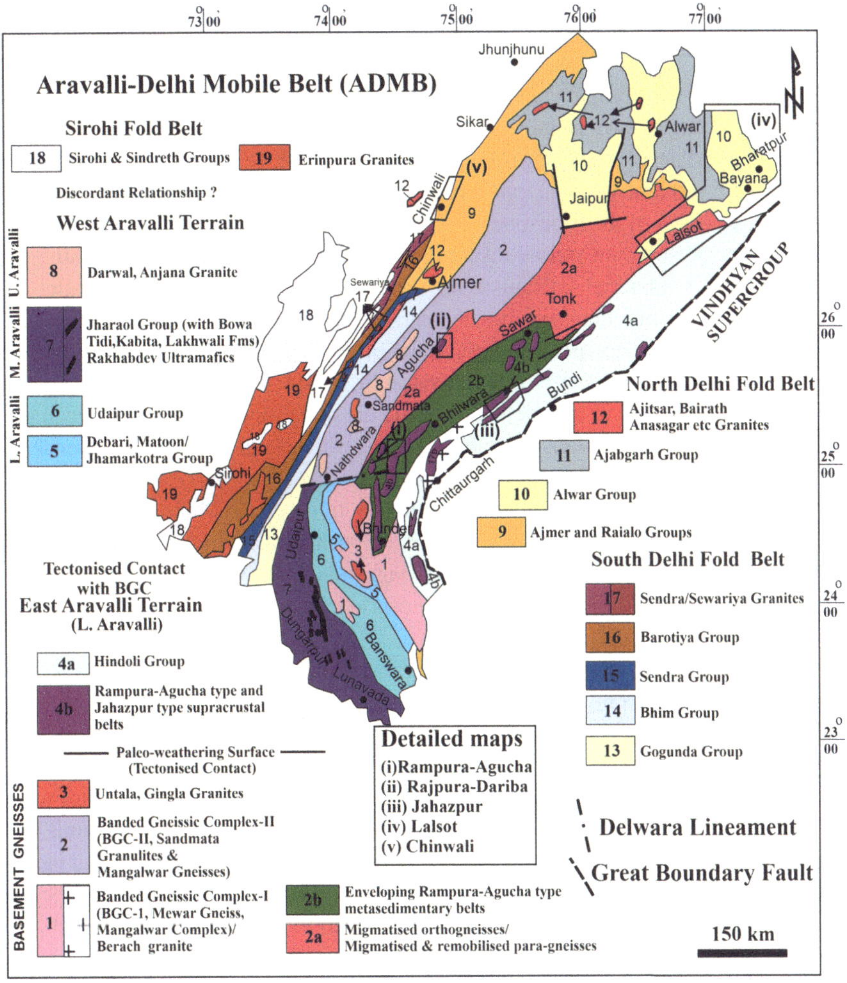

Fig. 4.2 Geological map of the Aravalli-Delhi Mobile Belt and its relation with the Aravalli Craton, based on Geological Survey of India map 1989, 2004): by Gupta et al. (1980a, b, 1997) and modifications by Fareeduddin and Banerjee (2020). Detailed geological maps of smaller (i) Rampura-Agucha area (Gandhi et al. 1984), (ii) Rajpura-Dariba belt (Gupta et al. 1997), (iii) Jahazpur area (Malhotra and Pandit 2000), (iv) Lalsot-Bayana area (Singh 1988), and (v) Chinwali granulite belt (Fareeduddin et al. 1995) areas are in text

southern belts (Fig. 4.2). The Proterozoic rocks of Aravalli and Delhi supergroups belong to a single mountain chain which displays several geo-tectonic and sedimentation domains (Naqvi and Rogers 1987) but share several common deformational and metamorphic histories. Two major tectonothermal events, one in the Late Paleoproterozoic and the other in the Late Mesoproterozoic–Early Neoproterozoic, have affected the evolution of these belts (Deb et al. 1989, 2001; Buick et al. 2006; Bhowmik et al. 2010; Meert et al. 2010).

4.2.2 Regional Geological Setup

The end phase of the Mangalwar Complex activity is marked by extensive volcanism as a result of rifting of the basement and outpouring of effusive lava on the peneplaned surfaces. Ahmad and Tarney (1994) and Ahmad et al. (2018a) interpreted these late Archean mafic volcanic lavas on top of the gneissic basement to be high-Mg basalt and suggested extension of the lithosphere and decompression melting of the underlying upper mantle. Absence of pillow lava and similar submarine lava features suggest the possibility of subaerial volcanism on land. It pre-dated the aluminous paleosol horizon of pyrophylite–talc–sericite–chlorite–kyanite schist, located at the base of the arkose–conglomerate–quartz arenite units of the basal Aravalli succession (Roy and Paliwal 1981; Banerjee 1996; Sreenivas et al. 2001a; Pandit et al. 2008; deWall et al. 2012). This paleosol unit demarcates the erosional unconformity between the gneissic basement and the clastic units of the Aravalli succession. Consequence to this observation, the volcanics with thin veneers of clean quartz arenite is being considered as a part of the last phase of igneous activity of the Mangalwar Complex and does not form a lithostratigraphic unit of the basal Aravalli succession (Delwara Group). In a similar tectonic scenario at Bagrunda area, west of Udaipur, basic metavolcanics surrounding the basement gneiss are mapped as part of the basement rocks (Sharma et al. 1988).

Geochemically, these lavas around Delwara (ca. 2.4 Ga) and basic volcanics occurring within the BGC (Mangalwar Complex) are directly comparable (Raza and Khan 1993). It is therefore inferred that the rifting of the basement resulted in the eruption and consolidation of mafic lava as a distinct end phase of the Mangalwar Complex. This was followed by a prolonged tectonic quiescence during which the aluminous soils developed on the coastal plains. Reactivation and repeated rifting around this time resulted into development of a geomorphology with small mountains and incipient Aravalli coastal plains, characterized by pebbly fluvial and deltaic sediments, which got tossed, winnowed, scoured and agitated to form large cross-bedded basal arkosic–quartz arenite units of the Aravallis.

4.2.3 Aravalli Mobile Belt

Coulson (1933), Gupta (1934), Gupta and Mukherjee (1938) and Heron (1935; 1953) identified the Aravalli 'System' of low-grade metamorphosed quartzite, phyllite and carbonates, several generations of granites and basic volcanics, and distinguished these from the overlying metamorphosed Delhi 'Group'. The Aravalli sediments were deposited on a peneplaned, metamorphosed and deformed basement of the Banded Gneissic Complex (BGC). Like most complexly folded Proterozoic fold belts in other parts of the world, the basement–cover relationship in the Aravalli Mobile belt also suffers from serious inconsistencies. Heron's (1953) original correlation of the Girwa valley sequences was subsequently revised (Mathur 1964; Poddar and Mathur

1965; Banerjee 1971a, b; Dutta 1980; Banerjee and Bhattacharya 1989, 1994; Gupta et al. 1980a, b, 1997; Roy and Paliwal 1981; Roy and Jakhar 2002 and others). Naha and Mohanty (1990) suggested that the gnesssic terrain of the BGC to be a product of granitization of the Aravalli sediments. This contention proved to be scientifically inconsistent after additional geological and geochemical evidences (Sharma 1988; Roy 1988; Gupta et al. 1997). The BGC and the Aravallis demonstrated unconformable relationship marked by extensive occurrence of aluminous paleosols at the base of the Aravalli succession (Banerjee 1996; Pandit et al. 2008). This boundary is prominently displayed through a reactivated dextral strike-slip boundary fault which separates the Hindoli belt from the basement gneisses in the north and the Berach Granite Gneiss in the south (Fareeduddin and Banerjee 2020). The status of BGC as the basement for the supracrustals of this region is widely accepted (Fig. 4.3).

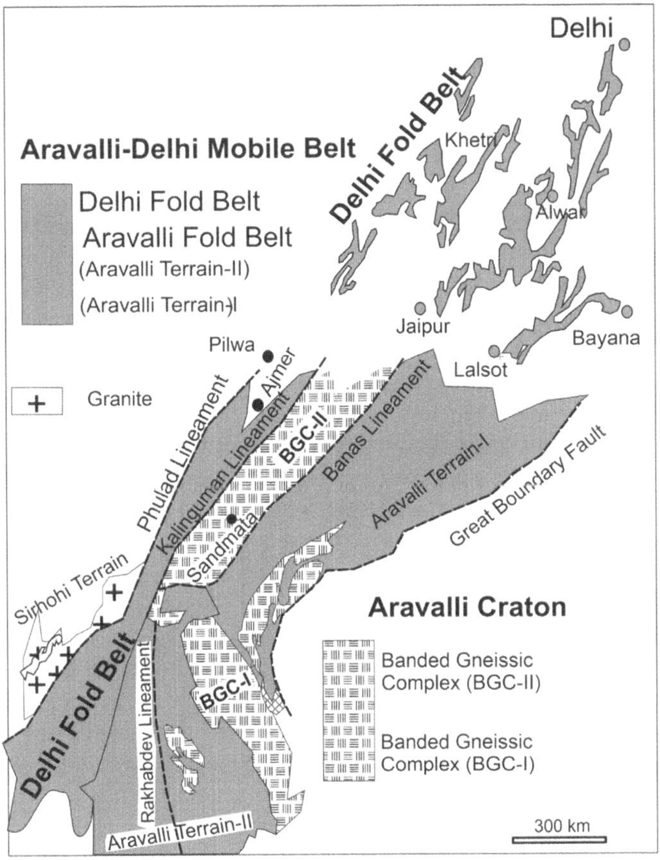

Fig. 4.3 Distribution of the Aravalli and Delhi Mobile Belts and their relations with the Aravalli Craton. Redrawn after Fareeduddin and Banerjee (2020)

4.2.3.1 Some Contentious Issues

Objections have been raised against the currently followed order of superposition in the Aravalli Supergroup. Some of these objections are: (i) stratigraphic position and correlation of arkose-conglomerate units overlying the BGC: do they occur as the basal lithological stratigraphic unit or occupy a higher position in the stratigraphic column? (ii) nature and status of several pink aplitic granite outrcops in and around Udaipur region: do they form part of the basement complex or constitute younger intrusives? (iii) correlation and stratigraphic status of basic volcanic rock suite (Delwara volcanics) overlying the basement: do they belong to the Aravallis or need clubbing with the basement complex? (iv) stratigraphic position of the sericite–pyrophyllite–kyanite-bearing paleosol at the base of the Aravalli sequence, (v) correlation and status of various carbonate and phosphate rock units in the entire Girwa Valley: do they belong to single stratigraphic order or reflect two different sedimentation domains of different ages? and (vi) maximum and minimum ages of the Aravalli sedimentary succession: how do we decide?

4.2.3.2 Aravalli Supergroup

The Aravalli Supergroup outcrops over a large expanse in the east-central and south-eastern parts of the Aravalli Mountain between south of Nathdwara-Bhilwara-Hindoli in the north and northeast and Udaipur–Lonavada belt in the south and southeast. As mentioned above, the two terrains come in contact along the Delwara Dislocation Zone (Sinha-Roy et al. 1998). Extensive mapping by Geologists of the Western Region of GSI (Raja Rao 1976; Raja Rao et al. 1971) indicates that the Bhilwara-Hindoli metasedimentary belt forms the older cycle and was named as Bhilwara Group [later elevated to Supergroup status by Gupta et al. (1997)] while rocks succession of the Nathdwara-Lunavada basin was assigned to a younger frame of Aravalli of the type area. Since the cover sequences are separated only in space and not necessarily in time, Fareeduddin and Banerjee (2020) referred the Aravalli supracrustal belts of the Nathdwara-Bhilwara-Hindoli area as the East Aravalli Terrain (EAT) and those between Nathdwara-Udaipur-Lunavada areas ('Aravalli type area') as South-Central Aravalli Terrain (SCAT).

The Nathdwara-Bhilwara-Hindoli sector in the EAT holds the key to the stratigraphic puzzles emanating from this region. In the Rajpura-Dariba area, the metasediments of the Rampura-Agucha belt occur as stretched, attenuated strips of medium to high-grade metasediments which are engulfed in the migmatite country (Fig. 4.4). Being deeply affected by longitudinal shears valid internal stratigraphic order could not be properly established. These rock units also occur as ~400 km long low-grade metamorphic belts with well defined litho-stratigraphic order towards eastern extremity of the ADMB and referred to as the Hindoli-Jahazpur belts.

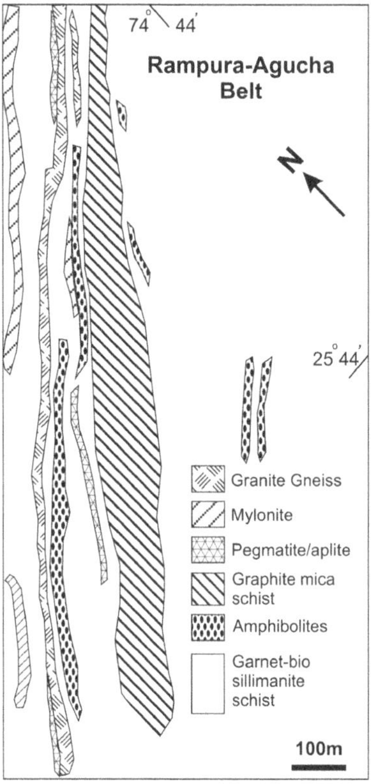

Fig. 4.4 Geological map of Rampura-Agucha area. After Gandhi et al. (1984)

Within the domain of EAT, significant Pb–Zn mineralized terrain, known as the Rampura-Agucha Belt, exposes pyroxene granulite, garnetiferous feldspathic sillimanite gneiss, garnetiferous amphibolite, metapelite, calc-silicates, and thick sulphide ore-bearing zones (Fig. 4.4; Gandhi et al. 1984). Pb-isotope model ages of 1.8 Ga (Deb et al. 1989) suggest its age marginally older than the base of the Aravalli Supergroup.

Another significant terrain known as the Bethumni-Sindeswar-Rajpura-Dariba belt north-northeast of Udaipur, is surrounded by migmatized gneiss and represented by quartzite, dolomite, calcareous, garnetiferous-staurolite-kyanite bearing phyllite and schist (Fig. 4.5). A co-axially folded isolated Sawar basin, located 50 km east of Agucha, is a repetitive sequence of meta-carbonate and schistose rocks occurring on top of the BGC. Southeast of Ajmer, the Pur Banera, Sawar-Bajta belt exposes medium-high grade metamorphic assemblages of sillimanite-bearing metapelites and diopsidic marble that are affected by three phases of folding where: (i) refolding (F_2) of the F_1 folds results in plunge inversions, and (ii) late ductile shear zones truncate the late F_3 folds (Ray 1988).

The Jahajpur-Hindoli Belt occurs towards the north-eastern part of the gneissic terrain near Bundi (Sinha-Roy and Malhotra 1988). These rocks are distributed along

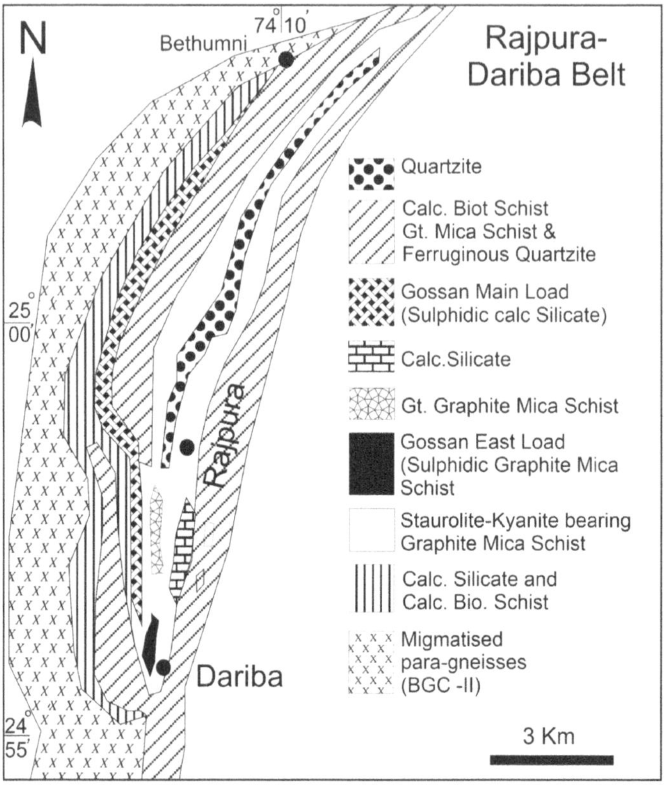

Fig. 4.5 Detailed geological map of the Rajpura-Dariba belt. After Gupta et al. (1997)

western and eastern belts and comprise of conglomerate, quartzite phyllite, dolomite, BIF and bimodal volcanics (Fig. 4.6; Malhotra and Pandit 2000). At places, conglomerate unit directly overlies the Jahajpur granite. These rocks exhibit amphibolite facies metamorphism except the rocks of Jahajpur-Hindoli belt which exhibit greenschist facies metamorphism. The western margin of this sequence in the Bhilwara sector is marked by a discontinuous shear zone, which demarcates the boundary between the granulitized Sandmata Complex and the Mangalwar Complex, the erstwhile Bhilwara Group. As mentioned before this shear zone was named Delwara Dislocation Zone by Sinha-Roy et al. (1998). The eastern margin of the this group is represented by a low-grade metamorphic belt at Jahajpur-Hindoli, sector, considered to be pre-Aravalli by Gupta et al. (1980a, b) and as part of the Aravalli succession (Sinha-Roy and Malhotra 1989; Sinha-Roy et al. (1998)). Perceptible and progressive increase in the metamorphism from east to west was noted in this sector. Roy and Rathore (1999) interpreted the high-grade rocks occurring in the Bhilwara belt as pop-up sequences during the phase of Aravalli basin inversion but later revised their opinion by invoking exhumation of these rocks along multiple shear zones (Roy and Jakhar 2002).

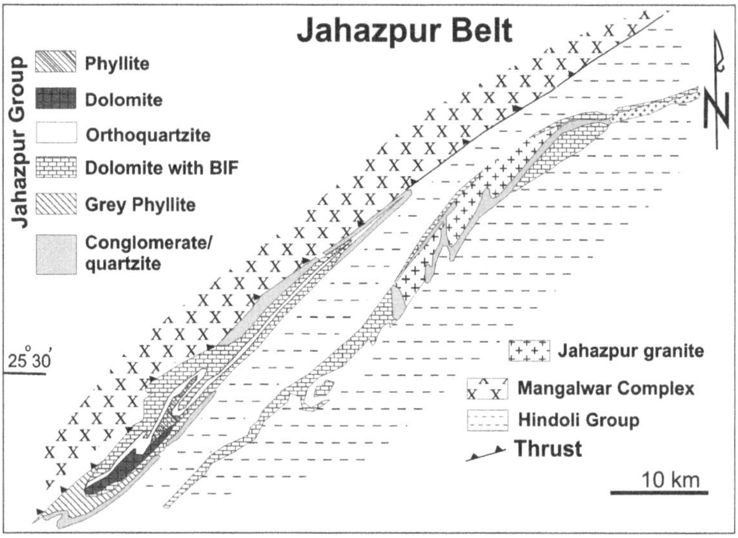

Fig. 4.6 Detailed geological map of the Jahazpur area. After Malhotra and Pandit (2000)

The South-Central Aravalli Terrain (SCAT) represents the type section of the Aravalli Supergroup and is spread from Nathdwara to Udaipur, Dungarpur, Banswara and Lonavada regions. These rocks are represented by mildly metamorphosed quartz arenite, limestone, dolomite, greywacke, black-grey metapelite and interlayered foliated metavolcanics. The Berach Granite initially considered basement for the Aravalli supracrustals was shown to display intrusive relationship with the eastern tracts of the Aravalli Supergroup (Raja Rao et al. 1971). In Udaipur area, the Ahar River Granite acted as the floor for the Aravalli sedimentaries (Dutta 1980). Stratigraphic status of the Lonavada and Champaner Groups is still poorly defined due to complex folding affecting these strata. Although invariably described as the youngest component of the Aravalli Supergroup, their relationship with the type Aravalli Section is not well established. Suggestions have been made that altogether separate marine embayment may have existed in which the Aravalli-like rocks were deposited in southern Rajasthan and northern Gujarat where these may or may not be time contemporaneous. No radiometric dates are available for the rocks of this sector.

4.2.3.3 Aravalli Type Area

Heron (1953) identified six major litho-marker units within the Aravalli 'System' of this sector: (i) amygdaloids and epidiorites, (ii) arkose and conglomerate, (iii) phyllite and biotite schist, (iv) quartzite, (v) carbonaceous band east of Udaipur city, and (vi) limestones. The bulk of the Aravalli succession represents two sedimentary packages of marine shelf and deepwater facies that facilitated its division into lower and upper Aravallis respectively with the Rakhabdeo Lineament marking a tectonic contact between the two. The Aravalli succession has been divided into smaller units in several ways. Mathur (1964) and Poddar and Mathur (1967) broadly distinguished a marine shallow-water shelf facies from deep-water facies within the Aravalli succession, though no distinctions were made at the formational level until aerial photo mapping of the Girwa (Udaipur) valley was initiated (Banerjee 1970). Variation in stromatolite morphology and orientation of primary sedimentary structures helped in establishing an east–to–west younging succession with re-designation of the Aravalli System as the Aravalli Group, subsequently elevated to Supergroup with three major 'Groups' (Gupta et al. 1980a, b, 1997). Banerjee (1971a, b) proposed the stratigraphy given in Table 4.1.

Banerjee and Bhattacharya (1994) distinguished an additional stratigraphic unit of pebbly conglomerate and arkosic sandstones between the Udaipur and Jharol Groups and named it the Sajjangarh Formation.

The Aravalli Supergroup was further subdivided by many contemporary workers including Gupta et al. (1980a, b) into six lithostratigraphic units accompanied by a detailed annotated map. This was followed by numerous stratigraphic schemes proposed by Roy et al. (1988), Nagori (1988) and Paliwal (1988) (see Banerjee and Bhattacharya 1989 for discussion). Gupta et al. (1997) proposed a modified version of their 1980 classification. Roy et al. (1993) and Sinha-Roy et al. (1993, 1998) subdivided the Aravalli Supergroup into three groups in ascending order: Delwara (meta-volcanics), Debari (shelf facies sediments) and Jharol (deep water facies sediments). This scheme was elaborated by (Roy and Kataria 1999; Roy and Jakhar (2002) as follows, and is mostly quoted (Table 4.2).

Stratigraphy of Roy and Jakhar (2002) was further revised by Purohit (2015) in which he added a couple of new stratigraphic units. Of late, Wang et al. (2018) published a stratigraphic scheme for the type Aravallis based on Roy and Jakhar (2002) and Absar and Srinivas (2015) and is the latest scheme available in the contemporary literature. Figure 4.7 shows a comparison of this stratigraphic scheme with the one proposed for the first time (Banerjee 1971a, b) and is now being contested (Fareeduddin and Banerjee 2020).

Table 4.1 Stratigraphic classification if the Aravalli Supergroup (Banerjee 1971a, b)

Jharol group	Phyllites, thin gray arenites, chlorite scihsts, proximal turbidite	Deep water facies
Udaipur formation	Graywackes, phylite, cobble beds, chlorite phyllite	Deeper shelf facies
Matoon formation	Black Shale, dolomite, Stromatolitic phosphorite	Shallow shelf facies
Debari formation	Pyrophyllite-sericite bed, arkose, pebbly conglomerate, quartz arenite, interbedded amphibolite and schist	
Erosional unconformity		
Basement gneissic complex (BGC)		

Table 4.2 Stratigraphic classification of the Aravalli Supergroup (Roy and Jakhar 2002)

Lakhawali formation	Upper Aravalli	Phyllite, dolomite, quartzite
Debari formation		Sandstone
Unconformity		
Bowa formation	Middle Aravalli	Sandstone, phyllite, Pb–Zn-Ag–rich dolomite
Udaipur formation		Conglomerate and sandstone. Phyllite
Unconformity		
Jhamarkotra formation	Lower Aravalli	Stromatolitic dolomite, phosphorite, carbon-phyllite, quartzite
Delwara formation		Basic metavolcanics and tuffs
Unconformity (Paleosol)		
Banded gneissic complex (BGC)		Tonalite-Trondjemite gneiss (3.3–2.9 Ga). Intruded by 2.66–2.44 Ga Granite

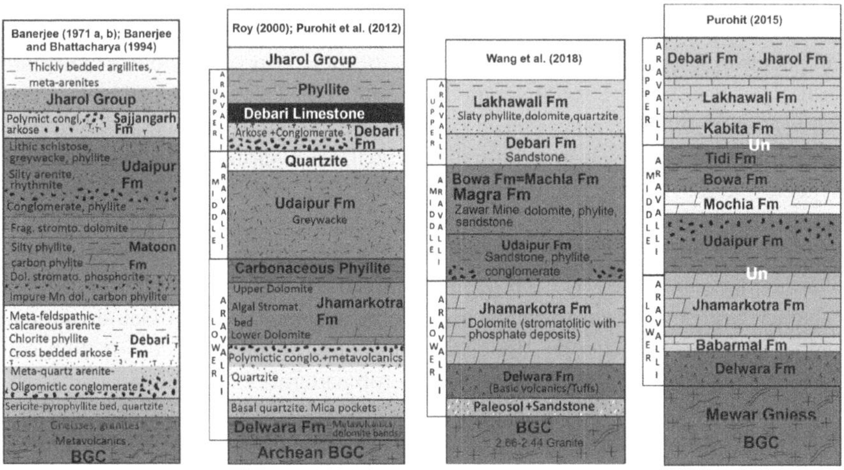

Fig. 4.7 Comparative stratigraphic schemes of the Aravalli Supergroup between Nathdwara and Lunavada in the type Area of Aravallis in Rajasthan. Standard geological symbols have been used. Wherever more than one lithology is present the rock types have been mentioned. After Fareeduddin and Banerjee (2020)

In the absence of measured standard sections a plethora of stratigraphic schemes (Fig. 4.7) created confusion and contradictions related to assignment of age, actual aerial extent of the groups and formations, and basement-cover relationship. Using the same geographical name for the rock-suites that are dissimilar in their litho-contents caused further correlation problems. Fareeduddin and Banerjee (2020) summarized these contentious issues and stated "irrespective of (i) lithostratigraphic level accorded (group vs formation) (ii) conflicting internal organization (e.g. Debari appearing at different stratigraphic levels) (iii) inclusion of the linear belt of mafic volcanics within the BGC (versus within Aravalli), (iv) nomenclature and their semantics (Matoon vs. Jhamarkotra) and differentiation of two carbonate rock assemblages (heavy vs. light carbon isotopes) are some of the most conspicuous features of the SCAT (see comments: Sharma 2009a, b).

4.2.3.4 Structure and Tectonics

The Aravalli Mobile Belt has undergone several tectono-magmatic events although the metamorphic imprints are not pervasive as demonstrated by well–preservation sedimentary structures. In the SCAT, three phases of folding (F1, F2, F3) affected the Aravalli succession although folding did not destroy the primary sedimentary structures, both in the clastic and carbonate rocks (Roy 1988 and many others). Gravity–induced small–scale recumbent folding has also developed in these rocks. The first tectonic episode produced east/west plunging tight isoclinal to reclined folds and slaty cleavage. The second generation of open to tight folds controls the map pattern of various lithounits. The hinges of these two generations of folds are parallel and resulted in coaxial geometry at some places, e.g. east of Udaipur city (Paliwal 1988), while in other places these are orthogonal to each other. Third generation of folds is open warps with sub-vertical axial surfaces that trend athwart the pervasive schistosity (Sinha-Roy et al. 1998). Roy and Jhakhar (2002) summarized deformation history of the Aravalli metasediments and its basement rocks in the SEAT domain wherein they opined that orientation of basins and the direction of shortening controlled style and geometry of superposed folds. Accordingly, belt is characterized by curved lineation over the fold hinges between Nathdwara and Udaipur, folding of extremely stretched pebbles due to multiple deformation and fold interference patterns controlling the outcrop pattern.

The reversal of dips of the basal quartzite-arkose-conglomerate unit in many parts of the SCAT is interpreted as the result of listric faulting (Roy 1988). These are very well documented in the Debari sector east of Udaipur (Roy 1988). Intense mylonitization in the uppermost parts of the BGC basement suggest strike-slip movement along shear zones where steeply plunging lineations also indicate involvement of a vertical component (Sinha-Roy 1988). The Banas Dislocation Zone (BDZ) separates the Aravalli fold belt and the BGC and therefore was interpreted to represent a reactivated boundary fault of the Aravalli rift basin (Sinha-Roy et al. 1993).

One of most talked regional scale polyphase folding is that of a 'hammer-head syncline' in the Kankroli-Rajnagar-Nathdwara sector north of Udaipur (transition area between EAT and SCAT) which is also the type area for unraveling the Raialo-Aravalli relationship. Extensive work by Naha and Halyburton (1974) revealed two generations of large-scale folding affecting the Railos as well as Aravallis (summarized in Naha and Mohanty 1990) These are west-trending isoclinal reclined folds which are refolded into upright folds on N- to NE-striking axial planes. The first set of conjugate sub-horizontal AF3 fold in the schistose rocks is accompanied by isoclinal AF_2 folds. The conjugate AF4 folds show vertical axial planes with NNW–SSE, and E–W strikes. The orientation of the principal stress axes from the late phase conjugate folds indicates that the AF_3 folds got induced by the maximum vertical compression and the AF_3 resulted from a NE–SW horizontal maximum compression parallel to the schistosity.

In the deeper southern domain of the belt beyond the Aravalli type sections, extensive Pb–Zn mineralization is noted, in the Zawar-Balaria-Mochia mine sectors where intense deformation was recorded along a narrow belt of shear zone between basement blocks of Sarara and Babarmal.

4.2.3.5 Metamorphism

Within the EAT the metasediments occurring on the margins of the BGC are relatively more metamorphosed than those disposed away from it. In the Rampura-Agucha lead-zinc mine, Deb and Sehgal (1997) reported peak geothermobarometric values of 650 °C and 6 kb for garnet-biotite and garnet-hornblende pairs, as well as sphalerite-hexagonal pyrrhotite-pyrite and garnet-plagioclase-sillimanite-quartz assemblages. Fareeduddin et al. (2014) also reported values of 677 °C and 4.01 kb for the garnet-biotite thermometer and the garnet-sillimanite-quartz-plagioclase from rocks of the same area. Mishra et al. (2006), recorded a temperature range of 580–600 °C and corresponding pressure of 5.5–6.1 kb inside kyanite field in the Bethumni-Sindeswar belt.

4.2.3.6 Geochronology

It is commonly assumed that the Aravalli sedimentation was initiated by 2.1–1.9 Ga only based on Rb–Sr isochron age of the Darwal Granite (Choudhary et al. 1984) near Nathdwara, which is seen intruding the migmatite country. These migmatites were considered the original Aravalli metasediments. Several studies over the years have shown that these migmatites are integral component of the BGC and do not represent Aravallis sediments. Deb and Thorpe (2004) did not accept the validity of argument that this age can be used for estimating the minimum age of the Aravalli succession.

Ahmad et al. (2008) recorded whole-rock Sm–Nd model ages of 2.3 and 1.8 Ga for the mafic volcanic within the Mangalwar Complex, representing the last phase of basement rifting. The age ranges, shown by these volcanics, support the proposal discussed here whereby these volcanics are considered part of the BGC. The above-mentioned ages, therefore, represent the end phase of BGC and not the beginning of the Aravalli sequence and can not be used in fixing the age of the Aravalli succession. A large array of Pb–Pb model ages of 1.8–1.7 Ga suggest the age for the lower parts of the Aravalli metasediments which overlies the metavolcanics of the Mangalwar Complex (Deb et al. 1989; Deb and Sarkar 1990; Deb and Thorpe 2004). Russell (1993) and Banerjee and Russell (1993) determined Pb–Pb isochron age of 1712 ±

34 Ma for the stromatolitic phosphatic dolomite of Matoon-Kanpur areas of SCAT. Significance of this age is discussed further in this article.

Phosphorites and dolomites of the Jhamarkotra–Matoon–Kanpur areas are assigned 2.2–2.0 Ga ages (Sreenivas et al. 2001a, b; Roy and Jakhar 2002; Maheshwari et al. 1999, 2002, 2010; Papineau et al. 2009; Purohit et al. 2010) indirectly by correlating these rocks with 2.3–2.05 Ga old carbonates from Greenland which exhibit anomalous carbon isotope values and reflect the Lomagundi–Jatuli isotope event (Schidlowski et al. 1976; Melezhik et al. 2007) considered to be of a global nature. Such indirect age assignment is considered scientifically inappropriate.

Recently, Wang et al. (2018) obtained consistent detrital zircon ages from the Aravallis although the stratigraphic interpretations of the data are debatable. These authors never mapped the area, and therefore used the stratigraphy column given in Roy and Jakhar's (2002) for assigning stratigraphic position of their samples for analysis. The zircon Sample DLW 1415 with 2560 Ma age peak appears to be collected from the volcanoclastic assemblage on the uppermost layers of the BGC and, therefore, shows the age compatible with the age of the basement. Sample DLW 1455 with 1880 Ma age peak certainly does not belong to the Middle Aravalli, instead reflect the age of basal conglomerate and quartzite of the Debari Formation of Banerjee (1971a, b) and not the greywacke–phyllite assemblage of the Udaipur valley. This age is consistent with the placement of arkose–conglomerate–quartzite of Debari village as the bottommost unit of the Aravalli succession. Sample DLW 1450 with 1780 Ma tag is rightly placed and reflect the age of the greywacke–quartzite–phylite assemblage of the Udaipur Formation (Banerjee 1971a, b), re-named as the Bowa Formation by Roy and Jakhar (2002). Samples DLW 1438 and DLW 1442 give the age peaks at 1760 Ma, which is consistent with the stratigraphic records. These conform to the ages presented by McKenzie et al. (2013), who determined a large population of detrital zircon of 1.9–1.7 Ga grains (~1772 Ma age peak) with the youngest single grain with $^{207}Pb/^{206}Pb$ age of 1762 ± 9 Ma from the arkosic-quartzite unit just below the stromatolitic phosphatic dolomite at Jhamarkotra. The sandstone from the Girwa valley greywacke-phyllite assemblage (Absar and Sreenivas 2015), overlying the stromatolitic carbonates contains a broad population of 1.8–1.6 Ga detrital zircon with abundant ~2.5 Ga grains and the single youngest zircon with $^{207}Pb/^{206}Pb$ age of 1576 ± 7 Ma. Presence of a large population of 1.9–1.7 Ga zircon grains in dominant phosphorite–dolomite with clastic intervals nullify the assertions of 2.2–2.0 Ga depositional ages for the basal Aravalli succession. The dates obtained by Absar and Sreenivas (2015) is consistent with the dates given by McKenzie et al. (2013) and supports ~1.7 Ga Pb–Pb model ages of Deb et al. (1989) and ~1700 Pb–Pb isochron ages of stromatolitic phosphorite (Banerjee and Russell 1993). The ~1.7 Ga maximum depositional age for this formation nullifies the interpretation that this unit was deposited during a ~2.0 Ga global phosphogenic event following the Great Oxidation Event (Papineau et al. 2009, 2013). The 1921 ± 67 Ma Pb–Pb age of dolomite from the Ghasiar area of Rama-Nathdwara sector north of Udaipur (Saranagi et al. 2006), displaying heavy carbon isotopic values, is consistent with the antiquity of these dolomites in comparison to the stromatolitic phosphorite and dolomite of the Kanpur-Jhamarkotra areas.

4.2.3.7 Depositional History

As stated above, the Aravalli succession started with a well-defined paleosol horizon, which is overlain by granite–derived cross-bedded arkose–conglomeratic quartzite and quartz arenite of the Debari Formation with the overlying Matoon Formation containing mineralized black shales, dolomites, bands of multi-component shales, stromatolitic dolomites and phosphorites (Banerjee 1971a, b) or the Jhamarkotra Formation (Roy and Jakhar 2002). The metagraywackes, meta-argillites, thin bands of shallow-water Alpine wild-flysch type conglomerates represent a more deeper shelf facies of the Udaipur Formation (Banerjee 1971a).

In some sections these marginally deeper Udaipur Formation is observed in contact with deep water carbonate sediments such as those found in the Nathdwara-Rama sectors. These micrtic dolomites are invariably recrystallized and metamorphosed into marble. No shallow water features are found in these carbonates. The contact relationship of these two rock units is poorly defined and appears to be tectonic. Based on presence or absence of phosphorite in the carbonate rock, Papineau et al. (2009, 2013) postulated the existence of two penecontemporaneous sub-basins within the epeiric Aravalli sea. The two carbonate units one in the Jhamarkotra-Matoon-Kanpur-Dakankotra sector show prominent presence of stromatolite and phosphorite (Fig. 4.8). Fine grained carbonates of the Nathdwara-Rama sector do not show any stromatolite nor do they contain phosphorite or any other mineralization. They also do not show the same chemistry or mineralogy. These isolated basins are not affected by faults or basement onlap of strata. Lithostratigraphic dissimilarity between the proposed "sub-basin" indicate that these sedimentary units were formed in different depositional environment and apparently at two different times.

Roy and Jakhar (2002) assigned all the carbonate rocks of Jhamarkotra-Matoon–Kanpur-Rama-Nathdwara areas (Jhamarkotra Formation) to 2.2–2.0 Ga bracket based on anomalously $\delta^{13}C$-enriched carbonates of the Nathdwara-Rama sector which reflects the Lomagundi–Jatuli isotope excursion of 2.3 and 2.05 Ga (Schidlowski et al. 1976; Melezhik et al. 2007). It is worthwhile pointing out that dolomite-marble rocks of the Nathdwara-Rama sector with positive carbon isotopic signatures are never seen in physical connect (Fig. 4.4) with the phosphorite and dolomite of the Jhamarkotra-Matoon-Kanpur areas displaying near zero per mil values for the carbonate carbon and heavy values for the organic carbon (Banerjee et al. 1986; Banerjee et al. 1992; Sreenivas et al. 2001a, b; Wang et al. 2018). The 1.9–1.7 Ga detrital zircon ages of the lower Aravalli phosphorite-dolomite units of the Matoon/Jhamarkotra Group (Fig. 4.2; McKenzie et al. 2013) negates the claims of correlation of $\delta^{13}C$ enriched non-phosphatic, non-stromatolitic dolomite and marbles of the Nathdwara-Rama-Ghasiar regions with the ~2.1 Ga Lomagundi–Jatuli event. The two carbonate successions are distinctly different, petrologically and isotopically. Papineau et al. (2009) postulated two penecontemporaneous sub-basins of the same age in which micritic dolostones and marbles of Rama-Nathdwara sector and the stromatolitic dolostones of Jhamarkotra-Kanpur-Matoon sector formed within the same epeiric Aravalli sea. This suggestion implies contemporaneous deposition of all carbonate-phosphate strata and also sets the age of all the carbonates in the region including

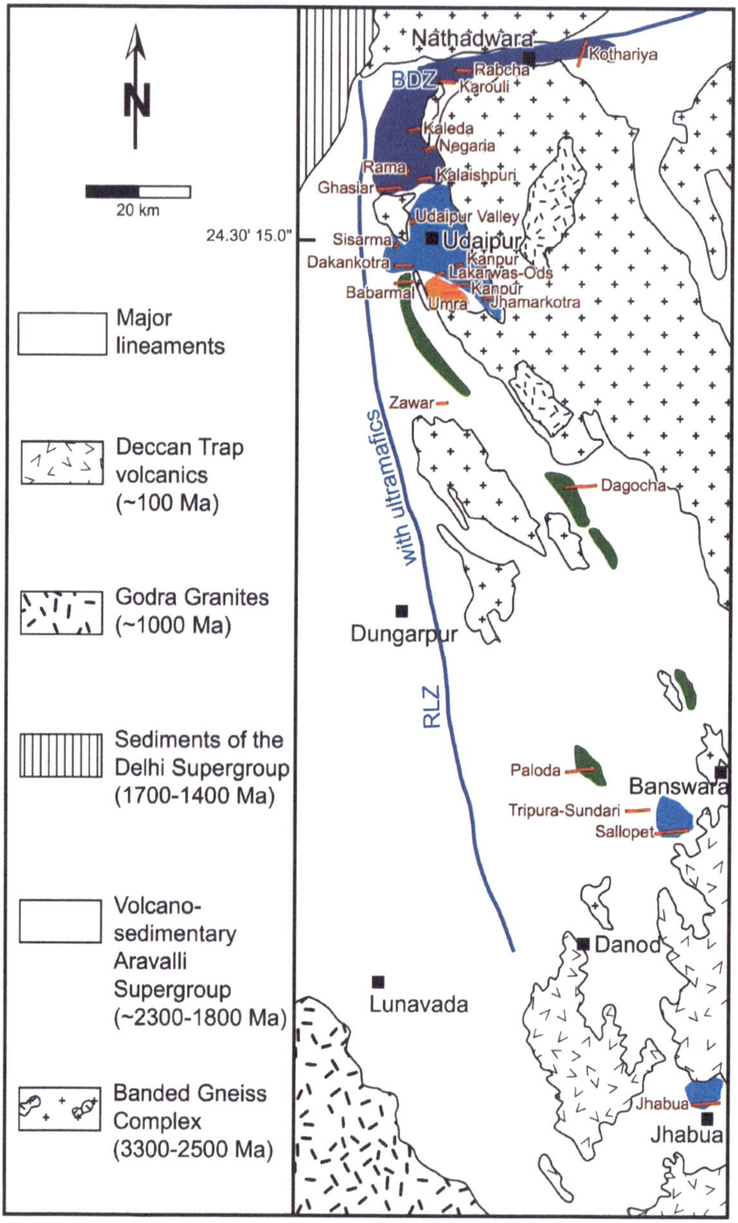

Fig. 4.8 Aravalli Basin with two contrasting carbonate groups with distinct carbon isotopic signatures. Deep Blue: Heavy $\delta^{13}C$; Light Blue: $\delta^{13}C$ close to zero per mil. After Roy and Purohit (2015)

those which are $\delta\ ^{13}C$-rich at ~2.2 Ga. The absence of any synsedimentary fault or evidence of any basement on lap of the strata does not support such a postulation. Marked lithostratigraphic, petrological and geochemical dissimilarities between carbonates of the proposed "sub-basins" clearly indicate that these sedimentary units were either laterally discontinuous due to facies change or that each "sub-basin" was a separate depositional system and were active at different periods of the Proterozoic history. McKenzie et al. (2013) contended that the Jhamarkotra Formation comprising phosphorite-dolomite beds and non-stromatolitic carbonates might have hosted a previously unidentified late Paleoproterozoic positive $\delta^{13}C$ excursion, however it is more appropriate to assign two different ages to these two isotopically different carbonates. Comparison of the isotopically heavy carbon dolostone-marble with the pink marble-dolostone unit stratigraphically and physically below the stromatolitic dolomite (Babarmal Formation ≡ Gupta et al. 1997; Purohit et al. 2012) is significant and supports an older age for the isotopically heavy carbonate unit. Purohit et al. (2012) distinguished a new depositional unit, the Babarmal Formation and interpreted it as the oldest sedimentation unit within the Jhamarkotra Formation (cf. Roy and Jakhar 2002). Carbonate rocks of Zawar mines are assigned to the Bowa Formation by Roy and Jakhar (2002), although their correlation with the similar dolostones of Nathdwara-Rama sector appears more logical.

Litho-association of the youngest monotonous metapelites and dirty sandstones of the Jharol Group was possibly deposited on continental slope and abyssal plain of the Aravalli Sea as deep-sea turbidites containing slivers of ultramafics and serpentinite as detached linear belt. Similar serpentinite bodies along the Rakhabdeo Lineament have been interpreted to demarcate the boundary between the shelf and deep-water facies (Fig. 4.8; Purohit 2015; Purohit et al. 2015), while Sinha-Roy et al. (1998) considered the Rakhabdeo ultramafics to be the result of complete rupturing and asthenospheric upwelling leading to formation of oceanic crust.

4.2.3.8 Salumbar-Bhukia Belt

Due to presence of significant gold and base metal mineralization in the Aravalli succession in the Salumber-Bhukia-Ghatol sections, this area has attracted the attention of the stratigraphers as well. Located ESE of Udaipur this belt is the extension of the type section of the Girwa valley. With the disposition of high-Mg tholeiitic basalts and komatiites on top of the basement gneisses, the succession includes Debari Group of quartzite/conglomerate that directly overlies the gneissic rocks of the BGC. The overlying dolomites, phyllites/carbon phyllites with intercalations of mica schist are placed under locally designated formations of Mukundpura and Jagpura (Shekhawat et al. 2007). The Jagpura Formation consists of meta-pelites (mainly quartz mica schist and garnet-biotite schist) and dolomitic marble.

4.2.4 *Delhi Mobile Belt*

4.2.4.1 Location

Metasedimentary rocks in the NE Rajasthan extending up to Gujarat were conjectured to have formed in spatially disconnected basins (Heron 1953) and possibly belong to a single contemporaneous lithostratigraphic domain of the "Delhi System." In later years two chronologically distinct intrusive magmatic events were identified with a gap ~800 Ma (Choudhury et al. 1984) thus separating the northern from the southern basin (Gupta et al. 1980a, b; Sinha-Roy et al. 1984) Thus, Fareeduddin and Banerjee (2020) suggested that the Delhi basins evolved diachronously in (i) northeastern Rajasthan with multiple depocenters, referred to as the North Delhi Fold Belt (NDFB), and (ii) all along the central and western edifice of the Aravalli Mountain Ranges (Main Delhi Synclinorium of Heron 1953) representing younger basin referred to as South Delhi Fold Belt (SDFB). The ENE–WSW to E–W trending Jaipur-Dausa-Alwar transcurrent fault possibly represent the contact between the NDFB and the SDFB (Sinha-Roy 1988). Physically and tectonically overlying the Aravalli Mobile Belt, folded and metamorphosed belt of Delhi Supergroup extends from Ajmer to Delhi in the north and between Ajmer to Ambaji in the south. Gupta and Bose (2000) have indicated the occurrence of migmatitic basement gneissic complex in intervening areas between NDFB and SDFB and even in the interbasinal areas of the NDFB.

4.2.4.2 Regional Geological Setting of the DFB

The Delhi Fold Belt (DFB) constitutes a narrow linear belt in central Rajasthan and is distinguished as the North Delhi Fold Belt (NDFB) in Delhi–Alwar–Jaipur sector and the South Delhi Fold Belt (SDFB) south of Ajmer extending to northern Gujarat. This division is primarily based on contrasting radiometric ages of intrusive granitoids within these two belts (Sinha-Roy 1984; Sinha-Roy et al. 1998). The NDFB is inferred to have formed in fossil grabens and identified as near-independent Lalsot-Bayana, Alwar and Khetri basins. The SDFB has an eastern Bhim Basin and western Sendra Basin, which are separated by the basement gneisses (Ramakrishnan and Vaidyanadhan 2008). U–Pb zircon dating of the NDFB and SDFB clearly suggest that the detrital zircon U–Pb age data underline a predominant Palaeoproterozoic source for detritus from the NDFB units and limit the maximum depositional age to c. 1720 Ma for the uppermost Ajabgarh Group (Wang et al. 2018). On the other hand, widespread Mesoproterozoic detritus in the SDFB constrains the initial sedimentation of these units to younger than 1055 Ma and does not allow any temporal correlation between the two belts. The two may have formed in independent basins and thereby display distinctly different depositional environments. Complex configuration of the DFB has resulted in variable nature of contact with different units of the Aravalli Craton. In the west, the contact with the Marwar Craton (hypothetical craton) is tectonized and partly marked by the ophiolite-bearing Phulad Lineament along this margin. In most part of this belt, the Delhis directly override the BGC (Mangalwar Complex), while the Kaliguman Lineament restricts its spatial distribution in the eastern parts and brings it in direct contact with the Sandmata Complex. The same lineament marks the tectonized/unconformable contact with the Aravalli Mobile Belt.

4.2.4.3 Stratigraphy

(i) **North Delhi Fold Belt (NDFB)**: In the Khetri, Lalsot-Bayana and Alwar basins, this belt has similar stratigraphy with the Raialo, Alwar and Ajabgarh Groups in ascending order and predominance of meta-arenites, carbonates and volcanics. The Khetri basin occurs in the western part of the NDFB near Sikar and Jhunjunu areas and represents a fault-bound graben having abundant interbasinal faults and synsedimentary dislocation zones. Several granites have intruded these metasedimentary sequences at different times. The Khetri Basin with copper mineralization seems to have formed in two sub-basins of the North and South Khetri Belts with quartzites, amphibolites, marbles, garnetiferous schists, andalusite–biotite schists, metagraywackes and felsic volcanics; an E–W trending Kantli Fault divides this basin into the North Khetri Belt and South Khetri Belt. In the NDFB around Kishangarh, a suite of alkaline rocks–the nepheline syenites are found at the contact with the Sandmata Complex. The intrusion is deformed into an elongated flattened dome and intruded synkinematically with the Delhi Orogeny. A carbonatite body at Newania was formed as an intrusive body within the Archean Untala Granite.

The belt shows variable folding from monocline to three phase of deformation trending NW–SE and superposed by NE–SW and NNE folds. Sinha-Roy (1988) proposed that the NDFB prograded from east through Alwar to the Khetri as pull-apart basin. The Khetri and Shyamgarh groups are eqivalents of Alwar and Ajabgarhs groups in these basins; the latter consists of meta-pelitic schist, pebbly quartzite, felsic volcanic, carbon phyllite, quartzite/conglomerates, carbonates, BIF and greywacke. Granites and dolerites intrude the sedimentary sequences at different places and assume local names.

Structures in Khetri Belt: Roy-Chowdhury and Dasgupta (1964) deciphered three phases of folding in the South Khetri belt. Ray (1974) suggested first two folds to be part of single deformation phase. The first generation of folds formed by the flexural mechanism through NW–SE directed horizontal compression and are associated with a strong axial planar foliation. Superposed second folds are coaxial with the first folds and are associated with shearing. The cross folds represent the third phase formed by NE–SW directed compressive strain on a regional scale. A similar study in the North Khetri Belt by Naha et al. (1988) supported Ray's (1974) observations who also recognized a set of folds with the subhorizontal axial plane as the third folds due to the gravitational load. In Alwar and Lalsot-Bayana basins, polyphase folds resulted in complex outcrop patterns with abundant structural duplications (Fig. 4.9).

More intense deformation, metamorphism and occurrence of base metals in this basin gives Khetri Belt a unique status among other basins of the NDFB (Mehdi et al. 2015). Recent petrogenetic studies by Kaur et al. (2018) had provided new age data for these rocks. The Alwar Group of rocks show large aerial extent while the Raialo Group of rocks include carbonates of Makrana areas with sandstones and volcanic rocks and inferred to have deposited in tectonically uplifted basin (Singh 1988). The Lalsot-Bayana Basin near Bharatpur has Raialo Group with well-sorted conglomerates, basic volcanics (Jahaj-Govindpur volcanic, tholeiitic to high-K basalts) and quartzite (Fig. 4.8). Deposition of quartzose Alwar Group and phyllitic Ajabgarh Groups constitute two major rock units of this region. Interspersed basic volcanics and granite bodies like Bairat, Ajitgarh, Barodia, and Harsora granites are intrusive in their disposition.

Delhi Rocks of Ajmer-Sambhar Lake Sectors: The area between Ajmer and Sambhar Lake has only few rock outcrops which appear similar to the rock of either Alwar (Heron 1953) or Ajabgarh (Kumbalgarh) Group (Gupta et al. 1997) while exact correlation has not be possible (see Fareeduddin et al. 1995) Some of these rocks appear to be similar to the Pre-Aravalli granite-gneisses of Beawar area, Aravalli sequence in the Sambhar Lake, and to the post-Delhi magmatic rocks. Recent workers indicate that (i) these sporadic outcrops of metasedimentatry do not compare with any of the

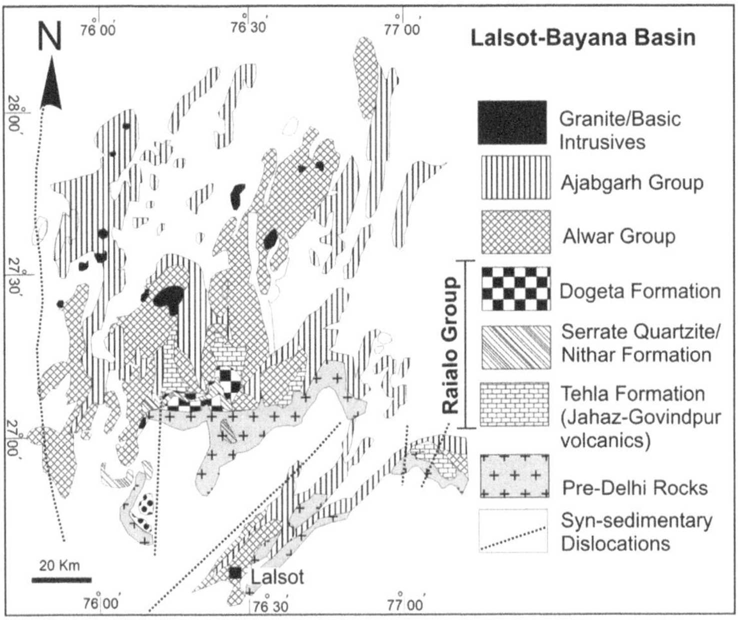

Fig. 4.9 Geological map of the North Delhi Mobile Belt (NDFB) around Lalsot-Bayana area (Singh 1988). Distinct fold patterns are even observed on Google Earth satellite imageries

constituent units of the Delhi Supergroup (Gupta and Bose 2000). Around Beawar, Heron (1953) observed structural continuity of the Anasagar granite-gneiss with the presumed Pre-Delhi inliers.

On the nothwest margin of the NDFB an interesting sequence of granulite to upper amphibolite facies association of calc-silicate granulite, garnet-cordierite granulite (orthopyroxene-bearing), leptynite gneiss, garnetiferous charnockite, garnetiferous two-pyroxene granulite, and garnetiferous amphibolite (Fareeduddin et al. 1995) is exposed within the 'Delhi-type' rock types in the Pilwa-Chinwali areas (Fig. 4.10). These granulite facies rocks occur as tectonic slices in the intermontane areas at the northwestern fringe of the Delhi fold belt proximal to- or west of the main Aravalli mountain range. Fareeduddin et al. (1995) estimated P-T conditions of metapelites 9.5, 7.5, and 5.5 kb and temperatures of about 630 °C and retrogressive pressure

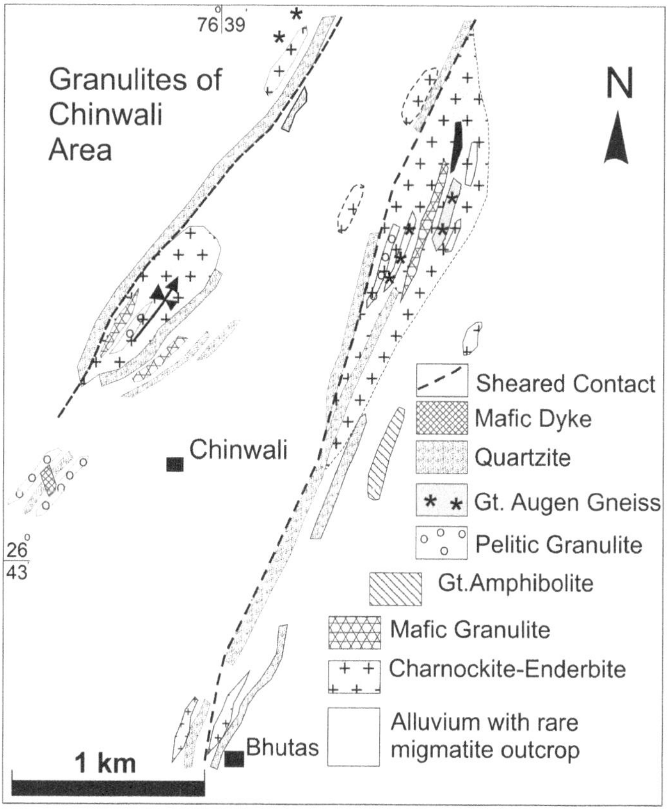

Fig. 4.10 Geological map of the Chinwali area showing granulite outcrops, tectonically juxtaposed with the Delhi quartzites. After Fareeduddin et al. (1995)

of 5.77 kb and 4.04 kb and temperatures of about 650 and 525 °C for charnockite's core and rim. Bhowmick et al. (2018) have indicated a medium-pressure granulite facies metamorphism (*P* between 4.9 and 6.8 kbar, $T > 760$–815 °C) along a counterclockwise P-T path.

$^{206}Pb/^{207}Pb$ data of igneous zircons from charnockite, intruding during peak granulite facies metamorphism, yielded an age of 1434.3 ± 0.6 Ma. Six detrital zircons from pelitic granulites yielded ages between 1522 and 1000 Ma (Fareeduddin and Kroner 1998). The latter becomes the maximum depositional age, hence gneissic fabric and second granulite facies metamorphism are distinctly younger and a post-1000 Ma event. Magmatic zircons from the Pilwa granite yielded an age of 1128 ± 0.7 Ma. Chemical dating of 21 monazite grains (Bhowmick et al. 2018) from six Pilwa-Chinwali granulites samples recorded three broad age domains: (i) 1305 ± 25 Ma, being the age of older metamorphism from detrital remnant in the pelites, (ii) between 1085 ± 10 Ma and 1010 ± 20 Ma as timing of granulite facies metamorphism, and (iii) 880 ± 35 Ma linked with the timing of Erinpura granite magmatism.

Mode of occurrences, lithologic assemblages, and metamorphic recrystallization events largely mimics the Sandmata terrain and, therefore, could be construed as the existence of Sandamata-like terrane across the Trans-Aravalli region.

Structures: Asymmetrical F2 folds in this terrain show alternate flat and steep overturned, limbs with easterly vergence (Mukhopadhaya et al. 2000) and do not conform to folding style of the Delhi rocks of the SDFB (Gupta et al. 1995). Chattopadhyay et al. (2012) identified thrusts associated with these easterly-vergent folds. These are refolded by coaxial D3 folds showing peak metamorphism (5.7 ± 1.5 kbar and temperature of 560 ± 50 °C) which do not compare with features observed in the rocks of the SDFB (Gupta et al. 1995). The folds of the two areas do not appear to be contemporaneous as indicated by two sets of synorogenic granites of widely different ages of 1600 Ma in the Ajmer area and 850 Ma in the SDFB (Choudhury et al. 1984). The U–Pb zircon dates suggest the crystallization age at 1849 ± 8 Ma. Deb and Thorpe (2004) obtained an age of 1745 Ma for the Sedex type Pb–Zn mineralization in Kayar, Ghugra, Lohagal areas. Using detrital zircon ages Mckenzie et al. (2013) inferred the quartzites of the SDFB strongly differ from the Alwar Group of the "North Delhi Belt".

(ii) **South Delhi Fold Belt (SDFB)**: The rocks of the SDFB can be visualized as part of a 'synclinorium' with a median anticline in the centre that exposes the basement rocks. West of Udaipur in the southern sector, the Aravalli Mobile Belt is in contact with the South Delhi Fold Belt (SDFB); the contact was described as an angular unconformity (Heron 1953), as sheared unconformity (Naha et al. 1984) or as a suture zone (Sinha-Roy 1988; Sugden and Windley 1984). On the other hand, Gupta and Bose (2000) inferred that the Aravalli and Delhi Supergroups are not juxtaposed as commonly believed but are separated by the basement complex. NNE–SSW to NE–SW trending SDFB extends from Ajmer in Rajasthan to Himmatnagar in Gujarat. Sen (1980) proposed that the Delhi rocks between Ajmer to Pindwara in the south are distributed in three longitudinal basins, viz. Bhim, Rajgarh and Barotiya with apparent age polarity towards west and the entire unit was shown to be overlain by a younger sequence in the west. Gupta et al. (1997) accepted the idea of westward younging of the rock sequence and proposed Gogunda Group and Kumbhalgarh Group corresponding to he Alwar and Ajabgarh Groups. Two sub-basins, namely the Rajgarh-Bhim and Barotiya-Sendra Basins are marked by prominent Barr Conglomerate at the base, followed upwards by quartzite–phyllite–dolomite sequences and at other places by bimodal volcanics. The Phulad Ophiolite is exposed in the western parts of the basin. Gupta et al. (1997) and Sinha-Roy (1988) considered the tectonized mafic-ultramafic volcanics and gabbroic bands as discontinuous lenses in the Barotiya-Sendra sub-basin as ophiolite strips and designated these as the Phulad ophiolite Belt. These rocks show close geochemical similarity to recent ophiolitic suites (Khan et al. 2005). Tectono-stratigraphically divisions of the Delhi Supergroup in the SDFB are given in Fig. 4.11 and Table 4.3 (Ramakrishnan and Vaidyanadhan 2008).

Structures: Four phases of mesoscopic level folding were recorded by Gupta et al. (1995) in the SDFB domain. The first F1 folds are rootless, flattened and geometrically isoclinal with axial planar schistosity and lineations. The F_2 folds are large and reclined with attendant mesoscopic folds which are upright, steeply inclined, vertical or reclined. These folds show strong axial planar cleavages. The F_3 folds are open or tight folds with occasional development of axial planar fabric. F_4 folds are cross warps. The complicated major structural pattern, resulting from 2nd to 3rd phases of folding and extensive shearing controls the map architecture of the SDFB rocks (Faredduddin and Banerjee 2020).

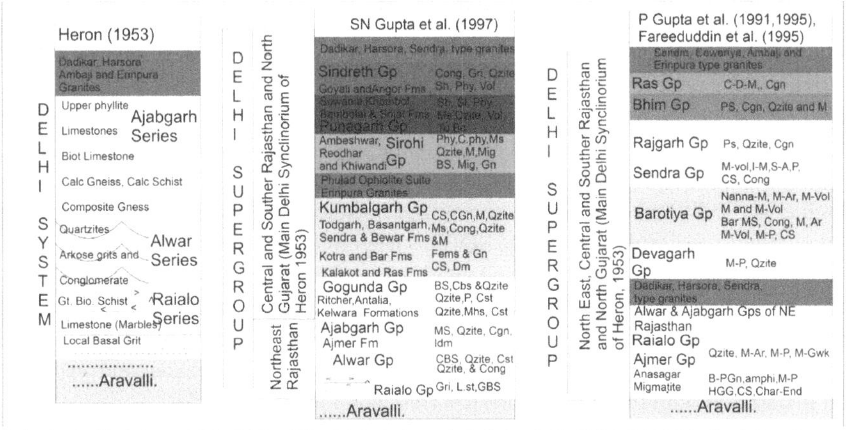

Explanations for abbreviations used: Bc-Bedded Chert, B-PGn- biotite bearing Para-gneiss, Bs-Biotite schist, Cbs-Cal biot schist, C-D-M-Calcitic dolomitic marble, C.Phy-carbon phyllite, CGn-Calc gneiss, Char- Charnockite,Cong-Conglomerate, Cs-Calc schist, Cst-Calc silicate, Dm-Dolomitic marble, End-Enderbites, Fems - Feldspathic mica schist, Gn-Gneiss, Gri-Grit, HGG-High grade granulites, Idm-Impure dolomitic marble, Lst -Limestone, M-Marble, M-Ar-Meta arkose, M-GWk-Meta greywake, Mhs- mica hornblende schist; M-P-Meta pelite,Mig-Migmatite, Ms-Mica schist, Phy-phyllite, Qaite-Quartzite, Sh-Shale,Sl-Slate,Tu-Tuff ,Vol-Volcanics

Fig. 4.11 Stratigrapy of Delhi Supergroup showing main schemes of classification. Abbreviations used for different rock types have been listed. After Fareeduddin and Banerjee (2020)

Table 4.3 Tectono-stratigraphic sub-division of the South Delhi Fold Belt (SDFB)

Bhim group	Peliteic schists, calc silicate, marble
Rajgarh group	Cross-bedded quartzite, pelitic schist, calc gneiss
Sendra group	Metabasalts, marble, arkose, amphibolites, pyroxene granulite, metagabbro and polymictic conglomerate
Barotiya group	Barr Conglomerate, mica schist, marble, arkose, bimodal volcanics
Basantgarh group	Metabasalts, metagabbro, calc schist, metapelites, quartzite, marble, metagraywacke

After Ramakrishnan and Vaidyanadhan (2008)

Metamorphism: The NKB and SKB of the Khetri Belt represent two metamorphic zones in the NDFB, i.e. andalusite-sillimanite and kyanite-sillimanite, respectively (Lal and Shukla 1975; Sharma 1988; Faredduddin and Banerjee 2020). Bhola (1989) studied metapelite rocks of Bayal area in Alwar basin and reported PT estimates to range from 540 to 600 °C and 4–5 kbar. Pant et al. (2000) recognized two sets of assemblages in sub-calcic amphibole-bearing rocks from the Bayal area wherein cordierite and gedrite have developed by consumption of chlorite and quartz and inferred the temperature of ~540 °C at pressure >4 kbar. Kundu et al. (2004) identified peak metamorphic assemblages in metapelites and metabasic rocks and estimated PT viz 5–7 kbar and 550–640 °C. On the other hand, Mehdi et al. (2015) recognized green schist facies rocks (~300 °C) in the Lalsot-Bayana basin. It is, therefore, possible to suggest a strong westward increase in the grade of metamorphism possibly due to tectono-metamorphic development of this basin later than the rocks exposed to the west. The Delhi Supergroup has undergone metamorphism and deformation like that suffered by basement rocks in the region.

In Ajmer-Sambhar Lake sector distinct metamorphic zonation can been seen from east to west: the staurolite-kyanite zone in the east and the sillimanite-muscovite zone in the center (Fareeduddin et al. 1995). Chattopadhyay et al. (2012) reported peak metamorphic conditions of 5.7 ± 1.5 kbar and 560 ± 50 °C, whereas Bose et al. (2017) recorded an increase in the metamorphic conditions to 7.7 ± 0.11 kbar and 592 ± 12 °C in metapelites towards eastern margins on the Ajmer-Shrinagar section.

In the Abu-Balaram-Kui-Surpagla-Kengora areas granulite were reported by Desai et al. (1978) which occur as tectonized lensoidal bodies within the low-grade rocks of the Ambaji sub-basin of the SDFB. The granulites comprise metapelite, calc-silicates/gneiss, basic granulites, and charnockite. Biswal et al. (1998a, b) included these as metamorphosed Delhi rocks and estimated P-T conditions at ~8 kbar, 900 °C. Singh et al. (2010) reported spinel–cordierite–garnet–sillimanite–quartz assemblage that has peak thermo-barometric condition at 5.5–6.8 kbar and $\geq$850 °C and juxtaposed with the Delhi quartzites (Fareeduddin et al. 1995).

Geochronology: In the NDFB the Delhi Supergroup is divided into the lower Alwar and upper Ajabgarh Groups. The initial comprehensive Rb–Sr dating brought to the fore widespread ages between 1340 and 1850 Ma (with relatively larger error-chron) for the granitic plutons of the NDFB (Choudhury et al. 1984). Revised and more precise zircon age data indicate 1.8–1.7 Ga age for the ADMB (Biju-Sekhar et al. 2003; Kaur et al. 2011) Detrital zircon from the Alwar Quartzite has yielded an age of 3671 ± 15 Ma. This, oldest zircon hitherto unreported from northwestern India, is explained as from the exposed basement in the region during the deposition in the North Delhi basin. Discordant U–Pb isotope age of ~2985 Ma represents the crystallization age of the granite below Raialo. Both basement phases currently remained unexposed and unidentified in this sector. Detrital zircons from the Raialo have maximum and minimum age range of 2510 Ma and 2309 Ma respectively. Zircons of youngest age in the Alwar Quartzites returned an age of 2110 ± 33 Ma. The identification of 3671 ± 15 Ma zircons ages adds a new dimension to the understanding of un-recognized Eoarchean to Neoarchean basement segments in this region. Biju Shekhar et al. (2003) considered them to represent the basement for the sediments of 1800–1700 Ma ages. Pant et al. (2008) suggested ~950 Ma for the metamorphism while Kaur et al. (2018) recognized at least four different ages of metamorphism (between 1.85 and 980 Ma and one phase of metasomatism (~900 to 850) based on zircon age data. The succession in the SDMB is apparently as young as ~1.0 Ga (Deb et al. 2001), based on Pb–Pb dating of Deri-Ambaji Pb–Zn ores. Different granitic bodies in the region show age range Rb/Sb age of 850 ± 50 Ma, while granites at Ambaji are 760 Ma old, which is almost same age that of zircon from the nearby charnockites. The volcanic rocks (1000–1100 Ma) and sediments (1240–860 Ma) indicate the age of the basin infill. The Rb–Sr age of 955 Ma to 740 Ma for the post-Delhi Pali-Sendra-Erinpura granites in the SDFB gets support from the precise zircon geochronology, which puts these rocks in the age bracket of 968 ± 1.2 Ma to 800 ± 2 Ma. Tiwari and Biswal (2019) observed that the Sendra Granite (~980 Ma) and Sewariya Granite (~860 Ma) got deformed by D1 deformation so that Delhi deformation cannot be older than 860 Ma. In the Mt Abu-Sirohi sector, Mt Abu granite is dated 764 ± 3 Ma to 779 ± 16 Ma. The granulite metamorphism and their exhumation have been dated at ~800 to 860 Ma and ~800 to 760 Ma respectively. In Ajmer-Sambhar Lake sector, U–Pb dates of zircon indicate that the precursor granite of the Anasagar migmatite within Ajmer town got emplaced within the supracrustal unit during an early folding event at approximately 1.85 Ga. On the other hand, monazite date of staurolite schist in this area yield a pooled age of 980 ± 22 Ma, which is assumed to be the age of peak metamorphism (Bose et al. 2017). The $^{206}Pb/^{207}Pb$ data for migmatitesis in the range of 1695.7–2258.1 Ma indicate a Paleo-Mesoproterozoic provenance for the sedimentary infills. The youngest age of ~1700 Ma constrain the closing time of this basin and is, therefore, contemporaneous to the Aravallis (Fareeduddin and Kröner 1998). The supracrustals of Sirohi and Sindreth Groups possibly represents the youngest unit of the Delhi Supergroup. Arora et al. (2017) observed that the Sirohi metasediments were deposited over a basement of 892 ± 10 Ma old Erinpura granites and got metamorphosed under low to medium grade metamorphism at 822 ± 29 Ma. The rock formations are disposed in NW–SE direction which that orthogonally impinges NE–SW trend of the South Delhi Fold. Since Sirohi sediments deposited over a post-Delhi Erinpura basement of ~890 Ma with structural discordance with Delhi structural fabric, Arora et al. (2017) interpret an existence of a younger Neoproterozoic fold belt deformed at 822 Ma.

4.2.5 Neoproterozoic Tectonics in the ADMB

Widespread Neoproterozoic magmatic bodies in the ADMB extend from the Tosham Volcanics (Haryana) in the north to gabbro–norite plutonic rocks of Balaram, Gujarat in the south. Table 4.4 summarizes the available Neoproterozoic U–Pb zircon and monazite ages from this belt. These intrusives are the Ojhar Granite (>960 Ma) in the Aravalli metasediments, carbonatite (955 ± 24 Ma) at Newania within pre-Aravallis, the Erinpura granite of 990 and 830 Ma age and the Mt. Abu granitoid (800–730 Ma) in the Sirohi Group west of the fold belt (Just et al. 2011; Van Lente et al. 2009; Pradhan et al. 2010; Ashwal et al. 2013). U–Pb zircon ages of 987 ± 6.4 and 986.3 ± 2.4 Ma from the Ambaji–Sendra rhyolites reveal the presence of an extensive arc terrane on the western flank of the Aravalli–Delhi orogenic belt in its southern and northern parts (Deb et al. 2001) and Deb and Thorpe (2004), while the same arc sequence is also established in the Punagarh Group from galena isotopic composition with a model age of ~940 Ma. In the southern part of the Ambaji–Sendra belt, a U–Pb zircon age of 836 +7/−5 Ma for the Siwaya gneissic granite is in accord with the previous age data for the felsic Erinpura plutons that have intruded the arc sequence (Deb et al. 2001). The other granitoids (Rahman and Mondal 2015) such as the Sendra–Ambaji granite (840–760 Ma), the Dudhi granite and Sendra granite (850–900 Ma (Choudhary et al. 1984; Tobish et al. 1994) are intrusives into the Delhi metasediments of the ADMB. Based on isotopic characters of the Sendra granitoid, Tobish et al. (1994) assigned its formation due to partial melting of underlying granite-gneiss crust. The Sewaria granitoid, occurring to the west of the ADMB, also exemplifies large-scale partial melting/migmatization of the Aravalli–Delhi metasediments. Thus, geochronological data from the ADMB on the sediments and intercalated volcanics between 1240 and 966 Ma (Deb et al. 2001; Deb and Sarkar 1990), plagiogranite intrusions at 1015 ± 4.4 Ma (Dharma Rao et al. 2013) and calc-alkaline granitoids at 968 ± 1 Ma (Pandit et al. 2003) record the evolution of a Neoproterozoic volcanic arc between the Marwar terrane and Aravalli Bundelkhand craton of the NW Indian shield. The supracrustals of Sirohi and Sindreth groups represents youngest sequences of Delhi Supergroup (Fig. 4.2). Fareeduddin and Banerjee (2020) have stated that the Sirohi metasediments were deposited over a basement of 892 ± 10 Ma old Erinpura granites and metamorphosed under low to medium grade metamorphism at 822 ± 29 Ma (Arora et al. 2017). It is worthwhile noting that the formational trends and the pervasive fabric in the metasediments are in NW–SE direction and corroborates well with the aero-magnetic signatures of this terrain (Gouda et al. 2015). The latter indicates the presence of a NW–SE trend that orthogonally impinged the NE–SW trend of the South Delhi Fold. Since the Sirohi sediments were deposited over a post-Delhi Erinpura basement of ~890 Ma with structural discordance with Delhi structural fabric, Arora et al. (2017) interpreted the existence of a Neoproterozoic fold belt which deformed at 822 Ma. Some workers do not agree to this interpretation and suggest they reflect the trend of the youngest generation of folds. Such trends are consistently observed in many other parts of the Delhi fold belts (Khetri belt of NDFB and the Todgarh areas in SDFB). In Sirohi

Table 4.4 Geochronology of Neoproterozoic magmatic events in Aravalli region, NW India

Ages (Ma)	Method	Rock type/locality	Event	References
Malani igneous suite				
751 ± 3	U–Pb, zircon	Rhyolite	MIS felsic volcanism	Torsvik et al. (2001)
765, 761 ± 16 767 ± 3	U–Pb zircon	Sindreth felsic volcanic and mafic volcanics	Sindreth volcanism	Dharma Rao et al. (2012), Van Lente et al. (2009)
771 ± 3	U–Pb zircon	Rhyolitic tuff, Jodhpur	MIS	Gregory et al. (2009)
Sirohi-Mt Abu event				
764 ± 3 767 ± 4 768 ± 3	U–Pb Zircon	Mt Abu Granite (augen gneiss; (deformed)	Mt Abu Batholith	Ashwal et al. (2013)
775 ± 26 779 ± 16	U–Pb–Th (mz) microprobe	Shear overprint and anatexis	Imprint of Malani magmatism	Just et al. (2011)
Delhi Orogeny and Erinpura magmatism				
800 ± 2 873 ± 3	U–Pb zircon	Tonalite/granodiorite basement	Erinpura Granite	Van Lente et al. (2009)
827.0 ± 8.8	U–Pb zircon	Harsani Granodiorite (Barmer)	Basement (Erinpura age) for MIS	Pradhan et al. (2010)
842, 846	U–Pb zircon	Inherited Zircons in Mirpur granite	Erinpura basement	Pandit (Pers. Comm.)
863 ± 23	U–Pb–Th (mz)	Granite Gneiss	Erinpura magmatic event (Sirohi region)	Just et al. (2011)
920 ± 0.8	^{207}Pb–^{206}Pb zircon evaporation	Veerwara migmatite	DFB basement rocks (Sirohi region)	Purohit et al. (2012)
968 ± 1.2	U–Pb zircon	Sendra Granite	Subduction and Calc–alkaline granite intrusion, DFB	Pandit et al. (2003)
987 ± 6.4 836 + 7/−5 986.3 ± 62.4	U–Pb Zr	Deri rhyolite, Siwaya Granite, Birantiya Khurd rhyolite	Ambaji-Sendra arc terrane of western Aravalli-Delhi fold belt	Deb et al. (2001)

area, the NW–SE trend appear to be dominant and controlled the map pattern. This is because of (i) open nature of the early fold, on which these are superposed.

4.2.6 Orogenies During Aravalli and Delhi

There is no ambiquity in recognizing polyphase deformation of the Aravalli and Delhi where a minimum of three regional folding phases have been identified (Naha et al. 1984; Roy 1988; Gupta et al. 1995). Naha and his co-workers in early 1984 established remarkable identity in the later phases of the deformational history of the Aravalli and Delhi rocks, with gravity-induced structures followed by conjugate folds due to longitudinal shortening (N–S in northeastern Rajasthan and NE–SW in central Rajasthan). High gravity of ~80 mGal short wavelength Bouguer anomaly caused by the density contrast of higher density rocks of the fold belt and the low density of the basement gneisses confirms the existence of major horst described by Fermor (1930). The earlier stages of the structural history of the Aravalli (Tewari and Rao 2003) and Delhi groups followed different paths. The E–W trending reclined folds of the first generation in the Aravalli rocks are not seen in the Delhi rocks. The NNE–SSW to NE–SW trending folds of the second generation in the Aravalli and older rocks are upright, whereas these structures in the Delhi rocks are of two phases—recumbent folds, followed by coaxial upright folds. This difference in the folding pattern suggest an angular unconformity between the Delhi and Aravalli and related rocks.

In the entire ADMB region, Roy (1988, p. 22) demonstrated temporally close DF_1–DF_2 folds, formed during Delhi orogeny being represented by AF_2 folds of the Aravallis whereas AF_1 folds are indicators of the Aravalli orogeny He further asserted that the DF_3 and AF_3 folding cannot be tied up with the Delhi deformation. Some late fold phases are also seen superimposed on the early Delhi folds (Gupta et al. 1995; Biswal et al. (1998a, b). The structural data, therefore, suggest an intense effect of the Aravalli and Delhi orogenies to generate the most prolific AF_1 folds in the Aravallis and $AF_2 + (DF_1 + DF_2)$ fold phases in the Aravalli–Delhi rocks. Postulation of identification of the Sirohi orogeny (Arora et al. 2017) or Champaner orogeny (Roy 1988, p. 22) is still conjectural.

Although Sharma (1988) accepted the fact that BGC and other pre-Aravalli metasediments have suffered multiple stages of metamorphism, he contended that both Aravalli and Delhi rocks are mono-metamorphic. As a result, he visualized only two orogenies affecting the south-central Rajasthan, the pre-Aravalli and Delhi orogeny. It is interesting to note that both the basement BGC rocks and overlying Delhis are more metamorphosed than the Aravallis. This prompted Sharma (1988) to argue that unless the two belts are subjected to only one regional metamorphic event, it can not explain the anomalous juxtaposition of two metamorphic belts.

Bhowmick et al. (2010) inferred that the M1 sillimanite-facies metamorphism and younger kyanite-facies HP metamorphic event caused re-heating and prograde burial of the M1 granulites which they attributed to crustal thickening of the entire ADMB domain during the ca. 1.7 Ga collisional orogeny. Another metamorphic recrystallization event recorded in the ADMB relates to the ca. 1.0 Ga collisional orogenic front (Bhowmik et al. 2018). This event encompasses high grade basement rocks and suprcrustals of highly mineralized Rampura-Agucha and Pur-Banera. Besides The 1.0 Ga metamorphic P-T paths move clockwise from Sandmata metamorphic complex and Anasagar migmatites and Shrinagar metapelites near Ajmer. A metamorphic imprint of ~1.0 Ga was recorded in the rocks of the Alwar basin (Pandey et al. 2013). Bhowmick et al. (2018) suggest that radiogenic heat production in a tectonically thickened continental collisional zone could be the dominant mechanism of middle to lower crustal heating in vast stretches of the ADMB at ca. 1.0 Ga. Mishra and Ravi Kumar (2014) have also recorded gravity high on ADMB with dipping crustal structures indicating uplifted lower crust with crustal thickening (~45 km) in the central part. The IGRF corrected magnetic data (Gouda et al. 2015) suggest much widespread existence of magnetic bodies similar to those seen in the Sandmata region in the alluvial covered tracts of the north and northwestern parts of Rajasthan implying pervasive presence of 'Sandmata-like' metamorphic event, the implications of which need to be further explored. Fareedudin and Banerjee (2020) have pointed out the significance of this data in identifying a Sandmata-like succession in the Trans-Aravalli region.

Seismic profile along Nagaur–Jhalawar transect displays opposing dips of reflectors along Phulad and Delwara Lineaments that marks the subduction–collision zones. Another reflector dipping NW marks the dislocation between the Bundelkhand and AravallCratons. Magnetotelluric studies clearly brought out this high reflector. Seismic reflection also imaged a mid-crustal reflector in the form of dome under Delhi Fold Belt and Sandmata Complex (Ramakrishnan and Vaidyanadhan 2008).

Recent studies draw correlation of the ADMB with chronologically and tectono-metamorphically similar basins in Rodinia continent assembly (for example, the Jiangnan Fold Belt in S. China (Zhao et al. 2018) and the Proterozoic mobile belts of central, eastern and southeastern India (Bhowmik and Dasgupta 2012). Such correlations, therefore, suggest that the ADMB constitute an important part of the growth and the assembly of the Greater Indian Landmass in Rodinia (Fareeduddin and Banerjee 2020; see also Mondal 2009).

4.2.7 *The Crustal Evolutionary Models*

4.2.7.1 Ensialic Rifting Model

One of the earliest tectonic models (post-geosynclinal concepts; Sharma 1995) proposes the ensialic rifting (at ca 2000–1900 Ma) as the reason for the formation of large Aravalli basin followed by subsidence and crustal shortening (assuming Aravalli orogeny at 1900–1800 Ma). Another ensialic rifting (1800–1700 Ma) favored Delhi sedimentation, followed by underplating by mantle magma, decoupling of mantle-lithosphere and subduction (ca 1800–1700 Ma). Deformation of Delhi sediments

(Delhi orogeny at ca 1700–1500 Ma) and a large period (ca 1500–900 Ma) of mantle lithospheric welding led to the tectonic exhumation of the granulites, followed by several episodes of Erinpura type granite magmatism.

4.2.7.2 Plate-Tectonic Model

Using trace element geochemistry and petrofacies analysis Banerjee and Bhattacharya (1989, 1994) applied plate tectonic concepts to assign each of the Aravalli clastic rocks to specific tectonic setting and interpreted the evolution of the Aravalli basin in a passive margin setting which changed to active margin tectonics towards the end phase (Fig. 4.12).

Sinha-Roy (2000) visualized Proterozoic development in the region involving Mewar/Bundelkhand craton and Marwar Craton towards east and west, respectively. The Aravalli basin formed during the Paleoproterozoic as an aulcogen that widened southwards from a triple junction near Nathdwara. A set of strike-slip faults from aulacogen triple point splayed and resulted in the formation of pull-apart basins with mineralized belts like Rajpura-Dariba, Agucha-Bhilwara, etc. According to him, the Aravalli aulacogen closed at 1800 Ma which may require revision in view of newly generated dates. A suture zone represented by the Antri-Rakhabdeo tectonic line was formed and synkinematic granites (Darwal, Anjana) were emplaced when the two cratonic blocks collided, and the oceanic basin floor thrust under the continental block. Another major strike-slip fault emanating from the triple junction opened the Delhi basins in two tectonic domains: (i) the NDFB evolved as fault-bound grabens and half-grabens as part of aborted rifts were intruded by granites at 1450–1650 Ma, and (ii) the SDFB as a linear belt formed as a transtensional rift that opened into a sea like the Red Sea, with eastern continental platform sediments (Bhim and Rajgarh groups) and the western volcanic suites of Barotiya and Sendra Groups consisting of basalt-andesite-rhyolite (island arc suite). This trough was closed by subduction, with the formation of a trench, a magmatic arc, and a marginal basin. The 1000–1100 Ma old Ranakpur gabbros and co-eval Phulad ophiolite suite/mélange were emplaced and was followed by Sendra-Ambaji and Erinpura granites (900 Ma) due to westward thrusting of the Delhi oceanic crust. Several new publications during recent years focus the development of individual segments of the ADMB under the overall ambit of the plate-tectonic models envisaged above (see Wang et al. 2018; Zhao et al. 2018).

Sharma (2009a, b, p. 166–170) has elaborated different stages of tectonic evolution of the Aravalli–Delhi Mobile Belt since Proterozoic (see also Verma and Greiling 1993) and visualized its evolution from the ductile stretching of the continental crust to produce initial rifted basins for the deposition of the sediments, subduction, metamorphism, magmatism and exhumation (Fig. 4.13).

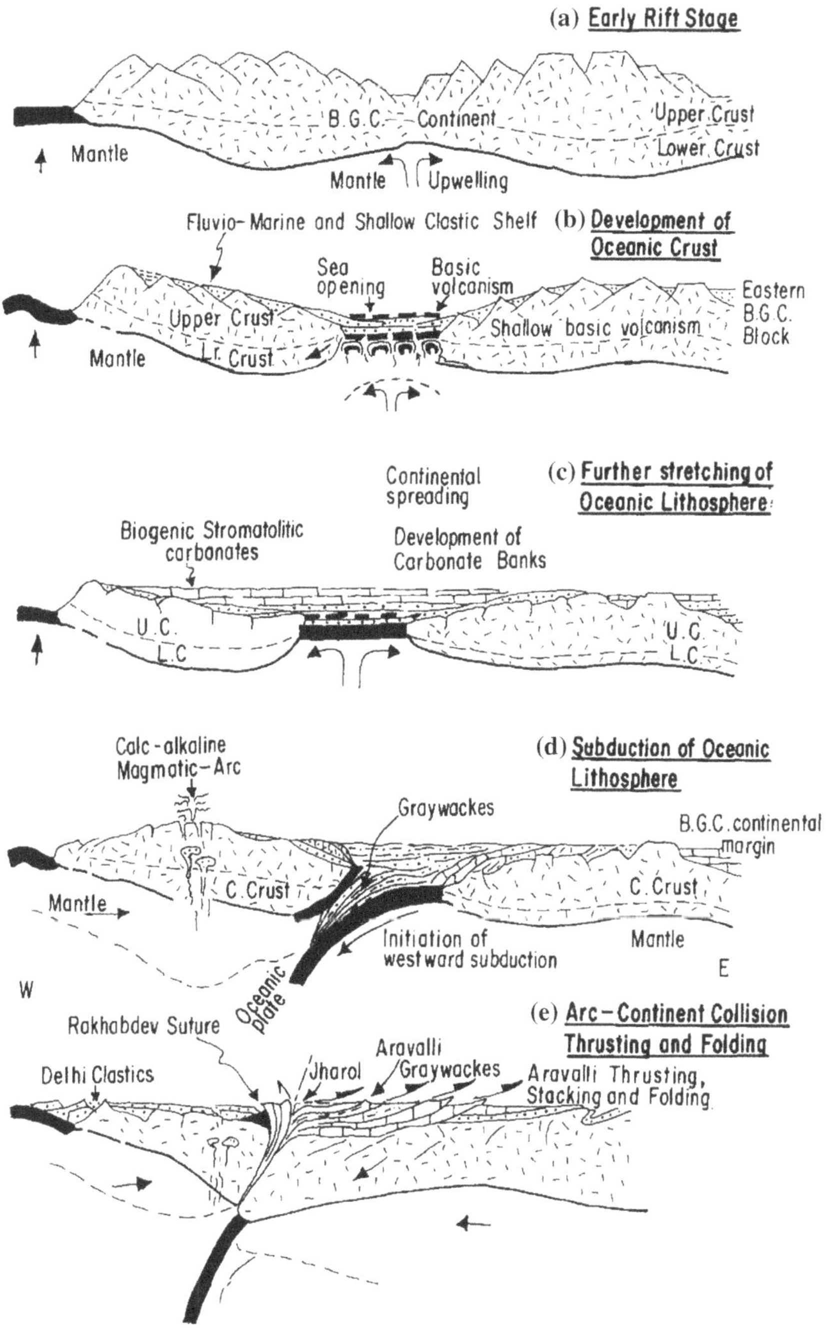

Fig. 4.12 Changing plate tectonic scenario from early basement rifting to final stabilization of the Aravalli sedimentation basin. After Banerjee and Bhattacharya (1994) modified after Sugden et al. (1990)

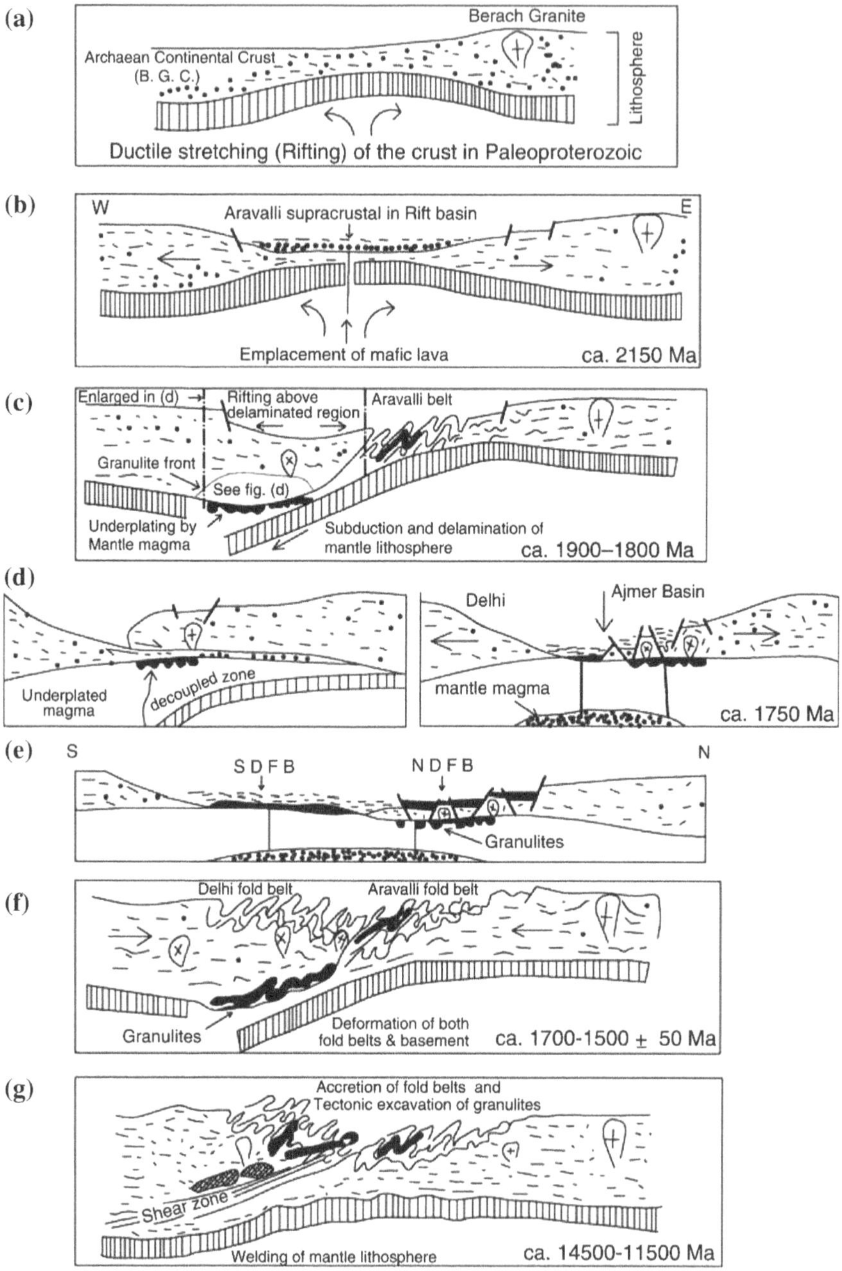
(a)
Berach Granite
Archaean Continental Crust
(B. G. C.)
Lithosphere
Ductile stretching (Rifting) of the crust in Paleoproterozoic
(b)
W
E
Aravalli supracrustal in Rift basin
Emplacement of mafic lava
ca. 2150 Ma
(c)
Enlarged in (d)
Rifting above delaminated region
Aravalli belt
Granulite front
See fig. (d)
Underplating by Mantle magma
Subduction and delamination of mantle lithosphere
ca. 1900–1800 Ma
(d)
Underplated magma
decoupled zone
Delhi
Ajmer Basin
mantle magma
ca. 1750 Ma
(e)
S
N
S D F B
N D F B
Granulites
(f)
Delhi fold belt
Aravalli fold belt
Granulites
Deformation of both fold belts & basement
ca. 1700-1500 ± 50 Ma
(g)
Accretion of fold belts and Tectonic excavation of granulites
Shear zone
Welding of mantle lithosphere
ca. 14500-11500 Ma

◀**Fig. 4.13** Tectonic evolution of the Aravalli-Delhi Mobile Belt, Rajasthan, NW India. **a**, **b** Plume-generated ductile stretching, development of rift basin on the Archaean crust and deposition of Aravalli supracrustals and Delwara volcanics. **c**, **d**, **e** Westward subduction, delamination of mantle lithosphere and magma underplating resulting in granulites formation and deposition of the Delhi supracrustals in a series of rift basins. **f**, **g** Convergence of different blocks, welding of the mantle lithosphere, accretion of the fold belts and tectonic excavation of granulites. After Sharma (2009a, b)

4.3 Central Indian Tectonic Zone (CITZ)

4.3.1 Location

The Precambrian crust of Central India consists of two Archaean cratonic domains, viz. Bundelkhand Craton in the north and Bastar Craton in the south, which were accreted along the ENE–WSW trending Central Indian Tectonic Zone (CITZ). Both the cratons are made up of gneisses–supracrustals–granite association and record independent evolutionary histories. The Bastar Craton with WNW–ESE to N–S trend is skewed from ENE–WSW to E–W trends. The limits of the CITZ is defined by the Son–Narmada North Fault (SNNF) in the north and Central Indian Shear Zone (CIS) in the south (Radhakrishna 1989; Acharyya and Roy 2000). In the east, the CITZ continues into the Chotanagpur Gneissic Complex (CGGC) and further northeast into the Shillong Plateau (Acharyya 2001) and Mikir Hills of the Meghalaya Craton (Fig. 4.14).

4.3.2 Distribution

The CITZ comprises Proterozoic supracrustal belts of varying metamorphic grade set in a largely undifferentiated gneiss and syn- to post-kinematic granite (Fig. 4.15). These can be differentiated into three prominent components:

(i) The supracrustal belts in the CITZ include **Mahakoshal, Betul and Sausar belts** distributed in the northern, central and southern parts. Dismembered small supracrustal slivers have also been recognized in Bilaspur and Sarguja districts of MP (Mishra and Ravi Kumar 1994).

(ii) The granulite belts in the CITZ are **Makrohar Granulite Belt (MG), Ramakona–Katangi Granulite Belt (RKG) and Balaghat–Bhandara Granulite Belt (BBG)**. The MG belt occurs south of Mahakoshal supracrustal belt, whereas RKG and BBG belts are exposed along the northern and southern margins of Sausar supracrustal belts, respectively. Granulite-facies BIF is reported from the northern part of Betul supracrustal belt.

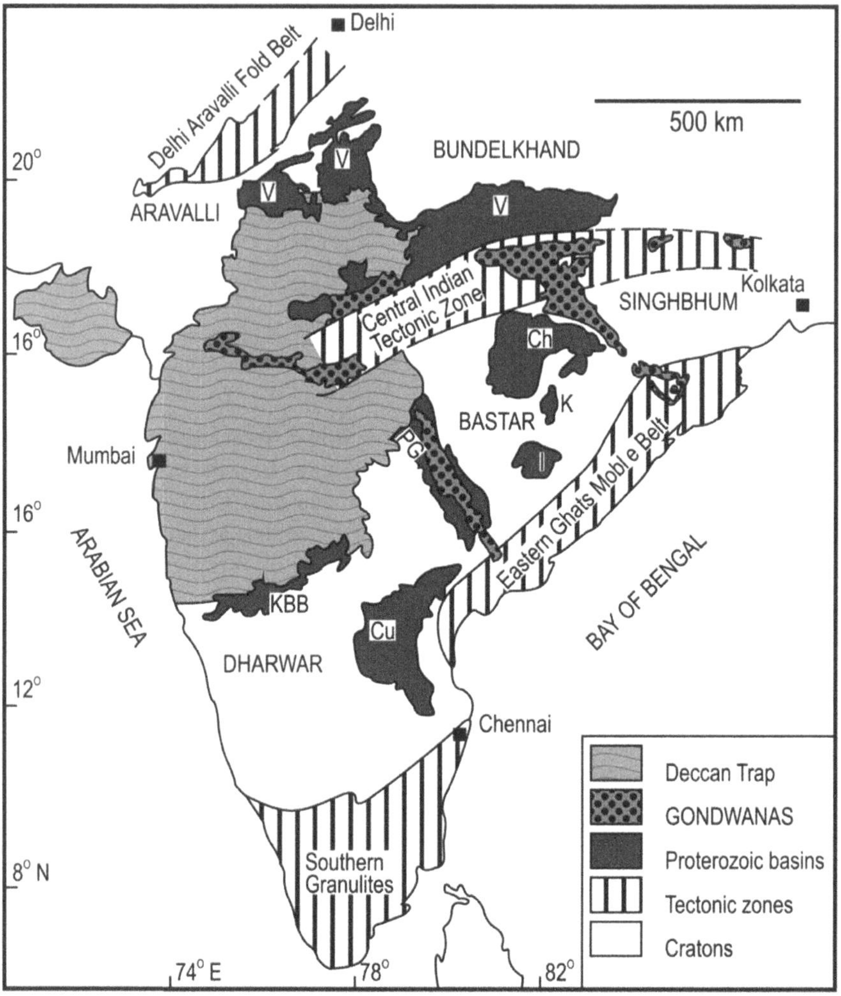

Fig. 4.14 Tectonic map of India showing distribution of major Proterozoic mobile belts/tectonic zones showing Aravalli–Delhi Mobile Belt (ADMB), Central Indian Tectonic Zone (CITZ), Eastern Ghats Mobile Belt (EGMB), and South Granulite Terrain (Pandyan Mobile Belt-PMB). Modified after Rao and Reddy (2002). Distribution of the Purana Basins, including Vindhyan (V), Chhattisgarh (Ch), Khariar (K), Indravati (I), Pranhita–Godavari (PG), Cuddapah (Cu) and Kaladgi–Bhima (KBB) basins are also indicated. After Chaudhuri et al. (2015)

(iii) **Son–Narmada South Fault** (SNSF), **Gavilgarh-Tan Shear** and **Central India Shear** are brittle ductile to ductile shear zones identified in this belt. The overall tectonic trend in the CITZ is ENE–WSW.

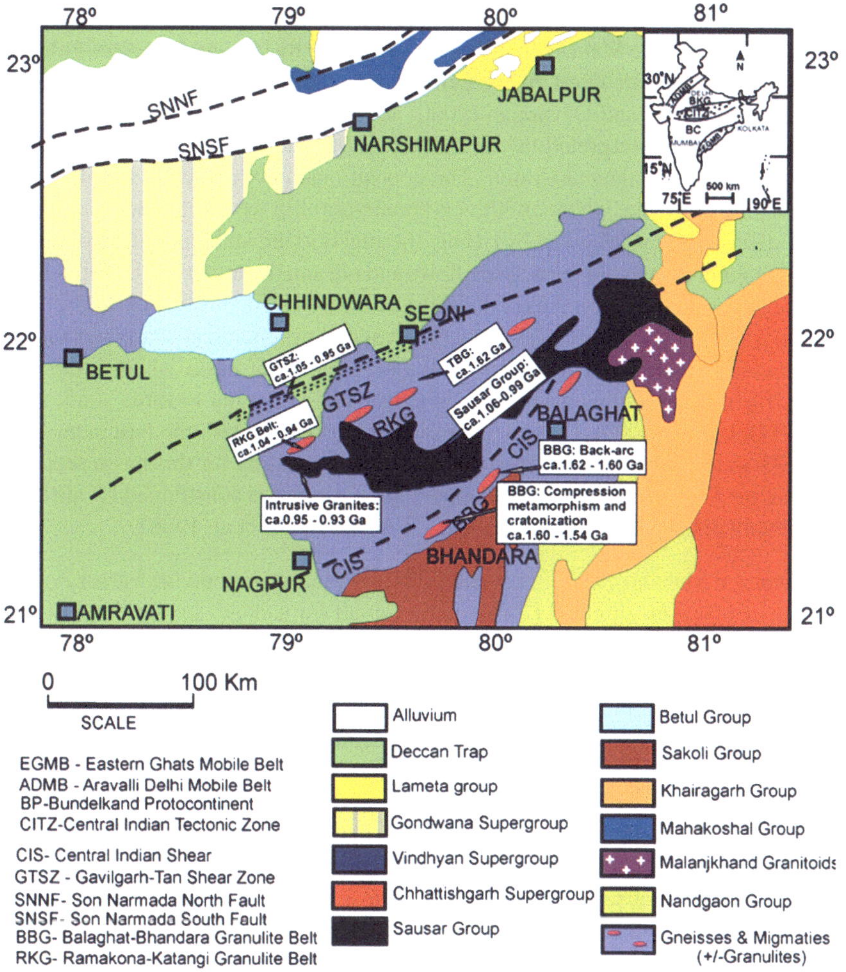

Fig. 4.15 Simplified geological map of the Central Indian Tectonic Zone (CITZ) showing the major supracrustal belts and tectonic lineaments. Geochronological data from Bhowmik et al. 2012) and Chattopadhyay et al. (2015a, b). Inset indicates positions of major Proterozoic mobile belts of India (e.g. EGMB, ADMB, CITZ). After Chattopadhyay et al. (2017)

4.3.2.1 Supracrustal Belts: Mahakoshal, Betul and Sausar Belts

(i) **Mahakoshal Supracrustal Belt (MSB)**: Northern parts of the CITZ are represented by ENE–WSW trending Mahakoshal supracrustal belt in a fault-bound asymmetrical rift basin (Roy and Bandyopadhyay 1990). The Son–Narmada North Fault (SNNF) in the north and the Son–Narmada South Fault (SNSF) in the south limits its outcrop exposures for over 600 km. The Archaean gneiss–granite–supracrustal complex of the Sidhi gneisses is exposed in the north, while Proterozoic granitic intrusive are abundant in the south. The lithological assemblage of the Mahakoshal belt is represented by low-grade metamorphosed quartzite-carbonate-chert-BIF-greywacke-argillite-mafic volcanics. Nair et al. (1995) and Roy and Devarajan (2000) suggested the stratigraphy of this belt; the latter have interpreted this belt as a pericratonic basin along southern margin of the Bundelkhand Craton. The depositional environment was inferred to be shallow marine, where arenite-carbonate-argillite were deposited along with the BIF. This was followed by rifting, thermal doming and outpouring of tholeiitic magma along with pyroclastic flows and ultramafic intrusives. Uplifted basin margins provided sediments to the debris, followed by deposition of the argillites and iron-rich sediments (BIF). Lithological and geochemical characters clearly indicate a continental rift setting for the Mahakoshal basin where lithofacies distribution highlights asymmetrical rifting and younging towards south (Roy and Devarajan 2000). Due to its intimate relationship with the basement Bundelkhand granites (Rb–Sr ages up to ca. 2.2 Ga), rifting for this basin seems to have post-dated ca. 2.2 Ga. The upper age of the belt is constrained by intrusive calc-alkaline granites, which yield ca. 1.8 Ga (Sarkar et al. 1998).

Within the Mahakoshal belt, several granitoid plutons intrude the Parsoi Formation, out of which an elliptical E–W trending small Jhirgadandi stock is largely metaluminous (I-type) to rarely peraluminous (S-type) granitoid. U–Pb SHRIMP zircon $^{206}Pb/^{238}U$ ages for microgranular enclaves (1758 ± 19 Ma) and host granitoid (1753 ± 9.1 Ma) from this pluton (Bora et al. 2013) support their coeval character. This pluton also points to the existence and role of Columbian continental component in the evolution of Mahakoshal Belt of the CITZ.

The supracrustal rocks here show upright to slightly overturned folding on the ENE–WSW trending and steep southerly dipping axial planes, in response to N and NNW-directed bulk sub-horizontal compression. During the deformation of the belt a prominent reverse-slip ductile shear zone along the SNSF helped channeling the granitic magma. The emplacement of granitic magma into the middle crust, in turn, increased the overall heat budget, leading to low-pressure, medium temperature regional metamorphic andalusite and cordierite assemblage (Roy et al. 2002b).

(ii) **Betul Supracrustal Belt (BSB)**: With Mahakoshal belt in the north and Sausar supracrustal belt in the south, the Betul supracrustal belt (BSB) is made up of ENE–WSW trending volcano-sedimentary rocks, mafic–ultramafic rocks and granites along with quartzite, garnetiferous schist, calc-silicates, BIF and bimodal volcanics. The presence of graded bedding and the nature of cross bedding indicates shallow marine environment of deposition. Metabasalts show pillow structures compatible with the bimodal composition of the lava. This character distinguishes it from the unimodal mafic volcanics of Mahakoshal from Sausar supracrustal belt and collision-related high-pressure granulite belt (Bhowmik et al. 1999, 2000).

The BSB exhibits metamorphic grade varying from lower- to middle amphibolite facies evident from the assemblages like garnet-biotite-staurolite-andalusite calcite/dolomite-antigorite-talc-tremolite epidote-diopside, quartz-grunerite-magnetite-actinolite-hornblende-plagioclase and quartz-K-feldspar-plagioclase-muscovite In the northern part, an ENE–WSW trending mafic–ultramafic belt occur as discreet sheets or plutons.

The mafic volcanic rocks of the bimodal suite are low-K tholeiites and contain geochemical signatures of arc magmatism. Copious granitic magmatism is also evident. The syn- to post-tectonic mafic–ultramafic rocks are interpreted to have been generated from an enriched mantle source, which is different from the mantle source for low-K tholeiites. Such a variable mantle source characters are also common in the arc environment. The lower age of volcano-sedimentary deposition in the Betul belt is not known. However, the ca. 1.5 Ga age of syn-tectonic granite constrains the upper age of the belt. Thus, the Betul belt is considered coeval with the Mahakoshal supracrustal belt.

(iii) **Sausar Supracrustal Belt (SSB)**: As a part of the larger CITZ approximately 300 km in length and 70 km in width, the Sausar Group of polymetamorphic sediments and manganese-bearing ores belt were once considered to be the oldest formations in central India. Roy et al. (2000) noted that the Sausar Belt was bounded on the north and south by granulite belts of different ages. On the southern margin of the CITZ Sausar supracrustals extend along a WNW–ESE to ENE–WSW belt, also as the Sausar Fold Belt (SFB) and is distinctly devoid of volcanic components. It is made up of arenite-carbonate-pelite-Mn-rich lithologies, resting over the gneisses, migmatites and granites (Narayanaswamy et al. 1963). Although these gneisses known as the Tirodi Biotite Gneiss (TBG), they clearly have the basement status relative to the Sausar metasediments (Chattopadhyay et al. 2015a, b). Mohanty (2010) considered these gneisses to have evolved through migmatization of the Sausar sediments and, therefore, does not qualify to be basement for the Sausar supracrustals. On the southern margin, Narayanaswami et al. (1963) also identified polymictic conglomerate containing pebbles of gneisses and granites with the deposition in a shallow marine, shelf environment. The lithofacies distribution in the belt indicate progressive deepening towards north, with source rocks lying in the southern parts of the belt (Chattopadhyay et al. 2001).

The Sausar supracrustal rocks have undergone polyphase deformation encompassing first phase of deformation of low angle thrusting, which led to the tectonic interleaving of basement and supracrustal rocks (Bhowmik et al. 1999; Chattopadhyay et al. 2001; Bandyopadhyay et al. 1995; 2001). It also resulted in E–W trending small-scale tight to isoclinal, recumbent to reclined folds and a mylonitic foliation (Roy et al. 2001a, b). The second phase of folding produced upright to steeply inclined plane non-cylindrical folds. The basement gneisses were co-folded with the supracrustal rocks. The deformation in the Sausar supracrustal belt is akin to thick-skinned fold-and-thrust mechanism, with hinterland towards north. The basement–cover relationships in the Sausar belt are mostly obliterated due to widespread tectonism. Four phases of deformation–an early southward thrusting phase and associated recumbent/reclined folds (D_1/F_1), followed by E–W trending upright/steeply inclined folding (F_2) of the thrust allochthon ('Deolapar Nappe') characterize the deformation of the Sausar Group. Minor F_3 folds, locally distorting the F_2 hinges, and late F_4 cross-folds are also observed (Chattopadhyay et al. 2003a, b). The Sausar Group of rocks have undergone Barrovian-type regional metamorphism, varying in grade from greenschist to amphibolite grades (T_{Max} ~ 675 °C at P ~ 7 kbar), exhibiting clockwise metamorphic P-T evolutionary path in a continental collisional set-up (Narayanaswamy et al. 1963; Bhowmick et al. 1999, 2012).

Study of tectonothermal events in the Sausar Fold belt has been in focus in the recent times. Chattopadhaya (2020) stated that late Mesoproterozoic to early Neoproterozoic (ca. 1063–993 Ma) ages can be assigned to the prograde and retrograde metamorphism in this belt. Metasediments of the Sausar Group and metapelitic granulites of the Ramakona–Katangi Granulite (RKG) Belt yielded ca. 1043 and 955 Ma EPMA monazite age peaks, while Bhowmick et al. (2012) obtained 938 Ma SHRIMP U–Pb zircon age of magmatic charnockite from the RKG domain. The EMP monazite ages of ca. 945 Ma from a syntectonic (syn-D_2) granite and ca. 928 Ma for a post-tectonic granites intrusive into the Sausar Group have been reported by Chattopadhaya et al. (2015). A little younger tectono-thermal history of the SSB than the RKG has been interpreted by these authors. The available radiometric dates confirm that the Sausar supracrustal rocks were deposited on a gneissic basement between ca. 1.5 and ca. 1.1 Ga. Post-kinematic granite yielded a Rb–Sr age 960 Ma which corroborates 950 Ma Ar–Ar age of cryptomelane in the Mn-formation (Lippolt and Hautmann 1994). Available radiometric data indicate that the Sausar Orogeny (ca. 1.1–0.9 Ga) may be correlatable with the global Grenvillian Orogeny.

In a recent review, Chattopadhayay (2020) has discussed some recent controversies which have been generated in the Sausar basin stratigraphy. A glaciogenic origin of the Sausar rocks has been proposed by Mohanty et al. (2015) who considered irregular rock fragments in the dolomitic marble to be diamictite and gave a 'dropstone' status to this lithological assemblage. The carbonate rocks associated with these "dropstone" like rocks are compared with the cap-carbonates commonly associated with glaciogenic sedimentary sequences. Highly negative $\delta^{13}C$ values of these dolomites are similar to many Proterozoic cold water carbonates and often mark the paleoenvironmental shifts at some stratigraphic boundaries. After suggesting a glacial origin of the Sausar rocks, Mohanty et al. (2015) proposed an early Paleoproterozoic to Mesoproterozoic deformation and metamorphism during in the time period of 1900–1800 Ma and 1500–1400 Ma. Chattopadhyay (2015) has noted that most of these radically new interpretations conflict with the primary outcrop level data. Radiometric age data for the Sausar Group of rocks and the associated granulites does not support any such interpretation (Chattopadhyay 2015; Bhowmik 2019).

The contact of the Sausar Group with the Tirodi Gneiss exposes some paleosols, according to Mohanty and Nanda (2016). The effects of pedogenesis and widespread weathering was interpreted on the strength of the nature of Ce and Eu anomalies, low ΣREE, and high $(La/Yb)_N$ ratios. This anomalous elemental behavior suggested deposition in a reducing sedimentary basin and is therefore believed to be linked to the oxygen-deficient Archaean-Paleoproterozoic transitional basins. Likewise, Sarangi et al. (2017) inferred anoxic depositional conditions of the Sausar basin and related it to the Paleoproterozoic interglacials using REE characteristics of the carbonates and enriched Fe, Mn, Zn and U values as proxy elements.

Yedekar et al. (1990) interpreted that the Sausar supracrustals were deposited on the passive margin of the Bundelkhand Craton, which implies provenance to the north of the basin. This contrasts with the presence of prominent conglomerate horizon along the southern margin and progressive deepening of the basin to the north. That would require a provenance to the south. Such an inference would also imply deposition of the Sausar supracrustal rocks on the Bastar Craton.

4.3.2.2 Granulite Belts in the CITZ

(i) **Makrohar Granulite Belt (MG)**: The Makrohar Granulite Belt occurs on the southern fringe of the Mahakoshal supracrustal belt where calc silicates, marble, BIF, metapelites, basic rocks, granites and intrusive gabbro-anorthosite represent this rock suite along with small rock bodies, metamorphosed in amphibolite and granulite facies (Pichai Muthu 1990). Dating of similar granitic rocks from the adjoining area yielded Rb–Sr ages in the range of ca. 1.7–1.5 Ga. A suite of dismembered granulite facies rocks described from the northern part of the Betul supracrustal belt are represented by BIF-calc silicate-marble along with intrusive mafic–ultramafic rocks and granites The 1.5 Ga syn-tectonic granite of the Betul supracrustal belt may constrain the upper age of granulite facies metamorphism.

(ii) **Ramakona–Katangi Granulite Belt (RKG)**: North of Sausar supracrustal belt occurs the Ramakona–Katangi Granulite (RKG) Belt with felsic migmatitic gneisses, mafic granulites, cordierite gneisses and garnetiferous metadolerite (Bhowmik et al. 1999). Ductile shear zones mark the contact between the granulites and Sausar supracrustal belt. The RKG belt is confined to the ENE–WSW trending shear zone. DSS profile shows a steep northerly dipping fault coinciding with the RKG belt (Mishra et al. 2000).

At least three stages of metamorphic evolution, marked by distinct mineralogical assemblages and P-T stability fields have been documented from this granulite belt (Bhowmik et al. 1999, 2000). These granulites and gneisses form the basement for the Sausar Group of rocks. Similarly, four phases of deformation and three phases of metamorphism have been established from metabasics and granulites of the RKG domain (Bhowmik and Roy 2003). The peak P-T condition reached ~9 to 10 kbar and ~750 to 800 °C, followed by isothermal decompression to ~7 kbar/775 °C and a later near-isobaric cooling at 6 kbar/650 °C, recording a clockwise P-T path. A similar P-T path was established from pelitic granulites (Bhowmik and Spiering 2004), indicating an overall collisional setting for the RKG granulites (Bhowmik 2019 and references therein). The Tirodi Biotite gneiss, believed to be product of peak metamorphism, yielded a Rb–Sr age of 1525 $\pm$ 70 Ma (Sarkar et al. 1986).

(iii) **Balaghat–Bhandara Granulite Belt (BBG)**: The southern margin of the CITZ is bordered by the ENE–WSW to NE–SW trending Balaghat–Bhandara Granulite (BBG) It consists of highly tectonized granulite facies mafic granulite, charnockite-enderbite, meta-ultramfites, cordierite granulite and quartzite, exposed as detached lenses along the CIS (Ramachandra and Roy 2001a, b; Bhowmick et al. 2005, 2014; Alam et al. 2017). The BBG belt is confined between two northerly dipping shear zones and shows polyphase folding (Ramachandra and Roy 2001a, b).

Two phases of granulite facies mineral assemblages are recorded from this belt. Melt generation is indicated during the first phase of metamorphism, resulted in development of migmatitic gneiss; mineral assemblages record P-T of 7 kbar and 900 °C (Bhowmik et al. 2000). During the second phase coronal and symplectitic garnets were developed and the P-T estimates reveal 8 kbar and 700 °C (Ramachandra 1999). Subsequently, the BBG belt is superimposed by amphibole facies metamorphism produced during the Sausar Orogeny. Based on this superimposition, the BBG belt is the basement for the Sausar Group of rocks (Ramachandra and Roy 2001a, b). Lower crustal, ultra-high-temperature (UHT) granulite facies metamorphic conditions at ca. 1.6 Ga in the central part of the BBG belt and medium temperature granulite facies metamorphism in the east and northeast were inferred from SHRIMP U–Pb zircon and EMPA monazite dating (Bhowmik 2019). In short, the UHT domain in the BBG belt is polyphase metamorphosed with three distinct cycles of metamorphism between 1.6 and 1.54 Ga. Application of a recently-developed tool, based on diffusion kinetics (Bhowmik and Chakraborty 2017) to the ultra-hot orogenic domain (i.e. BBG belt) has helped in reconstructing the pulses of colliding plates. Episodic mantle-scale thermal perturbations at 1.6–1.54 Ga is, therefore, a characteristic feature of the BBG belt, and distinguishes it from other metamorphic belts of the CITZ (Bhowmik et al. 2014; Bhowmik and Chakraborty 2017).

Yedekar et al. (1990) and Jain et al. (1991) interpreted the BBG belt to represents exhumed oceanic crust of the BC, which underwent granulite facies metamorphism during the collisional orogeny. The BBG belt therefore marks the suture zone between the Bundelkhand and the Bastar Cratons. On the other hand, Ramachandra and Roy (2001a, b) suggested an intra-continental tectonic setting based on the geochemical characters (Sarkar et al. 1993) of the mafic granulite protoliths. The lithological characters, metamorphism and tectonic setting of the BBG is very similar to the Archean granulite belts in the Bastar Craton (Ramachandra 1999). In other words, granulites of the BBG belt formed during Archaean period, were exhumed within the CITZ during Sausar Orogeny. The ~2.2 Ga old Rb–Sr age of the Amagaon Gneiss (Sarkar et al. 1981) provide a crude evidence. Several shear zones in the CITZ owe their origin to different deformational events which facilitated emplacement of granitic magma (Roy et al. 2002a; Roy and Hanuma Prasad 2001b).

4.3.3 Son–Narmada South Fault (SNSF)

Southern margin of Mahakoshal Supracrustal Belt is marked by a 700 km long E–W to ENE–WSW trending fault following the course of Narmada and Son Rivers in central India. This Paleoproterozoic shear zone dips to the south at high angle and is intimately related to the opening of the Mahakoshal Basin. Several phases of reactivations led to the emplacement of granitic plutons along the shear zone. The granitic rocks in some parts of this shear zone reflected both pre-full crystallization (PFC) and crystal-plastic deformational fabrics (CP) (Roy et al. 2002a). They were superimposed by mylonitic fabric, which trends E–W to ENE–WSW and dips steeply to the south. The shear deformation and syntectonic granitic magmatism appear to be coeval with the second phase of deformation in the adjoining MSB. Overall, a contractional tectonic regime is inferred for the shear zone deformation, which has deformed quartz-diorite, granodiorite, granite and adamellite (Sarkar et al. 1998). Contractional tectonic regime and associated calc-alkaline magmatism indicate a convergent margin, wherein mantle derived magmas interacted with the continental crust and resulted in wide spectrum of granitic rocks. These granitic rocks, which extends further south of SNSF, yield Rb–Sr ages in the range of ca. 1.8–ca. 1.5 Ga and represents the oldest crust forming event in the CITZ (Sarkar et al. 1998).

4.3.4 Gavilgarh-Tan Shear Zone (GTSZ)

In the central part of the CITZ, the Gavilgarh-Tan Shear Zone (GTSZ) is a 900 km long NE–SW to ENE–WSW trending belt, separating the Betul Supracrustal Belt (BSB) from the southerly-located Sausar Supracrustal Belt. The GTSZ is represented by intensely mylonitized granites and gneisses (Roy and Hanuma Prasad 2001; Golani et al. 2001). Like the SNSF, the GTSZ is also characterized by large-scale ENE–WSW trending concordant granitic plutons/sheets. Detailed studies of both the PFC and CP fabrics show that the granitic magma was channeled along the GTSZ and deformed subsequent to its solidification. An ENE–WSW trending, more than 300 km long and 2–3 km wide, brittle-ductile shear/fault zone runs through the unclassified gneisses exposed between the Betul and the Sausar Supracrustal belts (Fig. 4.15; Golani et al. 2001; Chattopadhyay and Khasdeo 2011). The eastern part of this linear structure is expressed as a ductile shear zone comprising sheared Precambrian gneisses and granitoids, while the western part is mainly observed as a brittle fault zone affecting the Gondwana sandstone and Deccan Trap basalt flows, within which slivers of the sheared gneissic basement rocks are also found. Repeated reactivation of the fault/shear zone from Proterozoic to Quaternary has been established by Chattopadhyay et al. (2008) and Bhattacharjee et al. (2016).

In recent years our knowledge about the kinematics of the Gavilgar-Tan shear zone has significantly increased. Even then, temporal tectonic correlation with other units of the CITZ (e.g. Sausar/Betul belts) is still poorly understood. Recently, Chattopadhyay et al. (2017) established that granitoids have intruded the shear zone syntectonically between ca.1.2 Ga and 0.95 Ga, using U–Pb zircon and U-Th_{total}-Pb monazite ages (also Roy and Hanumana Prasad 2003). Transpression and mylonitization was restricted between 1.05 and 0.95 Ga indicating early Neoproterozoic age for the granite magmatism in the GTSZ. This interpretation gives credence to the ca 1.06–0.93 Ga age of collision event in this fold belt (Chattopadhyay et al. 2017). Rb–Sr age of 1147 $\pm$ 16 Ma and initial Sr ratio of 0.7096, indicate their derivation from the crustal sources (Pandey et al. 1998). Granitic rocks along the GTSZ belong to granodiorite–granite–alkali granite suite, wherein last two phases represent bulk of the magmatism. They are peraluminuous, calc-alkaline–alkaline in nature and akin to collision-related granitoids. A suite of charnockite-mangerite group of rocks along the GTSZ near Mandla, exhibit similar PFC and CP fabrics as that of the adjoining granitoids (Bhowmik 2000).

4.3.5 Central Indian Shear Zone (CIS)

The northern margin of the Bastar Craton is delimited by the CIS which trends NE–SW in the western part and WNW–ESE in the east. The DSS profile reflects its south dipping nature although surface rocks dip sub-vertically. In the western part, the CIS separates the northern Betul Supracrustal Belt (BSB) from the southern gneissic complex of the BC on the east. It separates the Chattisgarh Group in the south from Bilaspur–Raigarh supracrustal belt. The syn-kinematic granites, which are prominent in the eastern part of the CIS, i.e. in Bilaspur district, include diffusely banded biotite-granite (s.l.), megacryst-bearing granite and coarse-grained homophanous granite intruding adjoining Bilaspur–Sarguja belt of granitic rocks belong to granodiorite-quartz monzonite-granite suite. These are peraluminous to metaluminous and belong to both I- and S-types. These granites shown extensive mylonitization. The mylonitic foliation strikes WNW–ESE and dips steeply to NNE. In the supracrustal belt near Betul, granitic rocks yielded an age of ca. 1.5 and 0.9 Ga, whereas ca. 1.2–0.9 Ga ages are common in the granites adjacent to the Sausar and Bilaspur–Raigarh belts. The available radiometric dates, though meager, indicate a broad polarity in the granitic magmatism, with successive younger phases encountered towards south.

4.3.6 Great Indian Proterozoic Fold Belt (GIPFOB)

The CITZ mainly documents four tectonothermal events of 1.9–2.2, 1.6–1.8, 0.9–1.0 and 0.5–0.75 Ga in granulite and amphibolite grade metamorphic rocks, and are intimately associated with crustal-scale shear zones and widespread felsic plutonism, acid and bimodal volcanism (Acharyya 2003a; Acharyya and Roy 2000; Bhowmik et al. 2012; Mishra et al. 2000; Radhakrishna 1989; Radhakrishna and

Naqvi 1986; Roy and Hanuma Prasad 2003; Yedekar et al. 1990; Rekha et al. 2011; Rekha and Bhattacharya 2013). It is visualized that 'garland' of uplifted major tectonic accretionary zone, called as the Great Indian Proterozoic Fold Belt (GIPFOB) (cf., Prabhakar et al. 2014; Mukherjee et al. 2019; Jain et al. 2020), was a highland mountain belt and eroded off around ~1.00 to 0.90 Ga to provide detrital zircon to the Great Himalayan Sequence (GHS) of the Himalaya (Fig. 4.16).

The accretionary Neoproterozoic phase (~1.0 Ga) of the Proterozoic fold belts in the Central India is marked by emplacements of acid volcanic effusive and tuff deposits between 0.9 and 1.0 Ga (Bhattacharya and Dhang 2011; Bickford et al. 2011; Das et al. 2009; Gopalan et al. 2013; Malone et al. 2008; Mukherjee et al. 2012; Patranabis-Deb et al. 2007), which is the closing period of the Upper Vindhyan, the Chhattisgarh/Khariar and the Indravati basins. Bickford et al. (2011) linked this acid volcanic episode to the collision between East Antarctica and India and suturing of North and South Indian cratonic blocks along the CITZ during the Rodinia assembly. Greenville-age granitoids, Neoproterozoic (1.0–0.90 Ga) prograde amphibolite-facies metamorphics and older anatectic gneisses in the southern part of this belt

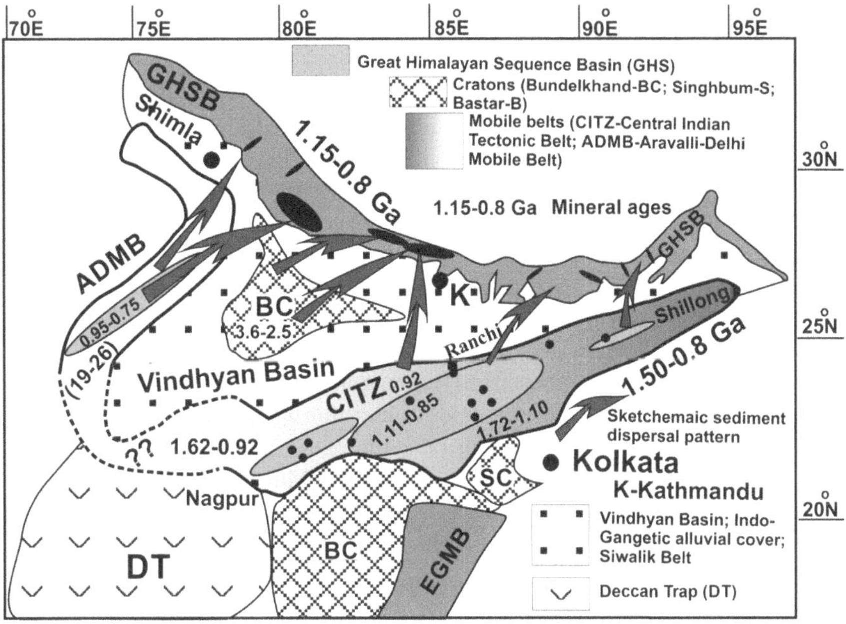

Fig. 4.16 Great Indian Proterozoic Fold Belt (GIPFOB) (cf. Prabhakar et al. 2014) of the Aravalli Delhi Fold Belt (ADFB) and the Central Indian Tectonic Zone (CITZ) as a postulated highland for sources of the Neoproterozoic detrital zircons in the Great Himalayan Sequence (GHS). Collision between South and North Indian Blocks caused uplift of this mountain, which was rapidly eroded off to provide detritus to the GHS basin from Himachal to Arunachal. Black dots: Approximate areas of mineral ages; ellipses indicate sections of the GHS with Neoproterozoic zircons; arrows indicate hypothetical sediment dispersal. Modified after Mukherjee et al. (2019) and Jain et al. (2020)

(Maji et al. 2008) appear to be linked to the Chotanagpur Granite Gneissic Complex (CGGC). The association of these Neoproterozoic granitoids and granites with high-grade Grenvillian metamorphic terranes in northeast and north–central CGC was noted by Chatterjee and Ghose (2011). This convergence with northward oceanic subduction appears to have led to the emplacement of such plutons (Rekha et al. 2011).

All these granite gneissic terranes with exposures of granitic bodies falls within the CITZ. Singh and Krishna (2009) obtained an Rb/Sr isochron age of 1005 ± 51 Ma for the grey granite and 815 ± 47 Ma from the younger pink granite. The CGGC yielded a range of Rb/Sr whole-rock ages, obtained from the granites of Jajawal (1100 ± 20 Ma), Ekma (1025 ± 11 Ma) in Gangpur Basin, Gumla (1048 ± 138 Ma), Marne (1065 ± 74 Ma) and, alkali syenite (1059 ± 104 Ma) and diorite (1138 ± 193 Ma) of Kailashnath Gufa (Pandey et al. 1986a, b, 1995; Mallick 1993; Krishna et al. 1996). Pb isotope date of 955–960 Ma is also in records. U–Pb zircon geochronology of numerous granitoids of this region yielded $^{207}Pb/^{206}Pb$ weighed mean ages between 1150 and 950 Ma, with an older phase around ca. 1750–1600 Ma and an younger early Paleozoic phases (Yin et al. 2010). Adjoining clastic sediments yielded detrital zircon peaks ca. 1600 Ma and 1200–1100 Ma indicating local erosion and derivation of detritus from the uplifted granitoid bodies (Yin et al. 2010).

Bhowmik et al. (2012) suggested that the pre-1.0 Ga Indian landmass consisted of at least three microcontinental blocks, the North Indian block, the South Indian Block and the Marwar block were united by c. 1.0 Ga. Peak and retrograde stages of metamorphism recorded in the schists from the central domain of the Central Indian Sausar Mobile Belt at 1062 ± 13 Ma and 993 ± 19 Ma monazite ages (Bhowmik et al. 2012). The Aravalli/Delhi region is also characterized by granitic intrusions with ages of ca. 1.0–1.1 Ga (Deb et al. 2001), which therefore makes the Sausar sedimentary sequence a time equivalent of the Aravalli–Delhi mobile belt of the South-Central Rajasthan in NW India. On the other hand, Stein et al. (2004) suggested that juxtaposition of northern and southern Indian cratonic nuclei along the CITZ took place took place during the earliest latest Archean and earliest Paleoproterozoic based on Re–Os ages (2490 ± 2 Ma) of granitoids from within the Sausar Belt (Malanjkhand granitoid batholith) and Cu-Mo-Ag mineralization ages (2446–2475 Ma). These ages are almost identical to U–Pb zircon ages of 2478 ± 9 Ma and 2477 ± 10 Ma for the same unit (Panigrahi et al. 1993).

4.3.7 Plate Tectonic Models

Yedekar et al. (1990) proposed a plate tectonic model for the evolution of the CITZ, in which a southerly dipping subduction of the Bundelkhand Craton was invoked below the Bastar Craton. According to them the subduction was initiated at ca. 2.4 Ga, which led to the development of rift basins and arc-related intrusions (e.g. Dongargarh and Malanjkhand granites) in the latter. This subduction system was culminated with the continent–continent collision at ca. 1.5 Ga, during which the Sausar sediments of passive margin were migmatized. This model considers BSB to be the obducted granulitic oceanic crust, which was exhumed during collisional orogeny.

A host of later workers reject this proposition as this model does not explain the evolution of the RKG belt, the MG belt, the Betul Supracrustal Belt and the vast expanse of Mesoproterozoic (ca. 1.8–1.0 Ga) granitic magmatism lying to the north of the CIS. It is now widely accepted that the Sausar basin progressively shows younging towards north and probably received bulk of the sediments from the Bastar Craton. The structural vergence in the Sausar Group of rocks is consistent with the model of thick-skinned fold-and-thrust belt with hinterland towards north.

Roy and Prasad (2003) proposed a new model invoking a north-directed subduction of the oceanic crust of the Bastar Craton below the Bundelkhand. Although the timing of the initiation of the subduction system is not yet constrained, it is assumed that the Mahakoshal rift basin (ca. 2.2) is a back-arc rift, related to this subduction. Thus, it is likely that the subduction system might have initiated as early as ca. 2.2 Ga. It was followed by closure of the Mahakoshal basin at ca. 1.8 Ga and accompanied by calc-alkaline granitic magmatism channeled through reverse-slip ductile shear zones in a contractional tectonic regime. Calc-alkaline magmatism, contractional tectonic regime and low-pressure metamorphism are akin to continental margin-arc setting. This magmatism continued up to ca. 1.5 Ga. The Betul Supracrustal Belt, which lies to the south of the Mahakoshal belt, developed as an intra-arc belt, filled with sediments and bimodal volcanic. Since, this basin was situated in an arc environment, it may be considered as coeval with the Mahakoshal basin of the back-arc. The basin closed at ca.1.5 Ga, as recorded by the syn-tectonic granitic rocks. This event was also accompanied by large-scale mantle melting, which resulted in copious hydrous ultramafic–mafic magmatism. The active subduction system is culminated at ca. 1.5 Ga with the continent–continent collision. This event is marked by the formation of collision related RKG belt, which marks the suture between the two cratons. The northerly dipping structural grain along the RKG belt also supports the south-directed thrusting during the collision. The amalgamated BC and BKC formed the basement for the Sausar supracrustal rocks of psammite–pelite–carbonate association, deposited in a continental shelf setting. The Sausar basin was closed perhaps at ca. 1.1 Ga, due to continued south-directed thrusting. The closing phase was also marked by the emplacement of profuse granitic rocks. The CIS, along which the BSB belt was exhumed, may have developed during or just before the Sausar deformation.

Based on deep seismic reflection and magnetotelluric imaging data acquired across the CITZ, Mall et al. (2008) observed reflectivity characteristics dipping towards each other and creating a domal structure across the diffused Central Indian Suture (CIS), while deepest set of reflections varies between 41 and 46 km and is imaged sporadically across the profile with the largest amplitude occurring in the northwest (Fig. 4.17). These data were interpreted as recording the presence of a mid-Proterozoic collision between two micro-continents, with the Satpura Mobile Belt being thrust over the Bastar craton (Fig. 4.18). Likewise, Azeez et al. (2017) interpreted the magnetotelluric (MT) data across the Central Indian Tectonic Zone (CITZ) as a suture along the Tan Shear Zone, which coincides with a north-dipping

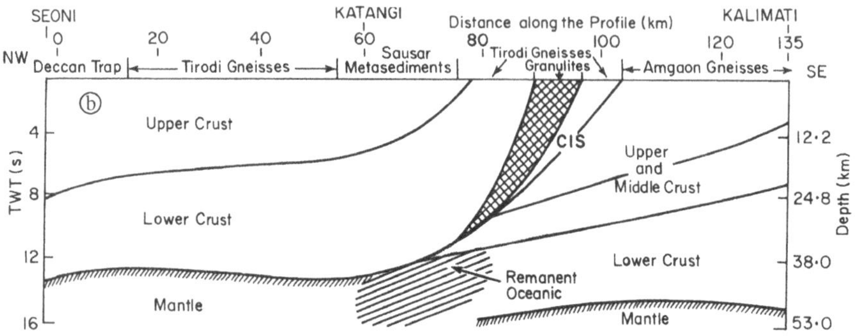

Fig. 4.17 Crustal seismic section across the CITZ and elements of the proposed tectonic model, showing main features of the crust, their relationship within the tectonic zone, collision of the two micro-continents and their suturing. After Mall et al. (2008)

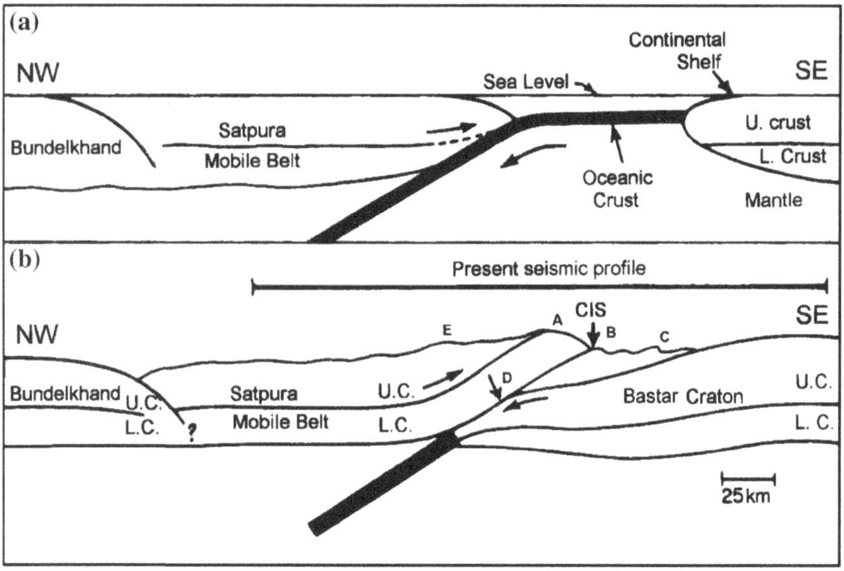

Fig. 4.18 Cartoons of tectonic interpretation of the seismic data. **a** Neoproterozoic collision between two micro-continents the present day Satpura Mobile Belt and the Bastar craton. **b** Postulated geometry produced by collision and suturing of the Satpura Mobile Belt and the Bastar Craton. After Mall et al. (2008)

conductive horizon from the upper crust to the Moho extends further into the upper mantle as a subvertical moderately conductive zone.

4.4 Singhbhum Mobile Belt (SMB)

4.4.1 Location

The Singhbhum Mobile Belt (SMB) is a geological term specific to the ensemble of folded, low to medium grade meta-sedimentary and meta-igneous rocks (1.0–2.4 Ga), sandwiched between the Archaean (>2.4 Ga) Singhbhum Craton in the south, and the Meso/Neo-Proterozoic (0.9–1.7 Ga) Chotanagpur Gneissic Complex in the north with north-dipping ductile shear zones (Fig. 4.19). In the south, the SMB and the Singhbhum Craton is separated by the Singhbhum Shear Zone (Copper Belt Thrust–Mahato et al. 2008). Based on stratigraphic correlation, sedimentological information, data on magma genesis, and chronologic information within the rock units flanking the SMB, several tectonic models have been proposed for this mobile belt; e.g. intraplate subduction (Sarkar and Saha 1983), microcontinental subduction, back-arc marginal setting (Bose and Chakrabarty 1981; Bose et al. 1989), intra-cratonic extension, rifting and ensialic orogenesis (Gupta et al. 1980a, b; Mukhopadhyay 1984; Sarkar et al. 1992).

In this chapter we have preferred the term 'Singhbhum Mobile Belt (SMB)' instead of using either the North Singhbhum Mobile Belt or North Singhbhum Orogen (NSO) since the southern component of this belt does not exist.

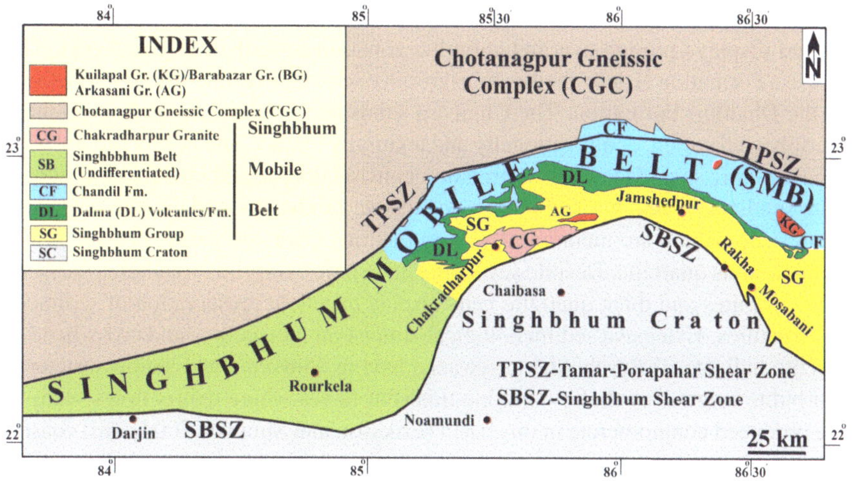

Fig. 4.19 Regional geological map of the Singhbhum Mobile Belt between the Singhbhum Craton and the Chhotanagpur Gneissic Complex. Redrawn after Saha (1999)

4.4.2 Gangpur and Kunjar Groups

A thick sequence of intricately folded and refolded metamorphosed argillite (locally carbonaceous), dolomite and limestone, profusely intruded by granite and basic rocks is found in an independent Gangpur Basin. Mica schist and manganese rich rocks (gondite) are found in this belt. These rocks are correlated with parts of Kolhan (Bhattacharya and Chatterjee 1964; Sahoo and Das 2015) and Koira Groups. Likewise, the Kunjar Group represents the Gangpur Group in the north and Singhbhum Group and Koira Group in other areas. A model Pb age of ~1.66 Ga (Viswakarma and Ulabhaje 1991), determined from the associated ore deposits (galena), has been interpreted as that of the syngenetic evolution of the ore and broadly representative of the age of Gangpur sedimentation (Acharyya 2003b).

4.4.3 Ghatsila Belt

A thick succession of meta-argillite, quartzite and effusive basic volcanics named as the Singhbhum Group, which is divided into a lower Chaibasa Formation (~2000 to 4000 m thick) and an upper Dhalbhum Formation (~400 m) (Sarkar and Saha 1983). Mazumdar (2005) favors a two-fold lithostratigraphic subdivision of the entire Dhalbhum–Dalma volcano-sedimentary assemblage with a lower Dhalbhum Formation and an upper Dalma Formation. Mica schist is the main rock type in the lower part and displays progressive metamorphic zonation towards the central part. The Chaibasa Formation is described as transgressive sequence resting over the basement and the Dhanjori Formation. The Chaibasa sandstone is very fine- to fine-grained, generally well sorted, compositionally and texturally mature, and is interbedded with the heterolithic shale facies. This formation consists of repeated intercalation of precursor sandstones, fine-grained sandstone/siltsone-mudstones and mudstone on dcm to mm scale, which are metamorphosed to lustrous mica and garnet-bearing schist and micaceous quartzite. Despite severe metamorphic overprints weathered surfaces of the laminites and thick quartzite beds display excellent preservation of sedimentary structures. Extensive sedimenological studies on metapelites and cross-bedded quartzite indicate (i) flyschoidal graywacke having convolute and slump structures of turbidite origin, (ii) shallow marine tidal-flat facies where debris flows seem to have produced conglomerate in this basin (Eriksson and Simpson 2004), (iii) coastal marine to fluviatile in origin.

The overlying Dhalbhum Formation is magnetite-bearing meta-argillites, quartzites and interstratified basic intrusives. These rocks display well preserved fluvial and aeolian structures in thick beds of fine and mature quartzites, outcropping in lenses of hundreds of meters in length and several tens of meters thickness, and intercalated with micaceous schists. These sediments are inferred to have formed in the tidal flat domain with telltale evidence of aeolian and intermittent fluvial episodes.

Table 4.5 Generalized comparative stratigraphy of Singhbhum Mobile Belt

Dunn and Dey (1942)		Sarkar and Saha (1983)		Gupta and Basu (2000)	
Iron ore series	Iron ore stage	Dalma Lavas		Dalma group	Up. Dalma Fm. Lr. Dalma Fm.
		Singhbhum Group	Dalbhum Fm. (Phyllite, quartzite, epidiorites sills) Chaibasa Fm. (Garnet-staurolite kyanite mica schist, hornblende schist, quartzite)	Chaibasa group	Up. Chaibasa Fm. Lr. Chaibasa Fm.

These rocks are overlain by the Dalma Formation containing volcanic rocks. A linear belt of granitic rocks is also found emplaced in the Chaibasa Formation near the Singhbhum Shear Zone. These granites with massive core and foliated rim are known as the Arkasani Granite. Rb–Sr age indicate 1000–1100 Ma time window for these granites. Another belt of low grade metapelite with rhythmic bands of chert and organic rich beds, similar to Ghatsila Formation, is identified between Chotanagpur Gneiss and Dalma Volcanics. This unit is known as Chandil Formation containing impure carbonates, carbonatite, felsic volcanics, syenite and granite. Thick beds of welded tuff with rocks of rhyolite and dacite compositions occur interbedded in this formation. The quartzite layers in the argillites preserve sedimentary features of subtidal to intertidal zones. Mazumdar (2005) contended that poor sediment sorting, lenticular geometry and unimodal cross-strata orientation in combination with compositional and textural immaturity suggest that the Chandil sandstones are fluvial deposits. Associated shales can therefore be inferred as of flood plain origin. Singh (1997, 1998) proposed an aeolian origin for the acidic tuffs associated with the Chandil sandstone. This succession also contains alkali syenite with nepheline and sodalite, mafic and ultra-mafic intrusives and metabasalts. Contact of this formation with the Chotanagpur Gneiss is sheared and is dotted with outcrops of granites, ranging in ages from 2100 to 2300 Ma (Rb–Sr age0 and Pb–Pb age of ~3100 Ma: Pb–Pb. In the east, around Ghatsila low pressure metamorphics are seen.

Various stratigraphic schemes for the Singhbhum Mobile Belt are compared in Table 4.5.

4.4.4 *Metasediments of Dhanjori Belt*

The Dhanjori Formation (2.1 Ga: Roy et al. 2002c) comprises siliciclastic sedimentary rocks interlayered with ultramafic to mafic (felsic at places) volcanic and volcaniclastic rocks, deformed and metamorphosed to greenschist facies An extensive polymict conglomerate-orthoquartzite belonging to the Dhanjori Basin is positioned between the Singhbhum Craton and Singhbhum Orogen. Gupta et al. (1985) divided Dhanjori Group into (i) a lower formation of pebbly metapelite with ultramafics and (ii) an upper formation of arkose, conglomerate and metapelites and volcanics of Dalma lavas. These conglomerates are considered submarine in origin and debris flow deposited them on the granitic sea floor. The basal beds are covered by metapelites, orthoquartzite, BIF, tuffs and ultramafics. Another view is that the entire sedimentary succession formed in a fluvial to tidal flat environments in which the sandstone composition varies from arkose to feldspathic arenite to lithic arenite/wacke, depending on relative proportion and dominant species of feldspar (Mazumder 2005). Mazumder and Sarkar (2004) presented a detailed account of the facies analysis and the mode of Dhanjori sequence building. Poor sediment sorting, compositional immaturity, lenticular body geometry and section-wide unimodal palaeocurrent pattern suggest that the Dhanjori sandstones are fluvial deposits (Eriksson et al. 1998; Mazumder and Sarkar 2004; Mazumder 2005).

In the west, this succession shows gradational contact with the underlying Koira Group of rocks. Dipping towards west and south these metasediments are intensely folded along the Singhbhum Shear Zone. A less deformed Simlipal Group of rocks in the south is correlatable to rocks of the Dhanjori Group. Since these rocks overlie Singhbhum Granite as well as metabasalts and tuffs (Badampahar Group or Gorumahisani Group), correlation of the surrounding metasediments with the Singhbhum Group was also suggested (Sinha et al. 1997). The Dhanjori lavas are basaltic to andesitic of tholeiitic series with Pb–Pb and Sm–Nd isochron ages of 2820 and 2790 Ma (Ramakrishnan and Vaidyanadhan 2008).

On the eastern margin of the Singhbhum Craton, the Dhanjori Group of metasediments is associated with a set of granitoids known as the Mayurbhanj Granite (Saha 1994). These granites have been compared with the Simlipal Granite by Iyenger and Murthy (1982). Notopahar, Nilgiri and Chakdar Pahar exposes similar granites in the region, with ages varying from 2084 $\pm$ 70 Ma (Rb–Sr systematics) to 3008 $\pm$ 8 Ma and 3092 $\pm$ 5 Ma (Pb–Pb zircon dates) like the ages of the Singhbhum Granite (Mishra 2006). The Sm–Nd isotopic analyses of basic-ultrabasic rocks of the Dhanjori Basin yield an isochron age of 2072 $\pm$ 106 Ma (MSWD = 1.56) indicating their age as early Proterozoic (Roy et al. 2002c).

4.4.5 Dalma Volcanic Formation

A synclinally folded extensive volcanic sequence commonly known as Dalma Volcanics is developed within the Singhbhum Mobile Belt. The northern limb of the syncline is overturned to the south and is restricted by the Dalma Thrust. Refolding is seen in the eastern and western ends. The Dalma Formation conformably overlies the Dhalbhum Formation and is represented by a thick sequence of mafic–ultramafic volcanic rocks with lenses of basic agglomerates (Gupta et al. 1980a, b;

Sengupta et al. 2000). These volcanic rocks are grouped into (i) a lower formation which overlies the Chaibasa Formation of the Sighbhum Group and are made up of volcanics intercalated with carbonaceous meta-argillites, tuff and quartzite, and (ii) an upper, extensively distributed high Mg-basalt and komatiite. Reworked pyroclasts are commonly associated with mafic lavas flow, which are amygdaloidal in character and pillowed. All the Dalma rocks show greenschist facies metamorphism. These rocks are sometimes described as ophiolites (Ramakrishnan and Vaidyanadhan 2008). Multiple deformation and eruption events make it difficult to ascertain the age of all the volcanic episodes. It is envisaged that the onset of basaltic volcanism in the Dalma marks a rift system that lead to the development of a local basin in which argillites accumulated together with occasional coarser grained, sandy slump deposits. The commonly associated black chert is probably of hydrothermal origin as it makes dykes in schist and quartzite. At places only coarse tuffs and agglomerates with large bombs are found along with carbonaceous phyllite and carbon-rich shale and assume the name of the Chandil Formation as described above (Ray et al. 1996; cf. Acharyya 2003b). Some workers prefer to retain these argillites under the Dalma formation. Geological and geophysical investigations of Bhattacharya and Bhattacharyya (1970), however, clearly show that the metasedimentary package lying north of the Dalma volcanic belt is younger than the Dalma volcanic rocks. A Rb–Sr age of 1487 $\pm$ 34 Ma has been determined recently for the acid tuffs of the Chandil Formation (cf. Sengupta et al. 2000; Sengupta and Mukhopadhaya 2000). Almost complete lack of sedimentary structures hints deposition of the sedimentary beds below wave/storm wave base. The oval cluster of granite exposures comprising the Kuilapal Granite (cf. Ghosh 1963; Ghosh and Saha 2018; Saha 1994) exposed within the Chandil supracrustal has yielded an Rb–Sr whole rock isochron age of 1638 $\pm$ 38 Ma (Sengupta et al. 1994), which according to Acharyya (2003a), fixes the upper age limit of the Chandil volcanics. The precise age of the Dalma volcanic rocks is unknown. A Rb–Sr whole rock isochron of the gabbro–pyroxenite intrusives into the Dalma volcanic rocks yields an age of 1619 $\pm$ 38 Ma (Roy et al. 2002b). Based on very similar trace element and rare earth element characteristics of these intrusives and the Dalma mafic volcanic rocks, both are inferred to be comagmatic (Acharyya 2003b). It has been proposed that the Dalma volcanic rocks were metamorphosed and intruded by comagmatic gabbro–pyroxenite rocks around 1600 Ma (Acharyya 2003a, b).

4.4.6 Chakradharpur Granite Gneiss (CGG)

An isolated granite body is exposed within the Singhbhum Mobile Belt around Chakradharpur amidst the supracrustal SG (Fig. 4.19; Sengupta et al. 2000). This granite gneiss forms the basement to the overlying Singhbhum Group, while its pegmatoidal phase intrudes both the older gneiss and the enveloping supracrustals. Geochemically, older tonalite gneiss of this granite is considered equivalent to the SBG-I/SBG-II, whereas pegmatoidal phase is considered equivalent to younger granite bodies (Saha 1994).

4.4.7 Singhbhum Shear Zone (SSZ)

Also known as the Copper Belt Thrust Zone, this tectonically disturbed belt forms an arcuate outcrop pattern of nearly 200 km in the Singhbhum region. It consists of a series of shear planes which at times turn into high angle fault. Extreme ductile shearing and complex deformational history mark this zone, having several episodes of metasomatism, migmatization and extensive mineralization of Cu, U, tungsten, apatite. Granite mylonite and quartz-mica phyllonites are found along the ductile zones. Stratigraphically, rocks of this shear zone overlie the Dhanjori and Koira (IOG) metapelites often marked by a sheared conglomerate and, in turn, overlain by the rocks of the Chaibasa Formation. This order of succession has remained undisturbed and makes it an integral part of the mobile belt. Rocks within the shear zone are represented by argillites and volcanoclastics with abundance of mafic and ultramafic intrusives. Within this belt the Chaibasa Formation show intermingling with the Arkasani Granite. At least three sets of inclined and sinistral folds and a late sub-horizontal fold affected these lithologies. The sense of shearing indicates a uniform up-dip movement during thrusting. Multi-episodic evolution of this shear zone has been interpreted at 2200, 1800, 1600 and 1000 Ma (Mishra 2006). Contrary to the earlier interpretation of Sarkar and Saha (1962) and Saha (1994) it has been recently been established that the SSZ does not mark the interface between the Singhbhum Archaean nucleus and Proterozoic SMB (cf. Blackburn and Srivastava 1994; Gupta and Basu 2000). In the north, the Dalma volcanics and the Chandil volcaniclastics were extruded along this belt and contemporaneously metamorphosed soon after ca. 1600 Ma which is the date of this prominent metamorphism (cf. Acharyya 2003a, b).

4.4.8 Tamar Porapahar Shear Zone (TPSZ)/South Purulia Shear Zone (SPSZ)

The southern margin of SMB is marked by Singhbhum Shear Zone (SSZ), whereas the northern limit is delineated by the South Purulia Shear Zone (SPSZ) or the Tamar-Porapahar Shear Zone (TPSZ)/lineament, where vast expanse of Chhotanagpur Granite Gneiss Complex (CGGC), an extension of the CITZ, abuts against the SMB. The SPSZ is approximately 150 km long and 4–6 km wide, roughly E–W trending zone and is known for the occurrence of alkali carbonatite complex with alkali syenite, apatite-magnetite rock, carbonatite, alkali granite and phlogopite-amphibolite. A number of small granitoids such as Biramdih, Beldih and Barabazar lie on or close to the SPSZ; the latter is a small oval shaped plutonic body emplaced into felsic volcanics and metapelites of the Singhbhum Group along the southern fringe of the SPSZ (Dwivedi et al. 2011). Geochemically, the Barabazar granite is predominantly peraluminous A-type granite with abundant alkali feldspar, perthite, and minor plagioclase, biotite and accessory minerals, and formed in 'within plate granite' tectonic set up. They obtained a Pb–Pb isochron age of ca. 1771 Ma as emplacement in rift related environs due to partial melting of stabilized lower/middle crust (initial 87Sr/86Sr $= 0.7302 \pm 0.0066$). The younger whole-rock Rb–Sr isotope age of c. 971 Ma from the same body indicates closure of this basin when the Singhbhum Craton finally juxtaposed against the Northern Indian Shield along CITZ during the global Grenvillian orogeny (Dwivedi et al. 2011).

4.5 Eastern Ghats Mobile Belt (EGMB)

4.5.1 Location

Almost running parallel to the western coast of Bay of Bengal, the Eastern Ghats ranges have discontinuous low-lying and NE–SW trending mountains of Precambrian rocks. These ranges of about 1000 km length extend from Brahmani River in Orissa to around Ongole in Andhra Pradesh, with maximum width of about 300 km in the north and tapers southwards to ~30 km. The largest dissected segment of these ranges is in the Dandakaranya region between the Mahanadi and Godavari Rivers, with highest peak of Arma Konda (1680 m). Still farther southwest and beyond the Krishna River, the Eastern Ghats appear as a series of low-lying ranges. The ranges are transversely dissected by the southeasterly-flowing rivers in their lower reaches in the coastal regions.

4.5.2 Geological and Tectonic Framework

The EGMB, named after the above mountains, is predominantly a granulite belt on the eastern margin of the Archean Dharwar–Bastar–Singhbhum Cratons, having a thrust contact (Fig. 4.20; Kovach et al. 2001; Ramakrishnan and Vaidanadhan 2008; Sharma 2009a, b; Dasgupta et al. 2013; Valdiya 2015). The belt is traversed by two Gondwana-bearing NW–SE trending Mahanadi and Godavari Grabens, while the WNW–ESE trending Rengali Belt limits the EGMB in the north, where its contact with the Singhbhum craton is marked by shear zones, mainly the Sukinda Thrust. The southern limits of the EGMB are vaguely defined in the Nallamalai Hills, where

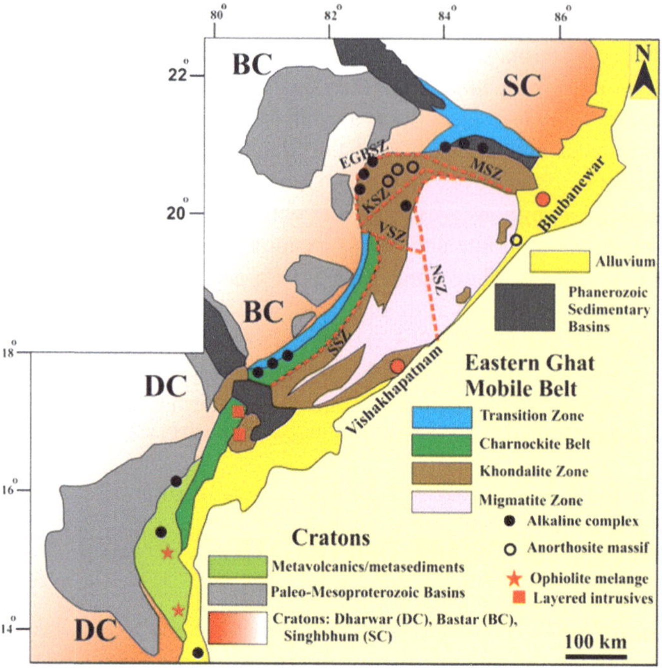

Fig. 4.20 Simplified geological map of the Eastern Ghats Mobile Belt (EGMB), showing distribution of major lithological units and important intrusive bodies. Major shear zones: Mahanadi Shear Zone (MSZ). Korapur-Sonepur Shear Zone (KSZ). Eastern Ghats Boundary Shear Zone (EGBSY). Vamshadhara Shear Zone (VSZ). Nagavalli Shear Zone (NSZ). Sileru Shear Zone (SSZ). After Ramakrishnan and Vaidanadhan (2008), Dasgupta et al. (2013), and other published sources

it possibly merges with the Southern Granulite Terrain. In order to establish its continuity further southwards, Chandrasekhar et al. (2018) observed similar deep magnetotelluric (MT) structure and subsurface high electrical resistivity to confirm its southward extension between Ongole and Nellore.

The EGMB has been extensively investigated for it structure, deformation, characteristic ultra-high temperature metamorphism, geochronology and tectonic modelling after the pioneer works by V. Ball, C. S. Middlemiss, L. L. Fermor, H. Crookshank and others. Walker (1902) classified the garnet-sillimanite-graphite gneiss as khondalite after the Khond tribe of the region. Subsequent geological investigations by the GSI finally produced a new geological map on 1:1000,000 scale (Ramakrishnan et al. 1998) with detailed record of geological setting and evolution (see Ramakrishnan and Vaidanadhan 2008). This terrane has been vital in the Rodinia reconstruction of the India and Antarctica (Upadhyay 2008; Dasgupta et al. 2013; Bose and Dasgupta 2018).

At present, different schemes of classification of the EGMB are prevalent based on its lithological, metamorphic and isotopic characters or a combination of these (see Bhattacharya 1996; Dasgupta et al. 2013 for reviews). Ramakrishnan et al.

(1998) proposed the following 4-fold longitudinal division of the EGMB based on distribution of lithogies (Fig. 4.20).

i. **Western Charnockite Zone (WCZ)** of charnockite and enderbite with lenses of mafic–ultramafics and minor metapelitic khondalite. Several generations of charnockites mark this belt; the oldest ones become deformed and garnet-bearing gneisses.
ii. **Western Khondalite Zone (WKZ)** of dominantly metapelitic khondalite, intercalated quartzite, calc-silicates, marble, and high Mg–Al granulites containing sapphirine; these are considered as supracrustals, which were deposited on the basement. Since both have suffered granulite facied metamorphism, it becomes difficult to distinguish these separately. Granulites are intruded by charnockite/enderbite and massif-type anorthosites.
iii. **Central Charnockite–Migmatite Zone (CMZ)** of migmatitic gneisses with minor amounts of high Mg–Al granulites and calc-silicates, which are intruded by charnockite–enderbite, pophyritic granitoid and massive anorthosite in the Chilka Lake region.
iv. **Eastern Khondalite zone (EKZ)**, with lithological similarity with WKZ, but lacking anorthosite. In the Chilka Lake area, leptynites are interlayered with khondalites, and represent granitoids, which were produced by dehydration melting of metapelites of different composition (Sen and Bhattacharya 1997).

Besides these important zones, Ramakrishnan et al. (1998) also recognized a Transition Zone along the western margin of the EGMB with the cratons to (Fig. 4.21a). Chetty and Murthy (1994) and Chetty (2001, 2017) recognized several major important ductile shear zones along the western margin and within the EGMB and classified it into 9 different terranes on the basis of presence of shear zones, lineation, fold patterns and their axial surfaces. Ricker et al. (2001) evolved another classification of the EGMB into Domain I to Domain IV on the basis of Nd-model ages, cutting across lithological boundaries of the EGMB (Fig. 4.21b). Dobmeier and Raith (2003) classified the EGMB into 12 crustal provinces and sub-provinces, having distinct lithology, structure, metamorphic grades and geological histories; main provinces are the Krishna Province in the extreme south, the Jeypore Province in the west, the Eastern Ghats Province in the center, and the Rengali Province in the north (Fig. 4.21c).

4.5.3 Deformation

4.5.3.1 Western Margin of the EGMB

The EGMB has been referred to as a convergent orogen which was evolved under NW–directed subhorizontal compression due to oblique collision of continental plates (Chetty and Murthy 1994; Ramakrishnan et al. 1998; Bhattacharya and Kar

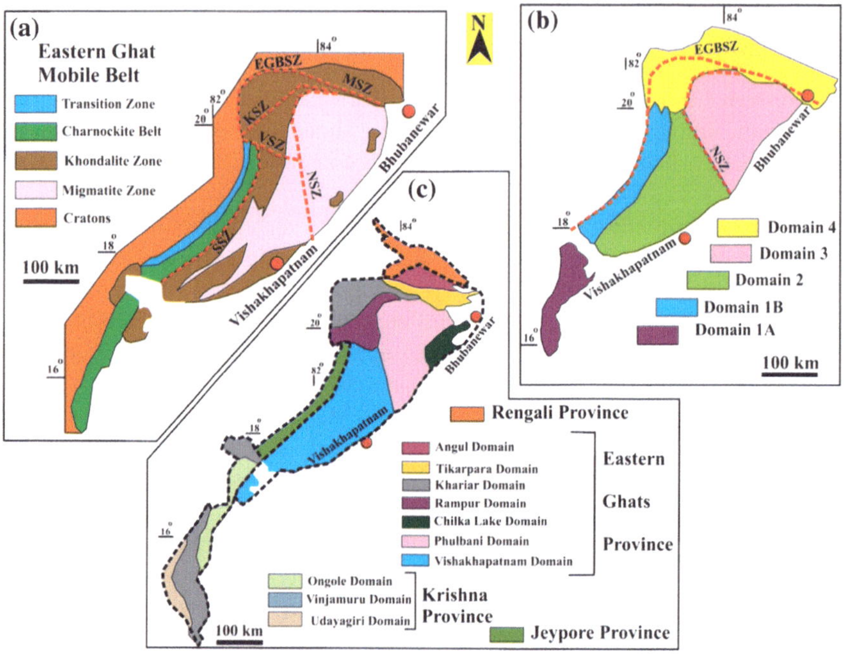

Fig. 4.21 Classification of the Eastern Ghats Mobile Belt (EGMB), based on **a** gross lithology (Ramakrishnan et al. 1998), **b** domains identified by isotopic characters–Nd-model ages (Rickers et al. 2001), **c** lithological, metamorphic and isotopic characters (Dobmeier and Raith 2003)

2002). Large-scale tectonics of the EGMB reveals west-directed thrusting of high-grade metamorphosed charnockite-khondalite-migmatite terrane over the Dharwar–Bastar craton along a ductile shear zone, possibly a result of collision from the Eastern Antarctica around Enderby Land (Fig. 4.22; Ramakrishnan and Vaidyanadhan 2008; Bhadra et al. 2004; Gupta et al. 2000). Further southwestern continuation of this belt is indicated from gravity and DSS profiles which reveal dense crust due to Moho upwarp. Western margin is remarkably marked by alkaline and carbonatite bodies which were possibly emplaced along a rift zone–an ancient suture zone and subsequently deformed during Eastern Ghats orogeny (Leelanandam et al. 2006).

Western margin of the EGMB with the Archean Craton has been called as the West Odisha Boundary Fault (Mahalik 1994), the Eastern Ghats Boundary Shear Zone/Fault (Ramakrishnan et al. 1998; Crowe et al. 2003), the Transition Zone (Ramakrishnan and Vaidyanadhan 2008), the Terrane Boundary Shear zone–TBSZ (Biswal et al. 2000; Biswal and Sinha 2003), and the Sileru Shear zone–SSZ (Chetty 1995, 2017). The Rangeli Province juxtaposes the EGMB in the north through various dextral shear zones like the Mahanadi Shear Zone (Mahapatro et al. 2009). The TBSZ is about 3–5 km wide listric thrust zone and dips moderately around 45 to 60° eastwards. It deforms both the lithologies of the Archean Craton and the EGMB, including numerous post-tectonic plutonic complexes. On the basis of unified 2-D

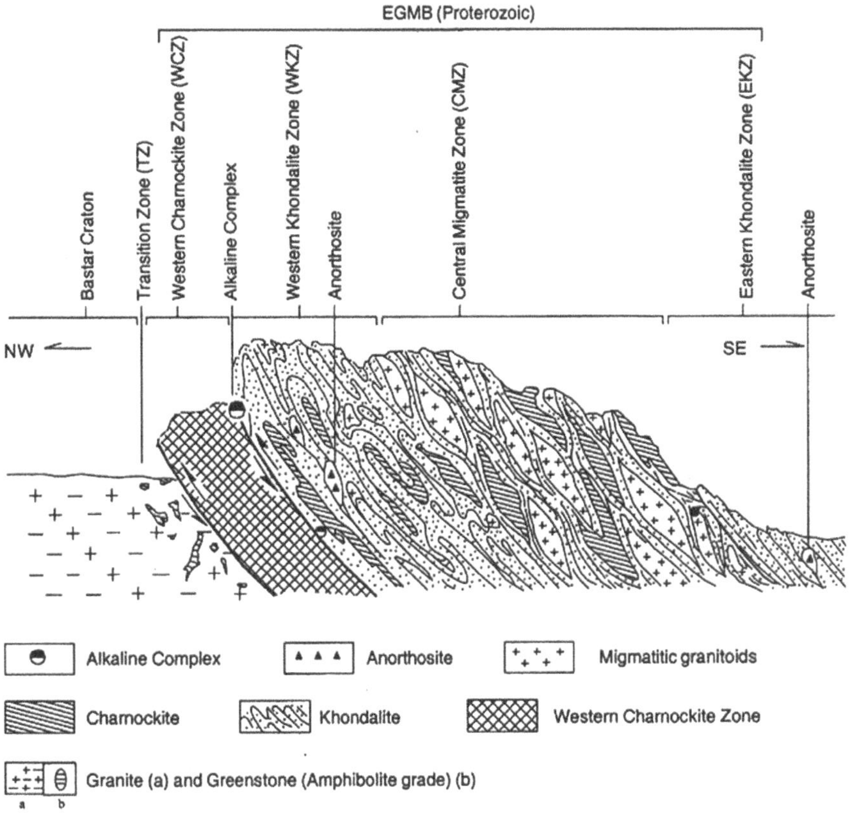

Fig. 4.22 Generalized cross-section of the EGMB and its tectonic contact with the Bastar Craton. Note intrusive alkaline plutons and anorthosites near the craton-mobile margin. After Ramakrishnan and Vaidanadhan (2008)

density modeling, Kumar et al. (2004) suggested an eastward-dipping ~38 to 40 km thick crust beneath the Indian Cratons against the 35 km thick EGMB crust with a steep gravity gradient along the margin of the two tectonic domains. Parallel gravity contours and structural grains of the EGMB along with narrowing of gravity contours along the EGMB boundary indicate that a faulted contact (TBSZ–Biswal et al. 2000) exists from the Cuddapah basin up to the Singhbhum craton, which juxtaposes crustal blocks of contrasting character (Subrahmanyam and Verma 1986).

Along its western margin, the UHT granulites and multiply-deformed migmatitic gneisses are thrust westward over hornblende granite and sedimentary rocks of the Bastar Craton, and preserve thrust-related shear fabrics (Gupta et al. 2000; Bhadra et al. 2003; 2004; Sinha et al. 2010). In the foreland craton, undeformed granites show eastwards progressive decrease in grain size, associated with low-temperature increasing penetrative foliation; it has produced orthogneiss near mylonitized contact zone with charnockite. Westward thrusting of migmatitic gneiss of the EGMB has juxtaposed these units against rather cold banded gneiss–granite of the Bastar Craton during later deformation history (Gupta et al. 2000).

Within the TBSZ, intense ductile shearing has imbricated different litho-units of the Indian Craton and the EGMB as 'mélange' within the TBSZ, whose structural geometry reveals its top-to-W/NW overthrust geometry (Mahalik 1994; Chetty 1995; Biswal et al. 2000, 2002; Bhadra et al. 2004; Sinha et al. 2010). Along the northwestern margin, contact between the Indian Craton (Bastar Craton) and the EGMB is retrogressed within the Lakhna Shear Zone with development of quartzo-feldspathic mylonite from the cratonic protoliths and a narrow-retrogressed zone from the charnockite gneisses (Biswal et al. 2000). This zone is marked by S-C fabric, asymmetric winged porphyroclasts, quartz ribbons and intragranular faults; the shear zone reveals 2.5–4.7 km NW-directed thrusting over the craton under simple shear. Northwesterly verging thrust nappes mark the western arcuate margin as salient part of a fold thrust belt, windows and klippen along with lateral ramps (Biswal et al. 2002). This margin also contains many synkinematically emplaced alkaline intrusives, constraining the age of thrusting ~1.4 Ga (Sarkar and Paul 1998). However, the thrusting might have continued up to the Neoproterozoic time when the EGMB was amalgamated with the Bastar and Dharwar cratons (Dobmeir and Raith 2003; Upadhyay et al. 2006a).

In low-temperature mylonites of the TBSZ, Sinha et al. (2010) observed distinct top-to-NW ductile shear indicators, including (i) contrasting ductile behaviour of quartz and brittle fracturing of feldspars, (ii) S-C shear fabric with C-planes paralleling the main shear zone, (iii) mantled microcline porphyroclast showing NW verging shearing, and (iv) microfractured feldspars along conjugate set out of which the main fracture trends parallel to S-fabric antithetically (Fig. 4.23b–d).

In the northwestern margin of the EGMB, garnet-sillimanite-gneisses are folded within calc-gneisses by second generation large-scale F2 folds around Lathore, Balangir district and bear imprints of polyphase deformation and ductile shearing (Biswal et al. 1998a, b). These folds are coaxial to the oldest F1 isoclinal and intrafolial folds, which possess penetrative axial planar foliation/gneissosity S1. The F2 folds are open to tight with axial plane foliations, striking NE–SW and dipping moderately to SE with variable plunging hinges. Fold interference of F1 and F2 produces Type 3 interference pattern, though it also sometimes develops sheath folds. Map pattern around Lothal reveals a large-scale southwesterly plunging overturned F2 fold. Third generation F3 folds are mostly gentle upright folds having NW–SE striking axial plane foliations, whose superposition on the F3 folds resulted in the domes and basins structures.

4.5.3.2 Deformation Within the EGMB

Detailed deformation patterns of the whole EGMB are investigated in limited areas like Angul in the north and Ongole in south. In Angul region, Halden et al. (1982) recognized four distinct deformation phases D1 to D4. Metamorphic differentiation banding S1 (D1 structure) is axial planar to F1 folds and cuts the S0 in fold hinges.

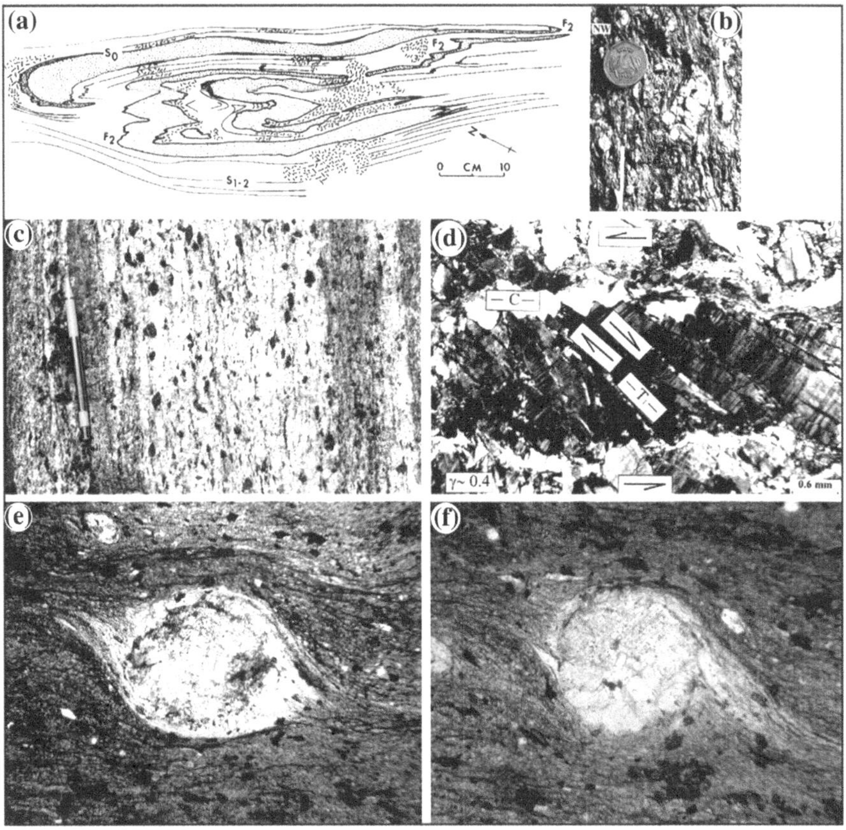

Fig. 4.23 Deformation within the EGMB and its margin within the Terrane Boundary Shear Zone (TBSZ). **a** Attenuated mafic granulite layer(s) within sillimanite-garnet gneiss, folded and boudinaged in D2 with quartz-feldspar-biotite neosome in boudin necks. **b** Sigmoidal feldspar porphyroclasts in granite mylonite. **c** Stretching lineation in the nepheline syenite emplaced along the same shear zone, Odisha. **d** Faults representing book-shelf gliding in feldspar porphyroclasts of granite mylonite. Photo rotated anti-clockwise. **e–f** σ- and δ-types porhyroclasts in mylonitized porhyroclasts in granite mylonites. All photographs are from the Terrane Boundary Shear Zone (TBSZ), Eastern Ghats Mobile Belt, Odisha. Courtesy Prof. TK Biswal, IIT Mumbai

Therefore, lithological layering S0 containing alternating basic granulites, khondalites, quartzo-feldspathic gneisses, representing the main metamorphic episode, are the earliest in the sequence of events. F2 isoclinal folds, their axial planar foliation S2 and fold hinges F2, including L2 rods (D2 phase), deform the earliest recognizable planar and linear structures. In limbs of the F2 folds, S2 forms a composite fabric with S1. Migmatization and emplacement of syntectonic granites occurred during the D2 event, where mineral segregation and compositional banding of migmatitic quartzo-feldspathic gneisses, metapelites, calc-granulites, mafic granulites and even Opx-bearing granitoids are taken as S1. D3 deformation produces widely spaced shears S3 and axial surfaces to large open to tight folds, which control disposition of rock units in the Angul district (Fig. 4.23a; Halden et al. 1982). Another phase of charnockitization is noticed within cores of F3 folds, while vertical S3 axial planar fabric within gneisses acts as conduits for quartzo-feldspathic neosomes, leading to the charnockitization of the pre-existing rock.

A four-phase deformation history is recorded in the structural architecture of the EGMB. The first three phases were dominated by folding, and the last one by shearing and fracturing. The NE–SW to N–S regional trend is defined by the first- and second-generation folds (Fig. 4.23a). The second-phase folding was coaxial with the first, and the fold interference produced hook-shaped structures in the khondalites. The third episode of deformation produced asymmetric folds that gently plunge northeastwards in metapelites and upright disharmonic structures in migmatites (Bhattacharya et al. 1994). Pieces of charnockite near the hinge of these third-generation folds represent discontinuous relicts of larger layers that were torn apart and displaced. The generally southeastward- to eastward-dipping rock successions all through their expanse are cut by several ductile and brittle shear zones developed along the early-generation folds. The shear zones are associated with mylonites and pseudotachylites (Chetty and Murthy 1994; Biswal et al. 1998a, b, 2000; Chetty et al. 2003). In some parts of the terrane, dome-and-basin structure of varying scale formed due to the third phase of folding (Natarajan and Nanda 1981).

4.5.4 Metamorphism

The EGMB is essentially a granulite–gneissic terrain, having the following rock types, some of which are distributed into distinct belts: (i) charnockite–enderbite (orthopyroxene–quartz–feldspar ± garnet), (ii) khondalite (garnet–sillimanite–perthite–quartz) and metapelitic granulites containing quartzite bands and Mg–Al granulites (sapphirine–spinel–cordierite–orthopyroxene–sillimanite–garnet ± quartz) lenses. Inclusions of these granulites are seen in mafic granulites and associated with garnet leptynites, (iii) leptynite (plagioclase–quartz– perthite ± garnet), (iv) calc-granulites (wollastonite–scapolite–calcite–plagioclase–garnet–clinopyroxene) as bands and lenses within metapelites, (v) mafic granulite (orthopyroxene–clinopyroxene–plagioclase–garnet), (vi) quartzo-feldspathic gneiss (quartz–feldspar–biotite ± garnet), and (vii) augen gneiss.

Various types of P-T paths for the ultrahigh temperature (UHT) metamorphosed Mg–Al granulites and other lithologies have been inferred from the EGMB from northern parts to extreme south, and are categorized as follows:

(i) the UHT thermal peak > 1000 °C at 10 kbar and subsequent high-P isobaric cooling (IBC), having clockwise or anti-clockwise path (Bhattacharya and Kar 2002; Bhattacharya et al. 2003; Sarkar et al. 2003).
(ii) High-P isobaric cooling (IBC) and subsequent isothermal decompression (ITD) up to 750 °C and 5 kbar (Mohan et al. 1997; Rickers et al. 2001), and

(iii) Isobaric heating–cooling path in the UHT granulites on either side of the Godavari graben, controlled by intrusive plutonic complexes, with thermal peak at 950–1000 °C and 6–8 kbar (Bose et al. 2000; Dasgupta and Sengupta 2003).

4.5.4.1 Characters

Northern segment: One of the most characteristic features of high-grade EGMB terrane in the northern parts is the presence of ultrahigh temperature (UHT) metamorphism, which is marked by the presence of sapphirine (Spr), quartz–spinel (Spl), and quartz–orthopyroxene (Opx)–sillimanite (Sill)–quartz mineral associations. It has been investigated from different localities and provided typical P-T trajectories (Fig. 4.24): (i) high-grade granulites from Kakanuru (Kamineni and Rao 1988), Anantgiri (Sengupta et al. 1990), Araku (Sengupta et al. 1990), Rajamundry (Dasgupta et al. 1993), Kondapalle (Sengupta et al. 1999; Dasgupta et al. 1995; Mohan et al. 1997), Paderu (Lal 1997), Rayagada (Bhattacharya and Kar 2002), Sunkarametta (Bose and Das 2007 and others), (ii) calc-silicate granulites from Rajamundri (Dasgupta et al. 1993), Borra (Bhowmick et al. 1995), Chilka Lake (Sen et al. 1995; Bhattacharya 2004; Raith et al. 2007; Sengupta et al. 2008), Koraput (Nanda et al. 2008), and (iii) mafic granulites and orthopyroxene-bearing quartzo-feldspathic granulites (Dasgupta et al. 1991, 1993; Sengupta et al. 2008; Mohan et al. 2003; Bhui et al. 2007; Das et al. 2008), and Madhuravada and Vizianagaram (Rao et al. 1995).

Petrological studies from Anantagiri (Sengupta et al. 1990), Madhuravada and Vizianagaram (Rao et al. 1995), Paderu (Lal 1997) and Rajamundry (Dasgupta et al. 1997) revealed an early M1 UHT metamorphism with thermal peak at 900–1000 °C/8–10 kbar, followed by high pressure isobaric cooling (IBC) to 750–800 °C, which is documented in coronas at thermal peak due to reactions involving Spr + Spl + Qz (Dasgupta and Sengupta 1995; Shaw and Arima 1996). Both clockwise (Lal 1997; Bhattacharya and Kar 2004) and anti-clockwise paths (Dasgupta et al. 1997; Sengupta et al. 1990) were deciphered for the northern segment. In Chilka Lake granulites, Sen et al. (1995) deduced three phases of isothermal decompression (ITD) paths with two intervening discontinuous isobaric cooling (IBC) paths from peak UHT metamorphism at 8–10 kbar/1000 °C down to 4.5 kbar/650 °C (Fig. 4.24j; also Rickers et al. 2001). In contrast, Mg–Al granulites from Visakhapatnam show a simple isobaric heating-cooling path (Bose et al. 2000). In Diguva Sonaba area of Vishakhapatnam district, Prakash et al. (2015) recorded textural evidences of prograde clockwise P-T path in Spr–Spl-bearing granulite, which had undergone peak metamorphic conditions of c.1000 °C/12 kbar, and an isothermal decompression path due to rapid uplift of a tectonically thickened crust. In northern EGMB, a new locality (Kaithapalli, NW of Chilka Lake) of Spr–Spl-granulite enclaves within khondalite having reaction textures and symplectites yielded an isothermal decompression path by about 2.5 kbar from ~10.5 kbar and ~950 °C (Fig. 4.24l; Prakash et al. 2019). From high Mg–Al granulites of Sunki in central EGMB, peak UHT metamorphism

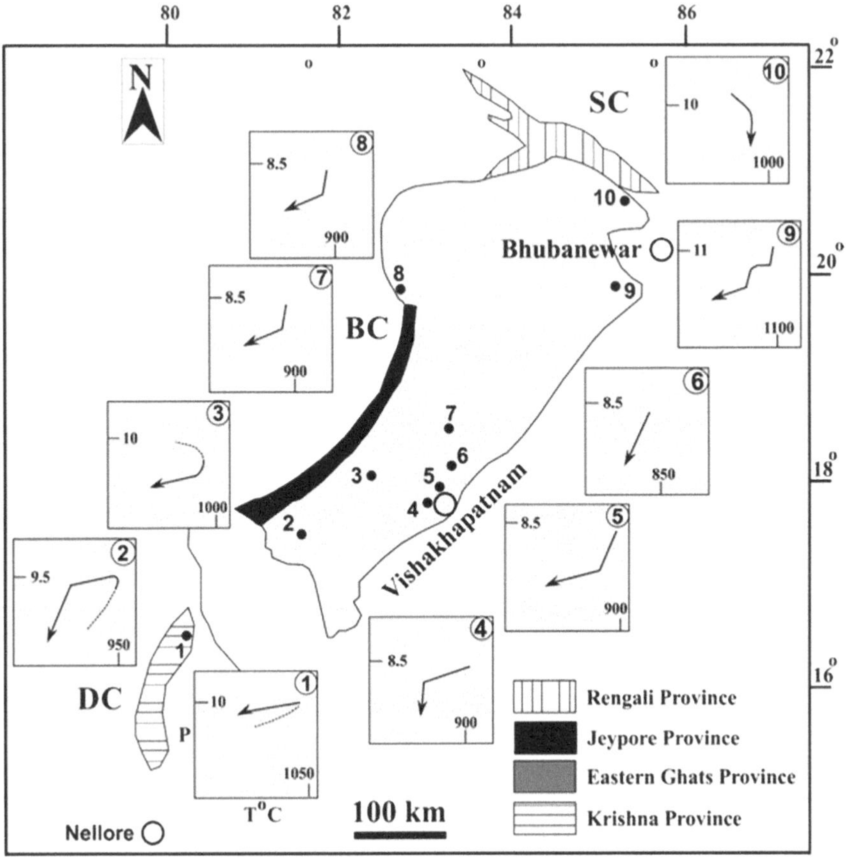

Fig. 4.24 Synoptic metamorphic history of the EGMB, shown by P-T paths at different places. 1–Kondapalle (Sengupta et al. 1999). 2–Rajamundri (Neogi et al. 1999). 3–Paderu (i. Bhattacharya and Kar 2002; ii. Prakash et al. 2019). 4–Anantagiri (Sengupta et al. 1990). 5–Anakapalle (Dasgupta et al. 1994). 6–Garbham (Dasgupta et al. 1992). 7–Rayagada (Shaw and Arima 1997). 8–Deobhog (Gupta et al. 2000). 9–Kaithapalli (Prakash et al. 2019). 10–Chilka (Sen et al. 1995). 11–Jeypore (Kar et al. 2003)

of >ca. 960 °C/9.7 kbar was recorded with decompression and minor cooling to ~7.5 kbar/~900 °C with a single long-lived protracted monazite growth inclusion in garnet and cordierite between ca. 1043 and 922 Ma during high-temperature retrograde evolution (Korhonen et al. 2011, 2013a). In contrast to a long-time gap between peak conditions elsewhere, the Sunki locality indicates a simple evolution of a UHT peak followed by decompression and cooling (Fig. 4.24).

Southern segment: In the segment south of the Godavari rift near Kondapalle and Chimakurthy, the EGMB exhibits heating to extreme ~1000 °C temperatures and cooling trajectories at different pressures. Textural relationships in spinel ±

sapphirine granulite lenses within leptynites near Gokavaram in the EGMB led documentation of several retrograde reactions, subsequent to prograde dehydration melting. These spinel granulites evolved through an anticlockwise P-T trajectory with peak metamorphism at 950 °C and >9 kbar, which was followed by near-isobaric cooling and superimposed event of near-isothermal decompression (Fig. 4.24). Similar results were obtained by Sengupta et al. (1999) in quartz– and corundum–bearing metapelitic granulites, intruded by layered gabbronorite–pyroxenite–anorthosite at Kondapalle, where an anticlockwise heating–cooling path is inferred from UHT metamorphism at 1000 °C and >8 kbar in the lower crust.

Two metamorphic events are recorded in the Ongole domain (Fig. 4.24a; Sarkar et al. 2014): (i) first stage UHT metamorphism at T > 950 °C, P = 6.5–7 kbar from spinel–quartz association, T = 950–1000 °C from orthopyroxene in garnet + cordierite–bearing metapelites, T = 900–1000 °C from mesoperthite and plagioclase pairs and other geothermometers, followed by near isobaric heating–cooling trajectory; (ii) second stage of higher pressure and lower temperature metamorphism at ca. 780 °C/9.5 kbar (thick garnet overgrowths over coarse-grained garnet porphyroblasts and orthopyroxene + sillimanite ± kyanite ± spinel symplectites (replacing cordierite) in metapelites and charno-enderbites, followed by its last stage isobaric cooling and finally the near-isothermal decompression to ca. 4 kbar.

4.5.5 Age of Metamorphism

Three distinct age groups of the UHT metamorphism are decipherable in the EGMB: the Archaean, Mesproterozoic and Neoproterozoic–Pan-African, though the Neoproterozoic event is the most prolific (Vinogradov et al. 1964; Paul et al. 1990; Shaw et al. 1997; Krause et al. 2001; Dobmeier and Simmat 2002). Paul et al. (1990) and Shaw et al. (1997) suggested a distinct Late Archaean 2.6–2.8 Ga age for basic magmatism and ensuing high temperature metamorphism. In addition, mesoscopic to microscopic structures suggest that the emplacement of charnockite massif was broadly syntectonic with the D1 deformation.

Melting and generation of charnockites under granulite facies around Jenapore, Orissa are dated ~3.0 Ga by Sm–Nd, WR Rb–Sr systematics and Pb–Pb zircon methods with Sm–Nd model dates between 3.4 and 3.5 Ga;—$Nd_{(0)}$ values provide evidences for an early Archaean continental crust in this high-grade terrane (Bhattacharya et al. 2001). Khondalite, associated gneiss and granulite from central belt are older than 1.4 Ga anorthosite and alkaline intrusives (Sarkar et al. 1981; Ramakrishnan et al. 1998), implying the granulite facies metamorphism during this period.

Mezger and Cosca (1999) opined that last high-grade metamorphism, superposed on several metamorphism and deformation episodes within central and eastern tectonic units of the northern EGMB, occurred at ca. 960 Ma (U–Pb ages on monazite, allanite, zircon and sphene). In addition, the Western Charnockite Zone reveals a major thermal event around 1.6 Ga. Discordant sphene ages along a reference line from ca. 935 to 504 Ma indicate a thermal disturbance during a Pan-African deformation phase, which is supported by concordant U–Pb zircon age of 516 ± 1 Ma in a vein and Pan-African hornblende $^{40}Ar/^{39}Ar$ plateau ages. These are similar to published ages for the Rayner Complex and Prydz Bay region of Eastern Antarctica, and provide evidences for these belts as parts of an extensive orogenic belt originated during the Grenvillian Orogeny ~1.0 Ga (Mezger and Cosca 1999). In contrast, high resolution geochronological data from central segment of the EGMB reveal sustained UHT conditions (T > 900 °C) for >50 My between ca 970 and 930 Ma, using $^{207}Pb/^{206}Pb$ zircon and monazite ages, and perhaps for as long as 200 Ma from ca 1130 to 930 Ma, during a single CCW tectono-metamorphic event (Korhonen et al. 2011, 2013b).

Two metamorphic events in the Ongole domain, dated by in situ U–Pb monazite, appear to be separated by 60–80 Ma when HT-LP and second metamorphisms occurred at ca. 1620 and 1540 Ma, respectively (Sarkar et al. 2014). In this domain, the UHT metamorphism is most likely caused by magmatic heat advection during arc growth, while second metamorphism was caused by crustal thickening during continental collision and subsequent rapid exhumation of the overthickened crust.

Available U–Pb mineral ages from Ongole and Kondapalle suggested the UHT metamorphism (granulite facies metamorphism) at 1.6 and 1.3 Ga, respectively, after which these did not experience any granulite facies event. However, $^{40}Ar/^{39}Ar$ amphibole ages gave ~1.0 Ga (Grenvillian age) in shear zones (Crowe et al. 2001).

4.5.6 Igneous Plutons

Numerous alkaline and mafic bodies mainly intrude along the western margin of the EGMB as nepheline syenite, hornblende syenite, syenite and quartz syenite plutons, like those of the Khariar Alkaline Complex in the west or the Prakasan alkaline province south of Godavari Graben (Leelanandam 1998). The Kharia rocks are deformed as lensoidal bodies, nearly parallelling the gneissic foliation. These bodies are either emplaced (i) in a paleo-rift at the Indian cratonic margin, which was subsequently deformed during collision of the EGMB (Sarkar and Schenk 2014), (ii) syntectonically during thrusting and shearing along the western and northern margin (Chetty 2001; Biswal et al. 2007), (iii) generated by mantle melting in the presence of CO_2 fluid (Banerjee et al. 2013), (iv) emplaced before the crustal reworking in the southern parts at ~1.5 to 1.2 Ga (U–Pb and Pb–Pb ages–Upadhyay and Raith 2006; Upadhyay et al. 2006a). The origin of Khariar alkaline complex is attributed to basaltic magma derived from partial melting of enriched lherzolite mantle source within the lithosphere. The basaltic magma fractionated within mantle and gave rise to nepheline syenite (Upadhyay et al. 2006b).

Massive anorthosites are mapped along its northern margin (Sarkar et al. 1981), while Leelanandam (1998) reported a layered igneous complex of anorthosite-gabbro–pyroxenite–chromitite from Kondapalle area in southern EGMB.

4.5.7 Tectonic Evolution and India–Antarctica Connection

Postulated Indo-Antarctic correlations reveal that the UHT metapelitic sediments of the EGMB were deposited on the Indian Shield along a passive intracontinental rifted margin, indicated by alkaline magmatism, during ~1.50 to 1.25 Ga, and were derived from the Indian Shield (Upadhyay 2008; Upadhyay et al. 2009; Ratre et al. 2010; Das et al. 2017). At many places, the UHT metamorphism of these sediments is characteristics of this belt at pressure up to 10 kbar/~1050 °C and at 1.17 Ga with a CCW P-T path, possibly due to burial by intracratonic collision before this metamorphism. Taking the most conservative temperature estimate of 1070 °C for the studied rocks, Dasgupta et al. (2017) estimated an abnormally high geothermal gradient of >110 °C $kbar^{-1}$, possibly due to large advective heat transport in a back–arc basin or asthenospheric/mantle upwelling, which has resulted in an extension or slab detachment. Voluminous calc-alkaline felsic magmas of arc affinity–protoliths of EGMB charnockites (Bose et al. 1997)—represent a change in tectonic regime from extensional to convergent-type with the subduction of East Antarctican oceanic crust beneath the EGMB (Das et al. 2017).

Finally, the India and east Antarctica collided at ~0.95 to 0.90 Ga and shaped the Rodinia supercontinent, resulting into steeply decompressive P-T path as a consequence of exhumation of overthickened crust. Amalgamation and separation of India and East Antarctica took place at least twice during the Mesoproterozoic–Neoproterozoic period, with the final docking at least at 0.95 Ga. Thus, the Eastern Ghats Mobile Belt (EGMB) was an integrated part of the Rodinia supercontinent, and that both the EGMB and East Antarctica shared common tectonothermal events at 0.95–0.90 Ga (Li et al. 2008; Chattopadhyay et al. 2015a, b; Dasgupta et al. 2017).

Far away across the Indian Ocean, the terrane of East Antarctica encompasses Napier, Enderby Land, Vestfold Hills and the Raynex Complex bearing some similarity with the EGMB (Upadhyay et al. 2006a). Veevers (2009) quantitatively reconstructed passive continental margins of India and Antarctica by continental and oceanic crust (COB) boundary by fitting to their pre-rift position by eliminating pre-drift extension (Fig. 4.25). As a consequence, the Lambert and Mahanadi Rifts are aligned in a Permian–Triassic rift system and the Napier salient fitting in the depression to the east of the Cuddapah Basin. Strong connections are indicated in: (i) Singhbhum Province–Prydz–Rauer–Vestfold Hills, (ii) Northern Eastern Ghats Mobile Belt–Bastar Craton–Kemp Land–MacRobertson Land, (iii) Southern Eastern Ghats Mobile Belt–Dharwar Craton–Napier, and (iv) Southern Granulite Terrain–Sri Lanka–Lützow–Holm Terrane–Rayner Complex. Some of the common events recorded in both the continents are the 1.3–0.9 Ga (Grenville) convergence in the Eastern Ghats–Rayner Mobile Belt in Rodinia, and 0.50–0.60 Ga (Pan-Gondwanaland) events that accreted the Southern Granulite Terrain and Sri Lanka to the Antarctic–Indian region. It appears the Gondwanaland fragmented around 0.14 Ga along the Eastern Ghats–Rayner Mobile Belt fabric and further along the Permian–Triassic rifts (Veevers 2009).

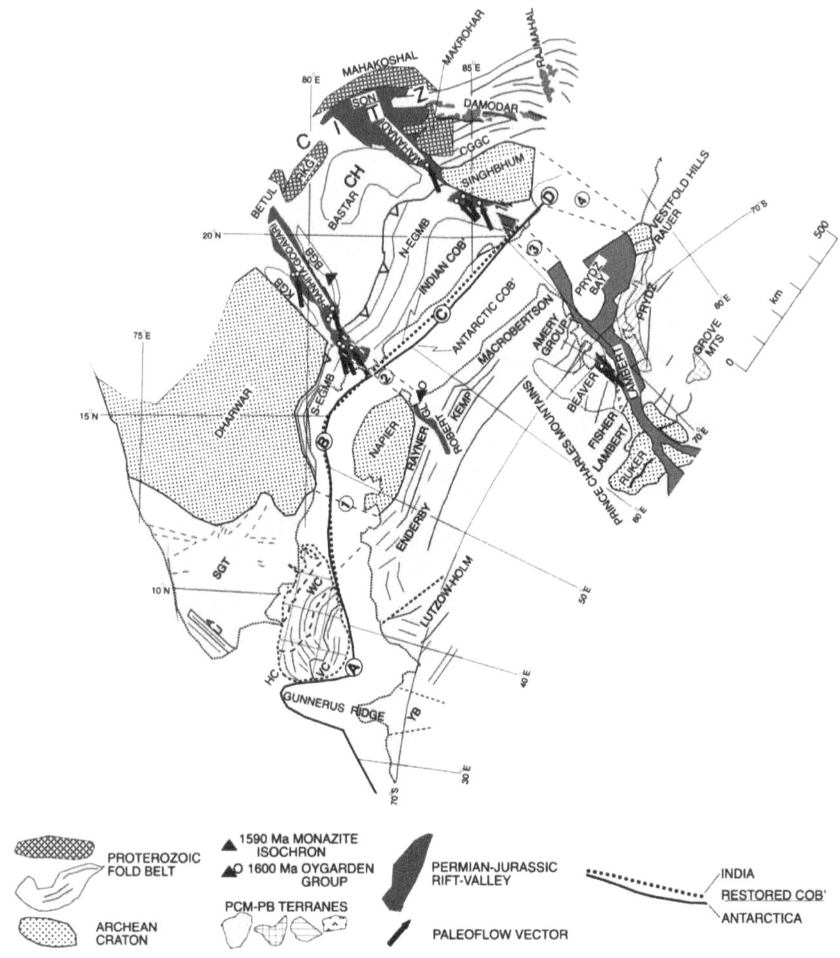

Fig. 4.25 Palinspastic restoration of India and Antarctica along continent–ocean boundaries (COB′–Antarctica full line, India broken line). Archean cratons and Proterozoic fold belts shown. Achankovil Unit (AU). Bhopalpatnam Granulite Belt (BGB). Chotanagpur Granite Gneiss Complex (CGGC). Central Indian Tectonic Zone (CITZ). Highland Complex (HC). Karimnagar Granulite Belt (KGB). Segmented Northern and Southern Eastern Ghats Mobile Belts (N-EGMB/S-EGMB). Prydz Bay (PB). Prince Charles Mountains (PCM). Ramakona–Katangi Granulite Belt (RKG). Restored continent–ocean boundary (COB'). Southern Granulite Terrain (SGT). Vijayan Complex (VC). Wanni Complex (WC). Yamato–Belgica Complex (YB). After Veevers (2009) and Fig. 7 therein for details

4.6 Karimnagar and Bhopalpatnam Granulite Belts

The Dharwar and Bastar cratons join each other along a NW–SE trending tectonic element with P–G valley basin in the intervening gap. Along these border regions are found two tectonostratigraphic units, the Karimnagar and the Bhopalpatnam granulites belts (KGB and BGB; Joy et al. 2018). These granulite enclaves and bands occur mixed with granites, granite gneisses and amphibolites (Fig. 4.26; Rajesham et al. 1993; Santosh et al. 2004; Vansutre et al. 2013).

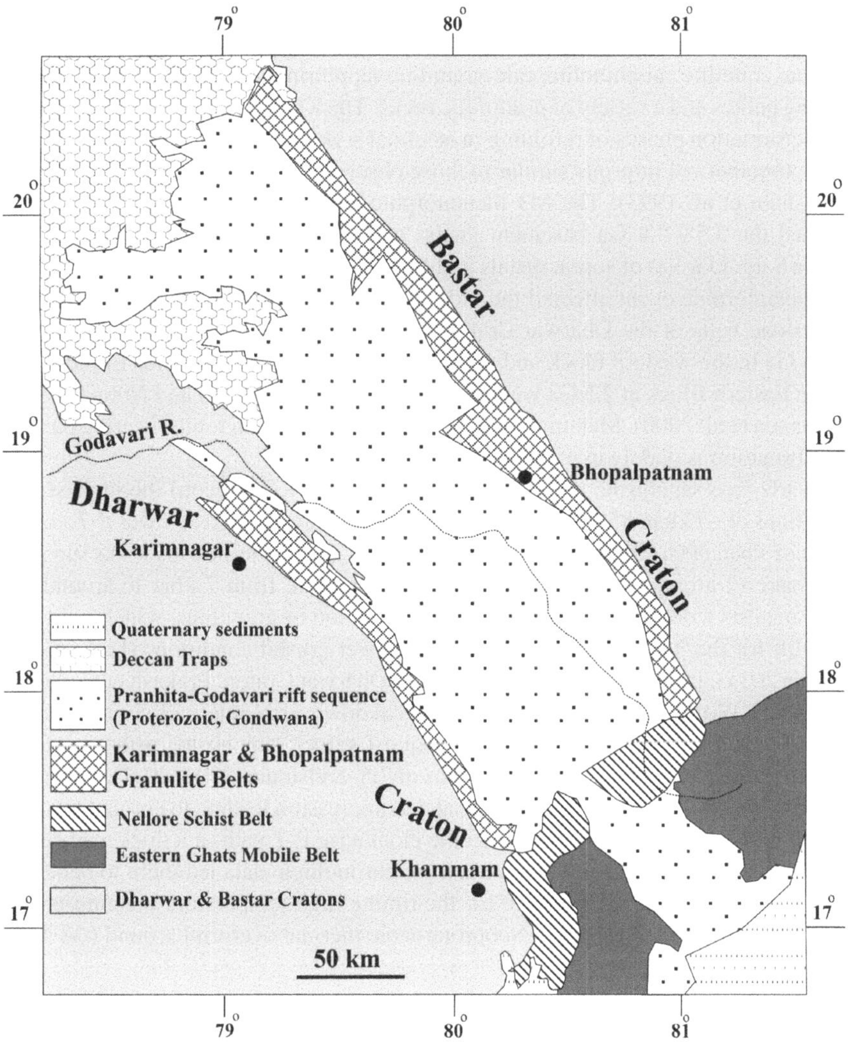

Fig. 4.26 Simplified geological map of Karimnagar and Bhpalpatnam granulite belts, Godavari valley. Redrawn after Rajesham et al. (1993) and Chaudhuri et al. (2012)

4.6.1 Karimnagar Granulite Belt (KGB)

The Karimnagar Granulite Belt (KGB) is an Archaean supracrustal-granite assemblage in a 150 × 20 km metamorphosed belt of granulite grade, which represents the suture zone between the Dharwar and Bastar cratons (Fig. 4.26; Rajesham et al. 1993). This belt contains widespread coarse-grained unfoliated charnockite containing wide variety of high-grade enclaves of charnockite-enderbite gneiss, ultrabasicbasic granulite, amphibolite, calc-granulite, sapphirine-orthopyroxene-cordierite-bearing pelites and a variety of psammitic rocks. The KGB is characterized by at least two deformation phases of refolding in isoclinal style with axial planar foliation and the tectonothermal imprints similar to those observed in the eastern Dharwar craton (Rajesham et al. 1993). The M1 metamorphic event is dated at ca 3.0 Ga which affected the 3.58–3.4 Ga basement gneiss ca 3.0 Ga Peninsular Gneiss and 3.2–3.1 Ga Sargur Group of supracrustals in the western block (Nutman et al. 1992). The M2 metamorphic event affected the older rocks as well as 2.7–2.6 Ga rocks of the greenstone belts of the Dharwar Craton. Emplacement of the Chitradurga Granite at 2.6 Ga in the western block and the Closepet granite and granitoids of the Kolar belt in Eastern Block at 2.5 Ga were the follow up events (Friend and Nutman 1991; Jayananda et al. 2000). Metamorphosed fine grained cummingtonite-bearing banded iron formation is widely exposed in the entire belt.

Quartz-free sapphirine and mafic granulites of the KGB record the highest *P-T* conditions of ~7 kbar, 850 °C, whereas the gneisses were formed at lower *P-T* conditions (~5 kbar, 800 °C). In addition, the presence of andalusite-bearing rocks suggests a pressure of around 2.5 kbar. This change in pressure from 7 kbar to around 2.5 kbar suggests a decompressive path for the evolution of granulites, which indicates an uplift for the granulite-facies rocks from lower crustal conditions (Prakash and Sharme 2011). In the northeastern parts of the Dharwar Craton, Prakash et al. (2017) observed NW-trending granulites having breakdown of biotite and corundum and formation of peak metamorphic assemblage of spinel, sapphirine, orthopyroxene, cordierite and K-feldspar at P-T conditions of 7.5–8 kbar and 800–840 °C. This thermal peak was followed by near isothermal decompression leading to coronal textures at ~4.3 to 4.5 kbar and 750–800 °C in the clockwise P-T path in a thickened continental crust. New SHRIMP U–Pb zircon geochronologic data led them to believe a Neoarchaean (2604 ± 25 Ma) age for the timing high temperature metamorphism and accretion of the terrane and Neoproterozoic thermal overprint around 638 Ma.

4.6.2 Bhopalpatanam Granulite Belt (KGB)

The 300-km long Bhopalpatanam granulite belt (BGP) lies on the western edge of the Bastar craton on the northeastern rim of the Pranhita-Godavari rift zone and is composed of two pyroxene granulites, ultramafics, quartzite, calc-silicate rocks, Mg–Al metapelites (Vansutre and Hari 2010). Santosh et al. (2004) reported zircon (EPMA) ages from the KGB with the cores recording ages of up to 3.1 Ga and rims with ages of 2.6 Ga. In contrast, the cores of zircons recovered from the BGB demonstrate core ages of 1.9 Ga and the rims of 1.6–1.7 Ga. The monazites from the BGB also record the 1600 ± 3 Ma age, interpreted as the most important tectonothermal event in the BGB as it is not recorded from the KGB nor from the Bastar craton.

4.7 Nellore–Khammam Schist Belt

The Nellore–Khammam schist belt (NKSB) is about 600 km long and 30–130 km wide schistose terrane and separates the marginally metamorphosed and deformed intracratonic Proterozoic Cuddapah basin to the west and the Ongole domain of the Eastern Ghats Mobile belt (Hari Prasad et al. 2000). In the northern sector it is called the Khammam schist belt (Ramam and Murty 1997; Hari Prasad et al. 2000; Okudaira et al. 2001; Saha et al. 2015) occurring between the EGMB and the Pranhita-Godavari valley sedimentaries. The NKSB is divided into two sedimentary stratigraphic entities namely: the Vinjamuru Group and the Udaigiri Group. These are intermingled with several igneous intrusive rocks of variable composition.

4.7.1 Vinjamuru Group

This group is made up of rocks of amphibolite facies showing characters of a volcano-sedimentary assemblage (Moeen 1998; Dobmeier and Raith 2003; Saha et al. 2015; Sain et al. 2017). The Vinjamuru gabbro of the NKSB has yielded Sm–Nd isotope whole rock age of 2654 ± 100 Ma and a 1911 ± 88 Ma isochron age (Vadlamani 2010), which are attributed to the intrusion of ~1.9 Ga MORB-type gabbro in the 2.7 Ga old NKSB (Vadlamani 2010) and also the Cuddapah succession. This intrusion was linked to a major extensional event along the East Dharwar craton margin. Geochemical and geochronological evidence of ~1.9 Ga volcanic event represents a large anorogenic igneous event in this sector that occurred during a major convergent orogenic event along the southeastern margin of the EDC. The andesites and rhyolites of the Vinjamuru Group are ~1868 and 1771–1791 Ma old as determined by the zircon Pb evaporation method (Vadlamani et al. 2012). The Vinukonda Granite intrusive in the Vinjamuru Group is dated as ~1590 Ma (U–Pb zircon TIMS; Dobmeier et al. 2006) and represents the minimum age of the Vinjamuru Group. On the other hand, fission track and K–Ar dates show pegmatite events at 1600, ~1000 and 600 ± 100 Ma (Ghosh et al. 2004) from the southern part of NKSB. The ~500 Ma event is also represented by the phengite Rb–Sr ages from mylonites of the Vinukonda granite (Dobmeier et al. 2006). The anorthosite assemblage of Chimalpahad in the NKSB has yielded Sm–Nd model age of ~1170 Ma (Dharma Rao et al. 2011b) and interpreted it as an accreted arc fragment within the NKSB. Yoshida et al. (1996) reported 1126 Ma isochron age of metapelite from the Khammam area. Based on an Sm–Nd mineral isochron of 824 ± 53 Ma from amphibolites of the Khammam schist belt, Okudaira et al. (2001), interpreted metamorphism of these rocks during tectonic accretion of the EGMB to the Dharwar–Bastar craton.

4.7.2 Udaigiri Group

This group is dominated by rocks showing greenschist facies metamorphism. The only age available from the Udaigiri Group is very poorly constrained at 1929 ± 130 Ma, based on a single grain xenotime analysis by Das et al. (2015).

4.7.3 Prakasam Alkaline Complex

The Prakasam alkaline plutons occurring along the boundary between the Vinjamuru and Ongole domains are dated 1242 and 1369 Ma and possibly represent the rifting during the breakup of Columbia (Upadhyay 2008).

4.7.4 Kandra Ophiolite Complex (KOC) and Kanigiri Ophiolitic Melange (KOM)

The Kandra ophiolite complex (Vijaya Kumar et al. 2010) in the south and the Kanigiri ophiolite melange (Dharma Rao et al. 2011a) in the central part have also been identified. All these rock assemblages have suffered multiple deformation, metamorphism and are associated with different phases of granites (Saha et al. 2015). The granulite facies rocks of the NKSB facing the Ongole domain were considered Late Paleoproterozoic by Dobmeier and Raith (2003). Saha et al. (2015) on the other hand, place them in the Late Neoarchaean to Mesoproterozoic period and infer that the NKSB evolved independent of the Eastern Ghats belt. This schist belt is also interpreted as the easternmost greenstone belt of the eastern Dharwar craton (Vadlamani 2010).

The 1850–1900 Ma SHRIMP U–Pb zircon ages (Vijaya Kumar et al. 2010) and similar Sm–Nd isochron age (Vadlamani 2010) for the Kandra ophiolite complex fixes the time frame of emplacement of these ophiolites. On the other hand, the Kanigiri complex is dated at around 1330 Ma (LA-ICP-MS U–Pb zircon ages by Dharma Rao et al. (2011a). There is a 500 My age difference between these two ophiolite complexes suggesting accretion along a craton margin over a prolonged period of convergence and proposed multiple cycles of ophiolite emplacement based on the divergent nature of thrusting in the area (Dharma Rao et al. 2011a).

4.8 Pandyan Mobile Belt (PMB)

4.8.1 Location and Terminology

A distinct domain of gneiss-granulite, called as the Pandyan Mobile Belt (PMB) after the legendary dynasty of South India or Southern Granulite Terrain (SGT), lies to the south of the Palghat–Cauvery Shear Zone (PCSZ) and the Achankovil Shear Zone (ASZ) (Fig. 4.27; Ramakrishnan 1993; GSI and ISRO 1994). Though extensive granulite massifs are located to the north of the PCSZ till the Fermor Line, the term PMB is now entrenched into the geological literature survey. Based on Sm–Nd isotopic signatures (Bhaskar Rao et al. 2003) and U–Pb zircon and monazite ages, it is suggested that the boundary of the PMB may be relocated along the newly recognized V-shaped Karur-Kambam-Painavu-Trichur Shear Zone (KKPTSZ) (Ghosh et al. 2004).

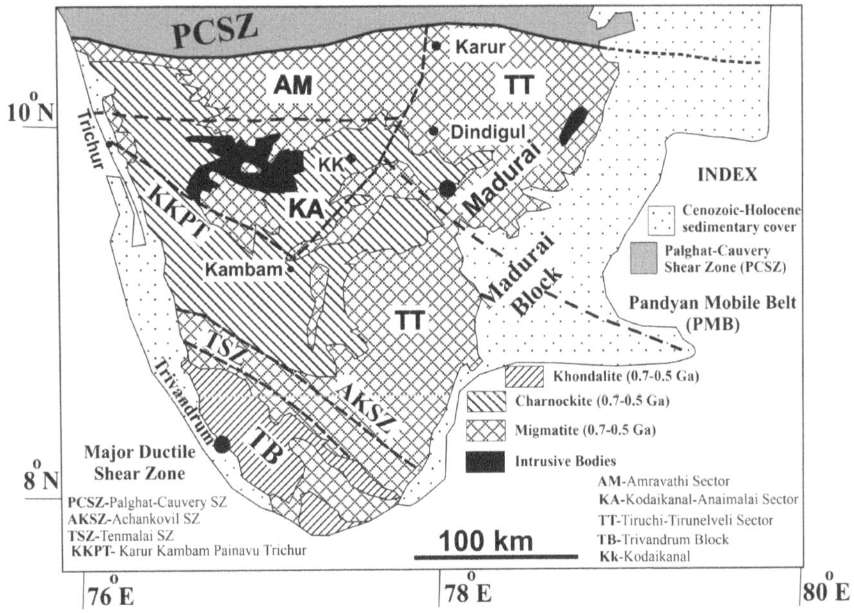

Fig. 4.27 Simplified geological map of the Pandyan Mobile Belt (PMB) showing distribution of different blocks and shear zones. Redrawn after Ramakrishnan and Vaidyanathan (2008). Location of the Karur Kambam Painavu Trichur (KKPT) Shear Zone is drawn from Ghosh et al. (2004)

4.8.2 Divisions

Ramakrishnan and Vaidyanathan (2008) divided the PMB into (i) Marginal Zone, (ii) Madurai Block or Central Zone, and (iii) Southern Zone (Trivandrum Block or Ponmudi Sub-Block (also known as the Kerala Khondalite Belt) and Nagercoil Sub-Block. Since the Marginal Zone lies much to the north of the PCSZ, it should not be considered as the part of the PMB.

4.8.2.1 Madurai Block

Bounded by the PCSZ in the north and the Achankovil Shear Zone (ASZ) in the south, the Madurai Block was further subdivided into the (i) Amravathi sector (AM), (ii) Kodaikanal-Anaimalai sector (KA), and (iii) Tiruchi-Tirunelveli Sector (TT) (Gopalkrishnan 2001).

Named after the Amravathi River, this sector contains enderbitic charnockite, hornblende-biotite gneiss, lenses of pyroxene granulite and ultramafics, besides grn-sill gneiss, calc-silicates, quartzite, pink granite and crystalline limestone. F1Sill-garnet schist from Madukkarai provided Sm–Nd–WR Pan-African age of 560 ± 17 Ma (Meißner et al. 2002). F1 isoclinal folds on bedding surfaces are coaxially folded by ENE-trending large-scale folds in this sector.

The adjoining Kodaikanal-Anaimalai sector (KA) contains migmatite gneiss with ultramafic and mafic enclaves and quartzite-carbonate-pelite suite. Sapphirine-orthopyroxene granulite, cordierite-garnet-spinel rocks and charnockites are associated together at different places; the latter is observed around Palani, Cardamom, Kodaikanal and Trichur. Orthogneiss is observed around the Annamalai Hills. Regional trends change from WNW to ENE and NE, becoming N–S between Dindigal and Karur.

Brown and Raith (1996) recorded sapphirine-bearing, orthopyroxene-sillimanite ± garnet granulite in the Palni Hill Ranges of the Madurai block as evidence of ultrahigh-temperature (UHT) metamorphism during which primary mineral assemblages were partially overprinted by reaction coronas and symplectites indicating decompression. Primary mineral assemblages indicate c. 12 kbar and c. 900–1000 °C P-T conditions, while reaction textures reveal decompression P-T conditions of c. 5 kbar and c. 850 °C in an inferred clockwise P-T path.

The most extensive Tiruchi-Tirunelveli (TT) Sector between the Periyar River and Achankovil SZ is a highly complex association of folded domes and basins, containing quartzite-carbonate-pelite suite (QCP) as the Khondalite Group. Highly metamorphosed carbonates contain calc-silicates, garnet-sillimanite-graphite gneiss, garnet-cordierite gneiss, cordierite-spinel gneiss, sapphirine gneiss, charnockite, garnet-quartz-feldspar gneiss (leptynites) are the main rock types. The QCP suite also contains bands of basic granulite and amphibolite.

Intercalated sapphirine-bearing lithologies are located at several places in the Madurai Block including Kodaikanal (Mohan et al. 1996; Prakash and Shastry 1999).

Tateishi et al. (2004) recently discovered equilibrium sapphirine + quartz assemblages enclosed in garnet in pelitic granulites from the southern part of the Madurai Block. Their results suggest that at least some parts of the block experienced T > 1050 °C (at 8–10 kbar) UHT metamorphism. In this block, UHT metamorphism has been verified using the calcite-graphite isotope exchange thermometer where sapphirine granulite yielded temperature estimates of around 1000 °C. Homogenous $\delta^{13}C$ and $\delta^{18}O$ values in calcite and graphite cores from marbles give highest metamorphic temperature ~1060 °C and retrograde cooling ~750 °C from the rims (Satish-Kumar 2000).

4.8.2.2 Trivandrum Block (TB)

The southernmost segment of the PMB is comprised of the NW–SE trending structural fabric to the south of the Achankovil Shear Zone (AKSZ) in the Trivandrum Block (Santosh 1996) and is also termed as the Kerala-Khondalite Belt (Chacko et al. 1987; 1996). It contains dominantly of metasedimentary migmatitic gneiss including garnet-bearing felsic gneiss (leptynite) and granulite facies garnet + spinel + cordierite + sillimanite metapelites (khondalite) with small amounts of interlayered charnockite, calc–silicates and pyroxene granulite. It is subdivided into the Ponmudi and Nagercoil Sub-blocks. The Ponmudi has similar lithologies like the TT Block, while the Nagercoil Block has mainly charnockite, gneiss with thin layers of anorthosite and norite.

Of interest is the development of incipient charnockite on leptynite (garnet-quartzo-feldspar gneiss), which were first recognized from Ponmudi (Ravidra Kumar et al. 1985) and subsequently from several localities in the Madurai and Trivandrum Blocks as well as from Gondwanic regions of Sri Lanka and Antarctica. In this variety, orthopyroxene develops after the breakdown reaction of biotite + quartz (+ garnet). Patches in these incipient charnockites are developed through carbonic metamorphism either by infiltrating CO2 fluids from the mantle, anatexis of crustal carbonates or drop in H2O pressure during regional metamorphism.

Within the TB, metamorphic conditions increased from 5 kbar/750 °C in the center to 8–9 kbar/< 1000 °C in the north (Collins et al. 2010), though Morimoto et al. (2004) and Tadokoro et al. (2008) suggested UHT granulite facies assemblage (spinel + quartz) of greater than 950 °C, at pressures up to 12 kbar. It is likely that there was no such progressive P-T gradient and khondalite suite was possibly subjected to HT and UHT metamorphism. In the Trivandrum Block, the Neoproterozoic-Cambrian metamorphism range from early estimates of 5.5–7.0 kbar/700–800 °C (Chacko et al. 1996) to 6.5–7.0 kbar/925 ± 20 °C (Braun and Bröcker 2004). Based on pseudo section analysis Cenki et al. (2002) obtained a clockwise P-T path culminating at 6–7 kbar/900–950 °C, which is consistent with that defined by Nandakumar and Harley (2018). Most recent estimates have confirmed peak metamorphic conditions of 830–925 °C/6–9 kbar, followed by suprasolidus decompression (Blereau et al. 2016). Garnet-bearing migmatite, associated leucogranite and leucosome, both classified as the S-type in khondalite of the Trivandrum Block, are uniformly peraluminous and formed through granulite facies dehydration melting of metasedimentary protoliths (Nandkumar and Harley 2018).

Dominant regional granulite-facies metamorphism and migmatization in the Trivandrum Block is the latest Neoproterozoic to Cambrian in age, though zircon in granitic and charnockitic orthogneiss suggests magmatic protolith ages of 2.1–2.0 Ga, 1.88–1.84 Ga and 1.76 Ga (Kröner et al. 2015). Migmatitic garnet-biotite-quartz-feldspar gneiss and khondalite record Archaean (3.3–2.7 Ga) to Paleoproterozoic (2.0 Ga) Nd model ages (Harris et al. 1994; Cenki et al. 2004).

Metamorphic ages defined from near concordant zircon rims in orthogneisses, migmatitic leucosomes and paragneisses lie between 570 and 515 Ma (Cenki et al. 2004; Collins et al. 2014; Clark et al. 2015; Kröner et al. 2015; Santosh et al. 2009; Shabeer et al. 2005; Taylor et al. 2014; Whitehouse et al. 2014), zircon and monazite in migmatite leucosome indicates cooling of the Trivandrum Block granulites to an elevated residual granite solidus by 523 ± 7 Ma (Taylor et al. 2014)

Harley and Nandakumar (2016) documented a relict 1.92 Ga high-T metamorphism, anatexis and melt segregation, proving that at least some of the metasediments in the Trivandrum Block are polymetamorphic. This Paleoproterozoic tectonothermal event was strongly overprinted by the main Neoproterozoic-Cambrian granulite facies tectonism which itself involved significant partial melting and formation of garnet-bearing assemblages at 0.65 GPa/>820–860°C. Monazite growth at ca. 565 Ma and its modification at ca. 520 Ma indicates that HT metamorphism in the Pan-African event was long-lived for about 45 Myr, indicating a long-lived collisional hot orogenic belt that welded Gondwana from ca. 580–510 Ma.

Some of the homogeneous cores of sub-rounded detrital monazite grains in khondalite yielded a maximum age between 610 and 569 Ma; these grains have undergone fluvial transport and deposition within this belt was as young as, or later than, this range (cf. Santosh et al. 2006a). U–Pb SIMS ages of detrital zircons reveal that original sediments from the Trivandrum Block gneisses were deposited between ~1900 and ~515 Ma (Collins et al. 2007). Monazite ages from khondalite indicate that it belongs to Late Proterozoic/Cambrian (ca. 560–520 Ma) (Santosh et al. 2006b), while U–Pb zircon metamorphic ages across this block are around 513 ± 6 Ma (Collins et al. 2007), like younger overprints in the Madurai Block and PCSZ. Thus, all these blocks were affected by the granulite facies metamorphic event during the final collisional stage of the Gondwana assembly.

Within the KKB, coarse-grained augen gneiss represents an early stage of the Pan-African granitic magmatism at ~590 Ma and formed from high-T dehydration–melting of the various KKB gneiss in granulite facies after the main stage of Pan-African ductile deformation (Braun 2006).

4.8.2.3 Nagercoil Block (NB)

The Nagercoil Block (NB), located to the SW of the Trivandrum Block, preserves massif type charnockites with emplacement ages of 2.1–2.0 Ga and Hf and Nd model ages ranging from Archaean to Palaeoproterozoic (2.8–2.2 Ga), identical to the protoliths of charnockites from within the Trivandrum Block itself (Ghosh et al. 2004; Cenki et al. 2004; Kröner et al. 2012, 2015; Johnson et al. 2015). The protolith, model age and metamorphic age record of the Nagercoil Block (NB) indicates that it is essentially contiguous with the Trivandrum Block (TB) (Kröner et al. 2015; Johnson et al. 2015).

4.8.3 *Deformation*

The Pandyan Moblie Belt (PMB) is dissected by three major large-scale shear zones: (i) V-shaped Karur-Kumbum-Painavu-Trichur Shear Zone (KKPTSZ), (ii) Achankovil Shear Zone–AKSZ, and (iii) Tenmala Shear Zone–TSZ.

4.8.3.1 Karur-Kumbum-Painavu-Trichur Shear Zone (KKPTSZ)

A NE–SW trending and V-shaped major shear zone extends from Karur to Kambam and turns sharply towards NW to Trichur (Ghosh et al. 2004). Structural discontinuity across the KKPTSZ is very pronounced in the eastern parts where structural trends are subparallel to the shear zone while complex fold interference patterns cannot be traced across this zone. Within the shear zone trends of lithological units are subparallel to it; east of Kotamangalam granitic mylonites are well developed with polyphase structures suggesting repeated deformation along this part of the shear zone.

New U–Pb zircon ages of syntectonic granite (567 ± 2 Ma) and sheared Oddhanchatram anorthosite (563 ± 9 Ma) date a major shearing event in the KKPTSZ between ~570 and 560 Ma. Large sections of mylonitic fabric along the KKPTSZ are recrystallized and overprinted by charnockite, hence (the KKPTSZ pre-dates this charnockitization which has occurred between 526 Ma and 548 Ma in the Kerala Khondalite Belt (Ghosh et al. 2004).

4.8.3.2 Achankovil and Tenmala Shear Zones (AKSZ/TSZ)

Along the southern edge of the Cardamom Hills charnockite massif a ~10 km wide shear zone, the Achankovil Shear Zone, extends along the Achankovil lineament due to changes in structural trends from N–S to NW–SE and subvertical to shallow dipping to across the lineament and contrasting lithologies from massive charnockite to intercalated charnockite and paragneiss (Drury and Holt 1980). Sinha-Roy et al. (1984) and Radhakrishna et al. (1990) disputed the presence of this shear zone, while Sacks et al. (1997) documented dextral strike-slip movement along a subparallel Tenmala Shear Zone, and Rajesh et al. (1998) reported structures indicating a sinistral sense of movement along the Tenmala Shear Zone.

In parts of the AKSZ and TSZ, Ghosh et al. (2004) observed similar lithologies, including leptynitic garnet-biotite gneiss, khondalite, cordierite-gneiss, calc-silicate, quartzite, and massive granite, which have undergone charnockitization to varying degrees and intensity. Regional charnockitization is post-tectonic and is superimposed on the regional structural trend of the paragneisses. Though isolated structures revealing both dextral and sinistral shear sense are present, a consistent of shear sense could not be verified in either the Achankovil or the Tenmala Shear Zone. In addition, structural trends along western parts of the SGT are transposed from E–W and NW–SE to NNW–SSE, and become parallel to the west coast of India, possibly due to the offshore dextral shear zone subparallel to the west coast—the West Coast Shear Zone'.

References

Absar N, Sreenivas B (2015) Petrology and geochemistry of greywackes of the ~1.6 Ga Middle Aravalli Supergroup, northwest India: Evidence for active margin processes. Intern Geol Rev 57:134–158

Acharyya SK (2001) Geodynamic setting of the Central Indian Tectonic Zone in central, eastern and northeastern India. Geol Surv India Spec Publ 64:17–35

Acharyya SK (2003a) The nature of Mesoproterozoic Central Indian Tectonic Zone with the exhumed and reworked older granulites. Gondwana Res 6:197–214

Acharyya SK (2003b) A plate tectonic model for Proterozoic Crustal evolution of Central Indian Tectonic Zone. Gondwana Geol Mag 7:9–31

Acharyya SK, Roy A (2000) Tectonothermal history of the Central Indian Tectonic Zone and reactivation of major faults/shear zones. J Geol Soc India 55:239–256

Ahmad T, Tarney J (1994) Geochemistry and petrogenesis of late Archaean Aravalli volcanics, basement enclaves and granitoids, Rajasthan. Precambr Res 65:1–23

Ahmad T, Dragusanu C, Tanaka T (2008) Provenance of Proterozoic Basal Aravalli mafic volcanic rocks from Rajasthan, Northwestern India: Nd isotopes evidence for enriched mantle reservoirs. Precambr Res 162:150–159

Ahmad I, Mondal MEA, Bhutani R, Satyanarayanan M (2018a) Geochemical evolution of the Mangalwar Complex, Aravalli Craton, NW India: insights from elemental and Nd-isotope geochemistry of the basement gneisses. Geosci Front 9:931–942. https://doi.org/10.1016/j.gsf.2017.07.003

Ahmad I, Mondal MEA, Satyanarayanan M (2018b) Geochemistry of the Archean Metabasaltic Rocks from the Aravalli Craton, Southern Rajasthan, NW India: implications for petrogenesis and geodynamic setting. Paper presented at the National Seminar on Dynamics of Surface and Subsurface Geological Processes held at Department of Earth Sciences, Pondicherry University Puducherry, 8–9 Feb 2018

Alam M, Choudhary AK, Mouri H, Ahmad T (2017) Geochemical characterization and petrogenesis of mafic granulites from the Central Indian Tectonic Zone (CITZ). Geol Soc London Spec Publ 449(1):207–229

Arora D, Pant NC, Fareeduddin, Sharma S, Raghuram, Sadiq M (2017) Inferring a Neoproterozoic orogeny preceding the Rodinia break-up in the Sirohi Group, NW India. In: Pant NC, Dasgupta S (eds) Crustal evolution of India and Antarctica: the supercontinent connection, 457. Geol Soc London Spec Publ. https://doi.org/10.1144/SP457.8

Ashwal LD, Solanki AM, Pandit MK, Corfu F, Hendriks BWH, Burke K, Torsvik TH (2013) Geochronology and geochemistry of Neoproterozoic Mt. Abu Granitoids, NW India: regional correlation and implications for Rodinia paleogeography. Precambr Res 236:265–281

Azeez KKA, Patro PK, Harinarayana T, Sarma SVS (2017) Magnetotelluric imaging across the tectonic structures in the eastern segment of the Central Indian Tectonic Zone: preserved imprints of polyphase tectonics and evidence for suture status of the Tan Shear. Precamb Res 298:325–340. https://doi.org/10.1016/j.precamres.2017.06.018

Bandyopadhyay BK, Roy A, Huin AK (1995) Structure and tectonics of a part of Central Indian shield. Mem Geol Soc India 31:433–467

Banerjee DM (1970) Geological mapping on aerial photographs in Girwa Valley, Udaipur, in relation to phosphorite search. Unpubl Geol Surv India, Western Region Field Season Report 1968–69 with maps and sketches, 59p

Banerjee DM (1971a) Aravallian stromatolites from Udaipur, Rajasthan. J Geol Soc India 12:349–355

Banerjee DM (1971b) Precambrian stromatolitic phosphorites of Udaipur, Rajasthan. Geol Soc Am Bull 82:2319–2330

Banerjee DM (1996) A lower proterozoic palaeosol at BGC-Aravalli Boundary in South-Central Rajasthan, India. J Geol Soc India 48:277–288

Banerjee DM, Bhattacharya P (1989) Petrofacies analysis of the clastic rocks in the Proterozoic Aravalli Basin, Udaipur district, south central Rajasthan. Indian Min (Spec IGCP 217 Issue) 43(3–4):194–225

Banerjee DM, Bhattacharya P (1994) Petrology and geochemistry of graywackes from the Aravalli Supergroup, Rajasthan, India and the tectonic evolution of a Proterozoic Sedimentary basin. Precambr Res 67:11–35

Banerjee DM, Russell J (1993) Pb–Pb dating of Vindhyan stromatolites and a comparison with stromatolitic dates from other Proterozoic terrains. Sem Recent Developments in Vindhyan Geology. Jadavpur Univ (Abstract) 29

Bandyopadhyay BK, Chattopadhyay A, Khan AS, Huin AK (2001) Assembly of the Rodinia supercontinent: evidence from the Sakoli and Sausar belts in Central India. Gondwana Res 4:569–570

Banerjee DM, Schidlowski M, Arneth JD (1986) Genesis of Upper Proterozoic-Cambrian phosphorite deposits of India: isotopic inferences from carbonate fluorapatite, carbonate and organic carbon. Precambr Res 33:239–253

Banerjee DM, Deb M, Strauss H (1992) Organic carbon isotopic composition of proterozoic sedimentary rocks from India. In: Schidlowski M, Golubic S, Kimberley MM, McKirdy DM, Trudinger PA (eds) Early organic evolution. Springer, Berlin, pp 232–240

Banerjee A, Bhattacharya S, Sajeev K, Santosh M (2013) Numerical simulations of CO_2 migration during charnockite genesis. Geology 41:743–746

Bhadra S, Banerjee M, Bhattacharya A (2003) Tectonic restoration of polychronous mobile belt-craton assembly: constraints from corridor study across the western margin of the Eastern Ghats Belt, India. Mem Geol Soc India 52:109–130

Bhadra S, Gupta S, Banerjee M (2004) Structural evolution across the Eastern Ghats Mobile Belt-Bastar craton boundary, India: hot over cold thrusting in an ancient collision zone. J Struct Geol 26:233–245

Bhaskar Rao YJ, Janardhan AS, Vijaya Kumar T, Narayana BL, Dayal AM, Taylor PN, Chetty TRK (2003) Sm-Nd model ages and Rb–Sr isotopic systematics of Charnockites Gneisses across the Cauvery Shear Zone, southern India: implications to the Archaean—Neoproterozoic terrane boundary in the Southern Granulite Terrain. In: Ramakrishnan M (ed) Tectonics of Southern Granulite Terrain Kuppam-Palani Geotransect. Mem Geol Soc India, vol 50, pp 297–317

Bhattacharjee D, Jain V, Chattopadhyay A, Biswas RH, Singhvi AK (2016) Geomorphic evidences and chronology of multiple neotectonic events in a cratonic area: results from the Gavilgarh Fault Zone, central India. Tectonophysics 677:199–217

Bhattacharya S (1996) Eastern Ghats granulite terrain of India: an overview. J. Southeast Asian Earth Sci 14:165–174
Bhattacharya S (2002) Nature of trijunction in the East Indian Precambrian: example from the Eastern Ghats granulite belt, Singhbhum craton and Bastar craton around Paikmal, Western Orissa. Gondwana Res 5:53–62
Bhattacharya S (2004) High-temperature crustal-scale shear zone at western margin of the Eastern Ghats granulite belt, India: implications for rapid exhumation. J Asian Earth Sci 24:281–290
Bhattacharya DS, Bhattacharyya TK (1970) Geological and geophysical investigations of a basaltic layer in Archaean terrain of eastern India. Geol Soc Amer Bull 81:3073–3078
Bhattacharya AK, Chatterjee BK (1964) Petrology of precambrian Kolhan formation of Jhinkapani, Singhbhum District, Bihar. Geol Rund 53:758–779
Bhattacharya S, Kar R (2002) High temperature dehydration melting and decompressive P-T path in a granulite complex from the Eastern Ghats, India. Contrib Min Petrol 143:175–191. https://doi.org/10.1007/s00410-001-0341-6
Bhattacharya S, Kar R (2004) Alkaline intrusion in a granulite ensemble in the Eastern Ghats Belt, India: shear zone pathways and pull-apart structure. Proc Indian Acad Sci (Earth Planet Sci) 113(1):37–48
Bhattacharya S, Sen SK, Acharyya A (1994) Structural setting of the Chilka Lake granulite-migmatite-anorthosite suite with emphasis on the time relation of charnockites. Precambr Res 66:393–409
Bhattacharya S, Kar R, Misra S, Teixeira W (2001) Early archaean continental crust in the Eastern Ghats Belt, India: isotopic evidence from the charnockite suite. Geol Mag 138:609–618
Bhattacharya S, Kar R, Teixeira W, Basei M (2003) High-temperature crustal anatexis in a clockwise P-T Path: isotopic evidence from a granulite-granitoid suite in the Eastern Ghats Belt, India. J Geol Soc London 160(1):39–46. https://doi.org/10.1144/0016-764902-063
Bhola AM (1989) Petrology of regionally metamorphosed pelitic schists, SSW of Narnaul, Mahendragarh district, Haryana, India. In: Saklani PS (ed) Metamorphism, ophiolites and orogenic belts. Curr Trends in Geol, vol XII, pp 85–102
Bhowmik SK (2000) Granulite–granite connection: a case study from the northern part of the Sausar mobile belt in Central India. Unpubl Rep Geol Surv India, pp 1–15
Bhowmik SK (2019) The current status of orogenesis in the Central Indian tectonic zone: a view from its Southern margin. Geol J. https://doi.org/10.1002/gj.3456
Bhowmik SK, Chakraborty S (2017) Sequential kinetic modelling: a new tool decodes pulsed tectonic patterns in early hot orogens of Earth. Earth Planet Sci Lett 460:171–17
Bhowmik SK, Dasgupta S (2012) Tectonothermal evolution of the BGC in central Rajasthan, NW India: present status and correlation. J Asian Earth Sci 49:339–348
Bhowmik SK, Roy A (2003) Garnetiferous metabasites from the Sausar Mobile Belt: petrology, P-T path and implications for the tectonothermal evolution of the Central Indian Tectonic Zone. J Petrol 44:387–420
Bhowmik SK, Spiering B (2004) Constraining the prograde and retrograde P-T paths of granulites using decomposition of initially zoned garnets: an example from the Central Indian Tectonic Zone. Contrib Min Petrol 147:581–603
Bhowmick SK, Dasgupta S, Hoernes S, Bhattacharya PK (1995) Extremely high temperature calcareous granulites from the Eastern Ghats, India: evidence for isobaric cooling, fluid buffering, and terminal channelized fluid flow. Eur J Miner 7:689–703
Bhowmik SK, Pal T, Roy A, Pant NC (1999) Evidence for Pre-Greenvillian high-pressure granulite metamorphism from the northern margin of the Sausar mobile belt in central India. J Geol Soc India 53:385–399
Bhowmik SK, Pal T, Pant NC, Shome S (2000) Implication of Ramakona cordierite gneiss in the crustal evolution of Sausar mobile belt in central India. In: Precambrianrian Crust in Eastern and Central India, vol 57. Geol Surv India, Spec Publ, pp 131–150

Bhowmik SK, Sarbadhikari AB, Spiering B, Raith MM (2005) Mesoproterozoic reworking of Palaeoproterozoic ultrahigh-temperature granulites in the Central Indian Tectonic Zone and its implications. J Petrology 46(6):1085–1119

Bhowmik SK, Saha L, Dasgupta S, Fukuoka M (2009) Metamorphic phase relations in orthopyroxene-bearing granitoids: implication for high-pressure metamorphism and prograde melting in the continental crust. J Metamor Geol 27:295–315

Bhowmik SK, Berhardt HJ, Dasgupta S (2010) Grenvillian age high pressure upper amphibolites-granulite metamorphism in the Aravalli–Delhi Mobile Belt, Northwestern India: new evidence from the monazite chemical ages of zircons. J Asian Earth Sci 22:189–207

Bhowmik SK, Wilde SA, Bhandari A, Pal T, Pant NC (2012) Growth of the Greater Indian Landmass and its assembly in Rodinia: geochronological evidence from the Central Indian Tectonic Zone. Gondwana Res 22:54–72

Bhowmik SK, Wilde SA, Bhandari Sarbadhikari AB (2014) Zoned monazite and zircon as monitors for the thermal history of granulite terranes: an example from the Central Indian Tectonic Zone. J Petrol 55(3):585–621

Bhowmik SK, Dasgupta S, Baruah S, Kalita D (2018) Thermal history of a late mesoproterozoic paired metamorphic belt (?) during rodinia assembly: new insight from medium-pressure granulites from the Aravalli-Delhi Mobile Belt, Northwestern India. Geosci Front 9:335–354

Bhui U, Sengupta P, Sengupta P (2007) Phase relations in mafic dykes and their host rocks from Kondapalle, Andhra Pradesh, India: implications for the time–depth trajectory of the Palaeoproterozoic (late Archean?) granulites from southern eastern Ghats Belt. Precamb Res 156:153–174

Bickford ME, Basu A, Mukherjee A, Hietpas J, Schieber J, Patranabis-Deb S, Ray RK, Guhey R, Bhattacharya P, Dhang PC (2011) New U-Pb SHRIMP zircon ages of the Dhamda Tuff in the Mesoproterozoic Chhattisgarh Basin, Peninsular India: stratigraphic implications and significance of a 1-Ga thermal-magmatic event. J Geol 119:535–548

Biju-Sekhar S, Yokoyama K, Pandit M, Okudaira T, Yoshida M, Santosh M (2003) Late Paleoproterozoic magmatism in Delhi Fold Belt, NW India and its implication: evidence from EPMA chemical ages of zircons. J Asian Earth Sci 22:189–207

Biswal TK, Sinha S (2003) Deformation history of the NW salient of the eastern Ghats mobile belt, India. J Asian Earth Sci 22:157–169

Biswal TK, Gyani KC, Parthasarathy R, Pant DR (1998a) Implications of the geochemistry of the pelitic granulites of the Delhi supergroup, Aravalli Mountain Belt, Northwestern India. Precambr Res 87:75–85

Biswal TK, Sanjeevan O, Nayak BP (1998b) Deformational history of Eastern Ghat mobile belt around Lathore, Balangir District, Orissa. J Geol Soc India 51:219–225

Biswal TK, Jena SK, Datta S, Das R, Khan K (2000) Deformation of the terrane boundary shear zone (Lakhna Shear Zone) between the Eastern Ghats Mobile belt and the Bastar craton, in Balangir and Kalahandi districts of Orissa. J Geol Soc India 55:367–380

Biswal TK, Biswal B, Mitra S, Roy Mulik M (2002) Deformation pattern of the NW terrane boundary of the Eastern Ghats Mobile Belt, India. A tectonic model and correlation with Antarctica. Gondwana Res 5:45–52

Biswal TK, Waele BD, Ahuja H (2007) Timing and dynamics of the juxtaposition of the Eastern Ghats Mobile Belt against the Bhandara Craton, India: a structural and zircon U-Pb SHRIMP study of the fold-thrust belt and associated nepheline syenite plutons. Tectonics 26:1–21

Blackburn WH, Srivastava DC (1994) Geochemistry and tectonic significance of the Ongarbira metavolcanic rocks, Singhbhum district, India. Precambr Res 67:181–206

Blereau E, Clark C, Taylor RJM, Johnson TE, Fitzsimons ICW, Santosh M (2016) Constraints on the timing and conditions of high-grade metamorphism, charnockite formation and fluid-rock interaction in the Trivandrum Block, southern India. J Metamor Geol 34:527–549

Bora S, Kumar S, Yi K, Kim N, Lee TH (2013) Geochemistry and U-Pb SHRIMP zircon chronology of granitoids and microgranular enclaves from Jhirgadandi Pluton of Mahakoshal Belt, Central India Tectonic Zone, India. J Asian Earth Sci 70–71:99–114

Bose MK, Chakrabarty MK (1981) Fossil marginal basin from the Indian shield: a model for the evolution of the Singhbhum precambrian belt, Eastern India. Geol Rund 70:504–518

Bose S, Das K (2007) Sapphirine + quartz assemblage in contrasting textural modes from the Eastern Ghats Belt, India: implications for stability relations in UHT metamorphism and retrograde processes. Gondwana Res 11:492–503

Bose S, Dasgupta S (2018) Eastern Ghats Belt, Grenvillian–age tectonics and the evolution of the Greater Indian Landmass: a critical perspective. J Indian Inst Sci Multi Rev J. ISSN: 0970-4140 Coden-JIISADA

Bose MK, Chakrabarty MK, Saunders AD (1989) Petrochemistry of lavas from Proterozoic volcanic belt, Singhbhum, Eastern India. Geol Rund 78:633–648

Bose PK, Mazumder R, Sarkar S (1997) Tidal sand waves and related storm deposits in the transgressive Protoproterozoic Chaibasa Formation, India. Precamb Res 84:63–81

Bose S, Fukuoka M, Sengupta S, Dasgupta S (2000) Evolution of high Mg–Al gran-ulites from Sunkarametta, Eastern Ghats, India: evidence for a lower crustal heating-cooling trajectory. J Metam Geol 18:223–240

Bose S, Seth P, Dasgupta N (2017) Meso-neoproterozoic mid-crustal metamorphic record from the Ajmer-Shrinagar section, Rajasthan, India and its implication to assembly of the Greater Indian landmass during the Grenvillian-age orogenesis. In: Pant NC, Dasgupta S (eds) Crustal evolution of India and Antarctica: the supercontinent connection, vol 457. Geol Soc of London, Spec Publ

Braun I (2006) Pan-African granitic magmatism in the Kerala Khondalite Belt, southern India. J Asian Earth Sci 28:38–45

Braun I, Bröcker M (2004) Geochronology of granitic gneisses and leucogranites from the Kerala Khondalite Belt, southern India: implications for late Proterozoic crustal evolution in east Gondwana. Int J Earth Sci 93:13–22

Brown M, Raith M (1996) First evidence of ultrahigh-temperature decompression from the granulite province of southern India. J Geol Soc 153:819–822

Buick I, Allen C, Pandit M, Rubatto D, Hermann J (2006) The proterozoic magmatic and metamorphic history of the banded gneiss complex, central Rajasthan, India: LA-ICP-MS U-Pb zircon constraints. Precambr Res 151:119–142

Cenki B, Kriegsman LM, Braun I (2002) Melt producing and melt consuming reactions in the Achankovil cordierite gneisses, southern India. J Metamor Geol 20:543–561

Cenki B, Braun I, Bröcker M (2004) Evolution of the continental crust in the Pan-African mobile belt of southernmost India: evidence from Nd isotope mapping combined with U-Pb and Rb–Sr geochronology. Precamb Res 134:275–292

Chacko T, Ravindra Kumar GR, Newton RC (1987) Metamorphic P-T conditions of the Kerala (South India) Khondalite belt: a granulite-facies supracrustal terrain. J Geol 96:343–358

Chacko T, Lamb M, Farquhar J (1996) Ultra-high temperature metamorphism in the Kerala Khondalite Belt. Gondwana Res Group Mem 3:157–165

Chakraborty PP, Paul S, Das A (2005) Facies development and depositional environment of the Mungra Sandstone, Kolhan Group, Eastern India. J Geol Soc India 65:753–757

Chandrasekhar E, Ramesh D, Biswal TK (2018) Magnetotelluric evidence on the southward extension of the Eastern Ghats mobile belt from Ongole, India. J Asian Earth Sci 164:23–32

Chatterjee BK, Bhattacharya AK (1969) Tectonics and sedimentation in a precambrian shallow epicontinental basin. J Sedi Petrol 39(4):1566–1572

Chatterjee N, Ghose NC (2011) Extensive early Neoproterozoic high-grade metamorphism in north Chotanagpur gneissic complex of the Central Indian Tectonic Zone. Gondwana Res 20:362–379. https://doi.org/10.1016/j.gr.2010.12.003

Chattopadhyay A (2015) Discussion on "Carbon and oxygen isotope systematic of a Paleoproterozoic cap-carbonate sequence from the Sausar Group, central India" by Mohanty S, Barik A, Sarangi S and Sarkar A (2015) published in Palaeogeography, Palaeoclimatology, Palaeoecology 417, 195–209. Palaeogeog, Palaeoclimat, Palaeoeco 433:156–157

Chattopadhyay A (2020) Proterozoic orogenesis and crustal evolution in the Central Indian Tectonic Zone: current understanding from the recent works (Status Report 2015–2019). Proc Indian Nat Sci Acad, New Delhi, Special Issue, INSA Report to IUGS

Chattopadhyay A, Khasdeo L (2011) Structural evolution of Gavilgarh-Tan Shear Zone, central India: a possible case of partitioned transpression during Mesoproterozoic oblique collision within Central Indian Tectonic Zone. Precamb Res 186:70–88

Chattopadhyay A, Huin AK, Khan AS (2003a) Structural framework of Deolapar area, central India and its implications for Proterozoic nappe tectonics. Gondwana Res 6:107–117

Chattopadhyay A, Khan AS, Huin AK, Bandyopadhyay BK (2003b) Reinterpretation of stratigraphy and structure of Sausar Group in Ramtek-Mansar-Kandri area, Maharashtra, Central India. J Geol Soc India 61:75–89

Chattopadhyay A, Khasdeo L, Holdsworth RE, Smith SAF (2008) Fault reactivation and pseudotachylite generation in the semi-brittle and brittle regimes: examples from the Gavilgarh-Tan Shear Zone, central India. Geol Mag 145:766–777

Chattopadhyay N, Mukhopadhyay D, Sengupta DP (2012) Reactivation of basement: example from the Anasagar granite gneiss complex, Rajasthan, Western India. In: Mazumder R, Saha D (eds) Proterozoic of India. Geol Soc London Spec Publ 366:219–245

Chattopadhyay A, Das K, Hayasak Y, Sarkar A (2015a) Syn-and post-tectonic granite plutonism in the Sausar Fold Belt, central India: age constraints and tectonic implications. J Asian Earth Sci 107:110–121

Chattopadhyay S, Upadhyay D, Nanda JK, Mezger K, Pruseth KL, Berndt J (2015b) Proto-India was a part of Rodinia: evidence from Grenville-age suturing of the Eastern Ghats Province with the Paleoarchean Singhbhum Craton. Precamb Res 266:506–529. https://doi.org/10.1016/j.precamres.2015.05.030

Chattopadhyay A, Chatterjee A, Das K, Sarkar A (2017) Neoproterozoic transpression and granite magmatism in the Gavilgarh-Tan Shear Zone, central India: tectonic significance of U-Pb zircon and U-Th-total Pb monazite ages. J Asian Earth Sci 147:485–501

Chaudhuri AK, Deb GK, Patranabis-Deb S, Sarkar S (2012) Paleogeographic and tectonic evolution of the Pranhita-Godavari Valley, Central India: a stratigraphic perspective Amer J Sci 312(7):766–815

Chaudhuri AK, Deb GK, Patranabis-Deb S (2015) Conflicts in stratigraphic classification of the Puranas of the Pranhita-Godavari Valley: Review, recommendations and status of the 'Penganga' Sequence. In: Mazumder R, Eriksson PG (eds) Precambrian Basins of India: Stratigraphic and Tectonic Context. Geol Soc London Mem 43:165–183. https://doi.org/10.1144/M43.12

Chetty TRK (1995) A correlation of proterozoic shear zones between Eastern Ghats Belt, India and Enderby Land, East Antarctica, based on LANDSAT imagery. Mem Geol Soc India 34:205–220

Chetty TRK (2001) The Eastern Ghat mobile belt, India; a collage of juxtaposed terranes. Gondwana Res 4:319–328

Chetty TRK (2017) Proterozoic orogens of India. A critical window to Gondwana, vol 3. Elsevier Inc

Chetty TRK, Murthy DSN (1994) Collisional tectonics in the late Precambrian Eastern Ghats Mobile Belt; Mesoscopic to satellite-scale structural observations. Terra Nova 6:72–81

Chetty TRK, Vijay P, Narayana BL, Girdhar GV (2003) Structure of the Nagavali shear zone, Eastern Ghats Mobile Belt, India: correlation in the East Gondwana reconstruction. Gondwna Res 6(2):215–219

Chhattopadhyay A, Bandyopadhyay K, Khan AS (2001) Geology and structure of the Sausar fold belt: a retrospection and some new thoughts. Geol Surv India, Spec Publ 64:251–264

Choudhary AK, Gopalan K, Sastry CA (1984) Present status of the geochronology of the Precambrian rocks of Rajasthan. Tectonophysics 105:131–140

Clark C, Healy D, Johnson TE, Collins AS, Taylor RJM, Santosh M (2015) Hot orogens and supercontinent amalgamation: a Gondwanan example using SHRIMP U-Pb geochronology, REE geochemistry, phase equilibria and numerical modelling. Gondwana Res 28:1310–1328

Collins AS, Santosh M, Braun I, Clark C (2007) Age and sedimentary provenance of the southern granulites, South India: U-Th–Pb SHRIMP secondary ion mass spectrometry. Precamb Res 155:125–138

Collins AS, Clark C, Chetty TRK, Santosh M (2010) Ediacaran-cambrian tectonic evolution of Southern India. Indian J Geol 80(1–4):23–40

Collins AS, Clark C, Plavsa D (2014) Peninsula India in Gondwana: the tectonothermal evolution of the Southern Granulite Terrain and its Gondwanan counterparts. Gondwana Res 25:190–203

Coulson AL (1933) The geology of the Sirohi State, Rajputana. Mem Geol Surv India 63:166

Crowe WA, Cosca MA, Harris LB (2001) $^{40}Ar/^{39}Ar$ geochronology and Neoproterozoic tectonics along the northern margin of the Eastern Ghats Belt, in north Orissa, India. Precambr Res 108:237–266

Crowe WA, Nash CR, Harris LB, Leeming PM, Rankin LR (2003) The geology of the Rengali Province: implications for the tectonic development of northern Orissa, India. J Asian Earth Sci 21:697–710

Das S, Sahoo SR (2015) The time-transgressive Kolhans in Fan Delta-Lacustrine environment. Paleogeographic and tectonic implications. Int J Earth Sci Eng 8(1):45–58

Das S, Nasipuri P, Bhattacharya A, Swaminathan S (2008) The thrust-contact between the Eastern Ghats Belt and the adjoining Bastar craton (Eastern India): evidence from mafic granulites and tectonic implications. Precambr Res 162:70–85

Das K, Yokoyama K, Chakraborty PP, Sarkar A (2009) Basal Tuffs and contemporaneity of the Chattisgarh and Khariar Basins based on new dates and geochemistry. J Geol 117:88–102

Das S, Shukla D, Bhattacharjee S, Mitra SK (2015) Age constraints of Udayagiri domain of Nellore schist belt by xenotime dating around Pamuru, Prakasam district, Andhra Pradesh. J Geol Soc India 85:289–298

Das E, Karmakar S, Dey A, Karmakar S, Sengupta P (2017) Reaction textures, pressure–temperature paths and chemical dates of monazite from a new suite of sapphirine–spinel granulites from parts of the Eastern Ghats Province, India: insights into the final amalgamation of India and East Antarctica during the formation of Rodinia. In: Pant NC, Dasgupta S (eds) Crustal evolution of India and Antarctica: the supercontinent connection, vol 457. Geol Soc, London, Spec Publ. https://doi.org/10.1144/SP457.12

Dasgupta S, Sengupta P (1995) Ultrametamorphism in Precambrian granulite terrains–evidence from Mg-Al granulites and calc-granulites of the Eastern Ghats, India. Geol J 30:307–318

Dasgupta S, Sengupta P (2003) Indo-Antarctic correlation: a perspective from the Eastern Ghats granulite belt, India. In: Yoshida M, Windley BF, Dasgupta S (eds) Proterozoic East Gondwana: supercontinent assembly and breakup, vol 206. Geol Soc London Spl Publ, pp 131–143

Dasgupta S, Sengupta P, Fukuoka M, Bhattacharya PK (1991) Mafic granulites from the Eastern Ghats, India: further evidence for extremely high temperature crustal metamorphism. J Geol 99:124–133

Dasgupta S, Sengupta P, Fukuoka M, Chakraborti S (1992) Dehydration melting, fluid buffering and decompressional P-T Path in a granulite complex from the Eastern Ghats, India. J Metamor Geol 10:777–788. https://doi.org/10.1111/j.1525-1314.1992.tb00122.x

Dasgupta S, Sengupta P, Mondal A, Fukuoka M (1993) Mineral chemistry and reaction textures in metabasites from the Eastern Ghats belt, India and their implications. Min Mag 57:113–120. https://doi.org/10.1180/minmag.1993.057.386.11

Dasgupta S, Sanyal S, Sengupta P, Fukuoka M (1994) Petrology of granulites from Anakapalle—evidence for Proterozoic decompression in the Eastern Ghats, India. J Petrol 35:433–459

Dasgupta S, Sengupta P, Ehl J, Raith M, Bardhan S (1995) Reaction textures in a suite of spinel granulites from the Eastern Ghats Belt, India: evidence for polymetamorphism, a partial petrogenetic grid in the system KFMASH and roles of ZnO and Fe_2O_3. J Petrol 36:435–461

Dasgupta S, Ehl J, Raith MM, Sengupta P (1997) Mid-crustal contact metamorphism around the Chimakurthy mafic-ultramafic complex, Eastern Ghats Belt, India. Contrib Min Petrol 129:182–197

Dasgupta S, Bose S, Das K (2013) Tectonic evolution of the Eastern Ghats Belt, India. Precambr Res 227:247–258

Dasgupta S, Bose S, Bhowmik SK, Sengupta P (2017) The Eastern Ghats Belt, India, in the context of supercontinent assembly. In. Pant NC, Dasgupta S (eds) Crustal evolution of India and Antarctica: the supercontinent connection, vol 457. Geol Soc London Spec Publ. https://doi.org/10.1144/SP457.5

De Wall H, Pandit MK, Chauhan NK (2012) Paleosol occurrences along the Archean-Proterozoic contact in the Aravalli Craton, NW India. Precambr Res 216:120–131

Deb M, Sarkar SC (1990) Proterozoic tectonic evolution and metallogenesis in the Aravalli-Delhi Orogenic Complex, northwestern India. Precambr Res 46:115–137

Deb M, Sehgal U (1997) Petrology, geothermobarometry and C-O-H-S fluid compositions in the environs of Rampura-Agucha Zn-(Pb) ore deposits, Bhilwara District, Rajasthan. Proc Indian Acad Sci (Earth Planet Sci) 106:343–356

Deb M, Thorpe RI (2004) Geochronological constraints in the precambrian geology of Rajasthan and their metallogenic implications: sediment-hosted lead-zinc sulphide deposit. Narosa Publishing House New Delhi, pp. 246–263

Deb M, Thorpe RI, Cumming GL, Wagner PA (1989) Age, source and stratigraphic implications of Pb isotope data for conformable, sediment-hosted, base metal deposits in the Proterozoic Aravalli-Delhi Orogenic Belt, Northwestern India. Precambr Res 43:1–22

Deb M, Thorpe RI, Kristic D, Corfu F, Davis DW (2001) Zircon U-Pb and galena Pb-isotope evidence for an approximate 1.0 Ga terrane constituting the western margin of the Aravalli-Delhi orogenic belt, northwestern India. Precambr Res 108:195–213

Desai SJ, Patel MP, Merh SS (1978) Polymetamorphites of balaram-abu road area, North Gujarat and Southwest Rajasthan. J Geol Soc India 19(9):383–394

Dharma Rao CV, Santosh M, Purohit R, Wang J, Jiang X, Kusky T (2011a) LA-ICP-MS U-Pb zircon age constraints on the Paleoproterozoic and Neoarchean history of the Sandmata Complex in Rajasthan within the NW Indian Plate. J Asian Earth Sci 42:286–305

Dharma Rao CV, Santosh M, Wu Y (2011b) Mesoproterozoic ophiolitic mélange from the SE periphery of the Indian plate: U-Pb zircon ages and tectonic implications. Gondwana Res 19:384–401

Dharma Rao CV, Windley BF, Choudhary AK (2011c) The Chimalpahad anorthosite complex and associated basaltic amphibolites, Nellore Schist Belt, India: magma chamber and roof of a Proterozoic island arc. J Asian Earth Sci 40:1027–1043

Dharma Rao CV, Santosh M, Kim S (2012) Cryogenian volcanic arc in the NW Indian Shield: Zircon SHRIMP U Pb geochronology of felsic tuffs and implications for Gondwana assembly. Gondwana Res 22:36–53

Dharma Rao CV, Santosh M, Kim SW, Li S (2013) Arc magmatism in the Delhi Fold Belt: SHRIMP U–Pb zircon ages of granitoids and implications for Neoproterozoic convergent margin tectonics in NW India. J Asian Earth Sci 78:83–99

Dobmeier C, Raith M (2003) Crustal architecture and evolution of the Eastern Ghats Belt and adjacent regions of India. In: Yoshida M, Windley BF, Dasgupta S (eds) Proterozoic east Gondwana: Supercontinent assembly and breakup, vol 206. Geol Soc London Spec Publ, pp 145–168

Dobmeier C, Simmat R (2002) Post-Grenvillian transpression in the Chilka Lake area, Eastern Ghats Belt—implications for the geological evolution of peninsular India. Precambr Res 113:243–268

Dobmeier C, Lutke S, Hammerschmidt K, Mezger K (2006) Emplacement and deformation of the Vinukonda meta-granite (Eastern Ghats, India)—Implications for the geological evolution of peninsular India and for Rodinia reconstructions. Precambr Res 146:165–178

Drury SA, Holt RW (1980) The tectonic framework of the south India craton: a reconnaissance involving Landsat imagery. Tectonophysics 65:T1–T15

Dunn JA, Dey AK (1942) The geology and petrology of eastern Singhbhum and surrounding areas. Mem Geol Surv India 69 2:281–450

Dutta MK (1980) Petrology of rocks around Jhamarkotra and its bearing on phosphorite formation. M.Sc. Dissertation Univ Delhi 51p evaluation. Q J Geol Min Metal Soc India 52(2):67–76

Dwivedi AK, Pandey UK, Murugana C, Bhatt AK, Ramesh Babu PV, Joshi M (2011) Geochemistry and Geochronology of A-type Barabazar Granite: implications on the geodynamics of South Purulia Shear Zone, Singhbhum Craton, Eastern India. J Geol Soc India77:527–538

Eriksson KA, Simpson EL (2004) Precambrian tidalites: recognition and significance. In: Eriksson PG, Altermann W, Nelson DR, Mueller WU, Cateneau O (eds) The Precambrian Earth: tempos and events. Developments in Precambrian Geology, vol 12. Elsevier Science, Amsterdam, pp 631–642

Eriksson PG, Condie KC, Tirsgaard H, Muller WU, Altermann W, Miall AD, Aspler LB, Catuneanu O, Chiarenzelli JR (1998) Precambrian clastic sedimentation systems. Sedim Geol 120:5–53

Fareeduddin, Banerjee DM (2020) Aravalli Craton and mobile belt: An update. In: Faredduddin et al (eds) Geodynamic evolution of Indian subcontinent. Episodes, 43(1):88–108, https://doi.org/10.18814/epiiugs/2020/20005

Fareeduddin, Kröner A (1998) Single zircon age constraints on the evolution of Rajasthan granulites. In: Paliwal BS (ed) Precambrian rocks of India, pp 547–556

Fareeduddin I, Kirmani TR, Srivastava BL, Reddy AB, Bhattacharya J (1995) Lamprophyre dykes in South Delhi fold belt near Pipela, district Sirohi, Rajasthan. J Geol Soc India 46(3):255–261

Fareeduddin BR, Venkatesh R, Hanumantha Golani PR, Sharma BB, Neogi S (2014) Petrology and stable isotope (S, C, O) studies of selected sediment-hosted basemetal ore deposits in Proterozoic Aravalli-Delhi Fold Belts, Rajasthan. J Geol Soc India 83:119–141

Fermor LL (1930) Age of the Aravalli Range. Geol Surv India, 62

Friend CRL, Nutman AP (1991) SHRIMP U-Pb geochronology of the Closepet Granite and peninsular gneiss, Karnataka, south India. J Geol Soc India 38:357–368

Gandhi SM, Plival HV, Bhanagar SN (1984) Geology and ore reserve estimates of Rampura-Agucha Zinc-Lead Deposit, Bhilwara District, Rajasthan. J Geol Soc India 25(11):689–705

Ghosh SK (1963) Structural, metamorphic and migmatitic history of the area around Kuilapal, eastern India. Q J Geol Min Metal Soc India 35:211–234

Ghosh NC, Saha A (2018) Vestiges of older greenstone in Mesoarchaean Chakradharpur granite gneiss, Singhbhum Craton, India: implications for plume-lithosphere interaction at rifted cratonic margin. Geol J 1–23. https://doi.org/10.1002/gj3271

Ghosh JG, De Wit MJ, Zartman RE (2004) Age and tectonic evolution of Neoproterozoic ductile shear zonesin the Southern Granullite Terrain of India, with implications for Gondwana studies. Tectonics 23:TC3006. https://doi.org/10.1029/2002tc001444

Golani PR, Bandyopadhyay BK, Gupta A (2001) Gavilgarh-Tan shear: a prominent ductile shear zone in central India with multiple reactivation history. Geol Surv India, Spec Publ 64:265–272

Gopalan K, Kumar A, Kumar S, Vijayagopal B (2013) Depositional history of the Upper Vindhyan succession, central India: time constraints from Pb–Pb isochron ages of its carbonate components. Precambr Res 233:108–117

Gopalkrishnan K (2001) A paleo-Andean type margin within South Granulite Terrain, India. Geol Surv India Spec Publ 55:85–96

Gouda HC, Fareeduddin, Singh RK, Rajaram H, Kumar R, Kesavan R, Sharma R (2015) Aeromagnetic anomaly maps of Aravalli Craton and their interpretations: some new insights on stratigraphy and metallogeny of the region. In: GSI (2011) Geology and Mineral Resources of Rajasthan. Misc Pub, 30(3):1–130

Gregory LC, Meert JG, Bingen BH, Pandit MK, Torsvik TH (2009) Palaeomagnetic and geochronological study of Malani Igneous Suite, NW India: implications for the configuration of Rodinia and the assembly of Gondwana. Precambr Res 170:13–26

GSI and ISRO (1994) Project Yasundhara: Generalised geological map. Geol Surv India and Indian Space Res Organ Publ

Gupta BC (1934) The geology of central Mewar. Mem Geol Surv India 65:107–169

Gupta A, Basu A (2000) North Singhbhum Proterozoic mobile belt Eastern India—a review. Geol Surv India Spec Publ 55:195–226

Gupta P, Bose U (2000) An update on geology of the Delhi supergroup in Rajasthan. Geol Surv India Spec Publ 55:287–366

Gupta BC, Mukherjee PN (1938) Geology of Gujarat and southern Rajputana. Rec Geol Surv India 73:103–208

Gupta A, Basu A, Ghosh PK (1980a) The Proterozoic ultramafic and mafic lavas and tuffs of the Dalma greenstone belt, Singhbhum, eastern India. Can J Earth Sci 17:210–231

Gupta SN, Arora YK, Mathur RK, Iqbaluddin, Prasad B, Sahai TN, Sharma SB, Sahai BP, Sharma SB (1980b) 1:1000,000 Lithostratigraphic map of Aravalli region, southern Rajasthan and northeastern Gujarat. Geol Surv India Calcutta–Hyderabad

Gupta A, Basu A, Singh SK (1985) Stratigraphy and petrochemistry of Dhanjori greenstone belt, eastern India. Jour Geol Min Met Soc India 57:248-263

Gupta P, Fareeduddin RMS, Mukhopadhyay K (1995) Stratigraphy and structure of Delhi Supergroup of rocks in central part of Aravalli Range. Rec Geol Surv India 120:12–26

Gupta SN, Arora YK, Mathur RK, Iqbaluddin Prasad B, Sahai TN, Sharma SB (1997) Precambrian geology of the Aravalli region, southern Rajasthan and northeastern Gujarat. Mem Geol Surv India 123:262

Gupta S, Bhattacharya A, Raith M, Nanda JK (2000) Contrasting pressure-temperature-deformation history across a Vestigal Craton-mobile belt boundary: the Western Margin of the Eastern Ghats Belt at Deobhog, India. J Metamor Geol 18:683–697. https://doi.org/10.1046/j.1525-1314.2000.00288.x

Halden NM, Bowes DR, Dash B (1982) Structural evolution of migmatites in granulite facies terrane: Precambrian crystalline complex of Angul, Orissa, India. Trans. R. Soc. Edinburgh Earth Sci 73:109–118

Hari Prasad B, Okudaira T, Hayasaka Y, Yoshida M, Divi RS (2000) Petrology and geochemistry of amphibolites from the Nellore-Khammam schist belt, SE India. J Geol Soc India 56:67–78

Harley SL, Nandakumar V (2016) New evidence for Palaeoproterozoic High grade metamorphism in the Trivandrum block, Southern India. Precambr Res 280:120–138. https://doi.org/10.1016/j.precamres.2016.04.018

Harris NBW, Santosh M, Taylor PN (1994) Crustal evolution in South India: constraints from Nd isotopes. J Geol 102:139–150

Heron AM (1935) Synopsis of pre-vindhyan geology of Rajputana. Trans Nat Inst Sci India 1:17–33

Heron AM (1953) The geology of central Rajputana. Mem Geol Surv India 79:1–389

Iyenger SVP, Murthy YGK (1982) The evolution of Archean Proterozoic crust in parts of Bihar and Orissa, eastern India. Rec Geol Surv India 112(3):1–5

Jain SC, Yedekar DB, Nair KKK (1991) Central Indian Shear Zone: a major Precambrian crustal boundary. J Geol Soc India 37:521–548

Jain AK (2020) Geological evolution of the himalayan mountains. Springer Nature Switzerland AG, pp. 363–393. https://doi.org/10.1007/978-3-030-15989-4_10

Jain AK, Mukherjee PK, Singhal S (2020) Terrane characterization in the Himalaya since Paleoproterozoic. Episodes 43(1):346–357. https://doi.org/10.18814/epiiugs/2020/020021

Jayananda M, Moyen JF, Martin H, Peucat JJ, Auvray B, Mahabaleshwar B (2000) Late Archean (2500-2520 Ma) juvenile magmatism in the eastern Dharwar craton, southern India: constraints from geochronology, Nd-Sr isotopes and whole rock geochemistry. Precam Res 99:225–254

Johnson T, Clark C, Taylor R, Santosh M, Collins AS (2015) Prograde and retrograde growth of monazite in migmatites: an example from the Nagercoil Block, southern India. Geosci Front 6:373–387

Joshi M, Thomas H, Sharma RS (1993) Granulite facies metamorphism in the Archaean gneiss complex from North-Central Rajasthan. Proc Nat Acad Sci India (Sect A) 63:167

Joy S, Linde GVD, Choudhuri AK, Deb GK, Tappe S (2018) Reassembly of the Dharwar and Bastar cratons at ca. 1 Ga: evidence from multiple tectonothermal events along the Karimnagar granulite belt and Khammam schist belt, southern India. J Earth Sys Sci 127:76. https://doi.org/10.1007/s12040-018-0988-2

Just J, Schulz B, de Wall H, Jourdan F, Pandit MK (2011) Monazite CHIME/EPMA dating of the Erinpura granitoid deformation: implications for neoproterozic tectonothermal evolution of NW India. Gondwana Res 19:402–412

Kamineni DC, Rao AT (1988) Sapphirine granulites from Kakanuru area, Eastern Ghats, India. am Min 73:692–700
Kar R, Bhattacharya S, Sheraton JW (2003) Hornblende-dehydration melting in Mafic Rocks and the Link between Massif-Type Charnockite and associated granulites, Eastern Ghats Granulite Belt, India. Contrib Min Petrol 145:707–729. https://doi.org/10.1007/s00410-003-0468-8
Kaur P, Zeh A, Chaudhri N, Gerdes A, Okrusch M (2011) Archaean to Palaeoproterozoic crustal evolution of the Aravalli mountain range, NW India, and its hinterland: the U-Pb and Hf isotope record of detrital zircon. Precambr Res 187:155–164
Kaur P, Choudhury N, Elyas N (2018) Origin of trondhjemite and albitite at the expanse of A-type granite, Aravalli Orogen, India: evidence for new metasomatic replacement fronts. Geosci Front. http://doi.org/10.1016/j.gsf.2018.09.019
Khan MS, Smith TE, Raza M, Huang J (2005) Geology, Geochemistry and tectonic significance of mafic-ultramafic rocks of Mesoproterozoic Phulad Ophiolite Suite of South Delhi Fold Belt, NW Indian Shield. Gondwana Res 8:553–566
Korhonen FJ, Saw AK, Clark C, Brown M, Bhattacharya S (2011) New constraints on UHT metamorphism in the Eastern Ghats Province through the application of phase equilibria modelling and in situ geochronology. Gondwana Res 20:764–781
Korhonen FJ, Brown M, Clark C, Bhattacharya S (2013a) Osumilite–melt inter-actions in ultrahigh temperature granulites: phase equilibria modelling and implications for the P-T–t evolution of the Eastern Ghats Province, India. J Metamorph Geol 31:881–907
Korhonen FJ, Clark C, Brown M, Bhattacharya S, Taylor R (2013b) How long-livedis ultrahigh temperature (UHT) metamorphism? Constraints from zircon and monazite geochronology in the Eastern Ghats orogenic belt, India. Precambr Res 234:322–350
Kovach VP, Simmat R, Rickers K, Berezhnaya NG, Salnikova EB, Dobmeier C, Raith MM, Yakovleva SZ, Kotov AB (2001) The Western charnockite zone of the Eastern Ghats Belt, India: an independent crustal province of Late Archaean (2.8 Ga) and Paleopro-terozoic (1.7–1.6 Ga) terrains. Gondwana Res 4:666–667
Krause O, Dobmeier C, Raith MM, Mezger K (2001) Age of emplacement of massif-type anorthosites in the Eastern Ghats Belt, India: constraints from U-Pb zircon dating and structural studies. Precambr Res 109:25–38
Krishna V, Prasad RN, Pandey UK, Chabria T, Saxena VP (1996) Rb–Sr geochronology of Chhotanagpur gneiss granulite complex around Kailashnathgufa area, Raigarh district, M.P. In: Proceedings National Symposium Mass Spectrom VIIth, Gwalior, Indian Soc Mass Spectrometry, Defence Res Development Establishment, pp 445–447
Kröner A, Santosh M, Wong J (2012) Zircon ages and Hf isotopic systematics reveal vestiges of Mesoproterozoic to Archaean crust within the late Neoproterozoic-Cambrian high- grade terrain of southernmost India. Gondwana Res 21:876–886
Kröner A, Santosh M, Hegner E, Shaji E, Geng H, Wong J, Xie H, Wan Y, Shang CK, Liu D, Sun M, Nandakumar V (2015) Palaeoproterozoic ancestry of Pan-African high-grade granitoids in southernmost India: implications for Gondwana reconstructions. Gondwana Res 27:1–37
Kumar N, Singh AP, Gupta SB, Mishra DC (2004) Gravity signature, crustal architecture and collision tectonics of the Eastern Ghats Mobile Belt. J Indian Geophys Union 8(2):97–106
Kundu A, Kamal Kazim, Sharma S (2004) Metamorphism in North Delhi Fold Belt—a case study from Southwest Haryana and adjacent parts of Rajasthan. Indian Min 58(1–2):17–26
Lal RK (1997) Internally consistent calibrations for geothermobarometry of high-grade g-Al rich rocks in the system MgO-Al_2O_3-SiO_2 and their application to sapphirine-spinel gran-ulites of Eastern Ghats, India and Enderby Land, Antarctica. Proc Indian Acad Sci (Earth Planet Sci) 103:91–113
Lal RK, Shukla RS (1975) Genesis of cordierite–gedrite cummingtonite rocks of the Mahendragarh district, Haryana-Implication on metamorphism of the Delhi Fold Belt. In: Deb M (ed) Crustal evolution and metallogeny in the Northwestern Indian shield. Narosa Publishing House, New Delhi, pp 138–157

Leelanandam C (1998) Alkaline magmatism in the Eastern Ghats Belt—a critique. Geol Surv India Spl Publ 44:170–179

Leelanandam C, Burke K, Ashwal LD, Webb SJ (2006) Proterozoic mountain building in Peninsular India: analysis based primarily on alkaline rock distribution. Geol Mag 143:195–212

Li ZX, Bogdanova SV et al (2008) Assembly, configuration, and break-up history of Rodinia: a synthesis. Precambr Res 160:179–210. https://doi.org/10.1016/j.precamres.2007.04.021

Lippolt HJ, Hautmann S (1994) Ar40/Ar39 ages of Precambrian manganese ore minerals from Sweden, India and Morocco. Min Depos 30:246–256

Mahalik NK (1994) Geology of the contact between the Eastern Ghats Belt and the north Orissa craton, India. J Geol Soc India 44:41–51

Mahapatro SN, Tripathy AK, Nanda J, Roy A (2009) Coexisting ultramylonite and pseudotachylyte from the eastern segment of the Mahanadi shear zone, eastern Ghats Mobile Belt. J Geol Soc India 74:679–689

Mahato S, Goona Bhattacharya A, Mishra B, Bernhardt H-J (2008) Thermo-tectonic evolution of the North Singhbhum Mobile Belt (eastern India): a view from the western part of the belt. Precambr Res 162:102–127

Maheshwari A, Sial AN, Chittora VK (1999) High $\delta^{13}C$ Paleoproterozoic carbonates from the Aravalli supergroup, Western India. Int Geol Rev 41:949–954

Maheshwari A, Sial AN, Chittora VK, Bhu H (2002) A positive $\delta^{13}C_{carb}$ anomaly in Paleoproterozoic carbonates of the Aravalli craton, Western India-support for a global isotopic excursion. J Asian Earth Sci 21:59–67

Maheshwari A, Sial AN, Gaucher C, Bossi J, Bekker A, Ferreira VP, Romano AW (2010) Global nature of the Paleoproterzoic Lomagundi carbon isotope excursion: a review of occurrences in Brazil, India and Uruguay. Precambr Res 182:274–299

Maji AK, Goon S, Bhattacharya A, Mishra B, Mahato S, Bernhardt H-J (2008) Proterozoic polyphase metamorphism in the Chhotanagpur Gneissic Complex (India), and implication for trans-continental Gondwanaland correlation. Precambr Res 162:385–402. https://doi.org/10.1016/j.precamres.2007.10.002

Malhotra G, Pandit MK (2000) Geology and mineralisation of Jahazpur belt, southeastern Rajasthan. In: Deb M (ed) Crustal evolution and metallogeny in Northwestern Indian Shield. Narosa Publ House, New Delhi, pp 115–125

Mall DM, Reddy PR, Mooney WD (2008) Collision tectonics of the Central Indian suture zone as inferred from a deep seismic sounding study. Tectonophysics 460:116–123

Mallick AK (1993) Dating of granite plutons in Bihar mica belt. Bihar. Rec Geol Surv India 126(2):27–29

Malone SL, Meert JG, Banerjee DM, Pandit MK, Tamrat E, Kamenov GD, Pradhan VR, Sohl LE (2008) Paleomagnetism and detrital zircon geochronology of the Upper Vindhyan sequence, Son Valley and Rajasthan, India: a c. 1000 Ma closure age for the Purana basins? Precambr Res 164:137–159

Mathur RK (1964) Systematic geological mapping in parts of Udaipur district, Rajasthan. Unpubl Progress Report Field Season 1963–64 Geol Surv India

Mazumder R (2005) Proterozoic sedimentation and volcanism in the Singhbhum crustal province, India and their implications. Sedim Geol 176:167–193

Mazumder R, Sarkar S (2004) Sedimentation history of the Palaeoproterozoic Dhanjori Formation, Singhbhum, eastern India. Precambr Res 130:269–289

McKenzie NR, Hughes NC, Myrow PM, Banerjee DM, Deb M, Planavasky NJ, Maheshwari A, Sial AN (2013) New age constraints for the Proterozoic Aravalli-Delhi successions of India and their implications. Precambr Res 238:120–128

Meert JG, Pandit MK, Pradhan VR, Banks J, Sirianni R, Stroud M, Newstead B, Gifford J (2010) Precambrian crustal evolution of Peninsular India: a 3.0-billion-year odyssey. J Asian Earth Sci 39(6):483–515

Mehdi M, Kumar Santosh, Pant NC (2015) Low-grade metamorphism in the Lalsot-Bayana Subbasin of the North Delhi Fold Belt and its tectonic implication. J Geol Soc India 85:397–410

Meißner B, Deters P, Srikantappa C, Kohler H (2002) A reappraisal of the pressure-temperature path of granulites from the Kerala Khondalite Belt, southern India. J Geol 108:687–703

Melezhik VA, Huhma H, Condon DJ, Fallick AE, Whitehouse MJ (2007) Temporal constraints on the Paleoproterozoic Lomagundi-Jatuli carbon isotopic event. Geology 35:655–658

Mezger K, Cosca MA (1999) The thermal history of the Eastern Ghats Belt (India), as revealed by U-Pb and $^{40}Ar/^{39}Ar$ dating of metamorphic and magmatic minerals: implications for the SWEAT correlation. Precambr Res 94:251–271

Mishra S (2006) Precambrian chronostratigraphic growth of Singhbhum-Orissa craton, Eastern Indian Shield: an alternative model. J Geol Soc India 67:356–378

Mishra BK, Ravi Kumar M (1994) Stratigraphy and structure of the Precambrian basement rocks and Gondwanawana sequences in Sarguja, Raigarh and Bilaspur districts of Madhya Pradesh. Rec Geol Surv India 127:219–230

Mishra DC, Ravi Kumar M (2014) Proterozoic orogenic belts and rifting of Indian cratons: geophysical constraints. Geosci Front 5:25–41

Mishra DC, Singh B, Tiwari VM, Gupta SB, Rao MBSV (2000) Two cases of continental collisions and related tectonics during the Proterozoic period in India insights from gravity modelling constrained by seismic and magnetotelluric studies. Precambr Res 99:149–169

Mishra B, Upadhyay D, Bernhardt HJ (2006) Metamorphism of the host and associated rocks at the Rajpura-Dariba massive sulfide deposit, Northwestern India. J Asian Earth Sci 26(1):21–37. https://doi.org/10.1016/j.jseaes.2004.09.012

Moeen S (1998) P-T estimates from the Nellore schist belt (India) and evidence for superimposed metamorphic events. Geol J 33:1–15

Mohan A, Prakash D, Motoyoshit Y (1996) Decompressional P-T history in sapphirine-bearing granulites from Kodaikanal, southern India. J Southeast Asian Earth Sci 14:231–243

Mohan A, Tripathi P, Motoyoshi M (1997) Reaction history of sapphirine granulites and a decompressional P-T path in a granulite complex from the Eastern Ghats. Proc Indian Acad Sci (Earth Planet Sci) 106:115–129

Mohan A, Singh PK, Sachan HK (2003) High density carbonic fluid inclusions in charnockites from Eastern Ghats, India: petrologic implications. J Asian Earth Sci 22:101–130

Mohanty S (2010) Tectonic evolution of the Satpura mountain belt: a critical evaluation and implication on supercontinent assembly. J Asian Earth Sci 39:516–526

Mohanty SP, Nanda S (2016) Geochemistry of a paleosol horizon at the base of the Sausar Group, central India: implications on atmospheric conditions at the Archean-Paleoproterozoic boundary. Geosci Front 7(5):759–773

Mohanty SP, Barik A, Sarangi S, Sarkar A (2015) Carbon and oxygen isotope systematics of a Paleoproterozoic cap-carbonate sequence from the Sausar Group, central India. Palaeogeog Palaeoclimaet Palaeoeco 417:195–209

Mondal M (2009) Was Bundelkhand-Aravalli nucleus part of Ur supercontinent? Curr Sci 96:33–35

Morimoto T, Santosh M, Tsunogae T, Yoshimura Y (2004) Spinel+Quartz association from the Kerala khondalites, southern India: evidence for ultrahigh-temperature metamorphism. J Mineral Petrol Sci 99(5):257–278. https://doi.org/10.2465/jmps.99.257

Mukherjee A, Bickford ME, Hietpas J, Schieber J, Basu A (2012) Implications of a Newly Dated c. 1000-Ma Rhyolitic Tuff in the Indravati Basin, Bastar Craton, India. J Geol 120:477–485

Mukherjee PK, Jain AK, Singhal S, Singha NB, Singh S, Kumud K, Seth P, Patel RC (2019) U-Pb zircon ages and Sm-Nd isotopic characteristics of the Lesser and Great Himalayan sequences, Uttarakhand Himalaya, and their regional tectonic implications. Gondwana Res 75:282–297. https://doi.org/10.1016/j.gr.2019.06.001

Mukhopadhyay D (1984) Precambrian plate tectonics in the Eastern Indian shield. In: Sychanthavong SPH (ed) Crustal evolution and metallogeny. Oxford & IBH Publ Co., New Delhi, pp 75–100

Mukhopadhyay D, Bhattacharya T, Chattopadhyay N, Lopez R, Tobisch OT (2000) Anasagar gneiss: a folded granitoid in the Proterozoic South Delhi fold belt, central Rajasthan. J Earth Sys Sci 109(1):22–38

Nagori DK (1988) Paleogeographic and depositional environment of the phosphorite bearing early Proterozoic carbonate rocks around Udaipur, Rajasthan. In: Roy AB (ed) Precambrian of Aravalli Mountain, Rajasthan, India. Mem Geol Soc India, vol 7, pp 139–152

Naha K, Halyburton RV (1974) Early Precambrian stratigraphy of central and southern Rajasthan, India. Precambr Res 1:55–73

Naha K, Mohanty S (1990) Structural studies in the Pre-Vindhyan rocks of Rajasthan: a summary of work of the last three decades. In: Naha K, Ghosh SK and Mukhopadhyay D (eds) Structure and tectonics: the Indian scene. Proc Indian Acad Sci Bangalore, vol 99, pp 279–290

Naha K, Mukhopadhyay DK, Mohanty R, Mitra SK, Biswal TK (1984) Significance of contrast in the early stages of the structural history of the Delhi and the pre-Delhi rock group in the Proterozoic of Rajasthan, western India. Tectonophysics 105:193–206

Naha K, Mukhopadhyay DK, Mohanty R (1988) Structural evolution of the rocks of the Delhi Group around Khetri, northeastern Rajasthan. Mem Geol Soc India 7:207–246

Nair KKK, Jain SC, Yedekar DB (1995) Stratigraphy, structure and geochemistry of the Mahakoshal greenstone. Mem Geol Soc India 37:403–432

Nanda J, Gupta S, Dobmeier CJ (2008) Metamorphism of the Koraput Alkaline Complex, Eastern Ghats Province, India–evidence for reworking of a granulite terrane. Precambr Res 165:153–168

Nandakumar V, Harley SL (2018) Geochemical signatures of mid-crustal melting processes and heat production in a hot orogen: the Kerala Khondalite Belt. Lithos, Southern India. https://doi.org/10.1016/j.lithos.2018.11.016

Naqvi SM, Rogers JJW (1987) Precambrian geology of India. Oxford University Press, New York, 223 p

Narayanaswamy S, Chakravarthy SC, Vemban NA, Shukla KD, Subramaniam MR, Venkatesh V, Rao GV, Anandalwar MA, Nagarajaiah RA (1963) The geology and manganese ore deposits of the manganese belt in Madhya Pradesh and adjoining parts of Maharastra. Part I: general introduction. Bull Geol Surv India A–22 (I)

Natarajan V, Nanda JK (1981) Large scale basin and dome structures in the high grade metamorphics near Visakhapatnam, South India. J Geol Soc India 22:584–592

Neogi S, Dasgupta S, Sengupta P, Das N (1999) Ultra-high temperature decompression in a suite of granulites from the Eastern Ghats Belt-Bastar Craton Contact and Its Significance. Mem Gondwana Res Group 5:115–138

Nutman AP, Chadwick B, Ramakrishnan M, Vishwanatha MN (1992) SHRIMP U-Pb ages of detrital zircon on Sargur supracrustal rocks in Western Karnataka, Southern India. J Geol Soc India 39:367–374

Okudaira T, Hamamoto T, Hari Prasad B, Kumar R (2001) Sm–Nd and Rb–Sr dating of amphibolite from the Nellore-Khammam schist belt, SE India: constraints on the collision of the Eastern Ghats terrane and Dharwar–Bastar craton. Geol Mag 138:495–498

Paliwal BS (1988) Deformation pattern in the rocks of the Aravalli Supergroup around Udaipur, Rajasthan. In: Roy AB (ed) Precambrian of the Aravalli Mountain, Rajasthan, India. Geol Soc India Mem, vol 7, pp 153–168

Pandey BK, Gupta JN, Lall Y (1986a) Whole rock and Rb–Sr isochron ages for the granites from Bihar mica belt of Hazaribagh, Bihar, India. Indian J Earth Sci 12:157–162

Pandey BK, Upadhyaya LD, Sinha KK (1986b) Geochronology of Jajawal–Binda–Nagnaha granitoids in relation to uranium mineralisation. Indian J Earth Sci 13:163–168

Pandey BK, Chabria T, Gupta JN (1995) Geochronological characterisation of the Proterozoic terrains of Peninsular India: relevance to the first order target selection for uranium exploration. Explor Res At Min 8:187–213

Pandey BK, Krishna V, Chabria T (1998) An overview of Chotanagpur gneiss–granulite complex and adjoining sedimentary sequences, eastern and central India. Int Sem Precamb Crust in Eastern and Central India. In: Abs vol UNESCO-IUGS-IGCP-368, pp 131–135

Pandey M, Pant NC, Kumar S (2013) Criteria to distinguish between regional and contact zone monazite—a case study from Proterozoic North Delhi Fold Belt (NDFB), India. Episodes 36(4):275–289

Pandit MK, Carter LM, Ashwal LD, Tucker RD, Torsvik TH, Jamtveit B, Bhushan SK (2003) Age, petrogenesis and significance of 1 Ga granitoids and related rocks from the Sendra area, Aravalli Craton, NW India. J Asian Earth Sci 22:363–381

Pandit MK, De Wall H, Chauhan NK (2008) Paleosol at the Archean-Proterozoic contact in NW India revisited: evidence for oxidizing condition during paleo-weathering? J Earth Syst Sci 117(3):201–209

Panigrahi MK, Mookherjee A, Pantulu GVC and Gopalan K (1993) Granitoids around Malanjkhand copper deposit: types and age relationship. Earth Planet Sci Lett 102:399–413

Pant NC, Sheo Prasad, Kundu A (2000) Subcalcic amphibole bearing rocks from northern portion of the Khetri copper belt, Rajasthan, India. Lithos 5:175–186

Pant NC, Kundu A, Joshi S (2008) Age of metamorphism of Delhi Supergroup rock selection microprobe ages from Mahendragarh district, Haryana. J Geol Soc India 72:365–372

Papineau D, Purohit R, Goldgerg T, Pi D, Shields DA, Bhu H, Steele A, Fogel ML (2009) High primary productivity and nitrogen cycling after the Paleoproterozoic phosphogenic event in the Aravalli supergroup, India. Precambr Res 171(1–4):37–56

Papineau D, Purohit R, Fogel ML, Shields-Zhou GA (2013) High phosphorus availability as a possible cause for massive cyanobacteria production of oxygen in the Paleoproterozoic atmosphere. Earth Planet Sci Lett 362:225–236

Patranabis Deb S, Bickford ME, Hill B, Chaudhari AK, Basu A (2007) SHRIMP ages of zircon in the uppermost tuff in Chattisgarh Basin in central India require up to 500 Ma adjustments in Indian Proterozoic stratigraphy. J Geol 115:407–416

Paul DK, Ray Barman T, McNaughton NJ, Fletscher IR, Potts PJ, Ramakrishnan M, Augustine PF (1990) Archaean-Proterozoic evolution of Indian charnockites: isotopic and geochemical evidence from granulites of the Eastern Ghats Belt. J Geol 98:253–263

Pichai Muthu R (1990) The occurrence of gabbroic anorthosites in Makrohar area, Sidhi district, Madhya Pradesh. Precambrian of Central India. Geol Surv India Spec Publ 28:320–331

Poddar B, Mathur R (1965) A note on the repetitious sequence of greywacke-slate-phyllite in the Aravalli system around Udaipur, Rajasthan. Bull Geol Soc India 2:83–87

Poddar BC, Mathur RK (1967) A note on the repetitious sequence of greywacke-slate-phyllite in the Aravalli system around Udaipur, Rajasthan. Bull Geol Soc India 2:83–87

Prabhakar N, Bhattacharya A, Sathyanarayanan M, Mukherjee PK (2014) Structural, petrological, and chronological constraints from eastern India and implications for the ~1.0 Ga assembly of greater India. J Geol 122(4):411–432

Pradhan VR, Meert JG, Pandit MK, Kamenov G, Gregory LC, Malone SJ (2010) India's changing place in global Proterozoic reconstructions: a review of geochronologic constraints and paleomagnetic poles from the Dharwar, Bundelkhand and Marwar cratons. J Geodyn 50:224–242

Prakash D, Sharma IN (2011) Metamorphic evolution of the Karimnagar granulite terrane, Eastern Dharwar Craton, south India. Geol Mag 148(1):112–132. https://doi.org/10.1017/S0016756810000488

Prakash D, Shastry A (1999) Geochemical studies on the sapphirine–granulites from Kodaikanal, south India. Gondwana Res 2:603–604

Prakash D, Deepak Singh PC, Singh CK, Tewari S, Arima M, Frimmel HE (2015) Reaction textures and metamorphic evolution of sapphirine–spinel-bearing and associated granulites from Diguva Sonaba, Eastern Ghats Mobile Belt, India. Geol Mag 152(2):316–340. https://doi.org/10.1017/S0016756814000399

Prakash D, Chandra Singh P, Tewari S, Joshi M, Frimmel HE, Hokada T, Rakotonandrasana T (2017) Petrology, pseudosection modelling and U-Pb geochronology of silica-deficient Mg–Al granulites from the Jagtiyal section of Karimnagar granulite terrane, northeastern Dharwar Craton, India. Precambr Res 299:177–194

Prakash D, Vishal B, Naik AS, Yadav R, Rai SK, Tewari S, Yadav MK, Tiwari S, Dash S, Pattnaik C (2019) New occurrence of sapphirine-spinel-bearing granulite from NW of Chilka Lake, Eastern Ghats Belt, Odisha. J Geol Soc India 93:153–156

Purohit R, Sanyal P, Roy AB, Bhattacharya SK, Roy AB (2010) ^{13}C enrichment in the Palaeoproterozoic carbonate rocks of the Aravalli Supergroup, northwest India: influence of depositional environment. Gondwana Res 18:538–546

Purohit R, Bhu H, Mehta P, Avadich PC (2012) Tectono-stratigraphic status of the Proterozoic babarmal pink marble from the Aravalli supergroup. J Geol Soc India 79:235–250

Purohit R, Bhu H, Sarkar A, Ram J (2015) Evolution of the ultramafic rocks of the Rakhabdeev and Jharol Belts in southeastern Rajasthan, India: new evidences from the imagery mapping, petro-mineralogical and OH stable isotope studies. J Geol Soc India 85(3):331–338

Radhakishna BP (1989) Suspect tectonostratigraphic terrain elements in the Indian subcontinent. J Geol Soc India 34:1–24

Radhakrishna BP, Naqvi SM (1986) Precambrian continental crust of India and its evolution. J Geol 94(2):145–166

Radhakrishna T, Mathai J, Yoshida M (1990) Geology and structure of the high-grade rocks from Punalur-Achankovil sector, south India. J Geol Soc India 35:263–272

Rahman MS, Mondal MEA (2015) Evolution of continental crust of the Aravalli craton, NW India, during the Neoarchaean–Palaeoproterozoic: evidence from geochemistry of granitoids. Int Geol Rev 57:1510–1525

Raith MM, Dobmeier C, Mouri H (2007) Origin and evolution of Fe–Al granulites in the thermal aureole of the Chilka Lake anorthosite, Eastern Ghats Province, India. Proc Geol Assoc 118:87–100

Raja Rao CS (1976) Precambrian sequences of Rajasthan. Misc Publ Geol Surv India 23:397–516

Raja Rao CS, Poddar BC, Basu KK, Dutta A (1971) Precambrian stratigraphy of Rajasthan—a review. Rec Geol Surv India 101:60–79

Rajesh HM, Santosh M, Yoshida M (1998) Dextral Pan-African shearing along the southwestern edge of the Achankovil Shear Belt, south India: constraints on Gondwana reconstruction: a discussion. J Geol 106:105–114

Rajesham T, Bhaskar Rao YJ, Murti KS (1993) The Karimnagar granulite terrane–a new sapphirine bearing granulite province, south India. J Geol Soc India 4:51–59

Ramachandra HM (1999) Petrology of Bhandara and Balaghat granulite belt in parts of Maharashtra and Madhya Pradesh. Unpubl Rep Geol Surv India, pp 1–55

Ramachandra HM, Roy A (2001a) Evolution of the Bhandara-Balaghat granulite belt along the southern margin of the Sausar mobile belt of Central India. J Earth Sys Sci 110(4):351

Ramachandra HM, Roy A (2001b) Evolution of the Bhandara-Balaghat granulite belt along the southern margin of the Sausar mobile belt of Central India. Proc Indian Acad Sci (Earth Plan Sci) 110:351–368

Ramakrishnan M (1993) Tectonic evolution of the granulite terranes of southern India. Mem Geol Soc India 25:35–44

Ramakrishnan M, Vaidyanadhan R (2008) Geology of India, vol 1. Geol Soc India, Bangalore, p 556

Ramakrishnan M, Nanda JK, Augustine PF (1998) Geological evolution of the Proterozoic Eastern Ghats mobile belt. Geol Surv India Spl Publ 44:1–21

Ramam PK, Murty VN (1997) Geology of Andhra Pradesh. Geol Soc India, Bangalore India

Rao AT, Kamineni DC, Arima M, Yoshida M (1995) Mineral chemistry and metamorphic P-T conditions of a new occurrence of sapphirine granulites near Madhuravada in the Eastern Ghats, India. In: Yoshida M, Santosh M, Rao AT (eds) India as a fragment of East Gondwana. Gondwana Res Group Mem, vol 2, pp 109–121

Rao VV, Reddy PR (2002) A Mesoproterozoic supercontinent: evidence from the Indian Shield. Gondwana Res 5:63–74

Ratre K, Waele BD, Biswal TK, Sinha S (2010) SHRIMP geochronology for the 1450 Ma Lakhna dyke swarm: its implication for the presence of Eoarchaean crust in the Bastar Craton and 1450–517 Ma depositional age for Purana basin (Khariar), Eastern Indian Peninsula. J Asian Earth Sci 39:565–577

Ravindra Kumar GP, Srikantappa C, Hansen EC (1985) Charnockite in the making at Ponmudi, Kerala, South India. Nature 313:207–209
Ray SK (1974) Structural history of the Saladipura pyrite-pyrrhotite deposit and associated rocks, Khetri Copper Belt, Rajasthan. J Geol Soc India 15:227–246
Ray SK (1988) Structural controls of copper mineralization near Bajta, Ajmer Dist., Rajasthan. Mem T, Geol Soc India, pp 363–372
Ray KK, Ghosh Roy AK, Sengupta S (1996) Acid volcanic rocks between the Dalma volcanic belt and the Chhotanagpur Gneissic Complex, East Singhbhum and Purulia districts of Bihar and West Bengal. Indian Min 50:1–8
Raza M, Khan MS (1993) Geochemical characterization of the Delwara volcanic: evidence for the origin of Aravalli basin by extensional rifting. In: Casshyap SM (ed) Rift basins and Aulacogens. Gyanodaya Prakashan, Nainital, India, pp 119–129
Rekha S, Bhattacharya A (2013) Growth, preservation of Paleoproterozoic-shear-zone-hosted monazite, north of the Western Dharwar Craton (India), and implications for Gondwanaland assembly. Contrib Min Petrol 166(4):1203–1222. https://doi.org/10.1007/s00410-013-0917-y
Rekha S, Upadhyay D, Bhattacharya A, Kooijman E, Goon S, Mahato S, Pant NC (2011) Lithostructural and chronological constraints for tectonic restoration of Proterozoic accretion in the Eastern Indian Precambrian shield. Precambr Res 187(3–4):313–333. https://doi.org/10.1016/j.precamres.2011.03.015
Rickers K, Mezger K, Raith MM (2001) Evolution of the continental crust in the Pro-terozoic Eastern Ghats Belt, India and new constraints for Rodinia reconstruction: implications from Sm-Nd, Rb-Sr and Pb–Pb isotopes. Precambr Res 112:183–212
Roy AB (1988) Stratigraphic and tectonic framework of the Aravalli Mountain range. Geol Soc India Mem 7:3–31
Roy A, Devarajan MK (2000) A reappraisal of the stratigraphy and tectonics of the Proterozoic Mahakoshal belt, Central India. Precambrian Crust in Eastern and Central India. UNESCO-IUGS-IGCP-368. Geol Surv India Spec Publ 17:79–97
Roy A, Hanuma Prasad M (2001) Tectonomagmatic history of Tan Shear Zone and its environs: a preliminary study. Geol Surv India Spec Publ 64:273–288
Roy A, Hanuma Prasad M (2003) Tectonothermal events in Central Indian Tectonic Zone (CITZ) and its implications in Rodinian crustal assembly. J Asian Earth Sci 22:115–129
Roy AB, Jakhar SR (2002) Geology of Rajasthan: precambrian to recent. Scientific Publisher Jodhpur, India, p 421
Roy AB, Kataria P (1999) Precambrian geology of the Aravalli Mountains and neighbourhood: analytical update of recent studies. In: Kataria P (ed) Proceedings of seminar on geology of Rajasthan: status and perspective, Udaipur, pp 1–54
Roy AB, Paliwal BS (1981) Evolution of the lower Proterozoic epicontinental deposits: stromatolite bearing Aravalli rocks of Udaipur, Rajasthan, India. Precambr Res 14:49–74
Roy AB, Purohit R (2015) Lithostratigraphic, geochronological and depositional framework of the Precambrian basin of the Aravalli Mountains and adjacent areas, Rajasthan, India. Geol Soc London Mem 43(1):555–565
Roy AB, Rathore S (1999) Tectonic setting and model of evolution of granulites in central part of the Aravalli mountain belt, Rajasthan. In: Murthy NGK, Ram Mohan V (eds) Charnockite and granulite facies rocks, vol 4. Geol Assoc TN, Spec Publ Geology Dept., Chennai, India, pp 111–126
Roy AB, Paliwal BS, Shekhawat SS, Nagori DK, Golani PR, Bejarniya BR (1988) Stratigraphy of the Aravalli Supergroup in the type area. Geol Soc India Mem 7:121–131
Roy AB, Sharma BL, Paliwal BS, Chauhan NK, Nagori DK, Golani PR, Bejarnia BR, Bhu H, Ali Sabah M (1993) Lithostratigraphy and tectonic evolution of the Aravalli Supergroup: a protogeosysnclinal sequence. In: Cassayap SM (ed) Rift basins and aulocogens. Ganodiya Prakashan, Nainital
Roy A, Ramachandra HM, Bandopadhyay BK (2000) Supracrustal belts and their significance in the crustal evolution of Central India. Geol Surv India Spec Publ 55:361–380

Roy A, Hanuma Prasad M, Bhowmik SK (2001a) Recognition of pre-Grenvillian and Grenvillian tectonothermal events in the Central Indian Tectonic Zones: implications on Rodinian crustal assembly. Gondwana Res 4:755–759
Roy AB, Kröner A, Laul V (2001b) Detrital zircons constraining basement age in a late Archaean greenstone belt of south-eastern Rajasthan, India. Curr Sci 81:407–410
Roy A, Devarajan MK, Hanuma Prasad M (2002a) Ductile shearing and syntectonic granite emplacement along the southern margin of the Palaeoproterozoic Mahakoshal supracrustal belt: evidences form Singrauli area, Madhya Pradesh. J Geol Soc India 59:9–21
Roy A, Hanuma Prasad M, Devarajan MK (2002b) Palaeoproterozoic low pressure metamorphism, deformation and syn-kinematic granite emplacement in the Proterozoic Mahakoshal supracrustal belt of Central India. Gondwana Res 5:489–500
Roy AS, Jeykumar S, Aggarwal SK, Ebihara M (2002c) Sm Nd age and mantle source characteristics of the Dhanjori volcanic rocks, Eastern India. Geochem Jour 36:503–518
Roy-Choudhury MK, Dasgupta SP (1964) On the geology and mineralization in the Khetri copper belt, Jhunjunu and Sikar Districts, Rajasthan. Ind Min 31(2):188–198
Russell J (1993) Pb–Pb dating of precambrian carbonate rocks. Unpubl Ph.D. thesis Univ Oxford England
Sacks PE, Nambiar CG, Walters LJ (1997) Dextral Pan-African shear along the southwestern edge of the Achankovil shear belt, south India: constraints on Gondwana reconstructions. J Geol 105:275–284
Saha AK (1994) Crustal evolution of Singhbhum, North Orissa, Eastern India. Mem Geol Soc India 27:341
Saha D, Sain A, Nandi P, Mazumder R, Kar R (2015) Tectonostratigraphic evolution of the Nellore schist belt, southern India, since the Neoarchaean. Geol Soc London Mem 43:269–282
Sahoo SR, Das S (2015) Kolhan sedimentation in Chaibasa-Noamundi Basin, Jharkhand, India: a critical observation through lithofacies analysis. Int J Earth Sci Eng 8(2):60–66
Sain A, Saha D, Joy S, Jelsma H, Armstrong R (2017) New SHRIMP age and microstructures from a deformed A-type granite, Kanigiri, Southern India: constraining the hiatus between orogenic closure and post-orogenic rifting. J Geol 125(2):241–259
Santosh M (1996) The Trivandrum and Nagercoil granulite blocks. Gondwana Res Group Mem 3:243–277
Santosh M, Yokoyama K, Acharyya SK (2004) Geochronology and tectonic evolution of Karimnagar and Bhopalpatnam Granulite Belts, Central India. Gondwana Res 7:501–518
Santosh M, Collins AS, Tamashiro I, Koshimoto S, Tsutsumi Y, Yokoyama K (2006a) The timing of ultrahigh temperature metamorphism in Southern India: U-Th–Pb electron microprobe ages from zircon and monazite in sapphirine-bearing granulites. Gondwana Res 10:128–155
Santosh M, Morimoto T, Tsutsumi Y (2006b) Geochronology of the khondalite belt of Trivandrum Block, southern India: electron probe ages and implications for Gondwana tectonics. Gondwana Res 9:261–278
Santosh M, Maruyama S, Sato K (2009) Anatomy of a Cambrian suture in Gondwana: implications for the timing of Gondwana assembly. Gondwana Res 16:321–341
Sarangi S, Gopalan K, Roy AB, Sreenivas B, Sharma SD (2006) Pb–Pb age of carbonates of Jhamarkotra Formation: constraints on the age of Aravalli Supergroup, Rajasthan. J Geol Soc India 67:442–446
Sarangi S, Mohanty SP, Barik A (2017) Rare earth element characteristics of Paleoproterozoic cap carbonates pertaining to the Sausar Group, central India: implications for ocean paleoredox conditions. J Asian Earth Sci 148:31–50
Sarkar A, Paul DK (1998) Geochronology of the Eastern Ghats Precambrian Mobile belt—a review. Geol Surv India Spl Publ 44:51–86
Sarkar SN, Saha AK (1983) Structure and tectonics of the Singhbhum-Orissa iron ore craton, eastern India. Recent Res Geol Hind Publ Corp Delhi 10:1–25

Sarkar T, Schenk V (2014) Two-stage granulite formation in a Proterozoic magmatic arc (Ongole domain of the Eastern Ghats Belt, India): part 1. Petrology and pressure–temperature evolution. Precambr Res 255:485–509

Sarkar A, Bhanumathi L, Balasubrahmanyan MN (1981) Petrology, geochemistry and geochronology of the Chilka Lake igneous complex, Orissa State, India. Lithos

Sarkar SN, Trivedi JR, Gopalan K (1986) Rb–Sr whole rock and mineral isochron age of the Tirodi gneiss, Sausar Group, Bhandara district, Maharashtra. J Geol Soc India 27:30–37

Sarkar SC, Gupta A, Basu A (1992) North Singhbhum Proterozoic mobile belt, Eastern India: its character, evolution and metallogeny. In: Sarkar SC (ed) Metallogeny related tectonics of the Proterozoic mobile belts. Oxford & IBH Publ Co., Calcutta, pp 271–305

Sarkar G, Corfu F, Paul DK, Mcnaughton NJ, Gupta SN, Bishui PK (1993) Early Archaean crust in Bastar craton, central India: a geochemical and isotopic study. Precambr Res 62:127–137

Sarkar A, Boda MS, Kundu HK, Mamgain VV, Ravishankar (1998) Geochronology and geochemistry of Mesoproterozoic intrusive plutonites from the eastern segment of the Mahakoshal greenstone belt, Central India. In: IGCP-368 Seminar on Precambrian Crust in Eastern and Central India, Bhubaneshwar, pp 82–85

Sarkar S, Santosh M, Dasgupta S, Fukuoka M (2003) Very high density CO_2 associated with ultrahigh temperature metamorphism in the Eastern Ghats granulite belt, India. Geology 31:51–54

Sarkar T, Schenk V, Appel P, Berndt J, Sengupta P (2014) Two-stage granulite formation in a Proterozoic magmatic arc (Ongole domain of the Eastern Ghats Belt, India): part 2. LA-ICP-MS zircon dating and texturally controlled in-situ monazite dating. Precamb Res. http://dx.doi.org/10.1016/j.precamres.2014.05.024

Satish-Kumar M (2000) Ultrahigh-temperature metamorphism in Madurai Granulites, Southern India: evidence from carbon isotope thermometry. J Geol 108:479–486

Schidlowski M, Eichmann R, Junge CE (1976) Carbon isotope geochemistry of the Precambrian Lomagundi carbonate province, Rhodesia. Geochim Cosmochim Acta 40:449–455

Sen S (1980) Precambrian stratigraphic sequence in a part of the Aravalli Range, Rajasthan: a re-evaluation. Q J Geol Min Metal Soc India 52(2):67–76

Sen SK, Bhattacharya S (1997) Dehydration melting of micas in the Chilka Lake Khondalites: the link between the metapelites and granitoids. Proc Indian Acad Sci (Earth Planet Sci) 106:277–297

Sen SK, Bhattacharya S, Acharyya A (1995) A multi-stage pressure-temperature record in the Chilka Lake Granulites: the epitome of the metamorphic evolution of the Eastern Ghats, India? J Metamor Geol 13:287–298. https://doi.org/10.1111/j.1525-1314.1995.tb00219.x

Sengupta S, Mukhopadhaya PK (2000) Sequence of Precambrian events in the eastern Indian craton. In: Proceedings of international seminar, Precambrian crust in eastern and central India, (UNESCO-IUGS-IGCP 368), vol 7. Geol Surv India, Bhubaneswar, pp 49–56

Sengupta P, Dasgupta S, Bhattacharya PK, Fukuoka M, Chakraborti S, Bhowmik S (1990) Petrotectonic imprints in the Sapphirine Granulites from Anantagiri, Eastern Ghats Mobile Belt, India. J Petrol 31:971–996

Sengupta S, Paul DK, Bishui PK, Gupta SN, Chakraborty R, Sen P (1994) A geochemical and Rb–Sr isotopic study of Kuilapal granite and Arkasoni granophyre from the eastern Indian craton. Indian Min 48:77–88

Sengupta P, Sen J, Dasgupta S, Raith M, Bhui UK, Ehl J (1999) Ultra-high temperature metamorphism of metapelitic granulites from Kondapalle, Eastern Ghats Belt: implications for the Indo-Antarctic correlation. J Petrol 40:1065–1087. https://doi.org/10.1093/petrology/40.7.1065

Sengupta S, Sarkar G, Ghosh-Roy AK, Bhaduri SK, Gupta SN, Mandal A (2000) Geochemistry and Rb–Sr Geochronology of acid tuffs from the northern fringe of the Singhbhum craton and their significance in the Precambrian evolution. Indian Min 54:43–56

Sengupta P, Dasgupta S, Dutta N, Raith MM (2008) Petrology across a calc–silicate–anorthosite interface from the Chilka Lake Complex, Orissa: implications for Neoproterozoic crustal evolution of the Eastern Ghats Belt. Precambr Res 162:40–58

Shabeer KP, Satish-Kumar M, Armstrong R, Buick IS (2005) Constraints on the timing of Pan-African granulite- facies metamorphism in the Kerala Khondalite Belt of Southern India: SHRIMP mineral ages and Nd isotope systematics. J Geol 113:95–106
Sharma R (1988) Patterns of metamorphism in the Precambrian rocks of the Aravalli mountain belt. Mem Geol Soc India 7:33–76
Sharma RS (1995) An evolutionary model for the Precambrian crust of Rajasthan: some petrological and geochronological considerations. Geol Soc India Mem 31:91–115
Sharma RS (2009a) Eastern Ghats mobile belt. In: Cratons and fold belts of India. Springer Int Publ Switzerland, p 304. https://doi.org/10.1007/978-3-642-01459-8
Sharma RS (2009b) Cratons and fold belts of India. Springer, Heidelberg, p 304
Sharma BL, Chauhan NK, Bhu H (1988) Structural geometry and deformation history of the Early Proterozoic Aravalli rocks from Bagdunda, district Udaipur, Rajasthan. In: Roy AB (Ed) Precambrian of Aravalli Mountain, Rajasthan, India, vol 7. Mem Geol Soc India, pp 169–191
Shaw RK, Arima M (1996) High-temperature metamorphic imprint from calc-granulites of Rayagada, Eastern Ghats, India: implications of isobaric cooling path. Contrib Min Petrol 126:169–180
Shaw RK, Arima M (1997) Retrograde pressure-temperature path for spinel-bearing metapelites in Rayagada, Eastern Ghats, India. Mineral Petrol 60:41–59
Shaw RK, Arima M, Kagami H, Fanning CM, Shiraishi K, Motoyoshi Y (1997) Proterozoic events in the Eastern Ghats granulite belt, India: evidence from Rb-Sr, Sm-Nd sys-tematics and SHRIMP dating. J Geol 105:645–656
Shekhawat LS, Pandi MK, Joshi DW (2007) Geology and geochemistry of Paleoproterozoic low-grade metabasic volcanic rocks from Salumber area, Aravalli Supergroup, NW India. J Earth Syst Sci 116:511–524
Singh SP (1988) Stratigraphy and sedimentation pattern in the Proterozoic Delhi Supergroup, Northwestern India. Mem Geol Soc India 7:193–206
Singh SP (1997) Geochemistry of acid volcanics of the Dalma Group, Singhbhum, Eastern India. J Geol Soc India 49:437–441
Singh SP (1998) Precambrian stratigraphy of Bihar. In: Paliwal BS (ed) The Indian Precambrian. Jodhpur, India, Scientific Publ pp. 376–408
Singh Y, Krishna V (2009) Rb-Sr geochronology and petrogenesis of granitoids from the Chhotanagpur Granite Gneiss Complex at Raikera-Kunkuri Region, Central India. J Geol Soc India 74:200–208
Sinha KK, Krishna Rao N, Shah VL, Sunil Kumar TS (1997) Stratigraphic succession of precambrian of singhbhum: evidence from quartz pebble conglomerate. J Geol Soc India 49:577–588
Singh YK, Waele BD, Karmakar S, Sarkar S, Biswal TK (2010) Tectonic setting of the Balaram-Kui-Surpagla-Kengora granulites of the south Delhi terrane of the Aravalli mobile belt, NW India and its implication on correlation with the east African Orogen in the Gondwana assembly. Precambr Res 183:669–688
Sinha S, Alsop GI, Biswal TK (2010) The evolution and significance of microfracturing within feldspars in low-grade granitic mylonites: a case study from the Eastern Ghats Mobile Belt, India. J Struct Geol 32:1417–1429
Sinha-Roy S (1984) Precambrian crustal crustal interaction in Rajasthan, NW India, Proc. Seminar on Crustal Evolution of Indian Shield and its bearing on metallogeny. Indian J Earth Sci 11:84–91
Sinha-Roy S (1988) Proterozoic Wilson cycles in Rajasthan. In: Roy AB (ed) Precambrian of the Aravalli mountain range. Mem Geol Soc India, vol 7, pp 95–108
Sinha-Roy S (2000) Precambrian terrain evolution in Rajasthan. Geol Surv India Spec Publ 55:275–286
Sinha-Roy S, Malhotra G (1989) Structural relations of the cover and its basement: an example from Jahajpur belt, Rajasthan. J Geol Soc India 34:233–244
Sinha-Roy S, Mathai J, Narayanaswamy (1984) Structure and metamorphic characteristic of cordierite-bearing gneiss of South Kerala. J Geol Soc India 25:231–244

Sinha-Roy S, Mohanty M, Guha DB (1993) Banas dislocation zone in Nathdwara-Khamor Area, Udaipur district, Rajasthan, and its significance on the basement cover relations in the Aravalli Fold Belt. Curr Sci 65(1):68–72

Sinha-Roy S, Malhotra G, Mohanty M (1998) Geology of Rajasthan. Geol Soc India, p 278

Sreenivas B, Roy AB, Srinivasan R (2001a) Geochemistry of sericite deposits at the base of the Paleoproterozoic Aravalli Supergroup, Rajasthan, India: evidence for metamorphosed and metasomatised Precambrian Paleosol. Proc Indian Acad Sci (Earth Planet Sci) 110(1):39–61

Sreenivas B, Sharma SD, Kumar B, Patil DJ, Roy AB, Srinivasan R (2001b) Positive $\delta^{13}C$ excursion in carbonate and organic fractions from the Paleoproterozoic Aravalli Supergroup, northwestern India. Precambr Res 106(3–4):277–290

Stein HJ, Hannah JL, Zimmerman A, Markey RJ, Sarkar SC, Pal AB (2004) A 2.5 Ga porphyry Cu–Mo–Au deposit at Malanjkhand, central India: implications for Late Archaean continental assembly. Precambr Res 134:189–226

Subrahmanyam C, Verma RK (1986) Gravity field, structure and tectonics of the Eastern Ghats. Tectonophysics 126:195–212

Sugden TJ, Windley BF (1984) Geotectonic framework of the early-mid Proterozoic Aravalli-Delhi orogenic belt, NW India. In: Abstracts of the Geological Association Canada Program, vol 9, p 109

Sugden TJ, Deb M, Windley BF (1990) The tectonic setting of mineralization in the Proterozoic Aravalli-Delhi orogenic belt, NW India. In: Naqvi SM (ed) Developments in Precambrian Geology. Elsevier, pp 367–390

Tadokoro H, Tsunogae T, Santosh M (2008) Metamorphic P-T path of the eastern Trivandrum Granulite Block, southern India: implications for regional correlation of lower crustal fragments. J Min Petrol Sci 103:279–284

Tateishi K, Tsunogae T, Santosh M, Janardhan AS (2004) First report of sapphirine + quartz assemblage from southern India: implications for ultrahigh-temperature metamorphism. Gondwana Res 7:899–912

Taylor RJM, Clark C, Fitzsimons ICW, Santosh M, Hand M, Evans N, McDonald B (2014) Post-peak, fluid-mediated modification of granulite facies zircon and monazite in the Trivandrum Block, southern India. Contrib Min Petrol 168:1–17

Tewari HC, Rao VV (2003) Structure and tectonics of the Proterozoic Aravalli-Delhi geological provinces, NW Indian peninsular shield. Mem Geol Soc India 53:57–78

Tiwari SK, Biswal TK (2019) Dynamics, EPMA Th-U-Total Pb monazite geochronology and tectonic implications of deformational fabric in the lower-middle crustal rocks: a case study of Ambaji Granulite, NW India. Tectonics. https://doi.org/10.1029/2017TC004891

Tobisch OT, Collerson KD, Bhattacharya T, Mukhopadhyay D (1994) Structural relationship and Sm-Nd isotopic systematics of polymetamorphic granite gneisses and granitic rocks from central Rajasthan: implications for the evolution of Aravalli craton. Precambr Res 65:319–339

Torsvik TH, Carter LM, Ashwal LD, Bhushan SK, Pandit MK, Jamtveit B (2001) Rodinia refined or obscured: paleomagnetism of the Malani Igneous Suite (NW India). Precambr Res 108:319–333

Upadhyay D (2008) Alkaline magmatism along the southeastern margin of the Indian shield: implications for regional geodynamics and constraints on craton-Eastern Ghats Belt suturing. Precambr Res 162:59–69

Upadhyay D, Raith MM (2006) Intrusion age, geochemistry and metamorphic conditions of a quartz-monzonite intrusionat the craton-Eastern Ghats Belt contact near Jojuru, India. Gondwana Res 10:267–276

Upadhyay D, Raith MM, Mezger K, Bhattacharya A, Kinny PD (2006a) Mesoproterozoic rifting and Pan-African continental collision in SE India: evidence from the Khariar alkaline complex. Contrib Min Petrol 151:434–456

Upadhyay D, Raith MM, Mezger K, Hammerschmidt K (2006b) Mesoproterozoic rift-related alkaline magmatism at Eluchuru, Prakasam alkaline complex, SE India. Lithos 89:447–477

Upadhyay D, Gerdes A, Raith MM (2009) Unraveling sedimentary provenance and tectonothermal history of high-temperature metapelites, using zircon and monazite chemistry: a case study from the Eastern Ghats Belt, India. J Geol 117:665–683. https://doi.org/10.1086/606036

Vadlamani R (2010) Palaeoproterozoic (1.9 Ga) extension and breakup along the eastern margin of the Eastern Dharwar Craton, SE India: New Sm–Nd isochron age constraints from anorogenic mafic magmatism in the Neoarchean Nellore greenstone belt. J Asian Earth Sci 37:67–81

Vadlamani R, Kŕoner A, Vasudevan D, Wendt I, Tobschall H, Chatterjee C (2012) Zircon evaporation ages and geochemistry of metamorphosed volcanic rocks from the Vinjamuru domain, Krishna Province: evidence for 1.78 Ga convergent tectonics along the southeastern margin of the Eastern Dharwar Craton. Geol J 48:293–309

Valdiya KS (2015) Mesozoic Eastern Ghat Mobile belt. In: The making of India geodynamic evolution. Springer Int Publ, p 924. https://doi.org/10.1007/978-3-319-25029-8-6

Van Lente B, Ashwal LD, Pandit MK, Bowring SA, Torsvik TH (2009) Neoproterozoic hydrothermally altered basaltic rocks from Rajasthan, northwest India: implications for late Precambrian tectonic evolution of the Aravalli craton. Precambr Res 170:202–222

Vansutre S, Hari KR (2010) Granulite belts of central India with special reference to the Bhopalpatnam Granulite Belt: significance in crustal evolution and implications for Columbia supercontinent. J Asian Earth Sci 39:794–803

Vansutre S, Hari KR, Vishwakarma N (2013) Implications of geochemistry in support of Palaeo-Proterozoic tectonothermal evolution of Bhopalpatnam Granulite Belt, Bastar Craton, Central India. J Geol Soc India 81:503–513

Veevers JJ (2009) Palinspastic (pre-rift and -drift) fit of India and conjugate Antarctica and geological connections across the suture. Gondwana Res 16:90–108

Verma PK, Greiling RO (1995) Tectonic evolution of the Aravalli Orogen (NW India): an inverted Proterozoic rift basin? Geol Rundsch 4(4):683–696

Vijaya Kumar K, Ernst WG, Leelanandam C, Wooden JL, Grove MJ (2010) First Paleoproterozoic ophiolite from Gondwana: Geochronologic–geochemical documentation of ancient oceanic crust from Kandra, SE India. Tectonophysics 487:22–32

Vinogradov A, Tugarinov A, Zhykov C, Stapnikova N, Bibikova E, Khorre K (1964) Geochronology of Indian Precambrian. In: 22nd International Geological Congress, Abstract vol 10. New Delhi, pp 553–567

Viswakarma RK, Ulabhaje AV (1991) Sargipalli galenas: unusual lead isotope data from eastern India. Min Deposita 26:26–29

Walker TL (1902) Geology of Kalahandi State, Central Provinces. Mem Geol Surv India 33(3):1–22

Wang W, Cawood P, Pandit MK, Mei-Fu Zhou (2018) Evolving passive and active margin tectonics of the Paleoproterozoic Aravalli basin. Geol Soc Am Bull, NW India. https://doi.org/10.1130/B35027.1

Whitehouse MJ, Ravindra-Kumar GR, Rimša A (2014) Behaviour of radiogenic Pb in zircon during ultrahigh- temperature metamorphism: an ion imaging and ion tomography case study from the Kerala Khondalite Belt, southern India. Contrib Min Petrol 168:1042–1059

Yedekar DB, Jain SC, Nair KKK, Dutta KK (1990) The Central Indian collision suture. Precambrian of Central India. Geol Surv India Spec Publ 28:1–37

Yin A, Dubey CS, Webb AAG, Kelty TK, Grove M, Gehrels GE, Burgess WP (2010) Geological correlation of the Himalayan orogen and Indian craton: Part 1. Structural geology, U-Pb zircon geochronology, and tectonic evolution of the Shillong Plateau and its neighboring regions in NE India. Geol Soc Am Bull 122:336–359. https://doi.org/10.1130/B26460.1

Yoshida M, Bindu RS, Kagami H, Rajesham T, Santosh M, Shirahata H (1996) Geochronologic constraints of granulite terranes of South India and their implications for the Precambrian assembly of Gondwana. J Southeast Asian Earth Sci 14:137–147

Zhao J-H, Pandit MK, Wang W, Xia X-P (2018) Neoproterozoic tectonothermal evolution of NW India: evidence from geochemistry and geochronology of granitoids. Lithos 316–317:330–347

Chapter 5
Proterozoic 'Purana' Basins

5.1 Introduction

The Banded Gneiss Complex attached to the Aravalli Craton in the west with coeval Bundelkhand Craton in the center, the Singhbhum and the Meghalaya Cratons in the east, and the Bastar and Dharwar cratons in the south constitute the Precambrian nuclei of the Peninsular India. An independent Marwar Craton has been added to this list. These Archaean cratons provided floor to the deposition of the supracrustal sequences of the early to middle Proterozoic mobile belts showing a variable degree of folding, shearing, metamorphism, and magmatism. These have been discussed in Chap. 4. Linked to each of these cratons and mobile belts are several intra-continental basins, described variously as cratonic, epicratonic, pericratonic or intercratonic depending upon their positions related to the craton margin. These mildly locally deformed and unmetamorphosed sedimentary basins expose sequences ranging in age from Paleoproterozoic to Neoproterozoic.

There is consensus about the lower age limit of these sedimentary basins being constrained by the radiometric age data of the granitic and gneissic crystalline basement rocks. However, the timeframe of the closure of the basin continues to remain a contentious issue. Some workers extend the age up to the base of the Paleozoic based on the occurrence of indigenous biotic elements in the Upper Vindhyan, Kurnool and Bhima basins (Sharma and Pandey 2012; Sharma and Shukla 2012; Kumar 2016; Sharma et al. 2016). On the other hand, based on paleomagnetic data, zircon geochronology of kimberlite in one of the basins, Gregory et al. (2006) and Malone et al. (2008) emphatically suggested the closure age of these basins at ~1000 Ma. This interpretation of an early closure of these Proterozoic sedimentary basins received support from the studies carried out in other basins in the peninsula (Patranabis-Deb et al. 2007, 2012; Bickford et al. 2013, 2017; Mukherjee et al. 2012; Rao et al. 2013;

A. K. Jain et al., *Tectonics of the Indian Subcontinent*, Society of Earth Scientists Series, https://doi.org/10.1007/978-3-030-42845-7_5

Basu and Bickford 2015; Saha et al. 2016a, b; Sahoo et al. 2017). It is, however, realized that defining the cause of the basinal closure of the Proterozoic basins may not be straight forward, due to inter- and intra-basinal tectonism, unpredictable erosion of the younger sedimentary cover, modification in the coastal conditions, and climate change effects on the land leading to variation in the erosion rate of the source terrain. Meert and Pandit (2015) proposed that the basin closure date should be defined by the time it stopped receiving the sediments from the hinterland and then speculate for why sedimentation ceased.

These little disturbed Proterozoic sedimentary successions in different basins with widely variable sedimentary environmental settings were clubbed under a sac term 'Purana Basin', derived from the local Indian word 'Purana' meaning old. Holland (1907) introduced this name primarily because none of these sedimentary successions fitted in any stratigraphic series identified in Europe at that time. The term Purana does not signify any lithology, age or tectonic setting, hence it does not fit into any modern stratigraphic nomenclature approved by the IUGS Subcommission on Precambrian Stratigraphy. The redundancy of the term 'Purana' was debated in several seminars and meetings (Banerjee 2011) and led Ramakrishna and Vaidyanadhan (2008) to highlight this aspect (p. 455). However, this term is deeply entrenched in the Indian geological literature and following the Law of Priority in the Stratigraphic Code, its usage in the Indian stratigraphy may continue. Attempt has been made to enlarge the scope of the term *Purana* by including sedimentary successions of the craton-related fold belts under this ambit thus destroying the basic idea of the *Purana* nomenclature. Meert and Pandit (2015) treated the term Purana as chronostratigraphic unit and identified Purana-I spanning through 2.5–1.6 Ga, Purana-II covering the period between 1.6 and 1.00 Ga and Purana-III representing the activities between 900 and 541 Ma. They also tried to correlate them with different stages of the supercontinent formation. This attempt to dilute the meaning of *Puranas* has so far not been followed by later workers. Although the original definition of *Puranas* broadly included the rocks of Mesoproterozoic-Neoproterozoic ages, in recent revision, rocks of undeformed Paleoproterozoic ages have been included in this domain. Kale and Phansalkar (1991) estimated that 20% of the Precambrian terrain of the Peninsular India is covered by these unconformity-bound shallow water transgressive–regressive sedimentary sequences with an aggregate area of ~350,000 km^2 (Ramakrishnan and Vaidyanadhan 2008). These independent basins and sub-basins derived their sediments from the basement granites (Srivastava and Banerjee 1987) and gneisses, hence carry the mineralogical and geochemical signatures of the parent material. Although largely undeformed, folding and thrusting along the margin of the Cuddapah, Vindhyan and Godavari basins, granitic activity in parts of Vindhyan basin in eastern end, emplacement of kimberlite at a number of places in the Vindhyan, Indravati and Cuddapah basins, reactivation of the Great Boundary Fault and Central Indian Tectonic Zone are manifestations of Earth's movements during this period of the Peninsular history.

In this book, we have identified several intracratonic undeformed basins and sub-basins attached to specific cratons and mobile belts (Fig. 5.1). These are Marwar Basin

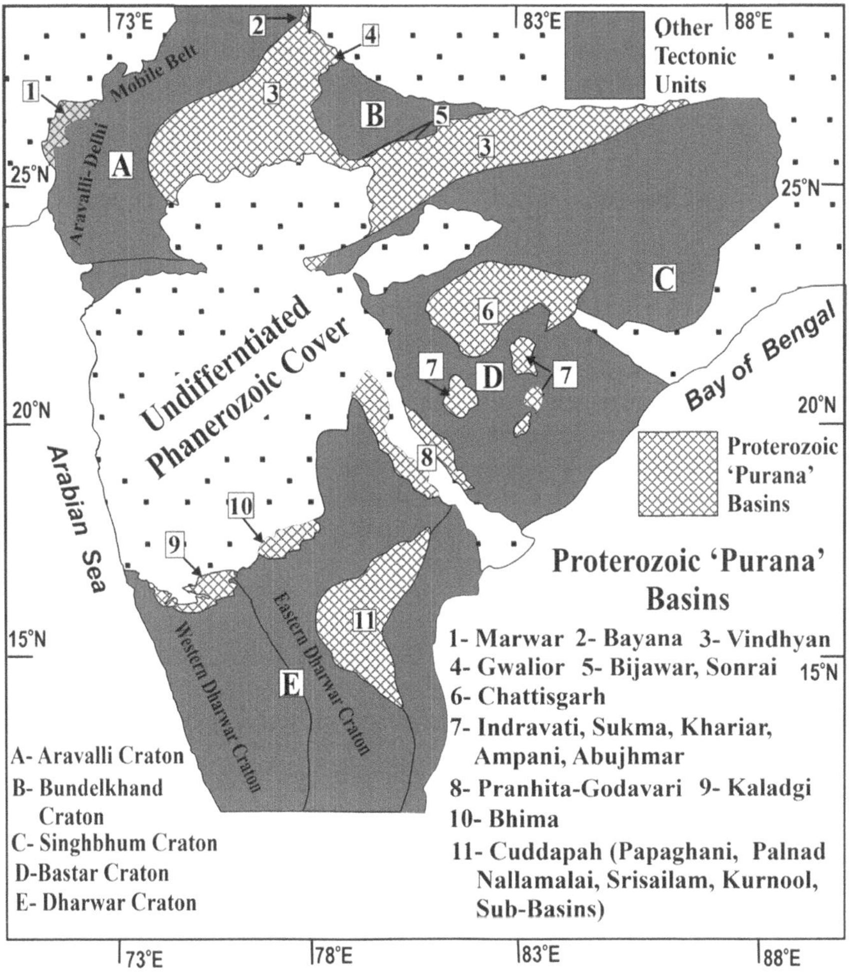

Fig. 5.1 Distribution of the Proterozoic 'Purana' sedimentary basins of the Indian Shield. Modified after original drawing provided by V. Kale

(Marwar Craton), Bayana and Vindhyan Basins (BGC–Aravalli–Bundelkhand Craton–Satpura Mobile Belt), Bijawar, Sonrai and Gwalior basins (Bundelkhand Craton), Chattisgarh, Khariar, Indravati, Sukma, Ampani, and Abujamar basins (Bastar Craton) and Pranhita–Godavari Basin (between Bastar and Dharwar Cratons), Cuddapah, Kaladgi, Bhima basins and Papaghani, Kurnool, Palnad, Nallamalai and Srisailam Sub-basins (Dharwar Craton and Eastern Ghats Mobile Belt). While all earlier inter-regional correlations were made based on similarities in lithology and depositional environment, new radiometric age data (McKenzie et al. 2013) have shown a new direction to the inter-regional correlations although this method of correlation is fraught with dangers of mismatch between the differences in the basinal settings with similarities of age. Also, radiometric dates available to date have very large uncertainties, hence these are not suitable for application to the shallow-water thin slices of sedimentary covers.

5.2 Marwar Basin

The metasediments of the Sirohi Group override the Erinpura Granite which, in turn, is overlain by 770–750 Ma Malani Igneous Suite (MIS) (~779–681 Ma) comprising felsic volcanics and intrusives along with minor basaltic dykes spread over 54,000 km^2 in SW Rajasthan, India (Rathore et al. 1999a; Bhushan 2000). An unmetamorphosed and undeformed sedimentary succession of the Marwar Supergroup (Marwar Basin) unconformably overlies the MIS. This basin is one of several Proterozoic intra-cratonic Purana basins in the Jodhpur–Nagaur–Bikaner–Ganganagar sectors of western Rajasthan (Fig. 5.2a; Pareekh 1981, 1984), and is composed of nearly 2000 m of unmetamorphosed and marginally deformed sedimentary units of sandstone, shale, carbonate and evaporites.

This clastic–carbonate package is divisible into two clastic groups: Jodhpur Group and Nagaur Group, and a carbonate–evaporite unit of the Bilara Group (Fig. 5.1b; Khan 1971; Pareek 1984; Cozzi et al. 2012). Based on similarity of lithology and subsurface borehole data, the Marwar Supergroup including the Birmania basin rocks have been broadly correlated with the Late Neoproterozoic to Early Cambrian rocks

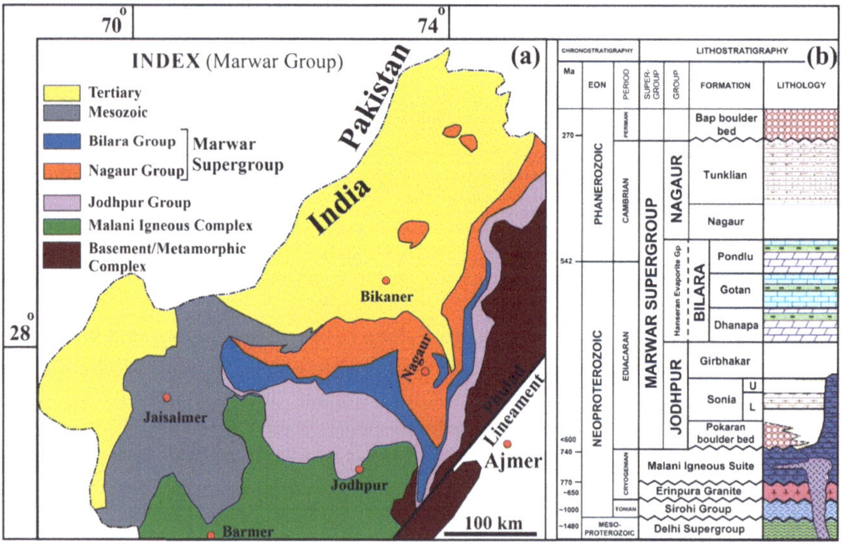

Fig. 5.2 **a** Simplified geological map of the Marwar Basin. Redrawn after Pareekh (1981). **b** Generalized stratigraphy of the Marwar Group after Cozzi et al. (2012)

of Salt Range (Husain and Banerjee 1986; Kumar 1999; Chauhan et al. 2004). The oldest Jodhpur Group consists predominantly of siliciclastic Sonia Sandstone, showing unconformable contact with the Neoproterozoic MIS. The unstratified rolled rhyolitic boulder bed overlying MIS and underlying the Sonia Sandstone known as Pokaran Boulder Bed was erroneously interpreted as a glacial till by Chauhan et al. (1991). The cross bedded Girbhakar sandstone formed in a fluvio-marine environment with some indications of lagoonal and marginal marine conditions. The carbonates and potash-rich cherty foetid limestone and dolomite of the Bilara Group conformably overlie the Jodhpur Group (Barman 1987) and reflect intertidal and lagoon calcareous facies with columnar and cabbage-like stromatolites. Reduced siliciclastic influx due to aridity in the provenance led to the precipitation of carbonates in the coastal warm waters. Based on drill holes data in the Bikaner-Ganganagar tract the evaporate bearing sub-surface carbonate rock was named as the Hanseran Evaporite Group (Kumar 1999; Dasgupta 1996). Seven evaporitic cycles of dolomite, anhydrite, halite, polyhalite and clay (Kumar 1999) represent this unit and considered coeval with low-dipping carbonates of the Bilara Group (Kumar 1999). Carbon and sulphur isotopic data (Mazumdar and Bhattacharya 2004; Mazumdar and Strauss 2006) lent support to this coeval hypothesis which was contested by Banerjee et al. (2007a, b) suggesting Hanseran evaporate formation to be younger than the Bilara carbonate rocks. Unconformably overlying the Hanseran Evaporite succession is a sequence of cross-bedded sandstone, siltstone and shale of the Nagaur Group yielding well-preserved trilobite traces in the Tunklian Sandstone which, in turn, is unconformably overlain by the Permo-Carboniferous Bap boulder bed (Pandey and Bahadur 2009). The cross-bedded clastics of the Nagaur Group (Pareek 1984) reflect low to high energy marginal marine to continental depositional environments. An isolated calcareous basin near Birmania is divide into Randha and Birmania Formations (Srikantia et al. 1969) where both the formations appear to overlie the MIS. Birmania Formation contains phosphorite and dolomite, while Randha is exclusively dolomitic (Husain and Banerjee 1986; Hughes et al. 2015). These rocks are considered equivalent of the Bilara–Nagaur segment of the Marwar basin, while these have also been correlated with Proterozoic–Cambrian transition level due to the occurrence of phosphorite (Maheshwari et al. 2003). Rocks of both the basins are considered Neoproterozoic-Early Cambrian in age. Virtual absence of body fossils in these sedimentary rocks has been attributed to excess of siliciclastic sediments, inhospitable chemical regimes and paleogeography. Pandey and Bahadur (2009) recorded trace fossils like *Gyrolithes* and Ediacaran fossils like *Arumberia*, indicates a late Neoproterozoic age for the Jodhpur Group (<570 Ma; Kumar and Pandey 2010 and references therein). Barman (1987) described many columnar and cabbage-like nodular stromatolites of the Neoproterozoic affinity. The Lower Cambrian trilobite traces like *Cruziana*, *Rusophycus* etc. in the Nagaur sandstone (Ahmad and Kumar 2014) is considered a significant stratigraphic breakthrough.

Discovery of Lower Cambrian trace fossil *Treptichnuspedum* (Srivastava 2014) is of greater significance in marking the stratigraphic boundary (Ahmad and Kumar 2014) and serve as the best indicator of the closure of the Marwar Basin. Carbon isotopes in the Bilara carbonates succession (Pandit et al. 2001) show dominance of negative $\delta^{13}C$ values with anomaly up to <−4.3‰PDB at the base of Dhanapa Formation and <6.5‰PDB in the overlying Gotan Formation. The isotopic profile shows a shift toward positive values, which has been related to the global end-Neoproterozoic-Early Cambrian carbon isotopic evolution curve. Mazumdar and Bhattacharya (2004) further suggested that the low $\delta^{13}C$ values indicate glacial climate.

Ansari et al. (2018) has recently added new carbon isotopic data for these carbonates and evaporates, while Banerjee et al. (1998), Strauss et al. (2001) and Mazumdar and Strauss (2006) determined sulphur isotopic compositions of trace sulphate (27.2–42.0‰, avg: 33.8 ± 3.1‰, n = 37) in the Bilara carbonates and Hanseran Evaporite Group (27.5–39.7‰, avg: 32.4 ± 3‰; n = 25) match well with globally recognized sulphur isotopic enrichment event at terminal Neoproterozoic. Organic geochemistry of soluble and insoluble fractions in the core sediment samples show normal alkanes range from C15 to C33 wherein C19 to C21 n-alkanes predominate and carbon preference index ranges from 1.12 to 1.19. These data helped in elucidating the paleoenvironment (Bhattacharya and Dutta 2015).

Paleomagnetic studies on hematite-bearing sandstone of the Jodhpur Group of the Marwar Basin support the interpretation of temporally distinct sedimentation episodes in the Marwar and Vindhyan basins (Malone et al. 2008; Davis et al. 2014). The Marwar Supergroup seems to have developed near the close of the Ediacaran Period. Similarity with the evaporite-bearing Lower Cambrian rocks of Salt Range and disposition on top over the Neoproterozoic rhyolite (779–681 Ma; Rathore et al. 1999a, b) are the only two criteria on which Marwar Basin rocks were assigned geological ages. TIMS U–Pb zircon ages are 770–750 Ma. U–Pb dates of detrital zircons from the Jodhpur sandstone yielded age peak which restricted the provenance to 800–900 Ma age bracket (Malone et al. 2008). Detrital zircon populations from the Nagaur sandstone and the Marwar Supergroup reveal a large concentration of grains between 700 and 1000 Ma and a small population of young grains with a peak at 540 Ma (McKenzie et al. 2011). Best estimate for the age of the Marwar Supergroup is between 521 and 570 Ma with the Ediacaran–Cambrian boundary (541 Ma), located within the Bilara Group. These sediments are believed to have deposited in a rift setting (Fig. 5.3; Mazumdar and Bhattacharya 2004; Cozzi et al. 2012). Correlations between the Marwar Supergroup and the Krol–Tal succession of the Lesser Himalaya is supported by nearly identical detrital zircon population variations (McKenzie et al. 2011; Hughes et al. 2015). Lithologically, the Marwar Supergroup, the Salt Range sequence, and the Ara Formation (Oman) show nearly identical cycles of carbonate–evaporite deposits (Banerjee et al. 2007a, b).

The Marwar Basin sediments were derived from the Erinpura and its equivalents rocks (George and Ray 2017). Some workers believe that the upper most unit of this basin coincided with the uppermost rock unit of the Vindhyan Basin in the type area. Since the Vindhyan sequence does not have zircon younger than 1.0 Ga, the Upper Vindhyan and Marwar sequences did not develop coevally (Malone et al. 2008). Correlation of the Bilara carbonates, Hanseran Evaporites, heavy hydrocarbons, phosphatic Birmania rocks (Husain and Banerjee 1986; Hughes et al. 2015) and Ara Formation (Huqf Supergroup) in south Oman Salt Basin (Banerjee et al. 2007a, b) suggests late Neoproterozoic-Early Cambrian age for these rocks (Mazumdar

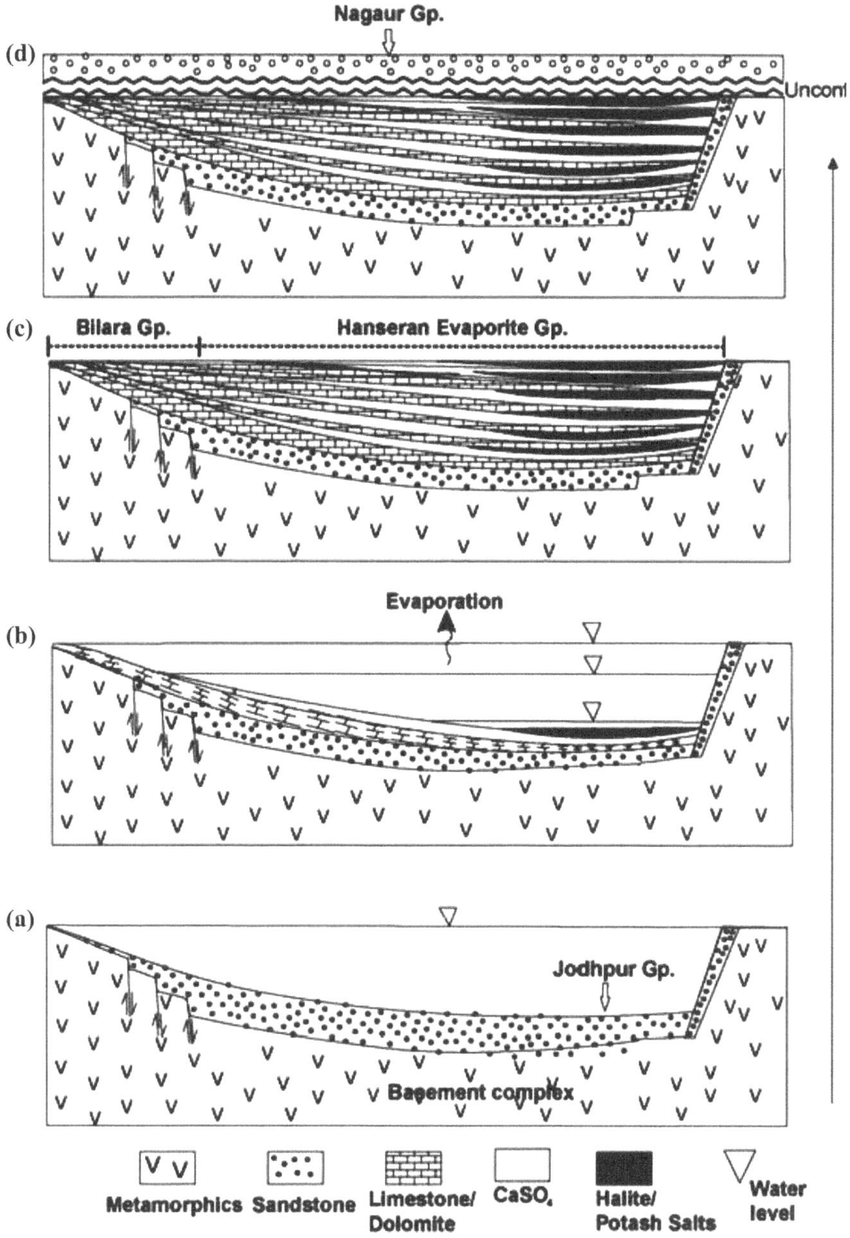

Fig. 5.3 Evolution of the Marwar Basin during the deposition of the Marwar Supergroup according to Mazumdar and Strauss (2006)

and Bhattacharya 2004). In this basin geophysical probes indicated vast subsurface expanse of these rocks in the Rajasthan Shelf (Pareek 1984; Sibal and Murty 2005; Cozzi et al. 2012). Low (<5°) westerly dips of the strata connect it with the eastern flank of the Indus shelf of the Indus-Baluchistan geosyncline or Indo-Arabian Geological Province (Shrivastava 1992).

5.3 Bayana Basin

Although traditional classification does not include this small unmetamorphosed and marginally deformed basin attached to the North Delhi Fold Belt, this ~1.8 Ga old NE–SW trending Bayana Basin deserves to be included as one of the intracratonic rift-related Purana basins as the easternmost limit of this belt (Raza et al. 2012). In this basin sandstone–conglomerate–volcanic succession represents depositional environments varying from fluvial to coastal marine tidal flat settings. Rapid alteration of facies types represents paleo-environmental shift between continental fluvial and shallow marine tidal/deltaic settings within dominant continental block with minor recycled orogen provenance for clastic detritus in the basin. Using standard provenance diagnostic Qm-F-Lt plots, major and trace element geochemical discrimination plots, Raza et al. (2012) inferred tectonically controlled sedimentation in this basin with major sedimentary fills coming from the gneisses and granites of the Banded Gneissic Complex (Aravalli Craton)/Bundelkhand Craton. On the other hand, paleocurrent analysis by Singh (1988) indicated the 'Dausa uplift' to be the sediment provenance. The ~1.8 Ga age of this sequence is highly tentative. Mixed assemblage of basic and acid volcanic rock derived sediments suggest a role of tectonics during sedimentation.

5.4 Gwalior Basin

Nearly E–W trending and ~80 km long unmetamorphosed sedimentary sequence the Gwalior Group was deposited along the northern margin of the Bundelkhand Craton (Fig. 5.4a) and is separated from the Bijawar Basin by a large gap. Although configuration wise both the basins show similar contact relationship with the Bundelkhand Craton and Vindhyan Basin, lithological package is not identical. Chakraborty et al. (2015a) recently wrote "despite being coeval recipients of detritus from a common provenance, that is the BGC, and having an analogous rift-related tectonic setting, the two Paleoproterozoic basins of Central India, the Gwalior and the Bijawar, vary in terms of sedimentation pattern", a point very explicitly explained by Pant and Banerjee (1990).

The lower part of the Gwalior group comprises of grit, conglomerate, glauconitic sandstone, stromatolitic limestone and is classified as the Par Formation (Fig. 5.4b). The siliceous, ferruginous shale, banded chert-jasper beds, volcanic tuff layers and

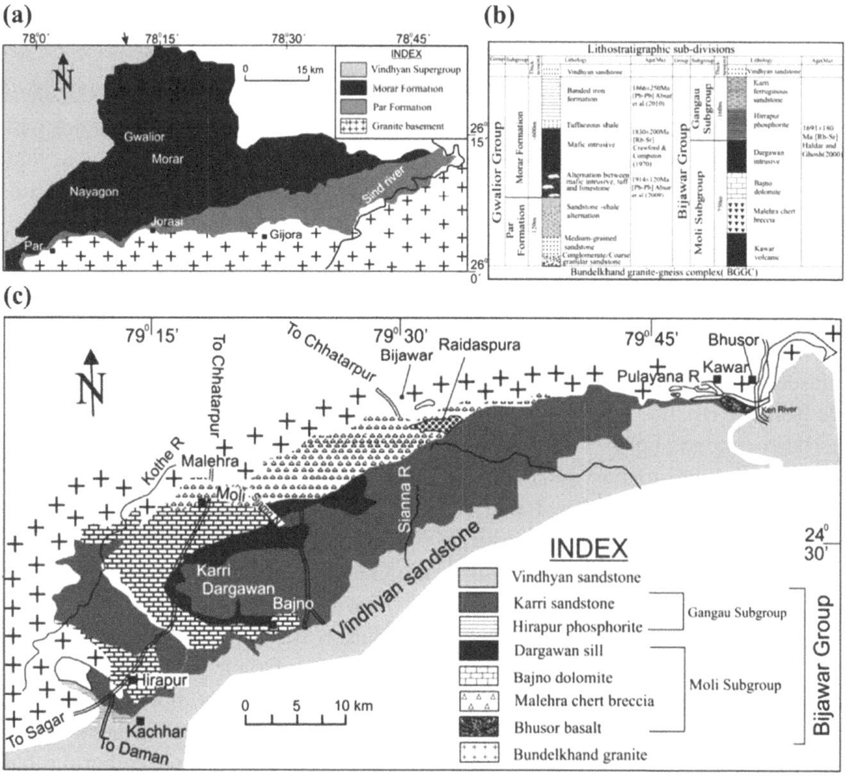

Fig. 5.4 **a** Simplified geological map of the Gwalior Group, Madya Pradesh. **b** Lithostratigraphy sub-divisions of the Gwalior and Bijawar Groups. **c** Simplified geological map of the Bijawar Basin. After Chakraborty et al. (2015b)

limestone constitute the overlying Morar Formation, which is intruded by several basic sills, resembling those of the Bijawar Basin. The Gwalior Basin displays clastic sedimentation in a variety of depositional settings, viz alluvial fan to braided fluvial veering to proximal shelf (e.g., Par Formation. Basic volcanics have yielded Rb-Sr isochron age of 1830 ± 200 Ma, while the sills in the region are dated as 1775–1790 Ma almost same as the mafic dykes seen in the Bundelkhand Craton. Absar et al. (2019) bracketed the Gwalior depositional history between 2000 and 1791 Ma. Crawford and Compston (1970) reported a Rb–Sr isochron age of 1830 ± 200 Ma for the mafic sill that intrudes the Morar Formation. Tholeiitic and sub-alkaline gabbroic sills yield mineral–whole-rock Sm–Nd isochron corresponding to an age of 2104 ± 23 Ma with model ages of 2.6–1.7 Ga, indicating that their mantle sources had a protracted evolution (Samom et al. 2017). Petrotectonic and geochemical studies of the mafic volcanic in the Gwalior basin reveal presence of mantle plume in an extensional intra-continental rift tectonic setting, which possibly facilitated emplacement of a Large Igneous Province (LIP) in the northern Indian Shield.

5.5 Bijawar Basin

In comparison to the Gwaliors, the Bijawar Group is characterized by chemical sediments, represented by Bajna Dolomite and Malhera Chert Breccia along with volcano-clastic sediments of the Kawar Formation. It is remarkable to note that, while Gwalior Basin indicates deposition of iron-rich sediments of 1.8 Ga BIF affinity, the Bijawar Basin was the repository of aluminous phosphorite (Banerjee et al. 1992) containing uraniferous and bituminous clays (Jha et al. 2012). These lithological and mineral associations have helped in explaining the paleooceanographic, paleoenvionmental and diagenetic history of these two basins.

On the southeastern margins of the Bundelkhand Craton, ENE–WSW trending metasedimentaries are exposed in about 100 km wide outcrops which are approachable from the Bijawar town, though these rocks are not exposed in the Bijawar town. The basin extends up to Sonrai in the west where its unconformable contact with the Vindhyan sandstone is well exposed. These rocks were mapped and investigated by Mathur and Mani (1970), Banerjee (1982), Pant and Banerjee (1990), Kumar et al. (1990) and Haldar and Ghosh (2000). Stratigraphically, it is subdivided into the Moli Subgroup which is overlain by the Gangau Subgroup. These are divided into Kawar Volcanics (Kumar and Singh 1978) with conglomerate at the base covered by a thick sequence of basaltic lava (Bhuson Basalt) and followed upward by well-developed grey, white, locally silicified Bajna Dolomite; the latter is profusely affected by doleritic/basaltic Dargawan sills. Some low-dipping sandstone beds, the Pukhra Sandstone and a prominent brecciated zone named as the Malhera Chert Breccia, occur towards the top of the subgroup. The Gangau Subgroup has two subdivisions, a lower Hirapur Phosphorite Formation and an upper Karri Ferruginous Sandstone Formation. The Hirapur Phosphorite Formation is made up of ferruginous shale, altered Fe-rich aluminous phosphorite rock units and numerous brecciated zones indicating frequent collapse and re-cementation during the sedimentation of these rock units (Banerjee et al. 1982). Although sedimentation took place in the shallow marine coastal waters, the present configuration and nature of the rock record is the result of tectonism and prolonged weathering.

Rb-Sr age of the Dargawan Sill is 1789 $\pm$ 71 Ma and of Kurrat Lava 1691 $\pm$ 180 Ma indicate 1700–1800 Ma as the time span of the Bijawar Group (Haldar and Ghosh 2000). U–Pb zircon concordia suggested an age of 1854 $\pm$ 7 Ma (Deb et al. 2002).

5.6 Sonrai Basin

The Sonrai Basin occurs as a detached basin located to the west of the Bijawar Basin. Rocks in this basin show large scale folding with WSW plunge. The base of the Sonrai Formation in this basin is marked by a prominent porphyritic tholeiitic basalt, known as the Kurrat Basalt. These are followed upward by the Solda Formation comprising induarated quartz arenite and ferruginous shales. Stratigraphic succession of this basin was proposed by Prakash et al. (1975) and Srivastava (1989). Grits, shale, ferruginous phosphorite and graphitic rocks are seen interbedded with dolomite (Banerjee 1982). Stromatolitic dolomites are found in an isolated inlier of the Bijawar Group within the Vindhyan succession and Deccan Traps (Joga area) in the Narmada Valley.

The type area of Bijawar Basin around Hirapur and detached Sonrai Basin contain phosphorite and uranium in different forms and phases. Uranium is either related to unconformities or controlled by the vein type hydrothermal fracture regime (Rawat et al. 2018). Some of the uranium mineralization is also linked to the bituminous shales. The mineralogy and chemistry of clay minerals associated with the uranium mineralization in the Sonrai Basin by has helped in understanding the genesis of such mineralization event (Jha et al. 2019). Phosphorite deposits in the Hirapur basin (Banerjee et al. 1982) and smaller occurrences in the Sonrai basin (Srivastava 1989) provided information about the origin of these mineralogical unique mineral assemblages.

High resolution heilborne time domain electromagnetic, magnetic and radiometric data sets have shown distinct changes in conductivity parameters that has outlined depth to resistivity basement, topography and the contrasts between the Bijawar, Bundelkhand and Vindhyans (Chaturvedi et al. 2012).

5.7 Vindhyan Basin

5.7.1 Configuration of the Vindhyan Basin

A thick pile of Meso-Neoproterozoic Proterozoic sedimentary rocks in the central India, designated as the Vindhyan Supergroup, is reflected on the geological map by a crescent shaped body, closely wrapping the exposures of the ~2.5 Ga Bundelkhand Granite Complex to its north. Two other Paleoproterozoic successions of the Bijawar and the Gwalior are seen underneath the Vindhyan rocks in the northern sector. The Vindhyan overlies the Mahakoshal Group in the south, the Chotanagpur gneisses in the east, the Banded Gneissic Complex, Aravallis and Delhis in the west. The Vindhyan succession is exposed in two geographical sectors, the Son valley in the east, and Rajasthan in the west. Direct contact of the two belts with near identical lithology and tectonics remains obscure and inferential due to large spread of the Cretaceous Deccan Volcanics, concealing the subsurface geology for any in situ observation (Fig. 5.5). Description of these igneous, metamorphic and sedimentary rock successions providing the floor for the Vindhyan sedimentation have been discussed in Chaps. 2 and 3.

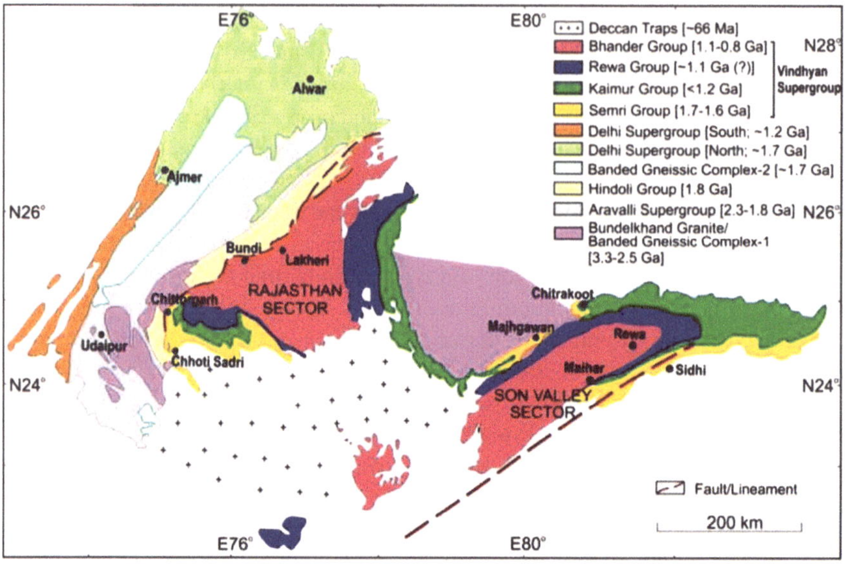

Fig. 5.5 Geological map of west-central India showing the Vindhyan Supergroup, basement complexes and Cretaceous Deccan Traps. Areal extent of the Semri, Kaimur, Rewa and Bhander Groups of the Vindhyan Supergroup are shown between the Great Boundary Fault and the Narmada Son Lineament along the western and eastern margins. After Shukla et al. (2019)

5.7.2 *Historical Perspective*

In the past several workers have contributed to our understanding of the Vindhyan System (now Supergroup); among which studies carried out by Auden (1933) are noteworthy. Regional mapping and interregional correlations of the widely distributed stratigraphic units was attempted by Prasad (1984), Sastry and Moitra (1984), Soni et al. (1987), Bhattacharya (1996 and references therein) and Bose et al. (2001). All these workers e.g. Chanda and Bhattacharya (1982), Soni et al. (1987) and Bose et al. (2001) simply fine-tuned the basic order of superposition and the stratigraphic framework established by Auden (1933) by adding local names to various subunits.

5.7.3 *Spatial Distribution*

The entire sedimentary units of the Vindhyan Supergroup show evidence of deposition in shallow marine environments with sporadic occurrences of deep-water facies. Refining the thickness estimate of Auden (1933), Bhattacharya (1996) estimated 5 km of clastic, volcanoclastic and carbonate sediments in the Vindhyan Basin. The basic subdivisions and their later nomenclature are given in Table 5.1.

Table 5.1 Lithostratigraphy of the Vindhyan Supergroup (after Ramakrishnan and Vaidyanadhan 2008)

Group	Formation	Alternative names
Bhander Group (1300–1500 m)	Maihar Sandstone	Upper Bhander Sst: Divided into Bhavpura Sh, Balwan Lst, Shikoda Sst
	Sirbu Shale	–
	Bundi Hill Sandstone	Lower Bhander Sst
	Lakheri Limestone	Bhander Lst, Nagod Lst
	Ganurgarn Shale	Simrawal Sh
Disconformity/gradational contact		
Rewa Group (100–300 m)	Govindgarh Sandstone	Upper Rewa Sst, Ghadra Sst
	Drummondganj Sandstone	–
	Jhiri Shale	Variegated Sh
	Asan Sandstone	Lower Rewa Sst, Itwa Sst, Kanar Sst
	Panna Shale	–
Normal contact/facies change		
Kaimur Group (400 m)	Dhandraul Quartzite	Upper Kaimur Sst, Scarp Sst
	Mangesar Formation	–
	Bijaigarh Shale	–
	Markundi Sandstone	Ghaghar Ss
	Ghurma Shale	Susnai Breccia
	Sasram Sandstone	Lower Kaimur Sst
Unconformity/normal contact		
Semri Group (3000–4000 m)	Suket Shale	BaghwarSh
	Rohtas Limestone	Nimbahera Lst: Divided into Bari Sh, Jiran Sst, Khori Malan Cong
	Chorhat Sandstone	Glauconite Bed, Rampur Sst, Basuhari Sst, Tirohan Breccia
	Bargawan Limestone	Salkhan Lst, Fawn Lst, Chorhat Lst, Tirohan Lst
	Kheinjua Shale	Olive Sh, Koldaha Sh, Binota Sh
	Chopan Porcellanite	Deonar Porc
	Kajrahat Limestone	Kuteshwar Lst, Bhagwanpura Lst, Tirohan Lst, Lohar Dol
	Arangi Shale	–
	Deoland Sandstone	Khardeola Sst, Pandwafall Sst
Angular unconformity/non-conformity		
Granites and supracrustals		

Sst sandstone; *Sh* shale; *Lst* limestone; *Dol* dolomite; *Porc* porcellanite; *Cong* conglomerate

In the Son Valley area at Maihar and its environs, the entire Vindhyan succession starting from the basal unit of the Semri Group (Lower Vindhyan) to the uppermost part of the Bhander Group (Upper Vindhyan) is exposed. The carbonate and volcanoclastic rich Semri Group of the Lower Vindhyan and the predominantly clastic succession of the Kaimur Group of the Upper Vindhyan are separated by disconformities, often characterized by conglomerate beds reflecting a hiatus in the sedimentation. Duration of this hiatus is conjectural (Bose et al. 2001). As mentioned above, the Semri Group was deposited on the variable basement rocks ranging from the Archean Bundelkhand Gneisses to Paleoproterozoic Bijwar and Gwalior Groups. Field relationship and recent radiometric dates confirms this contention. Bose et al. (2001) justified the subdivision of the Semri Group into five formations each designated by a local name and a distinctive lithology.

5.7.4 *Stratigraphy*

5.7.4.1 Lower Vindhyan

(i) **Semri Group: Basal Stage (Deoland Formation)**: The ~500 m thick Basal Stage (Conglomerate) of Auden (1933), resting over the Mahakoshal phyllite (~1750 Ma Ar/Ar dates of incipient mica; Unpublished data: Frank and Banerjee) in Rihand Dam area and Bijawar sandstones in other areas, has been named as the Deoland Formation by Sastry and Moitra (1984) which is followed by later workers. This locally well rounded conglomeratic pebbly sandstone with small carbonate lenses characterize the lithostratigraphic unit. Bedding patterns displayed include parallel beds and small-scale cross beds indicating a marginal coastal environment of deposition.

(ii) **Kajrahat Limestone**: The overlying ~300 m thick formation of shale and stromatolitic limestone is known as the Kajrahat Formation. The shale occupies the lower part of this formation while well-developed stromatolitic carbonate (dolomite/limestone) forms the upper part. The stromatolites of this formation have been correlated with those of the Calc Zone of Pithoragarh and Tejam in Kumaon (Valdiya 1969; Banerjee 1969, Kumar 1982, and many others). Attempts have been made to formalize the order of various carbonate units within this formation based on prevalence of particular lithological attribute. These flat-bedded and stromatolitic limestones represent fluctuating sedimentation milieu on the tidal flats of the ancient Vindhyan sea. Sporadic pyrite in these dark grey to grey to pink limestone and shales suggest entrapment of sediments in euxinic conditions within the open marine coastal tidal flats of the Kajrahat sea. Stromatolites of different form species, mainly the Conophyton sp. have been described from this formation (Valdiya 1969; Kumar 1982, 1999) and also Paleoproterozoic stromatolites (Kumar and Mishra 2007).

(iii) **Deonar Formation**: The overlying ~300 m thick Porcellanite Stage (cf. Auden 1933) is now known as the Porcellanite Formation (renamed as the

Deonar Formation) and is composed of volcanic tuffs, pyroclastics and horizontally bedded silicified shales. U/Pb dating of the volcanic zircons from these pyroclastics helped in fixing the age of these rocks while dating of detrital zircons provided a window to the timeframe in which these rocks existed.

(iv) **Olive Shale**: About ~100 m thick Olive Shale of Auden (1933), rechristened as Koldaha Shale or Kheinjua Shale, overlies the porcellanite-dominant beds indicating the end of volcanic activity in the basin. Alternating silt and shale with thin layers of graded sands as lenticular beds and friable nature characterize this member.

(v) **Salkhan Fawn Limestone**: The ~90 m-thick Fawn Limestone of Auden (1933) has been renamed as the Salkhan Limestone (Bargawan Limestone/Chorhat/Tirohan Limestone) after the Salkhan hill where it is best exposed. This fawn-coloured chert carbonate is more or less dolomitized and seen interlayered with minor calcareous shales. It is very hard, and its compact nature is attributed to extensive certification. Stromatolites of characteristic morphologies are preserved in these carbonate rocks.

(vi) **Rampur Formation**: Fairly thick (>100 m) glauconite-rich sandstone overlies the carbonate sequence (Fawn + Salkhan) of Chorhat Sandstone and is now known as the Rampur Formation. Some workers prefer to place horizon at the base of the overlying Rohtas Limestone Formation. This sedimentary unit shows a variety of primary sedimentary feature like hummocky cross bedding, sole marks etc. Various organic structures have been recorded from this lithological unit which were extensively used in interpreting the age of these Upper Vindhyan sedimentary sequences.

(vii) **Rohtas (Rohtasgarh) Limestone**: The rock unit overlying the glauconitic formation is predominantly calcareous and is represented by greyish blue and dark compact limestone with prolific development of stromatolitic bodies. Some of these large bun-like algal structures closely resemble the modern algal structures seen in the Shark Bay and Persian Gulf areas. This unit is known as the Rohtas Limestone and is overlain by a fairly continuous succession of structureless monotonous Bhagwar Shale, which is at places interbedded with discontinuous thin layers of porcellanite. In some sections, thick beds of organic-rich black shale are seen interbedded with the limestones. In the eastern Vindhyan basin these rocks are known as the Nimbahera Limestone.

(viii) **Bhagwar Formation**: Overlying a regional but diffused unconformity, the carbonate deposits of the Rohtas-Bhagwar Formation (Suket Shale) give way to the deposition of large-scale tidal-intertidal sandy beds. Although presence of mega-ripple beddings, large trough cross-beddings also suggest fluvial to partly aeolian environment, total absence of any overbank clayey flood deposits argues against such an interpretation.

5.7.4.2 Upper Vindhyan

(i) **Kaimur Group**: The Kaimur Group unconformaby overlies the Bhagwar Shale and contains exclusively the sandstone with sporadic beds of shale and a ferruginous conglomeratic bed at the base. These rocks are seen forming major scarps and were responsible for the present geomorphic configuration of the central Indian region. Lithostratigraphic subdivision of the Kaimur Group includes (Table 5.1):

The **Lower Kaimur Sandstone** is locally capped by a breccia horizon (**Susnai Breccia**), followed upwards by the **Ghagghar/Markundi Sandstone**. The **Bijaigarh Shale** is an interesting rock unit of the Lower Kaimurs containing excessive amount of kerogen-rich layers rendering these rocks black. These organic-rich layers have great potential for the source rocks for liquid/gas hydrocarbons in this Proterozoic terrain.

The **Upper Kaimur Sandstone** is also known as **Dhandraul Quartzite/Mangesar Formation**, exposed along a series of escarpments in the entire central Indian Vindhyan terrain.

(ii) **Rewa Group**: Nearly 150–200 m sandstone-shale succession overlies the highly indurated sandstone of the Kaimur Group and is designated as the Rewa Group. Two prominent buff coloured cross-bedded sandstones are separated and partly intercalated with silty, calcareous shales, designated as the **Panna Shale**, and are overlain by the **Lower Rewa Sandstone**. The bottom of this group is marked by diamondiferous conglomerate, derived from the kimberlite pipes cutting through the Kaimur sandstones. K-Ar dating of mica in the kimberlite pipes near Panna in the early sixties helped in constraining the age of the overlying Rewa and Bhander formations. The upper part of the Kaimur Group is represented by the **Upper Rewa Sandstone,** also known as the **Govindgarh Sandstone**.

(iii) **Bhander Group**: The Bhander Group occupies the youngest position in the Vindhyan stratigraphy. Successive shale-limestone-shale-sandstone succession represents the Bhander Group. The formational names, suggested by Sastry and Moitra (1984), are presently followed by all Vindhyan workers. It comprises of two units of shale (**Ganurgarh Shale** and **Sirbu Shale**), two units of sandstone (**Bundi Hill Sandstone** and **Maihar Sandstone**) and one unit of stromatolitic dolomite (**Lakheri Limestone**) (Chakraborty 2004); these define the stratigraphic units of this group. The Lakheri Limestone contain spectacular large stromatolites, alternating micritic-sparitic layers and ooidal beds. The sandstone units are medium to coarse grained with characteristic sedimentary features of a barrier island, and beach. Some indications of fluvial sedimentation have also been noticed. Provenance discrimination and tectonic setting interpretations were made (Banerjee and Banerjee 2010) based on modal analysis and whole rock geochemistry of the Bhander sandstone in Maihar–Nagod area.

5.7.5 *Sequence Stratigraphic Setting*

The Vindhyan Basin seems to have formed in an intracratonic-intracontinental tectonic setting. The Vindhyan sedimentary units have been also been studied in sequence stratigraphy perspective (Chakraborty 2006) where five distinct sequences have been identified with one or more system tracts. **Sequence 1** is represented by Deoland, Arangi and Kajrahat formations, **Sequence 2** is represented by Deonar, Koldaha, Chorhat, Salkhan, Rampur and Rohtas formations. **Sequence 3** with two system tracts primarily remains restricted to clastic facies of the Kaimur Group. **Sequence 4** is characterized by shale and sandstone of the Rewa Group with three system tracts, while **Sequence 5** has four system tracts and is represented by shales and carbonates of the Upper Vindhyan succession.

5.7.6 *Stable Isotopic Studies*

The carbonate facies of the Vindhyan Supergroup represents a wide spectrum of depositional facies of supratidal, intertidal, sub tidal, lagoonal, estuarine, inner and outer shelf. Carbon and oxygen isotopic studies on various carbonates and organic carbon in the entire Vindhyan succession provided inconsistent results primarily due to diagenetic resetting by isotopically lighter fluids, alteration due to meteoric diagenesis and difficulty in separating dolomite isotopic values from that of primary calcite (Banerjee et al. 1992, 2005, 2006; Friedman and Chakraborty 1997; Friedman et al. 1997; Kumar et al. 2005; Gileadeau et al. 2018). Among numerous published carbon and oxygen isotope data, values recorded for the dense micritic Rohtas Limestone of Semri Group only seems to be of near pristine nature so also the isotopic values of the organic carbon (Banerjee et al. 1992).

5.7.6.1 Organic Chemical Studies

Organic geochemistry of the carbonaceous laminae in the black shales of microbial mat origin, occurring in different stratigraphic levels in the Vindhyan succession, suggest a catagenetic stage of the early anthracite formation (Banerjee et al. 2006; Dayal et al. 2018), thereby reducing the chances of any large scale preservation of petroleum and gas.

5.7.7 *Depositional Environment*

It is generally agreed upon that the Vindhyan Basin is intracratonic in which entire sedimentary history remained restricted to shallow marine realm. This view has been perpetuated since the time of Auden (1933) and later endorsed by Valdiya (1969), Banerjee (1969), Mishra (1969), Mishra and Awasthi (1962), Singh (1973, 1980; 1985), Kumar (1982), Chanda and Bhattacharya (1982) and Chakraborty and Bhattacharya (1996b) and many others. Basumallick et al. (1996) identified tidal cyclicity of the strata. Opinion about the depositional environment veered around tidal flats, lagoon, sand-wave complexes, and more recently to fluvial (Bhattacharya and Morad 1993) and eolian (Chakraborty 1993, 1996; Bose and Chakraborty 1994; Bose et al. 2001). Bhattacharya (1996) proposed a fluvio-lacustrine model. Akhtar (1996) identified tell tale evidence of tidal flat-lagonal facies as the most common facies in the Vindhyan Basin, which occurs repeatedly throughout the entire sequence. Akhtar (1996) also described bar-shoal facies, delta facies and sporadic fluvial facies in some parts of the basinal deposits. Considerable work on the sedimentology and facies interpretations of the Vindhyan succession has been carried out in recent years by by Casshyap et al. (2001), Bose and his students at Jadavpur University (Bose et al. 1988, 2015), sedimentology group at Lucknow University under IB Singh and paleobiology-based interpretations by Scientists at the BSIP, Lucknow.

Early Vindhyan sedimentation represented by conglomerate of pebbly beds was initially interpreted as of glacial origin (Dubey and Chaudhary 1952; Ahmad 1955, 1981; Mathur 1954). This inference was rejected by later workers but once again gaining grounds through new age interpretations and paleogeographic and paleomagnetic reconstructions. The interpretations varied from glacial, fluvial, mass flow deposits to shallow shelf environments. Black organic shale are inferred to have formed in coastal lagoons with euxinic milieu. Profuse development of stromatolite in the Kajrahat Limestone indicated tidal flat (subtidal to supratidal) environment. Porcellanite Formation is pyroclastic in origin with views ranging from it being sub-aerial ash fall deposit to fissure and vent type submarine volcanism in the tidal flat environment. It was even considered a lagoonal deposits. Opinion varied about depositional setting of the Khenjua Shale ranging from lagoonal to tidal. During the formation of Koldaha Shale offshore to shoreface transition was interpreted. The Fawn Limestone was deposited in intertidal to supratidal environment while the Rampura Formation was inferred to have formed in storm dominated shelves. The Chorhat Sandstone formation was formed in shallow marine setting, while the Rohtas Formation interpreted to have formed in environments ranging from tidal flat with episodes of turbidite sequences reflecting slope deposits. Some workers attributed open shelf setting where rhythmical deposition was dominant. Shoaling upward trends were also recorded in the shelf setting.

5.7.8 Basin Configuration and Tectonics

The vast shallow water sedimentary Vindhyan succession reflects an intracratonic basin where penecontemporaneous slow subsidence provided space for the successive layers of sediments which got preserved in the geological records. From its present physical distribution pattern, it is amply clear that the basin existed as an E–W extending intra-cratonic rift basin, marginal to the Archean shields, with opening to the west (Bose et al. 1997, 2001).

Seismic studies suggested basement rifting (Kaila et al. 1989) with several extensional faults trending NW-SE that resulted because of dextral shear component (Ram 2005). It was proposed that the Vindhyan Basin originated as intracratonic half graben where all stages of rift phase could be distinguished. On the other hand, Bose et al. (1997) preferred rift basin origin of the Vindhyan Basin. Views regarding the origin of this basin varied from being rift basin (Kaila et al. 1989; Verma and Banerjee 1992, Ram et al. 1996), rift to sag transition (Bose et al. 2001), foreland basin (Auden 1933; Chakraborty and Bhattacharya 1996a, b; Raza and Casshyap 1996); intracratonic basin (Valdiya 1982), strike-slip basin (Chanda and Bhattacharya 1982; Crawford and Compston 1970) and synclise (Chanda and Bhattacharya 1982; Chaudhuri et al. 1999). There are doubts regarding the continuity and similarity of tectonic evolution of the Son-valley and Rajasthan Vindhyans. Some researchers believe there must have been two depocenters in a single basin with Bundelkhand granites acting as the extensive intrabasinal horst. Others infer that the Vindhyan Basin developed as a single basin, subsequently got separated during the Phanerozoic. High resolution deep seismic reflection studies (Prasad and Rao 2006), well-constrained by gravity, magnetic and magnetotelluric data suggested gradual thickening of the Vindhyan succession towards southeast from Bundi onwards. The velocities of the upper and lower Vindhyans were identified as 4.6–4.8 and 5.1–5.3 km/s. The NW limit of the Vindhyan Basin is demarcated by the Great Boundary Fault (GBF) that manifests as a 30 km wide NW-dipping thrust fault extending to a depth of 30 km. Bouguer gravity maps reveal a number of elongated highs and lows indicating probable horsts and grabens separating the Son Valley and Rajasthan Vindhyans.

Prasad and Rao (2006) identified a few deep-seated NW-dipping reflection bands which they interpreted as thrust faults developed in a compressional regime and collision of two crustal blocks, the Bundelkhand in the east and Mewar in the west. These thrusts extended to a depth of 30 km. Dimensions and depth extension of the thrust fault suggest that thin skinned tectonics were responsible for their evolution. It separates the Vindhyan sediments from the volcano-sedimentary formation of the Hindoli Group and represents the geologically mapped Great Boundary Fault of the region. The presence of Vindhyan only to the east of GBF suggest that the GBF was already in place when the Vindhyan Basin evolved. With ~1700 Ma ages now established for the lower Vindhyan sediments, the GBF should be older than that. The low Bouguer gravity under the Bundelkhand granite is accompanied by low gravity Gangetic Plain and high gravity Aravalli. A 110 km long gravity high extends from Jaipur to Ratlam cutting across the Aravalli and Vindhyan, which has been attributed to high density Bijawar rocks occurring below the Vindhyans (Qureshy and Warsi 1975; Verma and Banerjee 1992). It has been speculated that the rocks of Bijawar and equivalent Gwalior basin are likely to be present in subsurface covering a large spatial area than what it appears on the outcrop maps. A linear structure through Jhalawar-Jhansi to the Ganga Valley coincides with Asmara and Jhalawar lineaments. According to Ghosh (1981), the region between these two faults is the basement ridge with ENE–WSW orientation and possibly acted as the divide between the two Vindhyan sectors. This buried ridge seems to join the Faizabad ridge in the Ganga Plain. Airborne magnetic and ground studies reveal large magnetic anomalies representing the horst and graben structure of Paleoproterozoic basement Bijawar and Lower Vindhyan rocks. Geomorphotectonics of the basement in the Vindhyan Basin was carried out by Das et al. (1990) and revealed an asymmetric graben structure with the southern boundary fault between the Vindhyan and the Mahakoshal as the most prominent linear structure. Using magnetic and electrical methods, Singh and Ghosh (2003) mapped a prominent uplift in the basement to the north of Mirzamurad followed by steep slope to the south with thicker sediments of the Vindhyan s and Bijawar.

Recently, Khan (2013) observed that 160 m Khairmalia andesite in the Lower Vindhyan suggest the role of tectonic loading by andesitic volcanism in partial melting of the subducting slab in the formation of the Vindhyan Basin. It is also stated that although the Vindhyan Basin is an intra-cratonic basin, its genetic evolution suggests a back-arc basin due to continent-continent collision in Mesoproterozoic. It also experienced the Cenozoic tectono-thermal events that is envisaged from the relatively shallow Moho beneath the Central India (Kaila et al. 1985). DSS profiles also suggest that Vindhyan basin is riddled with deep-seated geofractures.

5.7.9 Paleocurrents

Geophysical studies have indicated westward to northwestward tilt of the basinal floor in which Vindhyan sediments were deposited. Consequently, depositional currents followed the basinal contours. The paleocurrent indicators like cross beddings and ripple marks in the sandy beds suggest supply of detritus from the source area in the south with currents directed towards west of NW (Chanda and Bhattacharya 1982). On the other hand, in the western flank of the Vindhyan Basin, the Upper Vindhyan succession shows cross laminations oriented across E–W paleo-shore line and represents on shore (south and offshore (north) dispersal of sediments (Akhtar 1996). The Vindhyan Basin is characterized by uniform unimodal westerly to northwesterly paleocurrent patterns suggesting a stable tectonics and persistence of paleoslope (Banerjee and Sengupta 1963). Chakraborty and Chaudhuri (1990) suggested that contemporaneous tectonism brought about a change in the paleoslope direction from the north during the early Rewa marine sedimentation phase to the west during the late Rewa fluvial phase.

5.7.10 Age of the Vindhyan Sedimentary Pile

The absolute age of the Vindhyan has remained a subject of discussion over a century. Summarizing the state of knowledge up to 1996, Vankatachala et al. (1996) concluded that the Vindhyan deposits range in age from 1400 to 550 Ma, in spite of a Pb–Pb age presented in 1993 at a Symposium on Vindhyans at Jadavpur University, where dates as old as ~1600 Ma were presented. The basement on which lower Vindhyan sediments were deposited varied from ~2500 Ma Bundelkhand Granite to ~1800 Ma old Bijawar and Gwalior groups. The controversy related to the age of Lower and Upper Vindhyan has remained an enigma for a very long time. Over the years several paleontological and isotopic techniques have been applied to constrain the age range of the Lower and Upper Vindhyan succession (Kumar et al. 1993; Ray et al. 2011). Glauconitic sandstone dated by Vinogradov et al. (1964) by K-Ar technique was later recalculated by Kreuzer et al. (1977) on samples from the Kheinjua Formation and suggested 1080 ± 40 Ma age for these rocks. It is worth noting that K-Ar dating technique gives highly erroneous dates for diagenetically altered rocks. This technique has an inherent limitation due to K and Argon's loss during diagnosis. Several Fission Track dates (Srivastava and Rajgopalan 1986, 1988, 1990) were also made available but due to large error bar have not been considered further in the geological literature. Rb-Sr isochron dates of the micas in the kimberlite pipes, intruding the lower part of the Kaimur sandstone, was reported to be 1140 ± 24 Ma (Crawford and Compston 1970) or 1067 ± 31 Ma (Kumar et al. 1993) remained in circulation in geological literature for several decades (Fig. 5.6).

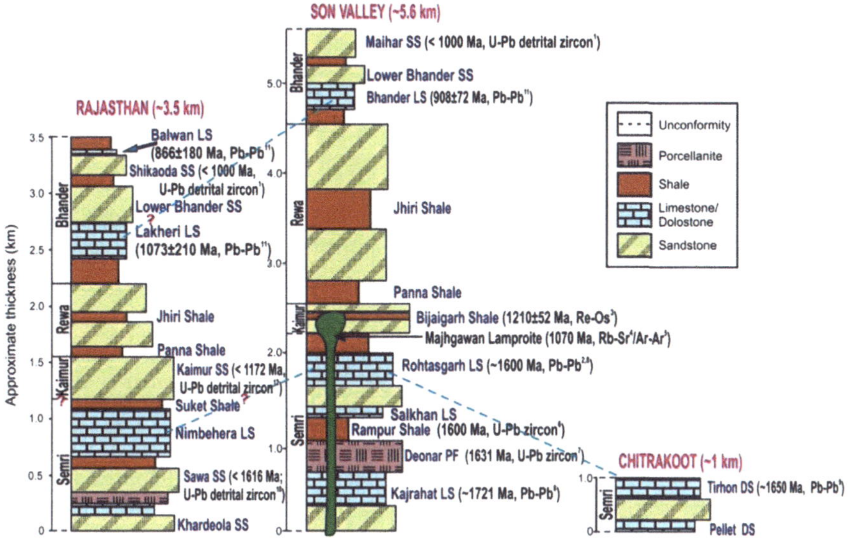

Fig. 5.6 Stratigraphic logs of the Vindhyan Supergroup in Rajasthan (West), Son Valley (Southeast), and Chitrakut (Northeast). Dashed lines connect correlatable limestone formations. Abbreviations: SS–Sandstone, LS–Limestone, DS–Dolostone, PF–Porcellanite Formation. After Shukla et al. (2019)

5.7.10.1 Seed of Age Controversy

Identification of metazoan trace fossils in the Chorhat Sandstones of the Semri Group (Sarkar et al. 1996; Seilacher et al. 1998), which at that time dated as ~1100 Ma old, triggered a world-wide controversy. Such an identification implied emergence of multicellular life forms much earlier than the globally accepted timeframe for such an evolutionary process. On scrutiny these structures were found to be inorganic and physically formed structures. This controversy was accentuated by erroneous identification of Ediacaran (Kathal et al. 2004) and Lower Cambrian Small Shelly Fauna (Azmi 1998) from the Semri sediments. In the same vain, premature interpretation of the Ar/Ar plateau age of 617 ± 4 Ma (Banerjee and Frank 2000) for the Lower Vindhyan micas was incorrectly cited to support an Ediacaran age for the basal Vindhyan. Extensive paleobiological studies (Sharma 1996a, b; Kumar 2016) involved studies on the microbial assemblages of at least 27 morphoforms, and numerous stromatolite form species, several trace fossils (Kumar and Pandey 2008), microbial mat induced structures (Sarkar et al. 2006), several forms of microfossils (Prasad and Asher 2016; Prasad et al. 2005; Kumar 1978; Kumar and Srivastava 1992a, b), some doubtful megafossils (Rai et al. 1997; Kumar et al. 2001; Hofmann 2005). Based on these paleobiological evidence the age interpretations remained at loggerhead (Azmi 2008) with those suggested based on precise U–Pb zircon ages. Although most paleobiologists (Sharma 2006a, b, p. 70) have now agreed to revise their age estimate of stromatolitic Lower Vindhyan carbonates to ~1600 Ma from ~1100 Ma, as stated earlier.

5.7.10.2 Isotopic Age Data

At a Vindhyan seminar at Jadavpur University, Banerjee and Russell (1993) demonstrated the antiquity of the Lower Vindhyans and ~1600–1650 Ma ages were presented for the Lower Vindhyan Rohtas Limestone, using the Pb/Pb isochrons. Ray et al. (2003) dated the Rohtas Limestone of the Semri Group as 1600 ± 130 Ma. Sarangi et al. (2004) obtained better results with Pb–Pb technique on similar carbonate rocks and reported 1599 ± 48 Ma age for these rocks (Gopalan et al. 2013; Kumar et al. 2001). Major change in the perception of the absolute ages of these rocks came when two papers simultaneously appeared in print (Rasmussen et al. 2002; Ray et al. 2002), both suggesting SHRIMP U–Pb dates of 1628 ± 8 Ma and 1631 ± 1 Ma respectively for the zircons from rhyolitic volcanic horizons of the Lower Vindhyan. These values have uncanny similarity with the Pb–Pb isochron ages of the Lower Vindhyan carbonates presented by Banerjee and Russell (1993) which got reconfirmed by Sarangi et al. (2004), Ray et al. (2003) and Bengtson et al. (2009). Rb/Sr model ages for glauconite (Kumar et al. 2001) provided added support for these ages. Unlike, robust nature of age constraints for the Lower Vindhyan rocks, the end phase of Upper Vindhyan, however remained mired with controversies. K-Ar dates of 741 ± 9 Ma for the Sirbu Shale of the Upper Vindhyan by Rathore et al. (1999b) and 650–750 Ma age suggested by Ray (2006) and Ray et al. (2003) are not well constrained and unreliable. At the same time, $^{87}Sr/^{86}Sr$ measurements on calcitic cements in stromatolitic Kajrahat Limestone provided some poor constraints (Ray et al. 2003). Tripathy and Singh (2015) provided Re-Os depositional ages for black shales of the Kaimur Group, while Chakrabarti et al. (2007) determined Nd isotopes to interpret the provenance of sediments.

5.8 Kolhan Group

Unconformably overlying several basement rocks like Sighbhum Granite, Jagannathpur volcanics and banded iron formation (Iron Ore Group) of the Singhbhum Craton, a thick pile of sedimentary rocks is exposed between Chaibasa and Noamandi as an intracratonic Proterozoic belt (Dunn 1940; Bhattacharya and Chatterjee 1964; Chatterjee and Bhattacharya 1969; Mukhopadhyay et al. 2006). The Kolhan Group, exposed in a NNE-SSW trending 60 km long Kolhan Basin, was divided into a basal 25 m thick sandstone-conglomerate (Mukhopadhyay 2001) unit of the **Mungra Sandstone** (Chakraborty et al. 2005), which is overlain by 20 m thick grey and purple stromatolitic carbonate rock assemblage and thin argillites designated as the **Jhinkpari Limestone**. These are topped by ~200 m thick shale and minor limestone known as the **Jetia Shale**. Structurally, this entire belt can be defined under three structural entities: (i) Domes and Basins in the east, (ii) homocline in the west with remnants of a thrust and (iii) a zone of inverted isoclinal folds. Conventionally, this basin is not described under the umbrella of the "Purana Basins" but its ~1600 Ma age allows undisputed correlation with other Proterozoic Basins of the Peninsular India.

The Kolhan Group is also recorded in two other belts: (i) **Nomira Basin** made up of low dipping beds of arkosic quartzite, sporadic pyrophyllite-sericite-talc rocks (paleosol?) on top of the Singhbhum Granite which show faulted contact with the volcanics, and (ii) **Mankarchua Basin** in Pala Lahara area where south dipping 50 m thick pebbly quartzite occur over the volcanic rocks of the region. A cross bedded chert bearing arenite known as the **Kamakhynagar Quartzite** is an equivalent rock unit.

Petrography, geochemistry and CIA values of the Kolhan siliciclastics, suggest an intensely weathered low-relief provenance dominantly composed of granitoid rocks. A warm and humid paleoclimate and a passive margin tectonic setting was inferred (Bandopadhyay and Sengupta 2004). On the other hand, linear outcrop pattern of the preserved sedimentary sequence with low matrix, lower feldspars and rock fragments presumably formed in an elongated trough, collectively convey the existence of a rift tectonic setting (Sahoo and Das 2015). They also suggested depositional framework under the time transgressive delta along the marine shelf with intermingled lacustrine (Das and Sahoo 2015) deposits. The rift setting of this basin was perhaps related to the fragmentation of the Rodinia supercontinent with the sediments derived from northerly and easterly source terrain (Bandopadhyay and Sengupta 2004).

5.9 Chattisgarh Basin

The Bastar Craton is dotted with numerous Proterozoic sedimentary basins, including the largest Chattisgarh Basin, besides Khariar, Ampani, Keskar, Singanpur, Abujhmar, Indravati and Sabari basins (Fig. 5.7; Ramakrishnan and Vaidyanadhan 2008). Opinions differ whether these were once connected and part of a single basin (Ramakrishnan 1987) or evolved individually (Kale and Phansalkar 1991).

The Chhattisgarh Supergroup of the Chhattisgarh Basin is exposed in the central India and occupies a large tract estimated to be 36,000 km^2. Total measured thickness of this supergroup is about 1500 m sedimentary layers of conglomerate, sandstone, quartzite, compositionally different types of shale, limestone, chert, dolomite and minor phosphorite (Dutt 1964; Naqvi and Rogers 1987). Earlier, the sedimentary sequence was divided into a basal Chandarpur Series and an upper Raipur Series

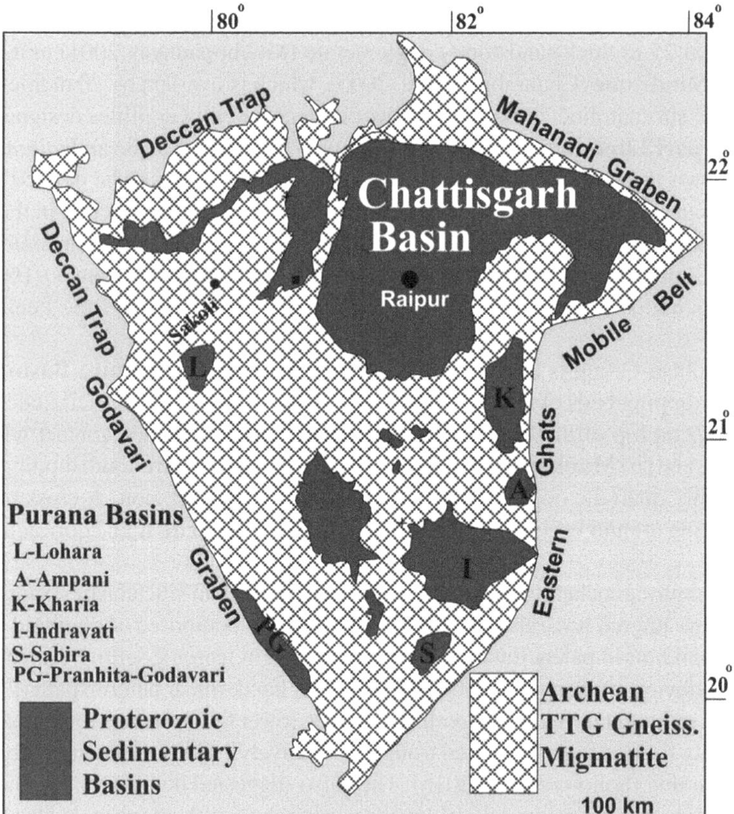

Fig. 5.7 Simplified geological map of the Proterozoic sedimentary basins developed on the Bastar Craton. Modified after Ramakrishnan and Vaidyanadhan (2008)

(Ball 1877; Dutt 1964; Schnitzer 1969). These nomenclatures were changed to lithostratigraphic term and got its new avatar as Supergroup and Groups (Murti 1987; Patranabis-Deb et al. 2007). After several revisions Murti (1987) finally established a lower Chandarpur Group (divided into Lohardih, Chaporadih and Kansa Pather Formations) containing conglomeratic, white and reddish brown arkosic sandstones. Environment of deposition has been interpreted differently ranging from a deltaic, storm-dominated shelf, subtidal, intertidal to high-energy shoreface (Chakraborty et al. 2002).

Chakraborty et al. (2015b) proposed an independent stratigraphic status of the Singhora Group in the Chhattisgarh stratigraphy based on structural data, AMS data and detrital zircon geochronology (Chakraborty and Barkat 2020). They suggested a four-tier stratigraphic subdivision of the basin with Singhora Group, Chandrapur Group, Raipur Group and Kharsiya Group in the eastern terrain and, Chandrapur, Raipur and Kharsiya Groups in the west (see Fig. 13.2 of Chakraborty et al. 2015b). Based on U–Pb detrital zircon geochronology, Nd isotope mapping and sediment geochemistry, Das et al. (2017) found similarities in the depositional history of the Singhora Group of the Chhattisgarh Supergroup with those of the Ampani and Khariar basins (Fig. 5.8).

The upper Raipur Group (divided into Charmuria, Gunderdehi, Chandi and Tarenga Formations), and separated by an unconformity with the Chandarpur Group

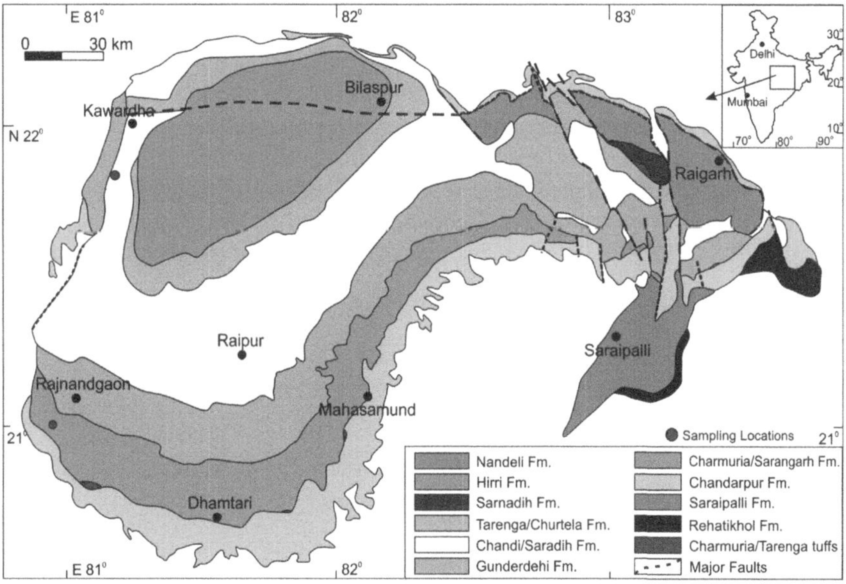

Fig. 5.8 Geological map of the Chattisgarh Supergroup, showing distribution of various stratigraphic units. Modified after Das et al. (1992)

is inferred to have deposited along the outer shelf and slope, characterized by limestone–shale-dominated sequence. The limestone–shale assemblage is the main component of the Raipur Group (Murti 1987; Jairaman and Banerjee 1978; Chaudhuri et al. 2002; Patranabis-Deb 2004). Conglomerate–shale–sandstone succession represents the proximal facies, while the distal facies is represented by limestone-shale assemblages. In some parts of the basin conglomerates of the Raipur Group rests unconformably and directly on the basement rocks and may be correlated with Chandarpur Group. Remapping of some of the sectors suggested subdivision of Chhattisgarh Supergroup into four groups: the Singhora, the Chandarpur, the Raipur, and the Kharsiya in temporally ascending order (Das et al. 1992; Saha and Patranabis-Deb 2014; Chakraborty et al. 2015b). Siliciclastics dominate the Singhora and Chandarpur groups. The sediments of the Chhattisgarh Supergroup were deposited in two subbasins, the Hirri in the west and the Bharadwar in the east, separated by the Sonakhan Greenstone Belt, and two proto-basins in the southeast: the Singhora and the Barapahar. There is no consensus in the tectonic framework in which the deposition took place. Three tectonic settings, an intracratonic sag, rift, and foreland were proposed at various points (Patranabis-Deb et al. 2007; Chakraborty et al. 2015a, b). Table 5.2 provides a glimpse of stratigraphy of the Chattisgarh Supergroup (Mohanty 2015), while Fig. 5.9 provides comparative picture of contemporary stratigraphic subdivisions of the Chattisgarh Basin along with a revised and modified version in the context of sequence stratigraphic framework (Chakraborty and Barkat 2020). Abrupt introduction of the fluvial sediments in the middle to distal marine shelf marks the Type 1 unconformity where several 'sequence' rank cycles appear stacked together. From paleocurrent, detrital zircon geochronology and Nd isotope mapping, unconformity formation across Saraipalli–Bhalukona Formation transition is supported by Das et al. (2017), while several other unconformities need further verification.

Two limestone formations of the Raipur Group, the Charmuria and Chandi limestones, have $\delta^{13}C$ in the range of 2.6–3.6‰, and 3.2–3.6‰, respectively which show variations from bottom to the top (George et al. 2018). The elevated $\delta^{13}C$ of the

Table 5.2 Chhattisgarh Supergroup sequence of the Bastar Craton (Mohanty 2015)

Stratigraphic units	Rock types	Age (in Ma)	Depositional/tectonic environment
Chhattisgarh Supergroup	Limestone, shale, sandstone and tuff	1011 ± 19	Peritidal carbonate platform
	Conglomerate, sandstone, shale	c. 1500	Fan delta, shore face shelf
	Unconformity		
	Conglomerate, porcellanite, sandstone, shale, limestone		Braid plain, alluvial fan, shallow-marine shelf
Unconformity			
Bastar craton			

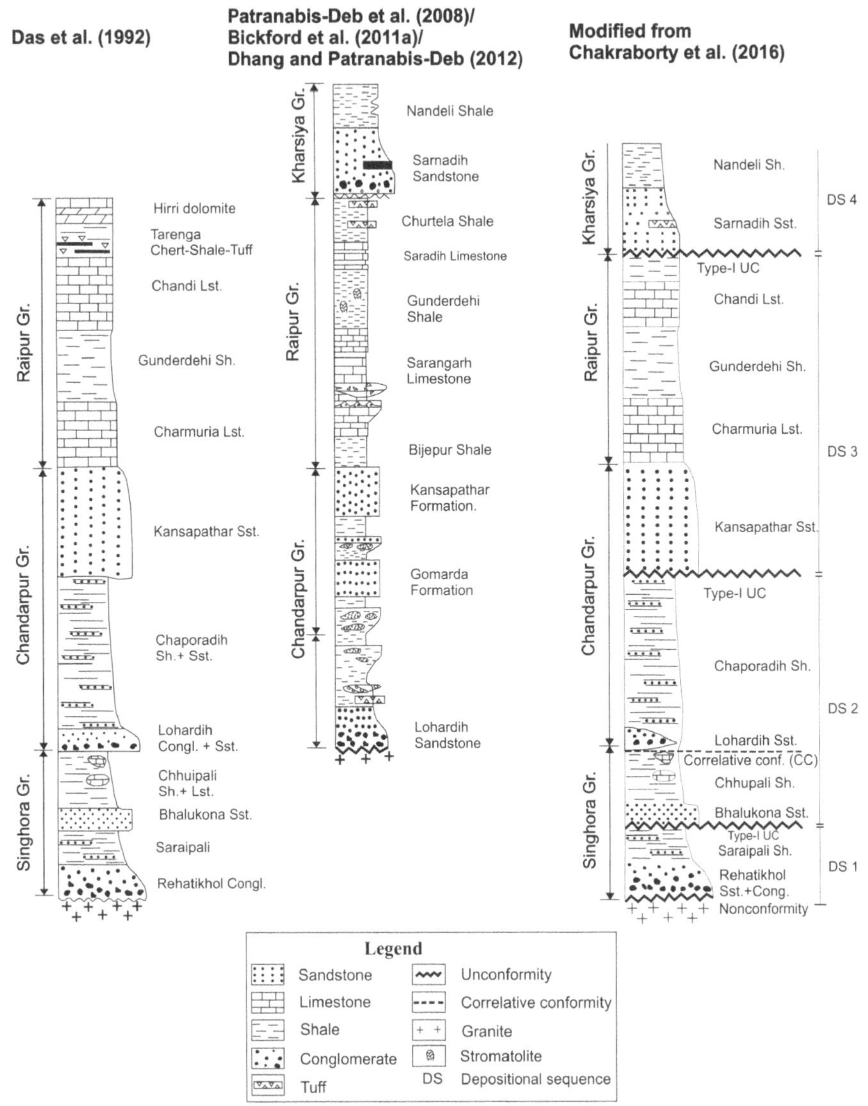

Fig. 5.9 Stratigraphic framework of the Chattisgarh Basin. After Chakraborty and Barkat (2020)

carbonates of the Raipur Group can be attributed to increased flux of organic carbon burial (Kump and Arthur 1999) consistent with the late Mesoproterozoic (1.2–1.0 Ga) $\delta^{13}C$ profiles in different parts of the world.

In the Charmuria Limestone placed in 'Sequence 3', $\delta^{34}S$ values (26.3 ± 0.9‰; n = 12) of pyrites indicated sulphidic deep marine environment (Sarkar et al. 2010). George et al. (2019) recorded positive $\delta^{13}C$ values like 2.6‰ and 3.6‰ and inferring an episode of higher organic carbon burial. An increase of $\delta^{13}C$ values from

1.4 to 4.3‰ in Charmuria and Chandi limestones is correlated with the global late Mesoproterozoic spike in the δ^{13}Cvalue in the contemporary oceans (George et al. 2019).

Authigenic glauconite of the Chapordih Formation yielded K-Ar age of 700–750 Ma (Kreuzer et al. 1977). Rocks of the Chattisgarh basin were correlated with 500 Ma geological event (Naqvi 2005) and assigned to late Proterozoic, based on stromatolitic assemblages (Jairaman and Banerjee 1978; Raha 1987). Dutt (1963), Kreuzer et al. (1977) and King (1981) correlated the Chattisgarh Supergroup with the Lower Vindhyan, Indravati, Kurnool and Bhima groups, based on lithological similarities and cyclic patterns of sedimentation.

Recent U–Pb SHRIMP ages of magmatic zircon from the rhyolitic Sukhda tuff on top of the Chhattisgarh sequence yielded 1011 ± 19 and 990 ± 23 Ma, and 1020 ± 15 Ma for the Sapos tuff (Patranabis-Deb et al. 2007; Bickford et al. 2011a, b; Mukherjee and Ray 2008, 2010). These authors, therefore, inferred that top of the Chattisgarh Supergroup could be up to 1000 Ma old and signifies the closing of the basin, with the base of the sequence (Singhora Group) around 1405 Ma (Saha and Patranabis-Deb 2014). Basal parts of stratigraphic successions in the Singhora and Khariar basins contain tuffaceous units with magmatic zircon crystallization ages as ~1500 Ma (Das et al. 2009) and 1405 ± 9 Ma (Bickford et al. 2011a) and 1455 ± 47 Ma (Das et al. 2009), respectively, which constrains the age of initiation of these basins (Das et al. 2009). Moreover, a diabase intrusive in the Saraipalli Formation has yielded a mineral-whole rock Sm–Nd isochron age of 1421 ± 23 Ma (Das et al. 2011). The convergence of all these data strongly support Mesoproterozoic <1600 Ma age of sedimentation in the basin, inferred from the detrital zircon age populations (Das et al. 2017).

5.10 Indravati and Sukma Basins

Adjacent to and closely similar to the Chattisgarh Basin, the Indravati Basin exposes an undeformed and unmetamorphosed ~500 m pile of shallow marine sediments in about 9000 km^2 over the Archean granites and supracrustals (Fig. 5.10). Several types of shale, limestone, stromatolitic dolomite, arkosic sandstone and conglomerate are recorded from this basin. Terrigenous rocks dominate the bottom of the sequence which are cross bedded, rippled and indicate shallow intertidal environment of deposition. These are followed upward by shale-limestone rhythmite up to the top of the sequence. Dutt (1963) gave the name Indravati Series for this Purana group of rocks in one of the corners of the Bastar craton and divided it into three Stages, namely Tirathgarh, Kanger and Jagdalpur. Schnitzer (1969) observed cyclicity of the succession in the Upper Beds and separated them from Lower Beds of Crookshank (1963) by a disconformity. The lithostratigraphic term, the Indravati Group was given by Sharma (1975). He identified Tiratgarh and Jagdalpur Formations which included Kanger limestone units. Ramakrishna (1978) redesignated these formations as subgroups, but reverted back to the identification of four formations in the ascending

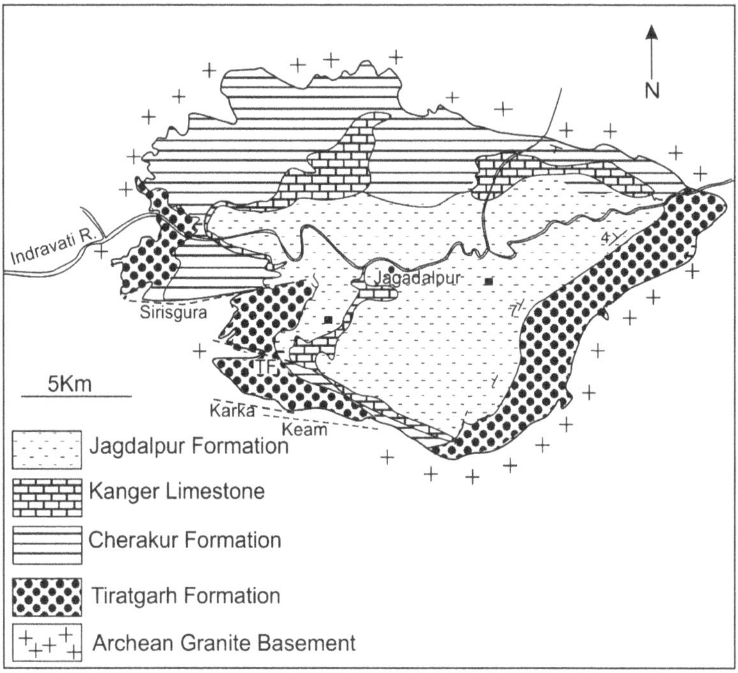

Fig. 5.10 Geological Map of the Indravati Basin. After Ramakrishna (1987)

order Tirathgarh, Cherakur, Kanger and Jagdalpur (Ramakrishna 1987). The rocks of this basin are nearly flat with undulating dips and is disposed of in a concentric manner. Crookshank (1963) recorded a few NE trending tight folds with dips up to 70° towards the eastern margin with a boundary fault. This sequence is partially affected by deformation in the Eastern Ghats Mobile Belt which appears to have resulted into a thrust. Decreasing gentle gradient of the regional gravity field towards the east was interpreted as due to downfolding and thickening of the crust in the mobile belt (Subrahmanyam 1980). Numerous normal faults with silicified surface and cataclastic features dissect the basin.

A triangular-shaped Sukma Basin of about 700 km^2 was identified by Ghosh of the GSI as an independent basin. Rocks of this basin closely resemble the Indravati sequence even up to the formation level with marginal variation at the member and bed level.

Although no rocks of Indravati Basin have been dated, their close correlation with the Chopardih Formation of the Chhattisgarh Basin puts it in the same age bracket for which K-Ar dates of glauconite (700–750 Ma: Kreuzer et al. 1977) are available. Recent U–Pb SHRIMP ages of zircon from the rhyolitic tuff from the top of the Chattisgarh/Indravati sequence (=Jagdalpur Formation) are 1011 ± 19 and 990 ± 23 Ma, indicating these rocks to be ~1000 Ma old (Bickford et al. 2011a, b). Both Indravati and Sukma Basins represent the faulted and eroded remnants of a single continuous Bastar-Chattisgarh superbasin (Ramakrishna 1987). Based on lithostratigraphy, mostly undeformed nature and absence of basic dykes, the Indravati Group was correlated with the lower Vindhyan and the Kurnool Group. Later, correlation with the Chattisgarh and connection through the outliers of Ampani, Keshkal and Khariar basins has been suggested over the years. Kimberlite pyroclastics, hosted within the Kanger Formation but below the Jagdalpur Formation (Mainkar et al. 2004), have yielded a LA-ICPMS U–Pb 620 ± 30 Ma age from autometasomatic titanite (Lehmann et al. 2007). U–Pb isotopic analyses of zircons from the Jagdalpur Formation tuffs point to the closure of the basin at 1001 ± 7 Ma (Mukherjee et al. 2012).

5.11 Khariar–Ampani Basins

Small independent sedimentary basins are also distinguished as the Khariar Basin (1500 km^2), the Ampani Basin (220 km^2) and the Abujmar Basin (3000 km^2). These basins have distinct peripheral boundary with the Eastern Ghats Mobile Belt and unconformably overlie the Bastar Craton and the 1466 Ma Lakhna dyke swarm (Ratre et al. 2010; Pisarevsky et al. 2013). Datta (1998) subdivided the Khariar succession into three informal units and correlated them with the Chandarpur Group of the Chattisgarh Basin. Das et al. (2001, 2009) classified these rocks under the Pairi Group which was divided into six formations. The succession is made up of terrigenous sediments of shallow shelves with a basal conglomerate containing clasts of basement granite and the Lakhna dykes. The Khariar Basin displays cluster of ages of 1455 ± 47 Ma for the monazite grains, derived from the tuff layers. Based on field disposition of the Kariar sediments with respect to dyke in the area, Das et al. (2011) estimated an age of 1450 Ma for this igneous activity, while Ratre et al. (2010) considered the sediments to be younger than 1460 Ma. The lower part of the Khariar Basin is considered correlative to the Singhora Group of the Chhattisgarh Basin.

The Ampani Basin fill succession spreads over 220 km^2 with folds (Walker 1902) and correlated with the Vindhyan and Cuddapah basins. Subsequent workers equated the Ampani with the Chattisgarh–Indravati master basin (Dutt 1964; Balakrishnan and Mahesh Babu 1987). The Ampani succession unconformably overlies the Bastar granite gneiss complex and shares a tectonic contact with the EGB.

5.12 Abujhmar Basin

Attached to the Bastar Craton the Abujhmar Basin is spread over ~3000 km^2 area is poorly studied basin till date (Mishra et al. 1987a, b). The Abjjhmar Group occurs on top of the Kotri-Dongargarh orogen. Its stratigraphic succession includes a lower Gundul Formation comprising polymictic conglomerate with clast of banded iron stones, rhyolites and quartzites. These unmetamophosed sandstones are covered by the amagdular tholeiitic Maspur Basalt along with gabbroic and doleritic sills and dykes (Ramakrishnan and Vaidyanadhan 2008).

5.13 Pranhita–Godavari (PG) Basin

NW–SE trending and about 450 km long Proterozoic Pranhita–Godavari (PG) Basin records ~6000 m thick sedimentary pile along the Pranhita–Godavari (PG) valley between the Bastar and Dharwar cratons, which are separated by wide outcrops of the Pakhal Basin along the Godavari valley (Fig. 5.11; Chaudhuri and Deb 2004; Chaudhuri et al. 2012). It is bounded by the Karimnagar Granulite Belt (KGB) in the southwest, while the Bhopalpatnam Granulite Belt (BGB) is exposed on the northeast margin (Santosh et al. 2004) The basin is girdled by the Khammam Schist Belt (KSB) and the Eastern Ghats Mobile Belt (EGMB) on the southeast. The basin constitutes the most important Proterozoic rift basin in the South Indian craton, though there are dissenting views (Sreenivasa-Rao 1987).

The PG Basin was earlier named as the Pakhal Basin (King 1881) with divisions into lower Pakhal Series and upper Sullavai Series, and record history of recurring rifting from Proterozoic onward. The Pre-Gondwana Proterozoic rifting record is preserved in this litho-package Godavari Supergroup, which is exposed in two linear Eastern and Western Belts. The Pakhal Supergroup includes shale, quartz arenite, dolomite, arkose and conglomerate. The lower part is dolomitic while the upper part is dominated by arenites. The Pakhals rest over the Peninsular gneissic complex and are divided into Mallamalai Group, Mulug Group, Albaka Sandstone and Penganga Group (divisible into 10 formations) and overlain by Sullavai Group with distinct disconformity/angular unconformity (Sreenivasa-Rao 1987). Subsequently, Chaudhuri (2003) designated the lithopackage as a 'mega-sequence' and subdivided it into three depositional cycles of '1st or 2nd order' affinity viz. Pakhal, Penganga and Sullavai, based on regional unconformities. A wide variation in the pattern of accommodation generation and spatio-temporal sedimentation motif is noticed between the depositional cycles. Smaller-order unconformities allowed further subdivision of each cycle as 'Subgroup', e.g. the bottommost cycle, represented by the Pakhal Group, subdivided into Mallampalli and Mulug subgroups; the Somanpalli Group is considered as coevally evolved deep water equivalent of shallow marine Mulug subgroup (Chaudhuri et al. 2012). Deposits belonging to shoreface and shelf depositional regimes are also described from rocks of the Somanpalli Group (Saha and Ghosh 1998). $^{40}Ar/^{39}Ar$ glauconite plateau ages from Mallampalli, Mulug and Somanpalli sandstones have yielded ages of 1686 ± 6 Ma, 1565 ± 6 Ma and 1620 ± 6 Ma, respectively. Considering possible loss of Ar on deep burial of sediments, an age older than 1600 Ma is suggested for the initiation of the Proterozoic sedimentation in the PG valley (Saha and Patranabis-Deb 2014).

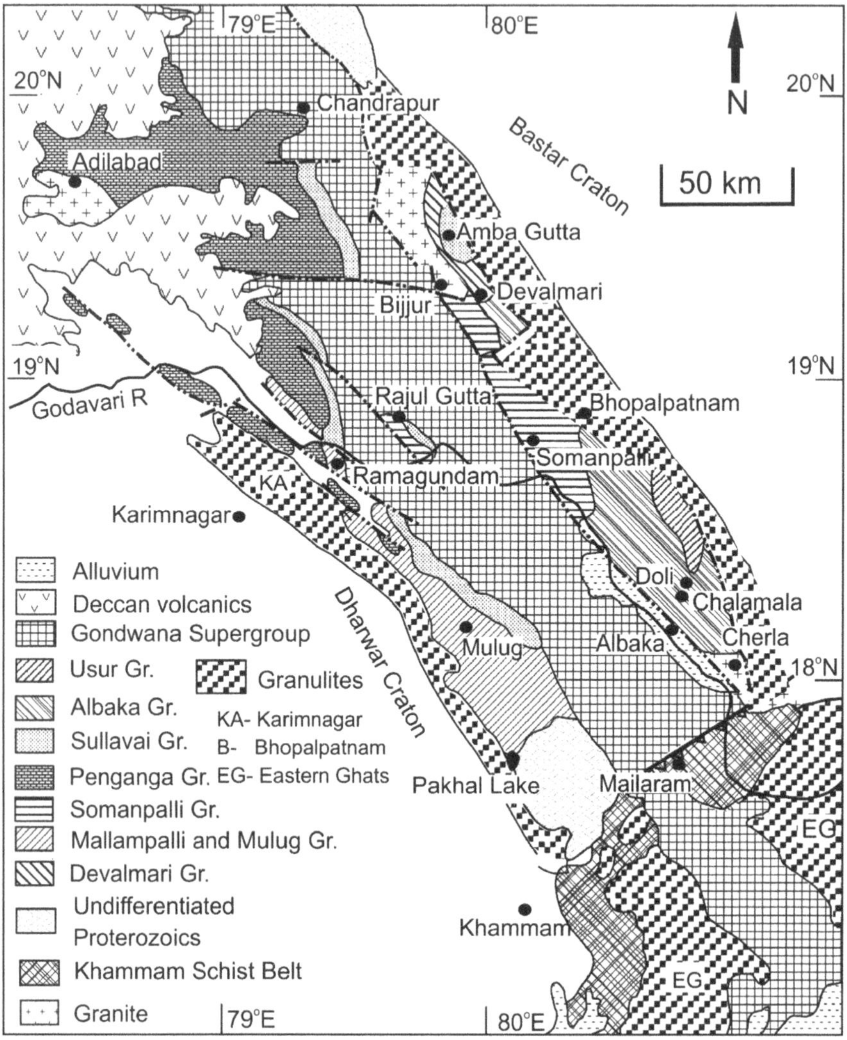

Fig. 5.11 Generalized geological map of the PG Valley, showing distribution of various lithounits of the PG Basin between the Karimnagar and Bhopalpatnam Granulite Belts. After Chaudhuri et al. (2012)

The second unconformity-bound cycle, exposed in the central and northern parts of the valley, is represented by the Penganga Group comprised of mixed carbonate–siliciclastic lithology. $^{40}Ar/^{39}Ar$ date of glauconite from carbonate allowed to assign a minimum age of 1200 Ma for the initiation of Penganga carbonate sedimentation. Based on age estimation, previous workers (Conrad et al. 2011; Saha and Patranabis-Deb 2014) have drawn correlation of the Penganga succession with the Chandarpur and Raipur successions of the Chhattisgarh Supergroup of Bastar Craton. From variations in stromatolite growth pattern in Penganga carbonate succession, Sarkar and Bose (1992) described a transition from basin plain to near shore subtidal across a reef.

Unconformably overlying rocks of the Penganga Group, Mulug subgroup or Mallampalli Subgroup, red feldspathic sandstones of the Sullavai Group, belonging to fluvial and eolian erg depositional systems, represents the third cycle. In this backdrop, correlation of the Albaka Group, unconformably overlying the Somanpalli Group and represented by siliciclastic tide-storm dominated lithopackage, remains tentative and has been correlated either with the Penganga Group or the Sullavai Group.

This sequence has been compared with lithostratigraphic units of Cuddapah Supergroup. Mathur (1982) reported 1276 ± 30 Ma, 1188 ± 14 Ma and 1142 ± 37 Ma and also one single age of 871 ± 21 Ma for the sandstone and arkosic units. Stromatolite assemblages show Mesoproterozoic affinity. Folding is seen locally with intensity of folding increasing from NE to SW. Rocks of the Pakhal Supergroup are largely unmetamorphosed, while mild metamorphic effects are seen in the SE sector.

Tectonic evolution of the PG Basin is envisaged as a rift-controlled structure and separation of the Dharwar and Bastar cratons from each other (Fig. 5.12; Chaudhuri et al. 2012). Unconformity-bound sequences provide control on episodic uplift, erosion of sub-strata and subsidence of the basin floor causing differential movements. The scale of differential subsidence and/or uplift, removal of an entire sequence beneath sequence boundaries, and sequence architecture consisting of thick clastic wedges provide the compelling evidence for tectonically driven sea-level changes and origin of the sequence bounding unconformities.

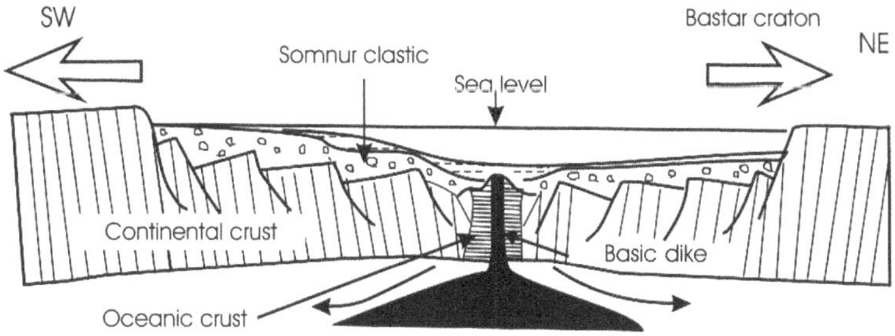

Fig. 5.12 Proterozoic evolution of the PG basin showing continental rift between the Dharwar and Bastar Cratons which were coalesced into a proto-continent. Deposition of shelf clastics, early stage of continental rupture and development of oceanic crust. After Chaudhuri et al. (2012)

5.14 Kaladgi Basin

Covering an area of nearly 8000 km^2 and extending for nearly 200 km, a ~4000 m thick sedimentary succession is given the name of Kaladgi Basin along the northern margin of the Dharwar Craton (Fig. 5.13; Jayaprakash et al. 1987; Kale and Phansalkar 1991; Dey 2015). Foote (1876) first described the stratigraphy of this entire basin following the earlier works by Christie, Newbold, and Carter in the early nineteenth century. This classification of the Kaladgis into the Lower and Upper was sustained until the mid-1960s (Pascoe 1965, p. 367). Nautiyal (1966) defined the Lower and Upper Series of the Kaladgi Basin as deposits of two separate but overlapping basins. Viswanathiah (1979) proposed a distinct angular unconformity separating these sequences. These rocks appear as small inliers with the Deccan Traps. Some part of this basin is concealed under the basaltic trap (Raha and Sastry 1982). Greenstone succession of Hungund Schist of the Dharwar Supergroup and granite gneisses of the Western Dharwar Craton provided floor for the Kaladgi sedimentation. The sedimentary beds of this basin are divided into a lower Kaladgi Group and an upper the Badami Group. Jayaprakash et al. (1987) suggested that the entire sequence can be clubbed into a Kaladgi Supergroup with older unit as the Bagalkot Group and the upper as the Badami Group.

5.14.1 *Bagalkot Group (Kaladgi Group)*

This group comprise of two sedimentary cycles, the lower Lokapur Subgroup and an upper Semikeri Subgroup (Dey 2015; Jayaprakash et al. 1987). Lokapur Subgroup occurs on top of the gneisses and granites of the Dharwar Craton with an angular

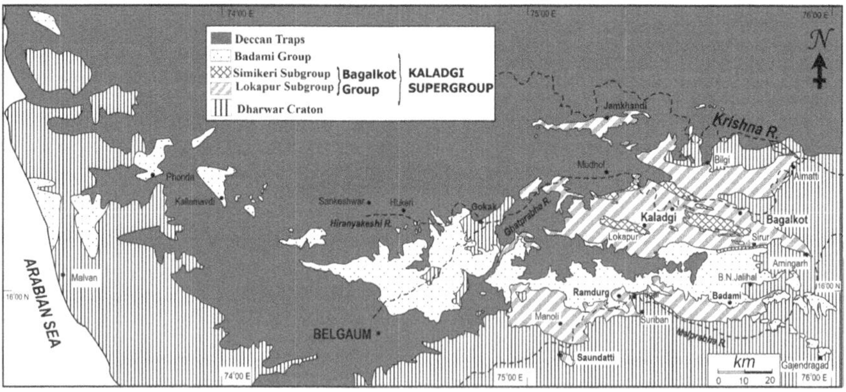

Fig. 5.13 Geological map of the Kaladgi Basin. Modified after Kale

unconformity. There is a marked disconformity between the Lokapur Subgroup and Simikeri Subgroup. Both the subgroups are divisible into ten formations characterized by distinct lithology. Broadly, three transgressive cycles of sedimentation can be recognized, each floored by the clastic suite of rocks, and grading into cyclic argillite-carbonate suites. These repeated marine transgressions on an episodically sinking, epicratonic shelf are responsible for the accumulation of more than 3500 m thick pile of sediments in the Kaladgi Basin. Stromatolites are recorded from the limestone and dolomite formations (Sharma and Pandey 2012). Rocks of this group are pervasively deformed. The deformation diminishes towards the basin margin. Tight, isoclinal to recumbent folding is seen in the central parts of the basin. (Jayaprakash et al. 1987). These are shear folds. In the axial sector of this basin, the Bagalkot Group has suffered cross-folding and multiple co-axial deformations in different phases.

Joy et al. (2019) have used the term Kaladgi Group instead of Bagalkot Group and changed some of the existing formation names (Jayaprakash et al. 1987). The Saundatti Quartzite of the old scheme renamed as Ramdurg Formation was used by Joy et al. (2019) for multipronged geochronological investigation in this basin. The lower part of this formation is characterized by poorly sorted, polymodal, very coarse to fine-grained quartz arenite, locally pebbly, and arkosic with lenticular bodies of conglomerates. These are petrologically the first-cycle arenite and confirm the erosional nature of the nonconformity below.

5.14.2 *Badami Group*

Unconformably overlying the Bagalkot/Kaladgi Group is the Kerur Formation which comes in sharp contact with the Katagiri Limestone of the Badami Group (Jayaprakash et al. 1987; Dey 2015; Saha et al. 2016a, b). The basal conglomerate as part of the Kerur Formation overlies the basement granitoids and schists as well as Bagalkot Group with a major angular unconformity. This formation comprises of mature silica-cemented medium-grained sandstone and shale. The lower part of the Kerur Formation comprises of poorly sorted and angular clasts. The matrix is ferruginous and clayey. The upper part is a medium to coarse-grained, moderately sorted subarkose to arenite with altered and fresh feldspars. The deposits occur as shallow, fining upward, wide channel fill bodies. The dominant palaeoflow is towards SW. The upper unit constituting the main body of the Kerur Formation consists of medium-grained, well-sorted, quartz arenite and feldspathic sandstone. Geochemistry of shales from the Kaladgi-Badami basin provide insight to the paleoweathering scenario at that period of time (Dey et al. 2008).

The Badami Group is mildly deformed, with large amplitudes of open folds, and gentle dips. Intraformational conglomerate in the Katagiri Limestone testifies to its penecontemporaneous deformation and basin floor instability. The equivalents of the Badami Group in the Konkan coastal strip display steep dips, but their deformation is attributed to a much younger (Tertiary) activity along the West Coast Fault.

5.15 Bhima Basin

A limestone-dominated succession of ~270 m thick and spread over an area of about 5200 km^2 is developed NE of Kaladgi/Bagalkot Basin in the Bhima Basin (Kale and Peshwa 1995). It overlies the granitic basement rocks of the Eastern Dharwar Craton (EDC) with a marked unconformity. Some parts of this basin are covered by Deccan basalt. Pioneering work in this basin has been done by many stalwarts, while recent studies by Dutt (1975), Janardhan Rao et al. (1975), Mishra et al. (1987a, b) and Kale et al. (1990) are noteworthy. The Bhima Group is divided into a lower Sedam Subgroup with two significant formations (Rabapalli Formation and Shahbad Limestone), and an upper Andola Subgroup of three formations (Halkal Formation, Katadevarhalli Limestone, and Harwal Shale). The succession begins with conglomerate and grit beds followed upward by quartz arenite, siltstone and olive, and purple shales. The overlying limestone is flaggy, grey, and cherty. Upwards, with a disconformity rests the limestone pebble beds grading into quartz arenite and, shale which gradually give way to siliceous limestone formation. These limestone formations show gradational contact with the topmost shale unit of the Bhima Basin. Rocks of this basin are horizontal with minor tectonic disturbances along the margins. Several microbiotas and Neoproterozoic organic remains are reported from the limestones and shales of this basin.

5.15.1 Age

A recent multipronged geochronological investigation by Joy et al. (2019) found U–Pb baddeleyite age of 1861 $\pm$ 4 Ma for the dolerite dike in the lower part of the Kaladgi/Bagalkot Group. This age suggests its equivalence to the Papaghani Group of the Cuddapah Basin which is the beginning of the Purana sedimentation at ~1900 Ma. Analysis of detrital zircon from the arenites of Kaladgi and Bhima Basins indicated variable sources of sediments, suggesting Kaladgi and Bhima to have different provenance. Paleocurrent analysis indicates a change in the provenance from S to SE to W or NW for the Kaladgi and Bhima sediments. U–Th–Pb and Rb–Sr dates indicate that limestone and glauconite bearing sandstones of the Bhima Group and Badami Group deposited around 800–900 Ma suggesting contemporaneity of the two successions. The unconformity between the Kaladgi Group and the overlying Badami Group, therefore, suggests that limestone and glauconite bearing sandstones of the Bhima Group and Badami Group deposited around 800–900 Ma. This suggests the contemporaneity of the two successions. The unconformity between the Kaladgi Group and the overlying Badami Group should, therefore, represent a time gap up to 1000 Myr.

5.16 Cuddapah Basin

Enormously thick Purana sedimentary sequence of the crescent-shaped Cuddapah Basin is exposed in the southeastern India in an area of about 45,000 km^2 in an epicratonic basin bordering the Dharwar Craton (Fig. 5.14; Kale and Phansalkar 1991; Kale 2016; Kale et al. 2020). The sequence is interbedded with volcanics, contains intrusive bodies, and shows complex evolutionary history. The sedimentary pile rests on the N-S to NNW-SSE trending Peninsular Gneiss-greenstone-schist of the Eastern Dharwar Craton, which are intruded by the Neoarchean (2.6–2.5 Ga) granitic batholiths and two phases of mafic dykes (~2.37 to ~1.89 Ga) during the Paleoproterozoic times (Chatterjee and Bhattacharji 2001; Ramakrishnan and Vaidyanadhan 2008; Kumar et al. 2015; Valdiya 2016; Mukherjee et al. 2018). The sediments from the Cuddapah Basin display an angular and erosional unconformity with the EDC basement all along its western boundary.

Earlier workers correlated the Cuddapah Basin with the Mesoproterozoic Eastern Ghats Mobile Belt (EGMB) or even Dharwarian Schist belts. However, new investigations link this basin with the Nellore Schist Belt/Nellore Fold Belt (Chetty 2001; Collins et al. 2015). Kale et al. (2020) proposed that the Cuddapah and Nellore Fold Belt constitute a combination of terrains/tectonic blocks that have a common, interlinked evolutionary history spanning the Late Paleoproterozoic to Neoproterozoic times (Sesha Sai 2013; Tripathy et al. 2019; Saha and Sain 2019).

The Nellore Schist Belt (NSB) is thrust upon the Cuddapah sediments along the Vellikonda Thrust (Nallamalai Thrust of King 1872) which marks the eastern boundary of the Cuddapah Basin (Fig. 5.14). The Udaigiri Group in south and the Vinjamuru Group in north show thrust contact with the Nallamalai Fold belt or a transpressive shear zone (Das et al. 2016). The NSB has a tectonic contact with the southern part of the Eastern Ghats Belt (EGB), referred to as the Ongole domain. Geophysical studies across the region surrounding the CB (Misra and Ravi Kumar 2013; Chandrakala et al. 2015) have been consistent in indicating that the continental crust below the CB and NSB is continuous, although dissected by the deep upper-crustal thrusts (Kale et al. 2020).

The metamorphics and intrusives from the Nellore Schist Belt occupy the eastern edge of this region and is replaced in the west by the folded low grade metasediments in the Nellore Fold Belt, which have suffered local low-grade (cataclasis dominated) metamorphism. This is bounded on east by the Vellikonda Thrust and on the west by the Maidukuru Thrust (=Rudravaram Line: Meijerink et al. 1984; Nagaraja Rao et al. 1987). Four undeformed and unmetamorphosed subbasins piggy back the Cuddapah Basin and host the sedimentary successions of the Neoproterozoic Palnad and Kurnool subbasins and Paleo to Mesoproterozoic Srisailam and Papaghni subbasins.

The Cuddapah Basin was studied by King (1872) who recognized the Cuddapah and Kurnool formations as two separate rock assemblages. Naraynaswami (1966), Meijerink et al. (1984), Nagaraja Rao et al. (1987), Mishra et al. (1987a, b), Prasad et al. (1987), Padmakumari and Dayal (1987) and a summary by Patranabis-Deb et al. (2012) and Kale et al. (2020) provide information about stratigraphy, tectonics

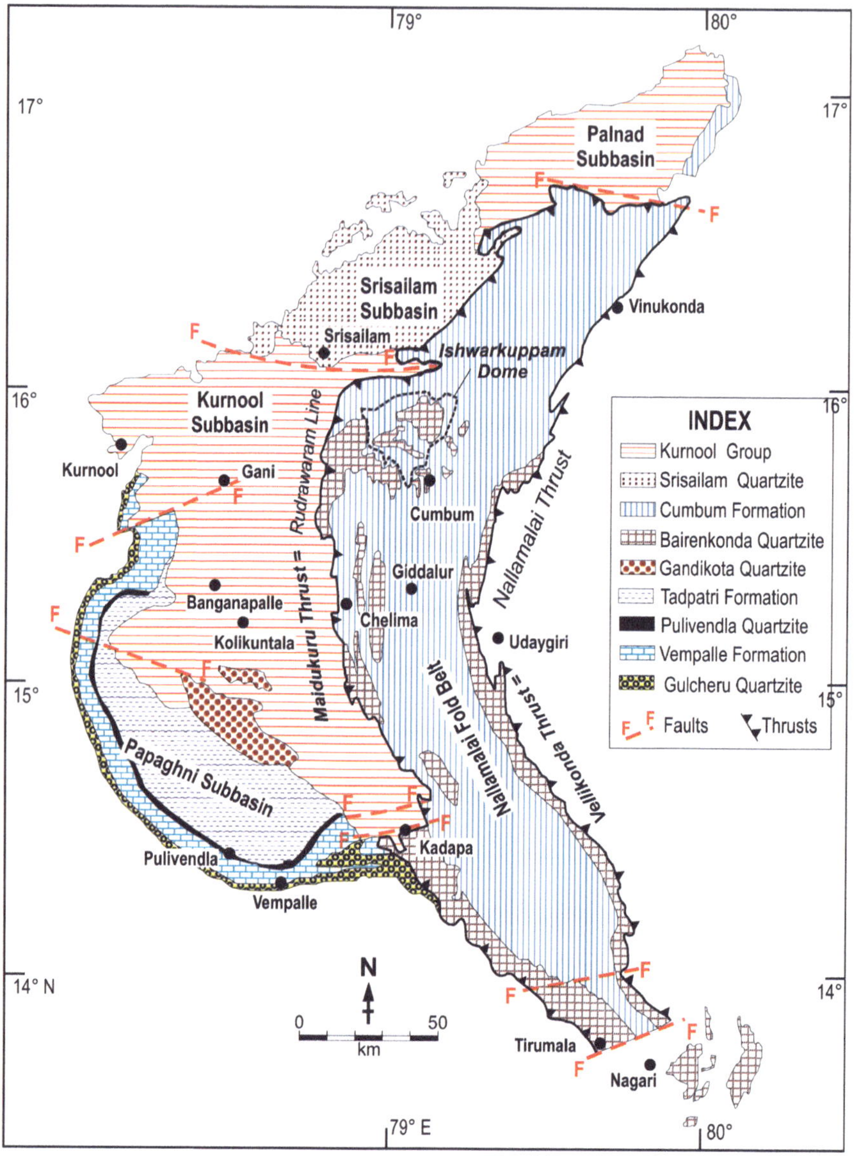

Fig. 5.14 Geological map of the Cuddapah (now officially spelled as "Kadapa") Basin and related adjoining Proterozoic sub-basins. After Kale et al. (2020)

and geochronology of this basin. The Kurnool Group was recognized as an independent unit by Nagaraja Rao et al. (1987) and was endorsed by Ramakrishnan and Vaidyanadhan (2008). The Cuddapah Supergroup is divided into the Papaghni, Chitravati and Nallamalai Groups in an ascending order (Nagaraja Rao et al. 1987; Saha et al. 2014); each separated by an unconformity. Recent workers have shown that deformed Nallamalai Group is bounded by major thrusts (Tripathy and Saha 2010; Saha and Tripathy 2012; Tripathy et al. 2019) and is not an autochthonous unit. Due to exclusion of the Nallamalai Group (Sesha Sai 2019) from the Cuddapah Supergroup (Saha et al. 2017), this 6000 m thick group has now been placed under the Nallamalai Fold Belt. The Cuddapah Supergroup now consists of the Papaghni Group and Chitravati Group and is unconformably overlain by quartzite of the Srisailam Formation. The Srisailam Formation, considered as the youngest unit of the Cuddapah Supergroup, rests unconformably and directly over the Peninsular gneiss and was named 'Kistna Series' by King (1872) after the River Krishna. It is capped unconformably by the Kurnool Group in the north, with local faulted contacts. Its contact with the Nallamalai Group is tectonic. The Cumbum Formation of the Nallamalai Group contains stromatolitic dolomites with phosphorite in Chelima-Pacherla area (Rao et al. 1997; Banerjee and Saigal 1988).

The Neoproterozoic Kurnool Group rests unconformably on other sequences of this basin as well as directly on the basement gneisses. The sediments of this group were deposited in two separate sub-basins, namely the Kurnool and Palnad subbasins.

Figure 5.15 summarizes the current stratigraphic status of the Cuddapah Basin and other subbasins along with their depositional environments (Kale et al. 2020).

5.16.1 Papaghni Subbasin

The **Papaghani Group** rests on the EDC basement complex with an erosional uncon formity. The lowest formation, the **Gulcheru Quartzite**, is made up of arkose, pebbly sandstone, glauconitic sandstone, conglomerate and mudstone and represents tectonically controlled alluvial fan facies accompanied by periodic basement uplift (Majumder et al. 2015). Ripple morphology suggest development in intertidal flats with passage to the carbonate platform of the Vempalle Formation (Saha et al. 2016a, b). Microbially-induced sedimentary structures (MISS) are found in its basal parts (Goswami et al. 2017). Geochemical indicator elements suggest provenance of these detritals to be active continental margin to continental arc settings (Chakrabarti et al. 2016). Sesha Sai et al. (2017) proposed that the iron-rich sediment possibly indicate the Paleoproterozoic oxygenation event (~2.1 Ga).

The overlying 1900 m thick **Vempalle Formation** consists of laminated cherty, stromatolitic and oolitic dolomite with thin intercalations of sandstone and mafic lava (Banerjee et al. 2019). The sandstones are profusely cross bedded with tepee structures, desiccation cracks, molar tooth structures and salt psedomorphs and indicate evaporitic depositional environment while carbonates indicate intertidal to subtidal environment (Phansalkar et al. 1991). The Vempalle dolomites show a variety of

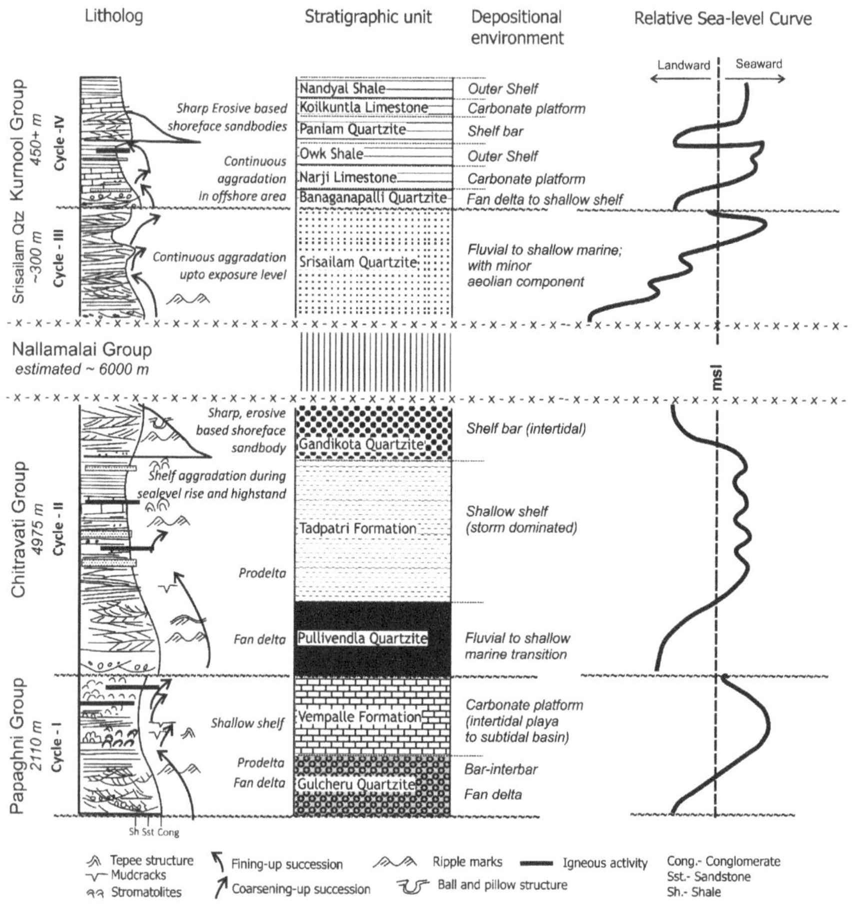

Fig. 5.15 Composite log (not to scale) of the Cuddapah Supergroup showing the major constituents, their depositional environments and the interpreted relative changes in the sea-level. After Kale et al. (2020)

stromatolites, reflecting depth partitioned growth of the microbial community (Bose et al. 2017; Patranabis-Deb et al. 2018). Soft sediment deformation structures including intraformational limestone breccia, large slumps, cracks and microdykes indicate Paleoproterozoic seismicity (Panja et al. 2018). Early Proterozoic tholeiitic lavas and sills (Kuppalapalle Volcanics) were emplaced during initial phase of extension of the intra-cratonic (Nagaraja Rao et al. 1987; Anand et al. 2003; French et al. 2008).

The **Chitravati Group** unconformably overlies the Papaghni Group and is subdivided into the Pulivendla Quartzite, Tadpatri Formation and Gandikota Quartzite in ascending order. Mafic sills and dykes are very widespread over the terrain. The ~75 m thick **Pulivendla Quartzite** consists of thickly bedded quartz arenite and laterally impersistent conglomerate, derived from the underlying Papaghni Group with a disconformity. Mass-flow mechanism is attributed to these alluvial fan deposits.

The quartz arenite beds with bipolar and bimodal cross stratification indicate deposition along the shore-line, at or very close to the sea-level. The Archean provenance has been inferred based on chemical characteristics of the sediments (Somasekhar et al. 2018).

Overlying the Pulivendla Quartzite with a sharp contact, the **Tadpatri Formation** of shale, sandstone and dolomite possesses dominant parallel planar lamination and indicate the deposition of shales in an outer shelf environment. The dolomitic units with algal laminae and stratiform stromatolites in upper part of the Tadpatri Formation suggest a supratidal to subtidal depositional environment (Mitra et al. 2018). Several dolerite sills, basaltic lava flows, ultramafic sill have been reported along with bimodal volcanics and tuffs (Anand et al. 2003; Goswami et al. 2018). The uppermost **Gandikota Quartzite** of the Chitravati Group has a gradational contact with the underlying Tadpatri Formation but unconformable relationship with the overlying Bairenkonda Quartzite of Nallamalai Group. The sandstone consists mostly of medium- to coarse-grained, well sorted and well-rounded quartz- and feldspathic arenite which was deposited in tidal and storm dominated open shelf environment.

Meijerink et al. (1984) identified an angular unconformity between these sandstones and the underlying Tadpatri shale (also Nagaraja Rao et al. 1987; Saha and Tripathy 2012; Patranabis-Deb et al. 2012). However, detrital zircon from Gandikota Formation suggests a Mesoproterozoic age (Collins et al. 2015) in contrast to the Paleoproterozoic age given to rest of the Chitravati Group.

5.16.2 Nallamalai Subbasin

The **Nallamalai Group** is a tectonically defined (rather than sedimentologically) segment of the Cuddapah Basin, bounded on either side by the Maidukuru and Vellikonda Thrusts (Fig. 5.14; Kale et al. 2020). The metasedimentary sequence within this subbasin is folded and underwent greenschist facies metamorphism, thus blurring the subdivisions within the stratigraphic profile. The contact between the **Chitravati Group** and the **Nagari Quartzite** and its lateral equivalent Bairenkonda Quartzites of the Nallamalai Group is defined by an angular unconformity. Lenses of pebbly conglomerates, gritty sandstones, thickly bedded quartz-rich sandstones, siltstones and interbedded shales are the constituent lithology. Saha et al. (2016a, b) interpreted a deltaic fan system, grading to a peritidal to inner shelf environment of deposition. Shales, dolomites and quartzites of **Cumbum Formation** includes folded and faulted lowly metamorphosed **Pullampet Slate**, **Giddalur Quartzite** and ferruginous shales and override the Bairenkonda Quartzite with gradational contact. Stromatolitic dolomites are phosphatic along with associated shales (Banerjee and Saigal 1988). Some of the 'porcellanite' and cherts interbedded within this shale-limestone sequence are ash-fall tuffs. Based on their overall characters, these sediments are attributed to an initial barrier beach environment that rapidly is replaced by tidal flats. Quartzose sandstones, red beds and shales are also present. Sedimentary structures suggest deposition by part of big river system, in dominantly arid environment (Patranabis-Deb et al. 2018).

5.16.3 Srisailam Subbasin

This basin is separated from the Papaghni subbasin (that hosts the older sequences in the CB) by the cap of younger sediments in the Kurnool subbasin. The **Srisailam Quartzite** is a 300 m thick sequence of supermature quartz arenite with a very high degree of mineralogical maturity along with ferruginous sandstones (red-beds) and minor shales. Large tabular cross strata indicate that they were deposited by large river system followed by aeolian deposition (Patranabis-Deb et al. 2012). The occurrence of red-beds within this sequence indicate deposited after the Great Oxidation Event (GOE) in an essential fluvial to marine environment. Basu et al. (2017) recognized five facies associations and suggested continental half-graben type basin with the black shale and sandstone attributed to a deep lacustrine and pro-delta environment.

5.16.4 Kurnool and Palnad Subbasins

The 450 m thick **Kurnool Group** with six constituent formations in Kurnool subbasin, overlies the Papaghni and Chitravati Groups and onlaps the gneissic basement of the EDC with a major unconformity. In Palnad area the Kurnool Group unconformably overlies the Peninsular Gneiss and the Srisailam Formation. The **Banganapalle Quartzite** rests upon the basement complex with an unconformity along its western edge of the Palnad subbasin and northwestern edge of the Kurnool subbasin. It also displays an angular and erosional unconformity with the underlying sediments from the Srisailam and Papaghni subbasins respectively. The basal polymictic conglomerate and coarse graded sandstone are overlain by multistorey bodies of trough cross-stratified coarse-grained sheet sandstones and are attributed to alluvial fan and braided fluvial plain deposits. Some of these sandstones were deposited as tidal bars. Laminated siltstone-shale in the uppermost **Banganapalle Quartzite** laterally passes into stromatolitic **Narji Limestone** which are widespread across the Kurnool and Palnad subbasins. Presence of pyrite in some of these beds indicate local anoxia. The ochre yellow coloured **Owk Shale** is laterally extensive across the Kurnool subbasin with a few interbedded fine sandy/silty beds, welded tuff and volcaniclastic sandstone. Carbonaceous organic remains and cynobacteria from this have been used for an upper Neoproterozoic age, correlatable to the biota of the Bhima, Vindhyan and Chattisgarh basins horizons (Sergeev et al. 2012; Sharma and Shukla 2012; Sharma et al. 2016; Sharma and Singh 2019). The succeeding **Paniam Quartzite** is marked by very well sorted medium- to fine-grained quartz arenite. King (1872) named these sandstones as "Plateau Quartzite". These texturally and mineralogically mature sandstone formed in open coastal environment. The buff coloured argillaceous **Koilkuntala Limestone** overlies the Paniam Quartzite with a gradational contact and is followed by brown thinly bedded **Nandyal Shale**. Storm laid, thin aeolian and normally graded silt/fine sand layers, often with fading ripple are preserved in thicker shale beds.

5.16.5 Tectonic Elements

Structural complexities of the CB–NSB block were worked out by Meijerink et al. (1984), Venkatakrishna and Dotiwala (1987), Geological Survey of India and the Atomic Minerals Directorate by using remote sensing analysis of the terrain. The following sequence of geologically and structurally distinct domains can be recognized along an E–W regional traverse across the Cuddapah basin as well as in the Deep Seismic Sounding profiles (Chetty 2011; Ram 2012; Kale et al. 2020) (Fig. 5.16).

(a) EDC: The deeply weathered and eroded cratonic block with the Archean gneisses and Paleoproterozoic mafic dyke swarms.
(b) Western subbasins: Unconformable capping of Paleo- to Mesoproterozoic sedimentary sequences of the Cuddapah Basin with interbedded igneous and subhorizontal Neoproterozoic sedimentary sequences of the Kurnool Group showing local folding.
(c) Nallamalai Fold Belt (NFB): N-S trending curvilinear regional folded Mesoproterozoic sediments that are thrust over the sedimentary successions of the western subbasins along its western edge.
(d) Nellore Schist Belt (NSB): Neoarchean metavolcanics of Mesoproterozoic accretionary sequences having collisional arc affinities, ophiolites with alkaline bodies; the belt is thrust over the NFB.

Geophysical studies across the CB since the 1980s by the NGRI (Kaila and Tewari 1985; Krishna Brahmam 1989; Kaila et al. 1987; Krishna Brahmam et al. 1986;

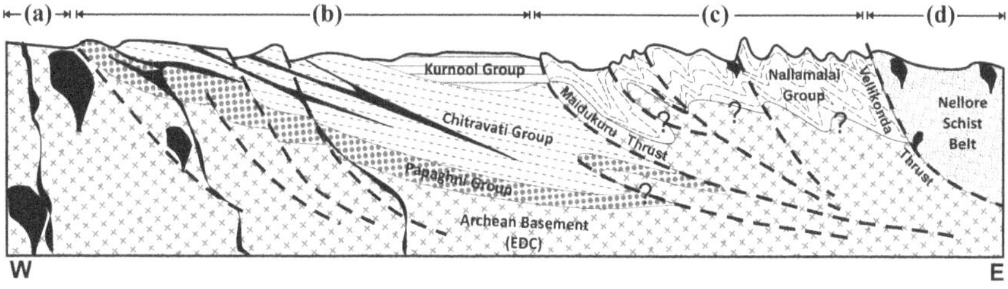

Fig. 5.16 Schematic E-W cross section (not to scale) across the Cuddapah Basin, based on geological and geophysical data. Tectonic terranes are **a** Basement Terrain of the EDC, **b** Western subbasins, **c** Nallamalai Fold Belt and **d** Nellore Schist Belt. Granitoid intrusives (red blobs with crosses), mafic dykes/sills–lavas (green). Faults are shown in thick lines. Modified after Johkan Ram (2012) and taken from Kale et al. (2020)

Ram Babu 1993; Kishore and Ram Rao 2004) and more recently by Chandrakala et al. (2013) reveal that these four juxtaposed terrains have mutually differing crustal characters. These studies also reaffirm the deep crustal nature of the Maidukuru and Vellikonda Thrusts that bound the NFB on either side.

5.16.6 Age

The interbedded volcanics from the Papaghni and Chitravati Groups were emplaced between 1.8 and 1.9 Ga (Sesha Sai et al. 2017; Sheppard et al. 2017). The ages of the mafic sills that intrude the Vempalle and Tadpatri Formations are also consistent with this data (Dhang et al. 2015). Two fresh sill samples, collected from more than 100 km apart at Pulivendla and Tadpatri, have given an identical age of 1899 ± 20 Ma as their emplacement age (Anand et al. 2003). This result is like a previously reported Rb/Sr age (1817 ± 24 Ma) for the Pulivendla sill (Bhaskar Rao et al. 1995). Both the ages imply that the deposition of the basal volcano-sedimentary Papaghni and Chitravati Groups in the Cuddapah Basin must have occurred prior to ~1900 Ma.

Zachariah et al. (1999) used $^{206}Pb/^{204}Pb$ systematics to date the U-mineralized horizons of the Vempalle and Tadpatri Formations at 1756 ± 29 Ma and interpreted it a minimum age for carbonate sedimentation and dolomitization within the Cuddapah Supergroup. Rai et al. (2015) had computed an age range of 1.9–2.0 Ga for dolomitization and uranium mineralisation in the same formation.

New U–Pb zircon ages from the mudstone-felsic tuff of the Owk Shale indicate that the detrital grains were derived from the 2.5 Ga Dharwar Craton with single 1880 Ma grain from 1890 Ma LIP (Bickford et al. 2013). Strict geochronological constraints make the Oak Shale younger than ca 1.9 Ga but its lower age is not fixed (Bickford et al. 2013). The detrital zircons from the siliciclastic sediments from the CB yielded U-Pb ages ranging from 2.54 to 0.91 Ga, which represent the ages of the provenance for the Papaghni and Chitravati Groups, possibly from the Dharwar craton (Collins et al. 2015). Sahoo et al. (2017) record an age of 2.53 Ga for magmatic zircons, while hydrothermally altered zircons provide concordia age of 2.32 and 2.12 Ga. U–Th–$Pb_{(total)}$ dating of monazite and uraninite from both the basement and cover in the Palnad Subbasin suggested that the Banganapalle Quartzite may have been deposited between 2.53 and 2.12 Ga. If these are taken as depositional ages, then these Kurnool Group rocks may belong to much older Gulcheru Quartzites.

Collins et al. (2015) and Sesha Sai et al. (2017) concluded the Nallamalai Group was deposited between 1.68 and 1.59 Ga, with inputs from the eastern tectonically active belt (EGB + NSB or collectively the Krishna Orogen) as the primary source of detritus. The Srisailam Formation may have been deposited at the same time, further westwards. The Kurnool Group may be reaffirmed as distinctly Neoproterozoic in age, with zircons dated at 913 ± 11 Ma from the Paniam Quartzite. To a large extent, this is consistent with the biotic remains reported from the Kurnool Group. Geochronology of the NSB has been constrained by the ages of granites and ophiolites that are present in the belt (Saha et al. 2017; Sain et al. 2017; Saha and Sain 2019).

The Vinjamuru Group hosts the 1.9 Ga Kandra ophiolite complex. This and the granulitic metamorphism (1.6–1.7 Ga) in the Ongole domain together suggest that the deposition of the Vinjamuru sediments may have occurred before 1.9 Ga in a deep marine environment on oceanic floor, while its metamorphism and tectonic upthrusting may have occurred around 1.6 Ga. The Udaigiri domain represents a younger arc that evolved and deformed around 1.3 Ga. The alkaline granitoid are younger events ~1.1 Ga.

Traditionally, the Cuddapah Supergroup (excluding the Kurnool Group) was recognized as the oldest 'Purana' sequence (Kale and Phansalkar 1991). Geochronological data has reaffirmed this to the extent that the Papaghni and Chitravati Groups are confirmed to be amongst the oldest of the Purana sediments, broadly coeval with the Bijawar and Gwalior Groups from the Bundelkhand Craton (Basu and Bickford 2015).

References

Absar N, Raza MQ, Augustine S, Managave S, Sarma DS, Balakrishnan S (2019) Trace, Rare-Earth Elements and C, O Isotope systematics of carbonate rocks of Proterozoic Bhima Group, Eastern Dharwar Craton, India: implications for the source of dissolved components, redox condition and biogeochemical cycling of mesoproterozoic ocean. In: Mondal MEA (ed) Geological Evolution of the Precambrian Indian Shield. Springer, Cham pp 297–326. https://doi.org/10.1007/978-3-319-89698-4_13

Akhtar K (1996) Facies, sedimentation processes and environments in the Proterozoic Vindhyan Basin, India. In: Bhattacharya A (ed) Recent advances in vindhyan geology. Mem Geol Soc India 36:127–136

Ahmad F (1955) Glaciation in the Vindhyan system. Curr Sci 24(7):231

Ahmad F (1981) Some problems in the paleogeography of the Vindhyan times in central India, vol 50. Geological Survey of India Miscellaneous Publications, pp 69–74

Ahmad S, Kumar S (2014) Trace Fossil assemblage from the Nagaur Group, western India. J Palaeontol Soc India 59(2):231–246

Anand M, Gibson SA, Subba Rao KV, Kelley SP, Dickin AP (2003) Early Proterozoic melt generation processes beneath the intra-cratonic Cuddapah basin, Southern India. J Petrol 44:2139–2171

Ansari AH, Pandey SK, Sharma M, Agrawal S, Kumar Y (2018) Carbon and oxygen isotope stratigraphy of the Ediacaran Bilara Group, Marwar Supergroup, India: evidence for high amplitude carbon isotopic negative excursions. Precambr Res 308:75–91. https://doi.org/10.1016/j.precamres.2018.02.002

Auden JB (1933) Vindhyan sedimentation in the Son Valley, Mirzapur District. Mem Geol Surv India 62(2):141–240

Azmi RJ (1998) Discovery of lower Cambrian small shelly fossils and Brachiopods from the lower Vindhyan of Son Valley, central India. J Geol Soc India 52:381–389

Azmi RJ (2008) A synoptic views of the Current discordant geo and biochronological ages of the Vindhyan Supergroup, central India. J Himalayan Geol 29(2):177–191

Balakrishnan P, Mahesh Babu M (1987) The geology of the Ampani outlier, Kalahandi, Koraput district, Orissa. In: Radhakrishna BP (ed) Purana basins of Peninsular India (Middle to Late Proterozoic), vol 6. Memoirs Geological Society of India, pp 281–286

Ball V (1877) Geology of the Mahanadi Basin and its vicinity. Rec Geol Surv India 10(4):167–186

Bandopadhyay BC, Sengupta S (2004) The paleoproterozoic supracrustal Kolhan group in Singhbhum craton, India and the Indo-African supercontinent. Gondwana Res 7(4):1228–1235
Banerjee DM (1969) A study of some stromatolites from the Calc-Zone of Sarju-Pungar Valley areas in Kumaun Himalaya and Vindhyans. J Paleontol Soc India 14:66–76
Banerjee DM (1982) Lithotectonics, phosphate mineralization and regional correlation of Bijawar Group of rocks of Central India. In: Valdiya KS, Bhatia SB, Gaur VK (eds) Geology of Vindhyachal. Hindustan Publishing Corporation (India), Delhi, pp 7–54
Banerjee DM (2011) Purana basins: stratigraphy in turmoil. The Annual General Meeting of the Geological Society of India, 23 Sept 2011, Department of Geology, Banaras Hindu University, Abstract 26–27
Banerjee A, Banerjee DM (2010) Modal analysis and geochemistry of two sandstones of Bhander Group (Late Neoproterozoic) in parts of the Central Indian Vindhyan Basin and their bearing on the provenance and tectonics. J Earth Syst Sci 119(6):1–15
Banerjee DM, Frank W (2000) How old is the Vindhyan Basin? In: International geological congress, Rio de Janeiro, Brazil, Volume of Abstracts 31 unpaginated
Banerjee DM, Russell JR (1993) New Pb/Pb dates for the carbonate rocks of Vindhyan Basin. In: Bhattacharya A (ed) Group discussion on geology of Vindhyans, Jadavpur University, Mar 1993. Abstract 29
Banerjee DM, Saigal N (1988) A study of Proterozoic phosphorite from Chelima-Pacherla area, Kurnool district, Andhra Pradesh, India. J Geol Soc India 32(1):32–39
Banerjee I, Sengupta S (1963) The Vindhyan basin—a regional reconnaissance of the eastern part. Q J Geol Min Metall Soc India 35:41–149
Banerjee DM, Khan MWY, Srivastava N, Saigal GC (1982) Precambrian phosphorites in the Bijawar rocks of Hirapur-Bassia areas, Sagar district, Madhya Pradesh, India. Miner Deposita 17:349–362
Banerjee DM, Deb M, Strauss H (1992) Organic carbon isotopic composition of Proterozoic Sedimentary rocks from India: preliminary results. In: Schidlowski M, Golubic S, Kimberley MM, McKirdy DM, Trudinger PA (eds) Early organic evolution: implications for mineral and energy resources. Springer, Berlin, pp 232–240
Banerjee DM, Strauss H, Bhattacharyya SK, Kumar V, Mazumdar A (1998) Isotopic composition of carbonates and sulphates, potash mineralization and basin architecture of the Nagaur-Ganganagar evaporate basin (north-western India) and their implications on the Neoproterozoic exogenic cycle. Miner Mag 62:106–107
Banerjee S, Bhattacharya SK, Sarkar S (2005) Facies, dissolution seams and stable isotope composition of the Rohtas Limestone, Vindhyan Supergroup in the Son Valley area, central India. Proc Indian Acad Sci (Earth Planet Sci) 114(1):87–96
Banerjee S, Dutta S, Paikary S, Mann U (2006) Stratigraphy, sedimentation and bulk organic geochemistry from the Proterozoic Vindhyan Supergroup (Central India). J Earth Sys Sci 45:37–47
Banerjee DM, Phukan P, Joachimski M, Kaufman AJ, Altermann W, Grover S, Mayr C, Suzane (2007a) Evaporite sequence of the Nagaur–Ganganagar basin, Western Rajasthan—new data and new interpretation (Late Extended Abstract). In: International conference on precambrian sediments, tectonics and second GPSS meeting, IIT, Bombay, pp 76–77
Banerjee S, Bhattacharya SK, Sarkar S (2007b) Carbon and oxygen isotopic variations in peritidal stromatolite cycles, Paleoproterozoic Kajrahat Limestone, Vindhyan basin of central India. J Asian Earth Sci 29:823–831. https://doi.org/10.1016/j.jseaes.2006.05.006
Banerjee A, Slowakewicz M, Majumdar T, Khan S, Patranabis-Deb S, Tucker M, Saha TA (2019) Paleoproterozoic dolomite (Vempalle Formation, Cuddapah basin, India) showing Phanerozoic type dolomitization. Precambr Res 328:9–26. https://doi.org/10.1016/j.precamres.2019.04.013
Barman G (1987) Stratigraphical position of the Marwar Supergroup in the light of stromatolite study. Geol Surv India Spec Publ 11:72–80
Basu A, Bickford ME (2015) An alternate perspective on the opening and closing of the intracratonic Purana Basins in Peninsular India. J Geol Soc India 85:5–25. https://doi.org/10.1007/s12594-015-0190-y

Basu H, Dandele PS, Kumar KR, Achara KK, Umamaheswara K (2017) Geochemistry of black shales from the Mesoproterozoic Srisailam Formation, Cuddapah basin, India: Implications for provenance, palaeoweathering, tectonics, and timing of Columbia breakup. Chem Erde. https://doi.org/10.1016/j.chemer.2017.10.002

Basumallick S, Sarkar BC, Banerjee S (1996) Tidal cyclicity in lower Bhander Sandstone, Maihar, Madhya Pradesh. J Geol Soc India 47(2):189–194

Bengtson S, Belivanova V, Rasmussen B, Whitehouse M (2009) The controversial "Cambrian" fossils of the Vindhyan are real but more than a billion years older. Proc Natl Acad Sci 106:7729–7734. https://doi.org/10.1073/pnas.0812460106

Bhaskar Rao YJ, Pantulu GVC, Reddy VD, Gopalan K (1995). Time of early sedimentation and volcanism in the Proterozoic Cuddapah basin, South India: evidence from the Rb-Sr age of Pulivendla mafic sill. In: Devaraju TC (ed) Dyke swarms of Peninsular India, vol 33. Memoirs Geological Society of India, pp 329–338

Bhattacharya A (1996) Recent advances in Vindhyan geology, vol 36. Memoirs Geological Society of India, p 331

Bhattacharya AK, Chatterjee BK (1964) Petrology of Precambrian Kolhan formation of Jhinkapani, Singhbhum District, Bihar. Geol Rundsch 53:758–779

Bhattacharya S, Dutta S (2015) Neoproterozoic-Early Cambrian biota and ancient niche: a synthesis from molecular markers and palynomorphs from Bikaner-Nagaur Basin, western India. Precambr Res 266:361–374

Bhattacharya A, Morad S (1993) Proterozoic braided ephemeral fluvial deposits: an example from the Dhandraul Sandstone Formation of the Kaimur Group, Son Valley, Central India. Sediment Geol 84:101–114

Bhushan SK (2000) Malani rhyolites—a review. Gondwana Res 3:65–77

Bickford ME, Basu A, Mukherjee A, Hietpas J, Schieber J, Patranabis-Deb S (2011a) New U-Pb SHRIMP zircon ages of the Dhamda Tuff in the Mesoproterozoic Chhattisgarh Basin, Peninsular India: stratigraphic implications and significance of a 1-Ga thermal-magmatic event. J Geol 119:535–548. https://doi.org/10.1086/661193

Bickford ME, Basu A, Patranabis-Deb S, Dhang PC, Schieber J (2011b) Depositional history of the Chhattisgarh basin, central India: constraints from new SHRIMP zircon ages. J Geol 119:33–50. https://doi.org/10.1086/657300

Bickford ME, Mishra M, Mueller PA, Kamenov GD, Schieber J, Basu A (2017) U–Pb age and Hf isotopic compositions of magmatic zircons from a rhyolite flow in the porcellanite formation in the Vindhyan Supergroup, Son Valley (India): implications for its tectonic significance. J Geol 125:367–379. https://doi.org/10.1086/691186

Bickford ME, Saha D, Schieber J, Kamenov G, Russell A, Basu A (2013) New U-Pb ages of zircons in the Owk Shale (Kurnool Group) with reflections on Proterozoic porcellanites in India. J Geol Soc India 82(3):207–216

Bose PK, Chakraborty PP (1994) Marine to fluvial transition: proterozoic upper Rewa sandstone, Maihar, India. Sediment Geol 89:285–301

Bose PK, Chaudhuri AK, Seth A (1988) Facies, flow and bedform patterns across a storm-dominated inner continental shelf: proterozoic Kaimur Formation, Rajasthan, India. Sediment Geol 59:275–293. https://doi.org/10.1016/0037-0738(88)90081-4

Bose PK, Banerjee S, Sarkar S (1997) Slope controlled seismic deformation and tectonic framework of deposition of Koldaha Shale, India. Tectonophysics 269:151–169

Bose PK, Sarkar S, Chakraborty S, Banerjee S (2001) Overview of the Mesoproterozoic-Neoproterozoic evolution of the Vindhyan basin (1.4–0.55 Ga). Sediment Geol 141:395–419

Bose PK, Sarkar S, Das NG, Mandal A, Chakraborty N (2015) Proterozoic Vindhyan Basin: configuration and evolution. In: Mazumdar R, Eriksson PG (eds) Precambrian basins of India: stratigraphic and tectonic context, vol 43. Geological Society London, Memoirs, pp 85–102. https://doi.org/10.1144/N43.6

Bose S, Chakrabarti G, Shome D (2017) Evolution of Palaeo-Mesoproterozoic open coast ramp-clues from stromatolite-bearing carbonate facies of Vempalle Formation, Cuddapah basin, India. Arab J Geosci 10:226. https://doi.org/10.1007/s12517-017-2988-y
Casshyap SM, Bhardwaj BD, Raza M, Singh A, Khan A (2001) Barrier inlet and associated facies of shore zone: an example from Khardeola Formation of Lower Vindhyan sequence in Chittaurgarh, Rajasthan. J Geol Soc India 58:97–111
Chakraborty C (1993) Morphology, internal structure and mechanics of small longitudinal (seif) dunes in an aeolian horizon of Proterozoic Dhandraul Quartzite, India. Sediment 40:79–85
Chakraborty C (1996) Sedimentary records of erg development over a braidplain: proterozoic dhandraul sandstone. Mem Geol Soc India 36:77–99
Chakrabarti R, Basu AR, Chakrabarti A (2007) Trace element and Nd-isotopic evidence for sediment sources in the mid-Proterozoic Vindhyan basin, central India. Precambr Res 159:260–274. https://doi.org/10.1016/j.precamres.2007.07.003
Chakrabarti G, Shome D, Kumar S, Stephens GM III, Kah LC (2016) Carbonate platform development in a Paleoproterozoic extensional basin, Vempalle formation, Cuddapah basin, India. J Asian Earth Sci 91:263–279
Chakraborty PP (2004) Facies architecture and sequence development in a Neoproterozoic carbonate ramp: Lakheri Limestone member, Vindhyan Supergroup, central India. Precambr Res 132:29–53. https://doi.org/10.1016/j.precamres.2004.02.004
Chakraborty PP (2006) Outcrop signatures of relative sea-level fall on a siliciclastic shelf: examples from the Rewa Group of Proterozoic Vindhyan basin. J Earth Syst Sci. 115(1):23–36
Chakraborty PP, Barkat R (2020) A status report on age, depositional motif and stratigraphy of Chhattisgarh, Indravati, Kurnool and Bhima basins, Peninsular India. In: Proceedings of the Indian National Science Academy (in Press)
Chakraborty C, Bhattacharya A (1996a) Fan delta sedimentation in the foreland moat: Deonar Formation, Vindhyan Supergroup, Son Valley. Mem Geol Soc India 36:27–48
Chakraborty C, Bhattacharya A (1996b) The Vindhyan basin: an overview in the light of current perspectives. Mem Geol Soc India 36:301–312
Chakraborty T, Chaudhuri AK (1990) Stratigraphy of the Late Proterozoic Rewa and paleogeography of the Vindhyan Basin in central India during Rewa sedimentation. J Geol Soc India 36:383–4012
Chakraborty PP, Sarkar A, Bhattacharya SK, Sanyal P (2002) Isotopic and sedimentological clues to productivity change in Late Riphean sea: a case study from two intracratonic basins in India. J Earth Sys Sci 111:379–390. https://doi.org/10.1007/BF02702051
Chakraborty PP, Paul S, Das A (2005) Facies development and depositional environment of the Mungra Sandstone, Kolhan Group, Eastern India. J Geol Soc India 65:753–757
Chakraborty PP, Pant NC, Paul PP (2015a) Controls on sedimentation in Indian Paleoproterozoic basins: clues from Gwalior and Bijawar basins, Central India. In: Mazumdar R, Eriksson P (eds) Precambrian basins of India: stratigraphic and tectonic context, vol 43. Geological Society, London, Memoirs, pp 67–83. https://doi.org/10.1144/M43.5
Chakraborty PP, Saha S, Das P (2015b) Geology of Mesoproterozoic Chhattisgarh Basin, central India: current status and future goals. Geol Soc Lond Mem 43:185–205. https://doi.org/10.1144/M43.13
Chakraborty K, Mukhopadhyay PK, Pankaj P (2016) Magmatism in western Cuddapahs: the mafic sills and lava flows of Vempalle and Tadpatri formations. J Geol Soc India 87:631–660
Chanda SK, Bhattacharya A (1982) Vindhyan: post Auden developments. In: Valdiya KS, Bhatia SB and Gaur VK (eds) Geology of Vindhyachal. Hindustan Publishing Corporation, New Delhi, pp 88–101
Chandrakala K, Mall DM, Sarkar D, Pandey OP (2013) Seismic imaging of the Proterozoic Cuddapah basin, South India and regional geodynamics. Precambr Res 213:277–289
Chandrakala K, Pandey OP, Prasad ASSSRS, Sain K (2015) Seismic imaging across the Eastern Ghats Belt-Cuddapah basin collisional zone, southern Indian Shield and possible geodynamic implications. Precambr Res 271:56–64

Chatterjee N, Bhattacharji S (2001) Petrology, geochemistry and tectonic setting of the mafic dykes and sills associated the evolution of the Proterozoic Cuddapah basin, South India. Proc Indian Acad Sci (Earth Planet Sci) 110:433–453

Chatterjee BK, Bhattacharya AK (1969) Tectonics and sedimentation in a Precambrian shallow epicontinental basin. J Sediment Petrol 39(4):1566–1572

Chaturvedi AK, Kora P, Pathak AK, Cerallos C, Rawat TPS, Miller R, Wiseman R, Parihar PS (2012) Geological minieu of the Bijawar Basin based on interpretation of geophysical data, Central India. In: 22nd proceedings of international geophysical conference and extension, Brisbane Australia, 1–4 Feb 2012

Chaudhuri AK (2003) Stratigraphy and palaeogeography of the Godavari Supergroup in the south-central Pranhita–Godavari Valley, south India. J Asian Earth Sci 21(6):595–611

Chaudhuri AK, Deb G (2004) Proterozoic rifting in the Pranhita–Godavari valley: implication on India-Antarctica linkage. Gondwana Res 7(2):301–312

Chaudhuri AK, Deb GK, Patranabis-Deb S, Sarkar S (2012) Paleogeographic and tectonic evolution of the Pranhita–Godavari Valley, Central India: a stratigraphic perspective. Am J Sci 312(7):766–815

Chaudhuri AK, Mukhopadhyay J, Deb SP, Chanda SK (1999) The Neoproterozoic cratonic successions of Peninsular India. Gondwana Res 2(2):213–225

Chauhan DC, Dubey JC, Ram B (1991) Geological analysis of part of Nagaur basin in the vicinity of Jodhpur city. In: Tandon SK, Pant CC, Casshyap SM (eds) Sedimentary basins of India: tectonic context. Gyanodaya Parkashan, Nainital, pp 3–73

Chauhan DC, Bhanwara R, Ram Narayan (2004) Jodhpur sandstone: a gift of ancient beaches to Western Rajasthan. J Geol Soc India 64:265–276

Chetty TRK (2001) The Eastern Ghats mobile belt, India: a collage of juxtaposed terranes (?). Gondwana Res 4(3):319–328. https://doi.org/10.1016/s1342-937x(05)70332-4

Chetty TRK (2011) Tectonics of Proterozoic Cuddapah basin, southern India: a conceptual model. J Geol Soc India 78:446–456

Collins AS, Patranabis-Deb S, Alexander E, Bertram CN, Falster GM, Gore RJ, Mackintosh J, Dhang PC, Saha D, Payne JL, Jourdan F, Backe G, Halverson GP, Wade BP (2015) Detrital mineral age, radiogenic isotopic stratigraphy and tectonic significance of the Cuddapah basin, India. Gondwana Res 28:1294–1309

Conrad JE, Hein JR, Chaudhuri AK, Patranabis-Deb S, Mukhopadhyay J, Deb GK, Beukes NJ (2011) Constraints on the development of Proterozoic basins in central India from Ar-40/Ar-39 analysis of authigenic glauconitic minerals. Geol Soc Ame Bull 123(1–2):158–167

Cozzi A, Rea G, Craig J (2012) From global geology to hydrocarbon exploration: Ediacaran-Early Cambrian petroleum plays of India, Pakistan and Oman. In: Bhat GM, Craig J, Thurow JW, Thusu B, Cozzi A (eds) Geology and hydrocarbon potential of Neoproterozoic–Cambrian Basins in Asia, vol 366, no. 1. Geological Society, London, Special Publications, pp 131–162. https://doi.org/10.1144/SP366.14

Crawford AR, Compston W (1970) The age of the Vindhyan system of Peninsular India. Q J Geol Soc Lond 25(499):351–371

Crookshank H (1963) Geology of the southern Bastar and Jeypore from the Bailadila Range to Eastern Ghats, vol 87. Memoirs Geological Survey of India, pp 149

Das Gupta SK (1996) Marwar Supergroup evaporates, Rajasthan. In: Bhattacharyya A (ed) Recent advances in Vindhyan geology, vol 36. Memoirs Geological Society of India, pp 49–58

Das S, Sahoo SR (2015) The time-transgressive Kolhans in fan Delta-Lacustrine environment: paleogeographic and tectonic implications. Int J Earth Sci Eng 8:45–58

Das LK, Mishra DC, Ghosh D, Banerjee B (1990) Geomorphotectonics of the basement in a part of the upper Son Valley of the Vindhyan basin. J Geol Soc India 35:445–458

Das DP, Kundu A, Das N, Dutta DR, Kumaran K, Ramamurthy S (1992) Lithostratigraphy and sedimentation of Chhattisgarh basin. Indian Miner 46:271–288

Das N, Dutta DRR, Das DP (2001) Proterozoic cover sediments of southeastern Chhattisgarh state and adjoining parts of Orissa. Geol Surv India Spec Publ 55:237–262

Das K, Yokoyama K, Chakraborty PP, Sarkar A (2009) EPMA U–Th–Pb Monazite dating and trace element geochemistry of tuff units from the basal part of Chhattisgarh and Khariar basin, central India: implications for basin initiation and depositional contemporaneity. J Geol 117:88–102

Das K, Bose S, Karmakar S, Dunkley DJ, Dasgupta S (2011) Multiple tectonometamorphic imprints in the lower crust: first evidence of ca. 950 Ma (zircon U-Pb SHRIMP) compressional reworking of UHT aluminous granulites from the Eastern Ghats Belt, India. Geol J 46(2–3):217–239

Das S, Shukla D, Mitra SK (2016) Evidences of inclined transpression at the contact between Vinjamuru group and Udayagiri group of Nellore Schist Belt, Andhra Pradesh, India. J Earth Syst Sci 125:1007–1020

Das K, Chakraborty PP, Horie K, Tsutsumi Y, Saha S, Balakrishnan S (2017) Detrital zircon U-Pb geochronology, Nd isotope mapping, and sediment geochemistry from the Singhora Group, Central India: implications toward provenance, its shift, and regional stratigraphic correlation. In: Mazumdar R (ed) Sediment provenance. Elsevier, pp 403–451. https://doi.org/10.1016/b978-0-12-803386-9.00015-0

Datta B (1998) Stratigraphic and sedimentologic evolution of the Proterozoic siliciclastics in the southern part of Chhattisgarh and Khariar, central India. J Geol Soc India 51:345–360

Davis JK, Meert JG, Pandit MK (2014) Paleomagnetic analysis of the Marwar Supergroup, Rajasthan, India and proposed interbasinal correlations. J Asian Earth Sci 91:339–351

Dayal AM, Devleena M, Kavitha S, Kalpana MS, Patil IJ, Sharma M (2018) Organic geochemistry of the Vindhyan sediments: implication for hydrocarbons. J Asian Earth Sci 91:329–338. https://doi.org/10.1016/j.jseaes.2014.03.010

Deb M, Thorpe R, Krstic D (2002) Hindoli group of rocks in the eastern fringe of the Aravalli-Delhi orogenic belt-archean secondary greenstone belt or proterozoic supracrustals? Gondwana Res 5:879–883. https://doi.org/10.1016/S1342-937X(05)70922-9

Dey S (2015) Geological history of the Kaladgi-Badami-Bhima basins, south India: sedimentation in a Proterozoic intracratonic set up. In: Mazumdar R, Eriksson PG (eds) Precambrian basins of India: stratigraphic and tectonic context, vol 43. Geological Society, London, Memoirs, pp 283–296

Dey S, Gajapati Rao R, Veerbhaskar D, Chaki A, Baidya TK (2008) Geochemistry of the shales from the Proterozoic Intracratonic Kaladgi-Badami Basins, Karnataka, Southern India as an indicator of paleoweathering and evolution of the Dharwar Craton. J Geol Soc India 71(4):483–501

Dhang PC, Saha D, Payne JL, Jourdan F, Backe G, Halverson GP, Wade BP (2015) Detrital mineral age, radiogenic isotopic stratigraphy and tectonic significance of the Cuddapah basin, India. Gondwana Res 28:1294–1309

Dubey VS, Chaudhary MS (1952) Late Precambrian glaciations in central India. Curr Sci 21(12):331–332

Dunn JA (1940) The stratigraphy of south Singhbhum, vol 63. Geological Survey of India Memoirs, pp 303–369

Dutt NVBS (1963) Stratigraphy and correlation of Indravati series (Purana Group) Bastar District, M.P. J Geol Soc India 4:35–49

Dutt NVBS (1964) A suggested succession of the Purana formations of the southern part of Chattisgarh, M.P. Rec Geol Surv India 93(2):143–148

Dutt NVBS (1975) Geology of the Kurnool System of rocks in Cuddapah and southeastern part of Kurnool district, Andhra Pradesh. Rec Geol Surv India 87:540–604

Foote RB (1876) Geological features of the southern Mahratta Country and adjoining districts, vol 12. Memoir Geological Survey of India, pp 70–138

French JE, Heaman LM, Chacko T, Srivastava RK (2008) 1891–1883 Ma southern Bastar-Cuddapah mafic igneous events, India: a newly recognized large igneous province. Precambr Res 160:308–322

Friedman G, Chakraborty C (1997) Stable isotopes in marine carbonates: their implications in the paleoenvironment with special reference to the Proterozoic Vindhyan carbonates (central India). J Geol Soc India 50(2):131–159

George BG, Ray JS (2017) Provenance of sediments in the Marwar Supergroup, Rajasthan, India: implications for basin evolution and Neoproterozoic global events. J Asian Earth Sci 147:254–270. https://doi.org/10.1016/j.jseaes.2017.07.027

George BG, Ray JS, Kumar S (2018) Geochemistry of carbonate formations of the Chhattisgarh Supergroup, central India: implications for Mesoproterozoic global events. Can J Earth Sci 56:335–346. https://doi.org/10.1139/cjes-2018-0144

George BG, Ray JS, Kumar S (2019) Geochemistry of carbonate formations of the Chhattisgarh Supergroup, central India: implications for Mesoproterozoic global events. Canadian J Earth Sci 56(3):335–346

Ghosh DB (1981) The Vindhyan basin in Bundelkhand-Son Valley region, vol 50. Miscellaneous Publications of the Geological Survey of India, pp 75–88

Gileadeau GJ, Sahoo SK, Linda KH, Henderson MA, Kaufmann AJ (2018) Proterozoic carbonates of the Vindhyan Basin, India: chemostratigraphy and diagenesis. Gondwana Res 57:10–25

Gopalan K, Kumar A, Kumar S, Vijayagopal B (2013) Depositional history of the Upper Vindhyan succession, central India: time constraints from Pb–Pb isochron ages of its carbonate components. Precambr Res 233:108–117. https://doi.org/10.1016/j.precamres.2013.04.014

Goswami S, Bhagat S, Zakaulla S, Kumar S, Rai AK (2017) Role of organic matter in uranium mineralisation in Vempalle dolostone; Cuddapah basin, India. J Geol Soc India 89:145–154

Goswami S, Upadhyay PK, Bhagat S, Zakaulla S, Bhatt AK, Natarajan V, Dey S (2018) An approach of understanding acid volcanics and tuffaceous volcaniclastics from field studies: a case from Tadpatri Formation, Proterozoic Cuddapah basin, Andhra Pradesh, India. J Earth Syst Sci 127:20. https://doi.org/10.1007/s12040-018-0929-0

Gregory LC, Meert JG, Pradhan V, Pandit MK, Tamrat E, Malone SJ (2006) A paleomagnetic and geochronologic study of the Majhgawan kimberlite, India: implications for the age of the Upper Vindhyan Supergroup. Precambr Res 149:65–75. https://doi.org/10.1016/j.precamres.2006.05.005

Haldar D, Ghosh RN (2000) Eruption of Bijawar lava: an example of Precambrian volcanicity under stable cratonic conditions. Geol Surv India Spec Publ 57:151–170

Hofmann HJ (2005) Palaeoproterozoic dubiofossils from India revisited-Vindhyan triploblastic animal burrows or pseudofossils? J Paleontol Soc India 50:113–120

Hughes NC, Myrow PM, McKenzie NR, Xiao S, Banerjee DM, Stockli D, Tang Q (2015) Age and implications of the phosphatic Birmania Formation, Rajasthan, India. Precambr Res 267:164–173

Husain V, Banerjee DM (1986) Birmania phospphorite: petromineralogical characters and paleogeology implications. J Geol Soc India 27:450–455

Jairaman R, Banerjee DM (1978) Studies on the stromatolites from the Raipur area, Chattisgarh Basin. In: Proceedings workshop on stromatolites, vol 44. Geological Survey of India, Miscellaneous Publications, pp 57–67

Janardhana Rao LH, Srinivasa Rao C, Ramakrishna TL (1975) Reclassification of the rocks of the Bhima basin, Gulbarga District, Mysore State, vol 23, no. 1. Geological Survey of India, Miscellaneous Publications, pp 177–184

Jayaprakash AV, Sundaram V, Hans SK, Mishra RN (1987) Geology of the Kaladgi-Badami Basins, Karnataka, vol 6. Geological Survey of India Memoirs, pp 201–225

Jha SK, Srivastava JP, Bhairam CL (2012) Clay mineralogical studies on Bijawar of the Sonrai Basin—Palaeoenvironmental implications and inferences on the Uranium mineralization. J Geol Soc India 79:117–134

Jha SK, Sharma R, Srivastava JP (2019) Trace element signatures and mineral chemistry of clays associated with the alteration of halos of Paleoproterozoic U mineralization in Bijawars of the Sonrai Basin, Central India. Acta Geochem. https://doi.org/10.1007/s11631-019-00363-9

Joy S, Patranabis-Deb S, Saha D, Jelsma H, Maas R, Soderlund U, Tappa S, Linde GV, Banerjee A, Krishnan U (2019) Depositional history of cratonic Purana basins in Southern India: a multipronged geochronology approach to the Proterozoic Kaladgi and Bhima Basins. J Geol 1–13. https://doi.org/10.1002/gj.3415

Kaila KL, Tewari HC (1985) Structural trends in the Cuddapah basin from Deep Seismic Soundings (DSS) and their tectonic implications. Tectonophysics 115:69–86
Kaila KL, Reddy PRK, Dixit MA, Koteshwara RP (1985) Crustal structure across the Narmada-Son lineament, central India from deep seismic soundings. J Geol Soc India 26(7):465–480
Kaila KL, Tewari HC, Roy Chowdhury K, Rao VK, Sridhar AR, Mall DM (1987) Crustal structure of the northern part of the Proterozoic Cuddapah basin of India from deep seismic soundings and gravity data. Tectonophysics 140: l–12
Kaila KL, Murthy PRK, Mall DM, Dixit MM (1989) The evolution of the Vindhyan Basin vis-à-vis Narmada-Son lineament, central India from deep seismic sounding. Tectonophysics 162:277–289
Kale VS (2016) Proterozoic basins of Peninsular India: status within the global Proterozoic systems. Proc Indian Natl Sci Acad 82(3):461–477
Kale VS, Peshwa VV (1995) Bhima Basin. Bangalore, Geological Survey of India, p 142
Kale VS, Phansalkar VG (1991) Purana basins of Peninsular India: a review. Basin Res 3:1–36
Kale VS, Mudholkar AV, Phansalkar VG, Peswa VV (1990) Stratigraphy of the Bhima Group. J Palaeontol Soc India 35:91–103
Kale VS, Saha D, Patranabis-Deb S, Sesha Sai VV, Tripathy V, Pillai SP. (2020) Cuddapah basin, India: a collage of Proterozoic sub-basins and Terranes. Proc Indian Natl Sci Acad Spec Issue (In Press)
Kathal PK, Patel DR, Alexander PO (2004) Age discrepancies on the Lower Vindhyan: impact of the recent fossil discoveries. Recent Res Econ Geol 2:53–62
Khan EA (1971) Geological mapping in parts of Jodhpur and Nagaur districts, Rajasthan. Field Session 1969–1970. Geological Survey of India (Unpublished Report)
Khan AA (2013) Paleogeography of the Peninsular India vis-à-vis geodynamic Petrotectonic significance of the Vindhyan Basin with special reference to Neo-Meso Proterozoic. J Indian Geol Congr 5(3):65–76
King W (1872) On the Kadapah and Karnűl Formations formations in the Madras Presidency, vol 8. Geological Survey of India Memoirs, pp 1–293
King W (1881) The geology of the Pranhita–Godavari valley, vol 18, no. 3. Geological Survey of India Memoirs, pp 151–311
Kishore RK, Ram Rao Ch (2004) Basement characteristics of the Papaghni basin of Eastern Dharwar craton (India)—an inference from aeromagnetic study. J Indian Geophys Union 4:311–317
Kreuzer H, Karre W, Kurssten M, Schnitzer WA, Srivastava NK (1977) K/Ar dates of two glauconites from the Chanderpur Series, Chattisgarh India. On the stratigraphic status of the Late Precambrian Basins in Central India. Geol Jahrb B 28:23–36
Krishna Brahmam N (1989) Gravity and seismicity of the Cuddapah basin and surrounding places. J Geol Soc India 34:373–384
Krishna Brahmam N, Sarma JR, Aravamadhu PS, Subbarao DV (1986) Explanatory brochure on Bouger gravity anomaly map (NGRI/GPH-6) of Cuddapah basin, India with scale 1: 250,000 NGRI, Hyderabad p 36
Kumar A, Padma Kumari VM, Dayal AM, Murthy DSN, Gopalan K (1993) Rb-Sr ages of Proterozoic kimberlites of India: Evidence for contemporaneous emplacement. Precambr Res 62:227–237. https://doi.org/10.1016/0301-9268(93)90023-U
Kumar S (1978) Stromatolites and environment of deposition of the Vindhyan Supergroup of Central India. J Paleontol Soc India 21–22:96–101
Kumar S (1982) Vindhyan stromatolites and their stratigraphic testimony. In: Valdiya, KS, Bhatia SB, Gaur VK (eds) Geology of Vindhyachal. Hindustan Publication Corporation, New Delhi, pp 102–112
Kumar A, Gopalan K, Rajagopalan G (2001) Age of the lower vindhyan sediments, Central India. Curr Sci 81:4
Kumar A, Parasharumulu V, Nagaraju E (2015) A 2082 Ma radiating dyke swarm in the Eastern Dharwar craton, southern India and its implications to Cuddapah basin formation. Precambr Res 266:490–505

Kumar S (1991) Lithostratigraphy, mineralogy and geochemistry of the Bhander Limestone (upper Vindhyan Supergroup), Satna area, Madhya Pradesh. Indian Mineral 45(3):139–148

Kumar S (2016) Megafossils from the Vindhyan Basin, Central India: an overview. J Palaeontol Soc India 61:273–286

Kumar S, Mishra Y (2007) Paleoproterozoic stromatolite Group Missassinia from the Khenjua Formation, Semri Group, Chopan area, Sonbhadra District, Uttar Pradesh. J Paleontol Soc India 52(1):97–102

Kumar S, Pandey SK (2008) Arumberia and associated fossils from the Neoproterozoic Maihar Sandstone, Vindhyan Supergroup, Central India. J Palaeontol Soc India 530:83–97

Kumar S, Pandey SK (2010) Trace fossils from the Nagaur Sandstone, Marwar Supergroup, Dulmera area, Bikaner district, Rajasthan, India. J Asian Earth Sci 38:77–85

Kumar S, Singh MP (1978) Pahoehoe toes in the basic volcanic flow of the Bijawar Group, Kurrat area, Lalitpur district, U.P. Curr Sci 47:773

Kumar S, Srivastava P (1992a) Microfossils from the black chert of Bhagwanpura Limestone (Middle Proterozoic) Vindhyan Supergroup, Chittorgarh area, Rajasthan, West India. Curr Sci 62(4):371–374

Kumar S, Srivastava P (1992b) Microfossils from the nonstromatolitic Middle Proterozoic Vindhyan chert, Chitrakut area, Uttar Pradesh. Comment. J Geol Soc India 39:446–448

Kumar B, Srivastava RK, Jha DK, Pant NC, Bhandaru BK (1990) A revised stratigraphy of rocks of type area of the Bijawar Group in Central India. Indian Miner 44:303–314

Kumar S, Schidlowski M, Joachimski MM (2005) Carbon isotope stratigraphy of the Paleo-Neoproterozoic Vindhyan Supergroup, central India: implication for basin evolution and intrabasinal correlation. J Paleontol Soc India 50(1):65–81

Kumar V (1999) Evolution and geological set up of the Nagaur-Ganganagar Basin, North Western Rajasthan. In: Paliwal BS (ed) Geological evolution of Northwestern India. Scientific Publishers, India, pp 34–60

Kump LR, Arthur MA (1999) Interpreting carbon-isotope excursions: carbonates and organic matter. Chem Geol 161:181–198. https://doi.org/10.1016/S0009-2541(99)00086

Lehmann B, Storey C, Mainkar D, Jeffries T (2007) In situ U–Pb dating of titanite in the Tokapal-Bejripadar kimberlite system, Central India. J Geol Soc India 69:553–556

Maheshwari A, Sial AN, Mathur SC (2003) Carbon and oxygen isotope profiles from the Terminal Pre-Cambrian Marwar Supergroup, Rajasthan, India. Carbonates Evaporites 18(1):268–276

Mainkar D, Lehmann B, Haggerty SE (2004) The crater-facies kimberlite system of Tokapal, Bastar district, Chhatisgarh, India. Lithos 76:201–217

Majumder T, Patranabis-Deb S, Nemec W (2015) Palaeoproterozoic sedimentation in the Cuddapah basin of south-eastern India. In: 31st IAS meeting of sedimentology, Kraków, Poland, 22–25 June 2015

Malone SJ, Meert JG, Banerjee DM, Pandit MK, Tamrat E, Kamenov G, Pradhan VR, Sohl LE (2008) Paleomagnetism and detrital zircon geochronology of the Upper Vindhyan sequence, Son Valley and Rajasthan, India: a 1000 Ma closure age for the Purana basins? Precambr Res 164:137–159

Mathur SM (1954) Late Precambrian glaciations in Central India—A rejoinder. Curr Sci 23(1):7–8

Mathur SM (1982) Precambrian sedimentary sequence of India: their geochronology and correlation. Precambr Res 18(1–2):139–144

Mathur SM, Mani G (1970) Geology of the Bijawar Group in the type area, Madhya Pradesh. Proc Symp Purana Formations of Peninsular India Univ Sagar, 313–320

Mazumdar A, Bhattacharya SK (2004) Stable isotopic study of late Neoproterozoic-early Cambrian (?) sediments from Nagaur-Ganganagar basin, western India: possible signatures of global and regional C-isotopic events. Geochem J 38:163–175

Mazumdar A, Strauss H (2006) Sulfur and strontium isotopic compositions of carbonate and evaporite rocks from the late Neoproterozoic-Early Cambrian Bilara Group (Nagaur-Ganganagar Basin, India): constraints on intrabasinal correlation and global sulfur cycle. Precambr Res 149(3–4):217–230

McKenzie NR, Hughes NC, Myrow PM, Xiao S, Sharma M (2011) Correlation of Precambrian-Cambrian sedimentary successions across Northern India and the utility of isotopic signatures of Himalayan lithotectonic zones. Earth Planet Sci Lett 31:471–483. https://doi.org/10.1016/j.epsl.2011.10.027
McKenzie NR, Hughes NC, Myrow PM, Banerjee DM, Deb M, Planavsky NJ (2013) New age constraints for the Proterozoic Aravalli-Delhi successions of India and their implications. Precambr Res 238:120–128. https://doi.org/10.1016/j.precamres.2013.10.006
Meert JG, Pandit MK (2015) The Archean and Proterozoic history of Peninsular India: tectonic framework for Precambrian Sedimentary basins in India. In: Mazumdar R, Eriksson PG (eds) Precambrian Basins of India: stratigraphic and tectonic context, Chap. 3, vol 43. Geological Society, London, Memoirs, pp 29–54
Meijerink AMJ, Rao DP, Rupke J (1984) Stratigraphic and structural development of the Precambrian Cuddapah basin, SE India. Precambr Res 26:57–104
Mishra RC (1969) The Vindhyan system. Presidential Address. In: 56th Indian Science Congress, Bombay, vol 2, pp 111–142
Mishra RC, Awasthi N (1962) Sedimentary markings and other structures in the rocks of the Vindhyan Formation of the Son Valley and Maihar-Rewa area, India. J Sediment Petrol 32(4):764–775
Mishra RN, Jayaprakash AV, Hans SK, Sundaram V (1987a) Bhima group of Upper Proterozoic—a stratigraphic puzzle. Mem Geol Soc India 6:227–237
Mishra VP, Datta NK, Ramachandra HM, Bhattacharya D, Das DP, Kumaran K, Ramamurthy S, Lakshmana BK, Rao N (1987b) Geological history of the Abhujmar Basin, Bastar District, Madhya Pradesh, vol 6. Memoirs Geological Society of India, pp 189–199
Misra DC, Ravi Kumar M (2013) Proterozoic orogenic belts and rifting of Indian cratons: geophysical constraints. Geosci Front 5(1):25–41. https://doi.org/10.1016/j.gsf.2013.03.003
Mitra R, Chakrabarti G, Shome D (2018) Development of Stromatolites in a mixed siliciclastic carbonate succession: an example from Palaeo-Mesoproterozoic Tadpatri Formation, Cuddapah basin, India. Int J Res Ann Rev 5:1248–1269
Mohanty SP (2015) Palaeoproterozoic supracrustals of the bastar craton: dongargarh supergroup and sausar group In: Mazumder R, Eriksson PG (eds) Precambrian basins of India: stratigraphic and tectonic context, Geol Soc London Mem 43:151–164. https://doi.org/10.1144/M43.11
Mukherjee A, Ray RK (2008) SHRIMP Ages of zircon in the uppermost tuff in Chattisgarh basin in central India require ~500 Ma adjustment in Indian Proterozoic stratigraphy: a discussion. J Geol 116(5):537–539
Mukherjee A, Ray RK (2010) An alternative view on the stratigraphic position of the ~1 Ga Sukhda tuff vis-à-vis chronostratigraphy of the Precambrians of central Indian Craton. J Geol 118:325–332
Mukherjee A, Bickford ME, Hietpas J, Schieber J, Basu A (2012) Implications of a newly dated ca. 1000-Ma rhyolitic tuff in the Indravati Basin, Bastar Craton, India. J Geol 120:477–485
Mukherjee S, Goswami S, Mukherjee A (2018) Structures and their tectonic implications of the Southern Part of Cuddapah basin, Andhra Pradesh, India. Iran J Sci Technol 1–17. https://doi.org/10.1007/s40995-018-0566-0
Mukhopadhyay D (2001) The Archaean nucleus of Singhbhum: the present state of knowledge. Gondwana Res 4:307–318
Mukhopadhyay J, Ghosh G, Nandi AK, Chaudhuri AK (2006) Depositional setting of the Kolhan Group: its implications for the development of a Meso to Neoproterozoic deep-water basin on the South Indian craton. S Afr J Geol 109:183–192
Murti KS (1987) Stratigraphy and sedimentation in Chattisgarh Basin in Purana Basins of Peninsular India, vol 6, pp 239–260
Nagaraja Rao BK, Ramalingaswamy G, Rajurkar ST, Ravindra Babu B (1987) Stratigraphy, structure and evolution of the Cuddapah basin. In: Purana basins of Peninsular India (Middle to Late Proterozoic), vol 6. Memoirs Geological Society of India, pp 33–86
Naqvi SM (2005) Geology and the evolution of the India plate. Capital Publ Co, New Delhi 1–450

Naqvi SM, Rogers JJW (1987) Precambrian geology of India. Oxford Univ Press, New York 1–223
Narayanswami S (1966) Tectonics of Cuddapah basin. J Geol Soc India 7:33–50
Nautiyal SP (1966) Precambs of Mysore plateau. Presidential Address (geology and geography section). In: 53rd session Indian science congress, vol 11, p 194
Padmakumari VM, Dayal AM (1987) Geochronological studies of some mafic dykes around Cuddapah Basin, south India. Mem Geol Soc India 6:369–379
Pandey DK, Bahadur T (2009) A review of the stratigraphy of Marwar Supergroup of west-central Rajasthan. J Geol Soc India 73:747–758
Pandit MK, Sial AN, Jamrani SS, Ferreira VP (2001) Carbon isotopic profile across Bilara Group rocks of Trans-Aravalli Marwar Supergroup in western India: implications for Neoproterozoic-Cambrian transition. Gondwana Res 4:387–394
Panja M, Chakrabarti G, Shome D (2018) Earthquake induced soft sediment deformation structures in the Paleoproterozoic Vempalle formation (Cuddapah basin, India). Carbonates Evaporites 34:491–505. https://doi.org/10.1007/s13146-017-0412-z
Pant NC, Banerjee DM (1990) Pattern of Sedimentation in the type area Bijawar Basin, Central India. Geol Surv India Spec Publ 28:156–166
Pareek HS (1981) Basin configuration and sedimentary stratigraphy of Western Rajasthan. J Geol Soc India 22:517–527
Pareek HS (1984) Pre-quaternary geology and mineral resources of Northwestern Rajasthan, vol 115. Memoirs of the Geological Survey of India, pp 1–99
Pascoe EH (1965) A manual of geology of India and Burma. Government of India Press, Delhi
Patranabis-Deb S (2004) Lithostratigraphy of the neoproterozoic Chattisgarh sequence, its bearing on the tectonics and palaeogeography. Gondwana Res 7:323–337
Patranabis-Deb S, Bickford ME, Hill B, Chaudhuri AK, Basu A (2007) SHRIMP ages of zircon in the uppermost tuff in Chattisgarh Basin in Central India require ~500 Ma adjustment in Indian Proterozoic stratigraphy. J Geol 115:407–415. https://doi.org/10.1086/518049
Patranabis-Deb S, Saha D, Tripathy V (2012) Basin stratigraphy, sea-level fluctuations and their global tectonic connections–evidence from the Proterozoic Cuddapah basin. Geol J 47:263–283. https://doi.org/10.1002/gj.1347
Patranabis-Deb S, Majumdar T, Khan S (2018) Lifestyle of the Paleooproterozoic stromatolite builders in the Vampalle Sea, Cuddapah basin, India. J Asian Earth Sci 157:360–370. https://doi.org/10.1016/jjeacs201708022
Phansalkar VV, Kale AS, Karmalkar NR, Kale VS (1991) An unusual evaporate association from the Papaghni Group, Cuddapah basin. J Geol Soc India 37(1):75–79
Pisarevsky SA, Biswal TK, Xuan-Ce Wang, De Waele B, Ernst R, Soderland U, Tait JA, Ratre K, Singh YK, Cleve M (2013) Palaeomagnetic, geochronological and geochemical study of Mesoproterozoic Lakhna Dykes in the Bastar Craton, India: implications for the Mesoproterozoic supercontinent. Lithos 174:125–143
Prakash R, Swarup P, Srivastava RN (1975) Geology and mineralization in the southern part of Bundelkhand in Lalitpur district, Uttar Pradesh. J Geol Soc India 16:143–156
Prasad B (1984) Geology, sedimentation and palaeogeography of the Vindhyan Supergroup, southeastern Rajasthan, vol 111. Memoirs Geological Survey of India, pp 1–107
Prasad B, Asher R (2016) Record of Ediacaran complex acanthomorphic acritarchs from the Lower Vindhyan succession of the Chambal Valley (East Rajasthan), India and their biostratigraphic significance. J Palaeontol Soc India 61:29–62
Prasad BR, Rao BV (2006) Deep seismic reflection study over the Vindhyans of Rajasthan: implication of geophysical setting of the basin. J Earth Syst Sci 115(1):135–147
Prasad B, Uniyal SN, Asher R (2005) Organic-walled microfossils from the Proterozoic Vindhyan Supergroup of Son Valley, Madhya Pradesh, India. Palaeobotanist 54:13–60
Qureshy MN, Warsi WEK (1975) Role of regional gravity surveys in concept oriented programmes. J Geol Soc India 16:44–54

Raha PK (1987) Stromatolites and correlation of the Purana (Middle to Late Proterozoic) basins of Peninsular India. In: Purana basins of Peninsular India, vol 6. Memoirs Geological Society of India, pp 393–397

Raha PK, Sastry MVA (1982) Stromatolite and Precambrian stratigraphy of India. Precambr Res 18:203–218

Rai V, Shukla M, Gautam R (1997) Discovery of carbonaceous megafossils (Chuaria-Tawuia assemblage) from the Neoproterozoic Vindhyan succession (Rewa Group), Allahabad, India. Curr Sci 73:783–788

Rai AK, Pandey UK, Zakaulla S, Parihar PS (2015) New 1.9–2.0 Ga, Pb–Pb (PbSL), age of dolomites from Vempalle Formation, Lower Cuddapah Supergroup, Eastern Dharwar craton. India. J Geol Soc India 86(2):131–136

Ram J (2005) Hydrocarbon exploration on onland frontier basins of India: perspective and challenges. J Paleontol Soc India 50(1):1–16

Ram J (2012) Neoproterozoic successions in Peninsular India and their hydrocarbon prospectivity. Geol Soc Lond Spec Publ 366:59–73

Ram Babu HV (1993) Basement structure of Cuddapah basin from gravity anomalies. Tectonophysics 223(3–4):411–422

Ram J, Shukla SN, Pramanik AG, Varma BK, Chandra G, Murthy MSN (1996) Recent investigations in the Vindhyan basin: implications for basin tectonics, vol 36. Memoirs Geological Society of India, pp 267–286

Ramakrishna M (1978) Stratigraphy of the Tiratgarh Subgroup of the Indravati Group, Bastar District, M.P. In: Proceedings symposium Purana formations of Peninsular India, University of Saugar, pp 260–267

Ramakrishna M (1987) Stratigraphy, sedimentary environment and evolution of Late Proterozoic Indravati Basin, Central India. In: Purana basins of Peninsular India, vol 6. Memoirs Geological Society of India, pp 139–160

Ramakrishnan M, Vaidyanadhan R (2008) Geology of India, vol 1. Geological Society of India, Bangalore, p 556

Rao SA, Sood NK, Basu PC, Rajendran N, Venkata Dasu SP (1997) Phosphorite in the Cuddapah basin, Andhra Pradesh, vol 6. Memoirs Geological Society of India, pp 511–518

Rao YH, Venkateshwarlu M, Papanna G, Ramalingeshwara Rao B (2013) Palaeomagnetism of Khairmalia volcanics, south of Chittorgarh–implications related to the basal age of the Proterozoic Vindhyan Supergroup. Curr Sci 104:364–367

Rasmussen B, Bose PK, Sarkar S, Banerjee S, Fletcher IR, McNaughton NJ (2002) 1.6 Ga U-Pb zircon age for the Chorhat Sandstone, Lower Vindhyan, India: possible implications for early evolution of animals. Geology 30:103–106

Rathore SS, Venkatesan TR, Srivastava RK (1999a) Rb–Sr isotopic dating of Neoproterozoic (Malani Group) magmatism from southwest Rajasthan, India: evidence of younger Pan-African thermal event by 40Ar–39Ar studies. Gondwana Res 2:271–281

Rathore SS, Vijan AR, Krishna BN, Prabhu BN, Mishra KN (1999b) Dating of glauconite from Sirbu Shale of Vindhyan Supergroup, India. In: Proceedings of the 3rd international petroleum conference on exploration, Petrotech, vol 99, pp 191–196

Ratre K, De Waele B, Biswal TK, Sinha S (2010) SHRIMP geochronology for the1450 Ma Lakhna dyke swarm: its implication for the presence of Eoarchaean crust in the Bastar Craton and 1450–517 Ma depositional age for Purana basin (Khariar), Eastern Indian Peninsula. J Asian Earth Sci 39:565–577

Rawat TPS, Roy M, Joshi GB (2018) Hydrothermal fracture controlled vein type Uranium mineralization in the Paleoproterozoic Bijawar Group of rocks, Sonrai Basin, Lalitpur district, Uttar Pradesh—fresh findings from the subsurface borehole data. J Geol Soc India 91:25–31

Ray JS (2006) Age of the Vindhyan Supergroup: a review of recent findings. J Earth Syst Sci 115:149–160

Ray JS, Martin MW, Veizer J, Bowring SA (2002) U–Pb zircon dating and Sr isotope systematics of the Vindhyan Supergroup, India. Geology 30:131–134

Ray JS, Veizer J, Davis WJ (2003) C, O, Sr and Pb isotope systematics of carbonate sequences of the Vindhyan Supergroup, India: age, diagenesis, correlations and implications for global events. Precambr Res 121:103–140

Ray JS, Davis WJ, Shukla AD (2011) Age of lower Vindhyans of Rajasthan. In: International conference on Precambrian sedimentology and tectonics. Second GPSS Meeting, IIT Bombay, Mumbai, p 75 (Abstract volume)

Raza M, Casshyap SM (1996) A tectono-sedimentary model of evolution of Middle Proterozoic Vindhyan basin, vol 36. Memoirs Geological Society of India, pp 287–300

Raza M, Ahmad AHM, Shamim Khan M, Khan F (2012) Geochemistry and detrital modes of Proterozoic sedimentary rocks, Bayana Basin, north Delhi fold belt: implications for provenance and source-area weathering. Int Geol Rev 54(1):111–129

Saha D, Ghosh G (1998) Lithostratigraphy of deformed Proterozoic rocks from around the confluence of the Godavari and Indravati rivers, South India. Indian J Geol 70(3):217–230

Saha D, Sain A (2019) Multiple convergences along an Archean craton margin: clues from Proterozoic ophiolite remnants, granites and granulite domains along the SE margin of India. J Geodyn 129:44–58. https://doi.org/10.1016/jjog.2018.04.004

Saha D, Patranabis-Deb S (2014) Proterozoic evolution of eastern dharwar and bastar cratons, India–an overview of the intracratonic basins, craton margins and mobile belts. J Asian Earth Sci 91:230–251

Saha D, Tripathy V (2012) Palaeoproterozoic sedimentation in the Cuddapah basin south India and regional tectonics—a review. In: Mazumder R, Saha D (eds) Paleoproterozoic of India, vol 365. Geological Society, London, Special Publications, pp 159–182

Saha D, Bhowmik SK, Bose S, Sajeev K (2016a) Proterozoic tectonics and trans-Indian mobile belts: a status report. Proc Indian Natl Sci Acad 82:445–460

Saha D, Patranabis-Deb S, Collins A (2016b) Proterozoic stratigraphhy of the southern India: cratons and global context. In: Stratigraphy and time scale, vol 9. Elsevier, pp 1–59, https://doi.org/10.1016/bs.sats.2016.10.003

Saha D, Mazumder R, Kar R (2017) Shallow marine to pelagic sediments from a dismembered ophiolite, Kandra, southern India—glimpses of ancient subduction zone related sedimentation. Gondwana Res 49:21–41

Sahoo SR, Das S (2015) Kolhan sedimentation in Chaibasa-Noamundi Basin, Jharkhand, India: a critical observation through lithofacies analysis. Int J Earth Sci Eng 8:60–66

Sahoo D, Pruseth KL, Upadhyay D, Ranjan S, Pal DC, Banerjee R, Gupta S (2017) New constraints from zircon, monazite and uraninite dating on the commencement of sedimentation in the Cuddapah basin, India. Geol Mag 147:1230–1246. https://doi.org/10.1017/S0016756817000140

Sain A, Saha D, Joy S, Jelsma H, Armstrong R (2017) New SHRIMP age and microstructures from a deformed A-type granite, Kanigiri, southern India: constraining the hiatus between orogenic closure and post-orogenic rifting. J Geol 125:241–259

Samom JD, Ahmad T, Choudhary AK (2017) Geochemical and Sm–Nd isotopic constraints on the petrogenesis and tectonic setting of the Proterozoic mafic magmatism of the Gwalior Basin, central India: the influence of Large Igneous Provinces on Proterozoic crustal evolution. In: Sensarma S, Storey BC (eds) Large igneous provinces from Gondwana and adjacent regions. Geological Society, London, Special Publications, p 463. https://doi.org/10.1144/SP463.10

Santosh M, Yokoyama K, Accharya SK (2004) Geochronology of tectonic evolution of Karimnagar–Bhopalpatnam granulite belts, Central India. Gondwana Res 7:501–518

Sarangi S, Gopalan K, Kumar S (2004) Pb–Pb age of earliest megascopic, eukaryotic alga bearing Rohtas Formation, Vindhyan Supergroup, India: implications for Precambrian atmospheric oxygen evolution. Precambr Res 132:107–121

Sarkar S, Banerjee S, Bose PK (1996) Trace fossils in Mesoproterozoic Koldaha Shale, central India and their implications. Neues Jahrb Geol Plaeontol Monatsh 7:425–438

Sarkar S, Bose PK (1992) Variation in late Proterozoic stromatolites over a transition from basin plain to nearshore subtidal zone. Precambr Res 551(1–2):139-157

Sarkar A, Chakraborty PP, Mishra B, Bera K, Sanyal P, Paul S (2010) Mesoproterozoic sulphidic ocean, delayed oxygenation and evolution of early life: sulphur isotope clues from Indian Proterozoic basins. Geol Mag 147(2):206–218
Sarkar S, Banerjee S, Samanta P, Jeevankumar S (2006) Microbial mat induced sedimentary structures and their implications: example from Chorhat Sandstone, M.P., India. Indian J Earth Sci 115:49–60
Sastry MVA, Moitra A (1984) Vindhyan stratigraphy—a review. Geol Surv India Mem 116(2):109–148
Schnitzer WA (1969) Zur stratigraphic and Lithologie des nordlochen Chattisgarh-Becken (ZentralIndien) Unterbesonder Berucksichtigung von Algenrff. Complexendt Geol Ges. 118:290–295
Seilacher A, Bose PK, Pflüger F (1998) Triploblasticanimals more than 1 billion years ago: trace fossil evidence from India. Science 282:80–83
Sergeev VN, Sharma M, Shukla Y (2012) Proterozoic fossil cyanobacteria. Palaeobotanist 61:189–358
Sesha Sai VV (2013) Proterozoic granite magmatism along the terrane boundary tectonic zone to the East of Cuddapah basin, Andhra Pradesh–petrotectonic implications for Precambrian crustal growth in Nellore Schist Belt of Eastern Dharwar Craton. J Geol Soc India 81:167–182
Sesha Sai VV (2019) Proterozoic magnesian feldspathic magmatism in the Nallamalai fold belt, Southern India–petrogenesis of Vellaturu–Ipuru–Nakerikally (VIN) arc magmatic granites, Eastern Dharwar craton. J Geol Soc India 93:409–418
Sesha Sai VV, Tripathy V, Bhattacharjee S, Khanna TC (2017) Paleoproterozoic magmatism in the Cuddapah basin, India. J Indian Geophys Union 21:516–525
Sharma VP (1975) A note on the stratigraphic classification of the Purana rock of Bastar District, Madhya Pradesh, vol 25, no. 1. Geological Survey of India Miscellaneous Publication, pp 171–175
Sharma M (1996a) Microbialites (stromatolites) from the Mesoproterozoic Salkhan Limestone, Semri Group, Rohtas, Bihar: their systematic and significance. Mem Geol Soc India 36:167–196
Sharma M (1996b) Paleobiology of Mesoproterozoic Salkhan Limestone, Semri Group, Rohtas, Bihar, India. J Earth Syst Sci 115(1):67–98
Sharma M (2006a) Late palaeoproterozoic (Statherian) carbonaceous films from the Olive Shale (Koldaha Shale), Semri Group, Vindhyan Supergroup, India. J Palaeont Soc India 51(2):27–35
Sharma M (2006b) Palaeobiology of mesoproterozoic salkhan limestone, Semri Group, Rohtas, Bihar, India; systematics and significance. J Earth System Sci 115(1):67–98
Sharma M, Pandey SK (2012) Stromatolites of the Kaladgis Basin, Karnataka, India, their systematic, biostratigraphy and age implications. Paleobotanist 61:103–121
Sharma M, Shukla Y (2012) OcCurrentence of helically coiled microfossil Obruchevellain the Owk Shale of the Kurnool Group and its significance. J Earth Syst Sci 121:755–768
Sharma M, Singh VK (2019) Megascopic carbonaceous remains from Proterozoic basins of India. In: Mondal MEA (ed) Geological evolution of the Precambrian Indian Shield. Society of earth scientists series. Springer, Berlin, pp 725–749. https://doi.org/10.1007/978-3-319-89698-4_27
Sharma M, Tiwari M, Ahmad S, Shukla R, Shukla B, Singh VK, Pandey SK, Ansari AH, Shukla Y, Kumar S (2016) Palaeobiology of Indian Proterozoic and early Cambrian successions–Recent Developments. Proc Indian Natl Sci Acad 82:559–579
Sheppard S, Rasmussen B, Zi JW Soma, Sekhar V, Srinivasa Sarma D, Ram Mohan M, Krapez B (2017) U-Pb dating of metamorphic monazite establishes a Pan-African age for tectonism in the Nallamalai Fold Belt, India. J Geol Soc Lond 174(6):1062–1069. https://doi.org/10.1144/jgs2017-051
Shrivastava BP (1992) Significance fourth dimensional stratigraphic markers in Paleozoic sediments of West Central Rajasthan-Palaeogeographic implications—petroleum habitat. Indian J Petrol Geol 1(2):224–244
Shukla AD, George BV, Ray JS (2019) Evolution of the Proterozoic Vindhyan Basin, Rajasthan, India: insights from geochemical provenance of siliciclastic sediments. Int Geol Rev 62:153–167. https://doi.org/10.1080/00206814.2019.1594412

Sibal VK, Murty GSSN (2005) Challenges in exploration for hydrocarbons in sedimentary basins of Rajasthan (a keynote address). In: Proceedings of natural oil gas lignite scenario with special reference to Rajasthan, pp 35–51
Singh IB (1973) Depositional environment of the Vindhyan sediments in the Son Valley area. In: Recent research in geology, vol 1. Hindustan Publication Corporation, New Delhi, pp 140–152
Singh IB (1980) Precambrian sedimentary sequences of India: their peculiarities and comparison with modern sediments. Precambr Res 12:411–436
Singh IB (1985) Paleogeography of the Vindhyan basin and its relationship with other Late Proterozoic basins of India. J Paleontol Soc India 30:35–41
Singh SP (1988) Stratigraphy and sedimentation pattern in the Proterozoic Delhi Supergroup, northwestern India, vol 7. Memoirs Geological Society of India, pp 193–206
Singh CL, Ghosh GK (2003) Study of Vindhyan sedimentation, crystalline basement and associated structures below the alluvial tracts of Varanasi region, Uttar Pradesh, by magnetic and electrical methods. J Geol Soc India 62:205–214
Somasekhar V, Ramanaiah S, Srinivasa Sarma D (2018) Geochemical characterisation of the siliciclastic rocks of Chitravati Group, Cuddapah Supergroup: implications for provenance and depositional environments. J Earth Syst Sci 127:54. https://doi.org/10.1007/s12040-018-0955-y
Soni MK, Chakraborty S, Jain VK (1987) Vindhyan Supergroup—a review. Mem Geol Soc India 6:87–138
Srikantia B, Sant VN, Sharma SB (1969) Geology and preliminary assessment of Birmania phosphorite deposit, Jaisalmer District. J Mines Metals Fuels 17:107–113
Srinivasa-Rao T (1987) The Pakhal Basin—a perspective. Geol Soc India Mem 6:161–187
Srivastava RN (1989) Bijawar phosphorite at Sonrai–Geology, sedimentation, exploration strategy and origin, vol 13. Memoirs Geological Society of India, pp 47–59 (Banerjee DM: Guest editor)
Srivastava P (2014) Largest Ediacaran discs from the Jodhpur Sandstone, Marwar Supergroup, India: their palaeobiological significance. Geosci Front 5:183–191
Srivastava N, Banerjee DM (1987) Proterozoic phosphorites of India: updated information part I and II. Mem Geol Soc India 6:471–487
Srivastava AP, Rajgopalan G (1986) F-T dating of Precambrian deposits of Vindhyan Group at Chitrakut, Banda district (UP-MP). In: Sharma KK (ed) Nuclear tracts, pp 41–52
Srivastava AP, Rajgopalan G (1988) FT ages of Vindhyan gluconitic sandstone beds exposed around Rawatbhata area, Rajasthan. J Geol Soc India 32:527–529
Srivastava AP, Rajgopalan G (1990) Glauconite ages of Vindhyan sediments in Rajasthan. Ind J Phy 64S(5):358–364
Strauss H, Banerjee DM, Kumar V (2001) The sulfur isotopic composition of Neoproterozoic to early Cambrian seawater—evidence from the cyclic Hanseran evaporates, NW India. Chem Geol 175:17–28
Subrahmanyam C (1980) Regional gravity and geological structure in Bastar Province, India. Geophy Res Bull 18:75–84
Tripathy V, Saha D (2010) Structure and low grade metamorphism of the east central part of the Proterozoic Nallamalai fold belt, south India–thrust stacking and discontinuous metamorphic gradients along eastern margin of East Dharwar craton. Ind J Geol 80(1–4):173–188
Tripathy GR, Singh SK (2015) Re–Os depositional age for black shales from the Kaimur Group, Upper Vindhyan, India. Chem Geol 413:63–72
Tripathy V, Satyapal, Mitra SK, Sesha Sai VV (2019) Fold-Thrust Belt Architecture and Structural Evolution of the Northern Part of the Nallamalai Fold Belt, Cuddapah basin, Andhra Pradesh, India. In: Mukherjee S (ed) Tectonics and structural geology: Indian context. Springer Geology, pp 219–252
Valdiya KS (1969) Stromatolites of the Lesser Himalayan carbonate formations and the Vindhyans. J Geol Soc India 10(1):1–25
Valdiya KS (2016) The making of India: geodynamic evolution. Society of Earth Scientists Series, 2nd edn. Springer, p 924

Vankatachala BS, Sharma M, Shukla M (1996) Age and life of the Vindhyans: facts and conjectures. Mem Geol Soc India 36:137–165

Venkatakrishna R, Dotiwala FE (1987) The Cuddapah salient: a tectonic model for the Cuddpah basin based on Landsat image interpretation. Tectonophysics 136:447–477

Verma RK, Banerjee P (1992) Nature of continental crust along the Narmada-Son Lineament inferred from gravity and deep seismic sounding data. Tectonophysics 202:375–395

Vinogradov AP, Tugarinov AI, Zhikov CI, Stanikova NI, Bibikova EV, Khorre K (1964) Geochronology of the Indian Precambrian. Report 22nd international geological congress, New Delhi, vol 10, pp 553–567

Viswanathiah MN (1979) Lithostratigraphy of the Kaladgi and Badami Groups, Karnataka, Indian. Mineral, Special Issue on Kaladgi, Badami, Bhima and Cuddapah basins. 18:122–132

Walker TL (1902) The geology of Kalahandi state, central provinces, vol 33 Part 3. Geological Survey of India Memoirs

Zachariah JK, Rao YB, Srinivasan R, Gopalan K (1999) Pb, Sr and Nd isotope systematics of uranium mineralized stromatolitic dolomites from the Proterozoic Cuddapah Supergroup, south India: constraints on age and provenance. Chem Geol 162(1):49–64

Chapter 6
Tectonics of the Himalaya

6.1 Introduction

This chapter provides summarized geological and geophysical account of the youngest and highest mountain belt to elucidate the role of the continental lithospheric subduction of the Indian Plate in the evolution of the Himalaya during the Cenozoic.

Paleogeographic reconstructions of the India-Asia domains during the Mesozoic postulate that the Tethyan Ocean separated the northern parts of the Indian continental lithosphere (ICL) from the southern Asian Plate (Fig. 6.1a; Stampfli and Borel 2002). These regions now constitute the Himalaya, Trans-Himalayan Ladakh-Karakoram Mountains and Tibet Plateau. As the Indian Plate converged anti-clockwise and northwards towards Asia, the Neo-Tethyan oceanic floor underwent intra-oceanic subduction to produce the Shyok-Dras Volcanic Arc (Fig. 6.1b; Searle et al. 1987; Thakur 1993; Rolland et al. 2000; Jain and Singh 2009; Copley et al. 2010). Between this arc and Asian Plate, a thick Paleo-Mesozoic Karakoram sedimentary sequence was deposited on its southern margin. This ocean closed along the Shyok Suture Zone (SSZ) in the north and subsequently along the Indus-Tsangpo Suture Zones (ITSZ) in south; two sutures demarcate contact between these plates (Fig. 6.1). These sutures preserve evidences of (i) initial Late Mesozoic subduction of the Neo-Tethys oceanic lithosphere along the SSZ during the Early Cretaceous–Lower Eocene with intervening intra-oceanic Shyok-Dras Volcanic Arc, (ii) emplacement of the younger calc-alkaline Trans-Himalayan plutons, and (iii) final closure of the Neo-Tethys along the ITSZ (Honegger et al. 1982; Searle et al. 1987; Hodges 2000; Rolland et al. 2000; Yin 2006; Jain and Singh 2009).

The ICL, its Proterozoic–Paleozoic supracrustals and Paleo-Mesozoic Tethyan sedimentary cover, were remobilized south of the ITSZ, and underwent Cenozoic deformation and metamorphism. It formed the Himalayan Metamorphic Belt (HMB) whose leading edge in the Tso Morari suffered ultra-high pressure (UHP) metamorphism at 53.3 ± 0.7 Ma at >120 km depth beneath the Trans-Himalaya (Leech et al.

A. K. Jain et al., *Tectonics of the Indian Subcontinent*, Society of Earth Scientists Series, https://doi.org/10.1007/978-3-030-42845-7_6

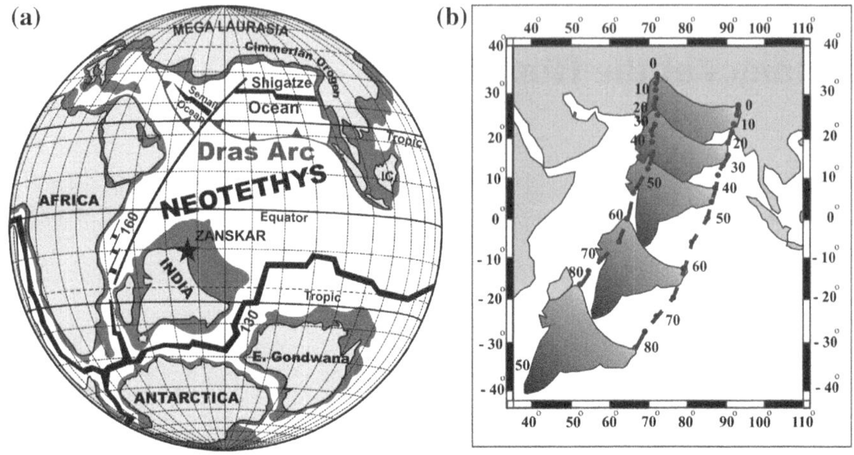

Fig. 6.1 **a** Paleogeographic reconstruction of the Neo-Tethys domain after the fragmentation of the Gondwanaland and movement of the Indian Plate. Of interest is the Mesozoic passive platform on the northern Indian margin, spread of the Neo-Tethys and development of the intra-oceanic Dras Volcanic Arc within the subduction zone. Redrawn after Stampfli and Borel (2002). **b** Movement of the Indian Plate with reference to Asia since 80 Ma and two points in the Himalayan syntaxes. Redrawn after Copley et al. (2010)

2005). Various thrusts such as the Main Central Thrust (MCT) and Main Boundary Thrust (MBT) caused extensive crustal shortening during the Himalayan tectonics (Fig. 6.1; Thakur 1993; Yin 2006).

Two contrasting hypotheses attempted to explain the geological evolution of the Himalaya: (i) continental lithospheric collision between the Indian and Asian Plates along the ITSZ (Dewey and Bird 1970; Thakur 1993; Hodges 2000; Searle et al. 1987; Yin 2006; Valdiya 2016), and (ii) continental lithospheric subduction beneath the Himalaya and Tibet till Kun Lun (Argand 1924; Powell and Conaghan 1973, 1975; Seeber and Armbruster 1981; Ni and Barazangi 1984; Zhao et al. 1993; Nelson et al. 1996; Nábělek et al. 2009; Jain 2014, 2017; Guo et al. 2017).

6.2 The Himalaya

The name ‘Himalaya’ is coined by linking two Sanskrit words: ‘Hima’ (snow)–‘Laya’ (dwelling), i.e. ‘Abode of snow’. The E–W trending arcuate mountain chain of the Extra-Peninsula runs NW–WNW to E–ESE for about 2400 km. It has its convexity towards the Indian Shield and is surrounded by low-lying Indo-Gangetic Plain in the south and the Tibetan Plateau in the north (Fig. 6.2). Width of the mountain is about 350 km in Pakistan and about 150 km in the east in Arunachal Himalaya. Physiographically, the Himalaya terminates at the Nanga Parbat (8130 m) in the west, and the Namcha Barwa (7782 m)–Gyala Peri (7234 m) in the east inside the Great

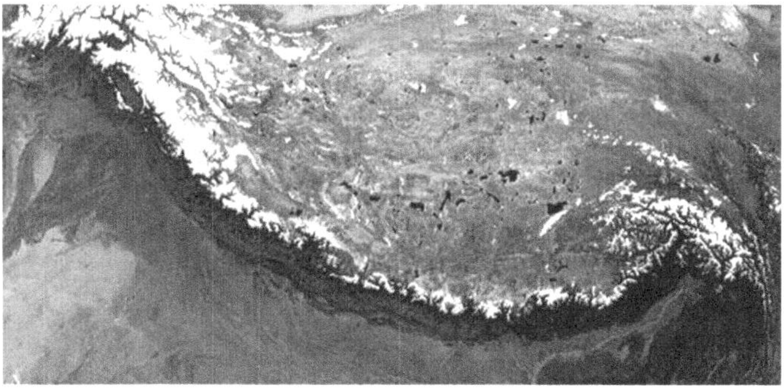

Fig. 6.2 Satellite imagery of the Himalaya mountain range, forming a convex arc towards the Indian Shield. The chain runs through Pakistan in the west, northern India, Nepal, Bhutan, with Central Asian high plateau in the north. *Source* Landsat GeoCover 2000

Bend of the Yarlung–Tsangpo River. Steepest gorges mark both the terminations of this range where the Indus River dissects the 4500–5200 m gorge in the west, and the Tsangpo River cuts through 6009 m between the mountains in the east (Fig. 6.2).

The Himalaya is drained by some major river systems like the Indus, the Ganga and the Brahmaputra, rising from the highest ranges; their combined drainage area hosts about 600 million people of different faiths. It has profound climate effects and keeps the Indian monsoon down south, while limiting the Artic winds in the north. Drained by some 19 major rivers, the Himalaya does not control all their watersheds, hence many rivers originate in Tibet and flow southwards, e.g. the Indus, Sutlej, Subansiri, Lohit and Brahmaputra. The Indus and the Brahmaputra Rivers are the largest among these; each having catchment basins of about 260,000 km^2 in mountains. The Indus system has the Indus, Jhelum, the Chenab, the Ravi, the Beas, and the Sutlej, and collectively define the vast system flowing into the Arabian Sea. The Ganga system has the Ganga, the Yamuna, the Ramganga, the Kali (Kali Gandaki), the Karnali, the Rapti, the Gandak, the Baghmati, and the Kosi Rivers, and flow into the Bay of Bengal along with the Brahmaputra system—the Tista, the Raidak, the Manas, the Subansiri, the Siang, and the Dibang–Lohit–Tirap Rivers.

6.3 Sub-divisions

The Extra-Peninsular Himalaya is bounded by various smaller mountain ranges in the west, e.g. the Pamir, Hindukush, Karakoram, Ladakh and Zanskar; the latter three mountains are loosely linked and combined with the Himalaya. The main Himalaya Mountain is comprised of at least four almost continuous ranges with distinct physiography, geology and tectonics due to southward Cenozoic convergence (Heim and Gansser 1939; Thakur 1993; Hodges 2000; Yin 2006; Jain et al. 2014,

2019; Valdiya 2016). The following units are tectonically juxtaposed from south to north: (Figs. 6.3 and 6.4).

(i) Indus-Ganga–Brahmaputra Plain between the Indian Craton/Proterozoic Sedimentary Basins and the Himalayan Orogen as low-lying alluvial Plain, drained by vast rivers and their thick late Quaternary to Recent alluvium. More commonly called as the Indo-Gangetic Plain, it abuts against the Sub-Himalayan (SH) Belt along the Himalayan Frontal Thrust (HFT) and is terminated in eastern parts of Arunachal Pradesh.
(ii) Sub-Himalayan (Outer Himalayan) ranges with maximum width of 100 km in Himachal Pradesh and elevation between 275 and 1000 m, having steeper southern slopes and flat-floored structural valleys. The Siwalik Range extends for almost entire length of the Himalaya with intermontane valleys, called the duns

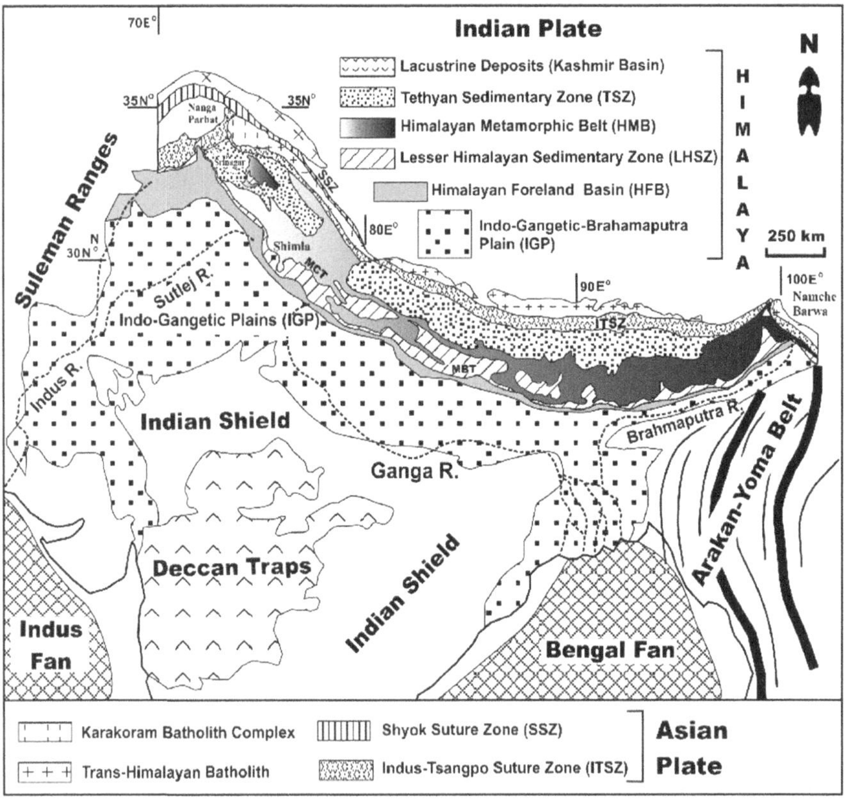

Fig. 6.3 Simplified regional geological map of the Himalaya and adjoining mountains in the plate tectonic framework. Also shown are the Cenozoic Sub-Himalayan foreland basin (HFB), Recent Indo-Gangetic Plains, and Indus-Bengal Fans, receiving sediments from the eroding mountains. SSZ—Shyok Suture Zone; ITSZ—Indus Tsangpo Suture Zone; MCT—Main Central Thrust; MBT—Main Boundary Thrust. Compiled from published data

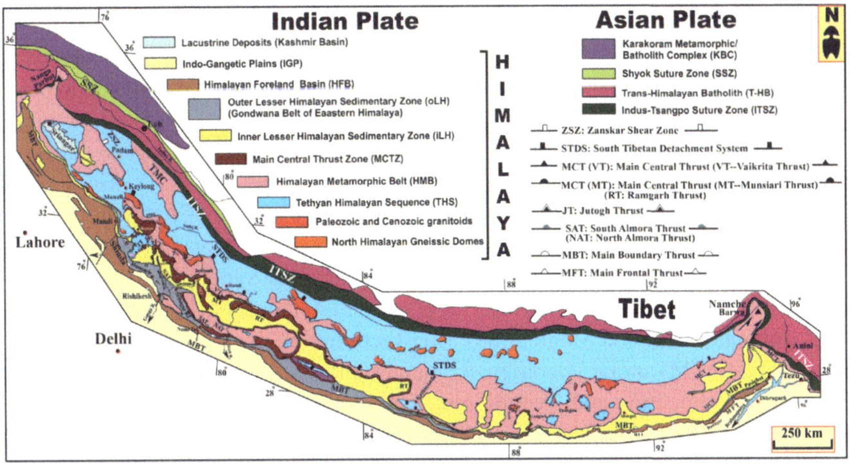

Fig. 6.4 Simplified regional geological map of the Himalaya from Kashmir to northeastern parts of Arunachal Pradesh, Trans-Himalaya and Karakoram Mountains. Impotant tectonic boundaries divide the Himalaya into distinct belts. Compiled from published data

(Dehra Dun, Uttarakhand). This belt is overthrust by the northern-hading Lesser Himalaya (LH) sedimentary belt along the Main Boundary Thrust (MBT).

(iii) Lesser Himalayan (Lower Himalayan) ranges are nearly ~80 km in width with valleys at nearly 1000 m and peaks going up to 4500 m; these exhibit dissected landscape. The Nag Tibba, the Dhauladhar and Pir Panjal Ranges branch off from the Great Himalayan Range. The LH Belt is overlain by regionally north-dipping metamorphic belt—the Himalayan Metamorphic Belt (HMB) along the Main Central Thrust Zone (MCTZ).

(iv) Great Himalayan (Higher Himalaya/Himadri) Range possesses elevations reaching above 8000 m and are covered by perpetual snow. It houses the highest peaks like the Mount Everest, the Nanga Parbat, the Dhaulagiri, the Nanda Devi, the Manaslu, the Kanchenjunga etc. It is mostly made up of the Himalayan Metamorphic Bely (HMB), constituting the core of the Himalaya and exhibits highest uplift. Further north, it is separated from the Tethyan Himalayan Sequence (THS) by the northerly-hading extensional South Tibetan Detachment System (STDS).

(v) Tethys Himalaya (Tibetan Himalaya) to the north of the Great Himalayan Range with ranges like Zanskar.

To the south of the Extra-Peninsular region, the Indo–Gangetic–Brahmaputra Plain, containing enormously thick eroded sediments from the rising Himalaya, separate the Himalaya from the Indian Shield.

Vast literature on the geology, tectonics, geochemistry, geophysics and geochronology of the Himalaya now exists that it would be extremely difficult to summarize. The readers are advised to refer to the books and monographs on (i) On the Geology of Krol Belt (Auden 1934), (ii) Central Himalaya (Heim and Gansser 1939),

(iii) Geology of Himalaya (Gansser 1964) and Geology of Bhutan (Gansser 1983), (iv) Geology of Nepal Himalaya (Hashimoto et al. 1973; Dhital 2015), (v) Geology of Kumaun Lesser Himalaya (Valdiya 1980a), (vi) Geology of Higher Himalaya (Sinha 1989), (vii) Himalayan Tectonics (Treloar and Searle 1993), (viii) Geology of Himachal Pradesh (Srikantia and Bhargava 1998), (ix) Geodynamics and collision tectonics of the Himalaya (Jain and Manickavasagam 1999, 2002), (x) Geological account of the Bhutan Himalaya (Bhargava 1995), (xi) Colliding Continents (Searle 2013), to name a few. There are a other special publications dealing with the Himalaya and adjoining mountains: (i) Tectonics of the Nanga Parbat Syntaxis and the Western Himalayas by Khan et al. (2000), (ii) Geology and Tectonics of Southeast Ladakh and Karakoram (Jain and Singh 2009), (iii) Tectonics of the Himalaya (Mukherjee et al. 2015), (iv) Myanmar: Geology, Resources and Tectonics (Barber et al. 2017), (v) Crustal Architecture and Evolution of the Himalaya–Karakoram–Tibet Orogen (Sharma et al. 2019), (vi) Himalayan Tectonics: A Modern Synthesis (Treloar and Searle 2019).

This chapter incorporates the geology and tectonics of the (i) Indus-Ganga–Brahmaputra Plain and (ii) Himalayan Orogen.

6.4 Tectonics of the Indus–Ganga–Brahmaputra Plain (IGBP)

6.4.1 Setting

Largest active foreland basin—the Indo-Gangetic (IGP) in the western parts, and the Brahmaputra and Bengal Basins in the east—are collectively called as the Indus-Ganga–Brahmaputra Plain (IGBP). The basin extends 3200 km from Rann of Kutch to Assam. The maximum width spans ~550 km in Panjab and a minimum of about 90 km in the extreme east, encompassing approximately 2.5-million km^2 of the fertile land of the northern Indian subcontinent. It accumulates an enormous quantity of eroded sediments and dissolved materials by large rivers, which also carry these into mega-fans of the Arabian Sea and the Bay of Bengal. Three major river systems have their tributaries. The Indus is in the extreme west and flows SSW. The Ganga system in the middle drain to the east. A large part of this plain is drained by the Brahmaputra system, which flows westward, and controls the depositional patterns of the IGBP (Fig. 6.5). The basin is restricted by the Himalayan Orogen in the north, while the Meghalaya Craton and Arakan–Yoma belt limits it in the east. Subsurface Archean ridges, basement faults, and uneven Vindhyan terrain dissect these Quaternary basins into deeper depocenters. The sedimentary succession of the Siwalik Supergroup is nearly 6.0 km thick, located near the Himalayan foothills (see Singh 1996, 1999; Ojha 2012; Clift 2017 for details).

The IGBP is an amalgamation of three large sub-basins, which are divided by subtle sub-surface ridges. These ridges control the drainage of (i) the Indus Basin,

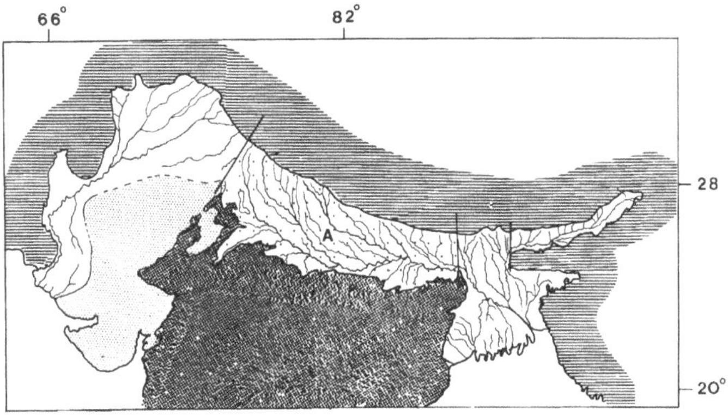

Fig. 6.5 Generalized setting of the Indus-Ganga–Brahmaputra Plain (IGBP), with the Ganga Plain and its tributaries in central parts, the Indus Plain (and tributaries) to its left and the Brahmaputra Plain (and tributaries) to its right. Dotted hatched area: Thar Desert, Dark area: Indian Peninsula, Horizontal hatched region: Himalaya including the Cenozoic Foreland Basin. Firm lines indicate approximate limits of basins. After Singh (1996)

(ii) the Ganga Basin, and (iii) the Brahmaputra Plain (also called as the Assam Basin). Thrust belts confine these basins in the north, southeast, and west. It exhibits all significant components of a foreland basin, namely an orogen (the Himalaya), deformed and uplifted foreland basin deposits adjacent to the orogen (Siwalik hills), a depositional bowl (IGBP) and peripheral cratonic bulge (Aravalli–Bundelkhand–Vindhyan Plateau–Meghalaya).

The Indus River originates from Mount Kailash and drains a large segment of the Tibetan Plateau, Trans-Himalayan and Karakoram Ranges before entering the Himalayan region of the Nanga Parbat in Pakistan. The main tributaries contributing the detritus to the Indus Basin are the Chenab, the Jhelum, the Zanskar, the Beas, and the Sutlej. The central Ganga Basin trends almost E–W and occupies the largest segment of the IGBP, before changing its orientation to almost N–S. The Brahmaputra River is fed by tributaries like Siang, Dibang, Lohit, Subansiri, Teesta, Burhi, Dihing, and Kopili. Together, they contribute to the sediment budget of the Brahmaputra Plains.

6.4.2 *The Ganga Plain*

The Ganga Plain (synonymous with the Ganga Basin) extends from the Aravalli Hills in the west up to the Rajmahal Hills in the east and occupies an area of about 250,000 km^2 to the south of the Himalayan foothills. A network of rivers, originating in the Himalaya, drains its different tectonic units. They flow south to southwest and then gradually swing to the southeast and meet the main trunk river, Ganga. Mountain-fed tributaries like Yamuna, Ramganga, Ghagra, Gandak, and Kosi show

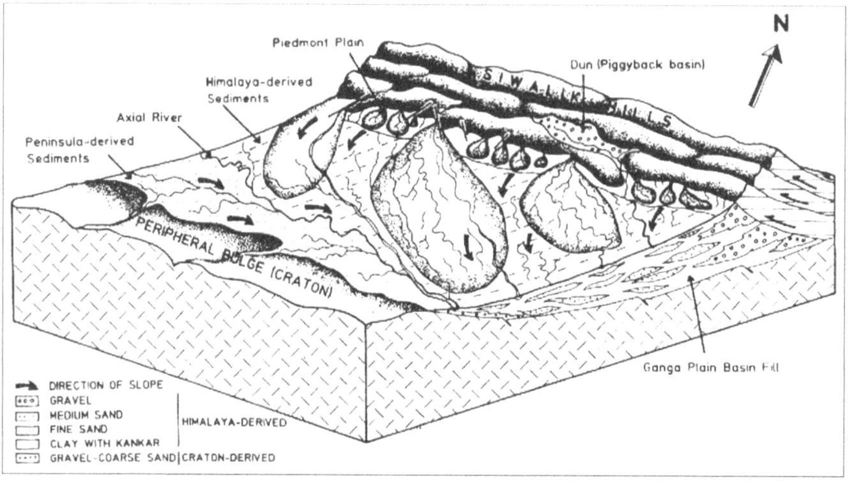

Fig. 6.6 Ganga Plain and its divisions into (i) Piedmont Fan Surface, (ii) Megafan Surface or Central Alluvial Plain in the middle, and (iii) Marginal Plain Upland Surface (MP). After Singh (1996)

multi-channel braided configurations. They have an almost similar geometry. In contrast, the Chambal and Son rivers and its distributaries drain the craton-region and flow NE/ENE (Fig. 6.5). Many small rivers also originate within the Ganga Plain and join the main trunk river at different points. Beyond Farakka, in the Bengal region, the river enters to form the Ganga Delta Plain.

Three geomorphological zones within the Ganga Plain are (i) south-sloping Piedmont Fan Surface (PF) in the north comprising Bhabhar and Terai zones, (ii) Megafan Surface (MF) or Central Alluvial Plain in the middle, and (iii) north-sloping Marginal Plain Upland Surface (MP) in the south (Singh 1992, 1996; Srivastava et al. 2003). Further, Active Flood Plain Surface (T0), River Valley Terrace Surface (T1), and Upland Interfluve Surface (T2) can also be delineated within the central zone of Megafan Surface (Fig. 6.6). Based on the rate of sedimentation and subsidence history, Thomas et al. (2002) classified the Indo–Gangetic Plain into Upper, Middle, and Lower units. The entire Ganga Plain has been segmented into numerous tectonic blocks of varying dimensions by several longitudinal and transverse faults developed as a result of neotectonic activities.

6.4.3 Precambrian Basement, Faults, and Depressions

Extensive geophysical survey and drilling by the Oil and Natural Gas Corporation indicated that the Ganga Plain has a complex evolutionary history. The Plain is underlain by the basement rocks of Proterozoic Aravalli, Delhi, and Vindhyan Supergroups and the Archean Bundelkhand Complex (Sastri et al. 1971; Rao 1973; Raiverman et al. 1983; Singh 1996; Ojha 2012). The basement rocks gently slope to the north and are found at ~4000 to 6000 m depth near the Himalayan foothills. One of the prominent subsurface basement ridges is the Delhi–Sargodha Ridge in the extreme west. In the central part, the Delhi–Hardwar Ridge, the Faizabad Ridge, the Mirzapur–Ghazipur Ridge, the Monghyr–Saharsa Ridge are of significance. Smaller ridges are Raxaul, Bahraich, and Puranpur. Several basement faults like Moradabad fault, Bareilly fault, Lucknow fault, West and East Patna faults and Malda fault (Sastri et al. 1971; Rao 1973) are well delineated (Fig. 6.6); some of these are seismically active.

Being a polycyclic basin, the Ganga Plain encompasses sub-surface northern continuation of the Vindhyan Basin and the overlying Himalayan foredeep basin (Cenozoic). Average alluvial fill is a thin veneer over the peninsular margin and basement ridges in the south and increases to the Himalayan foothill (5–6 km) in the north (Sastri et al. 1971; Singh and Singh 1971). Major rivers, i.e., the Yamuna, the Ganga, the Ghaghara, the Rapti, the Gandak, and the Kosi appears to be fault controlled. Several generations of faults and subsurface ruptures, as evident from the geophysical studies, suggest extensive neotectonic deformation of the region. These deformations were responsible for the development of the present configuration of the basin. The studies have indicated that various topographic breaks produced by these tectonic planes were responsible for the development of inland terminal fans as geomorphic markers on the down-thrown blocks (Fig. 6.7).

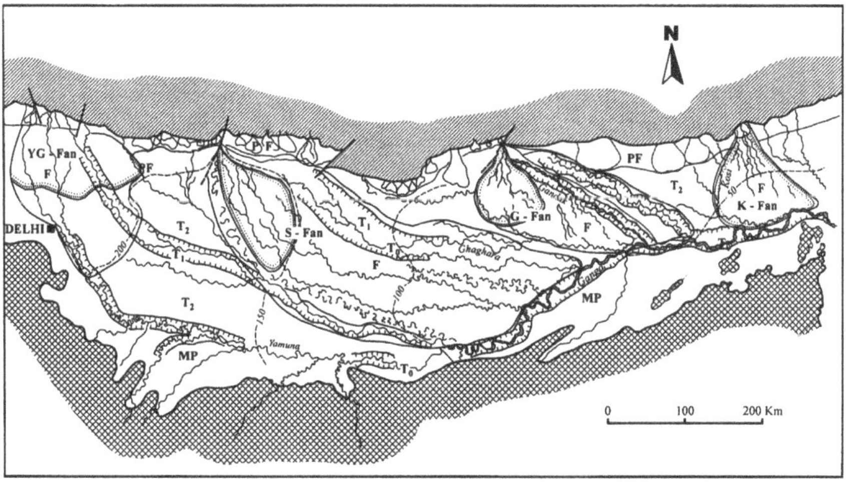

Fig. 6.7 Regional geomorphic surfaces of the Ganga Plain. PF—Piedmont Fan Surface in the north near the Himalayan Foothills. F—Megafan Surface in the middle (Yamuna–Ganga Megafan—YG, Sarda Megafan—S, Gandak Megafan—G, Kosi Megafan—K). T—Terrace Surfaces (T0—Active Flood Plain Surface. T1—River Valley Terrace Surface. T2—Upland Terrace Surface). MP—Marginal Plain Upland Surface in the south near the Indian Peninsula. After Srivastava et al. (2003)

Five basinal depressions are recognizable in the Ganga Valley based on gravity anomalies. These are the Sahaspur Depression, the Rampur Low, the Sarda Depression, the Gandak Depression, and the Madhubani Depression (Figs. 6.8 and 6.9; Ojha 2012). Of these, the Sahaspur Depression and Rampur Low probably developed during the Neoproterozoic Delhi Orogeny. The Sarda Depression is a large E–W trending depression with an accumulated sediment thickness of more than 7.0 km, having Puranpur Low and the Bahraich Low sub-depressions. The Gandak Depression is a significant depression located east of the Sarda Depression, with a sediment thickness of about 7.0–8.0 km. The Madhubani Depression developed as a narrow graben during the Middle Mesoproterozoic (1400 Ma) due to a transtensional rift, where and sedimentary accumulation of more than 7.0 km thick is recorded. The deepest well of the Ganga Valley (Madhubani-1) is in this depression reaching a depth of 5957 m. Northeast to the southwest geological cross sections through Peninsular India into the Ganga Valley, extending up to the deepest Madhubani depression reveal the sub-surface configuration of the Ganga Plain (Fig. 6.8). Integrated magnetotelluric, seismo-geological and borehole data further elaborate this configuration around Bahraich where maximum depth of the Archean basement is attained to more than 9 km possibly due to rift tectonics (Fig. 6.9d; Manglik et al. 2015).

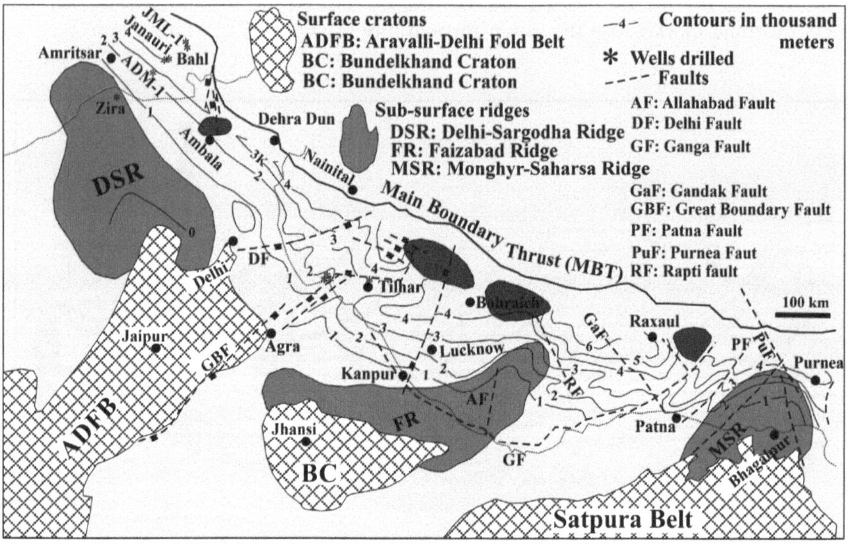

Fig. 6.8 Sub-surface configuration of the Ganga Plain showing prominent structural elements: the ridges, faults, depressions with northerly slope. Drawn and simplified by after Sastri et al. (1971), Rao (1973), Karunakaran and Rao (1979) and Singh (1996)

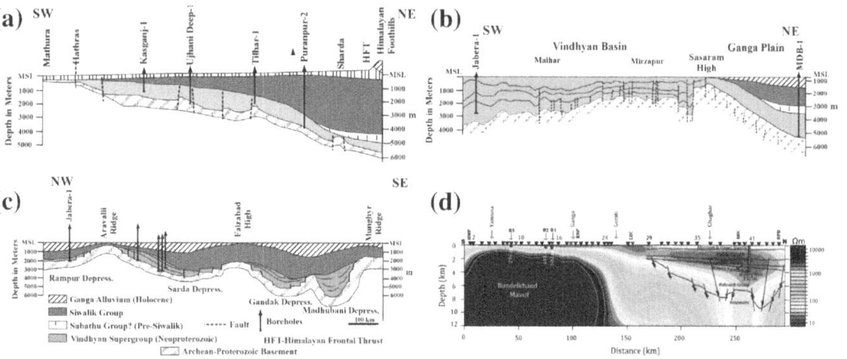

Fig. 6.9 Subsurface configuration of the Ganga Plain is showing the location of boreholes and basement configuration. Based on Sastri et al. (1971), Rao (1973), Karunakaran and Rao (1979), Ojha (2012)

6.4.4 *Neotectonism*

In the Ganga Plain, neotectonic deformation is widespread (Singh et al. 1997). Normal faults and thrusts on the surface in the unconsolidated Quaternary and Recent sediments produced small topographic breaks in the monotonous flat terrain with relief differences of 3–23 m. These were produced due to the change in compression directions (Parkash et al. 2000; Pati et al. 2011, 2012). Therefore, this plain is segmented into numerous tectonic blocks by several longitudinal and transverse faults. Parkash et al. (2000) studied the curvilinear nature of the longitudinal faults (through which the major rivers in the foreland basin flow) and suggested their development due to the compression parallel to sub-parallel with the Himalaya.

Due to the presence of NE–SW trending subsurface basement ridges, the E–W longitudinal faults become curvilinear towards the north. This compression results from the E–W extension, which leads to the development of NE–SW normal faults in the majority of the Plain. Due to a higher rate of convergence and upliftment rate, the compressional force is dominant, with the development of reverse faults. As the Plain is coupled and segmented with the Himalaya (Parkash et al. 2011), its geomorphic processes and tectonism of individual segments are controlled by the Himalayan tectonics (Pati et al. 2019). Neotectonic deformation in the Plain is manifest as surface ruptures. However, this relief difference is caused due to different tectonic events concerning these faults. This could be only a single event depending upon the number of tectonic events to which the concerned fault has responded. Along the fault, due to the erosion of the fault scarps, variable relief differences have been observed. The flow of these rivers is possibly controlled by active lineaments; some of these are identified as gravity faults (Singh 1996; Singh et al. 1996; Agarwal et al. 2002).

A 50 m-sediment core on the Ganga natural leevee near Kanpur has recorded the story of 100 ka of uninterrupted flood plain deposit. They show two pronounced geochemical excursions. One at ca. 70 ka and the other at 20 ka. They show changes in the $^{87}Sr/^{86}Sr$ ratio from 0.727 to 0.767 and εNd(0) values from −14.4 to −16.6 (Rahaman et al. 2009). These geochemical excursions coincide with decreased monsoon precipitation and an increase in glacial cover, which caused lower sediment supply from the Higher Himalaya.

6.4.5 Tectonic Blocks and Soil Chronology

On the basis of five soil chrono-associations of <500 years, >500 years, 2.5 ka, ~5 ka and ~8.5 ka, Parkash et al. (2000) recognized four tectonic blocks of the Piedmont, Upper Ganga–Yamuna, Lower Ganga–Yamuna and Ganga–Ramganga Blocks, while the Lower Ganga–Yamuna Block has Modinagar, Khurja and Etah sub-blocks. Uplift of the Lower Ganga–Yamuna, Ganga–Ramganga and Upper Ganga–Yamuna blocks took place at ~8.5 ka, ~5 ka, and 2.5 ka, respectively, while the Ganga–Ramganga block was tilted westwards around 1 ka, leading to shifting in the Ganga River course in the same direction. Smaller tectonic movements also took place during the recent past. The Upper Ganga–Yamuna Block and Modinagar sub-block have tilted to the east which resulted in the shifting of the Yamuna river towards east. The Khurja sub-block tilted very slightly to the west resulting in tightening of Yamuna river meanders. In this region, fault patterns resulted from compression from both the west and the southwest directions.

Since tectonically stable periods among successive faulting episodes promoted soil formation on the fan surface, Pati et al. (2011) suggested that the Upper Gangetic plain is relatively more stable than the Middle Gangetic Plain. They used the degree of soil development as one of the critical parameters. The Upper Gangetic Plain is marked by uplands with incised rivers and moderately to well-developed soils with salt efflorescence and calcrete. The Plain between Ghaghra, Rapti, Gandak, Baghmati, and Kosi forms a drainage network which, in turn, generated small inland streams. The Middle Gangetic Plain is marked by a relatively high rate of subsidence and by sedimentation in the mega-fans. They show weakly to moderately developed soils while the Lower Gangetic Plain is known or its deltaic sediments (Pati et al. 2015).

6.4.6 Tectonic Features of the Rapti–Gandak Plain

Using the GPS data across the Nepal Himalaya, Bürgmann et al. (1999) suggested that the Himalayan arc was divided into different segments, showing different convergence rates. Parkash et al. (2011) delineated the N–S trending fault along the Rapti River, dividing the Nepal Himalaya into western and eastern sectors. These faults can be distinguished by abrupt changes in the topography, convergence and upliftment rates. The Rapti and Gandak faults, controlling the Rapti and Gandak rivers respectively, divide the Plain into three tectonic blocks from west to east: Ghaghara–Rapti interfluves block, the Rapti–Gandak block, and the Gandak–Kosi interfan block (Pati et al. 2011). The Ghaghara–Rapti and the Rapti–Gandak blocks have been tectonically divided into sub-blocks by several NE–SW trending extensional normal faults. A similar segmentation of the Gandak–Kosi interfan block is due to several E–W trending thrust splays possibly off the HFT. Pati et al. (2015) picked up thrusts up to 15 m depth in the Gandak–Kosi block, using Ground Penetrating Radar (GPR).

6.4.7 Assam Plain

The Assam Plain is surrounded by hills and plateau. Structurally the basin is mainly bounded by thrust belts in the north, northeast, and southeast. The bulk of the Precambrian landmass of Meghalaya Craton lies to the south of the river Brahmaputra between longitudes 90° and 94° and controls the whole lower part of the river course. Compressional tectonism in northeast India is responsible for the present configuration of the Assam basin. Structures developed along the foothills of the Himalaya, and Naga hills show the migration of rock masses towards the basin. Besides, recent earthquakes have changed the basin morphology as well as river courses. It seems the basin is gradually becoming narrower and shallower under the influence of the present tectonism.

6.4.8 Genesis of the IGBP—A Foreland Basin

In the literature, the Indus-Ganga–Brahmaputra Plain (IGBP) is referred differently: Foredeep (Suess 1893–1909), Rift (Burrard 1915), Basin, Yoked Basin, Half Graben or Board Shallow basin (Rao 1973). The Indian–Asia Plate convergence produced the IGBP in Paleocene, and the basin was initiated as a foreland basin (Dewey and Bird 1970; Patriat and Achache 1984; Klootwijk et al. 1992; Beck et al. 1995). Burbank (1992) and Singh et al. (1996) suggested that overloading in the Himalaya was responsible for the creation of this basin due to the thrust-fold belt (Fig. 6.10). Gravity measurements across Ganga Plain and Himalaya show a deficit of mass below the Ganga Plain and excess of mass below Lesser Himalaya (Lyon-Caen and Molnar 1985).

Initiation of the IGBP took place around Early Miocene (ca. 20 Ma BP), when sediments, deposited during the early history of this basin (Middle Miocene-Middle Pleistocene: 16.0–0.5 Ma BP), are exposed in the Siwalik Hills. The present-day near-surface sediments of the basin are of Holocene age; the older Late Quaternary sediments and Siwalik successions are buried below these enormously thick younger sediments (Singh 1996, 1999; Tandon et al. 2008).

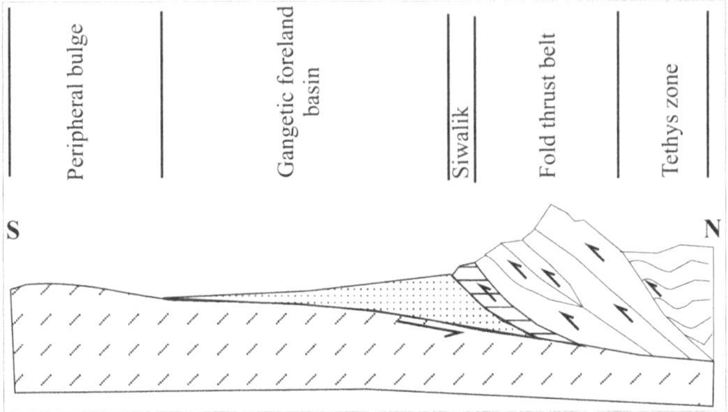

Fig. 6.10 Crustal evolution of the Ganga Plain due to overloading by the Himalayan Orogen. After Singh (1996)

6.5 Cenozoic Himalayan Foreland Basin (HFB)

The Himalayan Foreland Basin (HFB) is an elongated asymmetrical basin, which is located between the Himalayan Frontal Thrust (HFT) towards south and the Main Boundary Thrust (MBT) in the north. Sediment pile of this basin extends further south into the IGBP, where it overlaps the peripheral bulge of the Indian Craton (Fig. 6.11; Tandon 1991; Singh 1999; Kumar 2020). Many transverse basement ridges dissect the foreland basin into several sub-basins and control their sedimentation patterns and tectonics. It evolved with widespread Paleocene-Middle Eocene marine transgression, covering even parts of the Lesser Himalaya. This basin predominantly accumulated fluvial sediments, which were deposited by large southward-flowing

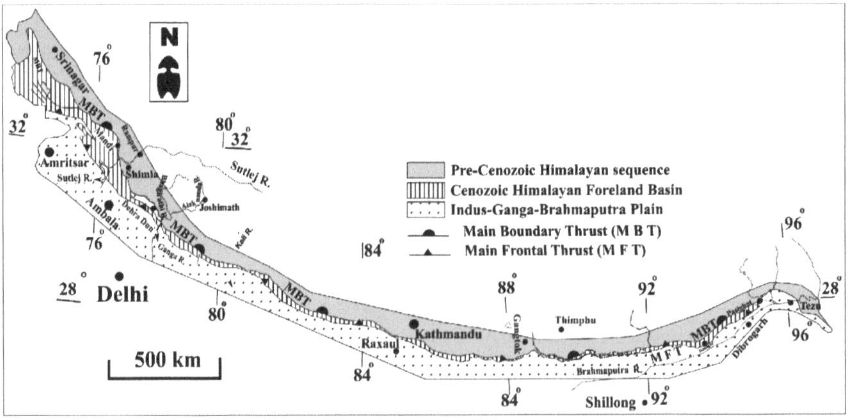

Fig. 6.11 Generalized geological map of the Cenozoic Himalayan Foreland Basin (HFB)

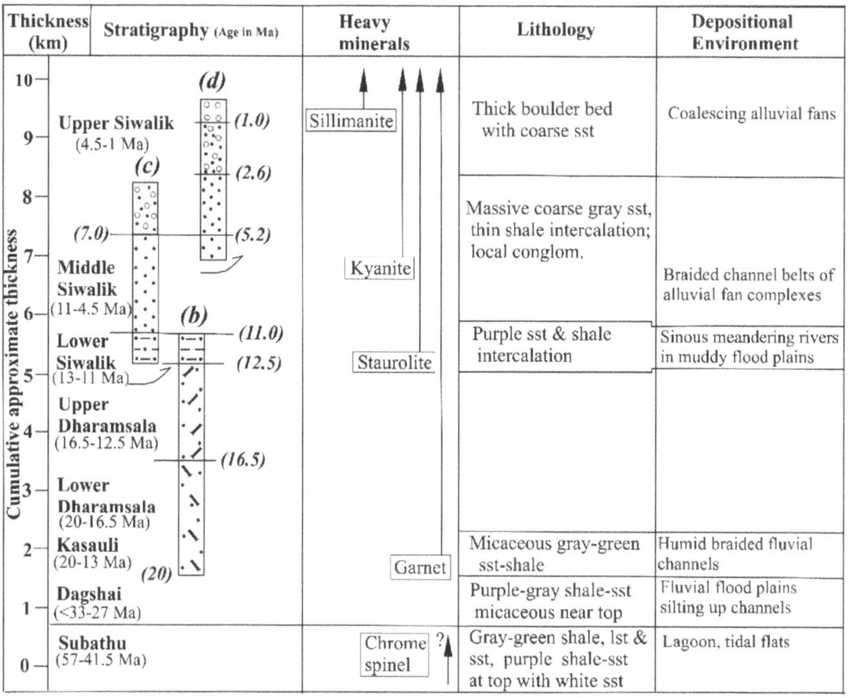

Fig. 6.12 Generalized stratigraphy of the Himalayan Foreland Basin (HFB). Compiled from various published sources. After Jain et al. (2009)

river systems from the rising Himalaya (Srikantia and Bhargava 1998). The following stratigraphic units characterize this belt (Fig. 6.12).

Between the Siwalik Group and the Precambrian rocks in the north along the Main Boundary Thrust the Paleogene sequence of the HFB extends interruptedly from Kashmir in the west to Arunachal Pradesh in the east. It is best developed in Himachal where the sequence is divisible into the Subathu, Dagshai (=Lower Dharamsala/Lower Murree) and Kasauli (=Upper Dharamsala/Upper Murree) Formations.

6.5.1 Subathu Formation

The Paleocene-Middle Eocene marine transgression commenced with the unconformable deposition of the Subathu Formation in the Proterozoic–Early Paleozoic Lesser Himalayan and Sub-Himalaya domains. The base of the formation is characterized by a Thanetian (~57.9 to 54.7 Ma) breccia due to growth faults, pisolitic impure bauxite or paleosoil horizon in Jammu region (Singh 2003) or volcanic ash beds (tonstein) in Shimla Hills; the latter contains fine-grained glass shards, kaolinite and angular β–quartz, indicating volcanic eruptions during initiation of the Subathu sedimentation (Siddaiah and Kumar 2007). Carbonaceous/coaly shale is overlain by dull green-grey fossiliferous splintery shale/siltstone, thin fossiliferous limestone and sandstone intercalations (Passage Beds). These rocks were deposited in euxinic and evaporitic lagoons and shallow tidal sea (Singh 1978) between Middle Paleocene and lower Middle Eocene (Mathur 1978). Clasts of felsic volcanics, chert, serpentine schist and high-Al and Cr spinel reflect a mixed source from the proto-Himalayan ITSZ (Najman and Garzanti 2000; Bhatia and Bhargava 2006). It is supported by ε_{Nd} values of (~−7.8 to −9) and $^{87}Sr/^{86}Sr$ ratios of ~0.710 to 0.715, which plot between fields of the ITSZ and Tethyan sedimentary cover. The sequence grades into variegated purple siltstone–shale alternations—the Passage bed sequence (Bhatia 2000) and, in turn, by dull white sandstone; the latter was deposited as coastal barrier in high-energy environment (Srivastava and Casshyap 1983). Further, the age of the Subathu Formation has been estimated to range between 61.5 and 43.7 Ma (Lakshami et al. 2000), while its uppermost part may be even 41.5 Ma, magnetostratigraphically (Sangode et al. 2005).

6.5.2 Dagshai Formation

The overlying unfossiliferous Dagshai Formation in Shimla Hills (=Dharamsala Formation) (Raiverman et al. 1983) is red siltstone and mudstone in lower parts, while sandstone and caliche appear in the upper parts. The arenites contain schistose clasts, zircon, garnet, tourmaline and epidote, which were derived from medium grade metamorphosed Himalayan rocks (Sinha 1970; Najman et al. 1997, 2004). ε_{Nd} and $^{87}Sr/^{86}Sr$ values range between ~−12.7 to −17.2 and ~0.753 to 0.775, respectively (Najman et al. 2000). It is likely that the Dagshai Formation was deposited either in an estuarine (Bhatia and Bhargava 2006) or meandering silting fluvial channels eroding the top of the Subathus (Bera et al. 2008).

The precise contact relationship between the Subathu and Dagshai Formations remains controversial since the transition was postulated from marine to fluvial environment (Bhatia 2000 and references therein; Bhatia and Bhargava 2006). On the contrary, an unconformity was proposed between these two sequences with a hiatus ranging from ~10 to <3 Myr (Najman et al. 1993, 2004; Bera et al. 2008). The age of the Dagshais is debatable: (i) 35.5 ± 6.7 Ma for the whole formation (Najman et al. 1994) or between <28 and 25 Ma from $^{40}Ar/^{39}Ar$ detrital micas (Najman et al. 1997), (ii) base younger than 31 ± 2 Ma from detrital-zircon FT ages (Najman et al. 2004) or 35.5 Ma (Lakshami et al. 2000; Najman 2006; Jain et al. 2009).

6.5.3 *Kasauli Formation*

The overlying Kasauli Formation contains about 2000 m-thick grey-green sandstone and alternating siltstone–mudstone; litharenites contain larger percentage of metamorphic fragments than the Subathu and Dagshai formations, and preserve isotopic characters like the latter. The formation appears to have been deposited in migratory braided river system under humid climate (Singh 1978; Najman et al. 1993). A Lower Miocene age (23–16 Ma) of this formation has been determined from floral and mammal remains (Arya et al. 2004). ^{40}Ar–^{39}Ar white mica ages constrain its age at <28 and 22 Ma (Najman et al. 2004) and 24.9 $\pm$ 2.1 to 20.7 $\pm$ 3.2 Ma from youngest ZFT peak for the uppermost part (Jain et al. 2009). As an equivalent of the Dharamsala Group, magnetostratigraphic age of the uppermost Kasauli is either 13 Ma (White et al. 2002) or 11.5 Ma (Lakshami et al. 2000).

6.5.4 *Siwalik Group*

Bulk of the HFB is comprised of over ~6000 m-thick Miocene-Pleistocene coarsening upwards Siwalik Group, which was deposited by the southerly-flowing fluvial system (Kumar et al. 2011). The sequence is extensively investigated regarding its depositional setting, biostratigraphy, faunal evolution and migratory routes, magnetostratigraphy, geochemistry, isotope characters, geochronology, exhumation, and tectonics (see Chirouze et al. 2013; Clift 2017; Kumar 2020; Mandal et al. 2019).

Many sub-basins like Kangra, Subathu, Dehra Dun mark the HFB due to presence of sub-subsurface ridges and transverse lineaments (Fig. 6.13a, b; Raiverman et al. 1983); litho-stratigraphy and sedimentology of each sub-basin are reviewed and dealt in detail by Kumar (2020). Stratigraphically, the Siwalik Supergroup is broadly subdivided into three Subgroup–Lower, Middle and Upper Siwalik, though their correlation and division are sometimes difficult along the entire length of the belt (Parkash et al. 1980; Tandon 1991; Kumar et al. 2011; Chirouze et al. 2013; Govin et al. 2018). The Kangra Sub-basin exposes all the three subgroups of the Siwalik, while the Lower Siwalik Subgroup is poorly exposed in the Subathu Sub-Basin, while it is totally absent in the Dehra Dun Sub-Basin.

6.5.4.1 Lower Siwalik Subgroup

The 1800 m-thick Lower Siwalik Subgroup of <13–11 Ma (White et al. 2001, 2002) of interbedded coarse purple-grey sandstone-brown shale was deposited by highly sinuous meandering rivers in broad muddy flood plains (Singh 1999; Parkash et al. 1980). Detritus was mainly derived from low-medium grade metamorphosed Himalayan sources between <13 and 9 Ma (cf. Clift 2017), as these source rocks are indicated

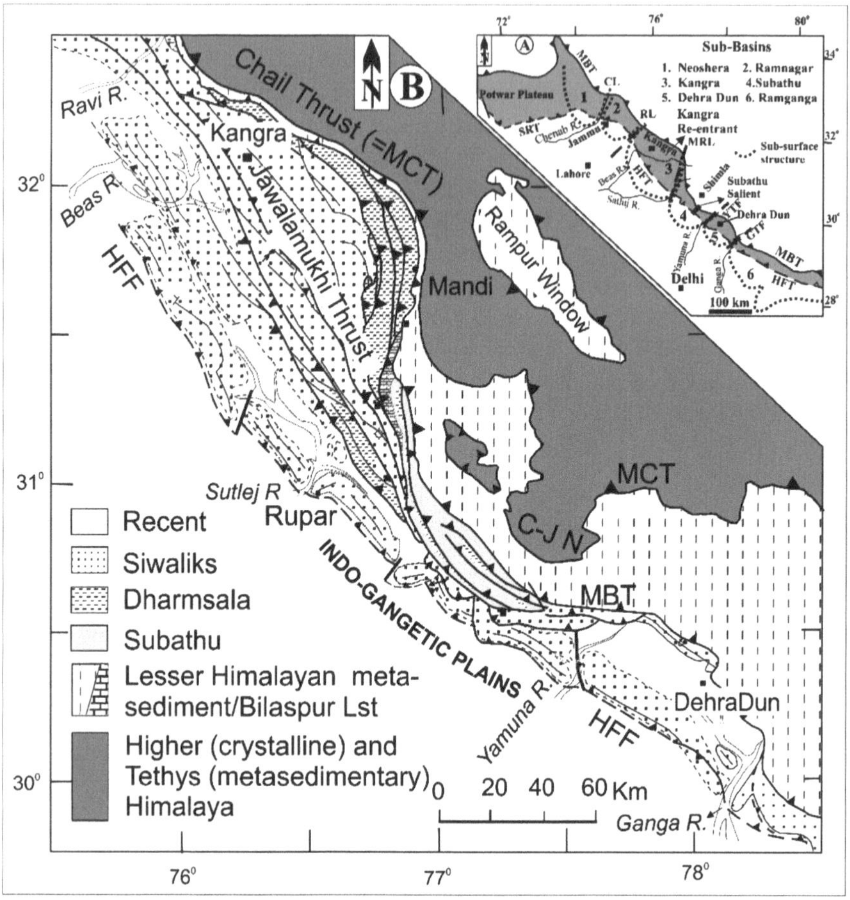

Fig. 6.13 **a** Simplified geological map of the Himalayan Foreland Basin (HFB) between Ravi and Ganga Rivers and its sub-basins marked by sub-surface ridges and lineaments. **b** Geological map of the northwestern parts of the HFB, its sub-basins and hinterland Lesser Himalaya. MCT—Main Central thrust; MBT—Main Boundary Thrust; HFF—Himalayan Frontal Fault; SRT—Salt Range Thrust; CL—Chanab lineament; RL—Ravi lineament; MRL—Manali-Rupar lineament; YTF—Yamuna Tear Fault; GTF—Ganga Tear Fault. Redrawn and modified Karunakaran and Ranga Rao (1979), Raiverman et al. (1983) and Kumar (2020)

by the presence of phyllite clasts, and epidote, garnet, tourmaline, zircon, rutile, chlorite and staurolite (Sinha 1970; White et al. 2002).

In the Subathu Sub-Basin between the Satluj and Yamuna Rivers, the Lower Siwalik Subgroup overrides the MS and US along the Nahan Thrust, while it is observed in the northern belt across the Jwalamukhi Thrust in the Kangra Sub-Basin (KSB). Fine-grained, dark grey sandstone and purple brown mudstone alternate each other with evidences of pedogenesis–calcrete, bioturbation and rootlets. Ribbon-shaped sandstone bodies display trough-cross stratification and ripple marks with mean paleoflow toward south (Kumar et al. 2004).

6.5.4.2 Middle Siwalik Subgroup (MS)

Over 2300 m-thick Middle Siwalik (MS) Subgroup contains cross-bedded, greyish, mica-rich medium to coarse multi-storied sandstone–siltstone–overbank mudstone. It was deposited between 11 and 4.5 Ma by major braided rivers with alluvial fan complexes. Coarsening-up greywacke beds have rock fragments ranging from 20 to 40%, with garnet, tourmaline, zircon, epidote, staurolite, zoisite, and kyanite as heavy minerals (Sinha 1970).

In the Kangra Sub-Basin, the Middle Siwalik Sub-Group contains sandstone with sheet-like geometry with low mudstone content, while conglomerate facies (~8.7 and 7.0 Ma) is well stratified, imbricated and fining upwards with rounded to well-rounded clasts of quartzite, basics, granite/gneiss and sandstone. Paleoflow pattern shows major change from S-SE (LS) to SW, parallelling the basin axis.

In the Subathu Sub-Basin, the MS (~6 and 5.5 Ma) consists of trough cross-stratified and thickly bedded (>40 m) multi-storey grey sandstone with minor mud-stone of brown to grey colour with calcrete and bioturbated paleosol. Paleoflow was nearly southeasterly.

Within the Dehra Dun Sub-Basin, the MS is sandstone-dominated multi-story complex with facies changing from sandstone–mudstone to sandstone and sand-stone–mudstone–conglomerate. These were deposited between 9.0 and 5.23 Ma (Kumar et al. 2003; Sangode et al. 1999). Mica-rich peppery grey sandstone is coarse to fine-medium grained with sheet geometry, having an overall southerly paleoflow direction.

Using major, trace elements, Sr–Nd isotope geochemistry and detrital zircon U–Pb geochronological records of 10-Myr long period of the MS and US sequences of the Dehra Dun re-entrant, Mandal et al. (2019) opined that these Late Miocene sediments were derived from the Tethyan Himalayan (THS), Greater Himalayan (GHS), inner Lesser Himalayan (iLH) and outer (oLH) sequences. Erosional unroofing of the LH Crystalline sequences (LHCS) was initiated ~6 Ma, possibly related to out-of-sequence movement of the Ramgarh-Munsiari Thrust as well as erosional recycling of older foreland basin deposits into younger Siwaliks since ~5.5 Ma; the latter may indicate the time of thrust propagation from the Lesser Himalaya into the foreland basin. Tectonic uplift and erosion of the orogenic wedge occurred in response to duplexing of the iLHS, concomitant high topography and rock uplift rate in the Greater and Tethyan Himalaya.

6.5.4.3 Upper Siwalik (US) Subgroup

The Upper Siwalik Subgroup of ~2.5 km thick conglomerate–sandstone–mudstone was laid down into coalescing alluvial fans between 4.5 and 1 Ma (Sangode et al. 1996). In the Kangra Sub-basin, the US succession has ~2000 m thick accumulation of conglomerate after 6 Ma with sandstone and rarely mudstone beds in the lower part and thickly-bedded conglomerate in the upper parts with increasing clasts size up to 50 cm in sheet-geometry (Kumar et al. 2004). These polymictic conglomerate includes clasts of white and pink quartzite, limestone, granite/gneiss, sandstone, basic volcanic etc. The paleoflow trend is towards SW.

In the Subathu Sub-Basin, the Upper Siwalik is dominantly exposed with the Tatrot, Pinjor and Bounder Conglomerate Formations after ~5.5 Ma. The Tatrot Formation (~5.5 to 2.58 Ma) contains variable grey sheet sandstone bodies, with a distinct conglomerate at 2.6 Ma (Tatrot-Pinjor formational boundary). In the Pinjor Formation, grey sheet sandstone is accompanied by pre-Tertiary clast-bearing stratified, imbricated conglomerate with clasts of quartzite, limestone, phyllite, slate, chert, granitoid, and rare basic igneous rocks. Paleoflow directions of grey and buff sandstone are almost perpendicular to each other and show SW and SE trends, respectively (Kumar et al. 1999). At the contact between the Pinjor and the Boulder Conglomerate Formations (~1.77 Ma), conglomerate clast composition changes from the Lesser Himalayan-derived quartzite to Sub-Himalayan-derived (Tertiary) sandstone. Conglomerates are stratified, imbricated and composed of subrounded to subangular clasts in buff sandy muddy matrix, yielding paleoflow directions dominantly towards S-SE, which changes to S-SW trend in the northwestern parts. Conformably and transitionally overlying Upper Siwalik (US) Sub-Group in the Dehra Dun Sub-basin starts with conglomerate, sandstone, mudstone facies, with gradually increasing, laterally and vertically amalgamated conglomerate sheets. Dominant clasts are of quartzite, with minor slate, phyllite limestone, granitoids and volcanics; their imbrication provide southwesterly paleo-drainage.

Since the HFB detritus was synchronously derived from the Himalaya, it provides an invaluable proxy for cooling and exhumation of the adjacent rising hinterlands. Detrital $^{40}Ar/^{39}Ar$ muscovite and ZFT youngest peaks become younger eastwards from 25–20 to 10–8 Ma; this trend is consistent with an eastward-migrating pulse of hinterland cooling (Clift 2017; Webb et al. 2017). Figure 6.14 conceptualizes chronology of depositional modifications and their related fluvial system between River Ravi and Ganga.

6.5.5 *Duns—Late Cenozoic–Holocene Tectonics*

The Cenozoic HFB is longitudinally and tectonically dissected by numerous small tectonic intermontane basins containing Late Cenozoic-Holocene sediments, lying over the Siwalik group. These are bordered by the outer Siwalik Hills on southern sides and the overthrust Lesser/Sub-Himalaya belt in the north along the MBT. Dehra

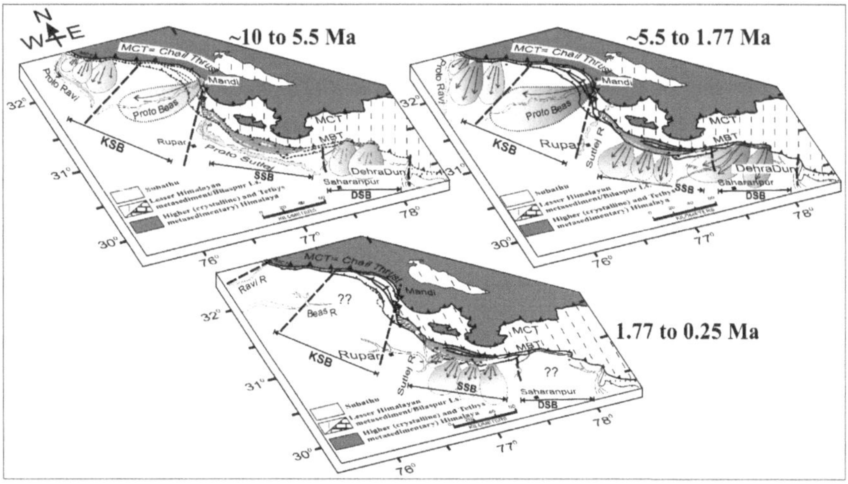

Fig. 6.14 Schematic depositional model illustrating chronologic-wise modifications and spatial relationships of fluvial system between Rivers Ravi and Ganga. Abbreviations as in Fig. 6.13b. Modified and redrawn after Kumar (2020) and Kumar et al. (2011)

Dun reveals a combined influence of basin margin regime and structures that transect the Dun where foreland propagating Himalayan fold-thrust system is largely controlled by the Main Boundary Thrust (MBT) and the Main Frontal Thrust (MFT) during Pleistocene. In Dehra Dun, fan sedimentation occurred in (i) oldest isolated hills of fan material (IHF) of debris flow deposits of ~43 to 33 ka, (ii) proximal fan (PF) sheet flow deposits of 34–21 ka, (iii) distal fan (DF) braided river deposits during 20–11 ka, and (iv) a minor phase of terrace deposition during 3–1 ka, all with abandoned intervening incision phases (Sinha and Sinha 2016). Though tectonism influenced the evolutionary history of the alluvial fans in the Dun, aggradation and entrenchment of fans were controlled by climatic perturbations.

6.5.6 *Detrital Sources in Foreland Basin*

6.5.6.1 Northwestern Himalayan Foreland Basin

Jain et al. (2009) investigated detrital-zircon fission-track (ZFT) characters of the Lower Cenozoic HFB sequence to determine exhumation of the Himalayan source rocks because of the India-Asia collision. Transgressive marine and oldest Paleocene–Eocene Subathu Formation (57.0–41.5 Ma) contains ~50 Ma detrital ZFT youngest peak P1, derived possibly from the Indus Tsangpo Suture Zone (ITSZ) and the Ladakh Batholith (LB) of the Asian Plate. A dominant older 302.4 ± 21.9 Ma peak (85% grains) with a few 520 Ma grains in this formation have been derived by erosion of the Zircon Partially Annealed Zone (ZPAZ) of 240–180 °C. As the first imprint of the Himalayan orogenesis, this zone affected the Himalayan Proterozoic basement and its Tethyan sedimentary cover. Since the detritus in the Subathu has been derived, both from the Indian and Asian Plates, possible suturing of these plates took place during the Subathu sedimentation (Fig. 6.15). A sudden change in the provenance is recorded in the detrital ZFT cooling ages in the Oligo-Miocene Dagshai and Kasauli formations, which have dominant 30 and 25 Ma youngest P1 peaks, respectively. We interpret a distinct unconformity spanning about 10 Myr between the Subathu and Dagshai formations. Since ~30 Ma, molassic sedimentation coincided with shifting of the source rocks to the Himalayan metamorphic belt. This

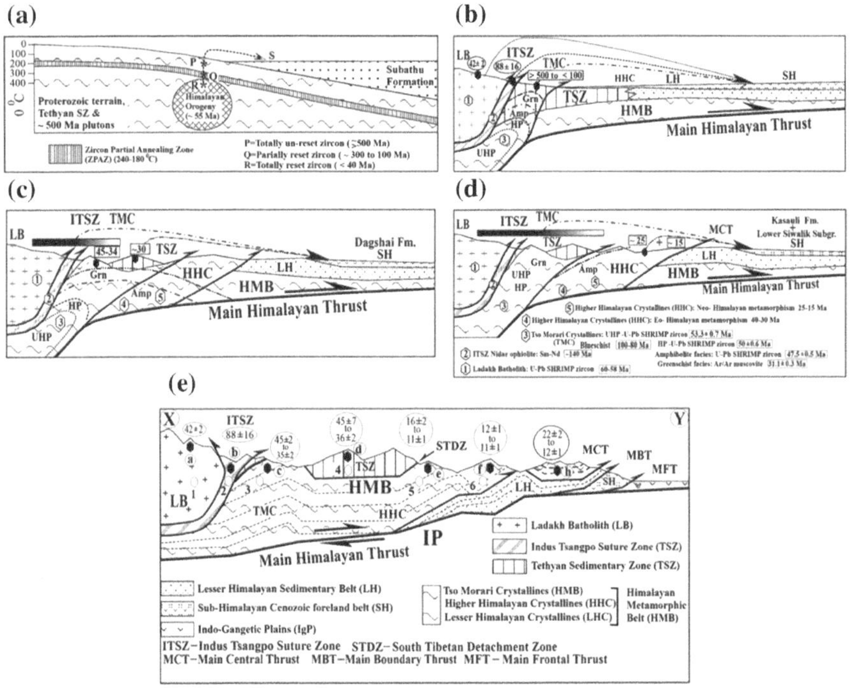

Fig. 6.15 Schematic sections of the possible source rocks for the Sub-Himalayan foreland basin. **a** Possible effects of the Himalayan orogeny on a Proterozoic terrain, its exhumation and hypothetical path of particles through different zones. Different source rocks at the deposition of the **b** Subathu Formation, **c** Dagshai Formation, **d** Kasauli Formation and Lower Siwalik Subgroup. **e** Geological cross-section of the NW Himalaya from the Indo-Gangetic Plains to the Ladakh Batholith, showing progressive unroofing and present-day measured zircon FT ages in the provenance. Filled polygons: Present-day distribution of zircon FT ages (Ma) across the Himalaya. Open polygons represent events across the Himalayan units: 1—U–Pb SHRIMP zircon (60.1 ± 0.9 Ma) from the Ladakh Batholith as age of emplacement. 2—^{40}Ar/^{39}Ar glaucophane age (90 ± 10 Ma) indicating blueschist metamorphism. 3—U–Pb SHRIMP zircon age (53–50 Ma) indicating UHP and HP metamorphism. 4—Low amphibolite facies metamorphism in the Tethyan Sedimentary Zone at ~47 Ma. 5—U–Pb zircon/monazite and U–Th–Pb monazite ages between 40 and 25 Ma due to the Eo–Himalayan metamorphism and 6-Neo-Himalayan metamorphism between 25 and 15 Ma. *Data Sources* See Jain et al. (2009)

belt has sequentially undergone three distinct cooling and exhumation pulses after the UHP–HP metamorphism (53–50 Ma) in the extreme north and two widespread M1 and M2 metamorphisms (40–30 and 25–15 Ma) in the middle parts. These appear to be largely responsible for incoming of the ~30 Ma old zircon Himalayan Peak (HP), ~25 and 15 Ma Young Himalayan Peaks (YHP), respectively; the latter appears within the Lower Siwalik Subgroup (13–11 Ma). During the Lower Siwalik deposition, pre-Himalayan Peaks (PHP) gradually decreased with the intensification of the Himalayan events in source rocks. Despite the uninterrupted fluvial sedimentation in the Dagshai–Kasauli–Lower Siwalik sequences since 30 Ma, breaks of ~5 to 7 Myr in zircon FT ages reveal pulsative cooling and exhumation in well-identified source areas. Though cooling and exhumation of the Himalayan source rocks remained almost uniform during the Eocene, source heterogeneity is reflected in fluvial sedimentation since 37 Ma from Pakistan to Nepal in response to the India-Asia collision.

6.5.6.2 Arunachal Himalaya

Along the Kameng River (western Arunachal Pradesh), Miocene–Pliocene Siwalik Group sediments were mainly derived from the Higher Himalayan source rocks during the period 13–7 Ma and <2.6 Ma, while during ca. 7 and 3 Ma time span these were sourced from the Gangdese Batholith and ITSZ (far less negative εNd and εHf values) through the Brahmaputra River system (Chirouze et al. 2013). Detrital zircon fission-track data indicate that Himalayan-derived sediments had ~1.8 km/m.y. rates in the fastest-exhuming areas, which were constant since ca. 13 Ma.

6.6 Lesser Himalaya

Lying between the Sub-Himalayan (Outer Himalaya) and the Great Himalaya in the north, the Lesser Himalaya, also called as the Lower Himalaya (Himachal), is 60–80 km wide belt trending NW–SE to E–W and WSW–ENE for nearly 2500 km (Fig. 6.16). Most of the terrain is highly dissected by southerly-flowing high-energy rivers with elevation from 600 to 4500 m, including occasional snow covers. Some of the southward-flowing rivers are Ravi, Beas, Sutlej, Yamuna–Tons, Alaknanda–Bhagirathi–Ganga, Kali, Karnali, Gandaki, Kosi, Arun, Tista, Manas, Kameng, Subansiri, Siang, etc. along with their tributaries, and dissect the Lesser Himalaya into highly rugged and mountainous region.

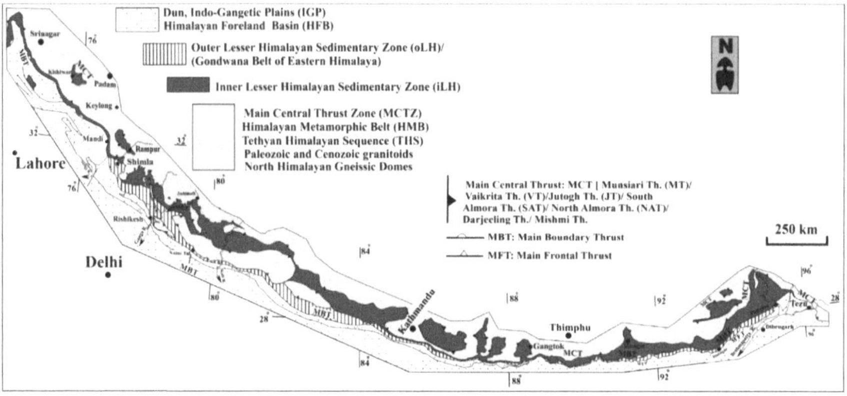

Fig. 6.16 Distribution of the Outer and Inner Lesser Himalayan Sedimentary Belts between Main Boundary Thrust (MBT) and the Munsiari Thrust (MT)/Main Central Thrust (MCT)

6.6.1 *Sub-divisions*

In general, three distinct belts characterize the Lesser Himalaya between the Main Boundary Thrust (MBT) and the Munsiari Thrust (MT)/Main Central Thrust (MCT), with the Lesser Himalayan Crystalline (LHC) nappe dividing the two belts in Uttarakhand–Nepal (Fig. 6.17; Jain 2020).

(i) Southernmost Neoproterozoic–Early Paleozoic Outer Lesser Himalayan (oLH) sedimentary belt encompassing the Shimla–Jaunsar–Blaini–Krol–Tal Groups and Subathu Formation between the MBT and Tons Thrust (TT)/North Almora Thrust (NAT) in the NW Himalaya, In the central sector and Sikkim–Bhutan–Arunachal Pradesh, it also comprises the Gondwana, Buxa and Manas Groups.

(ii) Northernmost Paleoproterozoic–Mesoproterozoic Inner Lesser Himalayan (iLH) sedimentary belt of the Shali–Rautgara–Gangolihat–Deoban–Berinag–Kuncha–Daling–Shumar Groups between the TT/NAT and MCT in the west (Valdiya 1980a, b; Jain et al. 2014). Besides the Shali Belt in contact with the Sub-Himalaya, the iLH is exposed beneath the overthrust belt of the Higher Himalayan Crystallines (HHC) in the Kishtwar and Kulu–Rampur windows in Himachal, Ghuttu–Garhwal windows in Uttarakhand, Arun window in Nepal, Ranjit window in Sikkim, Kuru Chu window in Bhutan, and Nacho-Menga windows in Arunachal.

(iii) The crystalline rocks of the Higher Himalaya thrust southwards and eroded into several remnants of synformal klippen of the Lesser Himalayan Crystalline (LHC) nappe (Garhwal–Ramgarh–Almora–Baijnath–Askot–Chiplakot and others) overriding the Proterozoic sedimentary belt of both the Inner and Outer Lesser Himalaya.

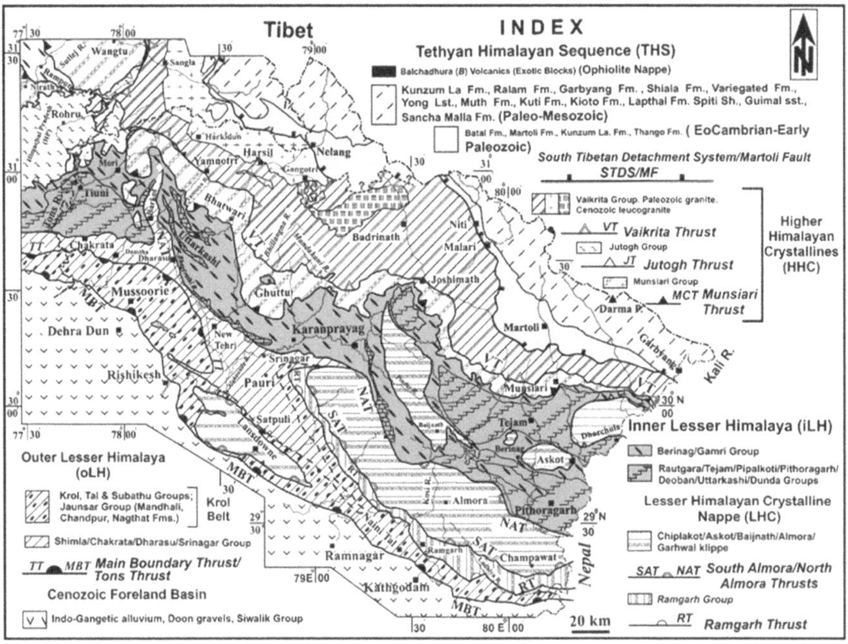

Fig. 6.17 Simplified geological map of the Uttarakhand Himalaya showing distribution of various litho-tectonic units including the Outer and Inner Lesser Himalayan (oLH, iLH) Sedimentary Belts. Compiled from various published sources. See Jain (2020) for details

6.6.2 *Outer Lesser Himalayan (oLH) Sedimentary Belt (NW Himalaya)*

The oLH Belt of the Neoproterozoic–Paleozoic–Mesozoic Shimla–Jaunsar and Blaini–Krol–Tal Groups is exposed between the MBT and Tons Thrust (TT)/North Almora Thrust (NAT). The oldest Shimla Group (≡Morar–Chakrata–Dharasu–Srinagar–Pauri Groups in different sectors) contains intercalated and alternating greywacke–slate–limestone–dolomite sequences at the base (Fig. 6.17; Table 6.1; Srikantia and Bhargava 1998). Valdiya (1970) interpreted the Simla Group of Himachal Pradesh and Garhwal as shaly flysch, deposited primarily by NNE-trending lateral turbidity currents originating from the Aravalli-Delhi System, while the overlying Blaini Boulder Beds were considered of this group as slide conglomerates from steep structural slopes toward the end of the Simla time. On the contrary, detailed sedimentological observations by Singh and Merajuddin (1978) indicated shelf-mud transition zone of a tidal flat complex for the deposition of the Simla Group.

Further south, within the Krol Belt (e.g., Krol Hill in Himachal Pradesh), the Neoproterozoic is represented by the Jaunsar Group comprising the Mandhali Formation (conglomerate-limestone-carbonaceous-slate-phyllite), the Chandpur Formation (phyllite-siltstone alternations-metavolcanics) and the Nagthat Formation

Table 6.1 Composite tectono-stratigraphy of the Lesser Himalayan (LH) sequence, Himachal-Uttarakhand

Belts		Group	Formation	Lithology	U–Pb Zr age (Ga)*	$\varepsilon_{Nd(t=0)}$
Garhwal Nappe		Amri/Bijni		Phy-sch-qz		
Garhwal Thrust						
Upper Cretaceous-Eocene			Subathu Fm	Ss-sh-ls		
			Singtali Ls	Arenool ls		
Unconformity (span: ~0.4 Ga)						
Cambrian		Tal Group	Upper Tal Fm	Qtz aren		
			Lower Tal m	Phos-Cht-Sh	0.525	−17.65
Neoproterozoic	Outer Lesser Himalaya (oLH)	Krol Group	Krol-E Fm	Gry Ls Frac		
			Krol-D Fm	Cht-Ls-Sh		
			Krol-C Fm	Bl Gr Ls, Frac		
			Krol-B Fm	Purp Gn Sh-Ls		
			Krol-A Fm	Dol Ls-Cht-Sl		
			Infra-Krol Fm	Sh-Sl-Qtz		
		Blain Group	Limestone-Sh Diamictite Quartzite Diamictite	Pk Microxln-Ls Gr Gn Diamict-Sh Qtzt Diamict	0.70	
		Jaunsar Group	Nagthat Fm	Wh Sst-Ark-Qtzt Cgl-Slt	0.82	−14 to −19
			Chandpur Fm	Gn-Qtzt-Phyl	0.88	
			Mandhali Fm	Qtzt-Ls-Mbl-Cgl-Slt-Phyl	0.95	

(continued)

Table 6.1 (continued)

Belts		Group	Formation	Lithology	U–Pb Zr age (Ga)*	$\varepsilon_{Nd(t=0)}$
		Simla/Chakrata/Dharasu/Srinagar/Pauri Group	Sanjaui Fm	Cgl-Sh-Qtzt-Gwke-Sh-Slt-St		
			Chaosa Fm	Gn-Purp sh-Slt-St-Gwke		
			Kunihar Fm	Ls-Dol-Sh-Slt		
			Basantpur Fm	Wh Qtz-Cgl-Sh-Slt-St-Ls		
Unconformity (span: ~1.0 Ga)						
Paleoproterozoic	Inner Lesser Himalaya (iLH)	Garhwal/Pithoragarh	Berinag Gr (Gamri, Rampur, Manikaran)	Pk-Wh Qz Aren-Qz Sch-Amph	2.0–1.7	−27.7 to −23.4
			Gangolihat/Deoban/Shali/Tatapani Gr	Dol-Ls-Stromlt Ls-Slt		
			Rautgara Fm	Gwke-Sl-Amph		
		Base not exposed				

[a]Maximum depositional age

Abbreviations: *Pk* pink; *Wh* white; *Gn* green; *Purp* purple; *Bl Gr* Bluish grey; *Sh* shale; *St* siltstone; *Sst* sandstone; *Ark* arkose; *Cgl* conglomerate; *Gwke* greywacke; *Diamict* diamictite; *Aren* arenite; *Qz* quartz; *Ls* limestone; *Dol* dolomite; *Mbl* marble; *Stromlt* stromatolite; *Areoo* arenaceous oolitic; *Microxln* microcrystalline; *Cht* chert; *Frac* fractured; *Phos* phosphatic; *Sl* slate; *Phy* phyllite; *Sch* schist; *Amph* amphibolite; *Fm* formation; *Gr* group (quart arenite–slate) in ascending order (Auden 1934). These rocks crop out in a series of doubly plunging synclines along nearly 280 km long belt between Solan in the NW and Nainital in the SE (Bhargava 1976; Shanker et al. 1997). In Kumaon, the oLH belt unconformably overlies the Bhowali Group containing quartzite–dolomitic limestone–slate–metabasalt–metarhyodacite.

The Nagthat siliciclastics reveal extensive development of multistoried medium-to thick-bedded sandstone, conglomerate and siltstone-shale, deposited under shoreface to proximal inner shelf tidal and stormy influence with coarsening upwards prograde stratigraphic sequence in gradual shallowing basin (Ghosh 1991). These contain sublithic arenite and subarkose with dominant sources from low to high grade metamorphic terrain and minor contribution from sedimentary and plutonic igneous rocks from the erosion of Aravalli-Delhi mountain chain and Bundelkhand granite-gneiss massif during Neoproterozoic.

The Nagthat Formation is unevenly covered by 2000 m thick glacial diamictite, siltstone and sandstone succession of the Blaini Group, representing the Marinoan glaciation of the Cryogenian period with at least two thick and regionally extensive diamictite units, separated by siliciclastics and argillites (Jain and Varadaraj 1978; Jain 1981). Clast composition variation, pebble imbrication and grain size variation indicate distinct control by glaciers, while carbonates were not exposed during sedimentation of the lower diamictite but were eroded off subsequently. Many clasts contain diagenetic and deformation features indicating an older orogenic cycle. A glaciogenic origin is supported by the presence of relatively abundant striated clasts and the local preservation of polished and striated pavement on underlying Simla Group clastics. The cap dolostone is characterized by lighter isotopes of 13 and 18 °C.

Nearly 400 m of shale, siltstone and minor sandstone overlie the Blaini diamictites as Neoproterozoic Infra-Krol Formation. These are followed upwards by a predominantly carbonate suite of the Krol Group. The Krol Sandstone as part of the Infra-Krol Formation is restricted to eastern and western ends of the E–W extending Krol Basin around Solan and Nainital. The overlying Krol Group is divided into Lower (Krol A), Middle (Krol B) and Upper (Krol C, D, E) (Auden 1934). This informal classification has survived all these years as the most suitable lithostratigraphic order of superposition. New stratigraphic names were proposed (Shanker and Mathur 1992) but did not find acceptance by other workers in this basin.

The Krol A formation is represented by argillaceous limestone interbedded with greenish grey calcareous shale. The Krol B is characteristically greyish-reddish shale with siltstone and thin layers of dolomite; the Krol C is bluish grey limestone and brecciated dolomite; the Krol D is dolomitic limestone, stromatolite-bearing cherty dolomite, shale, siltstone and thin layers of sandstone. The Krol E is thinly bedded limestone interlayered with calcareous shale, siltstone and dolomite. The entire Krol Group is traditionally interpreted as a pack of conformable lithologies (Auden 1934). However, detailed sequence stratigraphic studies identified several disconformable surfaces (Jiang et al. 2002) indicating interruptions in largely tidal to intertidal (Singh 1980) sedimentation in a shallow NW-sloping marine Krol Basin, with minor phases of deep-water environment (Fig. 6.18; Jiang et al. 2003).

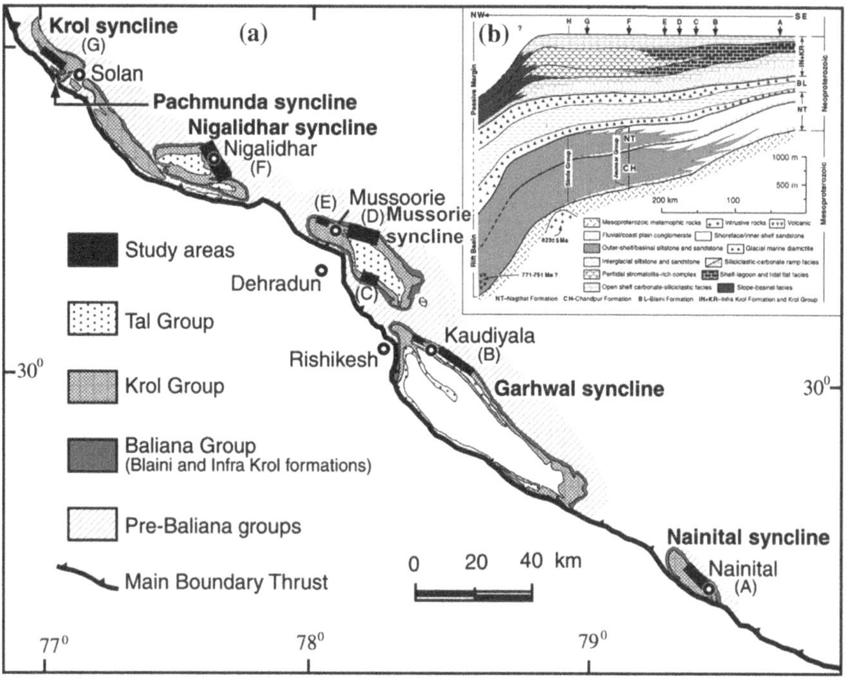

Fig. 6.18 **a** Simplified geological map of the Krol Belt in the Outer Lesser Himalaya (oLH) showing distribution of different synclines and measured sections. After Singh and Rai (1983). **b** Neoproterozoic stratigraphy and basin evolution of the Lesser Himalaya, northern India. Measured sections (A–G) located in various synclines of the Krol Belt. After Jiang et al. (2003)

At the end of the Krol E formation, carbonate deposition was interrupted by the development of ~150 m thick sequence in euxinic basin in which black shale, chert and organic matter rich phosphorites of the Lower Tal Formation were deposited. Hollow phosphatic tubes of small shelly fauna *Anabarites, Protohertzina,* and *Maldeotaia* of Tommotian age (Bhatt et al. 1985, 1989; Brasier and Singh 1987), pyrite-rich stromatolites (Singh and Rai 1977, 1983) and trilobite traces (Banerjee and Narain 1976; Rai and Singh 1983) dominated the depositional scene. This Lower Cambrian strata define the Proterozoic–Cambrian boundary (Banerjee et al. 1986, 1997; Shanker and Mathur 1992) in this part of the Himalaya. These rocks provide valuable information regarding the assumed existence of a stratified ocean during the terminal Neoproterozoic and beginning of earliest Cambrian (Nemakit-Daldynian). However, a few crucial Lower Cambrian stages are missing in this succession and, therefore, this boundary was not accepted as global Pc-C stratotype by the Cambrian Stratigraphic Subcommittee.

Up in the sequence, the Tal basin received about ~2500 m thick deposits of sandstone and shale under tidal to intertidal environments. Abundant animal trails, synsedimentary primary structures in sandstone and interbedded reddish brown shale represent the upper parts of the Lower Cambrian. Uppermost part of the Tal Group shows clean quartz arenite with sedimentary features indicating very shallow water coastal environment of lagoons, tidal flats and shoals (Singh 1979; Singh and Rai 1983). An unconformity at this level is marked by the deposition of Cretaceous Shelly limestone which, in turn, is overlain by the Paleocene Subathu Formation.

6.6.3 Outer Lesser Himalayan (oLH) Sedimentary Belt (Sikkim, Bhutan, Arunachal Himalaya)

The major lithotectonic units of the Lesser Himalaya in Bhutan are summarized in Table 6.2 (Bhargava 1995).

Table 6.2 Major lithotectonic units of the Eastern Bhutan

Tethyan Himalayan Sequence (THS)		Conglomerate, shale, siltstone, sandstone, limestone and volcanics
South Tibetan Detachment System (STDS)		
Great Himalayan Sequence (Higher Himalayan Crystallines-HHC)	Thimphu Group	Gneiss, migmatite, amphibolite and high grade metamorphics
Jaishidanda		Biotite-garnet-staurolite schist with granite gneiss
Main Central Thrust (MCT)/Jaishidanda Thrust		
Lesser Himalayan (LH) Sedimentary Belt		Shumar–Quartzite, phyllite, chlorite mica schist, limestone and tectonic slices of granite gneiss
		Shumar Thrust
		Buxa–Gritty feldspathic quartzite, limestone, dolomite, quartzite, conglomerate, slate, phyllite
		Buxa Thrust
		Duiri–Diamictite, phyllite, quartzite
		Damuda–Sandstone, siltstone, shale, coal beds
Main Boundary Thrust (MBT)		
Sub-Himalaya	Siwalik Group	Sandstone, siltstone, shale, clay, conglomerate
Recent Deposits		Sand, gravel and boulders, and deposits of the Brahmaputra Plains

6.6.3.1 Gondwana Group

The outermost LH Belt contains the Damuda Formation of the Gondwana Group of a lower succession of coal-bearing feldspathic sandstone, siltstone, carbonaceous shale, coal lenses/seams with plant fossils and an upper non-coal bearing succession, having characteristic lower Gondwana flora (Gokul 1983; Lakshminarayan and Singh 1995). In southeast the sequence occurs as an isolated and truncated narrow belt that extends into the Arunachal Pradesh. Like the other Gondwana sequences, Lower Permian marine fossils are associated with an intercalated succession of sandstone, siltstone and shale in upper parts of the Setikhola Formation (Joshi 1995).

6.6.3.2 Duiri Formation

Further to the north of the Gondwana sequence, the Duiri Formation is named after the Diu Ri valley, and mainly consists of quartzite and boulder slate (diamictite) (Long et al. 2011). Greyish boulder slate contains sub-rounded to rounded pebbles and boulders of quartzite, phyllite, slate, dolomite and gneiss in a sandy argillaceous slaty matrix, whereas pebbleless slate is carbonaceous and grey to dark grey.

Tectonically, the Diuri Formation is bounded by the Buxa Group in the north and the Damudas in the south and resembles the Blaini diamictite and Rangit Pebble slate n composition; the latter are of glacial origin and correlated with Permian Talchir Boulder Bed.

6.6.3.3 Buxa Group

Mallet (1875) grouped a thick sequence of dolomite, quartzite and shale into the 'Baxa Series' in western Duars after the famous Buxa Fort. The main repository of dolomite and limestone in Bhutan, the metasedimentary belt has been mapped along the entire foothills with thickness ranging between 1.5 and 6 km (Lahiri 1941). Separated from the Siwalik Group/Setikhol Formation by the Main Boundary Thrust (MBT), the group has been classified into four formations in ascending order: the Jainti Formation, the Manas Formation (including Thungsing Formation; cf. Jangpangi 1974), the Phuntsholing Formation and the Pangsari Formation (Tangri 1995). The upper contact of this group is defined by the Shumar Thrust.

The lowermost Jainti Formation (Member A) includes maroon-purple, green-grey phyllite, metasiltstone and quartzite, and is overlain by pink-white medium grained quartzite (Member B). The overlying Manas Formation contains intercalated sequence of white grey gritty quartzite, creamy dolomite, phyllitic slate and impersistent limestone, and is best exposed along the Manas River. The Phuntsholing Formation incorporates alternating sequence of white-grey, quartzite, phyllite and occasional limestone. The youngest Pangsari Formation contains white-cream arenaceous dolomite, phyllite and quartzite.

6.6.3.4 Thungsing Quartzite

A succession containing white, feldspathic, medium to coarse-grained quartzite, occasionally pebbly and conglomeratic have been included in the Thungsing Quartzite with minor slate and phyllite intercalations (Jangpangi 1974). This quartzite overlies the Buxa Group dolomite in Narphung region and is thrust over by the Shumar Nappe. As this sequence is also seen underlying the Diuri Boulder Slate, Jangpangi (1974) considered it coeval to the Gondwana, while Tangri and Pande (1995) opined it as a part of quartzite of the Manas Formation. Jangpangi (1974) has shown this quartzite separated from the Shumar quartzite on the western side by a cross fault and extended it into the Buxa-type dolomite and slate along the strike near Narphung. This quartzite is a part of the Thungsing Formation and constitutes subthrust part of the Shumar Thrust Sheet (Ray 1995).

Further east in Arunachal Pradesh, Immediately to the northwest of the Sub-Himalayan Siwalik Belt, the southern front of the Lesser Himalaya is composed of Permian Gondwana Group of the Outer Lesser Himalayan unit, and consists of carbonaceous shale, sandstone and a few coal beds belonging to the Garu (marine) and Khelong (fluvial) sequences (Acharya et al. 1975; Jain et al. 1974; Kumar 1997).

6.6.4 oLH Carbon and Oxygen Isotopes

Spatial and temporal variation of the carbon and oxygen isotopic composition in the Blaini dolomite (Etienne et al. 2011), Krol carbonate succession (Aharon et al. 1987; Kaufman et al. 2006) and Tal phosphorite (Banerjee et al 1986, 1997; Mazumdar and Banerjee 2001; Mazumdar et al. 1999) indicated great fluctuations. Kaufman et al. (2006) and Etienne et al. (2011) observed that the Blaini cap dolostone is depleted in both $\delta^{13}C$ and $\delta^{18}O$ isotopes, showing strong co-variation. $\delta^{13}C$ values range from -0.44 to -6.12, and $\delta^{18}O$ values vary between -5.49 and -13.27 with a pronounced negative excursion to $-5‰$ $\delta^{13}C$ followed by a recovery towards less negative values. In the Krol sequence, the $\delta^{13}C$ values mostly fall in negative field close to the sea water values but sharp excursions to positive values could be seen at the Krol B–Krol C boundary, top of the Krol C and along the Krol D and Krol E transition (Fig. 6.19; Kaufman et al. 2006). On the other hand, Krol E and lower Tal transition is characterized by siliceous sponge spicules from the Chert-Phosphate Member and a sharp change to positive $\delta^{13}C$ value, coinciding with the Pc-C boundary and carbonate-black shale transition (Banerjee et al 1986, 1997; Mazumdar and Banerjee 1998, 2001). Trace element and REE distributions of the Tal phosphorites gave deep insight into the Lower Cambrian phosphatic strata (Banerjee and McArthur 1989). These isotopic compositions of carbonate and organic carbon in the transition profile can be best interpreted as reflecting an oceanic anoxic event (OAE) prior to the phase of early Cambrian in shallow marine to restricted lagoonal environment where black shale and phosphorite could form (Mazumdar et al. 1999). In the Eastern Himalaya, the Buxa dolomite yielded $^{\delta 13}C$ values between $+3‰$ and $+6‰$,

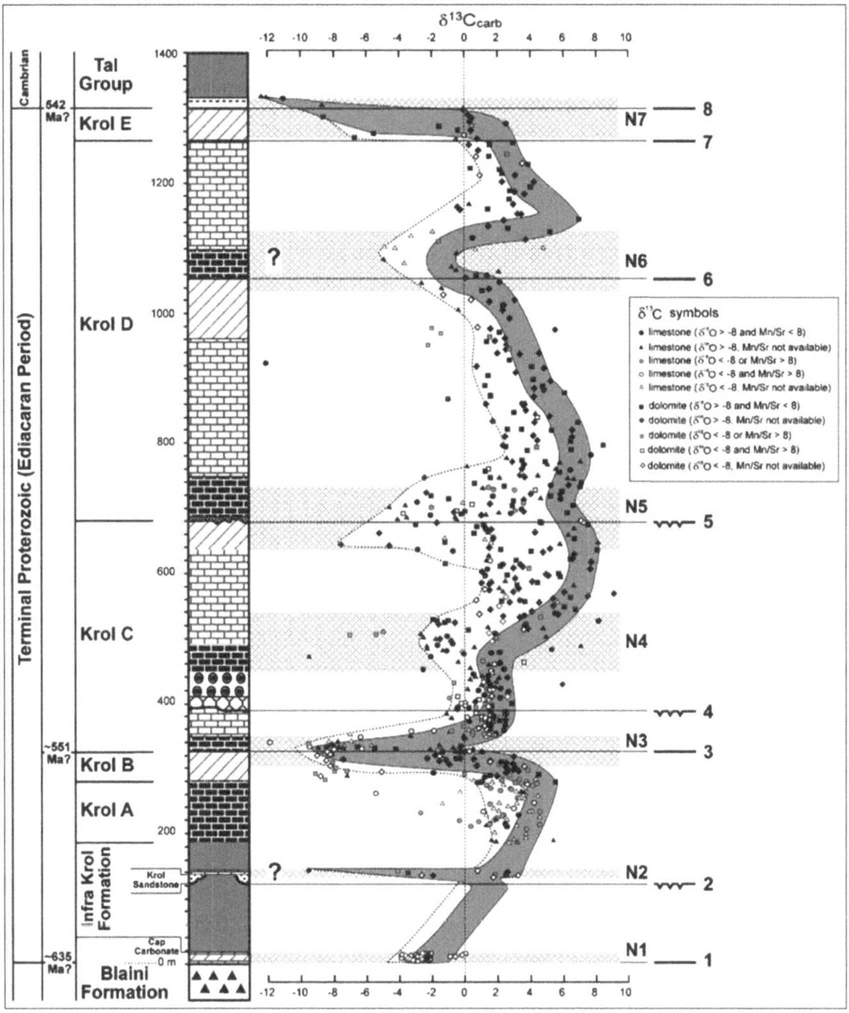

Fig. 6.19 Composite δ^{13}C curve for the terminal Neoproterozoic Ediacaran Period from the Krol Belt. After Kaufman et al. (2006)

suggesting deposition during an Early Cambrian positive δ^{13}C excursion (Long et al. 2011).

6.6.5 Lesser Himalayan Crystalline (LHC) Nappe

Synformal klippen of the Lesser Himalayan Crystalline (LHC) nappe (Garhwal, Ramgarh, Almora, Baijnath, Askot, Chiplakot, others) are thrust over the LH sedimentary belt (Valdiya 1980a, b). The lowermost Ramgarh Nappe of the Ramgarh–Debguru–Ulleri-type augen mylonite (ca. 1.85 Ga) is overlain by the Nathuakhan Formation containing quartzite-phyllite with 0.80 Ga youngest detrital zircons (Célérier et al. 2009; Mandal et al. 2015). The Almora Nappe overrides this unit with ca. 1.85 Ga mylonitized granite gneiss along the base, garnetiferous quartzite–schist alternations with 0.85–0.58 Ga youngest detrital zircons and 0.55 Ga intrusive granitoids (Trivedi et al. 1984).

6.6.6 Inner Lesser Himalayan (iLH) Sedimentary Belt

Lying between the North Almora Thrust (NAT) in the south and the Munsiari Thrust (MT)/Vaikrita Thrust (VT)/Main Central Thrust (MCT) in the north, vast Proterozoic sequence of the Inner Lesser Himalaya (iLH) extends as a linear belt between the Tons and Kali valleys. Because of its unfossiliferous character, facies variations and intricate tectonics, stratigraphic correlations are controversial. This belt extends both sides along the strike into Himachal, Jammu and Kashmir in the west and Nepal, Sikkim, Bhutan and Arunachal in the east.

In Jammu–Kashmir and southern Himachal Pradesh, outermost frontal Paleoproterozoic iLH Shali Belt (Srikantia 1977) contains quartzite–shale (Sundernagar Formation), mafic volcanics–slate (Mandi–Darla Volcanics) and quartzite–salt–marls–shale–dolomite (Shali Group). These are considered NW extension of the Rautgara of the Pithoragarh region. All over this belt, these rocks are intimately interbedded with vesicular basalts and tuffs, exposed from Mandi-Darla in Himachal, Alaknanada valley in Garhwal (the Rudraprayag volcanics) and Bhimtal in Kumaun.

This imbricated belt dips regionally NE and is thrust southwestward over the SH belt in the Punjab Re-entrant along the MBT (Table 6.3; Srikantia and Bhargava 1998). It contains the Tatapani limestone and Deoban Group carbonates within the Kulu-Rampur and other smaller windows along the Sutlej–Tons valleys beneath the Jutogh nappe and MCT zone (Fig. 6.17).

Further southeast in Uttarakhand, vast ~10 km thick extensive iLH Paleoproterozoic sedimentary sequence is developed beneath the overthrust metamorphic nappes. In eastern Kumaon, in immediate vicinity of the NAT, stratigraphically oldest and northerly-dipping Rautgara Formation starts with subgreywacke–slate–conglomerate–mafic flows (Valdiya 1980a, b). Although he postulated these sediments as deep sea turbidite, detailed analysis of sedimentary structures points to a shallow environment of deposition. This formation imperceptibly grades into the overlying Thalkedar Dolomite (stromatolite-bearing dolomitic limestone), the Sor Slate (green-brown-black slate), the Gangolihat Formation (another sequence of stromatolite-bearing

Table 6.3 Litho-tectonic sequence of the ILH belt, Eastern Kumaon

Tectonic unit	Formation	Lithology
	Askot/Dharamgarh/Baijnath Crystallines	
Inner	*Dharamgarh/Askot/Baijnath Thrust*	
	Berinag Formation	• Intercalated metavolcanics, chlorite schist, amphibolite • Sericite-bearing fawn-buff-white coarse to fine-grained quartz arenite and foliated quartzite; massive in character. Fuchsite-bearing quartzite
Lesser	Gangolihat Dolomite (Deoban Formation/Calc Zones of Tejam/Badolisera (Heim and Gansser 1939)	• Thalkedar Limestone: Massive blue limestone, dolomite and marls • Sor Slate: Interbedded grey-brown slate, marls and bedded limestone. Carbonaceous and pyritiferous grey dolomitic limestone • Blue-grey limestone • Deep purple dolomite with alternating chert • Blue-grey, cherty dolomite with stromatolites, intraformational conglomerate, lenticular magnesite schist and phyllite • Maroon slate and phyllite
Himalaya	Rautgara Formation	• Deep chocolate maroon slate-phyllite • Olive green grey slate-quartzite • Black ash grey fine-grained dense quartzite-black slate. Brown greywacke. Cross-beds, ripple marks
	Ramgarh Thrust/North Ramgarh Thrust	
Almora Nappe	Ramgarh Mylonite	• Mylonite, ultralmylonite. Porphyritic granitoids as protolith. Mylonitized garnet mica schist
	North Almora Thrust	

carbonates with shale, phosphorite and magnesite, dolomite and dolomitic limestone), finally into the Berinag/Garhwal Groups (mature quartz arenite–mafic flows) (Valdiya 1962; Kumar 1980). Likewise, the Deoban Limestone (dolostone) displays prolific development of algal biostromes. Distribution of primary sedimentary structures in the Rautgara, Gangolihat and Berinag Formations suggest subtidal, supratidal and coastal beach complex depositional environment, respectively, with northerly paleocurrents (Tewari 1994). Similar stromatolite-bearing carbonates with shale, phosphorite and magnesite are known as the Tejam Dolomite (Misra and Banerjee 1968). These carbonates are correlatable with the Shali Formation in Himachal, the Jammu Limestone and the Sirban Limestone in the Jammu-Kashmir sector. In Himachal, the Berinag-type quartz arenite and metavolcanics are identified as the Rampur Quartzite in the Sutlej Valley, the Manikaran Quartzite in the Beas Valley, the Shalimar Quartzite in the Chenab Valley and the Sauna Formation in Jammu. In the Eastern Himalaya, the Daling Group (Auden 1935) in Darjeeling area, Samchi–Shumar Group in Bhutan and Miri–Siang Group in Arunachal Pradesh correlate with this group of rocks.

6.6.6.1 iLH in Sikkim, Bhutan, Arunachal Himalaya

(i) **Shumar Group**: Nautiyal et al. (1964) named a group of metasedimentary rocks consisting of phyllite, slate, limestone, flaggy and massive quartzite in Nangkur-Shumar area of eastern Bhutan as the Shumar 'Series' immediately to the south of the crystalline thrust sheet. The Shumar Formation (Jangpangi 1974) extends throughout the length of the Lesser Himalaya in Bhutan and was classified as the Daling Shumar Group (Gansser 1983) or the Daling of the Darjeeling Hills (Sen Gupta and Raina 1978). This sequence has been found to be synonymous with the Samchi 'Series' (Phuntsholing Formation) (Nautiyal et al. 1964; Guha Sarkar and Singh Gaur 1963), exposed between the Manas and the Jaldhaka rivers.

In the type section of Shumar-Kharungla area or N–S trending Kuru half window of the eastern Bhutan, the Shumar Formation can broadly be divided into three parts: (a) lower part dominated by grey to greenish phyllite and muscovite-chlorite schist with intercalation of quartzite and occasional calcareous zones, (b) middle part mainly represented by white to greyish white, fine grained, thin to medium bedded and well jointed and fractured quartzite with minor mica schist intercalations and basic sills, and (c) upper part mainly comprising the mica schist with quartzite and thinly bedded grey to greyish white coloured limestone and marble beds. Concordant mylonitized granite gneiss sheets are present at different levels within the sequence, which have been considered as thrust slices of the Precambrian basement (Ray et al. 1989). At places, granite gneiss has intrusive contact with quartzite, and attained tectonized lensoidal form (Dasgupta 1995).

(ii) **Jaishidanda Formation**: An almost persistent package of metasedimentary rocks is exposed on the subthrust side of the Main Central Thrust (MCT) and is overlain by the Thimphu Group. This sequence has been classified as the Jaishidanda Formation in central and eastern Bhutan (Dasgupta 1995). It is also exposed in different windows in western Bhutan along the Wang Chu and Amo Chu valleys. This formation contains medium to coarse grained, light to dark grey and well foliated garnetiferous mica schist, intercalations of micaceous quartzite, carbonate bands, occasional lenses of carbonaceous schist and concordant sheet-like bodies of mylonitized granite gneiss. It is highly sheared to phyllonite with development of ductile shear fabric.

Precise contact relationship of this sequence with the overlying Thimphu Group and the underlying Shumar Formation is controversial, hence this litho-package is considered as a part of the overlying Thimphu Group (Jangpangi 1974; Sen Gupta and Raina 1978; Trichal and Jayaram 1989), while others included this sequence within the underlying Shumar Formation (Ray et al. 1989).

There is a sharp change in the grade of metamorphism across the Jaishidanda-Thimphu contact where its mineral assemblage reveals considerably lower grade of metamorphism in comparison to the immediately overlying Thimphu Group. All these evidences point to a thrust nature of the contact between the Jaishidanda Formation and the Thimphu Group.

In recent works by Davidson et al. (1997) and Daniel et al. (2003), the trace of the MCT separates thick Shumar quartzite and phyllite/schist of the Jaishidanda Formation from the biotite-rich schist/gneiss of the Great Himalayan sequence, thus corresponding to both protolith and metamorphic boundary.

In Arunachal, the iLH belt is either thrust over the Siwalik or the Gondwana Belt along the NNW-dipping Main Boundary Thrust (MBT) and, in turn, overthrust by the Khetbari-Miri Group-Abor Volcanics of the Lesser Himalaya (Jain and Balasubramanian 1981). The Lesser Himalayan metasediments are separated from the Himalayan Metamorphic Belt (HMB), locally called as the Siang Group, along the folded Main Central Thrust (MCT). The southernmost exposure of the MCT is locally named as the Tamen Thrust. In the upper reaches of the Subansiri River, the folded HMB is eroded off to expose the underlying Inner Lesser Himalayan sequence in the Menga and Nacho Windows (Singh and Jain 2007). The iLH Belt is folded into NNW-trending very large Siang Dome whose northeastern limb extends into the Mishmi Hills.

6.7 Himalayan Metamorphic Belt (HMB)

The Himalayan metamorphism represents one of the most prominent manifestations of the Himalayan orogeny along the entire strike length of the mountain belt and, therefore, is most well studied. Core of the Himalayan orogen encompasses highest peaks rising to more than 7000 m and is mostly comprised of high-grade metamorphic complex with a cover of the Tethyan Himalayan Sequence (THS). It constitutes the "Central Crystalline Axis" of Himadri/Great Himalaya belonging to the Himalayan Metamorphic Belt (HMB). Folded thrust nappe of the HMB extensively covers the LH sedimentary belt and is exposed into the (i) Lesser Himalayan Crystallines (LHC), (ii) Higher Himalayan Crystallines (HHC), and the (iii) Tso Morari Crystallines (TMC) from south to the north (Figs. 6.20 and 6.21). An extensional Zanskar Shear Zone (ZSZ)/South Tibetan Detachment System (STDS) separate this belt from the THS.

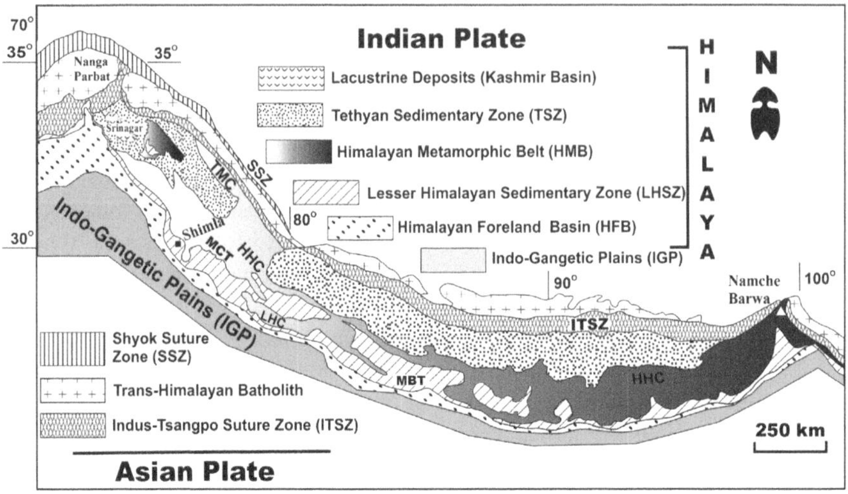

Fig. 6.20 Distribution of the Himalayan Metamorphic Belt (HMB) in the Himalaya, where it is overlain by the Tethyan Himalayan Sequence (THS) in the north and thrust southwards along the Main Central Thrust (MCT)

6.7.1 LHC Belt

Numerous klippen of low-medium grade metamorphosed LHC—the Salkhala Nappe, Garhwal and Almora Nappe etc. contain highly deformed alternating phyllite and schist. The lowermost Ramgarh Nappe of mylonite orthogneiss (~ca. 1.85 Ga), belonging to the MCTZ, is overlain by quartzite-phyllite of the Nathuakhan Formation, having ~0.80 Ga detrital zircons (Mandal et al. 2015). The Almora Nappe overrides this unit with mylonitized granite gneiss (~1.85 Ga) along the base and garnetiferous quartzite-schist alternations with 0.85–0.58 Ga youngest detrital zircons and 0.55 Ga intrusive granitoids (Fig. 6.21; Mandal et al. 2015).

As a part of the Almora Nappe, regional metamorphism within interbanded pelite-psammite sequence of the Askot Klippe in the north exhibit chlorite-biotite-muscovite-garnet-quartz, biotite-muscovite-garnet-quartz, biotite-muscovite-garnet-plagioclase-quartz and K-feldspar-sillimanite-garnet-biotite-plagioclase ± cordierite-quartz assemblages, which attained granulite facies metamorphism (Das et al. 2019). Disequilibrium textures along garnet-biotite contact indicate beginning of granulite facies metamorphism, which is substantiated by estimated temperatures of ~776 °C and 6.6 kbar for garnet rims.

Within the LHC Belt of the Alaknanda valley near Joshimath, Spencer et al. (2012) recorded P-T conditions of ~5 kbar and ~550 °C below the MCT, while a sudden jump takes place across it to ~14 kbar and 850 °C.

An integrated P-T path of the LH belt comprises garnet-kyanite grade metapelites along the footwall of the MCT from the Arunachal Himalaya (Goswami-Banerjee et al. 2014). A near isobaric peak Barrovian metamorphism at P ~8 to 9 kbar with

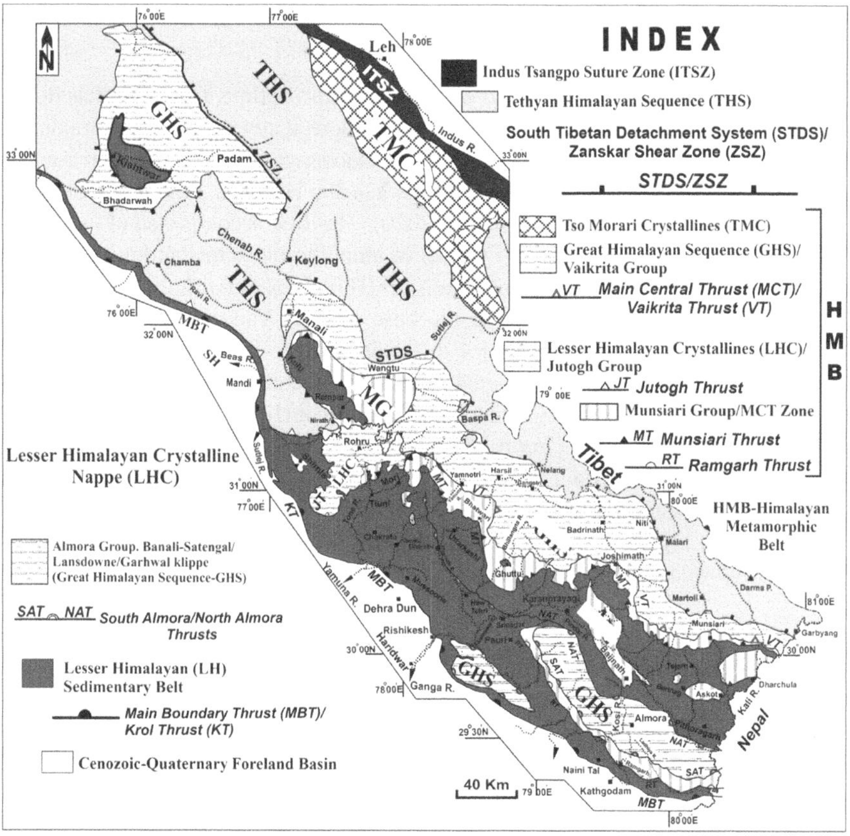

Fig. 6.21 Simplified geological map of the NW Himalaya showing distribution of metamorphic belts. Tso Morari Crystallines (TMC). Higher Himalayan Crystallines (HHC) with Munsiari Group at the base, and Great Himalayan Sequence (GHS) at the top, and Lesser Himalayan Crystalline (LHC) Belt

increase in T-maximum from ~560 °C in metagranite through ~590 to 600 °C in lower and middle garnet-zone to ~600 to 630 °C in upper garnet- and kyanite-zone rocks. The metamorphic sequence of the LH Belt additionally records a pre-Barrovian near isobaric thermal gradient in mid crust at ~6 kbar from ~515 °C in middle garnet zone to ~560 to 580 °C in the upper garnet- and kyanite zone, adjoining the MCT. A clockwise P-T path is proposed from the studied section of the Arunachal GHS with burial to lower crustal depths, prograde heating leading to kyanite-facies partial melting at T ~750 to 800 °C and steep, near isothermal decompression to mid-crustal depths and subsequent post-decompressional cooling.

6.7.2 HHC Belt

The HHC belt has been investigated along numerous sections from Zanskar in the west to Arunachal Himalaya in extreme east where it reveals classic Himalayan Inverted Metamorphism (HIM) (Jain and Manickavasagam 1993; Manickavasagam et al. 1999; Verma 1989; Harrison et al. 1998; Yin and Harrison 2000; Hodges 2000; Kohn 2014; Clarke et al. 2016; Pant et al. 2020). This belt is comprised of two litho-tectonic units: (i) a lower Munsiari Group of amphibolite, schist and augen gneiss (augen mylonite), delimited by the Munsiari Thrust (sensu stricto Main Central Thrust—Heim and Gansser 1939) at the base and the Vaikrita Thrust at the top (Fig. 6.21). Other common names for this unit are Lesser Himalayan Crystalline Sequence (LHCS), Main Central Thrust Zone (MCTz) etc. (ii) The upper main group of High Himalaya Crystalline Sequence (HHCS)/Great Himalayan Sequence (GHS) (also known as the Vaikrita Group in Uttarakhand), bounded by the Main Central Thrust (sensu lato)/Vaikrita Thrust at the base and the South Tibetan Detachment System (STDS) near the top (Jain et al. 2014)

There is vast literature available on metamorphic petrology of the HHC; some of these are reviewed here.

6.7.2.1 Deformation and Metamorphism

The HHC Belt in the NW Himalaya have suffered at least four superposed recognizable deformation (D1–D4) and associated metamorphic events, resulting into its complex history. These deformation and metamorphic events are summarized below.

(i) **D1 deformation and M1 metamorphism**: During this deformation, isolated, tight and appressed "flame" folds (F1) develop on lithological banding/metamorphic layering (S0; Fig. 6.22a). The axial-plane foliation S1 of these folds parallels this layering. During this deformation, most of the minerals crystallizing along the S1 foliation are chlorite, biotite, and quartz; hence it appears to have been subjected to lower greenschist facies M1 metamorphism.

(ii) **D2 deformation and M2 metamorphism**: D2 deformation was most widespread and produced penetrative foliation (S2) paralleling axial surfaces of close to isoclinal, reclined to recumbent F2 folds on S1 foliation or lithological and/or metamorphic layering (Fig. 6.22b). S2 foliation is essentially a ductile shear fabric, having regional dips between 20° and 40° NE. In schist the foliation is composed of S-surfaces that are progressively deflected and become subparallel to the ductile C-foliation of high strain on a scale of millimetres (Jain and Manickavasagam 1993). In various deformed granitoids, this typically resembles the S-C mylonitic foliation (Fig. 6.22c), and ubiquitously records top-to-southwest sense of ductile shearing, which is also evident from asymmetric quartz and feldspar augen, pressure shadows, rotational fabric within porphyroblasts and other shear criteria. Stretching mineral lineation (L2), coaxial to the F2 folds, plunges NE/SW almost down-dip, and was

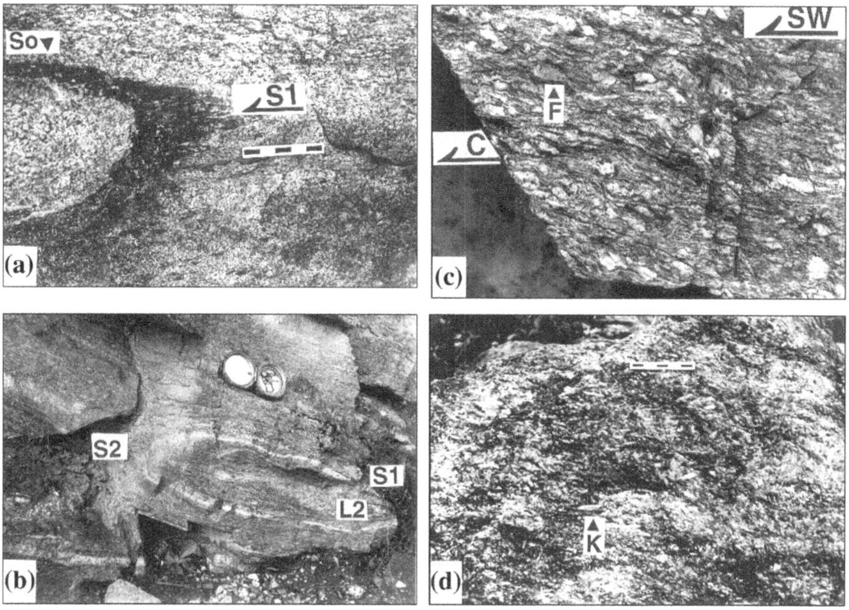

Fig. 6.22 Deformational structures from the Higher Himalayan Crystallines in western Himalaya. **a** F1 'flame-type' isoclinal fold having penetrative axial plane foliation S1 on lithological-metamorphic banding S0 in schist and gneiss. Location: Parkachik along the Suru valley, western Zanskar. Scale is 5 cm. **b** Reclined isoclinal F2 fold on S1 foliation in alternating schist and gneiss having rounded hinge zone and axial plane foliation S2. Note coaxial L2 mineral lineation on limbs. Location: About 5 km southeast from Sangla along the Baspa valley. **c** S-C foliation in augen gneiss, having asymmetric σ-type feldspar megacrysts (F) with top-to-southwest vergence. Location is near Karchham, Sutlej valley, Himachal. Scale is 2 cm. **d** Northeast-plunging L2 mineral lineation of long slender kyanite (K) blades in staurolite-kyanite schist. Location: Kundel along the Bhot Nala, southeast Kashmir. Scale is 5 cm

developed on composite S1-S2 and/C-foliation due to preferred orientation of mica, chlorite, tourmaline, amphibole, staurolite, kyanite, sillimanite, quartz, and feldspar (Fig. 6.22d). Many of these minerals grew along S and C surfaces during the M2 metamorphic event and were rotated due to the D2 ductile phase. Some of these minerals continued their growths on syntectonically grown mineral phases or developed separately over the foliation in the final stages of the M2 metamorphism in late to post-D2 deformation phase.

(iii) **D3 deformation and M3 metamorphism**: D3 phase produced isoclinal to tight F3a and superposed F3b folds that plunge gently either NW/SE. The F3b folds are open to closed and are inclined. Development of chlorite at the expense of garnet and biotite, and D3-related foliation have been observed during this deformation.

(iv) **D4 deformation and M4 metamorphism**: Discrete kinks, tensional gashes, and brittle shear zones represent the youngest D4 deformation event and occasional mineral growths.

Structures associated with extensional tectonics and ensuing exhumation of the HHC belt were produced during the D2 and subsequent deformation phases and are more penetratively developed toward its northern and upper boundary with the THS along the STDS. These are commonly associated with numerous melt-enhanced leucogranite veinlets in migmatite of higher metamorphic grades.

6.7.2.2 Textures and Index Minerals

The HHC metamorphics are foliated and gneissose from garnet through staurolite-kyanite to sillimanite—K-feldspar grades, and rarely granulite, having garnet, staurolite, kyanite and sillimanite porphyroblastic growths. Muscovite, biotite, quartz, and/or chlorite and chloritoid define the main foliation, which are also sometimes porphyroblastic in character. Staurolite is subordinate in the middle structural levels in southeast Kashmir and western Garhwal, possibly either due to restricted bulk composition or increased temperature. Chlorite is occasionally retrograde after garnet, staurolite, and biotite in staurolite- kyanite schist. Minor zircon, apatite, and ilmenite are present in the matrix, and as inclusions in biotite and garnet. The mineral assemblages include the following:

(i) Garnet grade: Chlorite–biotite–muscovite–chloritoid–garnet–plagioclase–quartz
(ii) Staurolite-kyanite grade: Biotite–muscovite–garnet–kyanite–quartz ± plagioclase ± K-feldspar ± staurolite ± chlorite
(iii) Sillimanite–muscovite grade: Garnet–biotite–muscovite–kyanite–sillimanite ± plagioclase–K-feldspar–quartz
(iv) Sillimanite–K-feldspar grade: Garnet–biotite–sillimanite–K-feldspar–plagioclase–quartz ± cordierite.

Garnet. Syntectonic garnet is very common in the HHC belt; though small post-tectonic idioblastic garnet is occasionally seen southeast Kashmir. Syntectonic garnet is subidioblastic to xenoblastic, and sometimes occurs as elongate clusters parallel to the foliation. It is poikiloblastic, having numerous inclusions of quartz, biotite, muscovite, carbon, ilmenite, rutile, plagioclase, kyanite, staurolite, and sillimanite in decreasing order of abundance. Some garnet is overgrown by an inclusion-free rim as post-tectonic to main foliation (Fig. 6.23a). Most of the syntectonic garnet deforms the main foliation, and is characterized by rotational fabric, indicated by numerous inclusions and asymmetrical pressure shadows (Fig. 6.23a); this suggests that it grew during the main metamorphism M2 and D2 deformation. Garnet is generally almandine-rich, having varying spessartite, pyrope, and grossular components.
Staurolite. Syntectonic staurolite poikiloblasts contain elongate quartz, and opaque and mica inclusions, that are either rotated like garnet or reveal typical sieve texture with Si ll Se (Fig. 6.23b). A few small subidioblastic staurolite grains, devoid of inclusions, are present in some samples, indicating their growth during late D2 deformation.

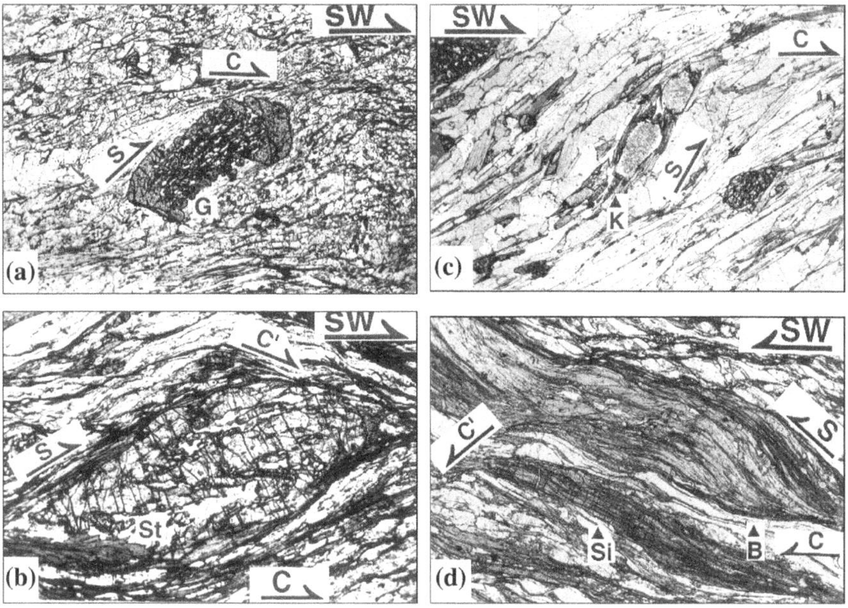

Fig. 6.23 Textural characteristics of the HHC, NW Himalaya. **a** Syntectonic garnet (G), riddled with quartz and biotite inclusions during its growth in M2 metamorphic episode along the S-C fabric. Note inclusion-free rim of subidioblastic garnet Location: About 5 km south of Kishtwar, SE Kashmir. **b** Growth of syntectonic staurolite (St) porphyroblast with quartz-biotite inclusions during M2 metamorphism along S-C foliation. Note distinct C′-shear band affecting staurolite tail and quartz-mica groundmass. Location: Junkal Gompa near Zanskar Shear Zone (ZSZ). **c** S-C shear fabric controlling porphyroblastic kyanite (K)-biotite. Near Masu, SE Kashmir. **d** Epitiaxally grown sillimanite (Si) over biotite (B) along top-to-southwest S-C fabric in sillimanite-biotite schist near ZSZ along Mulung Tokpo. Note SW-dipping extensional C′-shear band. Scale: Photographs 9.5 mm in width

Aluminosilicates. In the staurolite-kyanite grade, kyanite poikiloblasts grew mainly parallel to both S- and C-foliations and contain quartz inclusions having Si ll Se relationships (Fig. 6.23c). In higher grades, relict kyanite has reaction rims of fibrolite and/or cordierite. Some kyanite was folded by D3 deformation. Sillimanite commonly occurs along the main foliation and C-shear foliation and C′ shear bands (Fig. 6.23d), suggesting growth even during the intense ductile shearing of the D2 deformation. Fibrolite and/or sillimanite replaced biotite, muscovite, and kyanite during this deformation. Sillimanite–K-feldspar are broken and have corroded rims. Sillimanite needles are common, and fibrous sillimanite grows epitaxially in muscovite and biotite.

Feldspar. K-feldspar and plagioclase occur mostly as subidioblastic grains, having quartz and kyanite inclusions. Perthite is common in the higher grades. Plagioclase is more albitic (X_{An} = 0.12–0.29), but a few samples show an enriched anorthite content.

Chlorite. Chlorite occurs both in garnet and staurolite-kyanite grades; in the latter grade, chlorite is Mg-rich (Mg/Mg + Fe $\geq$ 0.5), and is associated with staurolite close to the hanging wall of the Vaikrita thrust in Sutlej valley.

6.7.2.3 Some Important Cross-Sections

(i) **Zanskar**: Along the Chenab River in Padar-Zanskar region, the HHC belt is thrust over the Lesser Himalayan Kishtwar window along its southwestern margin, while the Zanskar Shear Zone (ZSZ) marks its northeastern margin with the overlying Paleozoic-Mesozoic THS (Fig. 6.24; Herren 1987; Patel et al. 1993; Manickavasagam et al. 1999; Finch et al. 2014; Stüber et al. 2014). The Kishtwar window contains black carbonaceous phyllite, quartzite, sericite schist, metavolcanics, and granite gneiss. The overthrust HHC belt incorporates largely pelites, bands of marble and calc-silicate, quartzite, amphibolite-greenschist bands; pelites contain garnetiferous mica schist, staurolite-kyanite schist/gneiss, and sillimanite schist/gneiss. Migmatite and two-mica-bearing leucogranite are associated with highest grade metamorphic rocks. It is also intruded by Cambro-Ordovician granitoids that are now exposed as large doubly-plunging domes such as Sankoo, Shafat, Chisoti, and Bhuzun (Honegger et al. 1982; Kundig 1989).

In southeast Kashmir along the Chenab Valley, garnet grade within ~500 m of the MCT footwall zone of the iLH records rim 6.5 kbar P and 550 °C T, while hanging wall of the MCT in the staurolite-kyanite grade records garnet core T of 500–550 °C and P of 8.5 kbar; rim data indicate *T* of 600–650 °C and *P* of 8–9 kbar (Jain and Manickavasagam 1993; Manickavasagam et al. 1999). An increase of nearly 100 °C rise in T is seen with no significant change in P from the core. In sillimanite-muscovite grade, an increase in core T of about 170 °C is seen with no significant change in P, but rim data reveal a reduction of about 50 °C and 1–2 kbar. Garnet cores in sillimanite–K-feldspar grade reveal peak T and P of 780 °C and 10 kbar. However, for sillimanite-muscovite and sillimanite–K-feldspar grades, the rim data reveal a reduction of about 50 °C/1–2 kbar, and 100 °C and 4–6 kbar from core data, respectively. However, migmatite–leucogranite zone having sillimanite-bearing assemblages yield much reduced P-T values, possibly due to anatexis and melting; garnet is broken and shows resorbed margins. The Crystalline belt along the Padar–Zanskar section records a nearly uniform rim T for all the metamorphic grades and significant reduction in *P* from lower to higher grade rocks.

(ii) **Sutlej section**: Along the Beas, Parbati and Sutlej valleys in Himachal, the HHC Belt is thrust southwest over the LH Larji–Kulu–Rampur (LKR) Window along the folded Munsiari–Vaikrita (MCT)–Jutogh Thrust and their splays (Fig. 6.25). The Kulu–Rampur window is a NW–SE trending linear belt of the Rampur Volcanics, Manikaran Quartzite, carbonaceous phyllite, and concordant 1.8 Ga Bandal granitoid, belonging to the Rampur Group (Sharma 1977; Vannay et al. 2004; Caddick et al. 2007). The window is overridden by a thin sheet of the

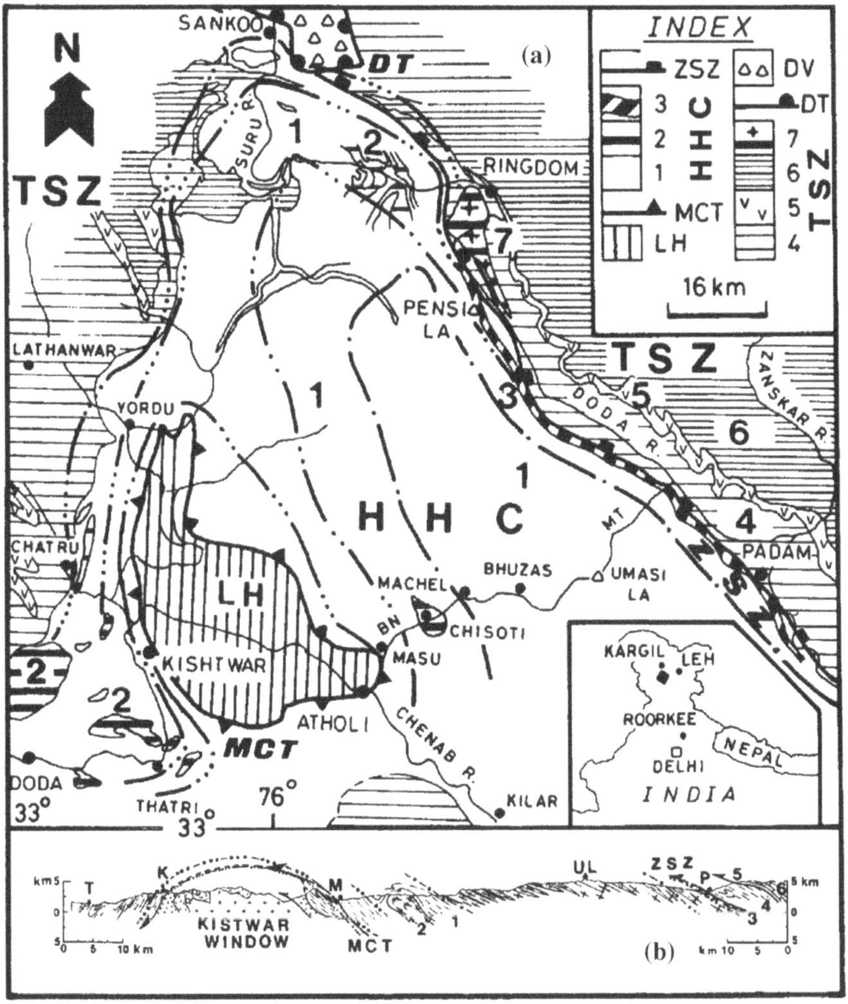

Fig. 6.24 **a** Simplified geological map of the HHC belt along Chenab River–Doda River, showing tectonic boundaries and metamorphic isograds. Lesser Himalayan (LH) Proterozoic sequence (Kishtwar Window). Higher Himalayan Crystalline (HHC) Belt: 1—schist, gneiss and migmatite, 2—granitoid intrusions, 3—leucogranite. Tethyan Himalayan Sequence/sedimentary zone (THS/TSZ): 4—Haimanta Group, 5—Phe volcanics, 6—Lilang Group, 7—granite. Dras Volcanics (DV). Main Central Thrust (MCT). Zanskar Shear Zone (ZSZ). Dras Thrust (DT). Bhot Nala (BN). Mulung Tokpo (MT). Dash-dot lines denote isograd boundaries: 4 dots–garnet, 3 dots–staurolite-kyanite, 2 dots–sillimanite-muscovite, 1 dot–sillimanite-K-feldspar. **b** Simplified geologic cross section along Thatri (T), Kishtwar (K), Masu (M), Umasi La (UL) and Padam (P), showing main structures and metamorphic isograds (dash-dot lines). After Jain and Manickavasagam (1993) and other published work

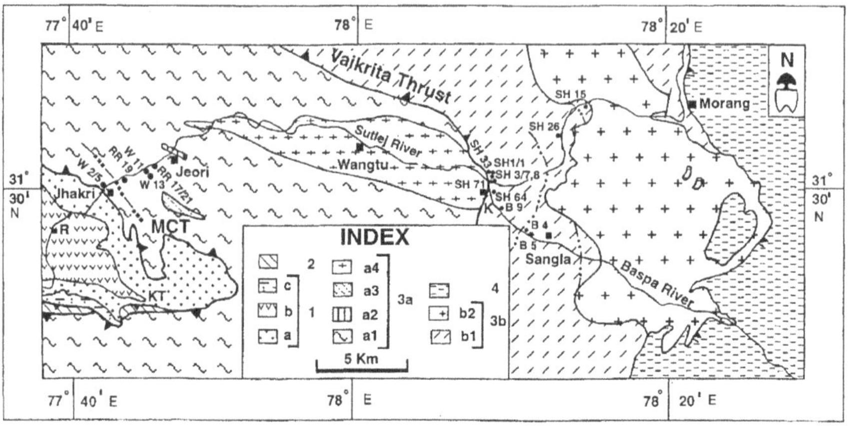

Fig. 6.25 Geologic map of the Sutlej-Baspa valleys, Himachal Pradesh. 1—Lesser Himalaya Rampur Group—(a) Manikaran Quartzite; (b) Rampur Volcanics; (c) carbonaceous phyllite. Kulu–Bajura–Munsiari nappe: 2. Augen mylonite. 3 (a) Jeori Group–staurolite/garnetiferous schist, banded biotite gneiss, augen gneiss (a1), amphibolite (a2), quartz mica schist (a3), Wangtu granite gneiss/granite/mylonite (a4). Vaikrita Group: 3 (b) Karchham Group–garnet/staurolite/kyanite/sillimanite schist/gneiss, calc-silicate, augen gneiss, and migmatite (b1), Akpa leucogranite (b2). Tethyan Himalayan Sequence (THS) (4) Abbreviations: KT—Kulu Thrust. R—Rampur; K—Karchham. Metamorphic isograd boundaries: 2 dots–sillimanite-muscovite grade; 3 dots–staurolite-kyanite grade. MCT hanging wall is in garnet–staurolite-kyanite transition zone. The Lesser Himalayan and Tethys Himalayan sediments are within biotite grade. Locations of samples used for pressure-temperature data are also indicated. MCT is written as the Munsiari Thrust, but the Vaikrita Thrust is actually MCT *sensu lato* Map after Singh (1993), and Manickavasagam et al. (1999)

dismembered Kulu–Bajura nappe (Frank et al. 1977) of highly mylonitized augen mylonite (≡Ramgarh mylonite) between the Ravi and the Sutlej valleys. In the Sutlej valley, the HHC Belt incorporates pelites, psammites, amphibolite and deformed 1.8-2.0 Ga Wangtu granite gneiss/mylonite (Singh 1993). This sequence is overridden by garnetiferous mica schist, staurolite-kyanite schist/gneiss, sillimanite schist/gneiss, and leucogranite along the Vaikrita Thrust (MCT).

Using overprinted contact metamorphic assemblages on the Paleoproterozoic Jeori–Wangtu granite gneiss/mylonite, Pant et al. (2006) demonstrated presence of pre-1800 Ma low-grade regional metamorphism in the Jeori–Wangtu sequence, while a sericite schist records a Paleoproterozoic soil horizon along the basement-cover contact (Bhargava et al. 2011a, b). Along the Sutlej valley, Manickavasagam et al. (1999) recorded garnet rim data between 490 and 500 °C for the garnet to staurolite-kyanite transition zone, and 570 and 590 °C and ~8 kbar for staurolite-kyanite grade, exposed between Jhakri and Wangtu in the lower structural levels on footwall of the Vaikrita Thrust. However, above hanging wall of the Vaikrita thrust to the STDS, the HHC records an increasing T of 575–620 °C and P of ~7 kbar in staurolite-kyanite grade, whereas sillimanite-muscovite grade exhibits 740–800 °C and ~9 kbar (see also Vannay et al. 2004; Caddick et al. 2007).

(iii) **Bhagirathi valley**: In the Garhwal Himalaya, the Munsiari Thrust separates the Lesser Himalayan sequence from the overthrust HHC Belt, which is subdivided into two distinct packages: (i) lower belt of Munsiari/Bhatwari Group containing quartzite, amphibolite, calc-silicate, garnet-biotite schist, and augen gneiss/mylonite on footwall of the Vaikrita Thrust (MCT), while its hanging wall contains staurolite-kyanite and sillimanite-bearing schist and/or gneiss, biotite granite gneiss, and Gangotri granite (Fig. 6.26).

On the hanging wall of the Munsiari Thrust along the Bhagirathi valley, Manickavasagam et al. (1999) noted rim P-T results as ~5.75 kbar and ~550 °C for the garnet grade in pelitic schist. Above the Vaikrita Thrust on its hanging wall, garnet cores record 500–545 °C in staurolite-kyanite grade and ~750 °C for sillimanite-muscovite grade in the GHS, garnet rim data reveal 8–9 kbar

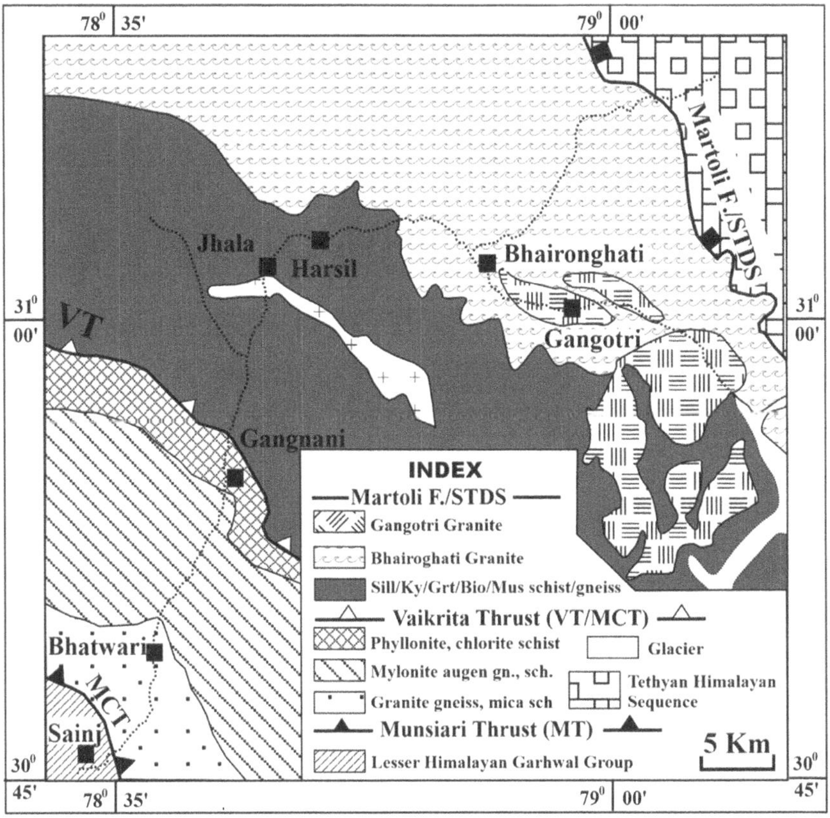

Fig. 6.26 Geologic map of the Bhagirathi valley, western Garhwal. After Manickavasagam et al. (1999) and Pêcher and Scaillet (1989)

and 600–650 °C for the staurolite-kyanite grade; and 9–11 kbar and ~700 °C for the sillimanite-muscovite grade (see also Metcalfe 1993).

(iv) **Alaknanda-Dhauli Ganga Valleys**: Along the Alaknanda valley, the Inner Lesser Himalayan (iLH) Sedimentary Belt is thrust southwards by the Munsiari Group along the Munsiari Thrust (MT), the earlier known Main central Thrust (MCT) (Fig. 6.27; Jain et al. 2014). The Munsiari Group, also known as the Lesser Himalayan Crystalline (LHC) Sequence/Main Central Thrust Zone (MCTZ), has a thrust contact with the structurally overlying Great Himalayan Sequence (GHS). It records first appearance of kyanite needles, and chemically homogeneous non-rotational garnet porphyroblast near the base (Thakur et al. 2015). Garnet from pelites yield 6.3–7.5 kbar and 550–582 °C in the MCTZ, and 8.0–10.0 kbar and 610–650 °C in basal part of the HHCS. The

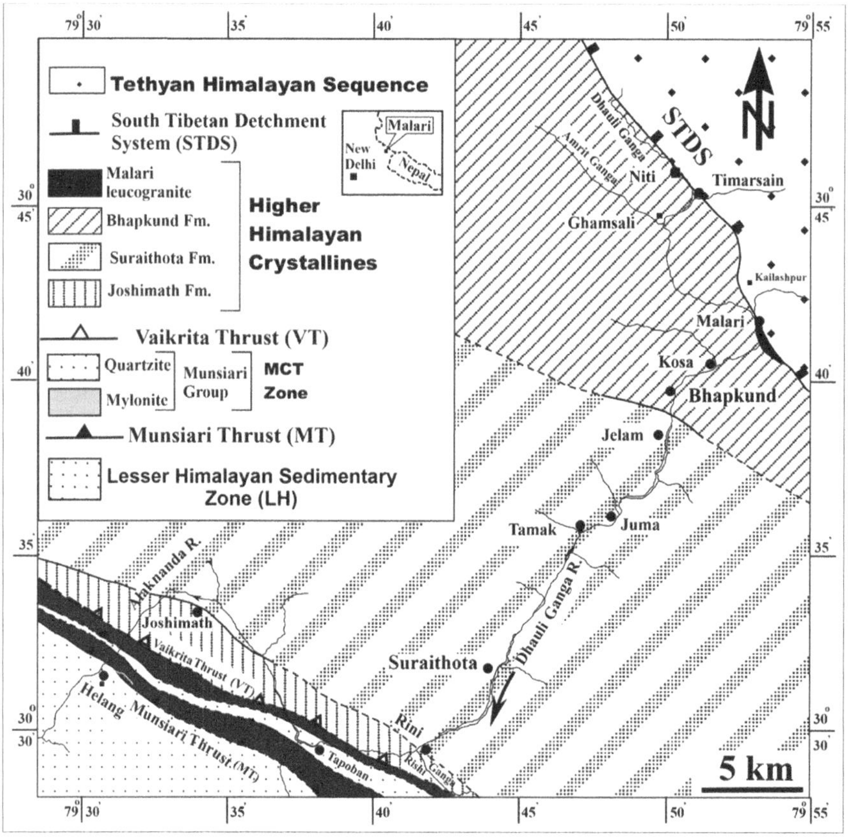

Fig. 6.27 Simplified geological map of the Alaknanda-Dhauli Ganga valleys showing distribution of major tectonic boundaries, e.g. the Munsiari Thrust (MT), the Vaikrita Thrust (VT) (now considered as the main Central Thrust–MCT) and the South Tibetan Detachment System (STDS). Modified after Jain et al. (2014) and Shreshtha et al. (2015)

results indicate continuity in the *P-T* field gradient across the contact between the MCTZ and GHS. The MCTZ shows an inverted metamorphic sequence from biotite to garnet zones in lower parts, while basal part of the GHS is in the kyanite zone with a continuous inverted metamorphic sequence. Earlier, Spencer et al. (2012) recorded dramatic P-T increase across the LHCS from ~5 kbar and ~550 °C to ~14 kbar and ~850 °C at ~3 km above the MCT in the GHS, though pressures then decrease up section to ~8 kbar in the GHS at nearly constant ~850 °C near the STD.

(v) **Darjeeling–Sikkim Himalaya**: One of the well-investigated regions of the Himalayan metamorphic belts is in Darjeeling–Sikkim, where progressive increase of both temperature and pressure is recorded with increasing grade and structural levels through the Lesser Himalaya, the MCTZ and the HHCS (Fig. 6.28; Neogi et al. 1998; Ganguly et al. 2000; Dasgupta et al. 2004). Within the Lesser Himalaya near Darjeeling, inverted Barrovian sequence in southern parts of the Lesser Himalaya (Darjeeling Hills) exhibits three deformation episodes D1–D3 and associated metamorphic phases M1a, M1b and M2 in which regional syntectonic low grade M1a–D1 phases and regional static post-tectonic M1b phase between D1 and D2 represent the prograde phases, while retrograde M2 phase took place between D2 and D3 episodes (Prakash et al. 2017). Following peak metamorphism at 5.8 kbar and 590 °C, the LH metamorphics underwent clockwise P-T-t path due to decompression of the crust (Prakash and Tewari 2016). The MCT zone near Darjeeling revealed polyphase deformation and Barrovian metamorphism with peak T of 570 and 780 °C in garnet and staurolite zones, respectively while peak P varied from 5.0 to 7.5 kbar (Tewari and Prakash 2016). In the overthrust HHC, migmatite has characteristic biotite ± garnet ± sillimanite-plagioclase-K-feldspar-quartz assemblage with evidence of extensive partial melting through biotite dehydration reaction (Tewari and Prakash 2017). Migmatite experienced peak P-T at 7.2 ± 0.5 kbar and 775 ± 20 °C, respectively with timing of crustal melting ~21 Ma.

In the MCT zone of upper parts of the Teesta River valley, Mukhopadhyay et al. (2017) opined that deformation mechanism differs from that of the underlying LH and overlying HHC sequence, with peak metamorphism of 750–800 °C before intense faulting at 20 Ma (450–700 °C) and isolated 11 Ma brittle phase.

Ganguly et al. (2000) calculated peak metamorphic conditions from the HHC sequence in the Sikkim–Darjeeling section as ~10.4 kbar, 800 °C temperature and extremely rapid (~15 mm/year) exhumation up to the depth of ~15 km, followed by a much slower process ~2 mm/year, up to at ~5 km depth. It is proposed that change of exhumation rate might reflect a process of tectonic thinning, followed by erosion and/or horizontal flow at shallow depth and suggests that the HHC exhumed from a depth of ~34 km within ~8 Ma.

Using in situ Th–Pb monazite ages, Mottram et al. (2014a, b) observed that peak P-T metamorphic conditions reached earliest in structurally highest part of inverted GHS above the MCT between ~37 and 16 Ma in southerly

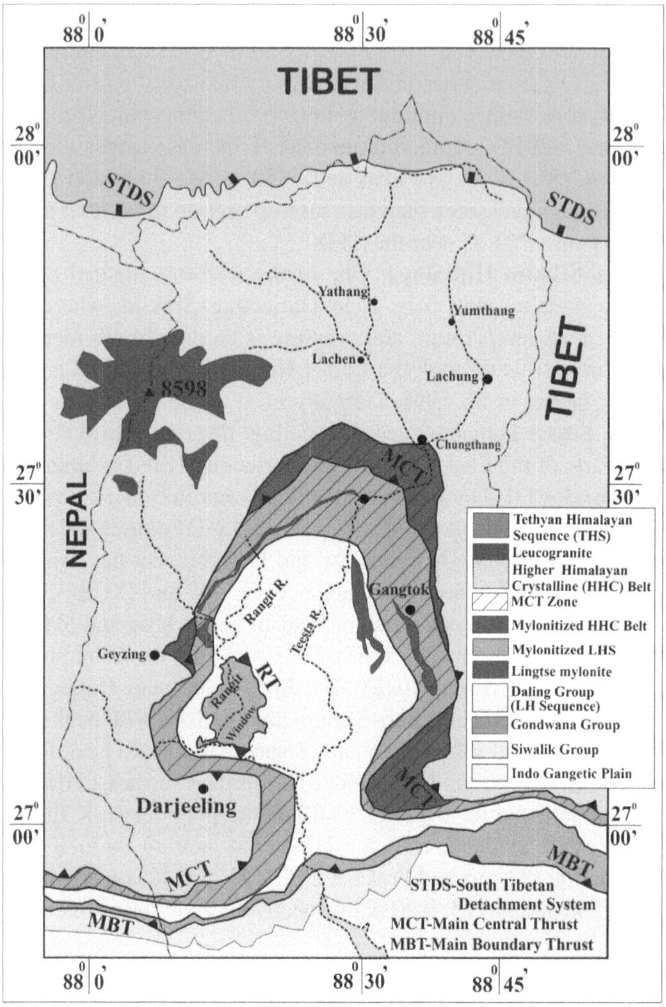

Fig. 6.28 Simplified geological map of the Darjeeling-Sikkim Hills adopted from various published sources. Main Central Thrust Zone (MCTZ) after Kellett et al. (2013) and Mottram et al. (2014a, b). Metamorphic isograds trend almost parallel to the MCT

leading-edge and between ~37 and 14.5 Ma in northern rear-edge of the MCT zone at peak P-T conditions of 10 kbar and ~790 °C. Likewise, upper LH footwall rocks attained P-T conditions of 9 kbar and ~655 °C between ~21 and 16 Ma in south transect and ~14.5 to 12 Ma in north. Lower LH footwall reached peak P-T of 8.5 kbar and ~580 °C at ~16 Ma in the south, and 9–10 Ma in north. Partial melting of the GHS hanging wall at ~23 to 19.5 Ma overlaps prograde metamorphism ~21 Ma in the LHS footwall. Progressive downwards-penetrating deformation and accretion of original LHS footwall material to the GHS hanging wall over a period of ~5 Ma is thus established, which may have occurred several times during the prolonged ductile evolution of the thrust along LHS–GHS protolith boundary (Mottram et al. 2014a, b).

In Sikkim Himalaya, LH protoliths (detrital zircon ages >1.8 Ga and εNd(0) of −27.7 to −23.4) are metamorphosed into continuous and systematic metamorphic field gradient of ~60 to 70 °C $kbar^{-1}$, having peak P-T increasing from ~5 kbar, 500 °C in garnet zone to ~8 kbar, 700 °C in sillimanite–muscovite zone with clockwise P-T paths with tight 'hairpin' shape with prograde slope of 20–30 °C $kbar^{-1}$ (Chakraborty et al. 2016). Garnet Lu–Hf chronometry indicates its growth from 16.8 to 11–10.5 Ma in sillimanite–K-feldspar to garnet zones, respectively, and peak metamorphism in all grades between 13 and 11 Ma simultaneously. Duration of melt formation and segregation is of the order of a few 0.1 Ma. Different protoliths of the GHS (youngest detrital zircons of 800 Ma) have undergone extensive migmatization in higher grades of ~8 to 12 kbar, 750–850 °C in at least two blocks with distinct, complex and multistage P-T-t histories with an older age of >20 Ma. These have different clockwise P-T path with isothermal decompression and near isobaric cooling (Chakraborty et al. 2016), possibly due to both channel flow at higher temperatures and critical taper/wedge tectonics at lower temperatures.

(vi) **Arunachal Himalaya**: Lithologically, the HHC belt in the Arunachal Himalaya is mainly comprised of high-grade sillimanite-bearing migmatitic gneiss, equivalent to those described from similar structural levels along orogenic strike (Yin et al. 2006, 2009; Warren et al. 2014; Goswami-Banerjee et al. 2014).

Metasediments of the Lumla Formation (LF) are exposed in a thrust-bound window through GHS (Yin et al. 2009). To the north of LF, garnet grade schist and gneiss are in contact with garnet free-sillimanite gneisses along the 'Zimithang Thrust'. This thrust is correlated with the Kakhtang thrust in Bhutan by many workers (Grujic et al. 1996; Yin et al. 2006, 2009). Metamorphic conditions along the Zimithang Thrust reveal peak temperature as ~525 to 650 °C in the hanging and footwalls (Warren et al. 2014). $^{40}Ar/^{39}Ar$ muscovite data from samples from either side of the thrust yield ages of ~7 Ma. Monazite grains in biotite–sillimanite gneiss in the hanging wall yielded ages of 16 ± 0.2–12.7 ± 0.4 Ma, while those on footwall yielded an older monazite ages of 27.3 ± 0.6–17.1 ± 0.2 (Wallis et al. 2014; Warren et al. 2014).

In eastern Arunachal Pradesh along the Siyom Valley, the LH sequence displays telescoped inverted greenschist–upper amphibolite facies pile, which is tectonically overlain by the GHS along the MCT by decreasing granulite to lower amphibolite facies; metamorphic isograds trend parallel to contemporaneous penetrative NW-dipping S2 foliation (Clarke et al. 2016). On footwall of the MCT, growths of garnet, staurolite and kyanite commenced at >8 kbar and ~550 °C, before syn- to post-S2 heating to ~640 °C at ~8.5 kbar probably at c. 18.5 Ma. However, on its hanging wall kyanite-garnet migmatite records peak conditions of ~10 kbar and ~750 °C and c. 21.5 Ma. Slip on the SE—propagating thrust was contemporaneous with early

Miocene metamorphism, which has initially controlled uplift of granulite to mid-crustal levels between 22 and 19 Ma, thermal relaxation within a disrupted LHS metamorphic profile inducing a post-S2 thermal peak in lower grade footwall rocks.

Near identical GHS metamorphic P-T paths were previously established from the adjoining Bhutan (Daniel et al. 2003) and Sikkim Himalaya (Ganguly et al. 2000; Harris et al. 2004; Neogi et al. 1998; Sorcar et al. 2014).

To summarize, the Himalayan metamorphism is characterized by

(i) Himalayan Inverted Metamorphism (HIM), both in the footwall and hanging wall of the Main Central Thrust (MCT).
(ii) It displays clockwise metamorphic path typical of underthrust block of continental crust thickened by converging continental plates (Fig. 6.29), with peak index metamorphic minerals (e.g. garnet, staurolite, kyanite, sillimanite) progressively growing along typical ductile shear fabric.
(iii) From the available dataset, Kohn (2014) has established that temperatures show sharp and continuous increase across the MCT from T ~550 °C in the LHS to T = 700–800 °C in the GHS, with common occurrence of kyanite isograd near the MCT. The belt experiences muscovite dehydration melting at ≥700 °C in the upper parts. The steepest temperature gradients are documented in Bhutan from central Kumaon to Langtang. Pressures show a steep increase in P in most transects with largest and most consistent step in the Kumaon to Langtang sector. Combined P-T conditions reveal that most rocks plot within kyanite stability field but at structurally high levels of the GHS, high T and low P increasingly prevail.

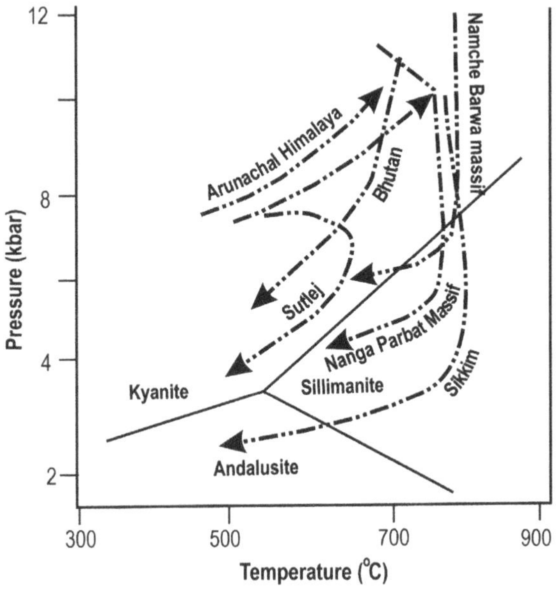

Fig. 6.29 Pressure-temperature paths across the Higher Himalayan Crystallines (HHC) Belt. Adopted from Pant et al. (2020)

(iv) Diachroneity of peak metamorphism and differential exhumation of the GHS with a 10 Ma age gap with an upper older slab exhuming along South Tibet Detachment (STD) at 27–25 Ma and lower slab along MCT at ~16 Ma (Wang et al. 2015). It is also evident that the exhumation rates along the Himalayan strike lengths are also unequal, being much higher within two Himalayan syntaxes (e.g. Zhang et al. 2015; King et al. 2016; Lang et al. 2016; Butler 2018). Thus, evolving character of the Himalayan metamorphism is well reflected in variable but younger dates from ~24 Ma onwards.

6.7.3 *Tso Morari Crystalline Belt (UHP Metamorphism)*

Eclogite was first reported from the inaccessible Rupshu district of Jammu and Kashmir (Berthelsen 1953), followed by their detailed regional geologic description (Srikantia and Bhargava 1978). Subsequently, the UHP rocks were observed from Kaghan, about 200 km west of Tso Morari (Pognante and Spencer 1991), adjacent Neelum valley, and later from Ama Drime, Nepal (Groppo et al. 2007). These UHPM belts have wide implications for angle of subduction, P-T-t path of the subducted block and recovery (Kaneko et al. 2003; Leech et al. 2005; Epard and Steck 2008; Singh et al. 2013).

In the NW Himalaya, the TMC is comprised of ~100 × 50 km long and NW–SE trending dome of quartzo-feldspathic gneiss, metamorphics and Paleozoic granitoids (Puga, Tanglang La, Polokongka La/Rupshu Granite, respectively), having dispersed small eclogite and garnet amphibolite bodies/lenses (Fig. 6.30; Thakur 1993; Jain et al. 2003; Leech et al. 2005). It is a UHP eclogite-gneiss-greenschist complex in southeastern Ladakh beneath the THS cover to the south of the ITSZ. Presence of coesite-bearing eclogite within the TMC and Kaghan (Pakistan) provides undisputed evidence that leading edge of the ICL subducted beneath the ITSZ to ~120 km depth in Early Eocene (Leech et al. 2005; Guillot et al. 2008).

The TMC attained peak P-T conditions for the UHP eclogite-gneiss at 750–850 °C and 2.7–3.9 GPa at ~55 Ma (de Sigoyer et al. 2000), and retrogressed to HP eclogite-facies (600 °C, 2.0 GPa), amphibolite-facies (600 °C, 1.3 GPa), greenschist-facies (350 °C, 0.4 GPa), and finally exhumed to the surface (Guillot et al. 1997). U–Pb SHRIMP zircon dates from the TMC pinpoint precise ages of peak UHP metamorphism (53.3 ± 0.7 Ma), HP metamorphism (50.0 ± 0.6 Ma) and amphibolite metamorphism (47.5 ± 0.5 Ma). ~30 Ma age of $^{40}Ar/^{39}Ar$ muscovite and biotite records last stage greenschist metamorphism (Leech et al. 2005, 2007). Present dataset reveals that final exhumation of the Tso Morari Complex, including the eclogite, was almost simultaneously with the HHC belt (Liou et al. 2004).

In contrast to the western Himalaya, the Arun valley eclogites reflect much younger age of UHP metamorphism at 20.7 ± 0.4 Ma in hanging wall of the MCT in immediate vicinity that may reflect either different origins or substantial time lags

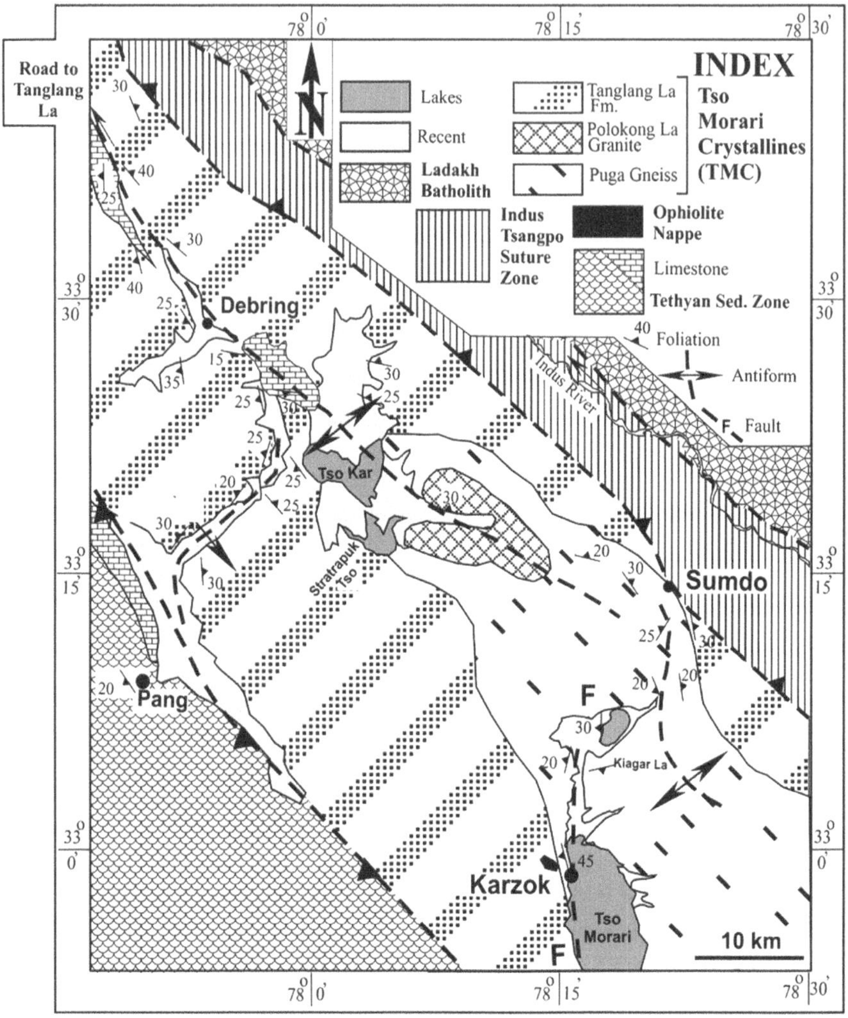

Fig. 6.30 Geological map of the Tso Morari Crystalline (TMC). Modified after Jain et al. (2003), and Jain and Singh (2009)

in tectonics along strike (Corrie et al. 2010). These have undergone three metamorphic episodes: (i) peak eclogite-facies metamorphism at ~670 °C and ≥15 kbar at 23–16 Ma, (ii) peak-T granulite episode at ~780 °C and 12 kbar, and (iii) late-stage amphibolite-facies metamorphism at ~675 °C and 6 kbar at ~14 Ma. Three models are postulated to explain observed P-T-t evolution: (i) the Main Himalayan Thrust (MHT) cuts deeper at Arun than elsewhere, (ii) along-strike diachroneity due to which subduction, slab breakoff, and ascent of India's leading edge occurred ~50 Ma in the west and ~25 Ma in Nepal, and still later in the east, and (iii) repeated loss of mantle lithosphere, first by slab breakoff at ~50 Ma, and again by delamination at ~25 Ma.

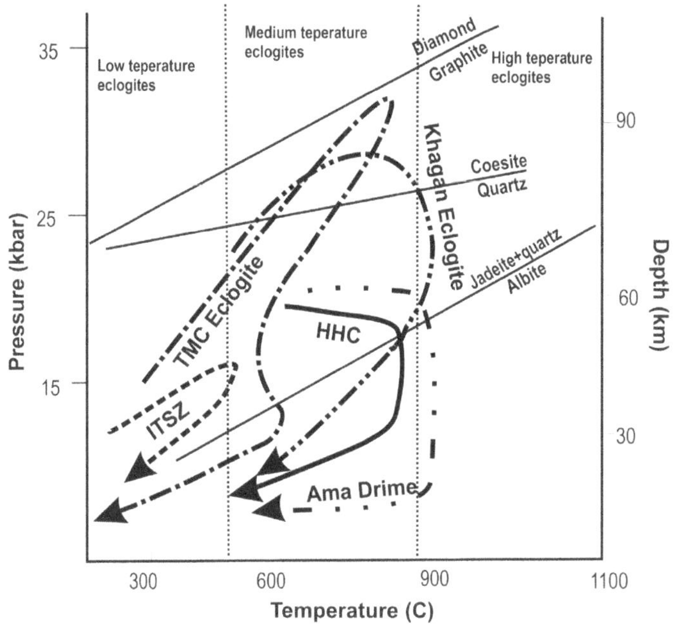

Fig. 6.31 Pressure-temperature paths from the Ultra-high pressure (UHP) rocks from the Himalaya. Adopted from Pant et al. (2020)

Eclogites from the Ama Drime Range further north of the Arun Valley localities are strongly overprinted by later metamorphism, where Groppo et al. (2007) proposed four metamorphic phases: (i) an initial eclogitic metamorphism (M1) at ~15 kbar and ~580 °C, (ii) an overprinted peak-T granulite-facies event (M2) at 8–10 kbar and ~750 °C, (iii) low-P granulitic event M3 at ~4 kbar and 750 °C, and (iv) a final cooling stage (M4) at 4 kbar and ~700 °C. U–Pb SHRIMP ages from low-U rims of zircon are 13–14 Ma during the low-P granulitic event during M3 metamorphism (Groppo et al. 2007). Cottle et al. (2009) interpreted monazite and xenotime ages of 13.2 ± 1.4 Ma as the timing of peak M2 granulite metamorphism, while Li et al. (2003) obtained zircon 206Pb/238U ages of 17.6 ± 0.3 Ma to represent granulite metamorphism; all these ages reflect the granulite-facies overprint, not the age of eclogitization.

Figure 6.31 summarizes the P-T paths from various eclogite bodies from the Himalaya.

6.8 Tethyan Himalayan Sequence (THS)

The Tethyan Himalayan Sequence (THS) is typically exposed beyond the Great Himalaya on northern passive margin of the Paleoproterozoic–Neoproterozoic Indian

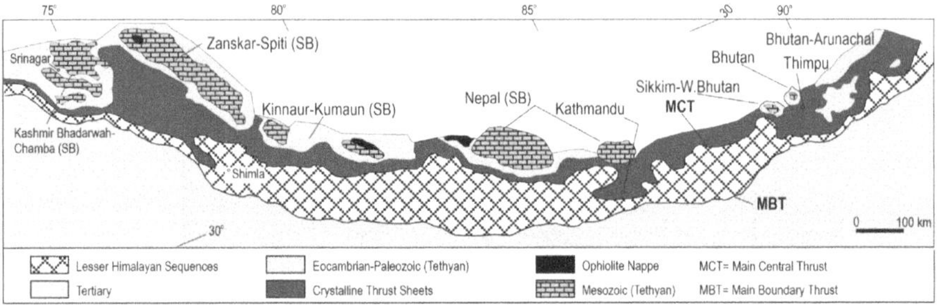

Fig. 6.32 Simplified geological map of the Himalaya showing the different sub-basins in the Tethyan domain. After Bhargava and Singh (2020)

Plate during Cambrian to Paleogene/Lower Eocene and represents a vast sedimentary ~10 km thick pile, deposited in the vast Tethyan Ocean (Fig. 6.32; Bhargava 2008a, b). Paleo–Mesoproterozoic sedimentary sequence commenced with eruptions of the Khewra Traps (Salt Range) and Singhi bi-modal Volcanics (Bhutan) possibly in a rift basin to the north representing the Tethys Himalaya. It rests over the Higher Himalayan Crystalline (HHC) Belt and is exposed in several disconnected basin traditionally identified by stratigraphic entities like Salkhala, Vaikrita, Kathmandu and Thimphu groups (Fig. 6.32).

Views about the stages of opening of this basin and spatial relationship with the Lesser Himalaya have been interpreted differently from time to time. Significant work on the Tethyan succession was by done by Middlemiss (1909, 1910, 1911) and Nakazawa et al. (1975) in Kashmir, Griesbach (1889), Stoliczka (1865), Hayden (1904), Srikantia (1981), Gaetani et al. (1986), Garzanti et al. (1998), Bhargava and Bassi (1998), Myrow et al. (2018), Hughes (2016) in Spiti, Heim and Gansser (1939), Valdiya and Gupta (1972), Banerjee (1974) and Kumar et al. (1972, 1977) in Kumaun and Garhwal, Bodenhausen et al. (1964), Fuchs (1967), Hagen (1968) and Bordet et al (1971) in Nepal and Gansser (1983) and Bhargava (1995) in Bhutan.

Since the THS is also observed within the Lesser Himalayan domain, Myrow et al. (2003) and Bhargava and Singh (2020) summarized three models to explains the relationship of the Tethyan basin with the Lesser Himalayan basin (Fig. 6.33). First model proposed a lofty rock barrier to separate two markedly different lithological assemblages and treated them as representing two separate tectonic domains. This old model still has few subscribers (Saxena, 1971; Bhargava 2008a, b, 2011). The second and the currently followed model envisages one single large basin in which both the Lesser as well as the Tethyan sequences were deposited 100 km apart (Brookfield 1993; Myrow et al. 2003, 2009, 2015; Hughes et al. 2005; Hughes 2016). Most workers now follow this model but it cannot explain the reason for sharp contrast in the facies of the Lower Cambrian rocks of the Lesser Himalaya with those of the Tethyan realm. The third model was proposed by DeCelles et al. (2000) in which the two basins must have formed at two different locations and brought together through large scale thrusting along the Main Central Thrust.

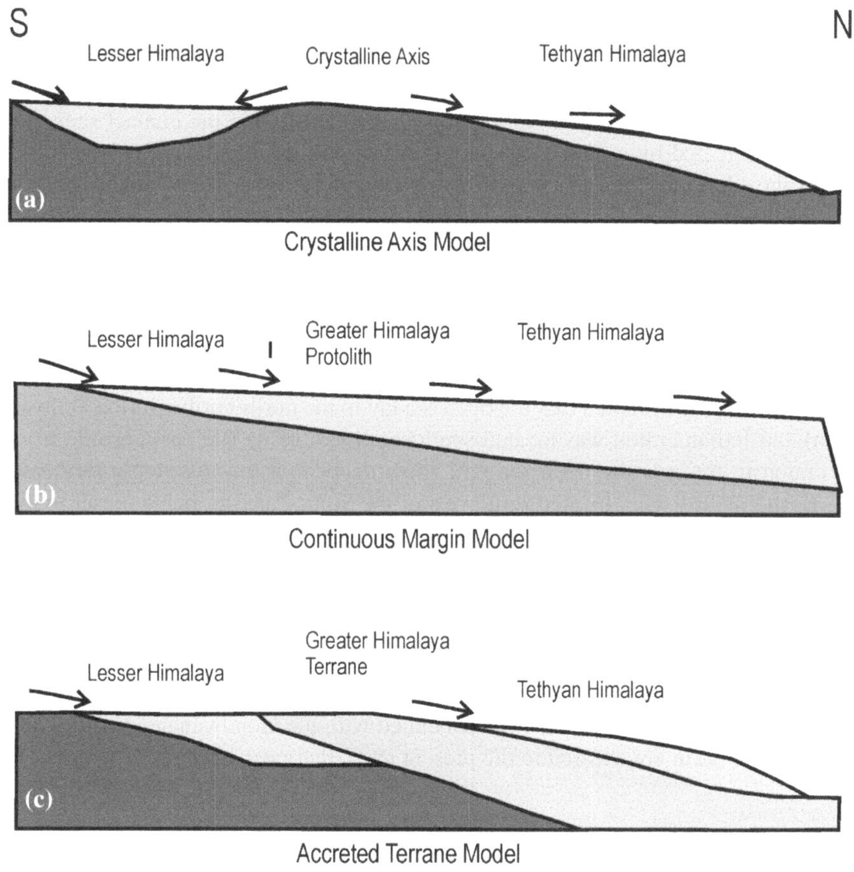

Fig. 6.33 Tectonic models explaining the relationship of the Tethyan basin *vis-à-vis* the Lesser Himalayan basin. Modified after Myrow et al. (2003)

Ganju and Khar (1984) interpreted the Tethyan Kashmir Basin to be a Neoproterozoic pericratonic basin which grew out of the movements along the deep-seated NW–SE trending Jhelum and other related collinear lineaments. Several volcanic rocks in Bhutan, sporadic mafic flows in Spiti, ferruginous tuffs in Kinnaur and basaltic rocks around Khewra in the Salt Range seem to represent the late Neoproterozoic rifting event, which resulted into the birth of the Tethyan Basin bordering the passive margin of the Indian plate (Bhargava 2015). Tethyan rocks have been primarily studied in four sectors namely the Kashmir, Spiti, Garhwal and Kumaun. Sporadic work in Nepal and Bhutan provided supporting evidence.

In Zanskar and Spiti Srikantia (1981) found a gradational contact between the crystalline basement rocks of Higher Himalaya and the Tethyan sediments. Gansser (1983) observed a normal gradational contact between the two contrasting rock assemblages in Bhutan. Occurrence of volcanics in some regions along the contact prompted some workers to interpret a nonconformity at this level. Burchfiel et al. (1992) inferred a normal fault along this contact and called it South Tibetan Detachment System (STDS). The STDS is identified in Bhutan along the margins of a klippe (Kellett et al. 2009). In Kumaun in Kali River section the contact seems to have been affected by a fault although no discernable prominent tectonic features are displayed. In Girthi Ganga and Dhauli Ganga sector in Malla Johar (Kumar et al. 1977) the contact is marked by widespread conglomerate. The Kashmir Basin located in the NW, adjacent to Spiti is largely represented by the shallow, marine facies of rocks which closely resembles the Cambrian-Early Permian lithofacies and biofacies of the Spiti and Zanskar region. On the other hand, post-Panjal Volcanics sedimentary sequences in this part reflect much shallower depositional conditions compared to the Spiti–Zanskar sector.

It is universally accepted that the open sea lay to the north/northeast (the Tethyan basin) and Indian craton was located south/southwest of it. The basin profile from the craton to towards the open sea will show deepening towards north/northeast. The Kashmir sub-basin, located southwest of the Spiti–Zanskar, being closer to the coast, is supposed have shallower litho- and biofacies. However, both the facies from the Cambrian to the lower Permian (Sakmarian) are like the corresponding sequences in Spiti–Zanskar. The post-Panjal Volcanic sequences in Kashmir, on the other hand, show shallower facies as compared to the Spiti–Zanskar. To explain such a distribution of the facies in space and time, Bhargava (2011) suggested that till early Permian, the Kashmir sub-basin formed lateral extension of the Spiti–Zanskar basin. During the Panjal volcanicity, associated with the Gondwanaland rifting, the Kashmir sub-basin got sheared to the present site (Bhargava 2011).

6.8.1 Cambrian

The Cambrian succession in the Kumaon (Kumar et al. 1972), and Spiti region was described by Srikantia and Bhargava (1998), Bhargava and Bassi (1998), Bhargava and Singh (2019), Myrow et al. (2009), Hughes (2016), Hughes et al. (2005, 2018) and many others in last three decades. Although the contact between the Higher Himalayan metamorphics and the Cambrian rocks is not clearly exposed due to glacial debris or tectonics, the conjectural boundary can be used for interpreting the overall biostratigraphy of the Cambrian succession.

In all contemporary literature, the Kunzam La Formation, a name coined by Srikantia and Bhargava (1998), is taken as the base of the Cambrian in the Spiti basin. These authors advocated the use of Kunzam La section as the type section for the Cambrians based on detailed biostratigraphic measurements. Recently, this contention has been contested due to incorrect estimation of thicknesses of stratigraphic units in the Kunzam La section (Hughes et al. 2018) and the Parahoo section was proposed as a more appropriate section for describing the Cambrian stratigraphy in the Spiti Himalaya. Srikantia and Bhargava (2018) refuted the evidence presented by Hughes et al. (2018) by stating that the Parahio section is truncated due to a

Table 6.4 Various lithostratigraphical schemes for the Cambrian and associated strata in the Spiti and Zanskar valleys of the Tethyan Himalaya. After Hughes et al. (2018)

Stoliczka 1865	Griesbach 1891	Hayden 1904	Reed 1910	Pascoe 1959	Nanda & Singh 1976; Garzanti et al. 1976	Srikantia et al. 1980	Fuchs 1982	Kumar et al. 1984	Myrow et al. 2006a, b	Singh et al. 2014	Hughes et al. 2018
Muth Series	Haimanta Series	Silurian System		Ordovician	Thaple Fm	Thango Fm	Shian Quartzite	Thango Fm	Thango Fm		Shian Formation
					Kurgiakh Fm						
					Karsha Fm: Teta Mbr.						
					Karsha Fm: Mauling Mbr.						
					Karsha Fm			Kumzum La Fm.: Parahio Mbr.			
Bhabeh Series	Vaikrita (Slates and Schist)	Cambrian System	Parahio Series	Parahio Series	Karsha Fm	Kumzum La Formation	Parahio Series	Kumzum La Fm.: Debsa Khad Mbr.	Parahio Formation	Kumzum La (=Parahio) Formation	Parahio Formation
		Haimanta Series		Haimanta Series	Phe Formation	Batal Formation	Haimanta Formation	Batal Formation	Batal ?? Formation	Batal Formation	Phe Formation

Dashed horizontal line—Stratigraphic unit extends beyond interval. Dashed slanting line—fault contact. Pecked line—Uncertain boundary. For full explanation see text. After Hughes et al. (2018)

fault and represents only late part of the Cambrian Series 2, Stage 4 to Wulian Stage, Miaolingian Series, while a complete succession is exposed in the Kunzam La section after which the formation was named. Reacting to the observation of Srikantia and Bhargava (2018), more detailed data and arguments based on contemporary Codes of Stratigraphic Nomenclature were presented by Hughes et al. (2019) to justify the selection of the Parahio section as more complete and better exposed section than the Kunzam La section (Table 6.4).

The Cambrian strata exposed in the Parahio Valley of Spiti provide the best-documented section of rocks belonging to the Cambrian System anywhere within the Himalaya. Building on earlier work in the last 150 years (e.g. Stoliczka 1865; Griesbach 1891; Hayden 1904; Reed 1910; Pascoe 1959; Srikantia et al. 1980; Srikantia 1981: Fuchs 1982; Kumar et al. 1984), Hughes et al. (2018) compiled a comprehensive taxonomic revision of all previously published available Cambrian fossil material from the subcontinent. The Parahio Formation is in the lower part of the Tethyan Himalayan succession which consists of primarily sedimentary and metasedimentary rocks bounded to the south by the South Tibetan Detachment System (STDS). In the Parahio Valley a tectonised lowermost unit of the Tethyan Himalaya, locally called the "Batal Formation", separates the Parahio Formation from the detachment fault. In Zanskar the STDS lies close to the base of the Parahio Formation (Myrow et al. 2006) which is conformably underlain by the Phe Formation. The Haimanta Group extends from the Chamba Formation at the base, through the Manjir and Phe Formations, and is capped by three fossiliferous formations namely the Parahio, Karsha and Kurgiakh formations (Bhargava and Bassi 1998). Reed (1910) introduced the term "Parahio Series" to describe the fossil-bearing succession, and Pascoe (1959) provided a comprehensive summary of the "Parahio Series". Stratigraphic and paleontological studies led to the use of a variety of stratigraphic nomenclatures for the upper part of the Cambrian succession in the Tethyan Himalaya. A ~2900 m thick trilobite- and brachiopod-bearing set of rocks with cyclic patterns of facies succession was named as Kunzam La Formation by Srikantia et al. (1980) occurring at the base of the overlying Thango Formation conglomerate (the equivalent of the Shian Formation). The name Kunzam La Formation has since been applied to the Parahio Valley by a number of workers (Bhargava et al. 1982; Bhargava and Srikantia 1985; Parcha 1996; Jell and Hughes 1997; Srikantia 1981). Pascoe (1959) described the lithostratigraphic properties of Parahio Series long before the alternative name Kunzam La was proposed. In terms of historical precedence, it is more than 100 years since the term "Parahio Series" was first proposed (Reed 1910, p. 62), whereas the term "Kunzam La Formation" has been in the literature for only four decades (Srikantia 1977). Singh et al. (2014, p. 82) wrote that the fossiliferous Cambrian rocks in the Parahio Valley (Spiti region) are well exposed and constitute the best studied Cambrian succession in the Himalaya. Detrital zircon analysis shows that rocks collected at Batal bridge, the type section of the underlying Batal Formation, contains grains as young as 520 Ma (Myrow et al. 2010). Hence these sheared rocks, interpreted to lie stratigraphically beneath the Parahio Formation in that section, may have been deposited after the lower parts of the Parahio Formation in the Parahio Valley section were laid down.

The Kurgiakh orogeny is the substantial tectonic event that took place along the northern Indian margin during the early Middle Ordovician and was also first recognized by Hayden in 1904 through the unconformity at the base of Ordovician sequence. In the Spiti as well as in the Zanskar regions of the Himalaya the base of the unit is strongly tectonised rather than being a straightforward stratigraphic one.

Parcha and Pandey (2016) described *Diplichnites*, *Dimorphichnus*, *Monomorphichnus*, *Palaeophycus*, *Planolites* and *Skolithos* from the Cambrian Kunzam La Formation, Chandra Tal section (Lahaul valley) and interpreted a shallow marine environment for the deposition. Singh et al. (2016) precisely demarcated the Hayden level 2 in the Kunzam La Formation in Spiti; it brackets the local stratigraphic range of *Oryctocephalus indicus*—a key taxon considered for the Global Stratotype and Point Section (GSSP) for the base of the Wuliaun Stage, Miaolingian Series (or Middle Cambrian) and possibly, the Cambrian Series 2—Wuliaun Stage, Miaolingian Series boundary. The taxon, *Eosoptychoparia (Danzhaina)* sp. was recorded for the first time from the Indian Himalaya.

Singh et al. (2017a) discovered the trilobite fauna *Oryctocephalus indicus*, *Pagetiasignificans*, *Kunmingaspispervulgata*, and *Bhargavia prakritika* from the Pin Valley (Spiti), which ranges from the top part of the Cambrian Series 2 (Stage 4) to the basal most part of the Series 3 (Stage 5). Singh et al. (2017b) described the Cambrian Series 3, Stage 5, trilobite fauna comprising *Peronopsis* sp., *Eosoptychoparia* cf. *spinosa*, *Gaotanaspis* cf. *pingzhaiensis* and *Gaotanaspis* cf. *transversa* immediately above the known *Oryctocephalus indicus* biozone (Cambrian Series 3, Stage 5) in the Parahio valley. Singh et al. (2017c) reported the first occurrence of trace fossils *Treptichnus pedum* and the *Treptichnus lublinensis* from the Cambrian succession of Spiti

(Tethyan Himalaya) as well as from Nigali Dhar Syncline in Himachal, respectively. The presence of these trace fossils only indicate that the beds are Cambrian or Early Ordovician and cannot be used for any paleontological zonation.

Storm-dominated, lower to upper shoreface of shallow-marine environment of a part of the Kunzam La Formation along the Shian section, Pin Valley was indicated from integrated ichnofabric and sedimentology by Chaubey et al. (2018). Yin et al. (2018) described low-diversity acritarch assemblages including a new species *Synsphaeridium parahioense* together with filamentous algae and cryptospore-like microfossils from Kunzam La Formation in the Parahio valley.

In Himachal, an integrated ichnology, sedimentary facies and sequence stratigraphy analyses of arenaceous succession of the Koti Dhaman Formation (Cambrian Series 2, Stage 4) of Tal Group, Nigali Dhar Syncline yielded trilobite traces of the Gondwana affinity along with seven ichnofossil assemblages, suggesting shallowing upward parasequences of a tidal flat complex (Singh et al. 2019).

The Lower Palaeozoic geology of the Kumaun–Garhwal regions remain considerably less well resolved than that of the Spiti region, and it may be a continuing challenge to apply terms that originated in the Kumaon to sections in the Pin/Parahio region. In Kumaun the deformed Cambrian rocks at the lower end of the succession show much stronger shearing than the underlying Ralam Formation itself and therefore Griesbach (1891) considered these slaty rocks to be part of the Vaikrita. This issue highlights difficulties with the use of the term Haimanta Group in the Cambrian stratigraphy. If the Ralam Formation in Gori Ganga and Dhauli Ganga sections formed the base of Griesbach's "Haimanta System", this original definition will imply that all of the rocks it contains are younger than the Parahio Formation (=Kunzam La Formation), whereas current usage applies the term Haimanta Group to sedimentary rocks in the Tethyan Himalaya that lie below the Ordovician unconformity (Bhargava and Bassi 1998).

In eastern Kumaun an unfossiliferous Lower Garbyang Formation (Banerjee 1974) is possibly Upper Cambrian in age as it underlies the Middle Ordovician gastropod (*Eccyliopteris kushanensis* Grabau) bearing Upper Garbyang Formation (=Changru formation of Valdiya et al. 1972). Trilobite fragment and liguilids have been report but never authenticated. The time gap between the Lower and Upper Garbyang Formation has been attributed to Late Cambrian Kurgiakh Orogeny (Spiti) by Srikantia et al. (1976). Detailed sedimentology of the rocks in the Kali River section bordering Nepal encompassing the Budhi Schist–Garbyang–Kalapani sections is given in Banerjee (1974). In the Gori Ganga–Girthi Ganga valleys Kumar et al. (1977) and Banerjee et al. (1975) identified Martoli Formation of meta-arenites and meta-argillites, overlain by a prominent Ralam Conglomerate. These two units, therefore, appear to be time equivalent of the Budhi Schist of the Kali River section in Kumaun Himalaya. The Garbyang Formation is divided into Lower, Middle and Upper units; each is characterized by near distinct lithology. A variety of trace fossils (*Aulichnites, Lavicyclus, Planolites, Phycodes*) from the Garbyang and Shiala Formations were described for the first time (Banerjee et al. 1975). Sinha and Verniers (2016) recorded chitinozoans *Belonechitinacapitata* from the Shiala Formation of Kali Valley.

6.8.2 Ordovician–Silurian

The Ordovician–middle Silurian succession of succeeding cycle of the Tethyan Himalaya began with a widespread transgression, and heralded sedimentation with a conglomerate, observed from Peshawar to Garhwal (Bhargava 2011). In Spiti there are two levels of conglomerate intervened by red-coloured quartz arenite succession. The lower conglomerate has clasts from the Cambrian and the older formations. The upper conglomerate contains pebbles of red coloured quartz arenite also, which suggests a pre-upper conglomerate sedimentologic cessation. The hiatus was long enough to form a horizon simulating ferricrete in Spiti (Bhargava and Bassi 1998). In Spiti Basin the Late Ordovician–Early Silurian is known as Takche Formation (=Yong Formation). It is arenaceous in the east and calcareous in the west, with tell-tale evidence of shallow depositional environment. The Takche Formation has an uneven development in Zanskar region due to pre-Muth erosion.

Pandey and Parcha (2018a) added to the existing record of Ordovician algae in the Himalaya by reporting calcareous algae *Dasyporellasilurica, Moniliporellamultipora*and *Vermiporella fragilis* from the Middle Ordovician (Dapingian to Darriwilian) Thango Formation in the Pin Valley. Chaubey et al. (2018) described warm-water Dasycladaceae algae *Mastopora* and *Cyclocrinites* from the Takche Formation (Upper Ordovician–lower Silurian), Parahio Valley, Spiti. Cryptostome bryozoan *Ptilotrypabapjaii*, reported from the lower part of the Yong Limestone in Kumaun, indicates an Upper Ordovician age. The Ordovician–Silurian boundary, thus, may lie in upper part of the Yong Limestone or in the overlying Variegated Formation (Swami et al. 2019). Using petrography, sedimentology and geochemistry of the Thango Formation siliciclastics a passive margin tectonic setting for these rocks was inferred (Rashid and Ganai 2018).

Myrow et al. (2018) reported a lower Silurian ozarkodinid fauna within upper member of the Takche Formation confirming a Llandovery age for its uppermost parts. Using high-resolution carbon isotope stratigraphy they suggested transition to ice house condition during the Ordovician to earliest Silurian.

The Ordovician sediments were extensively deposited along a long linear belt from Kashmir to borders of western Nepal. Bhargava (1995) opined that Ordovician rocks are not known from the Kumaun Himalaya. This view is a at variance with the observations of Valdiya et al. (1972) and Banerjee (1974) along the Kali River section, who described shale and carbonate overlying the Changru Formation (cf., Valdiya et al. 1972) or the Upper Garbyang (Banerjee 1974) as the Shiala Formation with Middle to Upper Ordovician fauna (*Rafinisquinaalternate, Sowerbyellarugos; Strophomena*). Shah and Sinha (1974) described many fossil genera and species from the Shiala Formation suggesting Middle Ordovician age. The Silurian to Lower Devonian Variegated Formation (*Favosites spitiensis, Orthis basalis, Polygnathids*) is a colourful shale, limestone, sandstone assemblage, overlain by white quartz arenite of the Middle to Upper Devonian Muth Quartzite Formation (*Schellwienellawillamis, Salopinalunata*), as a vast stable beach or barrier island (Bhargava and Bassi 1998; Draganits et al. 2003).

In Bhutan the Ordovician–Silurian is represented by thick carbonate buildup. The Silurian sequence shows syndepositional deformation (Kumar et al. 1977), representing paleoseismites.

6.8.3 Devonian–Carboniferous

In Kashmir, the Devonian–Carboniferous was described as the Agglomeratic Slates. These beds are known to yield *Eurydesma*. While the Gechang Formation in Spiti marks the end of marine sedimentation (Sakmarian-Asselian), in Kashmir ~290 Ma old Panjal Volcanic activity dominated the scene. Fresh water lakes formed at this time over the Kashmir region and plant debris got preserved in the (Kapoor and Maheshwari 1991; Kapoor et al. 1993, 2004). Cleal et al. (2016) reported rich fossil flora of Serpukhovian age from the Fenestella Shale of Manigam (Anantnag Kashmir). This assemblage of the northern margins of Gondwana, on the shores of the Paleo-Tethys, is dominated by remains of an eligulate, mainly monopodial lycopsid with persistent leaves.

Using a single zircon (U–Pb method), Agnihotri et al. (2018) assigned 329 ± 16 Ma age (late Viséan to early Serpukhovian) for the Fenestella Shale, the age tallies with the earlier observed age based on faunal assemblage. The palynoflora, pteridophytes, gymnosperms pollen grains and the trilete spores indicate a marine warm, temperate coastal depositional environment for the *Fenestella* Shale in the Banihal area region.

Deepening of the basin seems to have begun before the Givetian time. In Spiti, it ended up in deposition of thick gypsum deposits in the Lipak Formation (Late Tournaisian). The Po Formation/Fenestella Formation (Visean–Serpukhovian) with shale and sandstone is exposed in all the basins from Kashmir to Nepal.

In Spiti–Zanskar quartz arenite is prominently found in the Lipak Formation and its equivalent Kali Quartzite (Banerjee 1974) in Kali River section. It is dominantly limestone based in the Kashmir (Syringothyris Limestone). In Spiti–Zanskar the overlying rocks are diamictites (glacial/glacio-fluvial?) and are named the Ganmachidam Formation (=Permian Conglomerate). In Lahaul, Spiti and Kinnaur *Eurydesma* occurs in sandstone of the Gechang Formation overlying the diamictites. The equivalent of the Po Formation of Spiti is the Fenestella Shale in the Kali River section. A major transgression was identified following the volcanicity (Wuchiapingian transgression) which brought the Gungri Formation to rest over the Lipak Formation. The Gungri Formation is disconformably overlain by the Mikkin Formation (Induan-Olenkian). Late Changhsingian is absent in Spiti (Ghosh et al. 2016; Bhargava and Singh 2019) although Nakazawa et al. (1975) identified an uninterrupted succession. The Ganmachidam Formation (sensu stricto) of Spiti has not be specifically identified in the Kumaun Himalaya. However, a weathered bed with conglomeratic horizon mark a minor hiatus on top of the Fenestella Shale and underneath the Kringkrong Limestone (Kuling Series) (Banerjee 1974). No comparable rocks are found in Bhutan (Bhargava and Singh 2019).

Singh et al. (2017d) described a rich ichnofossils assemblage from the Po Formation (Visean–Serpukhovian), exposed at the base of the Ganmachidam Hill. Ganai et al. (2016) and Ganai and Rashid (2019) suggested that the Lipak, Po and Ganmachidam formations of Spiti had variable depositional environments from a warm Devonian, cooler in Middle Carboniferous and freezing at the Permo-Carboniferous intervals.

In Kumaun, uppermost part of the Muth Quartzite Formation is rusty brown due to weathering and contains beds of limestone and dolomite with diagnostic species of *Linoproductus and Chonetes, Canninophyllum and Caninia*, suggesting Lower Carboniferous age. These quartzites were therefore designated as a Kali Formation and are distinct from the Muth Quartzite. The overlying organic matter-rich shale contains *Fenestella plebei* along with *Proreteporaampala, Spirifer* and *Productus*, and compares well with the Middle Carboniferous *Fenestella* Shale Formation (=Po Series in Spiti) from other parts of the Tethyan Himalaya.

6.8.4 Permian–Triassic

Jasper et al. (2016) recorded fragments of tracheids that show homogenized cell walls, a characteristic feature of charcoal in the Late Permian Zewan Formation of the Guryul ravine which also records evidences of tsunamis (Brookfield et al. 2013). This feature is regarded an evidence of palaeo-wildfire, which permits reconstructing environment during the Late Permian Zewan Formation. Brosse et al. (2017) investigated the conodont biochronology and isotopic records of the Khunamuh Formation at the Guryul ravine which includes both the Permian-Triassic and the Griesbachian–Dienerian (Induan) boundaries; the latter is precisely constrained between beds GUR310 and GUR311, and corresponds to the replacement of segminiplanate (here *Clarkina* and *Neoclarkina*) by segminate (*Sweetospathodus* and *Neospathodus*) conodonts. This faunal turnover was possibly linked to a climate change at the Griesbachian–Dienerian transition, from a cool and dry to a hot and humid climate.

Petrographic, major and trace element geochemistry and rare earth elements studies of the late Permian and early Triassic sediments in the Guryul ravine revealed visible changes in lithostratigraphy from argillaceous–carbonaceous mudstone in the C Member of Zewan Formation, to fine grained argillaceous siltstone with quartz in the D Member (4 m below the Late Permian Event Horizon. Kumar et al. (2017) discussed geochemistry of the Permian-Triassic sequences of this section, and its implications for oceanic redox conditions. Table 6.5 presents the correlation of Tethyan Palaeozoic sequences of different sectors.

Two pronounced stages of oceanic oxygen deficiency are recorded in the latest Permian *Hindeoduspraeparvus–Clarkinameishanensis* Zone and the earliest Triassic *Isarcicellastaeschei* Zone (Huang et al. 2019). The first decline occurred within the uppermost Permian storm beds. The latter two biotic declines coincided with two marine anoxic events, as documented by pyrite framboid distributions. These anoxic

Table 6.5 Correlation of Tethyan Palaeozoic sequences of different sectors

Period		Salt Range/Raj.	Kashmir	Zanskar-Spiti-Kinnaur	Kumaon-Garhwal	Nepal	Bhutan	Peshawar
Permian	L	Zaluch Gr.	Khunamuh	Gungri	Kuling Sh.	Nenjung	Barishong	
	M		Panjal Volc.	Panjal Vol.				
	E	Nilawahan Gr. Bhadaura	Nishat Bagh & Mamal Beds	Gechang	Girthi	Senja		
Carboniferous	L		Pindabol	Ganmachidam		Thini Chu		
	M		Fenestella Sh.	Po		Ice Lake		Zaffar Kando
	E		Sytingothyris L.	Lipak			Ripakha	Khybar L.
Devonian	L		Wazura		Kali			
	M		Wazura-Muth	Lipak-Muth		Tilicho Pass		
	E		Muth	Muth	Muth			
Silurian	L							Nawshera
	M					Chandragiri		
	E		Gugaldhar	Takche	Shiala-Yong-Varigated	Chitlang Dark Band L.	Wachi La	Kandar
Ordovician	L							
	M							Panjpir
	E		Hallamulla	Thango	Ralam	Damgad		Misri Banda
Cambrian	L						Quartzite	
	M							
	E	Jhelum Gr. Nagaur Gr.	Hapatnar Gr.	Haimanta Gr.	Milam Gr.	?-?	Maneting	Amber

events are also recognized from the P/T Boundary beds in Barus Spur section in Kashmir. Two anoxic events at the Guryul Ravine correlate precisely with anoxic events in the Meishan GSSP.

In Spiti, Triassic sequence is identified as the Lilang Supergroup (Bhargava et al. 2004) and in the Kali River section of Kumaun Himalaya. Its lithostratigraphy and biostratigraphy remains consistent between Zanskar and Kumaun, although Kalapani Limestone and Kuti Shale are quite comparable. However, fauna of Spiti and Kumaun do not correlate well. The Tagling Formation, the youngest formation of the Lilang Supergroup is of Lias/Dogger age. With another submarine hiatus it is followed by the Spiti Formation of mid-shelf environment. Except for Kashmir, it is found in all other areas.

High resolution geochemical work (Ghosh et al. 2016) including stable isotope geochemistry across the P-Tr boundary in Atargu and Guling sections led to the interpretation of Late Permian shales of the Gungri Formations as anoxic sediments from deeper parts of the basin. Isotopic excursions in the range of 2.4‰ and 3.1‰ in these sections suggest P-Tr transition across a clayey and gypsiferous ferruginous layer. Original Pb isotope ratios and an increasing Pb abundance closer to the P-Tr boundary, distinguish the volcanic source of the Late Permian shales from continental crustal siliciclastic signature of the Early Triassic carbonates.

In Kumaun, marked by a zone of weathering possibly a discontinuity, the succession continues upward with a distinct shale unit designated as Castle Hill Shale (Banerjee 1974) which, in turn, grades imperceptibly into Middle to Upper Triassic Kalapani Limestone having coiled ammonites. These limestones are capped by another prominent shale horizon, the Kuti Shale, with abundant Upper Triassic fauna.

The topmost stratigraphic unit in this part of the Tethys Himalaya belongs to Belemnites and mollusk-bearing Upper Triassic to Lower Jurassic Kioto Limestone with beds of quartz arenite.

6.8.5 *Jurassic–Cretaceous*

The Spiti Formation in Spiti–Zanskar–Garhwal is conformable succeeded by the Guimal Formation (Valanginian/Early Hauterivian to Albian), which is interpreted both as proximal turbidites and also as shallow marine deposits (Bhargava 2008a, b). This was followed by sedimentation of the Chikkim Limestone and shales during Campanian/? Early Maastrichtian. A younger sequence of the Kelcha Formation was identified only in the Zanskar region (Spanboth Formation of Gaetani et al. 1983) of Thanetian to Early Ypresian age. The overlying lithological succession is Dumbur Formation (Chunglung La Formation of Gaetani et al. 1983) of ?Late Ypresian-Lutetian age. The Shilakong/Spongtong untramafic rocks occur as klippe over the Tethyan sequence in the Zanskar and Malla Johar areas in Uttarakhand (Bhargava and Singh 2020). These nappes represent squeezed up ophiolitic material from the deep-sea regime along the Indus Tectonic Zone, obducted to form the highest allochthon in the Himalaya. The faults in the area are high angled, though some of them show minor sinuous outcrop pattern. Such faults are considered as basement controlled listric-like folded faults (Bhargava and Bassi 1998, 2003).

The Jurassic–Cretaceous marine sediments represented by the Tagling (Early Jurassic), Spiti, Guimal (Early Cretaceous), and Chikkim (Late Cretaceous) formations crop out in the Spiti Valley. The Spiti and Giumal formations are rich in ammonoids. Based on planktonic foraminifera five planktonic foraminiferal assemblages in the Chikkim Formation could be identified. The litho-biostratigraphic evaluation of the Jurassic–Cretaceous succession in the Spiti Valley reveals the presence of the T-OAE of late Early Toarcian in the youngest few meters of the Tagling Formation, Late Valanginian Weissert OAE, latest Hauterivian Faraoni OAE and Early Aptian OAE 1a in the Giumal Formation, and probable an Early Albian OAE 1b, Late Albian OAE 1c and 1d, latest Cenomanian OAE 2 and Coniacian/Santonian OAE 3 in the Chikkim Formation (Pandey and Pathak 2016).

Pandey and Pathak (2016) reported the presence of the T-Oceanic Anoxic Event of late Early Toarcian in the youngest few meters of the Tagling Formation. Pandey and Pathak (2017) suggested that the Giumal Formation is ideal to carry out ammonoid-based biostratigraphic studies. The ammonoid assemblage in the Giumal Formation belongs to Berriasian to Early Aptian (Pandey and Pathak 2016, 2017; Pandey et al. 2018a). Pandey and Pathak (2017) recorded new species of the ammonoid genus *Olcostephanus* Neumayr from the Lower Cretaceous Giumal Formation which is helpful in intra-basinal to intercontinental biochronostratigraphic correlation.

The Chikkim Formation is post-Aptian (Pandey and Pathak 2016, 2017). Maceral analysis, Rock-Eval Pyrolysis, FTIR and NMR analyses of black carbonaceous shale

Table 6.6 Tethyan Mesozoic sequences of different areas

Age	Kashmir (Bhargava 2015)	Tandi (Srikantia & Bhargava 1979)	Zanskar-Spiti-Kinnaur (Bhargava and Bassi 1998)	Byans (Kumar et al. 1977)	Nepal (Dhittal 2017)	Bhutan (Bhargava 1995)
Cretaceous			Ophiolites	Ophiolitic exotic block	Ophiolites	
			Chikkim Fm. Giumal Fm.	Chikkim Fm. Giumal Fm.	Upper flysch Giumal Fm.	
Jurassic		Dilburi Fm.	Spiti Fm.	Spiti Fm.	Spiti Fm.	Sha Fm.
Triassic	Vihi Gr.	Gushal Fm.	Lilang S.group	Riwalibagar Gr.	Triassic	Lingshi

Table 6.7 Climatostratigraphy, sea level curve and important events recorded in the Spiti valley, which is the type section of the Tethyan Himalaya

Age	Formation	Climate	Important Geological Events
Cretaceous	Chikkim	Tropical	Deepening of basin, possibly coinciding with commencement of drift of Indian plate, folding and granite emplacement
	Giumal		Short lived shallowing (?), Tilt of the basin, proximal turbidites
	Spiti		Sudden submergence to shelf mud / slope and rise-basin margin (?)
Jurassic	Tagling	Warm-Humid	Transgression
Triassic	Rongtong-Para	Warm-humid-pluvial	Gradual shallowing, extensive patch-knoll reefs (middle Norian), Gradual shallowing
	Mikin-Chomule		Rapid deepening, partial rifting in Zanskar
Permian	Gungri	?Warm, humid	Extensive transgression, deepening of basin
	Gechang	Cold	Transgression, lag formation, regressioon or sudden submergence
Carboniferous	Ganmachidam	Cold	Rise in provenance area, regression, prograding cycles, coarser, restricted basin, some submarine highs become positive towards end (early Permian), rifting coinciding with Gondwana rifting
	Po	Warm-humid (pluvial)	Series of prograding cycles from shelf to shoreface
	Lipak	Arid	Deepening midway, hard ground formation, new areas inundated, local coral-algal reefs, Sabakha
Devonian	Muth	Uncertain, possibly a bit arid	Transgression on beach-mid shelf. Granitic activity
Silurian	Takche	Warm	Stablisation, coral-stromatoporoid-bryozoan-algal patch reef, shallowing and short lived break in sedimentation
Ordovician	Thango	Warm and arid	Sharp rise in provenance, formation of submarine highs, partial rifts in later part, granitic activity related to Kurgiakh orogeny
Cambrian	Kunzam La	-?Tropical, Warm	Extensive regression possibly a sequel to commencement of Kurgiakh orogeny, gradual shallowing
Eocambrian	Batal		Lithospheric activated rifting initially basic volcanism, acid volcanism at Ediacaran Cambrian

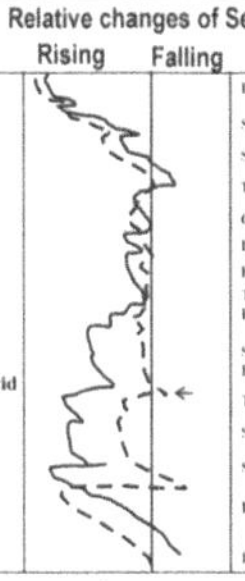

Subaerial break; Submarine break; Global (after Vail et al. 1977); Zanskar-Spiti-Kinnaur Basin (modified after Bhargava and Bassi 1998)

Modified after Bhargava and Bassi (1998), Bhargava and Singh (2020, in press)

in the Spiti and Chikkim formations of the Spiti valley revealed kerogen-III type of organic matter with TOC of 3.19% (Pandey et al. 2018b).

Table 6.6 summarizes the correlation of the Paleozoic and Mesozoic sequences of the Tethyan domain respectively, while important geological events are presented in Table 6.7.

6.9 Large-Scale Basin Configuration

Northern margin of the Indian Plate displayed variable tectonic configurations since Paleoproterozoic through Neoproterozoic and Phanerozoic during which large-scale basins evolved and deposited enormously thick sedimentary sequences. Of significance are the basins evolved during the (i) Paleoproterozoic and deposited sequences in the Inner Lesser Himalaya (iLH), (ii) Neoproterozoic with sedimentary pile in the Great Himalayan Sequence (GHS) and the Outer Lesser Himalaya (oLH) (iii)

Phanerozoic Tethyan Himalayan Sequence (THS) along the northern margin, and (iv) the Cenozoic Himalayan Foreland Basin.

6.9.1 Paleoproterozoic Inner Lesser Himalaya (iLH) Basin

An association of quartzite–limestone/dolomite–slate–metavolcanics of the Rautgara–Gangolihat–Deoban–Berinag–Kushma–Daling–Shumar sequences is exposed within the iLH beneath the overthrust Munsiari Thrust from Kashmir to Arunachal Pradesh (Valdiya 1980a, b; DeCelles et al. 2000; Kohn et al. 2010; Yin et al. 2009; Jain 2020). Detrital U–Pb zircon ages from this sequences—Rampur Group, Bhowali Group, Rautgara–Gangolihat–Berinag–Deoban–Damtha beneath the Munsiari Thrust, Daling Group in Sikkim, Shumar Group in Bhutan including metabasalt–metarhyodacite flows—reveal youngest U–Pb zircon peak between 2.0 and 1.7 Ga, and $\varepsilon_{Nd(0)}$ values between −27.7 and −23.4 (Fig. 6.34a–d; Mandal et al. 2016; Mukherjee et al. 2019; Jain et al. 2020). Along the Sutlej valley in the Kulu-Rampur window, the Rampur Group has youngest U–Pb DZ ages between 1.92 and 1.88 Ga (Fig. 6.34e; Richards et al. 2005; Webb et al. 2011), constraining its maximum depositional age. U–Pb zircon crystallization ages of 1800 ± 13 and 1840 ± 16 Ma from the Rampur metabasalt–metarhyodacite flows (Miller et al. 2000) and the intrusive Bandal granite of 1856 ± 3 Ma (Singh et al. 2009) further provide constrains to its depositional age. Beneath the Askot Nappe, the Berinag Group contains 1.86 Ga DZ peak (Fig. 6.34a; Mandal et al. 2016) and 1.87 Ga peak in its type area, while the lowermost Rautgara–Damtha formations yielded peaks of 1.75, 1.88 and 2.50 Ga (Fig. 6.34b–d; McKenzie et al. 2011). The Bhowali quartzite in Kumaon Himalaya has similar DZ population near the MBT (Mandal et al. 2015).

In Central and Eastern Himalaya, DZ ages are almost indistinguishable from each other in this belt; Nepal exhibits dominant DZ grain population between 2.0 and 1.7 Ga, minor peaks between 2.6 and 2.4 Ga (Fig. 6.34f, g; Parrish and Hodges 1996; DeCelles et al. 2000, 2001; Kohn et al. 2010; Gehrels et al. 2011 and references therein; Martin et al. 2005, 2011, 2017; Khanal et al. 2015). The Daling Group quartzite in Sikkim yielded U–Pb DZ peaks between 1.90 and 1.80 Ga and minor Neoarchean and lower Proterozoic components (Mottram et al. 2014a, b). In central Bhutan, the Daling–Shumar contact is marked by the DZ peak at 1.90–1.80 Ga and smaller peak around 2.50 Ga, and interstratified metarhyolite gave 1.89–1.79 Ga zircon, thus confirming the depositional age of these formations ~1.90 Ga (Fig. 6.34h–l; McQuarrie et al. 2008, 2013). Long et al. (2011) opined that the Daling-Shumar Group possesses youngest detrital zircon U–Pb peaks of ~1.8 to 1.9 Ga and intrusive orthogneiss bodies, thus these sequences had a Paleoproterozoic deposition age, like other sequences of the iLH in the NW Himalaya. It is, however, slightly younger (~1.78 Ga) in western Arunachal Pradesh (Fig. 6.34m; Webb et al. 2013).

Detrital zircon age data from nearly whole of the iLH domain, therefore, provide undisputed evidences that sheet-like matured Paleoproterozoic sedimentary

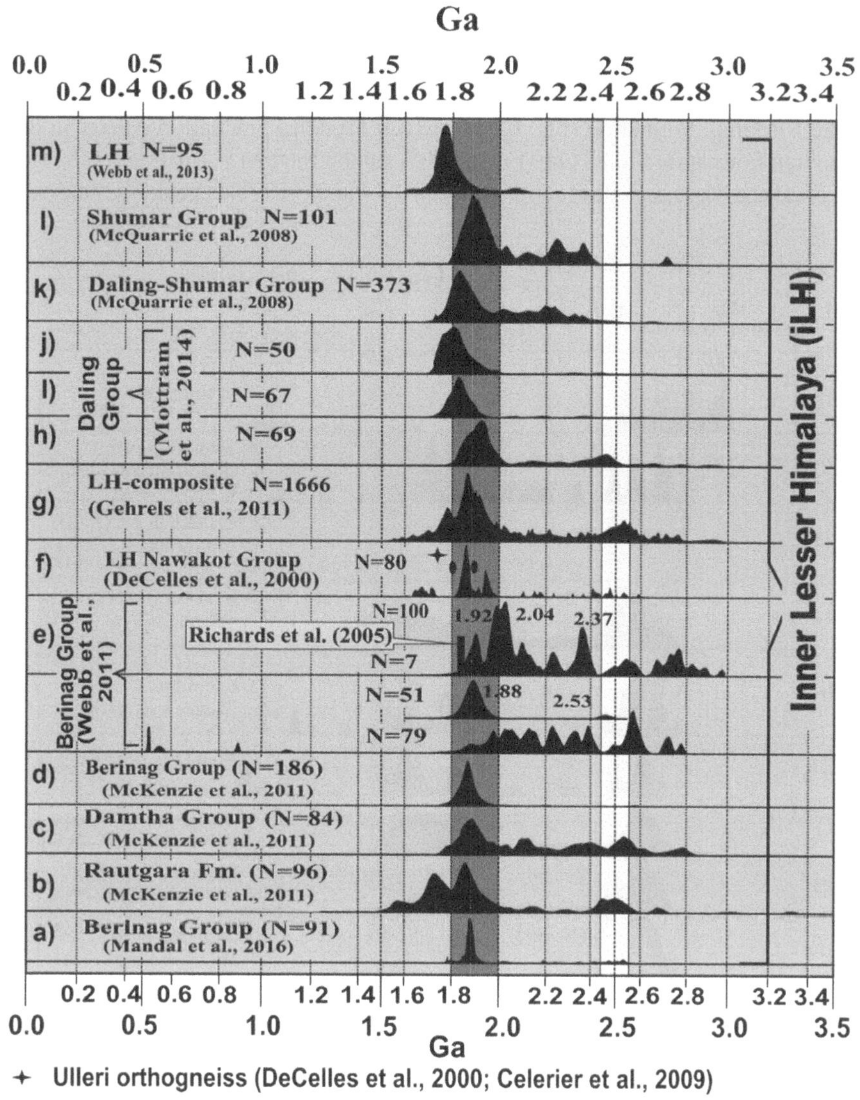

Fig. 6.34 Probability distribution plot of U–Pb detrital zircon ages from the Inner Lesser Himalayan (iLH) Sedimentary Belt. Two distinct zircon suites ~2.5 and 1.9–1.85 Ga characterize the iLH. Data sources in parenthesis (see references), while 'n' refers to number of the U–Pb zircon analyses

sequences were the deposited between ~1.90 and 1.80 Ga in a vast basin, having links with the Aravalli on one side and the Vindhyan sequence on the other.

6.9.2 oLH Detrital Zircon Pattern

Detrital zircon population from the Krol Belt has the oldest U–Pb peak of 0.95 Ga in the Mandhali Formation (Fig. 6.35a), which becomes gradually younger in the Chandpur Formation (~0.88 Ga) (Fig. 6.35b), and the Nagthat Formation (~0.82 Ga) (Fig. 6.35c, d); $\varepsilon_{Nd(t=0)}$ values range between −14 and −19. This peak becomes still

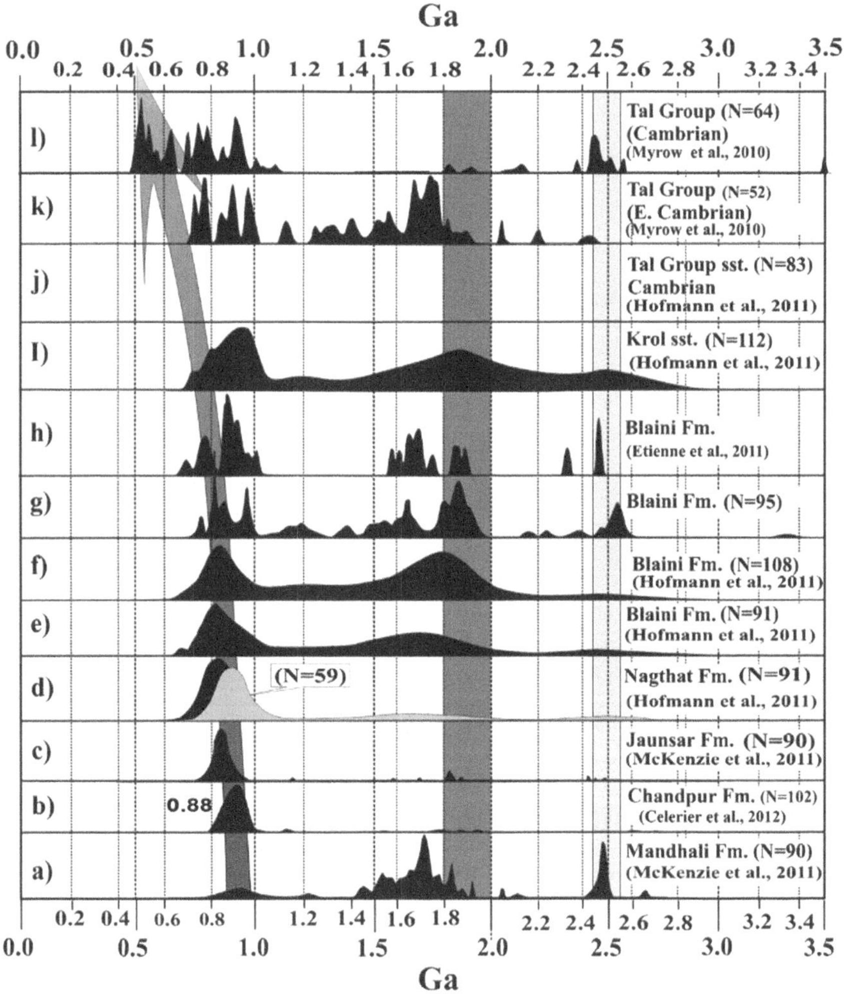

Fig. 6.35 Probability distribution plot of U–Pb detrital zircon ages from the Outer Lesser Himalayan (oLH) Sedimentary Belt. The oLH is distinct with appearance of the Neoproterozoic zircons with a younging pattern in Early Paleozoic zircons. Data sources in parenthesis (see references), while 'n' refers to number of the U–Pb zircon analyses Common distribution of ~2.5 Ga and 2.0–1.8 Ga zircons shown by vertical columns

younger with incoming of 0.70 Ga (692 ± 18 Ma) detrital zircon grains in the Blaini diamictite (Fig. 6.35e–h), ~0.70 Ga in the Krol sandstone (Fig. 6.35i) and youngest 0.525 Ga in trilobite-bearing Early Cambrian Lower Tal, respectively (Fig. 6.37j–l; Myrow et al. 2003, 2010; Etienne et al. 2011; Hofmann et al. 2011; McKenzie et al. 2011; Mukherjee et al. 2019; Jain et al. 2020).

In outermost LH Belt of Bhutan, the oldest Buxa Group quartzite yields ca. 0.9 Ga to ca. 520 Ma youngest detrital zircon peaks, indicating a Neoproterozoic–Cambrian(?) deposition age range (Long et al. 2011; McQuarrie et al. 2008). Above the Buxa Group, the Diuri Formation diamictite yielded a ca. 390 Ma youngest detrital zircon peak, indicating correlation with the late Paleozoic Gondwana supercontinent glaciation. In contrast, the Jaishidanda Formation in the MCT zone of central Bhutan yields youngest detrital zircon peaks ranging from ~0.8 to 1.0 Ga to ca. 475 Ma, indicating a Neoproterozoic–Ordovician(?) deposition age range (Long et al. 2011).

The above data further reveals a stratigraphic break of ~1.0 Ga between the iLH and oLH sedimentary basins throughout the Himalaya.

6.9.3 *Great Himalayan Sequence (GHS) Detrital Zircon Pattern*

This belt incorporates high grade metamorphosed Vaikrita Group/Tibetan slab in the north, associated metasediments of the Askot-Baijnath-Almora nappes, and Nathuakhan Formation (Ramgarh Nappe) in the south. These lithounits are characterized by a small Archean and Paleoproterozoic zircon fraction, and typical large Neoproterozoic population (1.05–0.80 Ga), having distinct age peaks at ~2.50 Ga, ~1.80 Ga and ~0.90 to 0.80 Ga (Fig. 6.36). Mukherjee et al. (2019) observed incoming of the Neoproterozoic DZ ~1.06 Ga in lowermost succession (Joshimath Formation) and decreasing age to ~0.86 Ga in the uppermost parts (Bhapkund Formation) Wherever good quality data is available, this pattern is characteristic of the GHS, e.g. in Bhutan. $\varepsilon_{Nd(0)}$ values fall between −12.5 and −19.2 (average −16.1) in the GHS. Presence of leucosome within orthogneiss of ~0.88 Ga and granitoid intrusives indicate that deposition in the GHS basin has probably ceased by 0.90 Ga (Mukherjee et al. 2019).

Such zircon distribution pattern is distinctly visible in the overthrust Neoproterozoic metasediments of the Garhwal, Ramgarh and Almora nappes. In the Nathuakhan Formation, U–Pb youngest DZ ages range between 0.88 and 0.81 Ga and $\varepsilon_{Nd(0)}$ values between −11.3 and −11.8 (Mandal et al. 2015).

The GHS belt is characterized by the following zircon patterns: (i) DZ ages between 0.77 and 2.60 Ga with peaks at ~0.82 and ~2.52 Ga above the Vaikrita Thrust along the Sutlej River (Fig. 6.36a, b; Webb et al. 2011; Spencer et al. 2012), (ii) peak ages of 0.8–1.2, 1.6–1.9 and 2.4–2.7 Ga with Neoproterozoic (540–750 Ma) grains making nearly half of these in Nepal (Fig. 6.36c; Gehrels et al. 2011) with an average $\varepsilon_{Nd(0)}$ value of −16 (Khanal et al. 2015; references therein), and an orthogneiss of

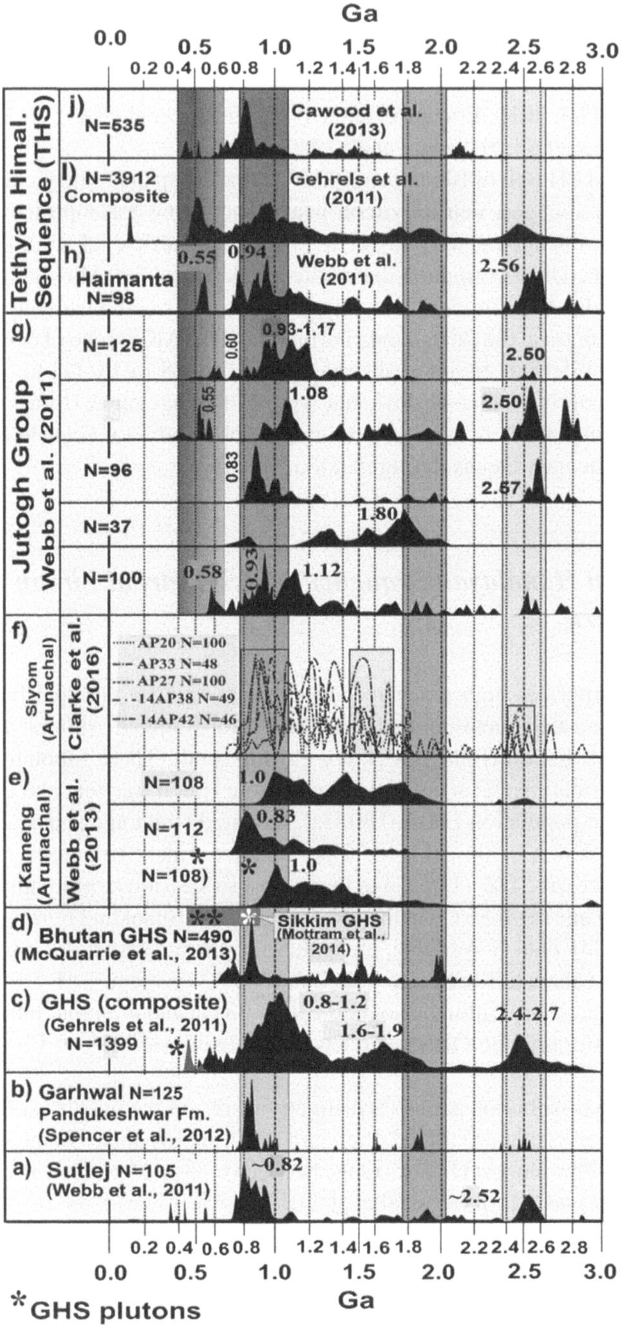
Ga
0.0 0.5 1.0 1.5 2.0 2.5 3.0
Tethyan Himal. Sequence (THS)
j) N=535
Cawood et al. (2013)
l) N=3912 Composite
Gehrels et al. (2011)
h) Haimanta N=98
Webb et al. (2011)
0.55 0.94 2.56
Jutogh Group Webb et al. (2011)
g) N=125
0.93-1.17
2.50
1.08
N=96
2.67
N=37
1.80
N=100
0.58 0.93 1.12
f) Siyom (Arunachal) Clarke et al. (2016)
AP20 N=100
AP33 N=48
AP27 N=100
14AP38 N=49
14AP42 N=46
e) Kameng (Arunachal) Webb et al. (2013)
N=108
1.0
N=112
0.83
N=108)
d) Bhutan GHS N=490 (McQuarrie et al., 2013)
Sikkim GHS (Mottram et al., 2014)
c) GHS (composite) (Gehrels et al., 2011) N=1399
0.8-1.2
1.6-1.9
2.4-2.7
b) Garhwal N=125 Pandukeshwar Fm. (Spencer et al., 2012)
a) Sutlej N=105 (Webb et al., 2011)
~0.82
~2.52
*GHS plutons

◀**Fig. 6.36** Probability distribution plots of U–Pb zircon ages from the Great Himalayan Sequence (GHS) and their comparison with the Jutogh Group and Tethyan Himalayan Sequence (THS). Common distribution of ~2.5 Ga, 2.0–1.8 and 1.0–0.85 Ga zircons are shown by vertical columns, while the Paleozoic zircons appear in the Jutogh Group and the THS. For data sources see references. 'n'—Number of the U–Pb zircon analyses

0.47–0.49 Ga (Cawood et al. 2007) thus constraining its age of deposition between 0.8 and 0.5 Ga, (iii) consistent younging DZ ages from 1.00 to 0.80 Ga from base to the top with minor Paleoproterozoic peaks of 2.45 and 1.9–1.75 Ga in central Bhutan with 0.50–0.48 Ga plutons intruding the lower parts; deposition of the Bhutanese GHS gets bracketed between ~0.50 and younger than 0.724 Ga (Fig. 6.36d; McQuarrie et al. 2013), (iv) youngest DZ ages between 1.0 and 0.83 Ga in western Kameng and Subansiri valleys of Arunachal with orthogneiss intrusives between 878 ± 12.6 and 836.9 ± 13.2 Ma and younger intrusive phase ca ~505 Ma (Yin et al. 2009), thus constraining maximum depositional age between 0.90 and 0.50 Ga (Fig. 6.36e; Webb et al. 2013), and (vi) DZ U-Pb ages along the Siyom valley (tributary of the Siang River) between 0.81 and 2.0 Ga having 809 ± 4.5 Ma orthogneiss; the Neoproterozoic (~810 Ma) maximum depositional age is constrained from adjoining 809 ± 4.5 Ma orthogneiss (Fig. 6.38f; Clarke et al. 2016).

Attempts are also made to correlate and discriminate the GHS (Vaikrita Group) from the Outer and Inner Lesser Himalayan (oLH and iLH) sedimentary sequences based on the DZ and $\varepsilon_{Nd(0)}$ characters (Ahmad et al. 2008; Myrow et al. 2003; Yin 2006; Célérier et al. 2009; Hofmann et al. 2011; McKenzie et al. 2011; Mandal et al. 2015). DZ from the oLH range from 0.95 Ga zircon peak in the oldest Mandhali Formation and becomes gradually younger to 0.70–0.525 Ga in the Marinoan Blaini diamictite and late Early Cambrian lower Tal trilobite-bearing strata; $\varepsilon_{Nd(t=0)}$ values range between −14 and −19 like the GHS (Ahmad et al. 2008; Myrow et al. 2003, 2010; Célérier et al. 2009; Hofmann et al. 2011; McKenzie et al. 2011). However, in spite of similarities in $\varepsilon_{Nd(0)}$ values with the GHS, the oLH sedimentary belt has distinct DZ population with incoming of younger 700 to 525 Ma grains in the Marinoan Blaini diamictite and late Early Cambrian lower Tal trilobite-bearing strata.

6.9.4 *Tethyan Himalayan Basin*

The GHS is followed by ~10 km thick Tethyan Himalayan Sequence (THS) to the north which were deposited in the open Tethyan Ocean and constitute a distinct sedimentary domain since late Precambrian (see Sect. 6.7). The Martoli and Ralam Formations from the lowermost parts of the THS in Uttarakhand contain youngest U–Pb DZ ages of ~0.57 and ~0.485 Ga, respectively. The older clusters are ~0.96 Ga and a few grains of 2.3 and 1.7 Ga, while some grains display Archean ages from 2.5 to 3.837 Ga. Distinct unconformity between the GHS and THS sequences is indicated by the presence of (i) ~0.55 to 0.50 Ga granitoids within the GHS and immediately beneath the THS, (ii) detrital zircons of the same age in lower parts of the THS, and (iii) conglomeratic Ralam Formation, with possible source from the GHS, besides other units. $\varepsilon_{Nd(t=0)}$ values indicate negative values similar to the GHS.

The Haimanta Formation of the THS in the Chandra Valley (Himachal Pradesh) yields similar DZ pattern between 0.55 and 3.0 Ga with additional peaks at ~0.94 and 2.56 Ga (Fig. 6.36h). Between Beas–Sutlej–Pabbar valleys, the Jutogh and Chail Formations have broadly same DZ age distribution with peaks at ~0.55, ~0.95, ~2.55 Ga and older grains (Fig. 6.36g; Webb et al. 2011). Distribution of DZ population bears strong similarities with the THS zircons (Fig. 6.36i, j; Gehrels et al. 2011; Cawood et al. 2013), and thus, provide a very powerful tool for correlation of different lithotectonic sequences.

In Nepal and Bhutan, Gehrels et al. (2011) noted that all the THS units from Cambrian to Cretaceous have almost similar U–Pb DZ age distribution with ages dominated by 480–570 Ma, 750–1200 and 2430–2560 Ma peaks (Fig. 6.36i). In central Bhutan, oldest Chekha Formation of the THS yielded 488–514 Ma youngest grain populations (McQuarrie et al. 2013).

6.9.5 Neoproterozoic (~1.10 to 0.85 Ga Zircons) Basin–Great Himalayan Sequence (GHS)—Rodinia Configuration

Arrival of the Neoproterozoic detrital zircons (1.1–0.85 Ga) in the GHS makes it distinct from the iLH across the MCT Zone, though migmatite and intrusive igneous bodies of the Paleozoic and Cenozoic ages are also recorded within the GHS. It is, therefore, imperative to ascertain source/s of these detrital zircons in the GHS, deposited along the northern passive margin of the Indian Craton. These Neoproterozoic zircons are omnipresent in the GHS along the entire length of the Great Himalayan belt, the LH metamorphic nappes, and the oLH; these units are either separated by the MCTZ or underlain by the iLH belt. Interestingly, the Neoproterozoic zircons predominate over the older zircons and mostly exhibit igneous characteristics with oscillatory zoning and low Th/U ratio. In view of its paleogeographic importance during the Neoproterozoic, sediment sources in this basin could either be from (i) the Neoproterozoic magmatic bodies from 'Within-basin', (ii) 'In-board' terranes from within the Indian Craton, or (iii) 'Out-board' external detritus supply from adjacent continents that came together during the Rodinia Supercontinent assembly (Fig. 6.37; Mukherjee et al. 2019; see Sect. 6.10 for more details).

6.9.5.1 'Within-Basin' Magmatic Bodies

~1.0 to 0.85 Ga DZ suite within the GHS and the Himalayan metamorphic nappes might have been sourced from intra-basinal, poorly developed, uplifted and eroded granite belt (Fig. 6.37a). Scanty geochronological data and poor recognition of such

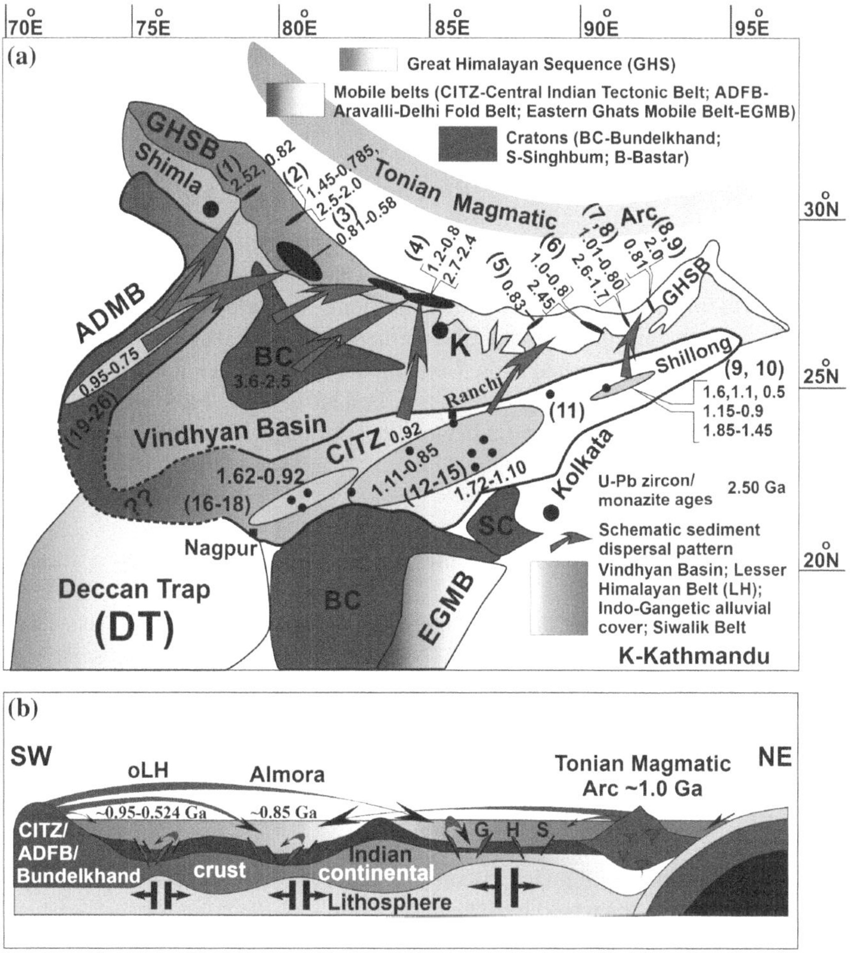

Fig. 6.37 Source for the Neoproterozoic detrital zircons in the Great Himalayan Sequence (GHS) and correlatable successions in the Lesser Himalaya. **a** 'In-board Indian cratonic sources from the Aravalli Delhi Fold Belt (ADFB), the Central Indian Tectonic Zone (CITZ), called together as garland-type Great Indian Proterozoic Fold Belt (GIPFOB) (cf. Prabhakar et al. 2014) and Tonian Magmatic Arc in north (cf. Spencer et al. 2018). The GIPFOB, extending into the Shillong-Meghalaya massifs, possibly remained highland during Neoproterozoic due to subduction-collision between South and north Indian Blocks and was rapidly eroded off to provide detritus to the GHS basin. Sections 1–9 yielded 1.1–0.85 youngest detrital zircons from Himachal to Arunachal in the GHS. **b** Schematic cross-section showing detrital sources in the GHS in north and 'Almora Basin'–oLH basins in south. Data Source: Mukherjee et al. (2019)

bodies in high grade metamorphosed GHS possibly explain non-prolific presence of ~850 Ma granitoids. Search for such bodies in high grade GHS will potentially reveal presence of such a source for the detrital zircons.

6.9.5.2 'In-Board' Indian Sources

Other possible sources, contributing the Neoproterozoic detrital zircons to the GHS basin, are the Aravalli-Delhi Fold Belt (ADFB) and Central Indian Tectonic Zone (CITZ), including Chotanagpur plateau and Shillong-Meghalaya massif (Fig. 6.37a). It is visualized that this 'garland' of the Neoproterozoic uplifted major tectonic accretionary zone, collectively called as the Great Indian Proterozoic Fold Belt (GIPFOB) between North and South Indian Blocks (cf., Prabhakar et al. 2014; see Mukherjee et al. 2019 and references therein for details), possibly was a highland mountain belt and eroded off around ~1.0 Ga (Fig. 6.37a).

6.9.5.3 External 'Out-Board' Terranes

This scenario visualizes that the Neoproterozoic sediments in the GHS basin were sourced from adjacent joined continents (Nubian-Arabia, Africa, Madagascar, eastern Antarctica, Australia) during the Rodinia Supercontinent assembly (DeCelles et al. 2000; Myrow et al. 2010; Spencer et al. 2012; see discussions and references therein). Such an off-board detrital source has limitations regarding the paleo-river systems that were required to transport sediments over 'extraordinary' distances (>2000 km) across the Indian craton. These paleo-rivers would have eroded dominantly Archean-Paleoproterozoic terranes with insignificant Neoproterozoic regions and carried sediments with mixed zircon population, in which zircon population of the Neoproterozoic ages would have been very less.

6.10 Tectonic Boundaries

Northern margin of the Indian continental lithosphere witnesses decreasing crustal shortening rates from ~15 to ~4 cm/year in the Himalaya (Jade et al. 2017) due to the India-Asia convergence since ~53 Ma. This convergence has produced the following south to north evolving orogen-scale main thrust systems (Fig. 6.38): (i) Quaternary Main Frontal Thrust (MFT) or Himalayan Frontal Thrust (HFT), (ii) Middle Miocene Main Boundary Thrust (MBT), and (iii) Early Miocene Main Central Thrust (MCT) (Thakur 1993; Meigs et al. 1995; Hodges 2000; Yin 2006; Robinson et al. 2006). In addition, an extension South Tibetan Detachment System (STDS) separates the northern margin of the Great Himalayan Sequence (GHS) from

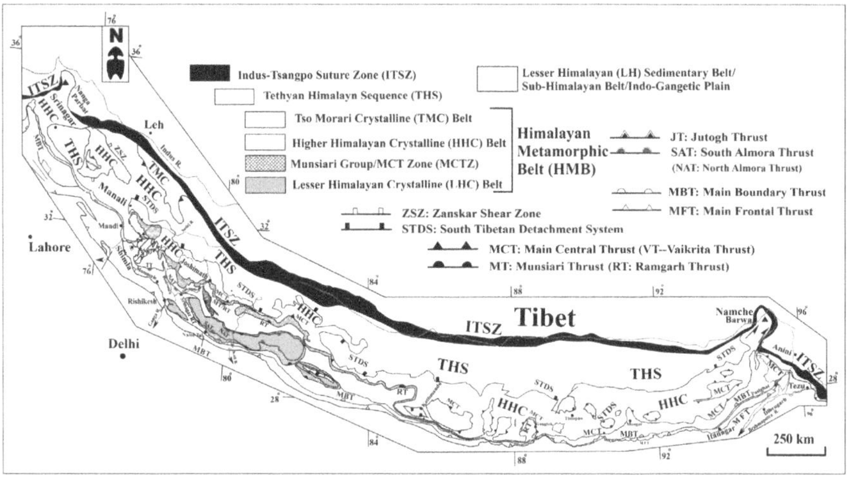

Fig. 6.38 Simplified geological map of the Himalaya showing major tectonic boundaries

the Tethyan Himalayan Sequence (Burchfiel and Royden 1985). These boundaries appear to merge with the blind Main Himalayan Thrust (MHT) (Zhou et al. 1993; Nabelek et al. 2009).

6.10.1 *Main Frontal Thrust (MFT)*

The Main Frontal Thrust (MFT) (≡Himalayan Frontal Thrust–HFT) is the youngest and outermost physiographic and tectonic boundary between the Cenozoic Sub-Himalayan (SH) Siwalik Belt and the southernmost Indus-Ganga-Brahmaputra Plains (Nakata 1972; Jayangondaperumal et al. 2018; Thakur et al. 2020), developed during the Quaternary. It generates and maintains active, modern Himalayan deformation and topographic front (Kumar et al. 2010; Burgess et al. 2012; references therein), and is the near-surface expression of the Main Himalayan Thrust (MHT) (Zhao et al. 1993). Thus, the MHT-MFT deformation boundary is the contemporary India-Asia plate interface (Thakur 2013) along which the Indian continental lithosphere subducts beneath Asia (Jain 2017). Since the youngest Siwalik Group has yielded a 0.5 Ma paleomagnetic age in the western Himalaya (Sangode et al. 1996), it is likely that the MFT has evolved progressively during later parts of Pleistocene-Holocene and caused deformation in the Sub-Himalaya. It has caused growth of longitudinal, front-parallel intermontane valleys (the Duns), and imbricated thrust system. The Duns occupy synclinal troughs in the evolving fold-thrust systems of the Sub-Himalaya and were filled by post-Siwalik piedmont sediments in piggy-back basins during the late Quaternary–Holocene period (Thakur et al. 2007; Srivastava et al. 2018).

The MHT-MFT fault system periodically releases strain during great (M > 8) earthquakes (Bilham et al. 2001) and can unlock to transfer strain during movements along these surfaces (Lavé and Avouac 2000), hence this deformation zone is significant for the Himalayan tectonics and natural hazards (Kumar et al. 2010; Malik et al. 2010; Jayangondaperumal et al. 2018 and references therein). Existing MFT fault zone deformation studies have been restricted to some outermost surface exposures of the Upper Siwalik (Srivastava and John 1999; Srivastava et al. 2018) and in trench excavations that typically access small fraction of the fault zone or only its minor splays (Kumar et al. 2010; Malik et al. 2010; Vassallo et al. 2015; and others).

6.10.1.1 Surface Deformation Associated with the Main Frontal Thrust (MFT)

Quaternary MFT motion has produced range-scale topography and folding (e.g. Karunakaran and Rao 1979; Raiverman et al. 1994; Malik et al. 2010; Barnes et al. 2011; Thakur 2013). The MFT has been interpreted as blind thrust (Mukhopadhyay and Mishra 2004), emergent (Powers et al. 1998), blind thrust in seismic sections and boreholes (Wesnousky et al. 1999), or simply a zone of intense brittle deformation (Srivastava and John 1999; Srivastava et al. 2018). This leads to variable shortening estimates within the MFT thrust sheet from ~27 to 17% in the Mohand Range (Powers et al. 1998; Mishra and Mukhopadhyay 2002), and fold kinematics such as fault-propagation folding (Lavé et al. 2005), fault-bend folding (Wesnousky et al. 1999), or generic growing anticline (Burgess et al. 2012).

The frontal Siwalik Ranges are marked by anticlinal ridges and in many places by synformal depressions of intermontane basins called duns, such as Soan dun and Dehradun in India and several duns in Nepal (Thakur et al. 2020). Some of these anticlines are located around Surin Mastgarh, Janauri, Chandigarh and Mohand (Fig. 6.39).

(i) **Surin Mastgarh Anticline (SMA)**: This symmetric anticline extends for nearly 250 km along its axial trace between Ravi and Chenab rivers with NE and SW dipping limbs with a subsurface decollement at ~6 km depth over the pre-Cenozoic sequence (Raiverman et al. 1994). The SMA forms the ridge topography of frontal Siwalik range where boundary between southern margin of the anticlinal ridge and Panjab plain is not sharp and linear due to onlap by alluvium, and fault-bend fold over the non-emergent HFT (Srinivasan and Khar 1996). Top part of the Upper Siwalik strata is assigned 0.2 Ma magnetostratigraphic age (Rao 1993), implying emergence of topography due to growth of fold post-dating ~200 ka.

(ii) **Janauri Anticline**: This anticline forms a 150 km long and 12–18 km wide ridge in frontal Siwalik range between the Beas and the Sutlej rivers, whose southern margin abuts against the MFT/Punjab alluvial plain, and northern margin is characterized by a back thrust (Fig. 6.40; Karunakaran and Rao 1979; Raiverman et al. 1994). On its back limb, intermontane Soan dun basin is

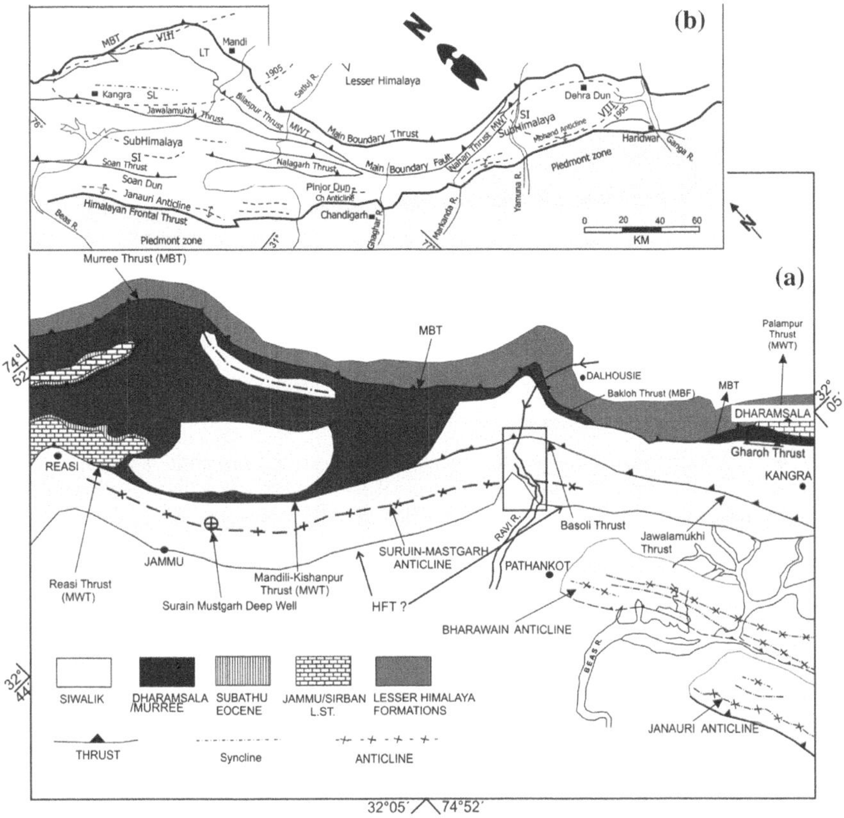

Fig. 6.39 **a** Tectonic map of the Sub-Himalaya between the MBT and the alluvial plain of Jammu region between the Riasi and the Kangra areas where the HFT is blind. Numerous frontal anticlines reveal their en echelon characters. **b** Sub-Himalaya between the MBT and the HFT between the Ganga and the Beas rivers, showing major frontal anticlines, dun (intermontane basins) and hinterland thrust faults. Dashed line ellipse is isoseismal VIII of 1905 Kangra earthquake. After Thakur et al. (2014)

filled with post-Siwalik sediments whose upper portion ranges between 36 and 10 ka (Suresh and Kumar 2009). Abandoned alluvial fan (AAF) on relatively flat central anticline ridge top are dated 42.9 ± 3.6 ka. With a dip of 30° of the HFT and vertical separation, shortening rates are 3.4 ± 0.2 mm/year and 6.9 ± 1.4 mm/year, respectively (Thakur et al. 2014).

(iii) **Chandigarh Anticline**: Frontal Upper Siwalik range exposes ~40 km long asymmetric Chandigarh Anticline with broad hinge zone over the HFT ramp as a fault-bend fold near Chandigarh (Fig. 6.39b; Thakur et al. 2009), with its southern margin marked by the HFT and the northern flank by a back-thrust whose back-limb is complemented by intermontane Pinjore dun basin. OSL-dated oldest alluvial fan sediments in this dun are 96–83 ka (Suresh et al. 2007),

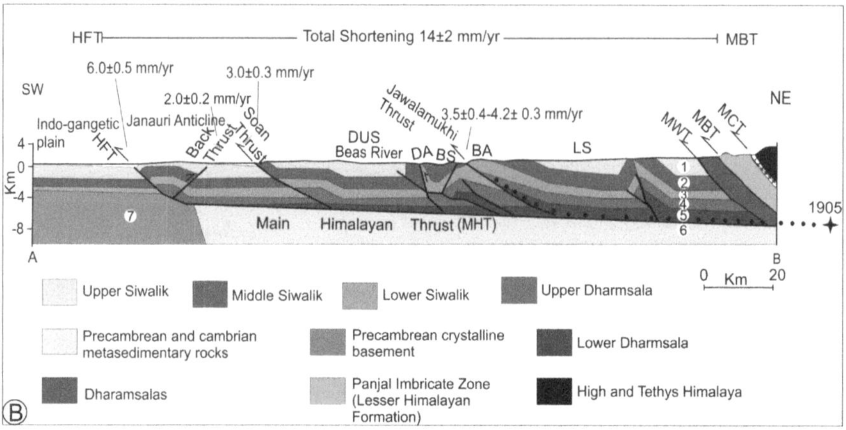

Fig. 6.40 Geological cross-section across the Panjab Re-entrant showing position of Janauri Anticline, the HFT and back-thrust. Other structures are: DUS—Dumskhar Syncline, BS—Balaru Syncline, BA—Balh Anticline, LS—Lambagaon Syncline, PA—Paror Anticline. Shortening across the section is also shown. After Thakur et al. (2020)

and top of the Upper Siwalik Boulder Conglomerate is 0.5 Ma in Haripur area (Sangode et al. 1996). Formation of the anticline and dun occurred between ~500 and 100 ka. Timing of formation of 40 km long Chandigarh Anticline is broadly correlatable with the assigned age of ~112 to 332 ka, derived from fault tip propagation rates and regional HFT slip rates (Barnes et al. 2011).

(iv) **Mohand Anticline**: In the frontal Siwalik Range, the Middle and Upper Siwalik strata are folded into an upright asymmetric Mohand Anticline between Mohand and Yamuna River in Uttarakhand, with its steeply dipping forelimb towards southwest and gently-dipping backlimb dipping towards northeast; it becomes subhorizontal broad syncline in Dehradun valley (Thakur et al. 2007) (Fig. 6.41). Boreholes around the Mohand Anticline struck pre-Tertiary basement at 4600 m depth (Karunakaran and Rao 1979). The HFT becomes almost horizontal towards north under the dun, forming part of the MHT (detachment),

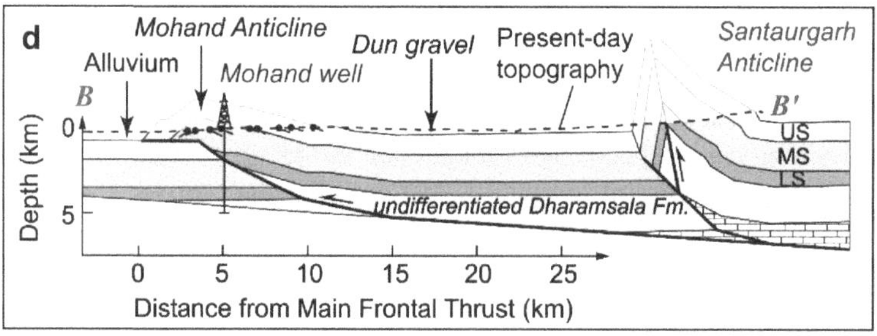

Fig. 6.41 Fold geometry of the Mohand Anticline, Dun Syncline and Santaurgarh Anticline, Sub-Himalaya, Mohand–Dehra Dun. After Mishra and Mukhopadhyay (2002) and Thakur et al. (2020)

implying the HFT to be a surface expression of the MHT. The HFT emerges at the Siwalik range front with steeper dip. A ramp is indicated over which the Mohand Anticline develops as a fault-bend fold (Wesnousky et al. 1999; Thakur et al. 2007). Strath terraces around 10–30 m high from the river bed provide estimated uplift, shortening and fault slip rates over the HFT as 6 mm/year, $\geq$11.9 $\pm$ 3.1 mm/year, and $\geq$13.8 $\pm$ 3.6 mm/year, respectively (Wesnousky et al. 1999) which is consistent with GPS-derived convergence rate (Banerjee and Burgman 2002).

Based on fault tip propagation rates and regional HFT slip rates, age of the ~80 km long Mohand Anticline is estimated as ~224 to 664 ka years (Barnes et al. 2011), which is nearly consistent with the earlier derived age of the Upper Siwalik Boulder Conglomerate and base of the Dun gravels between 500 and 200 ka years (Sangode et al. 1996, 1999). Barnes et al. (2011) also studied the erosion and relationship between faulting and base level changes in developing frontal fold topography. A significant erosion, ~15%, is estimated of the anticlinal topography, and that most (60–70%) of the relief and total catchment size is developed over a short (~5 km) distance from the fold tips. Srivastava et al. (2018) opined that a monocline structure in trishear fault propagation fold model is more suitable for deformation

(v) **Frontal anticlines in Nepal and eastern Himalaya**: In central Nepal south of Kathmandu, fault-bend folds are developed over a thrust system of the Main Frontal Thrust (MFT) ($\equiv$HFT), and Main Dun Thrust (MDT). The folds develop over the ramps rooting into the MHT at ~5 km depth. Using fluvial strath terraces across major fold along the Bagmati and Bakeya rivers, Lavé and Avouac (2000) estimated shortening over the MFT as 21 $\pm$ 1.5 mm/year during the late Holocene period.

In the foothills of Bhalukpong of western Kameng (Arunachal Pradesh), a 10 km wide active thrust zone of the MFT is mapped incorporating Tippi Thrust, Bhalukpong Thrust and Nimery Thrust (Burgess et al. 2012). Dating of fluvial terraces provided an estimated Holocene shortening of 23.4 $\pm$ 6.2 mm/year, accomodated over the MHT, while total shortening is distributed as ~8.4 mm/year, ~10 mm/year, and ~5 mm/year over the Bhalukpong Thrust, the Balipara Anticline, and the MFT, respectively.

6.10.2 Main Boundary Thrust (MBT)

The Main Boundary Thrust (MBT) is one of the major north-dipping Himalayan thrusts, developed during Middle Cenozoic, and forms the present-day structural and orographic boundary between the Sub- and Lesser Himalaya (Valdiya 1980a, b; 1992; Mugnier et al. 2004). Originally, it was defined as the fault contact between the older Tertiary sequence (Subatbu–Dagshai–Kasauli) of the Sub-Himalaya and the Siwalik, and called as the 'Main Boundary Fault' (Medlicott 1864; also Pilgrim and

West 1928; Auden 1934). At this level, a thrust gets identified within the NW Sub-Himalayan belt (Fig. 6.42) and is variously called as the (i) 'Boundary Fault' between the Murree and the Siwalik from NW syntaxis to Jammu–Poonch (Wadia 1928, 1931), (ii) Riasi Thrust and the Murree Thrust as the Main Boundary Thrust (MBT) (Karunakaran and Rao 1979; Raiverman et al. 1994) (iii) 'Main Boundary Fault' between Subathu, Murree and Jammu limestone inliers from the Siwalik (Wadia 1937), (iv) 'Palampur Thrust' between lower Tertiary Dharamsala from underlying Neogene Siwalik in Kangra (Karunakaran and Rao 1979; Raiverman et al. 1994), (v) 'Bilaspur Thrust' separating stromatolite-bearing Proterozoic limestone, lower Tertiary Subathu and Dharamsala from the Siwalik in Bilaspur area (Raiverman,

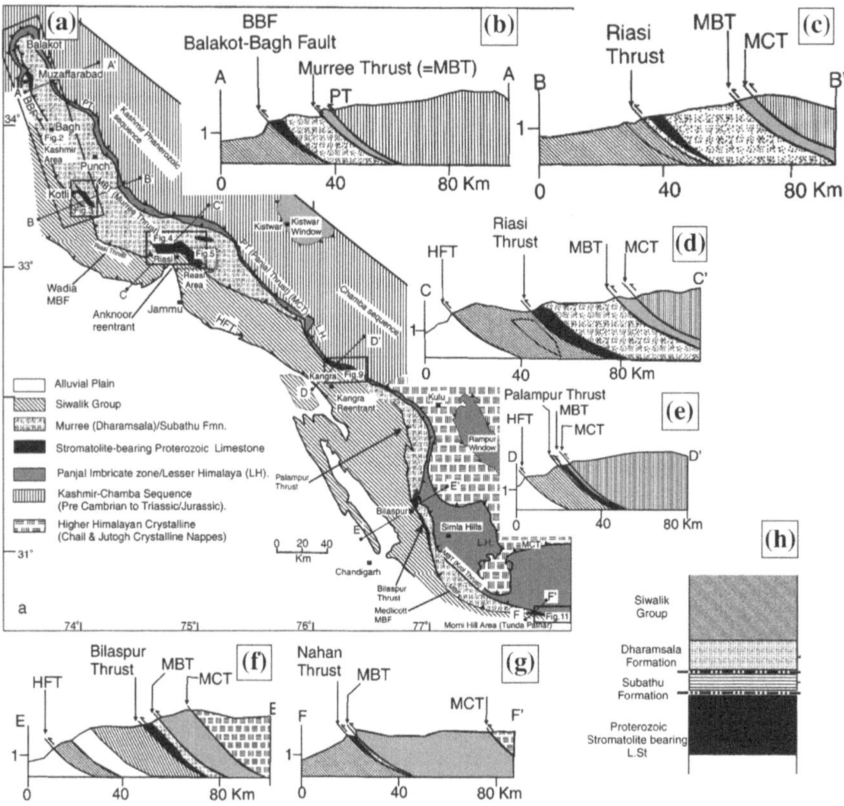

Fig. 6.42 **a** Outline tectonic map of the northwest Sub-Himalaya and adjoining regions showing regional geological and tectonic framework of the NW Sub-Himalaya and adjoining regions Lower Cenozoic sediments contain stromatolite-bearing Proterozoic limestone inliers zone. Cross-sections across Medlicott–Wadia Thrust (MWT) and Main boundary Thrust (MBT) through **b** Balakot (A–A′), **c** Kotli (B–B′), **d** Riasi (C–C′), **e** Kangra (D–D′), **f**) Bilaspur (E–E′), **g** Nahan (F–F′). **h** Composite stratigraphic column of the foreland basin showing Subathu unconformably overlying the Proterozoic limestone. After GSI map (1999), Thakur (2010)

2002), (vi) 'Nahan Thrust' between Lower Siwalik (Nahan Formation) and overlying lower Tertiary and stromatolite bearing-limestone around Nahan and Yamuna R.

In his map, Wadia (1931) showed the Boundary Fault extending NW–SE from northwest syntaxis to the river Ravi. The Murree Thrust was demarcated north of the Main Boundary Fault (MBF), separating the Murrees from Kashmir nappe sequence. Later workers (Karunakaran and Rao 1979; Raiverman et al. 1994) designated Wadia's MBF as the Riasi Thrust and the Murree Thrust as the Main Boundary Thrust (MBT) (Fig. 6.42). In Kangra region, the thrust separating the lower Cenozoic Dharamsala from the underlying Neogene Siwaliks was called the Palampur Thrust (Karunakaran and Rao 1979; Raiverman et al. 1994). Extension of the same thrust in Bilaspur area separating the stromatolite-bearing Proterozoic limestone, the lower Cenozoic Subathu and Dharamsala from the Siwalik was designated the Bilaspur Thrust (Raiverman 2002). Further east in the area between Simla hills and river Yamuna, the contact between the Lower Siwalik (Nahan Formation) and the overlying lower Tertiary and the stromatolite bearing-limestone was designated the Nahan Thrust or Medlicott's MBF. Post-Wadia, earlier studies were region specific lacking regional strike mapping that led to multiplicity of local names to the same formation and same fault at different places. Further southeast the Murree is overlapped by the Krol Thrust (Auden 1934), which brings Lesser Himalayan nappe against the Subathu and Dagshai or the Siwalik. From east of the Yamuna valley to the west of Gandak valley in central Nepal, the Krol nappe has completely concealed the analogue of the Palampur-Nahan Thrust. East of the Gandak valley up to the Kosi valley, metamorphic gneisses and schists of the Kathmandu nappe (Hagen 1958) of the Mahabharat Range override the Siwalik. Beyond the Kosi valley and the rest of the eastern Himalayan front, folded and imbricated succession of the Permian Rangit (=Garu) and Damuda (Gondwana) formations, sometimes including the basement (Jainti/Phuntsholing) is thrust against the Siwalik. Named as the Garu Thrust in the extreme east by Jain et al. (1974) this is the MBT in the Gondwana setting.

6.10.3 Medlicott Wadia Thrust (MWT)

Mapping by the Oil and Natural Gas Corporation (ONGC) in the entire Cenozoic Sub-Himalaya belt from Jammu region to Kali River revealed a tectonic boundary between this belt and pre-Tertiary Lesser Himalaya, and called it as the 'Main Boundary Thrust (MBT)', as revised nomenclature for the Murree Thrust and the Krol Thrust. Therefore, Thakur (2010) proposed a new name for the MBF as the Medlicott Wadia Thrust (MWT) and retained the present nomenclature of the MBT (Fig. 6.42). A brief account of the MWT in different sectors is given below to emphasize its independent status.

(i) **Balakot–Bagh Fault (BBF)**: On the western limb of the Hazara syntaxis (Wadia 1931), a belt of Pre-Cambrian Muzaffarabad Formation (stromatolite-bearing limestone and shale), extending NW from Muzaffarabad to Balakot, is overlain by Paleocene Patala and non-marine Murree formations (Figs. 6.42 and 6.43); the belt is faulted against the Miocene Murrees/Kamlial non-marine formation in the southwest. This fault was called as the Balakot–Bagh Fault (BBF) (Yeats et al. 2006; Hussain et al. 2009). This fault was ruptured during the 2005 Kashmir earthquake of magnitude 7.6 Mw with 3–7 m vertical separation in three segments of coseismic faulting (Kaneda et al. 2008). Further southeast, additional faults reveal right-step from the BBF and two en-echelon anticlines near Kotli (Hussain et al. 2009); SW faulted margin of the anticline extend into the Riasi Thrust in Jammu region with a connecting intervening Rawalkot fault (Fig. 6.43; Hussain et al. 2009).

(ii) **Riasi Thrust (Jammu)**: About 100 km long NW–SE trending mountain chain in Jammu-Riasi region of Neoproterozoic Jammu Limestone (Sirban Limestone) consists of interbedded cherty limestone-dolomite, quartzite and shale, and is unconformably overlain by bauxite, nummulitic limestone,

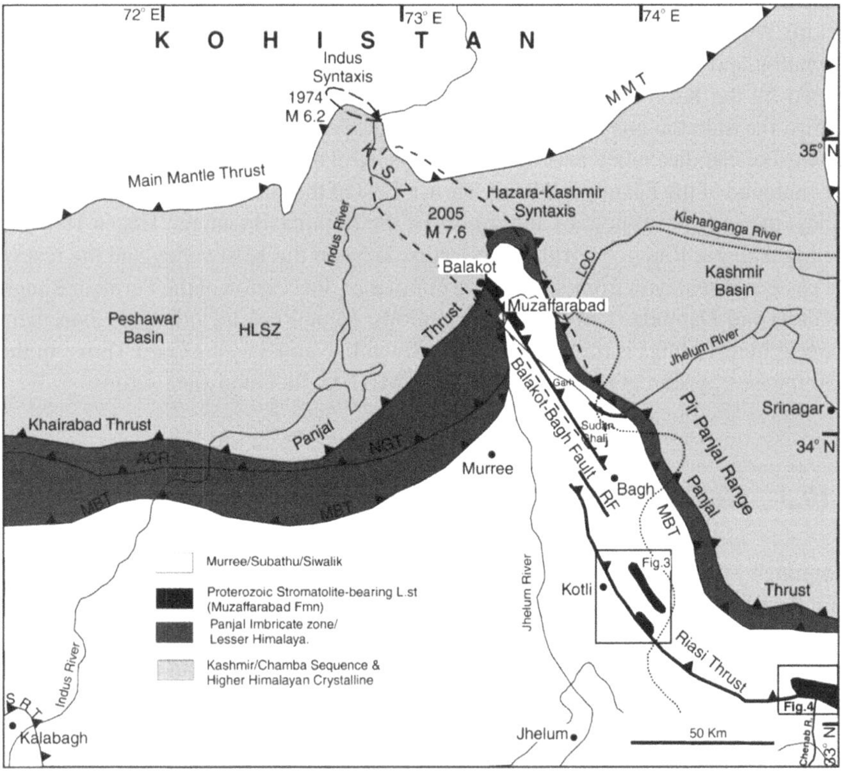

Fig. 6.43 Tectonic map of northwest Himalaya showing isoseismals of the 2005 Kashmir and 1974 Indus earthquakes. Balakot–Bagh Fault. Riasi Thrust. Rawalkot Fault (RF). Indus Kohistan Seismic Zone (IKSZ). Hazara Lower Seismic Zone (HLSZ). Nathia Gali Thrust (NGT) Attock Cherat Range (ACR). Main Boundary Thrust (MBT). After Hussain et al. (2009)

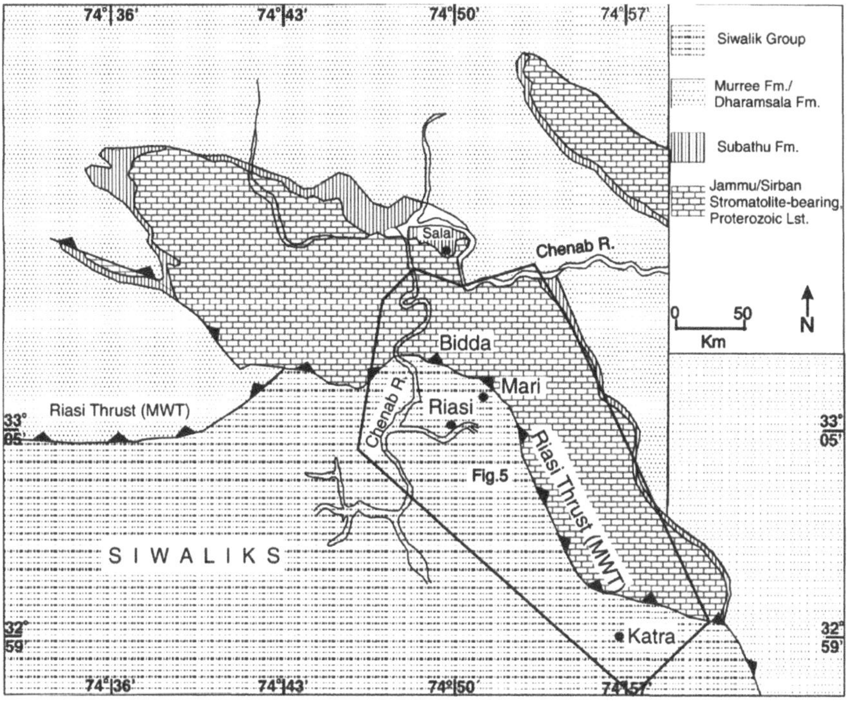

Fig. 6.44 Geological framework of the Riasi–Katra areas. Stromatolite-bearing Proterozoic Jammu/Sirban limestone inliers, overlain by the Subathu and Murree, are faulted against the Siwalik in the south along the Riasi Thrust. Medlicott–Wadia Thrust (MWT)

carbonaceous-ferruginous shale and coal seams of marine Paleocene–middle Eocene Subathu Formation, and non-marine Murree Formation in the north (Fig. 6.44). Southwestern margin of this limestone inlier makes a ~1000 m high fault scarp against the Middle–Upper Siwalik foothills of micaceous sandstone, red-orange shale and boulder conglomerate. This fault was initially named as the Boundary Fault (Wadia 1928) or Main Boundary Fault (Wadia 1931), and subsequently the Riasi Thrust (Karunakaran and Rao 1979). The fault witnessed Pleistocene and Holocene reactivation, as is evidenced by ~30 m scarp underlain by tilted fan surface dipping 20° NE (Awasthi et al. 1985). Other evidences of reactivation within the strath surfaces include presence of S-facing ~80 m high scarp, tilting of layered and consolidated gravels and folding due to fault-propagated fold over a low angle NE-dipping fault as a footwall imbricate of the Riasi Thrust.

(iii) **Riasi Thrust (Ravi River and Kangra reentrant)**: Separating the Murree from the Siwalik formations, the Riasi Thrust extends SE from the Riasi–Katra region towards the Ravi River, and is known as the Mandili–Kishanpur Thrust (Fig. 6.45) (Karunakaran and Rao 1979; Raiverman 2002). This thrust branches out into a northern branch as the Bakloh and Palampur Thrusts, while

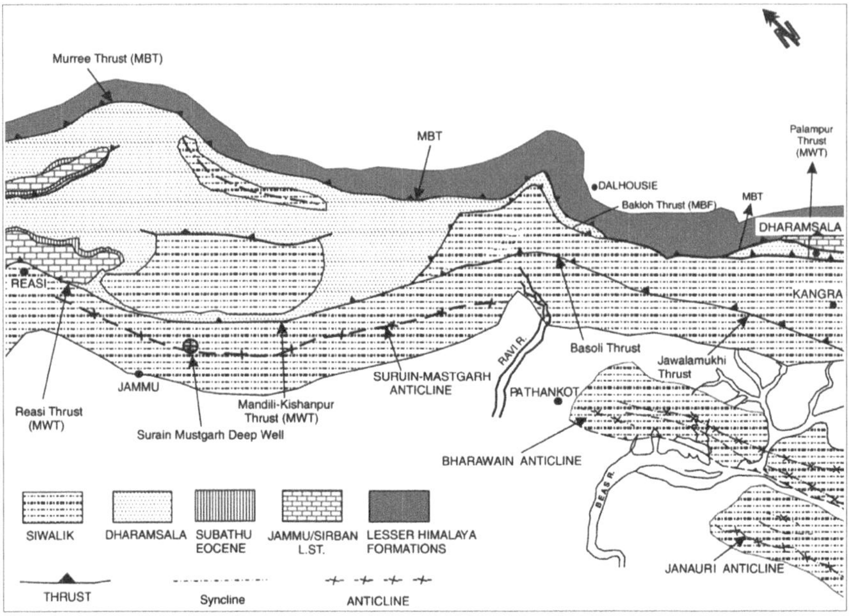

Fig. 6.45 Simplified geological map of the Sub-Himalaya between Riasi and Dharamsala showing extension of the MWT. Medlicott–Wadia Thrust (MWT). Modified after Karunakaran and Rao (1979) and Raiverman (2002) by Thakur et al. (2010, 2020)

southern branch extends southeast as the Basoli and Jawalamukhi Thrusts. The Dharamsala Formation (=Murrees) is truncated between Dalhousie and Dharamsala, hence stromatolite-bearing Proterozoic limestone, (Dharamkot limestone), occurs as thrust slice between the Dharamsala Formation and the Salkala Formation in Dharamsala area (Kumar and Mahajan 2001). The Palampur Thrust (MWT) between the Dharamsalas and Siwaliks represents eastern continuation of the Riasi Thrust and passes interruptedly in the Kangra reentrant due to concealment beneath the post-Siwalik late Quaternary–Holocene fan sediments. At places, late Quaternary–Holocene unconsolidated sediments of silt and sand with large boulders of the Dhauladhar granite are thrust over by the Upper Siwalik boulder conglomerate along an active footwall imbricate fault dipping NE at about 30° (Thakur et al. 2020).

(iv) **Bilaspur Thrust/Nahan Thrust (Bandla Range/Shimla Hills)**: The Palampur Thrust extends S and SE to Shimla Hills through intervening Bilaspur area, where stromatolite-bearing Proterozoic limestone and slates are overlain by the Subathu and Dharamsala in the Bandla range. These are thrust over by the Lesser Himalaya formations along the MBT, while southern margin of this belt is demarcated against the Siwalik by the Bilaspur Thrust as a southeastern extension of the Palampur Thrust (inset Fig. 6.42e) (Karunakaran and Rao 1979). The Bilaspur Thrust continues further SE in the Shimla Hills, where the reverse faulted contact between the Subathu Formation with the underlying Nahan Formation (Lower Siwalik) was designated as the Main Boundary Fault (MBF) by Medlicott (1864) and later followed by Pilgrim and West (1928, and others) (Fig. 6.46). Auden (1934) identified the Krol Thrust (=present MBT) as a boundary between the lower Tertiary of the Sub-Himalaya and the overthrust Krol Group of the Lesser Himalaya.

(v) **Nahan Thrust (Morni Hills/Sataun)**: Stromatolite-bearing large discontinuous bands and lenticular bodies of the Proterozoic massive limestone make a belt within lower Tertiary in Morni Hills and adjoining Tundapathar area of Sataun (Fig. 6.47) (Narain 1967). The limestone is unconformably overlain by the Subathu Formation and Dagshai Formation and are thrust over the Lower Siwalik along the Nahan Thrust (≡Bilaspur Thrust). The Nahan Thrust extends to Nahan and Sataun area with similar tectonostratigraphic setting. Along the Giri River, Proterozoic limestone occurs as discontinuous bodies between the underlying Lower Siwalik and the overlying Subathu Formation. The NE dipping Nahan Thrust brings the stromatolite-bearing limestone and the Subathus

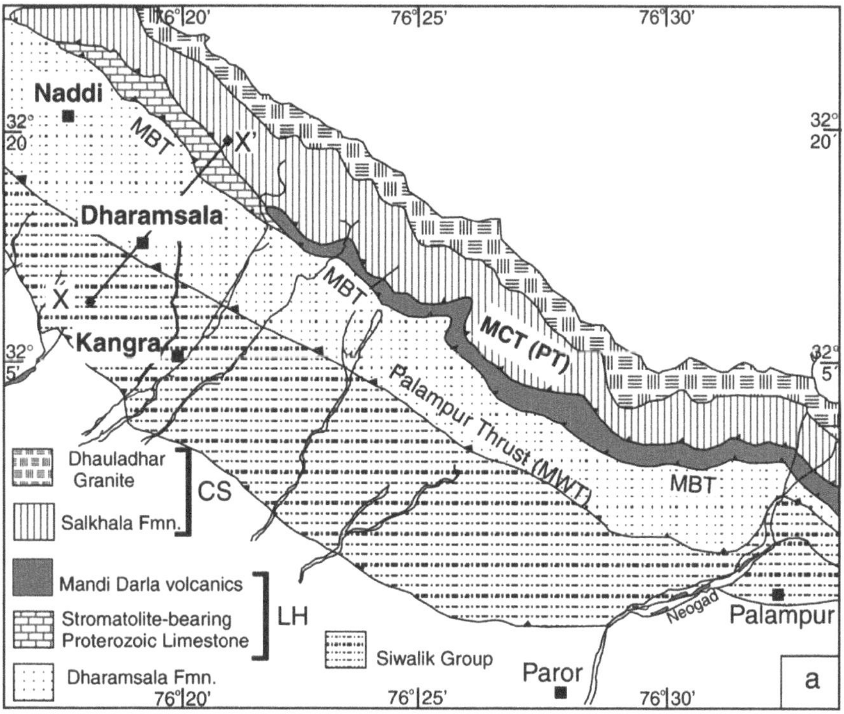

Fig. 6.46 Geological map of Dharamsala–Palampur area showing extension of the MWT and the MBT. Main Boundary Thrust (MBT), Palampur Thrust (PT), Main Central Thrust (MCT), Panjal Thrust (PT), Chamba sequence (CS), Lesser Himalaya (LH), Medlicott Wadia Thrust (MWT). After Kumar and Mahajan (1991)

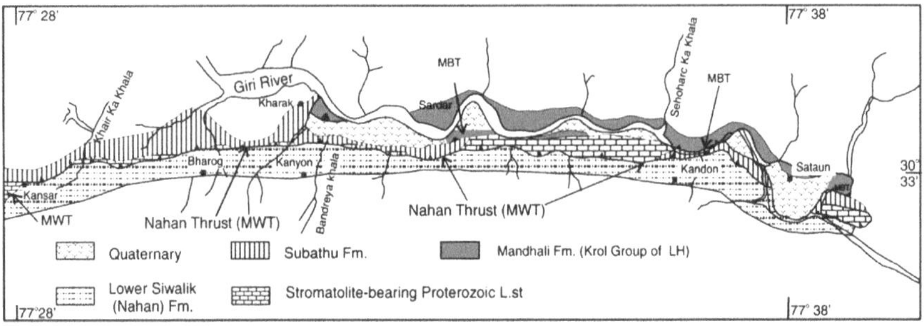

Fig. 6.47 Geological map of the Sub Himalayan MWT belt along Giri river near Sataun. Nahan Thrust/(MWT) defines a tectonic boundary between the Lower Siwalik (Nahan), and the overlying Subathus and Stromatolite-bearing Proterozoic limestone. From Thakur et al. (2010)

directly over the Lower Siwalik, and are thrust over by the Mandhali Formation (Lesser Himalaya) along the Krol Thrust (≡MBT). Further east, the Nahan Thrust extends into the Yamuna valley, where an active strike-slip fault cuts the MBT and the Nahan Thrust during late Holocene.

6.10.4 Reactivation of the Main Boundary Thrust (MBT)

In the NW Himalaya, the Main Boundary Thrust (MBT) have undergone initial displacement along its >1000 km segment prior to 10 Ma when foreland-basin fill underwent extensive changes in their depositional characteristics and accumulation rates between 11 and 9.5 Ma as a consequence of hinterland loading due to the formation of this thrust (Meigs et al. 1995). Combined with fission-track apatite ages from leading edge of the MBT between 8 and 10 Ma, rapid cooling followed the bedrock uplift and erosion about ~1 to 2 m.y. earlier. Thus, the MBT formed synchronously in the middle-late Miocene with a displacement rate of ~10 mm/year, that is coeval with late displacement on the Main Central Thrust (MCT).

After its formation ~11 Ma, the MBT exhibits neotectonic activation in certain localized areas (Valdiya 1992; Mugnier et al. 2004; Luirei et al. 2014; Philip et al. 2017; Thakur et al. 2020). In Dehra Dun, the MBT defines a physiographic and tectonic boundary between the Lesser Himalayan Krol Group of Mussoorie range, the Siwalik and the Dun gravel of the Dun valley (Fig. 6.41). The shear zone defining the MBT dips 25°–45° NE but, at places, it becomes steeper due to activation (Valdiya 1992). In the proximal part of Dun principal fan, the Lesser Himalayan rocks are thrust over the Dun gravels, whose OSL age of 31 ± 1 ka indicates late Quaternary activation of the MBT (Thakur et al. 2007). In eastern Kumaon (Gola River), the Lesser Himalayan Bhim Tal volcanics on hanging wall of the north dipping MBT abut against sandstone of the lower Tertiary Dharamsala/Lower Siwalik Formation in its footwall. OSL dating and paleoseismology trench investigation along a fault scarp paralleling the MBT within overlying alluvial fan reveals its age between 20 and 16.5 ka (Philip et al. 2017).

In Nepal Himalaya in Timari Khola, active faulting between the MFT and MBT is observed by Nakata (1989), where NW–SE active fault crosses the MFT showing vertical slip with downthrown to the north associated with right–lateral displacement. West of the Kosi River in central Nepal foothills ~120 km long Surkhet–Gorahi Fault produces a pressure ridge and associated linear depression having vertical displacement of 12–35 m (Nakata 1989).

6.10.5 Main Central Thrust (MCT)

6.10.5.1 Defining Main Central Thrust (MCT)

The Main Central Thrust (MCT) is one of the most significant and largest tectonic discontinuities as a ductile shear zones in the Himalaya, extending for nearly 2500 km along its entire length from Pakistan, western Zanskar, Nepal, Bhutan, Arunachal Pradesh, and even extends into Myanmar (Fig. 6.38; Thakur 1993; Hodges 2000; Searle et al. 2008; Yin 2006; Valdiya 2016). Heim and Gansser (1939) first introduced this term in Garhwal for 'a sharp contact where northerly-dipping crystalline rocks rest over the metamorphic limestone (p. 78)'. The crystalline rocks were called as the 'Crystalline Central Zone'. They noted that this thrust mass 'represents an enormous deep-rooted body of 10–15 km thick crystalline rocks, which must have been at a depth of 30 km below the surface at ~700 °C and thrust southwest after the deposition of the Cretaceous flysch (p. 225)'. It is therefore evident that there was only one thrust was recognized by Heim and Gansser (1939) on the regional scale, though they did not rule out possibilities of local repetition due to thrusting. Since then, the MCT has been investigated by various workers along its entire length (Fig. 6.38; Gansser 1964, 1983; Jain and Manickavasagam 1993; Hodges 2000; Searle et al. 2008, 2017; Carosi et al. 2018). Definition of the MCT has undergone numerous changes with time due to difficulties in its recognition in the field (see Searle et al. 2008, 2017; Jain et al. 2014, 2020; Martin 2017; Mukhopadhyay et al. 2017; Carosi et al. 2018). The following criteria are applied in its recognition, either independently or in combination.

(i) A lithological–structural–metamorphic characters (Heim and Gansser 1939), and identification of the MCT along a zone between gneiss of the HHC and LH sediments (Valdiya 1980a, b).
(ii) A high-strain reverse kinematic zone separating distinguishable protoliths (e.g. Martin et al. 2005; Martin 2017) with presence of ductile shear zones with different kinematics and its emplacement over the LH (Jain and Manickavasagam 1993; Davidson 1997).
(iii) A metamorphic–rheological approach (Searle et al. 2008) with distinct metamorphic breaks along the MCT (Vannay and Hodges 1996; Valdiya 1980a, b) or rheological (Gibson et al. 2016; Parsons et al. 2016a).

(iv) A chronological definition (Webb et al. 2013), having distinct isotopic signatures within the HHC (Ahmad et al. 2008), and U–Pb zircon ages and/or both (Jain et al. 2020; Mukherjee et al. 2019).
(v) Record of protracted deformation from ductile to brittle affecting several different lithologies along-strike (Carosi et al. 2018 and references therein).

6.10.5.2 Classification and Terminology

Two distinct MCT are now recognized throughout the Himalaya (Fig. 6.38): the Ramgarh/Munsiari and Vaikrita Thrusts in the NW Himalaya (Valdiya 1980a, b; Jain et al. 2014, 2020; Sheshrtha et al. 2015; Mukherjee et al. 2019), the MCT-1 and the MCT-2 in Nepal (Hashimoto et al. 1973; Arita 1983), bounding a package of rocks in the Main Central Thrust Zone (MCTZ). The latter terms are no more used in favor of the earlier ones (DeCelles et al. 2000; Robinson et al. 2001; Robinson 2008; Mukhopadhyay et al. 2017).

Valdiya et al. (1999 and references therein) identified an outermost lower crustal-scale mylonitized sheared package in the Uttarakhand Himalaya, bounded by the Munsiari Thrust (MT) along its base and the Vaikrita Thrust (VT) at its top. Valdiya et al. (1999) recognized the Vaikrita Thrust as the actual MCT (also Ahmad et al. 2008), though the Munsiari Thrust was the original MCT as per Heim and Gansser's (1939) definition.

The Main Central Thrust is recognized based on age, stratigraphy and metamorphism (Bordet 1961; Le Fort 1975; Gansser 1983; Parrish and Hodges 1996; Harrison et al. 1997; Ahmad et al. 2008; DeCelles et al. 2000; Catlos et al. 2001, 2002; Robinson et al. 2001; Daniel et al. 2003; Martin et al. 2005; Richards et al. 2005; Searle et al. 2008, 2017). However, none of these criteria can correctly and univocally identify the location of the MCT (Montemagni et al. 2016). Strain gradient is considered as the main criteria to define the MCT where a clear strain localization, evident from field and microstructural analyses, of deformation is present (Searle et al. 2008). The MCT roughly coincides with a wide zone characterized by Barrovian inverted metamorphism (Searle and Rex 1989; Davidson et al. 1997; Walker et al. 1999; Searle et al. 2008). Larson and Godin (2009) considered the Ramgarh Thrust (MCT I or Munsiari Thrust) as the MCT at the base of pervasively sheared rocks of the GHS, affected by inverted metamorphism (cf. Searle et al. 2008). Martin (2017) considered the MCT defining the protolith boundary where high-strain zone with thrust kinematics is present.

Within the MCTZ, P-T conditions of nearly peak metamorphism is bracketed between ~6 to 10 kbar and 520–650 °C (Hubbard 1989; Vannay and Grasemann 1998, 2001; Catlos et al. 2001; Vannay et al. 2004; Montemagni et al. 2016; Iaccarino et al. 2017). Interestingly, the MCTZ coincides with zone of inverse metamorphic grade ranging from biotite to kyanite/sillimanite zone in structurally higher levels (Searle et al. 2008; Kohn 2014; and references therein).

Such variable definitions of the MCT result in different geological maps regarding localition of the MCT itself and, hence, variations in thickness of the GHS and LHS. For example in the Kali Gandaki valley (Central Nepal), Le Fort (1975), Vannay and Hodges (1996) and Carosi et al. (2014) placed the MCT at the base of the garnet + kyanite-bearing gneiss of the GHS, and is nearly 15 km north to its position, recognized by Searle (2010) and Parsons et al. (2016a, b), who located the MCT in the Precambrian quartzites of the LHS unit where deformation attributed to the MCT vanishes to the south. Likewise, the MCT is shifted north in Sikkim towards the upper portion of the GHS at the base of sillimanite + K-feldspar zone within the migmatites of the upper GHS (Mottram et al. 2014a, b; Chakraborty et al. 2016, 2017). Such a variation in location of the MCT involves several kilometer thick parts of the LHS and lower parts of the GHS, including ages from ~23 to ~3 Ma (Godin et al. 2006; Catlos et al. 2002).

The GHS in Nepal also reveals a major structural-metamorphic-geochronological tectonic, which has been called as the Higher Himalayan Discontinuity (HHD) by Montomoli et al. (2013, 2015) and Carosi et al. (2018). The HHD is located much above the MCT with ages from ~28 to 17–16 Ma, which become distinctly younger to 17–16 Ma onward in the MCT zone (Montomoli et al. 2013; Carosi et al. 2018 and references therein). In the western Nepal, the Kalopani Shear Zone revealed first melting process during increasing metamorphic conditions in kyanite gneiss where two separated GHS parts along the shear zone moved between ~41 and 31 Ma (U–Th–Pb monazite age; Carosi et al. 2015). Its hanging wall was exhuming with footwall undergoing higher P and T up to ~27 to 26 Ma. Such process was repeated along the still lower level HHD and the MCT between 26 and 18 Ma, and subsequently between 17 and 13 Ma along the MCT and its footwall (Carosi et al. 2018).

6.10.5.3 Typical Cross-Section in Uttarakhand Himalaya

Figure 6.27 summarizes the litho-tectonics characters from one of the typical and complete cross-section through the Higher Himalayan crystalline (HHC) Belt along the Alaknanda–Dhauli Ganga valley in Uttarakhand Himalaya, where this belt is thrust over the Inner Lesser Himalayan (iLH) Sedimentary Belt along the Munsiari Thrust.

(i) **Munsiari Group**: Low- to medium-grade Munsiari Group is essentially made up of highly imbricated and mylonitized Paleoproterozoic rocks of dominant mylonitic augen mylonite (~1.85 Ga: Mukherjee et al. 2019), fine grained biotite-rich gneiss, garnetiferous mica schist, phyllonite, sheared amphibolite, and a persistent imbricated horizon of sericite-bearing foliated and sheared quartzite; the latter bears strong lithological resemblance with the LH Berinag Group quartzite. This group displays intense penetrative shearing with NE-dipping mylonitic S-C foliation and a consistent down-the-dip plunging lineation, though porphyritic granite protoliths sometimes escape intense shearing within the MCTZ (Fig. 6.48a). S-C-C′ foliation transects a well-developed

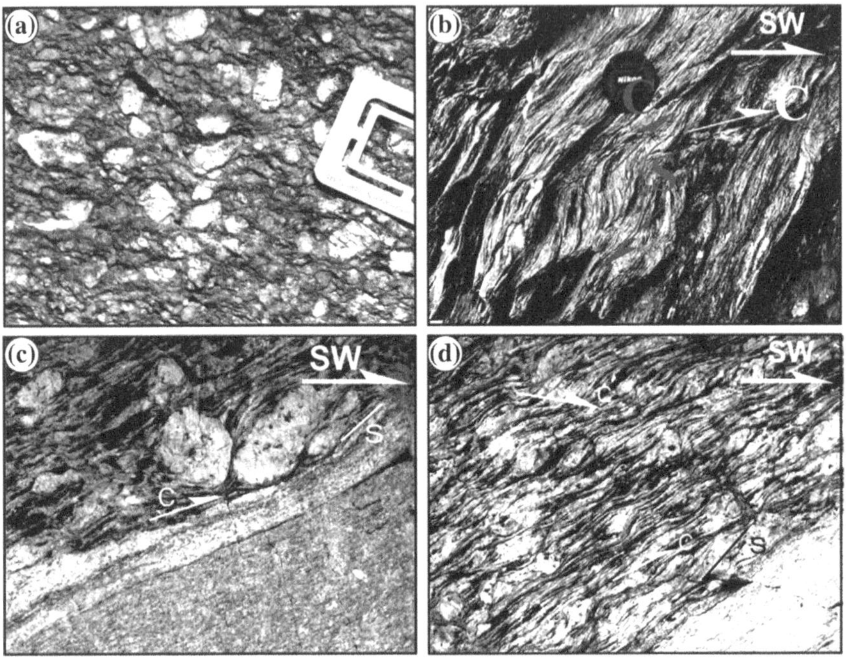

Fig. 6.48 Deformation and shearing within the MCT Zone along Bhagirathi and Alaknanda-Dhauli Ganga valleys. **a** Foliated megacrystic gneissose granite along the MCT, at Sainj, Bhagirathi valley showing minor ductile shearing not affecting the feldspar megacrysts. Scale: Sim card case. **b** S-C ductile shear fabric within mica schist having top-to-the-SW thrust shear sense, which is typical of the MCTZ. Loc.: Near Helang along Alaknanda valley. Lens cap for scale. **c**, **d** Typical augen mylonite from the MCT zone showing S-C-C′ ductile shear fabric and asymmetrical feldspar augen with top-to-the-SW thrust shear sense. Loc.: Dhak Nala near Tapovan, Dhauli Ganga valley

foliation S at a small angle; the latter becomes sigmoidal in character and gradually deflects near the C-surfaces paralleling the main shear zone (Fig. 6.48b). The belt is ubiquitously marked by small-scale asymmetrical structures like S-C and S-C′ foliation, porphyroclast (Fig. 6.48c, d) and porphyroblast, mineral fish, intrafolial folds, duplex structures, ductile—brittle shear zones, and asymmetric shear boudins. The LH quartzite is incorporated within the MCT zone due to imbrication. The pelitic samples within the Munsiari Group have quartz + biotite + muscovite + plagioclase ± garnet ± staurolite ± graphite ± andalusite (Spencer et al. 2012), while amphibolitic bodies have zoned calcic amphibole, plagioclase, quartz, biotite and minor epidote, opaque, chlorite and calcite. Peak P-T conditions, recorded by pelitic rocks, increase structurally upward from 0.5 GPa/500 °C up to c. 1.1 GPa/580 °C (Spencer et al. 2012), thus it displays inverted metamorphism. According to After peak metamorphism, the MCTZ rocks cooled to the white mica Ar–Ar closure temperature (350 ± 50 °C) at c. 9 Ma (Célérier et al. 2009).

(ii) **Vaikrita Group**: The overlying Vaikrita Group sequence of the HHC within the lower and upper amphibolite facies is divisible into three formations between the VT and the STDS: (i) the Joshimath Formation, (ii) the Suraithota Formation, and (iii) the Bhapkund Formation. The Joshimath Formation comprises garnet-biotite-muscovite schist and psammitic/pelitic gneiss, having (Spencer et al. 2012). The Suraithota Formation contains kyanite-garnct-biotite schist/psammitic gneiss and quartzites with thin amphibolite intercalations. Depositional features such as cross-beds are seen in less deformed rocks and indicate that many parts of this formation have been overturned. P-T conditions for this formation (≡Pandukeshwar Formation) are in the range of c. 14 kbar and c. 800–860 °C (Spencer et al. 2012). Topmost sequence of the Vaikrita Group is the Bhapkund Formation containing sillimanite (fibrolite)-garnet-biotite psammitic gneiss/schist, small tourmaline-rich leucogranite lenses/dykes and the Malari granite. Migmatization is pervasive within this formation. Geothermobarometric calculations by Spencer et al. (2012) reveal a pressure decrease from 10.5 to 8.5 kbar towards the STDS, while the temperature is roughly constant around 810–860 °C. A clear metamorphic break is present between the high-grade rocks of the Bhapkund Formation in the footwall of the STDS and the hanging wall low-grade to unmetamorphosed rocks of the Martoli Formation with a maximum T of c. 450 °C.

Overall, the HHC trends NW–SE in this valley with dips of the main foliation Sm towards NE- to E. Various shear sense indicators proliferate between Helang and Malari. Asymmetric structures such as S-C and S-C′ fabric, boudins, sigma structures, folds etc. provide valuable information regarding direction of tectonic transport from top-to-south or southwest ductile shears throughout the HHC and top-to-north or northeast shear sense in the middle to upper parts near the STDS (Shrestha et al. 2015).

6.10.5.4 Timing of the MCT

The activity time span of the MCT in different areas of the belt was estimated using mutually contrasting methods or criteria. The activity time span ranges from 23–20 to 15 Ma in different areas of the belt (see Godin et al. 2006; Montomoli et al. 2015 for an updated review) down to a reported c. 3 Ma in central Nepal (Catlos et al. 2001). In the Garhwal Himalaya, several researchers proposed preferred ages for the MCT activity based on different chronometers (K–Ar, Th–Pb and $^{39}Ar/^{40}Ar$) and on different non-isotopic sample characterizations. Metcalfe (1993) obtained K–Ar ages on biotite and muscovite from the Bhagirathi valley and proposed that it was active between 14 and 5.7 Ma. Catlos et al. (2002) extended their previous work in Nepal to western Garhwal beneath the Vaikrita Thrust and asserted that the Th–Pb ages of monazite constrain the age of the entire activity of the MCT in the central and western Himalaya to c. 6 Ma. Célérier et al. (2009) reported c. 9 Ma obtained using $^{39}Ar/^{40}Ar$ on muscovite from samples in the middle portion of the MCTZ near the village of Helang. Sen et al. (2015) obtained $^{40}Ar/^{39}Ar$ biotite ages of c. 10 Ma and interpreted them as ‘cooling ages’ correlated to the exhumation of the GHS caused by MCT thrusting at that time, while muscovite ages of c. 6 Ma were related to a late-stage deformation post-dating biotite cooling. Montemagni et al. (2019) recorded three different biotite–muscovite growths and recrystallization episodes along the Vaikrita Thrust in this valley viz., a relict mica-1 and mica-2 along the main mylonitic foliation; and mica-3 in coronitic garnet during its breakdown. Biotite step-ages between 8.6 and 16 Ma and muscovite step-ages between 3.6 and 7.8 Ma reflect sample-specific recrystallization markers, hence these ages were linked to the activity along the Vaikrita Thrust at least from 9 to 6 Ma at c. 600 °C and termination by 6 Ma.

6.10.6 South Tibetan Detachment System (STDS)

After the first recognition of an extensional fault system in southern Tibet within the contractional Himalayan orogen (Burg et al. 1984; Burchfiel and Royden 1985), the South Tibetan Detachment System (STDS) is now ubiquitously documented throughout the Himalaya from western parts, Nepal, Sikkim and Bhutan (Fig. 6.38; Herren 1987; Burchfiel et al. 1992; Patel et al. 1993; Grujic et al. 1996; Hodges et al. 1996; Carosi et al. 1998; Dèzes et al. 1999; Jain and Patel 1999; Jain and Manickavasagam 1999; Jain et al. 2005; Godin et al. 2006; Law et al. 2004; Shreshtha et al. 2015; Iaccarino et al. 2017).

Many north-dipping parallel low- to steep dipping normal faults of the STDS group exhibit top-to-the-north normal shear sense and juxtapose the Tethyan Himalayan Sequence (THS) over the Higher Himalayan Crystalline (HHC) belt/Tibetan Slab (Burchfiel et al. 1992). At places, these reveal brittle normal faults, detachment faults between the unmetamorphosed/poorly metamorphosed THS and underlying metamorphics, and ductile STDS-controlled shear zone at the top of the HHC (Leloup et al. 2010). In addition, the STDS also revealed an older top-to-the SW thrust shear sense, superposed by top-to-the-north normal faulting in Zanskar (Dèzes et al. 1999; Patel et al. 1993). Throughout the Himalaya, the STDS is now a well-established top-to-the-north-down structure, having ductile to brittle normal shear zones of variable thickness (Carosi et al. 1998; Searle 2010; Jain et al. 2005; Cottle et al. 2007, 2011; Kellett et al. 2010, 2013; Kellett and Grujic 2012; Finch et al. 2014; Shreshtha et al. 2015), separating medium- to high-grade GHS in the footwall from the THS in the hanging wall.

Recently, the STDS and a small intrusive Malari granite pluton in uppermost parts of the GHS are intensely investigated along the Alaknanda–Dhauli Ganga valley, Uttarakhand Himalaya (Sachan et al. 2010; Spencer et al. 2012; Jain et al. 2014; Sen et al. 2015; Shreshtha et al. 2015; Iaccarino et al. 2017; Montemagni et al. 2018). Megacrystic granite mylonite reveals an early top-to-the-SW ductile shear sense from the asymmetrical feldspar in uppermost parts of the HHC. It is superposed by shear bands, S-C shear fabric and asymmetrical rotated feldspar indicating younger top-to-the-NE ductile shearing (Fig. 6.49a). Extensional ductile shear zones reveal top-to-NE downward movements within migmatite of the upper parts of the Bhapkund Formation and contain leucogranite melt (Fig. 6.49b). Brittle-ductile shearing and development of intensely sheared granite at Malari produce cataclasite-related

Fig. 6.49 Ductile to brittle shearing along the South Tibetan Detachment System (STDS) along Dhauli Ganga Valley. **a** Megacrystic granite mylonite in uppermost parts of the HHC near Timarsain showing both top-to-the-SW and top-to-the-NE ductile shearing. Asymmetrical feldspar superposed by shear bands, S-C shear fabric and asymmetrical feldspar indicating. **b** Top-to-NE downward shear bands within migmatite, having leucogranite melt near Bhapkund. **c** Brittle-ductile shearing and development of intensely sheared granite at Malari. **d** Pseudotachylyte band within Malari Granite. **e** Steep NE-dipping brittle-ductile to brittle shear zones paralleling the STDS is shown by arrows near Timarsain. **f** Typical fault gouge of the Malari leucogranite

foliation (Fig. 6.49c). Occasionally, one also observes evidences of partially melting within zones of frictional heating, which has generated pseudotachylyte within Malari Granite (Fig. 6.49d). The STDS is marked by NE-dipping steeper brittle–ductile shear zones with top-to-NE sense of movement (normal) near the northernmost exposure of the HHC; geometry of these shear zones represents late phase of displacement along the STDS (Fig. 6.49e). End of the STDS activity is marked by the development of intensely brecciated leucogranite and fault gouge having large angular fragments (Fig. 6.49f).

Sachan et al. (2010) obtained U–Pb zircon age of 19.0 ± 0.5 Ma of this undeformed granite pluton which, according to them, has undergone only minor brittle deformation/alteration and intruded both the STDS and the overlying Tethyan Himalayan Sequence (THS). These data suggested that this leucogranite post-dates the ductile fabrics related to the STDS, hence ductile shearing along the STDS ceased before ~19.0 Ma in this valley. On the contrary, Jain et al. (2014), Shreshtha et al. (2015), Iaccarino et al. (2017) and Montemagni et al. (2018) opined that the Malari Granite has undergone ductile to brittle shearing controlled by the STDS, hence it is pre- to syn-tectonic to the ~19 Ma Malari granite.

In situ U–(Th)–Pb monazite ages of high-temperature mylonite below the Malari granite date the ductile shearing along the STDS as c. 19–20 Ma (Iaccarino et al. 2017). $^{40}Ar/^{39}Ar$ muscovite age of 16.0–16.5 Ma constrains the "waning stages" of ductile deformation of the STDS activity, while static muscovite grown after sillimanite at 14.3 Ma post-dates cessation of movement on the STDS (Montemagni et al. 2019). Based on $^{40}Ar/^{39}Ar$ muscovite dating and dynamic recrystallization in quartz and feldspar, Sen et al. (2015) proposed that the Malari leucogranite experienced ductile deformation until c. 15 Ma (Fig. 6.50).

Timing of the N–S extension along the STDS has been critically assessed and reviewed by Leloup et al. (2010) and Iaccarino et al. (2017), who observed that it was initiated at ~26 to 24 Ma and ceased at ~20 Ma in Zanskar (Finch et al. 2014). It became gradually younger eastwards between 15 and 16 Ma with the transition to brittle-ductile deformation at 13.9 Ma in western Bhutan (Montomoli et al. 2017). Hence, timing of the STDS was diachronous along the entire Himalayan belt unlike the proposition by Godin et al. (2006) that it remained uniform (Fig. 6.51).

6.10.7 Main Himalayan Thrust

During the receiver function surveys in Nepal, Schulte-Pelkum et al. (2005) and Nabelek et al. (2009) observed a detachment surface separating the underthrust Indian Plate and the Himalaya hanging wall wedge and called it as the Main Himalayan Thrust (MHT). An anisotropic zone was detected along this surface (Schulte-Pelkum et al. 2005), corresponding to the 'decollement' (Avouac 2003), where a northward dipping low velocity zone (LVZ) is caused by aqueous fluids resulting from dewatering of underthrust sediments (Nabelek et al. 2009).

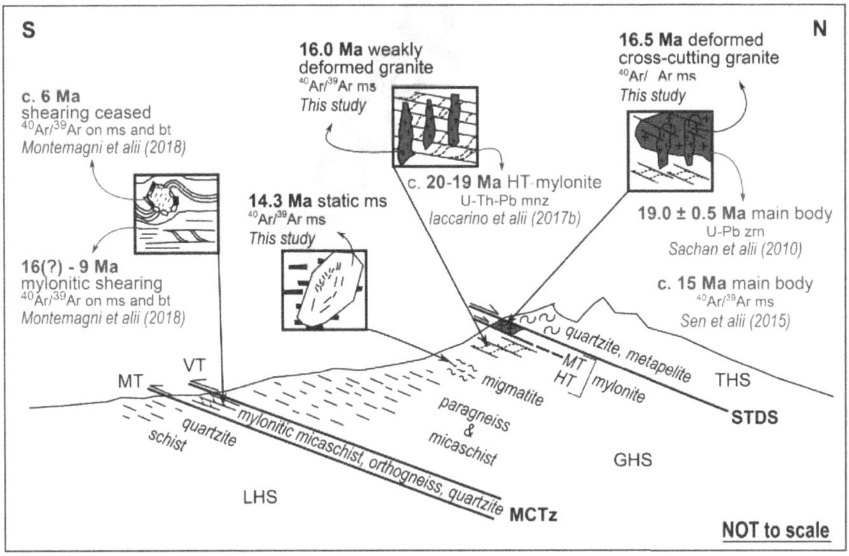

Fig. 6.50 Schematic geological transect along the Alaknaanda-Dhauil Ganga valley, showing available geochronological data, their mutual structural and tectonic interpretation. After Montemagni et al. (2018)

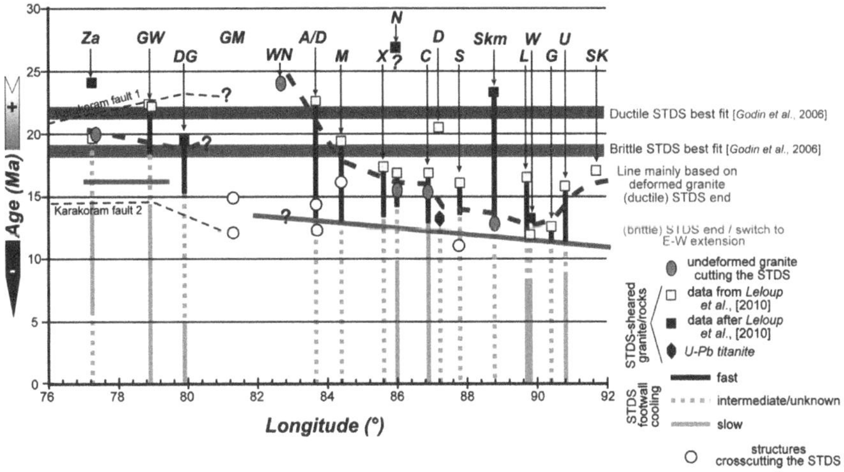

Fig. 6.51 Time chart along the South Tibetan Detachment System (STDS) from Zanskar to central Bhutan. After Iaccarino et al. (2017) and its Fig. 8 for detailed locations

Magnetotelluric (MT) profiles, seismic tomography, reflection profiles, earthquakes and their fault-plane solutions from the Himalayan arc and Tibet provided constraints on the present-day configuration of the Indian Plate and the MHT. Similar fluid-saturated fractured surface of very low <10 Ωm resistivity has also been picked up in the magnetotelluric (MT) profiles and seismic waves in intra-crustal low velocity layer (IC-LVL) from the NW Himalaya in Uttarakhand (Caldwell et al. 2013; Miglani et al. 2014; Rawat et al. 2014). This mid-crustal conductor of low resistivity zone is also indicated by partial melt in extreme northeast between Tso Morari and Leh (Ladakh), where its sub-horizontal geometry extends beneath high resistive LB and Karakoram Batholith (Gokarn et al. 2002; Arora et al. 2007). In NW-Himalaya, it defines the geometry of the subducting ICL even beyond Karakoram fault (Rai et al. 2006).

Using receiver function analysis of teleseismic earthquakes along the Sutlej valley (Himachal), Hazaika et al. (2017) imaged the Main Himalayan Thrust (MHT) as a gentle north dipping structure between the Sub- and Higher Himalaya with a ramp located further north, near the STDS in contrast to the reported ramp structure of the MHT beneath the MCT in Garhwal and Nepal Himalaya. Depth of the MHT varies from ~16 to 27 km across the Sub-, Lesser-, and Higher Himalaya, and increases to ~38 km beneath the STDS. In the Darjeeling–Sikkim region receiver functions also reveal a is horizontal geometry at a depth of ~10 to 15 km from 26.9°N to 27.45°N, and then northward dip at 15–20° and its ramp (Acton et al. 2010). Three flats and ramps are developed at a depth of 8–12 km beneath the Tista-Rangit double window (Bhattacharyya and Mitra 2009).

Majority of earthquakes define a narrow belt between north-dipping MBT and the MCT and are essentially controlled by a mid-crustal ramp within the MHT along which the Indian Plate underthrusts southern Tibet at about 10° (Seeber and Armbruster 1981; Ni and Barazangi 1984; Nelson et al. 1996). This surface dips at ~15° from about 10 to 20-km depth within the Himalaya and controls focal mechanisms not only of shallow ($\lesssim 30°$) earthquakes but also the great Himalayan earthquakes of M > 8 (Ni and Barazangi 1984). Outer seismic mid-crustal ramp extends into an aseismic 30–40 km deep seismic reflector, imaged in the INDEPTH experiment further north (Zhao et al. 1993). Seismic observations from the 25 April 2015, Mw 7.8 Gorkha earthquake in Nepal, their integration with geology and deep structures led Hubbard et al. (2016) to postulate that the slip patch closely matches an oval shaped, gently dipping MHT fault surface bounded on all sides by steeper ramps.

Further east in Bhutan structural, geophysical and thermokinematic models revealed the following geometry of the MHT:

(i) An essentially planar structure from surface trace of the MBT northward at a constant dip of 5–7° beneath the Bhutan Himalaya and southern Tibet (Robert et al. 2011),
(ii) Further north, below the Kangmar dome, the MHT is interpreted to dip northward at 35°–40° from ~42 to 45 km down to 70 km depth (Hauck et al. 1998). A balanced cross section at 89.55°E longitude suggests the MBT branches off

the MHT at ~7 km depth and the MHT dips uniformly northward at 4–5° until ~28°N (Tobgay et al. 2012).

Recent observations on thermochronological and cooling age patterns, uplift and exhumation rates in different parts of NW- and NE-Himalaya reveal their controls due to structural positions of dome/window/synform, klippen/nappe structures and thrusting/back-thrusting along different major faults (Sorkhabi et al. 1996; Jain et al. 2000; Vannay et al. 2004; Grujic et al. 2006; Patel and Carter 2009; Deekan et al. 2011; Adlakha et al. 2013, 2019; Coutand et al. 2014; DeCelles et al. 2016; Thiede et al. 2017; Gavillot et al. 2018). In many instances these surface structures appear to reflect the geometry and kinematics of the MHT with development of duplexes over the ramp of the MHT (Fig. 6.52). Patterns of exhumation are controlled by local tectonics that is dictated by the subsurface geometry of the MHT and its associated structures (Coutand et al. 2014; DeCelles et al. 2016; Thiede et al. 2017; Gavillot et al. 2018). Plots of the thermochronological ages over the published cross-section across various transects have been examined to understand how sub-surficial deformation i.e. development of duplex over the MHT ramp or surface breaking fault emerging over the MHT influenced the patterns of thermochronological ages and hence exhumation rates (Patel and ManMohan 2020a).

6.11 Himalayan Magmatism

Since Proterozoic the Himalayan Orogen witnessed episodic magmatism which can be classified into the following events (Fig. 6.53):

(i) Paleoproterozoic magmatic arc and associated granite-mafic bodies.
(ii) Neoproterozoic granitoids.
(iii) Cambro-Ordovician granitoids.
(iv) Permian mafic volcanism and related granitoids.
(v) Cenozoic granitoids.

6.11.1 Paleoproterozoic Magmatic Arc

The Paleoproterozoic magmatic bodies occur in far-travelled intensely mylonitized Main Central Thrust Zone (MCTZ), whose remnants are observed in the Munsiari Group, Askot–Baijnath–Chiplakot nappe, basal Almora Nappe mylonite and Ramgarh mylonite in Uttarakhand and Ramgarh thrust zone in Nepal and Eastern Himalaya (Fig. 6.54). Imbricated Baragaon mylonite belt of the Kulu–Bajura nappe in Himachal and the Munsiari Group in Uttarakhand represents the Late Paleoproterozoic magmatic arc (Kohn et al. 2010; Jain et al. 2015) between the iLH belt and the GHS from Ravi to Kali Rivers, and yielded $^{207}Pb/^{206}Pb$ ages between 1.95 and 1.89 Ga (Spencer et al. 2012; Mukherjee et al. 2019).

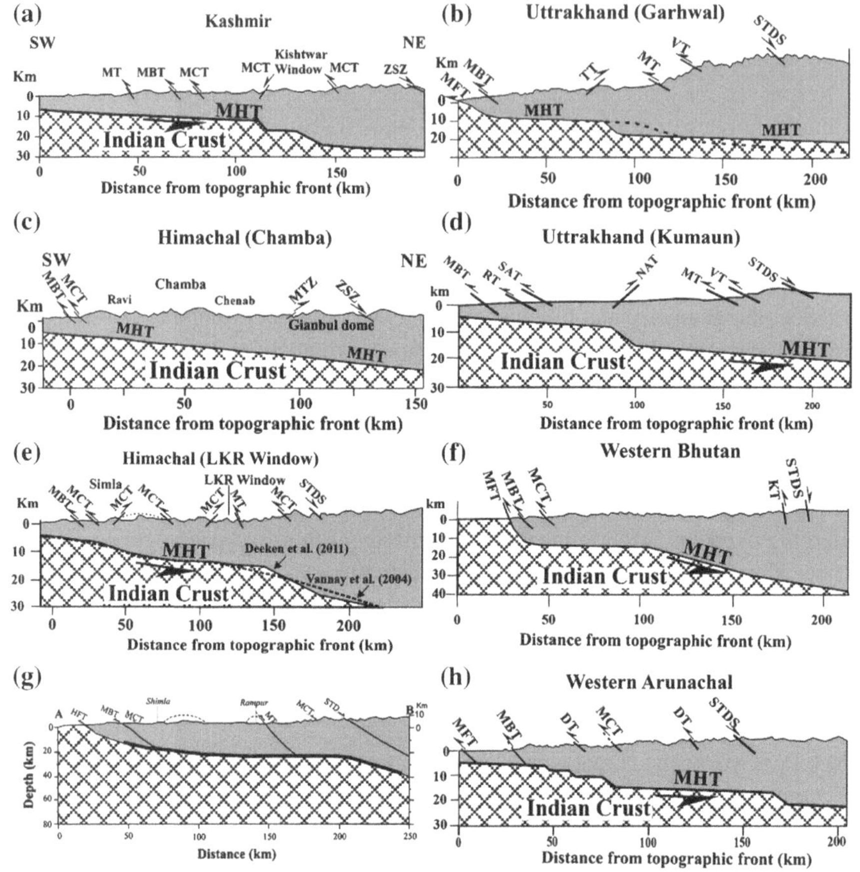

Fig. 6.52 Geometry of the Main Himalayan Thrust (MHT) along different sectors of the NW–Himalaya and NE–Himalaya. **a** Kashmir. After Gavillot et al. (2018). **b** Himachal–Chamba. After Deekan et al. (2011). **c** Himachal: Laji–Kullu–Rampur Window. After Vannay et al. (2004); Deekan et al. (2011). **d** Sutlej valley, Himachal. After Hazarika et al. (2017). **e** Uttarakhand Garhwal. Srivastava and Mitra (1994), Caldwell et al. (2013), Kanaujia et al. (2016). **f** Uttarakhand Kumaun. After Srivastava and Mitra (1994). **f** Western Bhutan. After Coutand et al. (2014). **g** Western Arunachal Pradesh. After DeCelles et al. (2106)

A belt of such mylonitized magmatic bodies to the north of the iLH represents the Late Paleoproterozoic magmatic arc based on texture, geochemistry and new zircon U–Pb age data from numerous magmatic bodies (Kohn et al. 2010; Rao and Sharma 2011; Mandal et al. 2016; Phukon et al. 2018; Jain et al. 2010; Mukherjee et al. 2019). The Paleoproterozoic Magmatic Arc (PMA) yielded $^{207}Pb/^{206}Pb$ ages between 1.95 and 1.87 Ga and average whole-rock $\varepsilon_{Nd(0)}$ −25 value (Mukherjee et al. 2019; Jain et al. 2020, in press). These ages replicated between 1.97 and 1.95 Ga along the Kali valley (Phukon et al. 2018) and 1.99–1.89 Ga in the Bhagirathi valley (Sen et al. 2018).

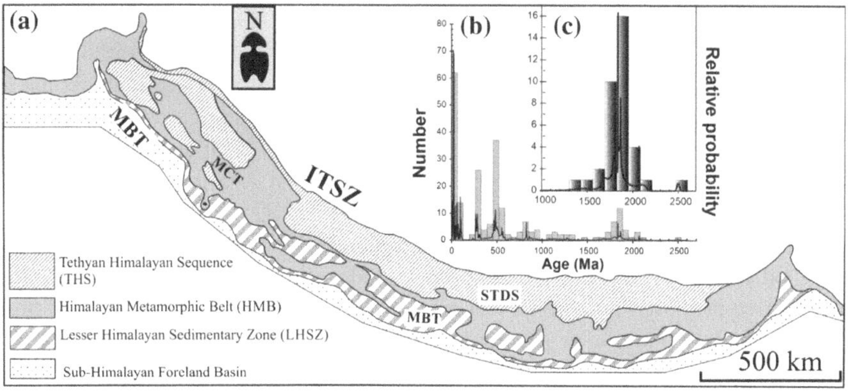

Fig. 6.53 **a** Simplified geological map of the Himalaya between the Main Boundary Thrust (MBT) and the Indus Tsangpo Suture Zone (ITSZ) showing distribution of main litho-tectonic units. **b** Histogram and relative probability curve of the published data set on Himalayan magmatic bodies around 1800 Ma, 1000–800 Ma, 500 Ma, 300 Ma and 45–15 Ma (Singh 2001, 2005, 2020; Singh et al. 2009; Dhiman and Singh 2020; Shellnutt 2020 and references therein). **c** Histogram and relative probability curve of the published data set on Paleoproterozoic magmatic bodies within the Main Central Thrust Zone

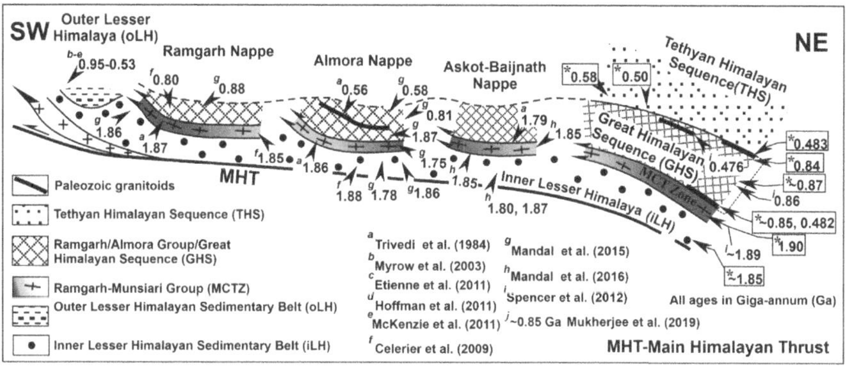

Fig. 6.54 Schematic NE–SW tectonic cross-section of the Uttarakhand Himalaya, showing distribution of different tectonic units and their U–Pb zircon ages. U–Pb crystallization ages of the Ramgarh-Munsiari groups (MCT Zone mylonite—part of the Paleoproterozoic magmatic arc) are shown along with youngest depositional ages in sediments of other units by their peak zircon ages. For simplicity, the Almora Nappe is shown distinct from the Ramgarh Nappe. Data Sources: a—Rb Sr WR ages Trivedi et al. (1984). b—Myrow et al. (2003). c—Etienne et al. (2011). d—Hofmann et al. (2011). e—McKenzie et al. (2011). f—Celerier et al. (2011). g—Mandal et al. (2015). h—Mandal et al. (2016). i—Spencer et al. (2012). j—Mukherjee et al. (2019)

Remnants of the LH crystalline nappe, representing southward extension of this arc, contain ~1.85 Ga zircon in the Chiplakot–Baijnath–Askot–Ramgarh–Almora Crystallines (Fig. 6.54). From the Chiplakot klippe, mylonitized augen gneiss has a 1.92 Ga age (Phukon et al. 2018). In extreme south, the Ramgarh orthomylonite has $^{207}Pb/^{206}Pb$ ~1.87 Ga zircon age (Célérier et al. 2009; Mandal et al. 2015, 2016); WR Rb–Sr 1820 ± 130 Ma isochron age from the basal Almora nappe granitoids (Trivedi et al. 1984) broadly resembles the U–Pb zircon age of this belt.

Based on texture, geochemistry and new zircon U–Pb ages from numerous magmatic bodies in innermost Lesser Himalaya, including the Ulleri-type mylonite, Kohn et al. (2010) proposed an active Paleoproterozoic magmatic arc between 1.88 and 1.78 Ga. The Ulleri mylonite granitoid intruded the Kuncha schist, which contained largest ~1.90 Ga DZ peak, thus constraining its maximum depositional age (Martin et al. 2011). Similar intrusive mylonitized augen gneiss provided ~1.75 Ga crystallization age along the Kodari section (Khanal et al. 2015), and 1.81 Ga Miltung augen gneiss in the Taplejung window (Sakai et al. 2013), and many others within the Munsiari Group sequence in Nepal (Kohn et al. 2010).

In Sikkim, sheared and imbricated Lingtse orthogneiss along the MCT zone possesses $^{207}Pb/^{206}Pb$ zircon crystallization age ~1.83 Ga (Mottram et al. 2014a, b), while in Central Bhutan quartz feldspar gneiss on footwall of the MCT yielded zircon ages between 1.76 and 1.84 Ga with magmatic monazites of 1.76 ± 0.076 Ga (Daniel et al. 2003). An orthogneiss below the MCT in Bhutan gave 1.844 ± 0.064 Ga with 1.90 Ga and older inherited grains (Long et al. 2011; McQuarrie et al. 2013). The ~1.75 Ga mylonite gneiss in western Kameng and Subansiri regions of Arunachal Pradesh possibly represents a somewhat younger phase of this arc (Yin et al. 2009).

6.11.2 Columbia Supercontinent and Paleoproterozoic Magmatic Arc

Globally, three basic configurations are visualized in positioning India in the Columbia Supercontinent configuration: (i) India having passive margin but away from North America (Rogers and Santosh 2002; Zhao et al. 2004), (ii) India directly adjacent to North America with subduction zone in the north (Hou et al. 2008), and (iii) rift-related Greater India domain with Paleoproterozoic magmatism, possessing 'within plate granite' characters (Singh et al. 2009; Larson et al. 2017; Martin 2017). Northern margin of the Indian Plate witnessed various tectonic setups since Paleoprotozoic; the oldest amongst these is the evolution of the Paleoprotozoic iLH basin with the Munsiari magmatic arc that separates the oLH and the GHS belts on either side.

This margin has been visualized either as a passive margin (Brookfield 1993; Myrow et al. 2003; Sakai et al. 2013), or epi-cratonic basin (Webb 2013) during the

Paleoproterozoic. However, Kohn et al. (2010) opined that it represents a Paleoprotezoic arc system and argued that linear belts of magmatic and clastic metasediments developed in a very short duration for about 100 Mya only.

On the contrary, a continental rift set-up was recently argued by Sakai et al. (2013), based on age and stratigraphic correlation of the Himalayan Paleoproterozoic assemblage. Martin et al. (2017) and Larson et al. (2017) also preferred a rift set-up based on compositions plots 'within plate granite' field in discrimination diagrams, and absence of intermediate felsic rocks and subduction-related accretionary sequences.

Mukherjee et al. (2019) and Jain et al. (2020) visualized magmatic arc set-up for the Munsiari Group and iLH sequences, and that granitoids of the Munsiari, Chiplakot, Askot, Ramgarh and basal Almora klippen were generated by subduction-related hydrous partial melting of the Paleoproterozoic mafic source and sediments (Rao and Sharma 2011; Mandal et al. 2016; Phukon et al. 2018). Thus, Paleoproterozoic scenario is similar an Andean-type magmatic arc set-up on northern margin of the Indian Plate (Kohn et al. 2010; Rao and Sharma 2011; Phukon et al. 2018) and is consistent with the Columbia Supercontinent set-up (Rogers and Santosh 2002; Hou et al. 2008).

This magmatic belt has almost similar zircon ages like the iLH sequences throughout the Himalaya and is visualized as the main Paleoproterozoic magmatic arc with the iLH as the back-arc; the latter derived its detritus, both from eroded magmatic arc in the north and the Indian craton along the southern margin.

Dominance of ~1.90 Ga U–Pb detrital zircons in sheet-like matured Paleoproterozoic sedimentary sequences of the iLH quartzite-volcanic association and the proposed Proterozoic magmatic arc establish an Andean-type arc magmatism (Kohn et al. 2010)—back-arc system during the Columbia period (cf., Rogers and Santosh 2002). This sedimentary sequence was deposited between ~1.95 and 1.87 Ga with $\varepsilon_{Nd(0)}$ values ~−25 throughout the Himalaya with its sedimentation terminating around end of the Orosirian period. The arc granitoids yield tightly clustered Paleoproterozoic upper intercept ages at 1.97 and 1.89 Ma, indicating almost contemporaneous sedimentation and crystallization at ~1.90 Ga. The magmatic arc–back-arc system belongs to one of the margins of the Columbia Supercontinent (Fig. 6.55a).

Out of different paleotectonic reconstructions for the Columbia Supercontinent, assembly, proposed by Rogers and Santosh (2002), Zhao et al. (2004) Hou et al. (2008), Kaur et al. (2013) and Mukherjee et al. (2019) are most relevant here (Fig. 6.55). An Andean-type magmatic arc system on edge of this continent involves North America (NA), India (I), Madagascar, Sri Lanka and North China (NC) during the Columbia 2.06–1.78 Ga period, when deposited sediments contained detrital zircons almost of the same age (Fig. 6.55a). The 'Paleoproterozoic Magmatic Arc (PMA)' of the Andean-type arc system is represented by the Munsiari Group—the MCT Zone (Mukherjee et al. 2019; Jain et al. 2020) defines one margin of this supercontinent. The volcano-sedimentary arc basin contains psammite schist and amphibolite, possibly belonging to its fore-arc and volcanic arc respectively, while intensely sheared augen mylonite represents its ~1.90 Ga plutonism. ~ca. 2.0–1.8 Ga iLH sedimentary sequence, incorporating the oldest Rautgara–Gangolihat–Berinag, their equivalents and interstratified mafic-felsic volcanics (cf., Miller et al. 2001;

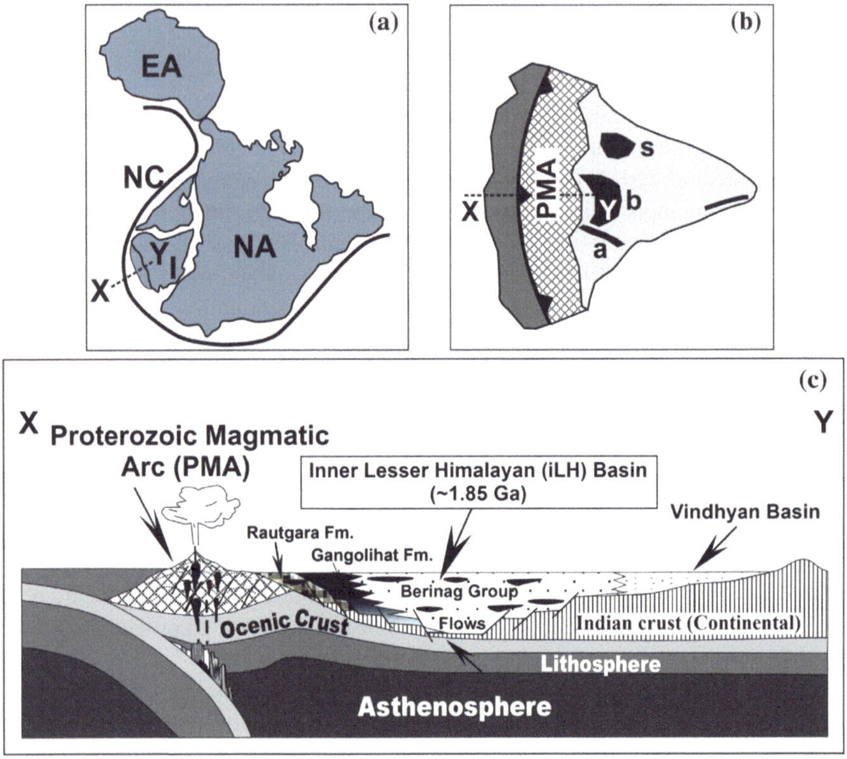

Fig. 6.55 Columbian reconstruction ca. 1.9 Ga with position of India and postulated magmatic arc. **a** Reconstruction after Rogers and Santosh (2002), Hou et al. (2008) and Kaur et al. (2013). NA—North America. EA—Eastern Antarctica. NC—North China. I—India. **b** Position of the Proterozoic magmatic arc, its configuration of fore-arc (purple), back-arc (yellow) and the position of India with the Aravalli (i), the Bundelkhand massif (ii), and the Singhbhum cratons (iii). Cross-section along XY shown. **c** Schematic reconstruction of the Paleoproterozoic back arc basins in relation to the Proterozoic magmatic arc during ~2.0 to 1.8 Ga. Subducting oceanic lithosphere caused its partial melting and emplacement of granitoids in the arc. Between the arc and Indian craton, back arc basin underwent rifting and received sediments from both the arc and the craton to be deposited within the iLH basin. After Mukherjee et al. (2019) and Jain et al. (2020, in press)

Richards et al. 2006), were contemporaneously deposited in the rifted back-arc basin (Fig. 6.55b). This evolving basin derived its detritus, both from the magmatic arc and the Indian Shield (Fig. 6.55c), and extended even into the Aravalli belt, where ~1.85 Ga subduction-related magmatism, erosion and sedimentation record contemporaneous yield of detrital zircons almost of the same age (Kaur et al. 2013; references therein). Additionally, ~3.5 Ga and ~2.7–2.5 Ga granitoids from the Aravalli and Bundelkhand cratons might have sourced a small component of the older zircons within the iLH.

Break-up of the Columbian supercontinent separated the Indian craton with its northern margin becoming passive as it drifted away and reassembled to form the

Rodinia Supercontinent during the Neoproterozoic. Thus, major part of this magmatic arc was lost and only a small part of the arc-crust remained with India. Extinct Proterozoic arc-crust and part of back-arc basin, represented now by the Munsiari Group and the iLH, respectively were subsequently transformed into passive marginal basement for deposition of the Neoproterozoic GHS.

6.11.3 Rodinia (Neoproterozoic) and Magmatism

During breakup of Rodinia, leading to amalgamation of Gondwanaland, magmatic rocks have played an important role in Greater India (Himalayan) domain. Geochemical characteristics of these magmatic bodies fall within plate granite (WPG) field and indicate a continuation of a passive northern margin of India even during Neoproterozoic time (Singh et al. 2002). Scanty geochronological data and poor recognition of such bodies in high grade metamorphosed Great Himalayan Sequence (GHS) possibly explain non-prolific presence of these ~0.85 Ga granitoids. Many ~0.90 to 0.80 Ga isolated granite intrusives are now recognized within metamorphic nappes (Figs. 6.37a and 6.56a, b; Mukherjee et al. 2019; Jain et al. 2020; Singh 2020): (i) 823 ± 2 Ma orthogneiss in Black Mountains–Mansehra pluton (Pakistan) with strongly discordant Pb–Pb ages between 1313 and 939 Ma (DiPietro and Pogue 2004), (ii) recently-recognized 1220–720 Ma zircon cores from the Dalhousie and Dhauladhar granite (Dhiman and Singh 2020), (iii) ca. 0.842 Ga mylonitic augen gneiss along southeastern margin of the Mandi Granite, (iv) ~0.831 Ga mylonitized augen gneiss near Chaupal near the MCT, (v) ca. 0.83 Ga mylonitic augen gneiss of the Audi shear zone, all in Himachal (iii–v) Webb et al. (2011), (vi) the Chor Granite pluton of the Jutogh Nappe with U–Pb zircon age of 823 ± 5 Ma (Singh

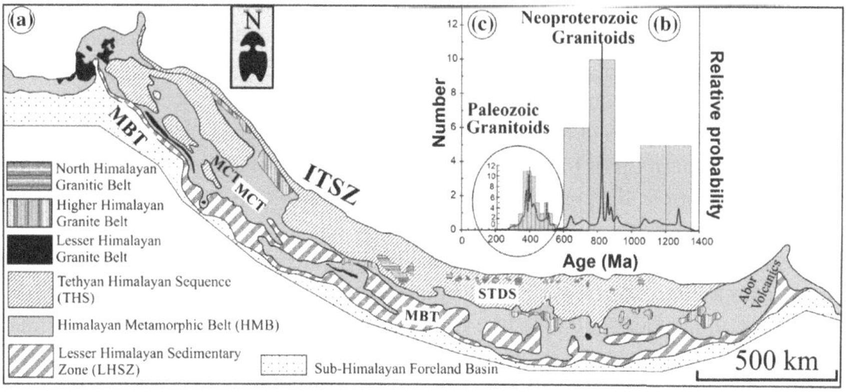

Fig. 6.56 **a** Simplified regional geological map of Himalaya with the main granitoid belts. **b**, **c** Histogram and relative probability curve of the published data set on Neoproterozoic and Cambro-Ordovician magmatic bodies, respectively. See text for data source. Modified after Singh (2020)

et al. 2002), (vii) inherited ~0.86 Ga zircon cores in the ~0.475 Ga orthogneiss in the Alaknanda valley (Spencer et al. 2012), (viii) an orthogneiss migmatite having U–Pb zircon ages between 1.00 and 0.90 Ga with its leucosome fraction between 0.95 and 0.875 Ga in the Dhauli Ganga valley (Mukherjee et al. 2019; Jain et al. 2020), (ix) 839 ± 28 Ma orthogneiss above the MCT, Sikkim (Mottram et al. 2014a, b), (x) 825 ± 9 Ma intrusive augen gneiss above the Kakhtang thrust, Kuru Chu, central Bhutan (Richards et al. 2006), (xi) Zimithang augen gneiss in Kameng and 878 ± 12.6 Ma mylonitized augen gneiss along the Zimithang shear zone in Kameng (Yin et al. 2009), (xii) 836.9 ± 13.2 Ma augen gneiss, Subansiri, western Arunachal (Yin et al. 2009), and (xiii) 809 ± 4.5 Ma orthogneiss, Subansiri (Clarke et al. 2016).

Tectonic setting of the Neoproterozoic magmatism within the GHS is possibly linked either to intracontinental rifting due to superplume beneath the Rodinia Supercontinent or crustal anataxis during syn- to post-collisional crustal thickening during the Neoproterozoic. It is apparent that a potentially extensive Neoproterozoic magmatic belt/arc existed in proximity of the GHS basin on its northern fringe of the Rodinia Supercontinent (Fig. 6.37b). Based on 860–840 Ma detrital zircon ages, enriched radiogenic Hf isotopic signatures (εHf = 10 to −20) and their comparisons with nearby potential sources, Spencer et al. (2018) postulated an active 'Out-Board' continental margin source for the detrital zircons of the GHS along the northern India margin. Though some relicts of this magmatism are observed within the GHS like those listed above (Mukherjee et al. 2019), it is likely that this source was tectonically removed during the Kurgiakh orogeny (Spencer et al. 2018) and prior to the accretion of the Asian continental margin.

6.11.4 Gondwana Amalgamation and Cambro-Ordovician Granitoids

The Cambro-Ordovician magmatism in the Himalaya is widely manifested as two-mica granite pluton emplacements within the LH Granitic Belt, the HHC Belt and near the South Tibetan Detachment System (STDS) because of the Kurgiakh Orogeny (Srikantia 1977)/Kuunga Orogeny (Merdith et al. 2017). These events are linked with collision of southern tip of India with Congo and northwest margin with Azania and northern Madagascar during the early Pan African orogeny during the Gondwana amalgamation (Merdith et al. 2017).

Because of such large-scale Pan-African geodynamic processes, numerous granite bodies occur as discontinuous gneissic culminations/domes within the THS north of the STDS as independent plutons like the Kaghan, Lhago-Kangri, Kangmar, Nyimaling, Tso Morari, Rupshu, Jispa, Kade, Kokhsar, to name a few (Fig. 6.56a; Gansser 1977; Frank et al. 1977; Wang et al. 1981; Spencer 1993; Girard and Bushy 1999; Leech et al. 2005). In addition, following evidences record the Pan-African major tectonic event within the Himalayan domain:

(i) Thin-skinned thrusting in the Himalaya, based on cross-cutting Cambro-Ordovician granites post-dating regional garnet-grade metamorphism and regional ductile folds and foliation in central Nepal (Stocklin and Bhattarai 1977; Stocklin 1980; Gehrels et al. 2003).

(ii) Deformation, metamorphism, upliftment, and erosion of large segments of the Himalaya and Indo-Gangetic Plains incorporating Great Vindhyan Basin during Cambro-Ordovician and subsequent renewed sedimentation in the Lesser and Tethys Himalayas in the Krol and Tethys Basins (Jain et al. 1980). Evidences of such deformation episodes are documented in Tons valley and Spiti (Himalchal–Bhargava et al. 2011a, b; Wiesmayr and Grasemann 2002).

(iii) Low-pressure contact metamorphism along with preservation of pre-Himalayan fabric around Mandi around Cambro–Ordovician granites (Baud et al. 1982; Guntli 1993; Purkayastha et al. 1999; Girard and Bussy 1999; Singh and Jain 2008); a mean $^{206}Pb/^{238}U$ zircon age of 469 ± 8 Ma (Mandi Granite having apophyses) and constraining this metamorphism (Fig. 6.57; Singh and Jain 2008).

(iv) Evidences of pronounced thermal/metamorphic event from coeval garnet ages of 485 ± 19 Ma and 473 ± 14 Ma (Chor mountain–Bhargava et al. 2016), 534 ± 24 Ma (Alaknanda valley–Argles et al. 1999), 467 ± 3 Ma (NW Himalaya–Foster 2000); 445 ± 16 Ma (Central Nepal–Catlos et al. 2000), 548 ± 17 Ma (Namche migmatites–Catlos et al. 2002), 436 ± 8 Ma (Barun Gneiss–Catlos et al. 2002), to name a few.

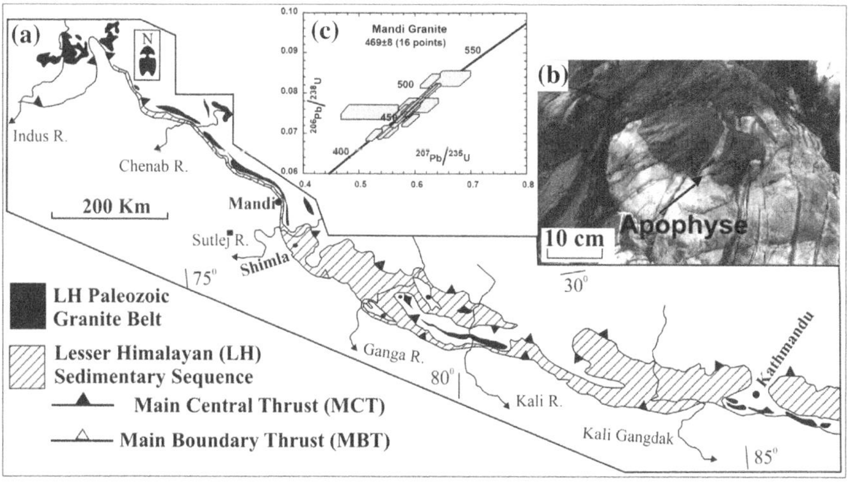

Fig. 6.57 **a** Simplified regional geological map of Himalaya with Cambro-Ordovician granitoid belts. **b** Arcuate apophyse from the Mandi Granite within the Jutogh Group. **c** Concordia diagram for the U–Pb zircon age of the mandi Granite. After Singh and Jain (2008)

6.11.5 Pangaea Supercontinent and Permian Magmatism

The Himalayan Orogen, in addition to extensive Cambro-Ordovician event, is affected by widespread Permian magmatism, which includes both plutons and mafic volcanism, and appears to be controlled by opening of the Neo-Tethyan Ocean between ~300 and 250 Ma in Pangaea Supercontinent (Zhao et al. 2018 and references therein). Some such granitoids in the NW Himalaya are Maklad granite, LoeSar Granite, Lower Swat, Parkachic and Sanko in Zanskar, Biotite granite below Shishma-Pangma Leucogranite and Yunam Granite in Upper Lahaul within NW Himalaya (Singh 2005; Shellnutt 2018 and references therein). The Early Permian (290 Ma) Panjal Traps (Phe Volcanics etc. in west and the Abor Volcanics in Eastern Himalaya) are the largest contiguous outcropping of volcanics (basaltic, andesitic and silicic) within the Himalaya that are associated with the Late Palaeozoic break-up of Gondwana. The Permian magmatic bodies occur as dykes or small plutons within the THS or close to it.

Covering an area of about 10,000 km^2 around Kashmir Valley, Pir Panjal and Zanskar ranges (Fig. 6.58a), the Traps are mostly basalt and minor basaltic andesite, andesite, rhyolite and dacite flows (Nakazawa and Kapoor 1973; Bhat et al. 1981; Shellnutt 2018). The Traps exhibit subaerial columnar joints and pillow structures indicating their eruptive and subaqueous environments. Intertrappean limestone and shale suggest intermittent pauses during volcanism. Maximum flow thickness of c. 3 km is greatest in western Kashmir and indicates the volcanic centre in that region (Nakazawa and Kapoor 1973). Nakazawa et al. (1975) suggested that the Panjal Traps are definitively constrained to the Early–Middle Permian (Sakmarian–Artinskian), which is consistent with a weighted mean ^{206}Pb/^{238}U zircon age of 289 ± 3 Ma from silicic traps near Srinagar (Shellnutt et al. 2011). Other Carboniferous–Permian granitic bodies in this region of western Himalaya are the Yunam A-type granite,

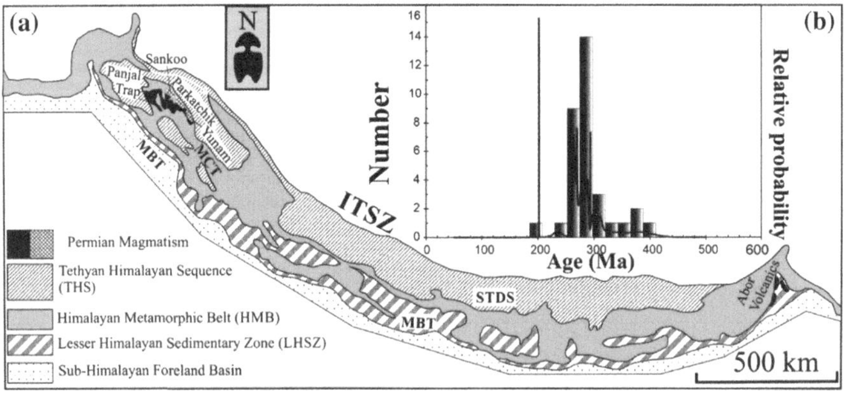

Fig. 6.58 **a** Simplified regional geological map of Himalayas with Permian magmatism. **b** Histogram and relative probability curve of the published data set on Permian magmatism (Table 1 of Shellnutt 2018 and references therein; Plot after Singh 2020)

Malakand, Ambela, Tarbela, Swat, Parkatchic and Sankoo and biotite granite below Shishma-Pangma Leucogranite; some of which are considered to be crustal melts associated with the injection of the mafic Panjal magma (see Table 1 in Shellnutt 2018; Fig. 6.58b).

Crustally contaminated basaltic Panjal Traps are continental tholeiite to ocean-floor basalt and possess Sr–Nd isotopes of $^{87}Sr/^{86}Sr_i = 0.7043–0.7073$ and εNd(t) = 0 ± 1) similar to a chondritic (subcontinental lithospheric mantle) source (Shellnutt 2018). Remaining basaltic rocks have a wide range of Nd εNd(t) = −6.1 to +4.3 and Sr ($^{87}Sr/^{86}Sr_i = 0.7051–0.7185$) isotopic values, with picritic primary melt composition of low-Ti Panjal Traps. The silicic volcanic rocks were derived by partial melting of the crust, whereas the andesitic rocks were derived by mingling between crustal and mantle melts (Fig. 6.59).

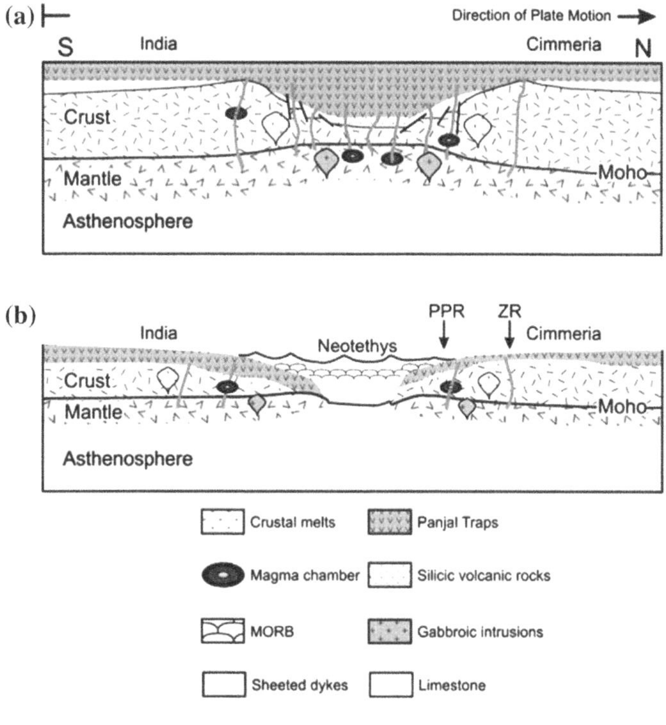

Fig. 6.59 Tectonic model for the evolution of the Panjal Traps. **a** Initial subaerial eruption followed by **b** subaqueous eruptions and eventual opening of the Neotethys Ocean. PPR—Pir Panjal. ZR—Zanskar Range the Panjal Traps initially erupted within a continental rift setting. The rift eventually transitioned into a nascent ocean basin that led to seafloor spreading and the formation of the Neo-Tethys Ocean and the ribbon-like continent Cimmeria. This magmatism is correlated with extensional tectonic setting (Shellnutt 2018)

6.11.6 Magmatism During Himalayan Orogenesis

Two stages of granitoid emplacements are associated with the Himalayan orogenesis: (i) subduction of the Neo-Tethyan oceanic plate during Cretaceous, followed by (ii) granitoids originated during Cenozoic India-Asia collision (Jain 2014). As a result of the northward subduction, Dras volcanic arc has formed with ages ranging from 105 to 65 Ma (Honegger et al. 1982; Schärer et al. 1984; Reuber 1989; Reuber et al. 1989). At the same time, the Kohistan arc also developed with calc-alkaline batholith emplacement between 120 and 85 Ma (for review see Singh 2001, 2005, and references therein) and development of Trans-Himalayan Batholith representing Andean-Type magmatism due to melting of north dipping Tethyan oceanic crust (Schärer et al. 1984). Both the batholiths have developed below an island arc located along the southern margin of Eurasia (Raz and Honegger 1989). The age of Trans-Himalayan bodies predate collision and represent subduction-related magmatism ranging from 102 to 54 Ma (for details see Chap. 7).

The subduction-related phase was followed by collision-related magmatism of smaller volume forming discontinuous small pluton known as the Higher Himalayan Leucogranites (HHL). These intrude closer to the STDS. These Syn-to post-Himalayan leucogranites have attained very large attention due to their association with collision (Le Fort 1986; Le Fort et al. 1987; Zeitler et al. 1993; Searle 1996; Harrison et al. 1997; Searle et al. 1999; Schneider et al. 1999, 2001; Singh 2001, 2005, 2020, in press and references therein; Verma 2003; Weinberg 2016). The Himalayan leucogranites are interpreted to have resulted either from (i) crustal anatectic melting due to fluid migration during intracontinental thrusting along the MCT (Le Fort et al. 1987; France-Lanord and Le Fort 1988), (ii) decompressional dehydration melting due to slip along the STDS controlling the emplacement of leucogranite plutons (Searle et al. 1992; Guillot and Le Fort 1995), or (iii) vapor-absent muscovite dehydration melting of metamorphic rocks due to shear heating along continuous active decollement (Harrison et al. 1997, 1999). However, irrespective of the mechanism of its generation, leucogranite belt was emplaced after 24 Ma after collision (Davidson et al. 1997; Searle et al. 1997, 1999; Guillot et al. 1999; Montomoli et al. 2017). U–Th–Pb ages of these HHL range between ~24 to 12 Ma and become younger eastwards (Searle and Godin 2003; Carosi et al. 2013 and references therein) and frequently occupy peaks of the Higher Himalaya. Crystallization ages from the HHL are younger ages to the peak metamorphic ages, e.g. 37–28 Ma in Zanskar (Vance and Harris 1999; Walker et al. 1999), 41–22 Ma in Himachal (Stübner et al. 2014), 44–26 Ma in Garhwal (Foster et al. 2000; Prince et al. 2000, 2001), 36 Ma (Hodges et al. 1996) and 36–28 Ma (Laccarino et al. 2015) in Annapurna (Nepal), and 36–34 Ma in Bhutan (Edward and Harrison 1997). $\varepsilon_{Nd(0)}$ values of some of the plutons range between -11 and -18, indicating absence of mantle components (Vigneress and Burg 2003; and references therein), and match well with the $\varepsilon_{Nd(0)}$ values (-13 to -19) of the Vaikrita Group (Ahmad et al. 2008; Mukherjee et al. 2019).

A few coevals in situ melts have also been reported from the migmatite zone (e.g., Harris et al. 2004; Rubatto et al. 2013; Jain et al. 2013; Singh 2019). Widespread partial melting of sillimanite–K-feldspar gneiss resulted in formation of migmatite and melt accumulation near the STDS along Dhauli Ganga Valley (Garhwal) during five Me1 to Me5 phases associated with various deformation events (Jain et al. 2013). Main migmatite Me1 phase parallels main foliation Sm as the stromatite layers and concordant leucogranite bands, while younger melt Me2, Me3 and Me5 phases are along small-scale ductile thrusts, extensional fabric and structureless patches, respectively. Large-scale melt migration is associated with cross-cutting Me4 irregular veins as possible conduits for larger leucogranite bodies, like the Malari leucogranite ~19.0 ± 0.5 Ma as small veinlets (cf. Sachan et al. 2010).

In the NW Himalaya (Bhagirathi valley), migmatite in sillimanite-muscovite zone exhibits flowage and in situ melt generation of tourmaline bearing leucogranite where zircon reveals protracted episodic growth from 46 to 20 Ma associated with peak metamorphism and HHL (Singh 2019).

6.11.6.1 Nanga Parbat-Harmosh Massif (NPHM)

In the NW Pakistan, the Nanga Parbat-Harmosh Massif (NPHM) of the western Himalayan Syntaxis represents the northernmost exposure of the HHC and is a complex mixture of ortho-and paragneiss (George et al. 1993). Leucogranite dykes from southern localities of Swat and Naran gave 35 and 50 Ma, respectively with older cores of ~1000, 1750 and >2600 Ma. Deepest parts of the NPHM contain leucogranites between 3 and 1 Ma (Zeitler et al. 1993, Schneider et al. 1999, 2001), however ages increase towards north between ~2.3 and 7 or 10 Ma (Schneider et al. 1999).

6.11.6.2 Zanskar

U–Pb dating of leucogranite bodies from the HHC in Zanskar revealed their crystallization ages between 21 and 19.5 Ma as timing of anatexis (Nobel and Searle 1995), which is younger to peak metamorphic ages of 37–28 Ma (Vance and Harris 1999; Walker et al. 1999). Migmatite from deepest structural levels gives U–Pb monazite ages of 20.6–19.5 Ma, while melt pods have magmatic monazite of 20.8 ± 0.3 Ma at higher levels. Inherited component of ~460 Ma reveals protolith of at least Ordovician age (Noble and Searle 1995). The Gumburanjon leucogranite on footwall of the STDS (Zanskar Shear Zone) has crystallization age of 21.4 ± 0.3 Ma with an upper intercept of 462 ± 38 Ma, interpreted as the age of source material (Walker et al. 1999). Based on U–Pb dating of monazite in leucogranite, Finch et al. (2014) opined that activity of the STDS started at ~26 to 24 Ma and ceased at ~20 Ma in Zanskar.

6.11.6.3 Garhwal

Deep incision by the Bhagirathi River in Garhwal provides exposures of the Gangotri–Badrinath leucogranite bodies at Shivling, Bhagirathi, Meru and Thalay Sagar peaks in the uppermost parts of the HHC (see Fig. 6.26). Whole-rock analyses of leucogranite have yielded a 5-point Rb–Sr isochron of 21.1 ± 0.9 Ma (Stern et al. 1989), while Th–Pb monazite dating provided an age of 22.4 ± 0.5 Ma for the Gangotri and 21.9 ± 0.5 Ma age for the Shivling leucogranites (Harrison et al. 1997); the latter has a 23.0 ± 0.2 Ma U–Pb monazite age (Searle et al. 1999).

In almost central parts of the HHC exhibiting sillimanite-muscovite grade of metamorphism, evidence of in situ partial melting are associated with migmatite, derived from metasediments, having detrital zircons with maximum $^{207}Pb/^{206}Pb$ dates between ~610 and 2745 Ma, and mostly clustering in the range of 737–1116 Ma (Singh 2019). Two samples of the tourmaline-bearing granite from the partial melt portion yielded evidences of zircon overgrowths; these rims possess Th/U ratios of <0.1 which is consisting with growth during metamorphism. SHRIMP dating of these zircon rims revealed their protracted growth for a time span of about 25 Mya between 46 and 20 Ma at around 850 ± 50 °C (Fig. 6.60; Singh 2019).

In the Dhauli Ganga valley, Sachan et al. (2010) obtained U–Pb zircon age of 19.0 ± 0.5 Ma of the Malari undeformed granite which has undergone only minor brittle deformation/alteration and intruded both the STDS and the overlying THS.

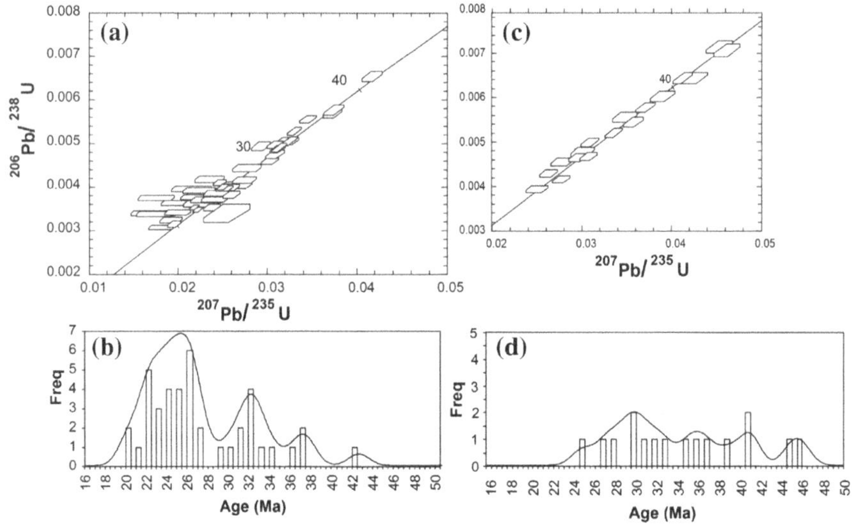

Fig. 6.60 Concordia diagrams and U–Pb zircon age histograms and cumulative probability distribution curves of zircon rims from in situ melt of tourmaline bearing leucogranite and migmatite from the Bhagirathi valley. Sample 1 with **a** concordia diagram and **b** histograms and cumulative probability distribution curve. Sample 2 with **c** concordia diagram and **d** histograms and cumulative probability distribution curve. After Singh (2020)

6.11.6.4 Nepal

Amongst the syn- to post-Himalayan Cenozoic plutons in Nepal, the Manaslu pluton is the most thoroughly investigated body, which cross cuts the STDS. Deniel et al. (1987) obtained a Rb–Sr isochron of 18.1 ± 0.5 Ma, using 11 WR samples with an initial $^{87}Sr/^{86}Sr$ ratio 0.7470 ± 0.0005 (MSWD = 7.3). An apatite–whole rock–K-feldspar–muscovite isochron from this pluton has given an age of 21.3 ± 0.4 Ma (Copeland et al. 1990), indicating that equilibrium has reached on the mineral scale. Ion-microprobe monazite analysis from the same body has yielded an age of 22.3 ± 0.5 Ma with ~600 Ma inherited cores (Harrison et al. 1995). For western Nepal, Carosi et al. (2013) reported the occurrence of an undeformed leucogranite cross-cutting the STDS at ~23 to 24 Ma. In addition, zircon and monazite dating of the Manaslu granite by conventional U–Pb method has yielded ages of 24.0 and 21.9 Ma, respectively (Scharer et al. 1986); different Miocene ages were caused due to crystallization of separate batches of magma by multiple injections within a span ~7 Ma (Deniel et al. 1987). Based on geothermal gradients, depth of intrusion was worked out to be ~8 to 15 km (Copeland et al. 1990).

Protracted melting is documented within upper parts of the GHS in the Annapurna and Nyalam regions of central Nepal (Kohn and Corrie 2011; Wang et al. 2015), east-central Nepal (Larson et al. 2011), the Mount Everest region (Cottle et al. 2009), and eastern Nepal (Imayama et al. 2012).

The Shisa Pangma (Xixabangma) massif is exposed along the Langtang Valley in Central Nepal and south Tibet just beneath the STDS, where zircon, uranite and monazite gave U–Pb age of 17.3 ± 0.2 Ma (Searle et al. 1997).

North of Mt. Everest (Qomolangma), the Rongbuk pluton in this valley cross-cuts the STDS and intrudes between 21 and 22 Ma (conventional U–Pb zircon and monazite ages) (Copeland et al. 1988), while single igneous zircon gave a concordant age of 19.5 ± 0.5 Ma, titanite 19.5 ± 1.3 Ma, and xenotime 20.6 ± 0.1 to 21.0 ± 0.1 Ma (Hodges et al. 1992). Cottle et al. (2015) reported U/Th–Pb monazite age of ~16.4 Ma from a folded leucogranite sills, whereas structurally higher sills and dykes that post-date the STDS fabric development are slightly younger at 15.6–15.4 Ma. The Everest leucogranite has true emplacement age of 20.6 Ma from xenotime (Parrish 1990). Brittle movements along the STDS took place between 16.37 and 16.67 Ma (U–Pb, $^{40}Ar/^{39}Ar$) with its extensional deformation spanning over at least 10 Ma (Hodges et al. 1998).

Along a long profile of the GHS in the Everest region, U–Th/Pb dating of zircon, monazite and xenotime constrained two phases of leucogranites at 20.8 ± 0.8 and 16.7 ± 0.4 Ma in the deeper part, while two phases of leucogranites are dated at 15.2 ± 0.2 Ma and 12.6 ± 0.2 Ma in northernmost exposed GHS outside the North Himalayan domes (Cottle et al. 2009).

6.11.6.5 Eastern Himalaya

In the extreme northern parts of Sikkim along the STDS, Kellett et al. (2013) obtained U–Pb zircon rim ages between 28.5 and 15.5 Ma, with a poorly defined cluster at 24.4 ± 0.7 Ma and 22.7–13.0 Ma age from cross-cutting dyke. One cordierite-bearing granite yields zircon rim ages of 19.8–13.4 Ma, with a poorly defined cluster at ca. 15.5 Ma.

In northwestern Bhutan it is speculated that muscovite-rich metapelites melting produced two-mica leucogranite between 17 and 14 Ma and possibly lasting till 12 Ma in the upper GHS. The first anatectic melts correspond to tourmaline leucogranite formed at ~20 Ma (Carosi et al. 2006) by dehydration melting of muscovite at relatively lower T while those leucogranite melts involving biotite formed at ~17 Ma. Brittle-ductile deformation along the STDS was active at 13.9 ± 06 Ma, hence large-scale antiforms and synforms mapped in NW Bhutan affected the GHS and STDS only after 14 Ma (Montomoli et al. 2017; also Cooper et al. 2015).

6.12 Exhumation

The Himalayan Orogen has experienced large-scale tectonics and undergone fast erosion due to fluvial, glaciers and mass wasting processes. Consequently, it has witnessed variable exhumation due to tectonics and/or coupled monsoon-controlled erosion (Jain et al. 2000, 2009; Wobus et al. 2005; Patel et al. 2009; Thiede et al. 2005, 2009; Thiede and Ehlers 2013; Clift 2017; Stübner et al. 2018; Patel and ManMohan 2020a, b). Exhumation patterns from the monsoon-affected NW– and NE Himalaya are compared with monsoon-deficient adjoining regions of the Trans-Himalayan (LB) and the Tso Morari Crystallines (TMC) to provide us inputs regarding role of precipitation and erosion in controlling exhumation.

6.12.1 NW Himalaya

In the NW Himalaya, the HHC and underlying LH sequences experienced differential exhumation rates during Miocene-Quaternary, and are modelled in tectonic framework of fast exhuming windows, thrusts and extensional faults, coupled with rapid erosion (Fig. 6.61a; Jain et al. 2000; Thiede et al. 2005, 2009, 2017; Gavillot et al. 2018; Stübner et al. 2018). Patel and ManMohan (2019; 2020b and references therein) relooked ~800 published $^{40}Ar/^{39}Ar$ (white mica/biotite), zircon fission track (ZFT), zircon U–Th/He (ZHe), apatite fission track (AFT), and apatite U–Th/He (AHe) from the HHC and the LH Crystalline klippen/nappes from both the NW and NE Himalaya to observe distinct variations in exhumation patterns. These patterns are controlled by the following structures.

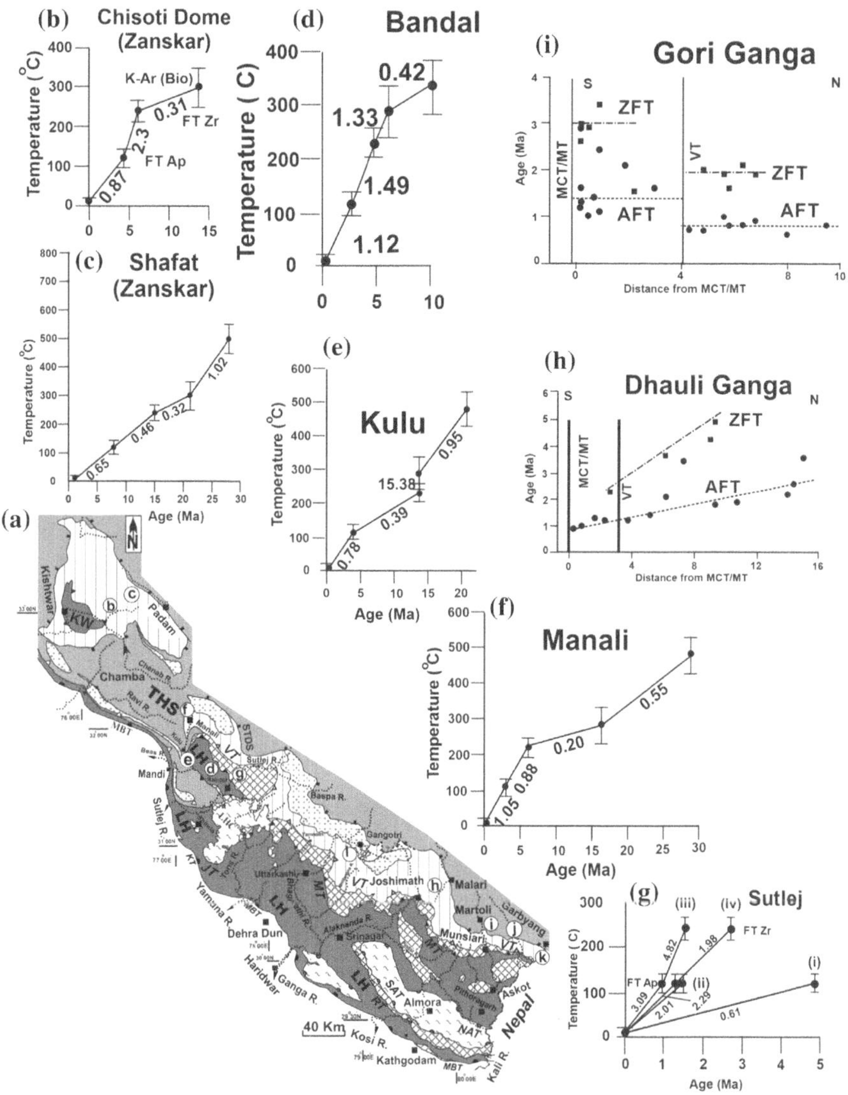

Fig. 6.61 Exhumation paths from the Himalayan Metamorphic Belt (HMB), NW Himalaya, **a** Simplified geological map NW-Himalaya showing locations of exhumation studies. **b**–**m** T-t curves from Chisoti Dome Zanskar, **c** Shafat, Zanskar, **d** Bandal, **e** Kulu, **f** Manali, **g** Sutlej valley, **h** Dhauli Ganga valley, **i** Gori Ganga valley, **j** Darma valley, **k** Kali valley, **l** Gangotri, Bhagirathi valley, **m** Tso Morari, **n** Ladakh Batholith. **o** Geological map of Bhutan and Arunachal Pradesh. Modified after Pebam et al. (2009) and Adhalka et al. (2013). T-t curves from **p** western Kameng, **q** Subansiri, and **r** Siang regions. Curves redrawn after Jain et al. (2000, 2009), de Sigoyer et al. (2000), Schlup et al. (2003), Leech et al. (2005, 2007), Patel and Carter (2009), Patel et al. (2009), Adlakha et al. (2013, 2018), Salvi et al. (2017), Kumar et al. (2018)

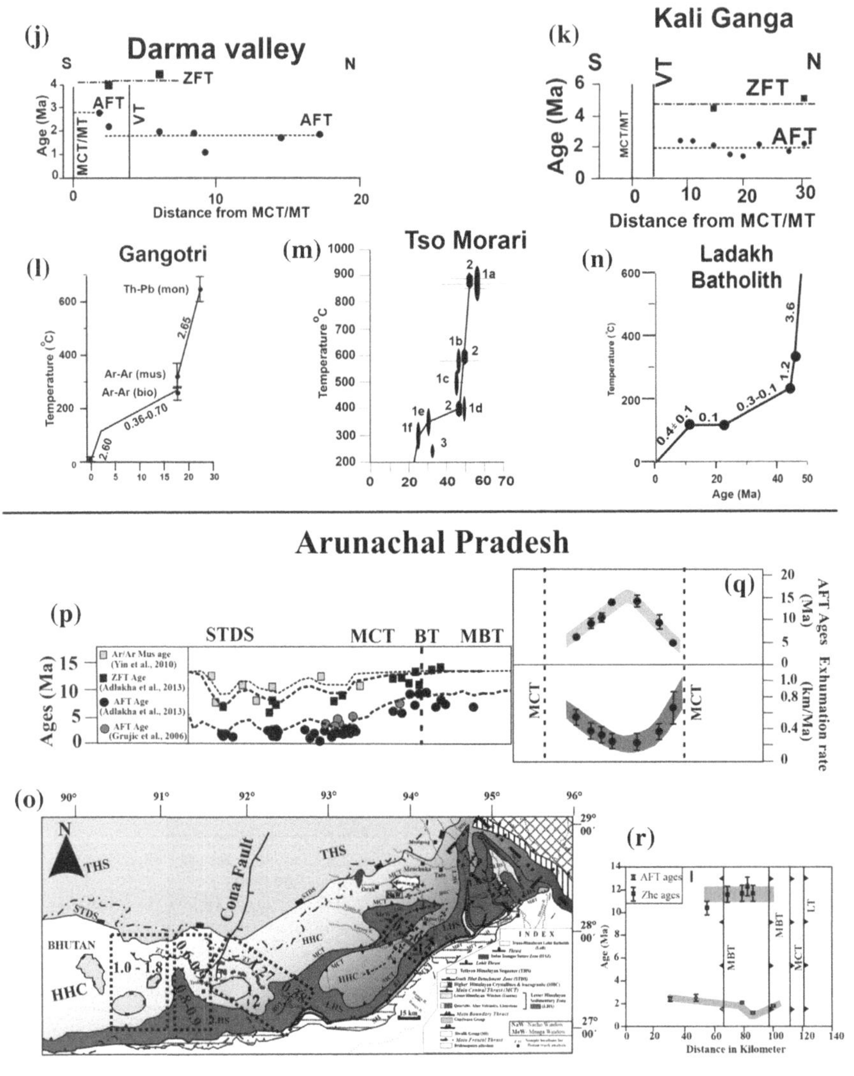

Fig. 6.61 (continued)

6.12.1.1 Domes/Windows

(i) **Western Himalayan syntaxis** witnessed rapid exhumation due to northerly-plunging dome and superposed faulting, which caused decrease in $^{40}Ar/^{39}Ar$ biotite ages from ~8.0 to 0.9 ± 0.1 Ma in south-central core. The ZFT-AFT ages indicated an enhanced exhumation rate of 5 mm/year during 2.0–0.5 Ma and its further acceleration to 9 mm/year since 0.5 Ma (Winslow et al. 1996).

(ii) **Suru/Chisoti Domes and Kishtwar/Kulu-Rampur Windows** have very young and fast exhumation rates in Jammu and Kashmir: 0.80 mm/year, 1.4 mm/year and 2.97 mm/year, respectively during 5–1 Ma (Kumar et al. 1995). In the Chisoti Dome, exhumation rates accelerated from 0.31 mm/year to 2.03 mm/year during 13.7–6.2 Ma and to 4.4 Ma, respectively. Subsequently, it slowed to 0.87 mm/year (Fig. 6.61b). Away from the Suru Dome, exhumation rates ranged from 0.32 to 0.65 mm/year since ~20 Ma (Fig. 6.61c). The Kishtwar Window indicated Quaternary folding around 1.0 Ma from its AFT age.

(iii) **Kulu-Rampur Window** exhumed relatively faster than its metamorphic cover along the Beas-Sutlej Valleys. Within the window, the Bandal Granite exhibited Late Miocene-Pliocene accelerated exhumation from 0.42 to 1.33 mm/year, as deciphered from $^{40}Ar/^{39}Ar$ sericite, Rb–Sr biotite and ZFT ages (Fig. 6.61d) (see Jain et al. 2000). The ZFT–AFT ages indicated that it finally exhumed at 1.49 ± 0.09 mm/year during 4.50 and 2.68 Ma before slightly slowing down to 1.12 ± 0.18 mm/year. On the southwestern flank of this window around Kulu, exhumation is much faster to 15.38 mm/year for a very short span around 13.7 Ma (Fig. 6.61e) and then much slower (0.55 mm/year) along the northwestern closure during 28.2 and 16.10 Ma (Fig. 6.61f). The Pliocene-Pleistocene accelerated exhumation rates between 1.49 and 1.12 mm/year, measured from the ZFT–AFT and AFT–surface, respectively surpass all rates within the window zone on limbs and along strike towards northwest (Fig. 6.61e, f). These range from 0.39 to 0.78 mm/year on southwestern flank around Kulu (Fig. 6.61e), between 0.88 and 1.05 mm/year towards northwest between Manali and Rohtang Pass (Fig. 6.5f) and much slower exhumation (0.61 ± 0.10 mm/year) on its northeastern limb since 4.9 Ma (Fig. 6.61g-curve i). All these rates reveal that the MCT has distinctly been active as an extensional normal fault after nappe emplacement on both flanks of the window and facilitated faster exhumation.

Thrusts

In the NW Himalaya, the FT ages vary across thrust boundaries, where these have controlled exhumation rates within the HHC in Sutlej Valley and Kumaon, e.g. the Chaura Thrust and the MCT. Along the former, footwall and hanging wall were differentially exhumed at 0.61 ± 0.10 mm/year from 4.9 ± 0.2 to 1.49 ± 0.07 Ma and 2.01 ± 0.35 mm/year from 1.49 ± 0.07 Ma to Present (Fig. 6.61g-curves i, ii). However, exhumation rate is 3.09 ± 0.71 mm/year since 0.97 Ma on its hanging wall. The character of the Vaikrita Thrust is uncertain within error limits of average AFT ages of 1.49 ± 0.07 Ma and 1.3 1 ± 0.22 Ma on both the walls, respectively. The ZFT–AFT ages from the Sutlej Valley reveal interplay of differentially moving crustal wedges both ways during Plio-Pleistocene (Jain et al. 2000).

Along the Dhauli Ganga and Gori Ganga Rivers in eastern Garhwal-Kumaon regions, the AFT ages systematically vary from 0.9 ± 0.3 to 3.6 ± 0.5 Ma from the MT to the north of VT (MCT) in Dhauli Ganga (Fig. 6.61h) in contrast to their step-wise variation across this thrust in the Gori Ganga from 1.6 ± 0.1 to 0.7 ± 0.04 Ma in the hanging wall (Fig. 6.61i). The ZFT ages of 1.8 ± 0.4 Ma remain constant in both the walls (Patel and Carter 2009). Further east along the Darma and Kali Ganga valleys, separated by ~40 km, respective AFT data-sets of 1.0 ± 0.1 to 2.8 ± 0.3 Ma and 1.4 ± 0.2 to 2.4 ± 0.3 Ma show very similar exhumation patterns like the Gori Ganga, and are controlled by Plio-Quaternary tectonic activity along the Vaikrita Thrust (VT-MCT) (Fig. 6.61j, k; Patel et al. 2009).

Though these valleys possess similar monsoonal conditions, spatio-temporal variations in the ZFT and AFT ages across the MCT zone and the HHC highlight role of thrusts during Pliocene-Pleistocene.

6.12.1.2 Extensional Fault and Erosion

An early Miocene rapid exhumation pulse around 18 Ma is recorded from the Gangotri leucogranite, by $^{40}Ar/^{39}Ar$ muscovite-biotite and ZFT–AFT ages due to the STDS (Fig. 6.61l). Together with Th–Pb monazite ~22.5 Ma age, it appears to have exhumed at a rate of 2.65 mm/year between 23.0 and 18.0 Ma before slowing down to 0.36 and 0.70 mm/year (ZFT–AFT). Another accelerated Late Pliocene-Quaternary exhumation followed at 2.60 mm/year since 1.67 Ma due to rapid erosion (Fig. 6.61l; Jain et al. 2009).

6.12.2 Tso Morari Crystalline (TMC) Belt

After attaining the UHP metamorphism around 53.3 ± 0.7 Ma due to its deep subduction (Leech et al. 2005, 2007), this metamorphic belt underwent a record exhumation till ~48 to 45 Ma from ~120 to 35 km through HP and amphibolite facies (Fig. 6.61m; de Sigoyer et al. 2000). Greenschist facies minerals grew at ~8 km between 45 ± 2 and 34 ± 2 Ma. Thus, the TMC witnessed record maximum exhumation rate of 17 mm/year during ~53 and 50 Ma and subsequent deceleration to 12 mm/year from 50 to 47 Ma, 0.3 mm/year till 34 ± 2 Ma (ZFT) and further lower rate when it attained ~120 °C at 24 ± 2 Ma (AFT) (Guillot et al. 2008).

6.12.3 Trans-Himalayan Batholith (LB)

The region of Trans-Himalaya lacks monsoon precipitation and exposes about 3-km elevation across the Ladakh Batholith and is ideal for determining the exhumation patterns in dry highlands. The batholith witnessed an Early-Middle Eocene very fast exhumation of 3.5 ± 0.9 mm/year between 50–45 and 48–45 Ma from closure temperatures of $^{40}Ar/^{39}Ar$ hornblende and Rb–Sr biotite, respectively and then to

1.2 ± 0.4 mm/year until 43–42 Ma (ZFT age) (Fig. 6.61n; Kumar et al. 2018). Exhumation rates finally decreased during Oligocene to a minimum of ~0.1 mm/a before a mild Late Miocene-Holocene acceleration (Kumar et al. 2007).

6.12.4 NE Himalaya

The NE Himalaya covers parts of the Bhutan Himalaya, Western and Eastern Arunachal Himalaya (Fig. 6.61o) from where thermochronological data are recently obtained (Grujic et al. 2006; Biswas et al. 2007; Adlakha et al. 2013, 2018; Coutand et al. 2014). In Bhutan Himalaya, exhumation rates are determined using ZFT, ZHe, AFT and AHe ages along two longitudinal regions of the western and eastern sectors. In the west, the AFT ages range between 1.4 ± 0.6 and 6.7 ± 0.8 Ma (mostly between 1.5 and 4.0 Ma), while these are between 3.0 ± 0.7 and 8.6 ± 0.8 Ma in the east (Grujic et al. 2006). Mean AFT age of 2.55 ± 0.15 Ma in the west and 5.08 ± 0.09 Ma in the east provide mean long-term exhumation rates of the order of 1.0–1.8 mm/year and 0.55–0.85 mm/year, respectively; the AFT ages are comparatively older than other part of the HHC in the NW Himalaya. Variations in AFT ages and exhumation rates are ascribed to the Shillong Plateau as an orographic barrier that has reduced precipitation in the interior of Himalayan orogen of the eastern Bhutan and, hence, lowered erosion rates (Grujic et al. 2006; Biswas et al. 2007). Therefore, Grujic et al. (2006) argued that exhumation and erosion variations in Bhutan are climatically driven and, in turn, are likely to influence region's recent tectonic evolution. Additional new ZFT, ZHe, AFT and AHe age data from Bhutan reveal similar age versus distance patterns from south to north with ages becoming older from the MBT across the MCT, progressively younging further north, and finally again becoming older northwards, both in the western and eastern Bhutan (Coutand et al. 2014; McQuarrie et al. 2014). 3-D thermokinematic modelling of age patterns indicated strong tectonic influence of variably dipping geometry of the Main Himalayan Thrust (MHT) between western and eastern Bhutan Himalaya and steady-state topography.

In western Arunachal Himalaya across the HHC, the AFT ages range from 1.4 ± 0.2 to 2.9 ± 0.3 Ma (mean ~2.1 Ma), while the ZFT ages range from 4.5 ± 1.0 to 8.9 ± 1.3 Ma (mean ~7.0 Ma); in the Lesser Himalayan sequence, the former are from 5.6 ± 0.6 to 12.4 ± 1.3 Ma (mean ~7.9 Ma) and the latter from 10.9 ± 0.6 to 14.1 ± 1.1 Ma (mean ~12.5 Ma) (Adlakha et al. 2013). Despite being located in the rain shadow of the Shillong plateau, the AFT and ZFT ages are as young as 1.3 ± 0.2 and 4.5 ± 1.0 Ma, respectively and do not correlate with spatial gradients in precipitation. Thermo-kinematic modelling of these data suggests that local exhumation patterns reflect gradients in rock uplift dictated by fault kinematics along the MCT and ultimately with the MHT (Fig. 6.61p; Adlakha et al. 2013). Further southeast in the Subansiri region along the strike, the AFT cooling ages within the synformally-folded HHC nappe, emplaced over the LH belt, range between 5.0 ± 0.8 and 14.4 ± 1.3 Ma, with slow exhumation rates varying from 0.25 ± 0.12

to 0.69 ± 0.25 mm/year since Middle to Late Miocene, but relative faster on either limbs of the synform and slow in the core (Fig. 6.61q; Adlakha et al. 2018). By contrast, the LH antiformal Menga Window zone from the adjacent Kurung Valley section has youngest AFT age of 2.1 ± 0.3 Ma and suggests fastest exhumation in the window whereas the AFT ages increase away from the window zone within the HHC (Pebam et al. 2008). Further east along the Siang valley, the AFT and U–Th/He zircon (ZHe) ages highlight the unique spatial erosion along the axial regions of the Siang antiform. All the ZHe and AFT cooling ages range between ~14 and 9 Ma, and 2.4 Ma to present, respectively (Salvi et al. 2017). Plot of the AFT and ZHe cooling ages versus distance across the Siang Antiform shows a horizontal linear trend of the ZHe ages across the antiform, whereas AFT ages show a concave upward trend (Fig. 6.61r), indicating that the antiform started developing post ~1.5 Ma (AFT age).

Older AFT cooling ages and slow exhumation in the frontal synformal nappe and fast exhumation in the LH window zone represent a décollement-type analogue model. Lack of any correlation with precipitation reveals absence of climate-driven tectonic deformation via focused erosion, while close correlation with the synform suggests strong tectonics–exhumation linkage.

It has been recently described that the thermochronological age pattern has compatibility with strong tectonic influence, involving a variably dipping MHT geometry (see Sect. 6.9). Duplex development due to horse emplacement above the MHT at depth controls the locus of exhumation. Horse emplacement at depth produces localized regions of relatively high uplift rates (Boyer and Elliott 1982; Molnar and Lyon-Caen 1988). It is a mechanism that explains young cooling ages and fault-related exhumation across structural culminations in the Nepalese hinterland (Herman et al. 2010; Robert et al. 2009) and in the Sikkim Himalaya hinterland (Abrahami et al. 2016; Landry et al. 2016).

Thermochronological age patterns have strong tectonic influence, involving a variable dipping MHT geometry (see Sect. 6.9). Duplex development due to horse emplacement above the MHT at depth controls the locus of exhumation. Horse emplacement at depth produces localized regions of relatively high uplift rates (Boyer and Elliott 1982; Molnar and Lyon-Caen 1988). It is a mechanism that explains young cooling ages and fault-related exhumation across structural culminations in the Nepalese hinterland (Bollinger et al. 2006; Brewer and Burbank 2006; Robert et al. 2009; Herman et al. 2010) and in the Sikkim Himalaya hinterland (Abrahami et al. 2016; Landry et al. 2016).

6.13 Large-Scale Himalayan Tectonics

Since the postulate of collision between India-Asia continents or large-scale underthrusting of India beneath Tibet, various models have been proposed for large-scale tectonics of the Himalaya (Fig. 6.62).

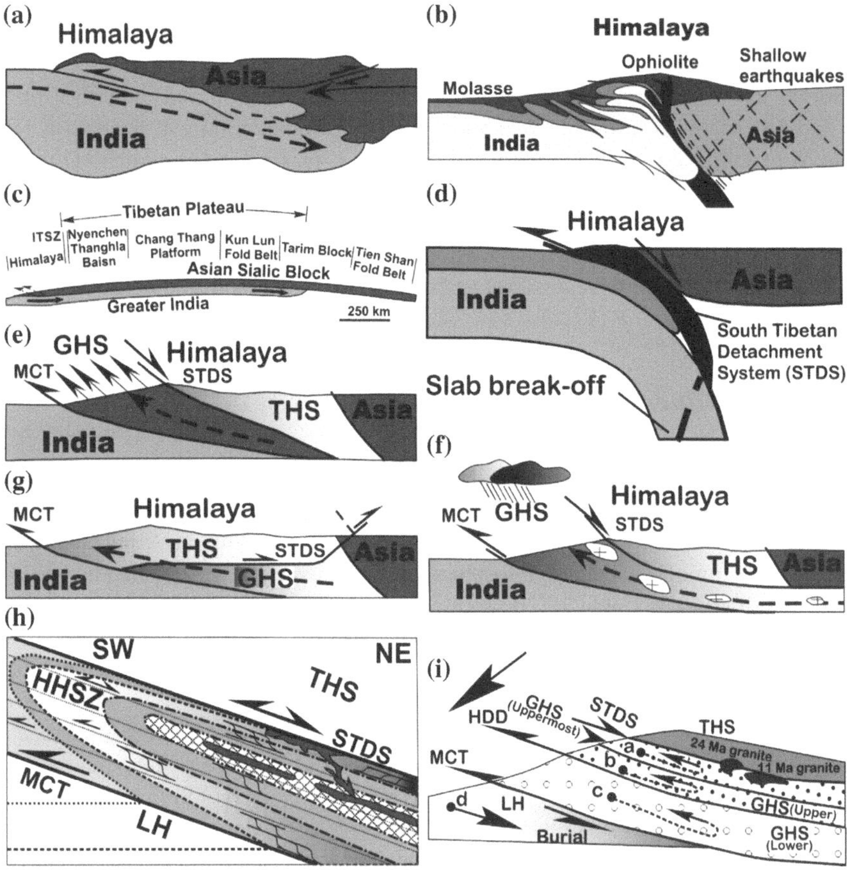

Fig. 6.62 Schematic diagrams showing large-scale tectonic evolution of the Himalaya. **a** Large-scale underthrusting of India showing its deformation and extension till Kun Lun (Argand 1924). **b** Continent-continent collision (Dewey and Bird 1970). **c** Underplating of the Indian crust beneath Asia as subhorizontal slab (Powell and Conaghan 1975). **d** Buoyant rise of partially subducted upper continental crustal slice to upper crustal levels, triggered by slab detachment (Chemenda et al. 2000). **e** Southward wedge extrusion of high-grade metamorphic complex between non-parallel southward-directed MCT and northward-directed STDS (Burchfiel and Royden 1985). **f** Channel flow model of southward migration of partially molten lower and/or middle crustal rocks in a channel and its forced extrusion due to intensified rainfall and resultant erosion (Nelson et al. 1996; Beaumont et al. 2001). **g** Tectonic wedging model involving emplacement of the high-grade metamorphic core at depth, and bound by thrust-backthrust system (Yin 2006; Webb et al. 2017). **h** Intra-continental ductile shear zone model involving high grade GHS sequence with distributed small-scale S-C penetrative ductile shear zones, causing inversion of metamorphic isograds (shown by thick lines), partial melting and leucogranite generation (Jain and Manickavasagam 1993; Jain et al. 2005). **i** In-sequence ductile shearing model involving GHS sequence with downward shifting shearing with time and development of Himalayan discontinuities (Carosi et al. 2018)

6.13.1 Large-Scale Underthrusting of India

Prior to the advent of Plate Tectonics, Argand (1924) illustrated underthrusting of India beneath Asia and its "collision" with the Asian continent beneath Kun-Lun Mountains, and "not" beneath the Himalaya (Fig. 6.62a). his model had three components: India, Asia and deposits of the Tethys between them. He also documented ductile deformation of the Indian continental crust in south, its doubling up due to underthrusting beneath the Himalaya and its rise. The underthrust Indian subcontinent extended beneath Tibet as far north as the Kunlun fold belt.

6.13.2 Continent-Continent Collision

As a follow-up of Plate Tectonics and its application to the mountain-building processes, Dewey and Bird (1970) formulated concepts of collision-type mountain belts for the Alpine-Himalaya. As India approached and collided with the Asian continent after subduction of its oceanic lithosphere, part of this segment obducted as uppermost ophiolite nappes, which rode southwards over the deformed metamorphosed Indian continent and its platform sediments (Fig. 6.62b). Buoyancy caused the uplift of the Himalaya, its extensive erosion, and deposition of molassic sediments into the frontal HFB.

Numerous other tectonic models for evolution of the Himalaya have since been postulated and summarized in Powell and Conaghan (1975) (e.g. Tapponier et al. 1982; Hodges 2000; Yin 2006): (i) Carey's symmetrical model of intra-crustal shortening and thickening by divergent penetrative deformation along the India-Asia margins, i.e. along the Indus-Tsangpo suture Zone, (ii) intra-crustal shortening within hot and ductile Asia during convergence of cold India leading to basement re-activation of the Tibetan Plateau, (iii) repeated southward collision and thrusting of small micro-continental blocks/plates within leading Asian edge throughout Phanerozoic, (iv) India and Tibet belonging to single Gondwanaland fragment with northern edge of the Indian Plate lying beyond the Tibetan plateau in Tien Shan, and the Himalaya evolving intra-continentally, and (v) indentation/extrusion tectonics by northward rigid India indenter push into softer plastic Asian mass, which caused a narrow Himalayan thrust belt, large-scale strike-slip faulting in Asia and rift zones.

6.13.3 Intra-crustal Shortening of Indian Subcontinent by Underthrusting

Geological settings of various Himalayan and Trans-Himalayan tectonic units made Powell and Conaghan (1973, 1975) to visualize that the Himalaya was produced by large-scale underthrusting of the Gondwanian Indian continent beneath Asia along sub-horizontal crustal fracture with its suturing with Asia along the ITSZ and not directly by continent-continent collision (Fig. 6.62c).

6.13.4 Subducting Indian Continental Crust

This model visualized the buoyant rise of detached subducted continental crustal slice to upper levels by compression and buoyancy along the STDS, possibly due to slab detachment at depth (Fig. 6.62d; Chemenda et al. 2000). This slice may be internally soft low-viscosity material, flowing in a channel.

6.13.5 Himalayan Wedge-Extrusion Model

The MCT and non-parallel north-dipping normal fault system caused southward extrusion of high-grade Himalayan metamorphic tapered core (Fig. 6.62e; Burchfiel and Royden 1985). Gravitational collapse of overthickened continental crust was proposed as a mechanism for development of the STDS and synchronous movements along the MCT.

6.13.6 Channel Flow Model

Southward extrusion of high-grade metamorphic rocks and granitic intrusions within the GHS was caused within a low viscosity mid-crustal channel, bounded by the MCT and STDS near the surface (Nelson et al. 1996; Beaumont et al. 2001; Gowdin et al. 2006). A pressure gradient between Tibet and India caused such movements, enhanced by focused precipitation, erosion and exhumation along southern margin of the Great Himalaya (Fig. 6.62f).

6.13.7 Tectonic Wedge Model

In Himachal, the THS is juxtaposed against the LH sequence along the MCT due to its termination against the STDS. Therefore, the GHS core remains at depth and subsequently forces itself towards the surface (Fig. 6.62g; Webb et al. 2007).

6.13.8 Ductile Shear Model

An inverted metamorphism at structurally higher levels across the GHS was earlier explained either by recumbent folding of isograds, structural juxtaposition/thrusting, downward heat transfer from hot plutons/metamorphic belt, shear frictional heating along the MCT, or transposition of isograds by simple/general shear. Jain and Manickavasagam (1993) and Jain et al. (2005) proposed that northward subduction of the ICL caused prograde regional metamorphism up to a maximum depth of 25–35 km, when it attained P-T conditions of 5–9 kb and 550°–780 °C. Millimetre-spaced C-foliation sigmoidally bent and transposed S-foliation on small-scale towards southwest within a ~20 km thick non-coaxial intra-continental ductile shear zone (Jain and Manickavasagam 1993). This model postulated that metamorphic isograd surfaces also underwent small-scale displacements by C-foliation with a cumulative displacement of around 80–120 km. Migmatite and leucogranite were produced during decompressional partial melting in sillimanite-muscovite and sillimanite-K-feldspar isograds in upper parts (Fig. 6.62h).

6.13.9 Southward-Propagating 'In-Sequence' Ductile Thrusting

The GHS was exhumed during ~25 to 17 Ma by contemporaneous shearing along the MCT and the STDS. Carosi et al. (2018) identified three tectono-metamorphic discontinuities/ductile shear zones since ~40 Ma, with top-to-the-S/SW sense of shear. These progressively exhumed the GHS from uppermost to the lowermost parts, with intervening Higher Himalayan Discontinuity (HHD) (Fig. 6.62i). In-sequence shear model brought more metamorphosed tectonic slices from deeper parts to uppermost structural levels; lowermost unit was juxtaposed at 17–13 Ma along the MCT.

6.14 Timing of India-Asia Convergence

Paleomagnetic and other evidences indicated that estimated timing of closure of the Neo-Tethys along the ITSZ and the India-Asia impingement/'collision' varied between 65 and 35 Ma (Leech et al. 2005; Bouilhol et al. 2013; Jain 2014, 2017). Paleomagnetic anomalies in the Indian Ocean point a slow-down in northward movement of the Indian Plate around 55 $\pm$ 1 Ma due to its impingement (Copley et al. 2010; references therein). Paleolatitude evidences reveal an India-Asia suturing of the Himalayan Tethyan succession with Lhasa terrane at 46 $\pm$ 8 Ma, when these terranes started overlapping at 22.8 $\pm$ 4.2°N paleolatitude (Dupont-Nivetet al. 2010). Stratigraphically, maximum age of initiation of India-Asia collision in the ITSZ (Ladakh) was deciphered at 56.5–54.9, and 50.5 Ma from termination of continuous marine sedimentation, and minimum age of closure of the Tethys, respectively (Garzanti and van Haver 1988). In this scenario, the ICL travelled to the ITSZ trench at 58 Ma (Garzanti et al. 1987). Renewed clastic supply within the Zanskar sequence revealed closure of the Tethys at ~56 Ma (Sciunnach and Garzanti 2012), though first arrival of Asian-derived detritus in uppermost Tethyan sediments indicated this convergence at ~50 Ma (Najman et al. 2010). Further east in southern Tibet and elsewhere, two-stage collision events at ~55 Ma and younger Oligocene are indicated between an island arc system with India (Event 1) and India-Asia continental collision (Event 2; Aitchison et al. 2007). Similar scenario has been visualized by van Hinsbergen et al. (2012): 50 Ma collision of micro-continent fragment and continental Asia, and 25 Ma hard collision.

This timing can, additionally, be resolved by comparing ages of continental subduction (UHP metamorphics–TMC), with oceanic subduction (Ladakh Batholith–LB). In the TMC, distinct metamorphic events are decipherable from peak UHP, HP eclogite and amphibolites facies at 53.1 $\pm$ 0.7, 50.0 $\pm$ 0.6 and 47.5 $\pm$ 0.5 Ma, respectively. Across the ITSZ, multiple pulsative crystallization and emplacement within the LB are recorded between ~100 and 41 Ma, with its peak crystallization at 57.9 $\pm$ 0.3 Ma (Jain 2014). Thus, it is evident that two contrasting deep crustal processes took place ~58 Ma across the ITSZ, leading to the India-Asia contact (Jain 2014).

6.15 Geological Evolution of the Himalaya

Once the Tethyan Oceanic lithosphere subducted in the Trans-Himalaya and closed the Tethyan Ocean (cf. Jain 2014, 2017), subsequent continental lithospheric subduction caused the convergence of the Indian and Asian Plates through various stages. Thus, the evolution of the Himalayan tectonic units can be visualized as follows.

6.15.1 First Stage of Continental Subduction–Tso Morari Crystallines (TMC)

The Indian Continental Lithosphere (ICL) touched the ITSZ trench at ~58 Ma, which coincided with peak plutonic growth of the LB. First stage of the continental lithospheric subduction in the Himalaya is, therefore, recorded along the leading edge of the Indian Plate in Tso Morari, where it subducted down to depth of ~120 km to produce carbonate-coesite-bearing UHP eclogite at >39 kb, >750 °C and 53.1 $\pm$ 0.7 Ma, and subsequent retrogression to HP and amphibolite facies between 50.0 and 48 Ma, respectively (Fig. 6.63; Guillot et al. 1997; Leech et al. 2005, 2007). $^{40}Ar/^{39}Ar$

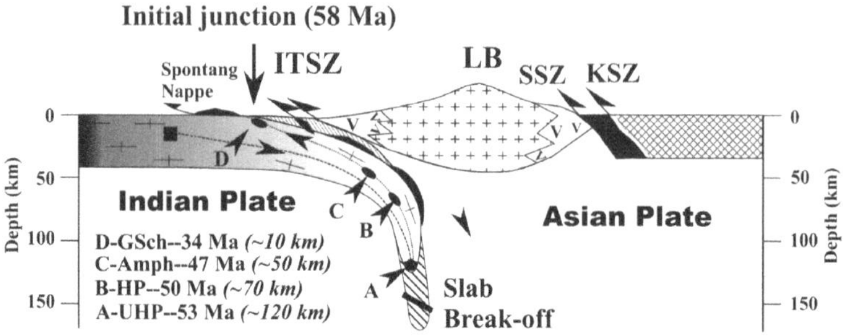

Fig. 6.63 Steep continental lithospheric subduction of the Indian Plate in the TMC, its UHP metamorphism and exhumation. Development of gneiss-hosted UHP eclogite at 53.1 ± 0.7 Ma. The TMC follows same exhumation and cooling paths through various metamorphic facies. See text for details. During the HP facies, steeply subducted continental lithosphere melted along southern margin of the Ladakh Batholith to produce crustal contaminations

mica and ZFT ages provide evidences of late stage exhumation and shallow crustal stabilization between 45.0 and 34 ± 2 Ma (ZFT) (Schlup et al. 2003).

6.15.2 First Rise of the Himalaya in the Tso Morari

After witnessing the first evidences of subduction of the ICL along the ITSZ, it had suffered the UHP to greenschist metamorphism during a period of ~53 to 34 Ma in the TMC terrane, which was then exhumed to the surface. Initially, this exhumation was very fast at a rate of ca. 1.7 cm/year between 53 and 50 Ma and 1.2 cm/year between 50 and 47 Ma and then slowed down to ca. 0.3 cm/year till 35 Ma (Guillot et al. 2008).

The Himalaya, therefore, first witnessed its rise and emergence of land mass in the Tso Morari region between 53 and 50 Ma when the continental crust was exhumed from a depth of ~120 km. The terrane was uplifted very fast rate to near-surface and eroded off to shed detritus into the HFB in the south and the ITSZ basin (Jain et al. 2009). This subducting lithospheric slab did not melt till its decompression around 50 Ma after the HP metamorphism. Very steep geometry of the Indian lithosphere and its melting along the ITSZ, thus, explain sharp isotopic changes in the overlying LB along its southern margin.

6.15.3 Second Stage of Continental Subduction–HHC

Folded HMB slab, underwent peak Late Eocene pre-MCT regional Barrovian metamorphism in upper amphibolite facies at 650–700 °C, 8–9 kb and around 45–35 Ma (Hodges 2000; Foster et al. 2000) (Fig. 6.64). It is likely that the ICL witnessed the

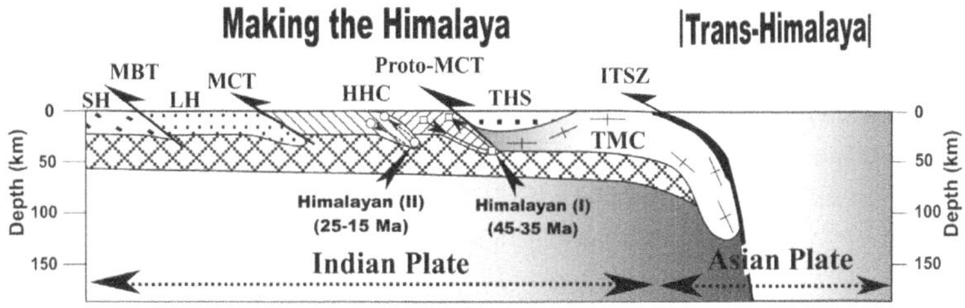

Fig. 6.64 Continental subduction in the NW-Himalaya. Shallower and younger continental subduction in the HHC produced peak Eocene (~48 to 45 Ma), and pre-MCT metamorphism in upper amphibolite facies. Younger Miocene ~25 Ma metamorphism, anataxis and leucogranite generation between 25 and 15 Ma. Major discontinuities within HHC caused its differential exhumation. See Carosi et al. (2018) for details. Thrusting and imbrication along the MBT and MFT during late Miocene-Pliocene-Pleistocene

development of the proto-MCT at around 45 Ma and underwent shallow continental subduction to a depth of 35–25 km after subduction in the Tso Morari region; both the regions witnessed almost simultaneous amphibolite metamorphism at nearly same time. $^{40}Ar/^{39}Ar$/K-Ar hornblende and muscovite ages gave its cooling through 500 $\pm$ 50 °C and 350 $\pm$ 50 °C between 40 and 30 Ma (Sorkhabi et al. 1999).

6.15.4 Third Stage of Continental Subduction–Within HHC

Within the HHC belt, younger Miocene ~25 Ma metamorphism and partial melting led to leucogranite generation between 25 and 15 Ma, though leucosome melt production also took place occasionally during the Eocene and Oligocene (33–23 Ma), as well (Carosi et al. 2018). These melts appear to have evolved in a southward extruding Himalayan orogenic channel, bounded by the MCT and STDS (Grujic 2006). Subsequent Miocene–Pleistocene exhumation is widespread in the HHC (Fig. 6.64), followed by its extensive erosion to produce detritus for the Cenozoic Himalayan foreland and Indo-Gangetic–Bengal basins (Hodges 2000; Yin 2006); these patterns are either controlled by tectonics, concomitant erosion or a combination of two processes (Jain et al. 2000).

6.15.5 Present-Day Configuration

During the past two decades, configuration of the Indian Plate beneath the Himalaya and adjoining regions has been imaged by magnetotelluric profiling, seismic tomography and focal plane mechanism of recent earthquakes.

6.15.5.1 Magnetotelluric (MT) Evidences

Magnetotelluric (MT) profiles from the NW-Himalaya image the present-day geometry of the Indian Plate (IP) (Fig. 6.65). Along Roorkee-Gangotri profile (Uttarakhand), the Precambrian basement (>1000 Ωm electrical resistivity) is overlain by a northerly gently-dipping and 7-km thick veneer of Miocene-Recent conducting sediments (<50 Ωm) within the Indo-Gangetic Plains (IGP) and Sub-Himalaya. Their contacts represent fluid-saturated fractured zone of the Main Himalayan Thrust (MHT) along which the IP subducts beneath the Himalaya (Fig. 6.65a; Israil et al. 2008; Miglani et al. 2014). A sharp discontinuity defines the MBT where resistivity increases in the Precambrian LH belt. Below the MHT-controlled conductor, high electrical resistive zone at ~30 km depth represents northerly-subducting IP continental crust (Miglani et al. 2014). Fluid-filled fractured MHT zone of very low <10 Ωm resistivity extends up to a depth of ~30 km within a ramp and is marked by earthquake cluster and heat flow. Further southeast along the Bijnor-Joshimath-Malari traverse (Uttarakhand), ~5 km thick low-angle and NE-dipping intra-crustal high conducting layer (IC-HCL) of very low resistivity is observed across the IGP and the Higher Himalaya with an intervening ramp (Rawat et al. 2014); the whole structure also controls velocities of seismic waves in intra-crustal low velocity layer (IC-LVL) (Caldwell et al. 2013). These layers are fluid-saturated due to partial melt or released water from metamorphic reactions.

This mid-crustal conductor of low resistivity zone may also indicate partial melt in extreme northeast between Tso Morari and beyond, where its sub-horizontal geometry extends beneath high resistive LB and Karakoram Batholith (Fig. 6.65b; Gokarn et al. 2002; Arora et al. 2007).

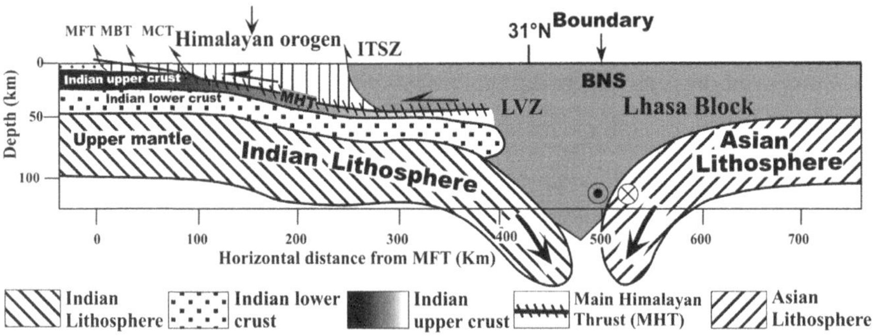

Fig. 6.65 Magnetotelluric sections from the NW-Himalaya to Karakoram. **a** MT section along Roorkee-Gangotri profile showing low resistivity (<10 Ωm) zone beneath MCT and cluster of small associated earthquakes (Miglani et al. 2014). **b** Deep resistivity structure of the subducting Indian lithosphere along Tso Morari and further northeast. NE-dipping low resistivity zone (~30 Ωm-white colour) is observed beneath THS, Tso Morari Dome (TMD), and beyond where is becomes subhorizontal beneath ITSZ and LB (Arora et al. 2007)

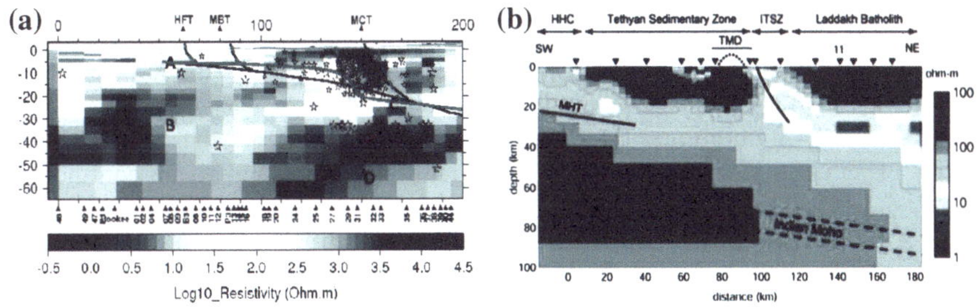

Fig. 6.66 Simplified geophysical cross-section of the India-Asia convergence beneath the Himalaya and Tibet from the Hi-CLIMB profile, depicting "collision" between two plates in the Bangong-Nujiang Suture (BNS). Large-scale tectonics of the Himalayan orogen is depicted as scrapped crustal wedge above the MHT and its limits by the ITSZ. Redrawn after Nábělek et al. (2009)

6.15.5.2 Seismological Evidences

Seismic tomography, reflection profiles, earthquakes and their fault-plane solutions from the Himalayan arc and Tibet provide constraints on the present-day configuration of the Indian Plate. In NW-Himalaya, the Moho runs almost parallel below low resistivity layer and defines geometry of the subducting ICL even beyond Karakoram fault (Rai et al. 2006). Majority of earthquakes are located within a narrow belt between north-dipping MBT and the MCT and are essentially controlled by a mid-crustal ramp within the MHT along which the Indian Plate underthrusts southern Tibet at about 10° (Seeber and Armbruster 1981; Ni and Barazangi 1984; Nelson et al. 1996). This surface dips at ~15° from about 10 to 20-km depth within the Himalaya and controls focal mechanisms not only of shallow ($\lesssim$30°) earthquakes but also the great Himalayan earthquakes of $M > 8$ (Ni and Barazangi 1984). Outer seismic mid-crustal ramp extends into an aseismic 30–40 km deep seismic reflector, imaged in the INDEPTH experiment further north (Zhao et al. 1993).

Subducted Indian lower crust does not stop at the ITSZ/IYS but it underthrusts and extends northwards beneath the partially molten Asian crust till Bangong-Nujiang Suture (BNS) where both the Indian and Asian lithospheric mantles subduct downwards due to flow (Fig. 6.66; Nábělek et al. 2009; Liang et al. 2012). Recent deep seismic reflection profiles by Guo et al. (2017) across ITSZ further strengthen a crustal-scale outline of the subducting Indian crust.

6.15.6 Present-Day Configuration: Continental Collision Versus Subduction

Geological setting of various Himalayan units clearly demonstrates that leading northernmost edge of the ICL was juxtaposed against the Trans-Himalayan ITSZ and LB, and not with any tectonic unit of the Asian continental lithosphere. It converged first with the intra-oceanic Shyok-Dras Volcanic Island Arc, which existed within the Tethyan Ocean, and then both of them moved towards the Indian continental margin (Stampfli and Borel 2002; Rolland et al. 2002; Maheo et al. 2004) or vice versa (Bouilhol et al. 2013). It is evident that Indian and Asian continental lithospheres, separated by Neo-Tethys, did not initially 'collide' directly to make the Himalaya (cf., Powell and Conaghan 1973).

As the Neo-Tethys finally closed along the ITSZ during Paleocene, timing of first India-Asia impingement is constrained at ~58 Ma by comparing ages of the Trans-Himalayan LB with the TMC. Both the LB and the TMC signify drastic geodynamic changes within ~4 to 5 Ma across the ITSZ.

At the time of India undergoing UHP metamorphism at ~53 Ma in the Tso Morari steep subduction zone, the ITSZ limited the Indian continental lithosphere. As the steep ICL took some time to reach depth of about ~120 km to have suffered UHP metamorphism from near surface, difference of ~4 to 5 Ma in these values is accounted for fast speed by which the plate underwent the subduction. This steep continental lithosphere subduction and its melting along the southern margin can account for younger pulses within the LB with more evolved isotopic signatures till 45 Ma without modifying the proposed age of the India-Asia contact (cf., Bouilhol et al. 2013; Jain 2014).

It is likely that the Himalaya first witnessed its rise and emergence during 53 and 50 Ma in the Tso Morari, when part of continental lithosphere exhumed after undergoing UHP metamorphism. Further imbrication and subduction of the ICL caused its two metamorphisms and associated exhumation episodes within the HHC during 45–35 and 25–15 Ma, respectively for the rise of the Himalayan Mountains (Fig. 6.67). It is clearly implied that original configuration of the ICL must have been very different: steeper and imbricated than the present-day geometry (O'Brien et al. 2001), which is attained by its northerly push and flattening of imbricated Himalayan packages, now lying sub-horizontally (Fig. 6.67).

Present-day geometry of the Indian continental lithosphere beneath the Himalaya, and beyond in Asia reflects its unique subducting near-horizontal configuration. Low

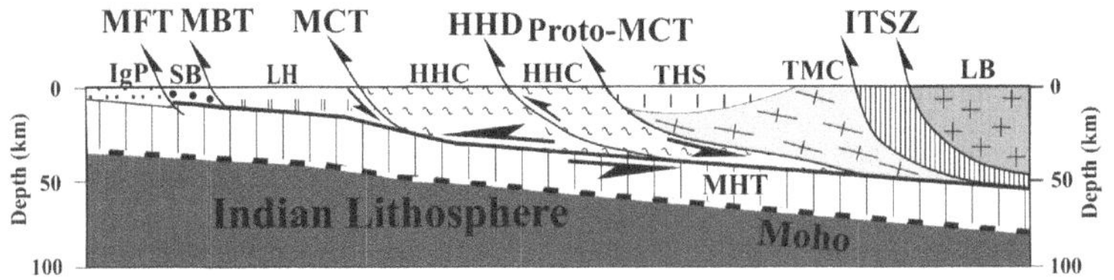

Fig. 6.67 Rotated imbricated slices of the Himalayan Metamorphic Belt (HMB) to gently dipping crustal wedges of the TMC, HHC including its Higher Himalayan Discontinuities (HHD) by northward push of the subducting Indian continental lithosphere along the MHT since post-15 Ma. Configuration is based upon the present-data magnetotelluric data from Garhwal and Ladakh-Karakoram. The MHT approximately follows upper–lower continental crust boundary (vertical lines). Indian Lithosphere shown by dark shade and Moho follows its upper contact

resistivity zone (5–10 Ωm) of partial melt/fluid-rich mid-crustal layer at 15–25 km above the MHT defines the top of the IP, which is underlain by relatively high electrical resistive ICL. This layer is observed throughout the Himalaya and beyond and is the first order structure in the northwest. A steeper northeast-dipping low resistivity zone beyond Tso Morari flattens out at depth beneath the LB and merges with this first order conductor (Arora et al. 2007).

Seismologically, majority of earthquakes and tomographic evidences delineate a basal MHT within the Indian Plate along which it subducts beneath the southern Tibet at about 10° or less (Nelson et al. 1996). In fact, signatures of the ICL extend northwards beneath the partially molten Asian crust till BNS, where both the Indian and Asian lithosphere mantles subduct northwards and southwards, respectively.

Thus, critical geological and geophysical evidences from the Himalaya and adjoining mountains lead us to postulate that the Indian continental lithosphere (ICL) subducted and imbricated sequentially in geological past since 58 Ma (Fig. 6.67). The overriding sequences were scrapped, thrust southward along the MCT-MBT systems, and sequentially deformed as crustal wedges in the rising Himalaya whose erosion caused deposition into the frontal Himalayan Foreland basin and the Indo-Gangetic Plains because of tectonics and monsoonal precipitation since Miocene. The Indian and Asian Plates interacted and 'collided' only in the Bangong-Nujiang Suture (BNS).

References

Abrahami R, van der Beek P, Huyghe H, Hardwick E, Carcaillet J (2016) Decoupling of long-term exhumation and short-term erosion rates in the Sikkim Himalaya. Earth Planet Sci Lett 433:76–88. https://doi.org/10.1016/j.epsl.2015.10.039

Acharya SK, Ghose SC, Ghosesn Shah SC (1975) The continental Gondwana Group and associated marine sediments of Arunachal Pradesh (NEFA), Eastern Himalaya. Himalayan Geol 5:60–82

Adlakha V, Lang KA, Patel RC, Lal N, Huntington KW (2013) Rapid long-term erosion in the rain shadow of the Shillong Plateau, eastern Himalaya. Tectonophysics 582:76–83

Adlakha V, Patel RC, Kumar A, Lal N (2018) Tectonic control over exhumation in the Arunachal Himalaya: new constraints from Apatite Fission Track Analysis. In: Sharma R, Villa IM, Kumar S (eds) Crustal architecture and evolution of the Himalaya–Karakoram–Tibet Orogen, vol 481. Geological Society, London, Special Publications. https://doi.org/10.1144/sp481.1

Ahmad T, Tanaka T, Sachan HK, Asahara Y, Islam R, Khanna PP (2008) Geochemical and isotopic constraints on the age and origin of the Nidar ophiolitic complex, Ladakh, India: implication for the Neo-Tethyan subduction along the Indus suture zone. Tectonophysics 451:206–224

Agarwal KK, Singh IB, Sharma M, Sharma S, Rajagopalan G (2002) Extensional tectonic activity in the craton ward parts (peripheral bulge) of the Ganga Plain foreland basin, India. Int J Earth Sci (Geol Rundsch) 91:897–905. https://doi.org/10.1007/s00531-002-0265-z

Agnihotri D, Pandita SK, Tewari R, Awatar R, Linnemann U, Pillai USK, A Joshi A, Gautam S, Kumar K (2018) Palynology and detrital zircon geochronology of the Carboniferous Fenestella Shale Formation of the Tethyan realm in Kashmir Himalaya: implications for global correlation and floristic evolution. J Asian Earth Sci. https://doi.org/10.1016/j.jseaes.(2017.09.003)

Aharon P, Schidlowsk M, Singh IB (1987) Chronostratigraphic markers in the end–Precambrian carbon isotope record of the Lesser Himalaya. Nature 327:699–702

Argand E (1924) La tectoniquedel' Asie. In: Proceeding 13th Internal Geology Congress 7:170–372

Arya R, Ambwani K, Sahni N, Sahni A (2004) First mammal and additional fossil flowers from the Kasauli Formation, Kasauli, Himachal Pradesh. J Geol Soc India 64:317–324
Auden JB (1934) The geology of the Krol belt. Rec Geol Surv India 67(4):357–454
Auden JB (1935) Traverses in the Himalaya. Geol Surv India 69:123–167
Awasthi SC, Prasad M, Srivastava VC (1985) Terraces and Neotectonism in Katra–Riasi, district Udhampur, Jammu and Kashmir. In: Merh SS, Vashi NM (eds) Proceeding quaternary episodes in India. Depart Geol, MS. University, Baroda, pp 39–48
Banerjee DM (1974) Stratigraphy and depositional characteristics of Tethyan sediments in Kuti–Kalapani area, Kumaun, India. Himalayan Geol 4:296–322
Banerjee DM, McArthur JM (1989) A note on the mineralogy and geochemistry of Mussorie phosphorites of the Lower Himalaya. J Geol Soc India 13:75–82
Banerjee DM, Narain MJ (1976) Trace fossils and the Environment of Deposition in the Lower Tal Formation of Mussoorie. J Sedimen Petrol 1:234–239
Banerjee DM, Tandon SK, Sinha AK (1975) Trace fossils in the Tethyan succession in a part of north–eastern Kumaun Himalaya: In: Proceedings of symposium on sediment, sedimentation and sedimentary environment, vol 9. University of Delhi, Theme, pp 117–185
Banerjee DM, Schidlowski M, Arneth JD (1986) Genesis of upper Proterozoic-Cambrian phosphorte deposits of India: isotopic inferences from carbonate fluorapatite, carbonate and organic carbon. Precamb Res 33:239–253
Banerjee DM, Schidlowski M, Siebert F, Brasier MD (1997) Geochemical changes across the Proterozoic–Cambrian transition in the Durmala phosphorite mine section, Mussoorie Hills, Garhwal Himalaya, India. Palaeogeog Palaeoclim Palaeoeco 132:183–194
Barber AJ, Zaw K, Crow MJ (eds) (2017) Myanmar: geology, resources and tectonics. Mem Geol Soc Lond 48
Barnes JB, Densmore AL, Mukul M, Sinha R, Jain V, Tandon SK (2011) Interplay between faulting and base level in the development of Himalayan frontal fold topography. J Geophys Res 116:1–19. https://doi.org/10.1029/2010JF001841. F0301
Beck RA, Burbank DW, Sercombe WJ, Riley GW, Barndt JK, Berry JR, Afzal J, Khan AM, Jurgen H, Metje J, Cheema A, Shafique NA, Lawrence RD, Khan MA (1995) Stratigraphic evidence for an early collision between northwest India and Asia. Nature 373:55–58
Bera MK, Sarkar A, Chakraborty PP, Loyal RS, Sanyal P (2008) Marine to continental transition in Himalayan foreland. Geol Soc Amer Bull 120:1214–1232. https://doi.org/10.1130/B26265.1
Berthelsen A (1953) On the Geol of the Rupshu District, N Himalaya. Medd Dan Geol Forening 12:350–415
Bhargava ON (1976) Geol of the Krol Belt and associated formations: a reappraisal. Mem Geol Surv India 106:106–234
Bhargava ON (1995) The Bhutan Himalaya: a geological account. Geol Surv India Spec Publ 39:1–245
Bhargava ON (2008a) An updated introduction of the Spiti geology. J Palaeont Soc India 53:113–129
Bhargava ON (2008b) Palaeozoic successions of the Indian Plate. Mem Geol Soc India 74:209–244
Bhargava ON (2011) Early Palaeozoic palaeogeography, basin configuration, paleoclimate and tectonics in the Indian plate. Mem Geol Soc India 78:69–99
Bhargava ON (2015) Evolution of the Tethyan and Karewa successions in Kashmir: a synthesis. J Palaeont Soc India 60(1):51–72
Bhargava ON, Bassi UK (1998) Geol of Spiti-Kinnaur Himachal Himalaya. Mem Geol Surv India 124:1–210
Bhargava ON, Singh BP (2019) A broad climatostratigraphy of the Himalaya. Himalayan Geol 40(2):220–238
Bhargava ON, Singh BP (2020) Geological evolution of the Tethys Himalaya. Episodes 43(1):404–416, https://doi.org/10.18814/epiiugs/2020/020024
Bhargava ON, Srikantia SV (1985) Trilobite and other trace fossils from the Kunzumla Formation, eastern Lahaul, Himachal Pradesh. J Geol Soc India 26:880–886

Bhargava ON, Kumar G, Gupta SS (1982) Cambrian trace fossils from the Spiti Valley, Himachal Himalaya. J Geol Soc India 23(4):183–191
Bhargava ON, Krystyn L, Balini M, Lein R, Nicora A (2004) Revised litho- and sequence stratigraphy of the Spiti Triassic. Albertiana 30:21–39
Bhargava ON, Frank W, Bertle R (2011a) Late Cambrian deformation in the Lesser Himalaya. J Asian Earth Sci 40:201–212
Bhargava ON, Kaur G, Deb M (2011b) A Paleoproterozoic paleosol horizon in the Lesser Himalaya and its regional implications. J Asian Earth Sci 42:1371–1380
Bhargava ON, Thoni M, Miller C (2016) Isotopic evidence of Early Palaeozoic metamorphism in the Lesser Himalaya (Jutogh Group), Himachal Pradesh, India: its implication. Himalayan Geol 37(2):73–84
Bhatia SB (2000) Faunal and floral diversity in the Subathu-Dgshai Passage Beds: a review. Himalayan Geol 21:87–97
Bhatia SB, Bhargava ON (2006) Biochronological continuity of the Paleogene sediments of the Himalayan Foreland Basin: paleontological and other evidences. J Asian Earth Sci 26:477–487
Bhatt DK (1989) Small shelly fossils, Tommotian and Meishucunian stages and the Precambrian-Cambrian boundary—implications of the recent studies in the Himalayan sequences. J Palaeont Soc India 34:55–68
Bhatt DK, Mamgain AK, Misra RS (1985) Small shelly fossils of Early Cambrian (Tommotian) age from Chert-Phosphorite Member, Tal Formation, Mussoorie syncline, Lesser Himalaya, India and their chronostratigraphic evaluation. J Palaeont Soc India 30:92–102
Biswas S, Coutand I, Grujic D, Hager C, Stockli D, Grasemann B (2007) Exhumation and uplift of the Shillong Plateau and its influence on the eastern Himalayas: new constraints from apatite and zircon (U-Th-[Sm])/He and apatite fission track analyses. Tectonics 26:TC6013. https://doi.org/10.1029/2007tc002125
Bodenhausen JWA, DeBooy T, Egelar CG, Nijhuis HJ (1964) On the geology of Central west Nepal—a preliminary note. In: 22nd International geological congress, vol 11, New Delhi. Special Publications, pp 101–122
Bordet P, Colchen M, Krummenacher D, Le FP, Mouterde R, Remmy M (1971) Recherches géologiques dans l'Himalaya du Nepal de la Thakkola. Coll Int Centre Nat Rech Sci Sci Terre 1–279
Boyer SE, Elliott D (1982) Thrust systems. Am Assoc Petrol Geol Bull 66:1196–1230
Brasier MD, Singh P (1987) Microfossils and Precambrian-Cambrian boundary stratigraphy at Maldeota, Lesser Himalaya. Geol Mag 124:323–345
Brookfield ME (1993) The Himalayan passive margin from Precambrian to the Cretaceous times. Sediment Geol 84:1–35
Brookfield ME, Algeo TJ, Hannigan R, Williams J, Bhat GM (2013) Shaken and Stirred: Seismites and Tsunamites at the Permian-Triassic Boundary, Guryul Ravine, Kashmir, India. Palaios 28:568–582
Brosse M, Baud A, Bhat GM, Bucher H, Leu M, Vennemann T, Goudemand N (2017) Conodont–based Griesbachian biochronology of the Guryul Ravine section (basal Triassic, Kashmir, India. Geobios 50(5–6):359–387. https://doi.org/10.1016/j.geobios.2017.10.001
Burbank DW (1992) Causes of recent Himalayan uplift deduced from deposited patterns in the Ganges Basin. Nature 357:680–682
Burchfiel BC, Royden LH (1985) North-south extension within the convergent Himalayan region. Geology 13:679–682
Burchfiel BC, Zhilang C, Hodges KV, Yuping L, Roden LH, Changong D, Jiene X (1992) The South Tibetan detachment system, Himalayan orogen: extension contemporaneous with parallel to shortening in collisional mountain belt. Geol Soc Am Bull 269:1–42
Bürgmann R, Larson KM, Bilham R (1999) Model inversion of GPS and leveling measurements across the Himalaya: implications of earthquake hazards and geodetic networks. Himalayan Geol 20:59–72

Burrard SG (1915) Origin of the Gangetic Trough, commonly called the Himalayan foredeep. Proc R Soc Lond 91:220–238

Butler RWH (2018) Tectonic evolution of the Himalayan syntaxes: the view from Nanga Parbat. Geol Soc Lond Spec Publ 483. https://doi.org/10.1144/sp483.5

Caddick M, Bickle M, Harris N, Holland T, Horstwood M, Parrish RR, Ahmad T (2007) Burial and exhumation history of a Lesser Himalayan schist: recording the formation of an inverted metamorphic sequence in NW India. Earth Planet Sci Lett 264:375–390

Carosi R, Lombardo B, Molli G, Musumeci G, Pertusati PC (1998) The south tibetan detachment system in the Rongbuk valley, Everest region: deformation features and geological implications. J Asian Earth Sci 16:299–311. https://doi.org/10.1016/S0743-9547(98)00014-2

Carosi R, Montomoli C, Laccarino S (2018) 20 years of geological mapping of the metamorphic core across Central and Eastern Himalayas. Earth Sci Rev 177:124–138

Cawood PA, Johnson MRW, Nemchin AA (2007) Early paleozoic orogenesis along the Indian margin of Gondwana: tectonic response to Gondwana assembly. Earth Planet Sci Lett 255:70–84

Cawood PA, Wang Y, Xu Y, Zhao G (2013) Locating South China in Rodinia and Gondwana: a fragment of greater India lithosphere? Geology 41(8):903–906. https://doi.org/10.1130/G34395.1

Célérier J, Harrison TM, Webb AAA, Yin A (2009) The Kumaun and Garhwal Lesser Himalaya, India. Part 1. Structure and stratigraphy. Geol Soc Am Bull 121:1262–1280

Chakraborty S, Anczkiewicz R, Gaidies F, Rubatto D, Sorcar N, Faak K, Mukhopadhyay DK, Dasgupta S (2016) A review of thermal history and timescales of tectonometamorphic processes in Sikkim Himalaya (NE India) and implications for rates of metamorphic processes. J Metamor Geol 34:785–803

Chaubey RS, Singh Birendra P, Mikuláš R, Bhargava ON, Kishore N, Prasad SK (2018) Integrated Ichnological and sedimentological analysis of the Cambrian Kunzam La (Parahio) Formation, Shian Section, Pin Valley, Spiti, Northwest Himalaya. Int J Stratig Geol Correl 26(7):709–721

Chemenda AI, Burg JP, Mattauer M (2000) Evolutionary model of the Himalaya–Tibet system: geopoem based on new modeling, geological and geophysical data. Earth Planet Sci Lett 174:397–409

Chirouze F, Huyghe P, van der Beek P, Chauvel C, Chakraborty T, Dupont-Nivet G, Bernet M (2013) Tectonics, exhumation, and drainage evolution of the eastern Himalaya since 13 Ma from detrital geochemistry and thermochronology, Kameng River Section, Arunachal Pradesh. Geol Soc Am Bull 125(3/4):523–538. https://doi.org/10.1130/B30697.1

Clarke GL, Bhowmik SK, Ireland TR, Aitchison JC, Chapman SL, Kent L (2016) Inverted Oligo-Miocene metamorphism in the Lesser Himalaya Sequence, Arunachal Pradesh, India; age and grade relationships. J Metamor Geol 34:805–820

Cleal CJ, Bhat GM, Singh KJ, Dar AM, Saxena A, Chandra S (2016) Spondylodendron pranabii—the dominant lycopsid of the late Mississippian vegetation of the Kashmir Himalaya. Alcheringa 40(4):443–455

Clift PD (2017) Cenozoic sedimentary records of climate–tectonic coupling in the Western Himalaya. Progr Earth Planet Sci 4:39. https://doi.org/10.1186/s40645-017-0151-8

Copley A, Avouac JP, Royer JY (2010) India-Asia collision and the Cenozoic slowdown of the Indian plate: implications for the forces driving plate motions. J Geoph Res 95: B03410. https://doi.org/10.1029/2009JB006634

Corrie SL, Kohn MJ, Vervoort JD (2010) Young eclogite from the Greater Himalayan Sequence, Arun Valley, eastern Nepal: P-T-t path and tectonic implications. Earth Planet Sci Lett 289:406–416

Cottle JM, Jessup MJ, Newell DL, Searle MP, Law RD, Horstwood MSA (2007) Structural insights into the early stages of exhumation along an orogen-scale detachment: the south tibetan detachment system, Dzakaa Chu section, Eastern Himalaya. J Struct Geol 29:1781–1797. https://doi.org/10.1016/j.jsg.2007.08.007

Cottle JM, Jessup MJ, Newell DL, Matthew S, Horstwood A, Noble SR, Parrish RR, Waters DJ, Searle MP (2009) Geochronology of granulitizedeclogite from the AmaDrime Massif: Implications for the tectonic evolution of the South Tibetan Himalaya. Tectonics 28:1–25 TC1002. https://doi.org/10.1029/2008tc002256

Cottle JM, Waters DJ, Riley D, Beyssac O, Jessup MJ (2011) Metamorphic history of the south tibetan detachment system, Mt. Everest region, revealed by RSCM thermometry and phase equilibria modelling. J Metamor Geol 29:561–582. https://doi.org/10.1111/j.1525-1314.2011.%2000930.x

Coutand I, Whipp DM, Grujic D, Bernet M, Fellin MG, Bookhagen B, Landry KR, Ghalley SK, Duncan C (2014) Geometry and kinematics of the Main Himalayan Thrust and Neogene crustal exhumation in the Bhutanese Himalaya derived from inversion of multithermochronologic data. J. Geophys Res (Solid Earth) 119(2):1446–1481. https://doi.org/10.1002/2013JB010891

Daniel CG, Hollister LS, Parrish RR, Grujic D (2003) Exhumation of the Main Central Thrust from Lower Crustal Depths, Eastern Bhutan Himalaya. J Metamor Geol 21:317–334

Das BP, Joshi M, Kumar A (2019) Tectonochemistry and P-T Conditions of Ramgarh and Almora Gneisses from Askot Klippe, Kumaun Lesser Himalaya. Acta Geol Sin 93(2):322–343

Dasgupta S (1995) Jaishidanda Formation. In: Bhargava ON (Ed) Bhutan Himalaya: a geological account. Geol Surv India Spec Publ 39:79–88

Dasgupta S, Ganguly J, Neogi S (2004) Inverted metamorphic sequence in the Sikkim Himalayas: crystallization history, P-T gradient and implications. J Metamor Geol 22:395–412

Davidson C, Grujic DE, Hollister LS, Schmid SM (1997) Metamorphic reactions related to decompression and synkinematic intrusion of leucogranite, High Himalayan crystallines, Bhutan. J Metamor Geol 15:593–612. https://doi.org/10.1111/j.1525-1314.1997.00044.x

de Sigoyer J, Chavagnac V, Blichert-Toft J, Villa IM, Luais B, Guillot S, Cosca M, Mascle G (2000) Dating the Indian continental subduction and collisional thickening in the northwest Himalaya: multichronology of the Tso Morari eclogites. Geology 28:487–490

DeCelles PG, Gehrels GE, Quade J, Lareau B, Spurlin M (2000) Tectonic implications of U-Pb zircon ages of the Himalayan orogenic belt in Nepal. Science 288:497–499

DeCelles PG, Robinson DM, Quade J, Ojha TP, Garzione CN, Copeland P, Upreti BN (2001) Stratigraphy, structure, and tectonic evolution of the Himalayan fold–thrust belt in western Nepal. Tectonics 20:487–509

Dewey JF, Bird JM (1970) Mountain belts and the new global tectonics. J Geophys Res 75:2625–2647

Dèzes PJ, Vannay JC, Steck A, Bussy F, Cosca M (1999) Synorogenic extension: quantitative constraints on the age and displacement of the Zanskar shear Zone. Geol Soc Amer Bull 111:364–374

Dhiman R, Singh S (2020) Neoproterozoic and Cambro-Ordovician magmatism: episodic growth and reworking of continental crust, Himachal Himalaya, India. Intern Geol Rev. https://doi.org/10.1080/00206814.2020.1716399

Dhital MR (2015) Geology of the Nepal Himalaya Regional Perspective of the Classic Collided Orogen. Springer, Berlin, pp 498

Draganits E, Grassemann B, Schmid HP (2003) Fluidization of pipes and spring pits in a Gondwana barrier–island environment: groundwater phenomenon, paleoseisimicity or a combination of both, In: Maltman AJ, Morley CK (eds) Subsurface mobilization, vol 216. Geol Soc of London, Spec Publ, pp 109–121

Epard JL, Steck A (2008) Structural development of the Tso Morari ultra-high pressure nappe of the Ladakh Himalaya. Tectonophysics 451:242–264

Etienne JL, Allen PA, le Guerroué E, Heaman L, Ghosh SK, Islam R (2011) The Blaini formation of the lesser Himalaya, NW India. In: Arnaud E, Halverson GP, Shields-Zhou G (eds) The geological record of neoproterozoic glaciations, vol 36. Mem Geol Soc London, pp 347–355. https://doi.org/10.1144/m36.31

Finch M, Hasalová P, Weinberg RF, Mark Fanning CM (2014) Switch from thrusting to normal shearing in the Zanskar shear zone, NW Himalaya: implications for channel flow. Geol Soc Am Bull 126(7–8):892–924. https://doi.org/10.1130/B30817.1
Frank W, Thoni M, Purtscheller F (1977) Geology and petrography of Kulu-South Lahul area. Ecologie geol l'Himalaya Paris 33:147–172
Fuchs G (1967) Zum Bau des Himalaya. Denksch Kaiserlichen Akad Wissensch/Math–Naturwissen Classe 113:1–211
Fuchs G (1982) The geology of the Pin valley in Spiti, Himachal Pradesh India. Neues Jahrb Geol Bundesanst 124:325–359
Gaetani M, Nicora A, Premoli SI, Fois E, Garzanti E, Tintori A (1983) Upper Cretaceous and Paleocene in Zanskar Range (NW Himalaya). Rivista Italiana Paleont Stratig 89:81–118
Gaetani M, Casnedi R, Fois E, Garzanti E, Jadoul F, Nicora A, Tintori A (1986) Stratigraphy of the Tethys Himalaya in Zanskar, Ladakh. Rivista Italiana Paleont Stratig 91:443–478
Ganai JA, Rashid SA (2019) Anoxia and fluctuating climate recorded from the Devonian-Carboniferous black shales, Tethys Himalaya, India: a multi-proxy approach. Int J Earth Sci 108:863–883
Ganai JA, Rashid SA, Masroor A (2016) Petrography and geochemical studies of Upper Paleozoic sandstones from the Tethys Himalaya, Spiti Valley, Himachal Pradesh. J Appl Geochem 18:3–14
Ganguly J, Dasgupta S, Cheng W, Neogi S (2000) Exhumation history of a section of the Sikkim Himalaya, India: records in the metamorphic mineral equilibria and compositional zoning of garnet. Earth Planet Sci Lett 183:471–486
Ganju JL, Khar BM (1984) Tectonics and Hydrocarbon prospect of Kashmir Valley-possible exploratory targets. Petrol Asian J 3:207–216
Gansser A (1964) Geology of the Himalayas. Lond Intersci Publ, pp 1–289
Gansser A (1983) Geology of the Bhutan Himalaya. Birkhäuser Verlag, Basel, pp 1–181
Garzanti E, Angiolini L, Brunton H, Sciunnach D, Balini M (1998) The bashkirian fenestella shales and the moscovian chaetetid shales of the Tethys Himalaya: south Tibet, Nepal and India. J Asian Earth Sci 16:119–141
Gavillot Y, Meigs AJ, Sousa FJ, Stockli D, Yule D, Malik M (2018) Late Cenozoic foreland-to-hinterland low-temperature exhumation history of the Kashmir Himalaya. Tectonics 37. https://doi.org/10.1029/2017tc004668
Gehrels GE, Kapp P, DeCelles P, Pullen A, Blakey R, Weislogel A, Ding L, Guynn J, Martin A, McQuarrie N, Yin A (2011) Detrital zircon geochronology of pre-Tertiary strata in the Tibetan-Himalayan orogen. Tectonics 30 TC5016. https://doi.org/10.1029/2011tc002868
Geological Survey of India Map (1999)
Ghosh SK (1991) Palaeoenvironmental analysis of late Proterozoic Nagthat Formation, NW Kumaun Lesser Himalaya, India. Sediment Geol 71:33–45
Ghosh NP, Basu AR, Bhargava ON, Shukla UK, Ghatak A, Garzione NC, Ahluwalia AD (2016) Catastrophic environmental transition at the Permian-Triassic Neo-Tethyan margin of Gondwanaland: Geochemical, isotopic and sedimentological evidence in the Spiti Valley, India. Gondwana Res 34:324–345
Godin D, Grujic D, Law RD, Searle MP (2006) Channel flow, ductile extrusion and exhumation in continental collision zones: an introduction. In: Law RD, Searle MP, and Godin L (eds) Channel flow, ductile extrusion and exhumation in continental collision zones. Geol Soc, London Spec Publ 268:1–23
Gokul AR (1983) Geological and Mineral Map of Bhutan: Scale 1:500,000, 1 sheet, Geol. Surv. India Map Printing Div Hyderabad
Goswami-Banerjee S, Bhowmik SK, Dasgupta S, Pant NC (2014) Burial of thermally perturbed Lesser Himalayan mid-crust: Evidence from petrochemistry and P-T estimation of the western Arunachal Himalaya, India. Lithos 208–209:298–311
Govin G, Najman Y, Dupont-Nivet G, Millar I, van Der Beek P, Huyghe P, Sullivan P, Mark C, Vögeli N (2018) The tectonics and paleo-drainage of the easternmost Himalaya (Arunachal Pradesh, India) recorded in the Siwalik rocks of the foreland basin. Am J Sci 318:764–798

Griesbach CL (1889) Geological notes—a sequence of formations in Spiti. Rec Geol Surv India 22:158–167

Griesbach CL (1891) Geology of the Central Himalayas. Mem Geol Surv India 23:1–232

Groppo C, Lombardo B, Rolfo F, Pertusati P (2007) Clockwise exhumation path of granulitized eclogites from the Ama Drime range (Eastern Himalaya). J Metamor Geol 25:51–75

Grujic D (2006) Channel flow and continental collision tectonics. In: Law RD, Searle MP, Godin L (eds) Channel flow, ductile extrusion and exhumation in continental collision zones, vol 268. Geol Soc Spec Publ, pp 25–37

Grujic D, Casey M, Davidson C, Hollister LS, Kündig R, Pavlis T, Schmid S (1996) Ductile extrusion of the Higher Himalayan Crystalline in Bhutan: evidence from quartz microfabrics. Tectonophysics 260:21–43

Guha Sarkar TK, Singh Gaur SK (1963) Geology of Hatisar–Kakulang river area, Tongsta, Bhutan. Geol Surv India (Unpubl Rept FS 1962–63)

Guillot S, de Sigoyer J, Lardeaux JM, Mascle G, Colchen M (1997) Eclogitic metasediments from the Tso Morari area (Ladakh Himalaya): evidence for continental subduction during India-Asia convergence. Contrib Mineral Petrol 128:197–212

Guillot S, Mahéo G, de Sigoyer J, Hattori KH, Pêcher A (2008) Tethyan and Indian subduction viewed from the Himalayan high-to ultrahigh-pressure metamorphic rocks. Tectonophysics 451:225–241

Guo X, Li W, Gao R, Xu X, Li H, Huang X, Ye Zh, Lu Zh, Klemper SL (2017) Nonuniform subduction of the Indian crust beneath the Himalayas. Sci Reports 7:12497. https://doi.org/10.1038/s41598-017-12908-0

Hagen T (1968) Report on the Geological survey of Nepal, 2. Geology of Thakkola including adjacent areas. Denkschr Schweiz naturfor Gesellsch 86:1–160

Harris NBW, Caddick M, Kosler J, Goswami S, Vance D, Tindle AG (2004) The pressure–temperature–time path of migmatites from the Sikkim Himalaya. J Metamor Geol 22:249–264

Harrison TM, Grove M, Lovera M, Catlos EJ (1998) A model for the origin of Himalayan anatexis and inverted metamorphism. J Geophys Res 103:27017–27032

Hashimoto H, Ohta Y and Akiba C (eds) (1973) Geology of Nepal Himalaya. Hokkaido Univ Sapporo Japan, pp 1–281

Hayden HH (1904) The geology of Spiti with parts of Bashahr and Rupshu. Mem Geol Surv India 36:1–121

Heim A, Gansser A (1939) Central Himalaya: geological observations of the Swiss expedition 1936. Mem Soc Helv Sci Nat 73:1–245

Herman F, Copeland P, Avouac JP, Bollinger L, Maheo G, Le Fort P (2010) Exhumation, crustal deformation, and thermal structure of the Nepal Himalaya derived from the inversion of thermochronological and thermobarometric data and modeling of the topography. J Geophys Res 115:B06407. https://doi.org/10.1029/2008JB006126

Herren E (1987) Zanskar Shear Zone: northeast-southwest extension within the Higher Himalaya (Ladakh, India). Geology 15:409–413

Hodges KV, Parrish RR, Searle MP (1996) Tectonics evolution of the central Annapurna Range, Nepalese Himalayas. Tectonics 15(6):1264–1291. https://doi.org/10.1029/96TC01791

Hodges KV (2000) Tectonics of the Himalaya and southern Tibet from two perspectives. Geol Soc Am Bull 112:324–350

Hofmann M, Linnemann U, Rai V, Becker S, Gärtner A, Sagawe A (2011) The India and South China cratons at the margin of Rodinia—Synchronous Neoproterozoic magmatism revealed by LA-ICP-MS zircon analyses. Lithos 123:176–187

Honegger K, Dietrich J, Frank W, Gansser A, Thoni M, Trommsdorff W (1982) Magmatism and metamorphism: the Indus Tsangpo Suture Zone. Earth Planet Sci Lett 60:253–292. https://doi.org/10.1007/s00410-013-0957-3

Hou G, Santosh M, Qian X, Lister GS, Li J (2008) Configuration of the Late Paleoproterozoic supercontinent Columbia: Insights from radiating mafic dyke swarms. Gondwana Res 14:395–409

Huang Y, Chena Z-Q, Algeo TJ, Zhao L, Baud A, Bhat GM, Zhang L, Guo Z (2019) Two-stage marine anoxia and biotic response during the Permian-Triassic transition in Kashmir, northern India: pyrite framboid evidence. Global Planet Change 172:124–139. https://doi.org/10.1016/j.gloplacha.2018.10.002
Hughes NC (2016) The Cambrian palaeontological record of the Indian subcontinent. Earth Sci Rev 159:428–461
Hughes NC, Peng S, Bhargava ON, Ahulwalia AD, Walia S, Myrow PM, Parcha SK (2005) The Cambrian biostratigraphy of the Tal Group, Lesser Himalaya, India, and early Tsanglangpuan (late early Cambrian) trilobites from the Nigali Dhar Syncline. Geol Mag 142:57–80
Hughes NC, Myrow PM, Peng S, Banerjee DM (2018) The Parahio Formation of the Tethyan Himalaya: The type section, thickness, lithostratigraphy and biostratigraphy of the best characterized Cambrian section in the Himalaya. J Palaeontol Soc India 63(1):1–18
Hughes NC, Myrow PM, Ghazi S, McKenzie NR, Stockli DF, DiPietro JA (2019) Cambrian geology of the Salt Range of Pakistan: linking the Himalayan margin to the Indian craton. Bull Geol Soc Am 131(7/8):1095–1114. https://doi.org/10.1130/B35092.1
Hussain A, Yeats RS, Lisa M (2009) Geological setting of the 8th October, 2005 Kashmir earthquake. J Seismol 13:315–325. https://doi.org/10.1007/s/10950-800-9101-7
Iaccarino S, Montomoli C, Carosi R, Montemagni C, Massonne H-J, Langone A, Jain AK, Visonà D (2017) Pressure-temperature-deformation-time constraints on the South Tibetan Detachment System in the Garhwal Himalaya (NW India). Tectonics 36. https://doi.org/10.1002/2017TC004566
Jade S, Shrungeshwara TS, Kumar K, Choudhury P, Dumka RK, Bhu H (2017) India plate angular velocity and contemporary deformation rates from continuous GPS measurements from 1996 to 2015. Scientific Rep 7:11439. https://doi.org/10.1038/s41598-017-11697-w
Jain AK, Goel RK, Nair NGK (1980) Implications of Pre-Mesozoic geological evolution of the Himalaya and Indo-Gangetic Plains. Tectonophysics 62:76–86
Jain AK (1981) Stratigraphy, petrography and paleogeography of the Late Paleozoic diamictites of the Lesser Himalaya. Sed Geol 30:43–78
Jain AK (2014) When did India-Asia collide and make the Himalaya? Curr Sci 106(2):254–266
Jain AK (2017) Continental subduction in the NW-Himalaya and Trans-Himalaya. Ital J Geosci 136(1):89–102. https://doi.org/10.3301/IJG.2015.43
Jain AK (2020) Geological evolution of the Himalayan Mountains. In: Gupta N, Tandon SK (eds) Geodynamics of the Indian Plate: evolutionary perspectives. Springer Geology pp 363–393. https://doi.org/10.1007/978-3-030-15989-4_10
Jain AK, Balasubramanian E (1981) Late Paleozoic diamictites of the eastern himalayan Gondwana belt, India. In: Hambrey MJ, Harland WB (eds) Earth's pre-pleistocene glacial record. Cambridge Univ Press pp 308–325
Jain AK, Manickavasagam RM (1993) Inverted metamorphism in the intracontinental ductile shear zone during Himalayan collision tectonics. Geology 21:407–410
Jain AK, Manickavasagam RM (eds) (1999) Geodynamics of the NW Himalaya. Gondwana Res Mem 6
Jain AK, Patel RC (1999) Structure of the higher himalayan crystallines along the Suru-Doda valleys (Zanskar), NW Himalaya. In: Jain AK, Manickavasagam RM (eds) Geodynamics of the NW Himalaya. Gondwana Res Group Mem 6:91–110
Jain AK, Singh S (2009) Geology and Tectonics of the Southeastern Ladakh and Karakoram. Geol Soc India Bangalore India 1–179
Jain AK, Varadaraj N (1978) Stratigraphy and provenance of Late Palaeozoic diamictites in parts of Garhwal Lesser Himalaya, India. Geol Rund 67:49–72
Jain AK, Thakur VC, Tandon SK (1974) Stratigraphy and structure of the Siang District of Arunachal Pradesh. Himalayan Geol 4:28–60
Jain AK, Singh S, Manickavasagam RM (2002) Himalayan Collision Tectonics. Gondwana Res Group Mem 7:114

Jain AK, Manickavasagam RM, Singh S, Mukherjee S (2005) Himalayan collision zone: new perspectives—its tectonic evolution in a combined ductile shear zone and channel flow model. Himalayan Geol 26(1):1–18

Jain AK, Kumar D, Singh S, Kumar A, Lal N (2000) Timing, quantification and tectonic modelling of Pliocene-Quaternary movements in the NW Himalaya: Evidence from fission track dating. Earth Planet Sci Lett 179:437–451

Jain AK, Singh S, Manickavasagam RM, Joshi M, Verma PK (2003) HIMPROBE Programme: integrated Studies on Geology, Petrology, Geochronology and Geophysics of the Trans-Himalaya and Karakoram. Mem Geol Soc India 53:1–56

Jain AK, Lal N, Sulemani B, Awasthi AK, Singh S, Kumar R, Kumar D (2009) Detrital-zircon fission-track ages from the Lower Cenozoic sediments, NW Himalayan foreland basin: clues for exhumation and denudation of the Himalaya during the India-Asia collision. Geol Soc Am Bull 121:519–535

Jain AK, Seth P, Shreshtha M, Mukherjee PK, Singh K (2013) Structurally-controlled melt accumulation: Himalayan migmatites and related deformation, Dhauli Ganga Valley, Garhwal Himalaya. J Geol Soc India 82:313–318

Jain AK, Shrestha M, Seth P, Kanyal L, Carosi R, Montomoli C, Iaccarino S, Mukherjee PK (2014) The Higher Himalayan Crystallines, Alaknanda–Dhauli Ganga Valleys, Garhwal Himalaya, India. In: Montomoli C, Carosi R, Law RD, Singh S, Rai SM (eds) Geological field trips in the Himalaya, Karakoram and Tibet. J Virtual Expl Electr Edit 47 paper 8, ISSN 1441–8142

Jain AK, Mukherjee PK, Singhal S (2020) Terrane characterization in the Himalaya since Paleoproterozoic. Episodes 43(1):346–357. https://doi.org/10.18814/epiiugs/2020/020021

Jangpangi BS (1974) Stratigraphy and tectonics of parts of eastern Bhutan. Himalayan Geol 4:117–136

Jasper A, Guerra SM, Abdalla MB, Hamad A, Bamford M, Bernardes-de-Oliveira MC, Tewari R, Uhl D (2016) The burning of Gondwana: Permian fires on the southern continent—a palaeobotanical approach. Gondwana Res 24:148–160

Jell PA, Hughes NC (1997) Himalayan Cambrian trilobites. Spec Pap Palaeont 58:1–113

Jiang G, Christie BN, Kaufman AJ, Banerjee DM, Rai V (2002) Sequence stratigraphy of the Neoproterozoic Infra Krol formation and Krol Group, Lesser Himalaya, India. J Sedim Res 72(4):524–542

Jiang G, Christie BN, Kaufman AJ, Banerjee DM, Rai V (2003) Carbonate platform growth and cyclicity at a terminal Proterozoic passive margin, Infra Krol Formation and Krol Group, Lesser Himalaya, India. Sediment 50:921–952

Joshi A (1995) Setikhola formation. In: Bhargava ON (ed) Bhutan Himalaya: a geological account, vol 39. Geol Surv India Spec Publ, pp 34–37

Kaneda H, Nakata T, Tsutsumi H, Kondos SN, Awata Y, Akhtar S, Majid A, Khatak W, Awan A, Yeats RS, Hussain A, Ashra M, Wesnousky SG, Kausar B (2008) Surface rupture of the 2005 Kashmir, Pakistan earthquake and its active tectonic implications. Bull Seism Soc Am 98:512–557

Kaneko Y, Katayama I, Yamamoto H, Misawa K, Ishikawa M, Rehman HU, Kausar AB, Shiraishi K (2003) Timing of Himalayan ultrahigh-pressure metamorphism: sinking rate and subduction angle of the Indian continental crust beneath Asia. J Metamor Geol 21(6):589–599

Kapoor HM, Maheshwari HK (1991) Early Permian paleogeography of the Peri-Gondwana in the Indian Segment. Curr Sci 61:648–653

Kapoor HM, Bajpai U, Maheshwari HK (1993) On a fossil cockroach from the Mamal Formation, Kashmir Himalaya. J Paleont Soc India 38:31–36

Kapoor HM, Maheshwari HK, Bajpai U (2004) Evolutionary history of the northern margin of Gondwana Supercontinent during the Late Palaeozoic. In: Srivastava PC (ed) Vistas in Palaeobotany and plant morphology: evolutionary and environmental perspective. Lucknow, pp 119–131

Karunakaran C, Rao R (1979) Status of hydrocarbon in the Himalayan region: contributions to stratigraphy and structure. Geol Sur India Misc Publ 41:1–67

Kaufman AJ, Jiang G, Christie BN, Banerjee DM, Rai V (2006) Stable isotope record of the terminal Neoproterozoic Krol platform in the Lesser Himalayas of northern India. Precamb Res 147:156–185. https://doi.org/10.1016/j.precamres.2006.02.007

Kaur P, Zeh A, Chaudhri N, Gerdes A, Okrusch M (2013) Nature of magmatism and sedimentation at a Columbia active margin: insights from combined U-Pb and Lu–Hf isotope data of detrital zircons from NW India. Gondwana Res 23(3):1040–1052. https://doi.org/10.1016/j.gr.2012.07.008

Kellett DA, Grujic D (2012) New insight into the south tibetan detachment system: not a single progressive deformation. Tectonics 31:TC2007. https://doi.org/10.1029/2011TC002957

Kellett DA, Grujic D, Erdmann S (2009) Miocene structural reorganization of the South Tibetan detachment, eastern Himalaya: implications for continental collision. Lithosphere 1:259–281

Kellett DA, Grujic D, Warren C, Cottle J, Jamieson R, Tenzin T (2010) Metamorphic history of a syn-convergent orogen-parallel detachment: the south tibetan detachment system, Bhutan Himalaya. J Metamor Geol 28:785–808. https://doi.org/10.1111/j.1525-1314.2010.00893.x

Kellett DA, Grujic D, Coutand I, Cottle J, Mukul M (2013) The South Tibetan detachment system facilitates ultra rapid cooling of granulite–facies rocks in Sikkim Himalaya. Tectonics 32:252–270. https://doi.org/10.1002/tect.(2014)

Khan MA, Treloar PJ, Searle MP, Jan MQ (eds) (2000) Tectonics of the Nanga Parbat Syntaxis and the Western Himalayas. Geol Soc London Spec Publ 170

Khanal S, Robinson DM, Mandal S, Simkhada P (2015) Structural, geochronological and geochemical evidence for two distinct thrust sheets in the "Main Central thrust zone", the Main Central thrust and Ramgarh–Munsiari thrust: implications for upper crustal shortening in the central Nepal. In: Mukherjee S, Carosi R, van der Beek PA, Mukherjee BK, Robinson DM (eds) Tectonics of the Himalaya, vol 412. Geol Soc London Spec Publ, pp 221–245. https://doi.org/10.1144/sp412.2

King GE, Herman F, Guralnik B (2016) Northward migration of the eastern Himalayan syntaxis revealed by OSL thermochronometry. Science 353(6301):800–804. https://doi.org/10.1126/science.aaf2637

Klootwijk CT, Gee JS, Peirce JW, Smith GM (1992) Neogene evolution of the Himalayan-Tibetan region: constraints from ODP site 758, northern Ninety East Ridge: bearing on climatic change. Palaeogeog Palaeoclim Palaeoeco 95:95–110

Kohn M (2014) Himalayan metamorphism and its tectonic implications. Ann Rev Earth Planet Sci 42:381–419

Kohn MJ, Paul SK, Corrie SL (2010) The lower Lesser Himalayan sequence: a Paleoproterozoic arc on the northern margin of the Indian plate. Geol Soc Am Bull 122:323–335

Kumar S (1980) Stromatolites and Indian biostratigraphy. J Paleont Soc India 23–24:166–183

Kumar G (1997) Geology of Arunachal Pradesh. Geol Soc India Bangalore pp 1−217

Kumar R (2020) Late Cenozoic Himalayan foreland basin: sedimentologic attributes. Episodes 43(1):417–428. https://doi.org/10.18814/epiiugs/2020/020026

Kumar S, Mahajan AK (2001) Seismotectonics of the Kangra region, Northwest Himalaya. Tectonophysics 331:359–371

Kumar G, Mehdi SH, Prakash G (1972) A review of stratigraphy of parts of Uttar Pradesh Tethys Himalaya. J Palaeont Soc India 15:86–89

Kumar S, Singh IB, Singh SK (1977) Lithostratigraphy, structure, depositional environment, palaeocurrent and trace fossils of the Tethyan sediments of Malla Johar area, Pithoragarh–Chamoli districts, Uttar Pradesh, India. J Palaeont Soc India 20:396–435

Kumar G, Raina BK, Bhargava ON, Maithy PK, Babu R (1984) The Precambrian-Cambrian boundary problem and its prospects, northwest Himalaya, India. Geol Mag 121:211–219

Kumar A, Lal N, Jain AK, Sorkhabi RB (1995) Late Cenozoic-Quaternary thermotectonic history of Higher Himalayan Crystallines (HHC) in Kishtwar-Padar-Zanskar region, NW Himalaya: Evidence from fission track ages. J Geol Soc India 45:375–391

Kumar R, Ghosh SK, Sangode SJ (1999) Role of thrusting in the evolution fluvial system, Himalayan Foreland Basin, India. In: Macfarlane A, Sorkhabi RB, Quade J (eds) Himalayan and Tibet: Mountain Roots to Mountain Tops, vol 328. Geol Soc Am Spec Publ, pp 239–256

Kumar R, Ghosh SK, Mazari RK, Sangode SJ (2003) Tectonic impact on fluvial deposits of Plio-Pleistocene Himalayan foreland basin, India. Sedim Geol 158:209–234

Kumar R, Ghosh SK, Sangode SJ (2004) Depositional environment of Mio-Pleistocene coarse clastic facies in the Himalayan foreland basin, India. Himalayan Geol 25:101–120

Kumar R, Ghosh SK, Sangode SJ (2011) Sedimentary architecture of late Cenozoic Himalayan foreland basin fill: an overview. Mem Geol Soc India 78:245–280

Kumar R, Jain AK, Nand Lal, Sandeep Singh (2018) Early-Middle Eocene exhumation of the Trans-Himalayan Ladakh Batholith, and the India-Asia convergence. Curr Sci 113:1090–1098

Kumar S, Wesnousky SG, Rockwell TK, Briggs RW, Thakur VC, Jayangondaperumal R (2006) Paleoseismic evidence of great surface rupture earthquakes along the Indian Himalaya. J Geophy Res 111:B03304. https://doi.org/10.1029/2004JB003309

Kundig R (1989) Domal structures and highgrade metamorphism in the Higher Himalayan Crystalline, Zanskar region, northwest Himalaya, India. J Metamor Geol 7:43–55

Lahiri A (1941) Geology of Buxa Duars. Quart J Geol Min Met Soc India 13(1):1–62

Lakshami KJP, Sudheer Kumar M, Bhalla MS, Rao GVSP (2000) Magnetostratigraphy of Himalayan sediments from Himachal Pradesh. J Indian Geophy Union 4(2):147–154

Lakshminarayan G, Singh B (1995) Siwalik Group. In: Bhargava ON (ed) Bhutan Himalaya: a geological account, vol 39. Geol Surv India Spec Publ, pp 23–28

Landry KR, Coutand I, Whipp DM Jr, Grujic D, Hourigan JK (2016) Late Neogene tectonically driven crustal exhumation of the Sikkim Himalaya: insights from inversion of multithermochronologic data. Tectonics 35. https://doi.org/10.1102/2015tc004102

Lang KA, Huntington KW, Burmester R, Housen B (2016) Rapid exhumation of the eastern Himalayan syntaxis since the late Miocene. Geol Soc Am Bull 128(9–10):1403–1422

Larson K, Piercey S, Cottle J (2017) Preservation of a Paleoproterozoic rifted margin in the Himalaya: Insight from the Ulleri-Phaplu-Melung orthogneiss. Geosci Front 1–13. https://doi.org/10.1016/j.gsf.2017.05.010

Lavé J, Avouac JP (2000) Active folding of fluvial terraces across the Siwalik Hills, the Himalayas of central Nepal. J Geophy Res 105(B3):5735–5770

Law RD, Searle MP, Simpson RL (2004) Strain, deformation temperatures and vorticity of flow at the top of the Greater Himalayan slab, Everest Massif, Tibet. J Geol Soc 161:305–320. https://doi.org/10.1144/0016-764903-047

Leech ML, Singh S, Jain AK, Klemperer SL, Manickavasagam RM (2005) The onset of India-Asia continental collision: early, steep subduction required by the timing of UHP metamorphism in W Himalaya. Earth Planet Sci Lett 234:83–97

Leech ML, Singh S, Jain AK (2007) Continuous metamorphic zircon growth and interpretation of U-Pb SHRIMP dating: an example from the Western Himalaya. Intl Geol Rev 49:313–328

Leloup PH, Mahéo G, Arnaud N, Kali E, Boutonnet E, Liu D, Xiaohan L' Haibing L (2010) The south tibet detachment shear zone in the Dinggye area. Time constraints on extrusion models of the Himalayas. Earth Planet Sci Lett 292:1–16. https://doi.org/10.1016/j.epsl.2009.12.035

Li D, Liao Q, Yuan Y, Wan Y, Liu D, Zhang X, Yi S, Cao S, Xie D (2003) SHRIMP U-Pb zircon geochronology of granulites at Rimana (southern Tibet) in the central segment of the Himalayan orogen. Chinese Sci Bull 48:2647–2650

Liou JG, Tsujimori T, Zhang RY, Katayama I, Maruyama S (2004) Global UHP magmatism and metamorphism: the Indus Tsangpo Suture Zone. Earth Planet Sci Lett 60:253–292

Long S, McQuarri N, Tobgay T, Rose C, Gehrels G, Grujic D (2011) Tectonostratigraphy of the Lesser Himalaya of Bhutan: implications for the along-strike stratigraphic continuity of the northern Indian margin. Geol Soc Am Bull 123(7/8):1406–1426. https://doi.org/10.1130/B30202.1

Luirei K, Bhakuni SS, Suresh N, Kothyari GC, Pant PD (2014) Tectonic geomorphology and morphometry of the frontal part of Kumaun Sub-Himalaya: appraisal of tectonic activity. Zeitsch Geomorph 58(4):435–458

Lyon-Caen H, Molnar P (1985) Gravity anomalies, flexure of the Indian plate and the structure, support, and evolution of the Himalaya and Ganga basin. Tectonics 4:513–538

Mallet FR (1875) On the geology and mineral resources of Darjeeling district and western Duars. Mem Geol Surv India 11:1–50
Mandal S, Robinson DM, Khanal S, Das O (2015) Redefining the tectonostratigraphic and structural architecture of the Almora klippe and the Ramgarh–Munsiari Thrust sheet in NW India. In: Mukherjee S, Carosi R, van der Beek PA, Mukherjee BK, Robinson DM (eds) Tectonics of the Himalaya, vol 412. Geol Soc London Spec Publ, pp 247–269
Mandal S, Robinson DM, Kohn MJ, Khanal S, Das O, Bose S (2016) Zircon U–Pb ages and Hf isotopes of the Askot klippe, Kumaun, northwest India: implications for Paleoproterozoic tectonics, basin evolution and associated metallogeny of the northern Indian cratonic margin. Tectonics 35:965–982. https://doi.org/10.1002/2015TC004064
Mandal SK, Scherler D, Romer RL, Burg JP, Guillong M, Schleicher AM (2019) Multiproxy isotopic and geochemical analysis of the Siwalik sediments in NW India: implication for the Late Cenozoic tectonic evolution of the Himalaya. Tectonics 38:120–143. https://doi.org/10.1029/2018TC005200
Manglik A, Adilakshmi L, Suresh M, Thiagarajan S (2015) Thick sedimentary sequence around Bahraich in the northern part of the central Ganga foreland basin. Tectonophy 653:33–40. http://doi.org/10.1016/j.tecto.2015.03.024
Manickavasagam RM, Jain AK, Singh S, Asokan A (1999) Metamorphic evolution of the NW–Himalaya, India: pressure-temperature data, inverted metamorphism, and exhumation in the Kashmir, Himachal, and Garhwal Himalaya. In: Macfarlane A, Sorkhabi RB, Quade J (eds) Himalaya and Tibet: Mountain Roots to Mountain Tops, vol 328. Geol Soc Am Spec Paper, pp 179–198
Martin AJ (2017) A review of Himalayan stratigraphy, magmatism, and structure. Gondwana Res 49:42–80. https://doi.org/10.1016/j.gr.2017.04.031
Martin AJ, DeCelles PG, Gehrels GE, Patchett PJ, Isachsen C (2005) Isotopic and structural constraints on the location of the Main Central thrust in the Annapurna Range, central Nepal Himalaya. Geol Soc Am Bull 117:926–944
Martin AJ, Burgy KD, Kaufman AJ, Gehrels GE (2011) Stratigraphic and tectonic implications of field and isotopic constraints on depositional ages of Proterozoic Lesser Himalayan rocks in central Nepal. Precamb Res 185:1–17
Mathur NS (1978) Biostratigraphical aspects of the Subathu Formation, Kumaun Himalaya. Recent Res Geol 5:96–112
Mazumdar A, Banerjee DM (1998) Siliceous sponge in the Early Cambrian chert-phosphorite member of the Lower Tal Formation, Krol Belt, Lesser Himalaya. Geology 26:899–902
Mazumdar A, Banerjee DM (2001) Regional variations in the carbon isotopic composition of phosphorite from the Early Cambrian Lower Tal Formation Mussoorie Hill, India. Chem Geol 175(1/2):5–16
Mazumdar A, Banerjee DM, Schidlowski M, Balaram V (1999) Rare-earth elements and stable isotope geochemistry of early Cambrian chert-phosphorite assemblages from the Lower Tal formation of the Krol Belt (Lesser Himalaya, India). Chem Geol 156:275–297
McKenzie NR, Hughes NC, Myrow PM, Xiao S, Sharma M (2011) Correlation of Precambrian-Cambrian sedimentary successions across northern India and the utility of isotopic signatures of Himalayan lithotectonic zones. Earth Planet Sci Lett 312(3):471–483. https://doi.org/10.1016/j.epsl.2011.10.027
McQuarrie N, Robinson D, Long S, Tobgay T, Grujic D, Gehrels G, Ducea M (2008) Preliminary stratigraphic and structural architecture of Bhutan: implications for the along strike architecture of the Himalayan system. Earth Planet Sci Lett 272(1–2):105–117
McQuarrie N, Long S, Tobgay T, Nesbit JN, Gehrels G, Ducea MN (2013) Documenting basin scale, geometry and provenance through detrital geochemical data: Lessons from the Neoproterozoic to Ordovician Lesser, Greater, and Tethyan Himalayan strata of Bhutan. Gondwana Res 23:1491–1510
McQuarrie N, Tobgay T, Long SP, Reiners PW, Cosca MA (2014) Variable exhumation rates and variable displacement rates: documenting a recent slowing of Himalayan shortening in western Bhutan. Earth Planet Sci Lett 286:161–174. https://doi.org/10.1016/j.epsl.2013.10.045

Medlicott HB (1864) On the geological structure and relations of the southern portion of the Himalayan ranges between rivers Ganges and the Ravi. Mem Geol Surv India 3(2):122

Meigs AJ, Burbank DW, Beck RA (1995) Middle-late Miocene (>10 Ma) formation of the Main Boundary thrust in the western Himalaya. Geology 23(5):423–426. https://doi.org/10.1130/0091-7613(1995)023%3c0423:mlmmfo%3e2.3.co;2

Metcalfe RP (1993) Pressure, temperature and time constraints on metamorphism across the Main Central thrust zone and High Himalayan slab in the Garhwal Himalaya. In: Treloar PJ, Searle MP (eds) Himalayan Tectonics, vol 74. Spec Publ Geol Soc London, pp 485–509

Middlemiss CS (1909) Gondwana and related marine sedimentary systems of Kashmir. Rec Geol Surv India 37:286–327

Middlemiss CS (1910) A revision of the Silurian-Triassic sequence in Kashmir. Rec Geol Surv India 40:206–260

Middlemiss CS (1911) Sections in Pir-Panjal Range and Sind Valley, Kashmir. Rec Geol Surv India 41:85–137

Miller C, Klotzli U, Frank W, Thoni M, Grasemann B (2000) Proterozoic crustal evolution in the NW Himalaya (India) as recorded by circa 1.80 Ga mafic and 1.84 Ga granitic magmatism. Precamb Res 103:191–206

Misra RC, Banerjee DM (1968) Stratigraphy, correlation and tectonics of Sarju-Pungar valley areas, district Almora-Pithoragarh. Pub Centre Adv Studies Geol Chandigarh 4:101–113

Molnar P, Lyon-Caen H (1988) Some simple physical aspects of the support, structure, and evolution of mountain belts. Processes in continental lithospheric deformation. Geol Soc Am Spec Pap 218:179–207

Montemagni C, Montomoli C, Iaccarino S, Carosi R, Jain AK, Massonne H-J, Villa IM (2018) Dating protracted fault activities: microstructures, microchemistry and geochronology of the Vaikrita Thrust, Main Central Thrust zone, Garhwal Himalaya, NW India. Geol Soc, London, Spec Publ SP481.3. https://doi.org/10.1144/sp481.3

Montomoli C, Carosi R, Rubatto D, Visonà D, Iaccarino S (2017) Tectonic activity along the inner margin of the South Tibetan Detachment constrained by syntectonic leucogranite emplacement in Western Bhutan. Ital J Geosc, 136:5–14. https://doi.org/10.3301/IJG.2015.26

Mottram CM, Argles TW, Harris NBW, Parrish RR, Horstwood MSA, Warren CJ, Gupta S (2014a) Tectonic interleaving along the Main Central Thrust, Sikkim Himalaya. J Geol Soc London 171:255–268. https://doi.org/10.1144/jgs2013-064

Mottram CM, Warren CJ, Regis D, Roberts NM, Harris NB, Argles TW, Parrish RR (2014b) Developing an inverted Barrovian sequence: insights from monazite petrochronology. Earth Planet Sci Lett 403:418–431

Mugnier JL, Huyghe P, Leturmy P, Jouanne F (2004) Episodicity and rates of thrust-sheet motion in the Himalayas (western Nepal). In: McClay KR (ed) Thrust tectonics and hydrocarbon systems, vol 82. Mem Am Assoc Petrol Geol, pp 91–114

Mukherjee S, Carosi R, van der Beek PA, Mukherjee BK, Robinson DM (eds) (2015) Tectonics of the Himalaya. Geol Soc Lond Spec Publ 412

Mukherjee PK, Jain AK, Singhal S, Singha NB, Singh S, Kumud K, Seth P, Patel RC (2019) U–Pb zircon ages and Sm–Nd isotopic characteristics of the Lesser and Great Himalayan sequences, Uttarakhand Himalaya, and their regional tectonic implications. Gondwana Res 75:282–297. https://doi.org/10.1016/j.gr.2019.06.001

Mukhopadhyay DK, Chakraborty S, Trepmann C, Rubatto D, Anczkiewicz R, Gaidies F, Dasgupta S, Chowdhury P (2017) The nature and evolution of the Main Central Thrust: Structural and geochronological constraints from the Sikkim Himalaya, NE India. Lithos 282:447–463

Myrow PM, Hughes NC, Paulsen TS, Williams IS, Parcha SK, Thompson KR, Bowring SA, Peng S, Ahluwalia AD (2003) Integrated tectonostratigraphic reconstruction of the Himalaya and implications for its tectonic reconstruction. Earth Planet Sci Lett 212:433–441

Myrow PM, Thompson KR, Hughes NC, Paulsen TS, Sell BK, Parcha SK (2006) Cambrian stratigraphy and depositional history of the northern Indian Himalayas, Spiti Valley, north-central India. Geol Soc Am Bull 118:491–510

Myrow PM, Hughes NC, Searle MP, Fanning CM, Peng S, Parcha SK (2009) Stratigraphic correlation of Cambrian-Ordovician deposits along the Himalaya: implications for the age and nature of rocks in the Mt. Everest region. Bull Geol Soc Am 120:323–332
Myrow PM, Hughes NC, Goodge JW, Fanning CM, Williams IS, Peng S, Bhargava ON, Parcha SK, Pogue KR (2010) Extraordinary transport and mixing of sediment across Himalayan central Ordovician-Gondwana during the Cambrian. Geol Soc Am Bull 122:1660–1670
Myrow PM, Hughes NC, Derry LA, McKenzie NR, Jiang G, Webb AAG, Banerjee DM, Paulsen TS, Singh BP (2015) Neogene marine isotopic evolution and the erosion of Lesser Himalayan strata: implications for Cenozoic tectonic history. Earth Planet Sci Lett 417:142–150
Myrow PM, Fike AD, Malmskog B, Leslie S, Zhang T, Singh BP, Chaubey RS, Prasad SK, (2018) Ordovician–Silurian boundary strata of the Indian Himalaya: record of the latest Ordovician Boda Event. Geol Soc Am Bull. https://doi.org/10.1130/b31860.1
Nábělek J, Hetenyi G, Vergne J, Sapkota S, Kafle B, Jiang M, Su H, Chen J, Huang B-S and the HI-CLIMB Team (2009) Underplating in the Himalaya-Tibet collision zone revealed by the Hi-CLIMB experiment. Science 325:1371–1374
Najman Y (2006) The sediment record of orogenesis: a review of approaches and techniques used in the Himalaya. Earth-Sci Rev 74:1–72
Najman Y, Garzanti E (2000) Reconstructing early Himalayan tectonic evolution and paleogeography from Tertiary foreland basin sedimentary rocks, northern India. Geol Soc Am Bull 112:435–449
Najman Y, Clift P, Johnson MRW, Robertson AHF (1993) Early stages of foreland basin evolution in the Lesser Himalaya. In: Treloar PJ, Searle MP (eds) Himalayan tectonics, vol 74. Geol Soc London Spec Publ, pp 525–540
Najman YMR, Enkin RJ, Johnson MRW, Robertson AHF, Baker J (1994) Paleomagnetic dating of the earliest continental Himalayan foredeep sediments: implications for Himalayan evolution. Earth Planet Sci Lett 128:713–718
Najman YMR, Pringle MS, Johnson MRW, Robertson AHF, Wijbrans JR (1997) Laser $^{40}Ar/^{39}Ar$ dating of single detrital muscovite grains from early foreland basin sediments in India: implications for early Himalayan evolution. Geology 25:535–538
Najman Y, Bickle M, Chapman H (2000) Early Himalayan exhumation; isotopic constraints from the Indian foreland basin. Terra Nova 12:28–34
Najman Y, Johnson K, White NM, Olivers G (2004) Evolution of Himalayan foreland basin, NW India. Basin Res 16:1–24
Nakata T (1989) Active faults of Himalaya, India and Nepal. Geol Soc Am Spec pap 232:243–264
Nakazawa K, Kapoor HM, Ishi K, Bando Y, Okimura Y, Tokuoka T (1975) The Upper Permian and the Lower Triassic in Kashmir, India. Mem Fac Sci Kyoto Univ Series Geol Mineral 42(1):1–106
Nautiyal SP, Jangpangi BS, Singh P, Guha Sarkar TK, Bhate VD, Raghavan MR, Sahai TN (1964) A preliminary note on the geology of Bhutan Himalaya. Rep 22nd Int Geol Cong New Delhi 11:1–14
Nelson KD, Zhao W, Brown LD, Indepth Team (1996) Partially molten middle crust beneath southern Tibet: synthesis of Project INDEPTH results. Science 274:1684–1688
Neogi S, Dasgupta S, Fukuoka M (1998) High P-T polymetamorphism, dehydration melting and generation of migmatites and granites in the Higher Himalayan Crystalline complex, Sikkim, India. J Petrol 39:61–99
Ni JF, Barazangi M (1984) Seismotectonics of the Himalayan collision zone: geometry of the underthrusting Indian plate beneath the Himalaya. J Geophys Res 89:947–963
Ojha PS (2012) Precambrian sedimentary basins of India: an appraisal of their petroleum potential. In: Bhat GM, Craig J, Thurow JW, Thusu B, Cozzi A (eds) Geology and Hydrocarbon Potential of Neoproterozoic–Cambrian Basins in Asia, vol 366. Geol Soc London Spec Publ, pp 19–58. https://doi.org/10.1144/SP366.11
Pandey S, Parcha SK (2018) Calcareous algae from the Ordovician succession (Thango Formation) of the Spiti Basin, Tethys Himalaya, India. Acta Palaeobotanica. https://doi.org/10.2478/acpa-2018-0009

Pandey B, Pathak DB (2016) The possibility of the Oceanic Anoxic Events (OAEs) study in the Indian marine Jurassic-Cretaceous outcrops. J Geol Soc India 87(3):261–267

Pandey B, Pathak DB (2017) Biostratigraphic implication of Olcostephanus Neumayr, (1875) (Ammonoidea) from the Lower Cretaceous Giumal Formation, Spiti Valley, Tethys Himalaya, India. Cret Res 70:244–251

Pandey B, Pathak DB, Jaitly AK (2018a) A new ammonite gen. nov. from the Lower Valanginian (Lower Geticeras Cretaceous) of the Spiti Valley, Tethys Himalaya, India. Himalayan Geol 39(1):115–120

Pandey B, Pathak DB, Mathur N, Jaitly AK, Singh AK, Singh PK (2018b) A Preliminary Evaluation on the Prospects of Hydrocarbon Potential in the Carbonaceous Shales of Spiti and Chikkim Formations, Tethys Himalaya, India. J Geol Soc India 92(4):427–434

Pant NC, Kundu A, Kumar R, Dorka BS, Prasher S (2006) Paleoproterozoic metamorphism in the Jeori-Wangtu Gneissic Crystallines (JWGC), Western Himalaya. J Asian Earth Sci 26:585–604

Pant NC, Singh P, Jain AK (2020) A Re-look at the Himalayan metamorphism. Episodes 43(1):369–380. https://doi.org/10.18814/epiiugs/2020/020023

Parcha SK (1996) Cambrian sequences in the Tethyan zone of Spiti Himalaya and its boundary problems. Newslett Strat 34:3–11

Parcha SK, Pandey S (2016) Trace fossils and microbially induced sedimentary structures from the early Cambrian successions of the Chandratal area, Spiti Basin, Tethys Himalaya. J Paleont Soc India 61(1):9–18

Parkash B, Sharma RP, Roy AK (1980) The Siwalik Group (molasse) sediments shed by collision of continental plates. Sedim Geol 25:127–159

Parkash B, Kumar S, Rao MS, Giri SC, Kumar CS, Gupta S, Srivastava P (2000) Holocene tectonic movements and stress field in the western Gangetic Plain. Curr Sci 79:438–449

Parkash B, Rathore RS, Pati P, Jakhmola RP, Singh S (2011) Convergence rates along the Himalayan Frontal Thrust inferred from terraces at Chandidevi Temple Hill, Hardwar, Northwestern Himalaya. Curr Sci 100(9):1426–1432

Parrish RR, Hodges KV (1996) Isotopic constraints on the age and provenance of the Lesser and Greater Himalayan sequences, Nepalese Himalaya. Geol Soc Am Bull 108:904–911. https://doi.org/10.1130/0016-7606(1996)108b0904:ICOTAAN2.3.CO;2

Pascoe EH (1959) A manual of the geology of India and Burma, 2nd edn. Gov India, pp 1–448

Patel RC, Carter A (2009) Exhumation history of the Higher Himalayan Crystalline along Dhauliganga-Goriganga river valleys, NW India: new constraints from fission-track analysis. Tectonics 28:TC3004. https://doi.org/10.1029/2008tc002373

Patel RC, ManMohan (2020a) Exhumation history of the Lesser Himalaya, NW-India. Proc Indian Sci Acad 86(2)

Patel RC, ManMohan (2020b) Mio-Pliocene Tectonics and Exhumation Histories of the NW- and NE-Himalaya. Episodes 43(1):381–403. https://doi.org/10.18814/epiiugs/2020/020023

Patel RC, Singh S, Asokan A, Manickavasagam RM, Jain AK (1993) Extensional tectonics in the collisional Zanskar Himalayan belt. In: Treloar PJ, Searle MP (eds) Himalayan tectonics, vol 74. Spec Publ Geol Soc London, pp 445–459

Patel RC, Adlakha V, Lal N, Singh P, Kumar Y (2009) Spatiotemporal variation in exhumation of the Crystallines in the NW-Himalaya, India: constraints from Fission Track dating analysis. Tectonophysics 504(1–4):1–13

Pathak DB (2007) Jurassic/Cretaceous boundary in the Spiti Himalaya, India. J Palaeont Soc India 52(1):51–57

Pathak M, Kumar S (2019) Petrology, geochemistry and zircon U-Pb-Lu-Hf isotopes of Paleoproterozoic granite gneiss from Bomdila in the western Arunachal Himalaya, Northeast India. Geol Soc Lond Spec Publ. https://doi.org/10.1144/sp481-2017-169

Pati P, Parkash B, Awasthi AK, Acharya V (2011) Holocene tectono-geomorphic evolution of parts of the Upper and Middle Gangetic Plain, India. Geomorph 128:148–170

Pati P, Parkash B, Awasthi AK, Jakhmol RP (2012) Spatial and temporal distribution of inland fans/terminal fans between the Ghaghara and Kosi rivers indicate eastward shift of neotectonic

activities along the Himalayan front. A study from parts of the upper and middle Gangetic Plain, India. Earth-Sci Rev 115:201–216
Pati P, Pradhan RM, Dash C, Parkash B, Awasthi AK (2015) Terminal fans and the Ganga plain tectonism: a study of neotectonism and segmentation episodes of the Indo-Gangetic foreland basin, India. Earth-Sci Rev 148:134–149
Pati P, Verma AK, Dash C, Patel NK, Gupta A, Sharma V, Jakhmola RP, Parkash B, Awasthi AK, Saraf AK (2019) Influence of neotectonism on geomorphology and depositional architecture of the Gandak megafan, middle Ganga plain, India. Geomorph 327:489–503
Patriat P, Achache J (1984) India-Euraisa collision chronology has implications for crustal shortening and driving mechanism of plates. Nature 311:615–621
Pebam J, Jain AK, Kumar R, Singh S, Lal N (2008) Tectonics v. erosion: evidences from apatite fission track and Rb–Sr (biotite and muscovite) thermochronology, Arunachal Himalaya. Himal J Sci 5:103–105
Philip G, Suresh NP, Jayangondaperumal R (2017) Late Pleistocene-Holocene strain release by normal faulting along the Main Boundary Thrust at Logar in the northwestern Kumaon Sub Himalaya, India. Quatern Int 462:50–64
Phukon P, Sen K, Srivastava HB, Singhal S, Sen A (2018) U–Pb geochronology and geochemistry from the Kumaon Himalaya, NW India, reveal Paleoproterozoic arc magmatism related to formation of the Columbia supercontinent. Geol Soc Am Bull. https://doi.org/10.1130/B31866.1
Pilgrim GE, West W (1928) The structure and correlation of the Simla rocks. Mem Geol Surv India 53:1–140
Pognante U, Spencer DA (1991) First report of eclogites from the Himalayan belt, Kaghan valley (Pakistan). Eur J Miner 3:613–618
Powell C McA, Conaghan PJ (1973) Plate tectonics and the Himalayas. Earth Planet Sci Lett 20(1):1–12. https://doi.org/10.1016/0012-821X(73)90134-9
Powell C McA, Conaghan PJ (1975) Tectonic models of the Tibetan plateau. Geology 3:727–731. https://doi.org/10.1130/0091-7613
Powers PM, Lillie RJ, Yeats RS (1998) Structure and shortening of the Kangra and Dehra Dun reentrants, Sub-Himalaya, India. Geol Soc Amer Bull 110(8):1010–1027. https://doi.org/10.1130/0016-7606
Prabhakar N, Bhattacharya A, Sathyanarayanan M, Mukherjee PK (2014) Structural, Petrological, and Chronological Constraints from Eastern India and Implications for the ~1.0 Ga Assembly of Greater India. J Geol 122:411–432
Prakash D, Tewari S (2016) Garnet-staurolite-mica schist from RangliRangliot, Eastern Himalaya: constraints from phase equilibria and Thermobarometry. J Geol Soc India 87:679–690
Prakash D, Tewari S, Singh PC (2017) Prograde Barrovian metamorphism along Darjeeling-Tista transect, Eastern Himalaya, India: constraints from textural relationship, phase equilibria and geothermobarometry. Geol J 53:243–281
Rahaman W, Singh SK, Sinha R, Tandon SK (2009) Climate control on erosion distribution over the Himalaya during the past ~100 ka. Geology 37(6):559–562. https://doi.org/10.1130/g25425a.1
Rai V, Singh IB (1983) Discovery of trilobite impression in the arenaceous member of Tal Formation. J Palaeont Soc India 28:114–117
Raiverman V (2002) Foreland sedimentation in Himalayan tectonic regime. Bishan Singh Mohendar Pal Singh Dehradun, pp 1–378
Raiverman V, Kunte SV, Mukherjee A (1983) Basin geometry, Cenozoic sedimentation and hydrocarbon prospects in northwestern Himalaya and Indo-Gangetic plains. Petrol Asia J 6:67–92
Raiverman V, Srivastava AK, Prasad DN (1994) Structural style in northwestern Himalayan foothills. Himalayan Geol 15:263–280
Rao MBR (1973) The subsurface geology of Indo-Gangetic Plains. J Geol Soc India 14:217–242
Rao DR, Sharma R (2011) Arc magmatism in eastern Kumaun Himalaya, India: a study based on geochemistry of granitoid rocks. Island Arc 20:500–519

Rashid SA, Ganai JA (2018) Depositional environments, provenance and paleoclimatic implications of Ordovician siliciclastic rocks of the Thango Formation, Spiti Valley, Tethys Himalaya, northern India. J Asian Earth Sci 157:371–386

Ray SK (1995) Lateral variations in geometry of thrust planes and its significance, as studies in the Shumar allochthon, Lesser Himalayas, eastern Bhutan. Tectonophysics 249:125–139

Ray SK, Bandopadhyay BK, Razdan RK (1989) Tectonics of a part of the Shumar allochthon in eastern Bhutan. Tectonophysics 169:51–58

Reed FRC (1910) The Cambrian fossils of Spiti. Paleont Indica 15:1–70

Richards A, Argles T, Harris N, Parrish R, Ahmad T, Darbyshire F, Draganits E (2005) Himalayan architecture constrained by isotopic tracers from clastic sediments. Earth Planet Sci Lett 236:773–796

Richards A, Parrish R, Harris N, Argles T, Zhang L (2006) Correlation of lithotectonic units across the eastern Himalaya, Bhutan. Geology 34:341–344

Robert X, van der Beek P, Braun J, Perry C, Dubille M, Mugnier J-L (2009) Assessing Quaternary reactivation of the Main Central thrust zone (central Nepal Himalaya): new thermochronologic data and numerical modelling. Geology 37(8):731–734. https://doi.org/10.1130/G25736A.1

Robinson, D.M., DeCelles, P.G., Patchett, P.J., and Garzione, C.N., 2001, The kinematic history of the Nepalese Himalaya interpreted from Nd isotopes: Earth Sci Lett 192, p. 507–521. https://doi.org/10.1016/S0012-821X(01)00451-4

Robinson DM, DeCelles PG, Copeland P (2006) Tectonic evolution of the Himalayan thrust belt in western Nepal: Implications for channel flow models. Geol Soc Amer Bull 118:865–885. https://doi.org/10.1130/B25911.1

Rogers JJW, Santosh M (2002) Configuration of Columbia, a Mesoproterozoic supercontinent. Gondwana Res 5:5–22

Rolland Y, Pêcher A, Picard C (2000) Middle Cretaceous back-arc formation and arc evolution along the Asian margin: the Shyok Suture Zone in northern Ladakh NW Himalaya. Tectonophysics 325:145–173

Rolland Y, Picard C, Pêcher A, Lapierre H, Bosch D, Keller F (2002) The cretaceous Ladakh arc of NW Himalaya—slab melting and melt–mantle interaction during fast northward drift of Indian Plate. Chem Geol 182:139–178

Sachan HK, Kohn MJ, Saxena A, Corrie SL (2010) The Malari leucogranite, Garhwal Himalaya, northern India: Chemistry, age, and tectonic implications. Geol Soc Amer Bull 122:1865–1876. https://doi.org/10.1130/B30153.1

Sakai H, Iwano H, Danhara T, Takigami Y, Rai SM, Upreti BN, Hirata T (2013) Rift related origin of the Paleoproterozoic Kuncha Formation, and cooling history of the Kunchanappe and Taplejung granites, eastern Nepal Lesser Himalaya: a multichronological approach. Island Arc 22:338–360. https://doi.org/10.1111/iar.12021

Salvi D, Mathew G, Kohn B (2017) Rapid exhumation of the upper Siang Valley, Arunachal Himalaya since the Pliocene. Geomorphology 284:238–249. https://doi.org/10.1016/j.geomorph.2016.09.032

Sangode SJ, Kumar R, Ghosh SK (1996) Magnetic polarity stratigraphy of the Siwalik sequence of Haripur (H.P.), NW Himalaya. J Geol Soc India 47:683–704

Sangode SJ, Kumar R, Ghosh SK (1999) Palaeomagnetic and rock magnetic perspectives on the post–collision continental sediments of the Himalaya, India. In: Radhakrishna T, Piper JDA (eds) The Indian subcontinent and Gondwana: a palaeomagnetic and rock magnetic perspective, vol 44. Mem Geol Soc India, pp 221–248

Sangode SJ, Kumar R, Siddaiah NS (2005) Magnetic polarity and rock magnetic studies across marine to continental transition (Subathu–Dagshai sequence) in the Himalayan foreland. Asia Oceania Geosci Soc 2nd Ann Meet SE30/3A–01–3/206

Sastri VV, Bhandari LL, Raju ATR, Datta AK (1971) Tectonics framework and subsurface stratigraphy of the Ganga basin. J Geol Soc India 12(3):222–233

Saxena MN (1971) The crystalline axis of the Himalaya, Indian shield and continental drift. Tectonophysics 12:433–447

Schlup M, Carter A, Cosca M, Steck A (2003) Exhumation history of eastern Ladakh revealed by $^{40}Ar/^{39}Ar$ and fission track ages: the Indus river-Tso Morari transect, NW Himalaya. J Geol Soc Lond 160:385–399
Schulte-Pelkum V, Monsalve G, Sheehan A, Pandey MR, Sapkota S, Bilham R, Wu F (2005) Imaging the Indian subcontinent beneath the Himalaya. Nature 435:1222–1225
Searle MP (2013) Colliding continents: a geological exploration of the Himalaya, Karakoram, and Tibet. Oxford Univ Press, Oxford, pp 1–464
Searle MP, Windley BF, Coward MP, Cooper DJW, Rex AJ, Rex D, Tingdong L, Xuchang X, Jan MQ, Thakur VC, Kumar S (1987) The closing of the Tethys and the tectonics of the Himalaya. Geol Soc Amer Bull 98:678–701
Searle MP, Crawford MB, Rex AJ (1992) Field relations, geochemistry, origin and emplacement of the Baltoro granite, central Karakoram. Trans Royal Soc Edin 83: 519–538
Searle MP, Parrish RR, Hodges KV, Hurford A, Ayres MW, Whitehouse MJ (1997) Shisha Pangma leucogranite, South Tibetan Himalaya: Field relations, geochemistry, age, origin and emplacement. J Geol 105:295–317
Searle MP, Noble SR, Hurford AJ, Rex DC (1999) Age of crustal melting, emplacement and exhumation history of the Shivling leucogranite, Garhwal Himalaya. Geol Mag 136 (5):513–525
Searle MP, Law RD, Godin L, Larson KP, Streule MJ, Cottle JM, Jessup MJ (2008) Defining the Himalayan Main Central Thrust in Nepal. J Geol Soc London 165:523–534
Searle MP, Avouac J-P, Elliott J, Dyck B (2017) Ductile shearing to brittle thrusting along the Nepal Himalaya: Linking Miocene channel flow and critical wedge tectonics to 25th April 2015 Gorkha earthquake. Tectonophysics 714–715:117–124
Seeber L, Armbruster JG (1981) Great Detachment Earthquakes Along the Himalayan Arc and Long-Term Forecasting. Earthquake Prediction:259–277 https://doi.org/10.1029/me004p0259
Sen Gupta S, Raina PL (1978) Geology of parts of Bhutan foothills adjacent of Darjeeling district. Indian J Earth Sci 5(1):20–33
Sen K, Chaudhury R, Pfänder J (2015) 40Ar–39Ar age constraint on deformation and brittle–ductile transition of the Main Central Thrust and the South Tibetan Detachment zone from Dhauliganga valley, Garhwal Himalaya, India. J Geody 88:1–13. https://doi.org/10.1016/j.jog.2015.04.004
Shah and Sinha (1974) Stratigraphy and tectonics of the Tethys zone in a part of western Kumaun Himalaya. Himalayan Geol 4:1–27
Shankar R, Kumar G, Mathur VK, Joshi A (1993) Stratigraphy of Blaini, Infra Krol, Krol and Tal succession, Krol belt, Lesser Himalaya, India. Indian J Petrol Geol 2:99–136
Shanker R, Mathur VK (1992) Precambrian-Cambrian sequence in Krol belt and additional Ediacaran fossils. Geophyt 22:27–39
Shanker R, Mathur VK, Kumar G, Srivastava MC (1997) Additional Ediacaran biota from the Krol Group, Lesser Himalaya, India and their significance. Geosci J 17:79–94
Sharma VP (1977) The stratigraphy and structure of parts of the Simla Himalaya. Mem Geol Surv India 106:1–488
Sharma R, Villa IM, Kumar S (eds) (2019) Crustal architecture and evolution of the Himalaya–Karakoram–Tibet Orogen. Geol Soc Lond Spec Publ 481
Shreshtha M, Jain AK, Singh S (2015) Shear sense analysis of the Higher Himalayan Crystalline (HHC) belt, and tectonics of the South Tibetan Detachment System (STDS), Alaknanda–Dhauli Ganga Valley, Uttarakhand Himalaya. Curr Sci 108(6):1107–1118
Siddaiah NS, Kumar K (2007) Discovery of volcanic ash bed from the basal Subathu Formation (Late Palaeocene–Middle Eocene) near Kalka, Solan District (Himachal Pradesh), Northwest Sub-Himalaya, India. Curr Sci 92(1):118–125
Singh IB (1976) Geological Evolution of Ganga Plain - An Overview. J Palaent Soc India 41:99–137
Singh IB (1978) On some sedimentological and palaeoecological aspects of the Subathu–Dagshai–Kasauli succession of Simla Hill. J Palaeont Soc India 21:19–28
Singh IB (1979) Some thoughts on evolution of Himalaya and northern limits of Indian Shield. Geol Rundsch 68:342–350
Singh IB (1980) Sedimentological evolution of the Krol Belt sediments. Himalayan Geol 8:657–683

Singh IB (1992) Geolgical evolution of Gangetic Plain: Present Status. In: Singh IB (ed) Gangetic Plain. Terra Incognita 1–14

Singh IB (1996a) Late Quaternary sedimentation of Ganga plain foreland basin. Proc Symp NW Himalaya and foredeep. Geol Surv India Spec Publ 21:161–172

Singh IB (1996b) Geological evolution of Ganga Plain—an overview. J Palaeont Soc India 41:99–137

Singh IB (1999) Tectonic control on sedimentation in Ganga Plain Foreland Basin: constraints on the Siwalik sedimentation models. In: Jain AK, Manickavasagam RM (eds) Geodynamics of the NW Himalaya. Gondwana Res Group Mem 6:3–37

Singh IB (2001) Proxy records of neotectonics, climate changes and anthropogenic activity in the Late Quaternary of Ganga Plain. Proc Nat Symp of Role of Earth Sciences in Integrated Developments and Related Societal Issues. Geol Surv India Spec Publ 65(1):xxxiii–1

Singh IB, Merajuddin (1978) Some sedimentological observations of the Chhaosa Formation (Simla Slates) in the Simla Hills. Himalayan Geol 8(2):683–702

Singh IB, Rai V (1977) On the occurrence of stromatolites in the Krol Formation of Nainital area and its implications on the age of Krol Formation. Curr Sci 46:736–738

Singh IB, Rai V (1983) Fauna and biogenic structures in Krol-Tal succession, Lesser Himalaya: their biostratigraphic and paleontological significance. J Paleont Soc India 28:67–94

Singh IB, Ansari AA, Chandel RS, Misra A (1996) Neotectonic control on Drainage System in Gangetic Plain, Uttar Pradesh. J Geol Soc India 47:599–609

Singh IB, Rajagopalan G, Agarwal KK, Srivastava P, Sharma M, Sharma S (1997) Evidence of Middle to Late Holocene neotectonic activity in the Ganga Plain. Curr Sci 73(12):1114–1117

Singh BP (2003) Evidence of growth fault and forebulge in the Late Paleocene (~57.9–54.7 Ma), western Himalayan foreland basin, India. Earth Planet Sci Lett 216:717–724

Singh BP, Virmani N, Bhargava ON, Kishore N, Gill A (2014) Yuehsienszella (Cambrian Series 2) trilobite from the Parahio Valley, Spiti region (Zanskar–Spiti Sub-basin), India and its biostratigraphic significance. J Palaeont Soc India 59:81–88

Singh BP, Virmani N, Bhargava ON, Negi RS, Gill A, Kishore N (2016) Trilobite fauna of basal Cambrian Series 3 (Stage 5) from the Parahio Valley (Spiti), Northwest Himalaya, India and its biostratigraphic significance. Annales Paleont 102:59–67

Singh BP, Chaubey RS, Bhargava ON, Prasad SK, Negi RS (2017a) The Cambrian trilobite fauna from the Shian (Saybang) section, Pin Valley (Spiti) and its biostratigraphic significance. Palaeoworld 26:25–36

Singh BP, Bhargava ON, Negi RS, Zhao Y, Yin L, Sharma CA (2017b) Additional trilobite fauna from the basal part of the Cambrian Series 3, Stage 5, Kunzam La (=Parahio) Formation, Parahio Valley, Spiti (Northwest Himalaya), India and its biostratigraphic significance. Annales Paleont 103:271–281

Singh BP, Bhargava ON, Sharma CA, Chaubey RS, Prasad SK, Negi RS, Kishore N (2017c) Treptichnus ichnogenus from the Cambrian of India and Bhutan: its relevance to the Precambrian-Cambrian boundary. J Paleont Soc India 62(1):39–51

Singh BP, Bhargava ON, Mikuláš R, Prasad SK, Singla G, Kaur R (2017d) Asteriacites and other trace fossils from the Po Formation (Visean–Serpukhovian), Ganmachidam Hill, Spiti Valley (Himalaya) and its paleoenvironmental significance. Geol Carapathica 68(5):464–478

Singh BP, Bhargava ON, Mikuláš R, Morrision S, Kaur R, Singla G, Kishore N, Kumar N, Kumar R, Moudgil S (2019) Integrated sedimentological, ichnological and sequence stratigraphical studies of the Koti Dhaman Formation (Tal Group), Nigali Dhar Syncline, Lesser Himalaya, India paleoenvironmental, paleoecological, paleogeographic significance. Ichnos. https://doi.org/10.1080/10420940.2019.1584560

Singh S (1993) Collision tectonics: metamorphic and geochronological constraints from parts of Himachal Pradesh, NW–Himalaya. Unpubl Ph.D. thesis, Univ Roorkee 289

Singh S (2005) A review of U-Pb ages from Himalayan Collisional Belt. J Himalayan Geol. 26(1):61–76

Singh S (2019) Protracted zircon growth in migmatites and In situ melt of Higher Himalayan Crystallines: U-Pb ages from Bhagirathi valley, NW Himalaya, India. Geosci Front 10:793–809. https://doi.org/10.1016/j.gsf.2017.12.014

Singh S (2020) Himalayan Magmatism through space and time. Episodes 43(1):358–368, doi.org/10.18814/epiiugs/2020/020021

Singh S, Jain AK (2007) Geology and tectonics of the Subansiri Corridor, Arunachal Himalaya. DCS–DST News Lett 17:21–24

Singh S, Jain AK (2008) Pan African Mandi Granite in the Lesser Himalaya: its field relationship and SHRIMP U-Pb zircon dating. In: Pan African Event: India and Antarctica. Geol Surv India Spec Publ 91:70–78

Singh P, Saikia A, Pant NC, Verma PK (2013) Insights into the P-T evolution path of Tso Morari eclogites of the north-western Himalaya: constraints on the geodynamic evolution of the region. J Earth Sys Sci 122(3):677–698

Singh RL, Singh KN (1971) Upper Ganga Plain. In: Singh RL (ed) India—a regional geography. Nat Geog Soc India Varanasi 922

Singh S, Barley ME, Brown SJ, Jain AK, Manickavasagam RM (2002) SHRIMP U–Pb in zircon geochronology of the Chor granitoid: Evidence for Neoproterozoic magmatism in the Lesser Himalayan granite belt of NW India Precamb Res 118:285–292. https://doi.org/10.1016/S0301-9268(02)00107-9

Singh S, Jain AK, Barley ME (2009) SHRIMP U–Pb c (1860) Ma anorogenic magmatic signatures from the NW Himalaya: implications for Palaeoproterozoic assembly of the Columbia Supercontinent. In: Reddy SM, Mazumdar R, Evans DAD, Collins AS (eds) Palaeoproterozoic supercontinents and global evolution, vol 323. Geol Soc London Spec Publ, pp 283–300

Sinha RN (1970) Heavy mineral investigation of the Siwaliks of Mohand, Dist. Saharanpur, U.P. J Geol Soc India 11:163–177

Sinha AK (1989) Geology of higher central Himalaya. Wiley, Chichester, pp 1–236

Sinha S, Sinha R (2016) Geomorphic evolution of Dehra Dun, NW Himalaya: tectonics and climatic coupling. Geomorph 266:20–32

Sinha HN, Verniers J (2016) Discovery of the chitinozoans Belonechitina capitata from the Shiala Formation of northeastern Garhwal-Kumaon Tethys Himalaya, Pithoragarh District, Uttrakhand, India. Geosci Front 7:859–864

Sorcar N, Hoppe U, Dasgupta S, Chakraborty S (2014) High-temperature cooling histories of migmatites from the high Himalayan crystallines in Sikkim, India: rapid cooling unrelated to exhumation? Contrib Mineral Petrol 167:1–34

Sorkhabi RB, Stump E, Foland K, Jain AK (1999) Tectonics and cooling history of the Garhwal Higher Himalaya (Bhagirathi Valley): constraints from thermochronological data. In: Jain AK, Manickavasagam RM (eds) Geodynamics of the NW Himalaya, vol 6. Gondwana Res Group Mem, pp 217–235

Spencer CJ, Harris RA, Dorais (2012) The metamorphism and exhumation of the Himalayan metamorphic core, eastern Garhwal region, India. Tectonics 31:TC1007. https://doi.org/10.1029/2010tc002853

Srikantia SV (1977) Sedimentary cycles in the Himalaya and their significance on the orogenic evolution of the mountain belt. Coll Int Centre Nat Recher Scient 268:395–408

Srikantia SV (1981) The lithostratigraphy, sedimentation and structure of Proterozoic–Phanerozoic formations of Spiti in the higher Himalaya of Himachal Pradesh, India. In: Sinha AK (eds) Contemporary geoscientific researches in Himalaya. Bishen Singh Mahendra PS Publ Dehradun, pp 31–48

Srikantia SV, Bhargava ON (1978) The Indus tectonic belts of Ladakh Himalaya: its geology, significance and evolution. In: Saklani PS (ed) Tectonic geology of the Himalaya. Today Tomorrow's Printers Publ, pp 43–62

Srikantia SV, Bhargava ON (1998) Geology of Himachal Pradesh. Mem Geol Soc India, pp 1–406

Srikantia SV, Bhargava ON (2018) Stratigraphic nomenclature of early Palaeozoics in the Spiti Himalaya: cobwebs cleared. J Palaeont Soc India 63(2):233–241

Srikantia SV, Ganesan TM, Sinha PK, Tirkey B (1976) Geology of Part of Zanskar Mountains, Ladakh Himalaya with special reference to late Caledonian "Kurgiakh Orogeny". "Geology, Mineral Resources and Natural Resources of Power Development of Himalayas with particular Reference to Kashmir". Geol Sur India, Kashmir Circle Abstracts, pp 1–3

Srikantia SV, Ganesan TM, Rao PN, Sinha PN, Tirkey B (1980) Geology of Zanskar area, Ladakh Himalaya. Himalayan Geol 8 Geology of Zanskar area, Ladakh Himalaya. Himalayan Geol 8 (for 1978):1009–1033

Srivastava DC, John G (1999) Deformation in the Himalayan Frontal Fault zone: evidence from small-scale structures in Mohand-Khara area, NW Himalaya. In: Jain AK, Mancikvasagam RM (eds) Geodynamics of the NW Himalaya. Gondwana Res Group Mem 6:273–284

Srivastava P, Parkash B, Sehgal J, Kumar S (1994) Role of neotectonics and climate in development of the Holocene geomorphology and soils of the Gangetic Plains between the Ramganga and Rapti rivers. Sediment Geol 94:129–151

Srivastava P, Singh IB, Sharma M, Singhvi AK (2003) Luminescence chronometry and Late Quaternary geomorphic history of the Ganga Plain, India. In: Gupta AK, Anderson DM, Malmgren BA (eds) Indian Ocean Monsoons: Land and Sea Record, vol 197. Palaeogeo Palaeoclim, Palaeoeco, pp 15–41. https://doi.org/10.1016/s0031-0182(03)00384-5

Srivastava P, Sinha R, Deep V, Singh A, Upreti N (2018) Micromorphology and Sequence Stratigraphy of the Interfluve Paleosols from the Ganga Plains: a Record of Alluvial Cyclicity and Paleoclimate During the Late Quaternary. J Sediment Res 88:105–128

Srivastava VK, Casshyap SM (1983) Evolution of the Pre-Siwalik Tertiary basin of Himachal Himalaya. J Geol Soc India 24:134–147

Stampfli GM, Borel GD (2002) A plate tectonic model for the Paleozoic and Mesozoic constrained by dynamic plate boundaries and restored synthetic oceanic isochrons. Earth Planet Sci Lett 196(1):17–33. https://doi.org/10.1016/S0012-821X(01)0058-X

Stoliczka F (1865) Geological section across the Himalayan mountain range from Wangtu Bridge on the river Sutlej to Sungdeo with an account of formations in Spiti accompanied by a revision of all known fossils from that district. Mem Geol Surv India 5:1–153

Stübner K, Grujic D, Parrish RR, Roberts NMW, Kronz A, Wooden J, Ahmad T (2014) Monazite geochronology unravels the timing of crustal thickening in NW Himalaya. Lithos 210–211:111–128. https://doi.org/10.1016/j.lithos.2014.09.024

Stübner K, Grujic D, Dunkl I, Thiede R, Eugster P (2018) Pliocene episodic exhumation and the significance of the Munsiari thrust in the northwestern Himalaya. Earth Planet Sci Lett 481:273–283

Suess E (1893–1909) Das Antlitz der Erde, 4 vols. Prague and Leipzig

Swami NK, Ernst A, Tripathi SC, Barman P, Bharti SK, Rana YP (2019) A new cryptostome bryozoan Ptilotrypa from the Upper Ordovician Yong Limestone Formation: tethyan sequence of Kumaun Higher Himalaya, India. J Paleont. https://doi.org/10.1017/jpa.2018.94

Tandon SK (1991) The Himalayan Foreland: focus on Siwalik Basin. In: Tandon SK, Pant CC, Casshyap SM (eds) Sedimentary Basins of India: tectonic context. Gyanodaya Prakashan, Nainital, pp 177–201

Tandon SK, Sinha R, Gibling MR, Dasgupta AS, Ghazanfari P (2008) Late quaternary evolution of the Ganga Plains: myths and misconceptions, recent developments and future directions. Mem Geol Soc India 66:1–41

Tangri SK (1995) Baxa Group. In: Bhargava ON (ed) Bhutan Himalaya: a geological account, vol 39. Geol Surv India Spec Publ, pp 38–58

Tangri SK, Pande AC (1995) Tethyan sequence. In: Bhargava ON (ed) Bhutan Himalaya: a geological account, vol 39. Geol Surv India Spec Publ, pp 109–142

Tewari VC (1994) Sedimentology of the rocks of Deoban Basin, Dhuraphat area, Saryu valley, Eastern Kumaon Lesser Himalaya. Geosci J 15(2):117–161

Tewari S, Prakash D (2016) Geothermobarometry and Barrovian Metamorphism of Darjeeling–Mangpu Region, Eastern Himalaya. In: Raju NJ (ed) Geostatistical and geospatial approaches for the characterization of natural resources in the environment. Springer, Berlin, pp 523–526

Tewari S, Prakash D (2017) Age of crustal melting in Higher Himalayan Crystalline Sequence (Darjeeling, Eastern Himalaya): constraints from SHRIMP U-Pb geochronology. Geol J 53:1516–1525. https://doi.org/10.1002/gj.2970
Thakur VC (1993) Geology of the Western Himalaya. Pergmon Press, Oxford, pp 1–355
Thakur VC (2013) Active tectonics of Himalayan Frontal Fault system. Int J Earth Sci 102 (7):1791–1810
Thakur VC, Pandey AK, Suresh N (2007) Late Quaternary–Holocene frontal fault zone of the Garhwal Sub Himalaya, NW India. J Asian Earth Sci 29(2/3):305–319
Thakur VC, Jayangondaperumal R, Malik MA (2010) Redefining Wadia–Medlicott's Main Boundary Fault from Jhelum to Yamuna: an active fault strand of the Main Boundary Thrust in Northwest Himalaya. Tectonophysics 489:29–42
Thakur SS, Patel SC, Singh AK (2015) A P-T pseudosection modelling approach to understand metamorphic evolution of the Main Central Thrust Zone in the Alaknanda valley, NW Himalaya. Contrib Mineral Petrol 170:2. https://doi.org/10.1007/s00410-015-1159-y
Thakur VC, Joshi M, Jayangondaperumal R (2020) Active tectonics of Himalayan frontal fault zone in the Sub Himalaya. In: Tandon SK, Gupta N (eds) Geodynamics of the Indian Plate: evolutionary perspectives. Springer Geology, Berlin
Thiede RC, Ehlers TA (2013) Large spatial and temporal variations in Himalayan denudation. Earth Planet Sci Lett 371–372:278–293
Thiede RC, Arrowsmith JR, Bookhagen B, Mcwilliams MO, Sobel ER, Strecker MR (2005) From tectonically to erosionally controlled development of the Himalayan orogen. Geology 33:689–692. https://doi.org/10.1130/G21483AR.1
Thiede RC, Ehlers TA, Bookhagen B, Strecker MR (2009) Erosional variability along the northwest Himalaya. J Geophys Res 114:F01015. https://doi.org/10.1029/2008JF001010
Thiede R, Robert X, Stübner K, Dey S, Faruhn J (2017) Sustained out-of-sequence shortening along a tectonically active segment of the Main Boundary thrust: the Dhauladhar Range in the northwestern Himalaya. Lithosphere. https://doi.org/10.1130/L630.1
Thomas JV, Parkash B, Mohindra R (2002) Lithofacies and paleosol analysis of the Middle and Upper Siwalik Groups (Plio–Pleistocene), Haripur-Kolar section, Himachal Pradesh, India. Sedim Geol 150:343–366
Treloar PJ, Searle MP (eds) (1993) Himalayan tectonics. Geol Soc Lond Spec Publ 74:630
Treloar PJ, Searle MP (eds) (2019) Himalayan tectonics: a modern synthesis. Geol Soc Lond Spec Publ 483
Trichal AK, Jayaram S (1989) Tungsten mineralization in skarn rocks, Dholpani-Bhurkhola areas, Geylegphug district, Bhutan. Geol Surv India Spec Publ 22:101–105
Trivedi JR, Gopalan K, Valdiya KS (1984) Rb–Sr ages of granitic rocks within the Lesser Himalayan Nappes, Kumaon, India. J Geol Soc India 25:641–654
Valdiya KS (1962) An outline of the stratigraphy and structure of the southern part of the Pithoragarh District, U.P. J Geol Soc India 3:27–48
Valdiya KS (1970) Simla Slates, the Precambrian flysch of Lesser Himalaya; its turbidites, sedimentary structures and palaeocurrents. Geol Soc Am Bull 81:451–468
Valdiya KS (1980a) Geology of the Kumaun Lesser Himalaya. Wadia Institute of Himalayan Geology, Dehra Dun India, pp 1–291
Valdiya KS (1980b) The two intracrustal boundary thrusts of the Himalaya. Tectonophysics 66:323–348
Valdiya KS (1992) The Main Boundary thrust zone of the Himalaya, India. In: Bucknam RC, Hancock PL (eds) Major active faults of the world: results of IGCP Project 206. Annales Tectonicae 6:54–84
Valdiya KS (2016) The making of India geodynamic evolution, 2nd edn. Springer, Berlin, pp 1–924
Valdiya KS, Gupta VJ (1972) A contribution to the geology of Northeastern Kumaun, with special reference to the Hercynian gap in Tethys Himalaya. Himalayan Geol 2:1–34
Valdiya KS, Gupta VJ, Powar KB, Banerjee DM (1972) Palaeozoic Biostratigraphy of the Tethys Himalaya in the North-Eastern Kumaun Himalaya. Nature 235:152–154

Valdiya KS, Paul SK, Tarachandra, Bhakuni SS, Upadhyay RC (1999) Tectonic and lithological characterization of Himadri (Great Himalaya) between Kali and Yamuna rivers, Central Himalaya. Himalayan Geol 10(2):1–17

Vannay JC, Grasemann B, Rahn M, Frank W, Carter A, Baudraz V, Cosca M (2004) Miocene to Holocene exhumation of metamorphic crustal wedges in the NW Himalaya: evidence for tectonic extrusion coupled to fluvial erosion. Tectonics 23:1–24

Verma PK (1989) The Himalayan Metamorphism. Geol Soc Spec Publ 43:377–383

Verma PK (2003) The Himalayan metamorphism and Generation of Melt. In: Proceedings of 90th Indian Sci Congr Part-II President Address, 34

Wadia DN (1928) Geology of the Punch state (Kashmir) and adjacent portions of the Northern Punjab. Mem Geol Surv India 51:253

Wadia DN (1931) The syntaxis of the northwest Himalaya and its rocks, tectonics and orogeny. Rec Geol Surv India 65:189–220

Wadia DN (1937) Permo Carboniferous limestone inliers in the Sub Himalayan Tertiary zone of Jammu, Kashmir Himalaya. Rec Geol Surv India 72:162–173

Wallis D, Phillips RJ, Lloyd GE (2014) Evolution of the Eastern Karakoram Metamorphic Complex, Ladakh, NW India, and its relationship to magmatism and regional tectonics. Tectonophysics 626:41–52. https://doi.org/10.1016/j.tecto.2014.03.023

Wang JM, Rubatto D, Jhang JJ (2015) Timing of partial melting and cooling across the Greater Himalayan Crystalline Comlex (Nyalam, Central Himalaya): in-sequence thrusting and its imlications. J Petrol 56:1677–1702

Warren CJ, Singh AK, Roberts NM, Regis D, Halton AM, Singh RB (2014) Timing and conditions of peak metamorphism and cooling across the Zimithang Thrust, Arunachal Pradesh, India. Lithos 200–201:94–110

Webb AAG, Yin A, Harrison TM, Celerier J, Burgess P (2007) The leading edge of the Greater Himalayan Crystalline complex revealed in the NW Indian Himalaya: Implications for the evolution of the Himalayan orogen. Geology 35:955–958. https://doi.org/10.1130/G23931A.1

Webb AAG, Yin A, Harrison TM, Celerier J, Gehrels GE, Manning CE, Grove M (2011) Cenozoic tectonic history of the Himachal Himalaya (NW India) and its constraints on the formation mechanism of the Himalayan orogen. Geosphere 7:1013–1061

Webb AAG, Yin A, Dubey CS (2013) U-Pb zircon geochronology of major lithologic units in the eastern Himalaya: Implications for the origin and assembly of Himalayan rocks. Geol Soc Am Bull 125:499–522. https://doi.org/10.1130/B30626.1

Webb AAG, Guo H, Clift PD, Husson L, Müller T, Costantino D, Yin A, Xu Z, Cao H, Wang Q (2017) The Himalaya in 3D: slab dynamics controlled mountain building and monsoon intensification. Lithosphere 9(4):637–651. ISSN 1941–8264. https://doi.org/10.1130/L636.1

Wesnousky GW, Kumar S, Mohindra R, Thakur VC (1999) Uplift and convergence along the Himalayan Frontal Thrust of India. Tectonics 18(6):967–976

White NM, Parrish RR, Bickle M, Najman YMR, Burbank D, Maithani A (2001) Metamorphism and exhumation of the NW Himalaya constrained by U-Th–Pb analyses of detrital monazite grains from early foreland basin sediments. J Geol Soc Lond 158:625–635

White NM, Pringle M, Garzanti E, Bickle M, Najman Y, Chapman H, Friend P (2002) Constraints on the exhumation and erosion of the High Himalayan Slab, NW India, from foreland basin deposits. Earth Planet Sci Lett 195:29–44

Winslow DM, Zeitler PK, Chamberlain CP, Williams IS (1996) Geochronologic constraints on syntaxial development in the Nanga Parbat region, Pakistan. Tectonics 15:1292–1308

Wobus C, Heimsath A, Whipple K, Hodges KV (2005) Active out-of-sequence thrust faulting in the central Nepalese Himalaya. Nature 434(7036):1008–1011. https://doi.org/10.1038/nature03499

Yeats RS, Kausar AB, Nakata T, Nakata T (2006) Conference examine deadly (2005) Kashmir earthquake. EOS 87:115

Yin A (2006) Cenozoic tectonic evolution of the Himalayan orogen as constrained by along-strike variation of structural geometry, exhumation history, and foreland sedimentation. Earth-Sci Rev 76:1–31

Yin A, Harrison TM (2000) Geologic evolution of the Himalayan-Tibetan orogen. Ann Rev Earth Sci 28:211–280

Yin A, Dubey CS, Kelty TK, Gehrels GE, Chou CY, Grove M, Lovera O (2006) Structural evolution of the Arunachal Himalaya and implications for asymmetric development of the Himalayan orogen. Curr Sci 90:195–206

Yin A, Dubey CS, Kelty TK, Webb AAG, Harrison TM, Chou CY, Célérier J (2009) Geologic correlation of the Himalayan orogen and Indian craton: part 2. Structural geology, geochronology, and tectonic evolution of the Eastern Himalaya. Geol Soc Am Bull 122:360–395. https://doi.org/10.1130/B26461.1

Yin Lei-Ming, Singh Birendra P, Bhargava ON, Zhao YL, Negi RS, Meng FW, Sharma CA (2018) Palynomorphs from the Cambrian Series 3, Parahio valley (Spiti), Northwest Himalaya. Palaeoworld 27:30–41. https://doi.org/10.1016/j.palwor.2017.05.004

Zhang YZ, Replumaz A, Wang GC, Leloup PH, Gautheron C, Bernet M, van der Beek P, Paquette JL, Wang A, Zhang KX, Chevalier ML (2015) Timing and rate of exhumation along the Litang fault system, implication for fault reorganization in Southeast Tibet. Tectonics 34:1219–1243

Zhao G, Wang Y, Huang B, Dong Y, Li S, Zhang G, Yu S (2018) Geological reconstructions of the East Asian blocks: From the breakup of Rodinia to the assembly of Pangaea. Earth-Sci Rev HYPERLINK https://www.sciencedirect.com/science/journal/00128252/186/supp/C 186: 262–286, https://doi.org/10.1016/j.earscirev.2018.10.003

Zhao GC, Sun M, Wilde SA, Li SZ (2004) A Paleo-Mesoproterozoic supercontinent: assembly, growth and breakup. Earth Sci Rev 67:91–123

Zhao W, Nelson KD, Che J, Quo J, Lu D, Wu C, Liu X. (1993) Deep seismic reflection evidence for continental underthrusting beneath southern Tibet. Nature 366:557–559

Chapter 7
Trans-Himalayan and Karakoram Ranges

7.1 Introduction

The Himalayan Mountains are bordered on the northwest by the Ladakh, Karakoram, Pamir and the Hindu Kush ranges, while the Tibetan Plateau in the central part controls its overall shape. A 50–60 km wide tectonic valley, the Indus–Tsangpo Suture Zone, separates these geographical units as a garland throughout the Himalayan ranges from extreme west to the east. These mountain ranges of the Trans-Himalayan and Karakoram Mountains are essentially drained by the Indus, Shyok, Nubra Rivers and their tributaries in the west, and Tsangpo-Brahmaputra drainage system in the east.

This chapter incorporates the geology and tectonics of the Trans-Himalayan and Karakoram Mountains.

7.2 Geology and Tectonics of the Trans-Himalayan

Paleogeographic reconstructions of the India-Asia domains during the Mesozoic postulate that the Tethyan Ocean separated the northern parts of the Indian continental lithosphere (ICL) from the southern Asian Plate (Stampfli and Borel 2002). These regions now constitute parts of the Himalaya, Trans-Himalayan Ladakh–Karakoram Mountains and Tibet Plateau. As the Indian Plate converged anti-clockwise and moved northwards towards Asia, the Neo-Tethyan oceanic floor underwent intra-oceanic subduction to produce the Shyok-Dras Volcanic Arc (Searle et al. 1987; Thakur 1993; Rolland et al. 2000; Jain and Singh 2009; Copley et al. 2010). Between this arc and Asian Plate, a thick Paleo-Mesozoic Karakoram sedimentary sequence was deposited on southern margin of Asian Plate in the Tethyan Ocean, which closed along the Shyok Suture Zone (SSZ) and subsequently along the Indus–Tsangpo

A. K. Jain et al., *Tectonics of the Indian Subcontinent*, Society of Earth Scientists Series, https://doi.org/10.1007/978-3-030-42845-7_7

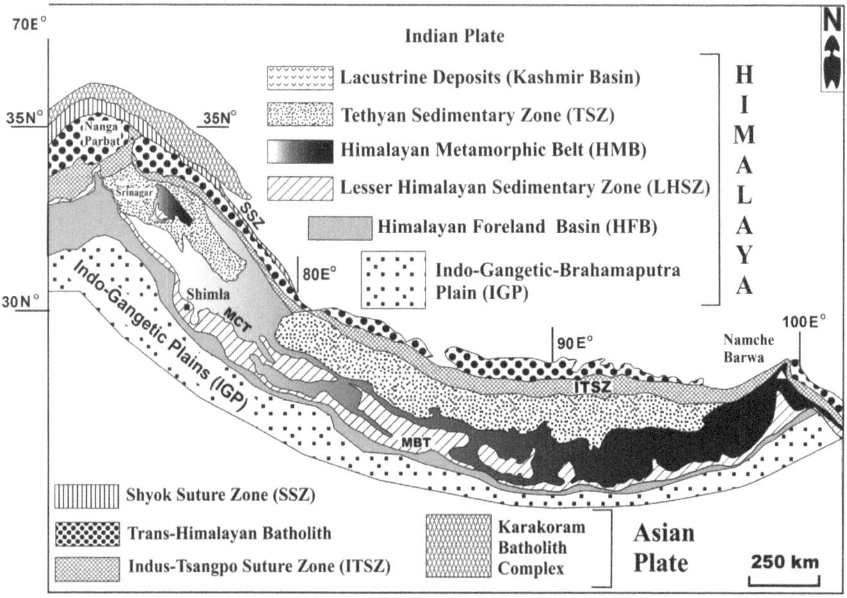

Fig. 7.1 Simplified geological map of the Trans-Himalaya and Karakoram Mountains showing distribution of main tectonic units

Suture Zones (ITSZ) in the south. These sutures demarcate contact between these plates (Fig. 7.1), and preserve evidences of (i) initial Late Mesozoic subduction of the Neo-Tethys oceanic lithosphere along the SSZ during the Early Cretaceous–Lower Eocene with intervening intra-oceanic Shyok–Dras Volcanic Arc, (ii) emplacement of the younger calc-alkaline Trans-Himalayan plutons, and (iii) final closure of the Neo-Tethys along the ITSZ (Honegger et al. 1982; Searle et al. 1987; Hodges 2000; Rolland et al. 2000; Yin 2006; Jain and Singh 2009). Salient characters of these geological units are summarized in Table 7.1.

The Himalaya is, therefore, delimited by km-wide ITSZ, while the SSZ is developed in the west. Both these suture zones bound the Trans-Himalayan Ladakh Range, while the Karakoram Range is developed further north of the region, drained by the Indus River and its tributaries. These ranges represent the cold deserts of Ladakh. In spite these having distinct geology than the Himalaya, many authors include these in the Himalaya, traditionally.

7.2.1 Indus–Tsangpo Suture Zone (ITSZ)

Nearly continuous ITSZ marks the junction of the Indian and Asian Plates (Dewey and Bird 1970; Gansser 1977; Shakelton 1981; Windley 1983) and defines the northern margin of the Himalaya. This junction of vital geodynamic significance for more

Table 7.1 Lithotectonic units of Trans-Himalaya and Karakoram

Tectonic unit	Formation	Lithology	Remarks
Karakoram Batholith Complex	Karakoram Batholith	Biotite granite, leucogranite, metamorphics	Remobilized southern edge of Asian Plate
		Granodiorite, diorite	
Karakoram Metamorphic Complex	Pangong Group Shear Zone Tangste Group	Slate, phyllite, gr-mica scist, calc silicates, marble, amphibolite, augen mylonite	Southern edge of Asian Plate
		Schist, migmatite, amphibolite, pegmatite, leucogranite	
Karakoram Shear Zone		Tirit granodiorite-diorite Mylonitized conglomerate, sandstone-slate, marls-limestone, volcanics, mylonite, serpentinite	Juxtaposition of Asian Plate over SSZ
Shyok Suture Zone (SSZ)	Saltoro Group Thrust Hundiri Fm. Thrust Shyok Volcanics	Ultramafics, gabbro, basalt, chert Conglomerate, sandstone, shale, limestone Basalt, andesite	Northernmost junction of Indian-Asia Plates
Intrusive/Thrust			
Trans-Himalayan Batholith	Khardung Volcanics Ladakh Batholith	Rhyolite, andesite, sandstone Dolerite dykes Granodiorite, diorite, mafic bodies, granite	Andean-type batholith
Thrust			
Indus-Tsangpo Suture Zone	Hemis Group	Conglomerate, sandstone	Southernmost Junction of Indian-Asia Plates
	Indus Group	Conglomerate, sandstone, siltstone, shale alternations	
	Nidar Ophiolite	Ultramafics, gabbro, basalt, dolerite dykes, chert, shale	

(continued)

Table 7.1 (continued)

Tectonic unit	Formation	Lithology	Remarks
Thrust			
Tso Morari Crystallines (TMC)	Polokongka La Granite	Porphyritic granite gneiss	Northern edge of Subducted Indian continental lithosphere
	Tanglang La Fm.	Carbon. schist, quartzite, staurolite-garnet schist, marble, eclogite	
	Puga Gneiss	Garnet gneiss, pelites, eclogite	

than 2000 km signifies the disappearance of the Neo-Tethys oceanic lithosphere by northward subduction of the Indian Plate during Aptian–Albian (120–110 Ma; Maheo et al. 2004; Pudsey 1986) or as early as 140 ± 32 Ma (Ahmad et al. 2008). Presence of the ophiolite complexes and associated sediments along the entire length of the ITSZ from Ladakh to southern Tibet signify disappearance of the vast oceanic lithosphere (Gansser 1980; Srikantia and Razdan 1980, 1981). Two main litho-tectonic associations characterize the ITSZ:

(i) Southward-obducted Spongtang Ophiolite Nappes over the Tethyan Himalayan Sequence (THS), defining the northern continental margin of the Indian Plate, and
(ii) Discontinuous steeply dipping ITSZ as the main suture zone (Gansser 1964; Honegger et al. 1982; Thakur 1993) containing (a) the Dras Volcanics (basalt, dacite, pillow lavas and cherts) defining the Upper Jurassic to Upper Cretaceous volcanic island arc, (b) dismembered ophiolite bodies and mélanges (Nidar ophiolite, Shergol ophiolite, Zildat ophiolite, Shigatse/Xigaze ophiolite, to name a few), (c) Lamayuru, Nindam and Indus Formations as flysch sediments, and (d) Indus molasse of essentially fluvial clastic sediments, mainly derived from the eroded sequences from nearby sequences (Fig. 7.2).

7.2.1.1 Spongtang Ophiolite Nappe

Numerous ophiolite klippen, known as the Kiogarh ophiolite nappe (Heim and Gansser 1939) in Uttarakhand, the Spongtang Nappe (Reuber et al. 1987) or the Shilakong ophiolite (Srikantia and Razdan 1981) in Zanskar, and the Karzok ophiolite of the Tso Morari (Berthelsen 1953), characterize southern margin of the ITSZ, where oceanic lithosphere is obducted onto the THS of the Indian platform margin (Heim and Gansser 1939; Gansser 1964). In Zanskar, gently dipping thrusts bind the Spongtang ophiolite nappe, making a NW–SE trending doubly plunging synform, and constrains the nature of oceanic lithosphere. This nappe is made up of Photang

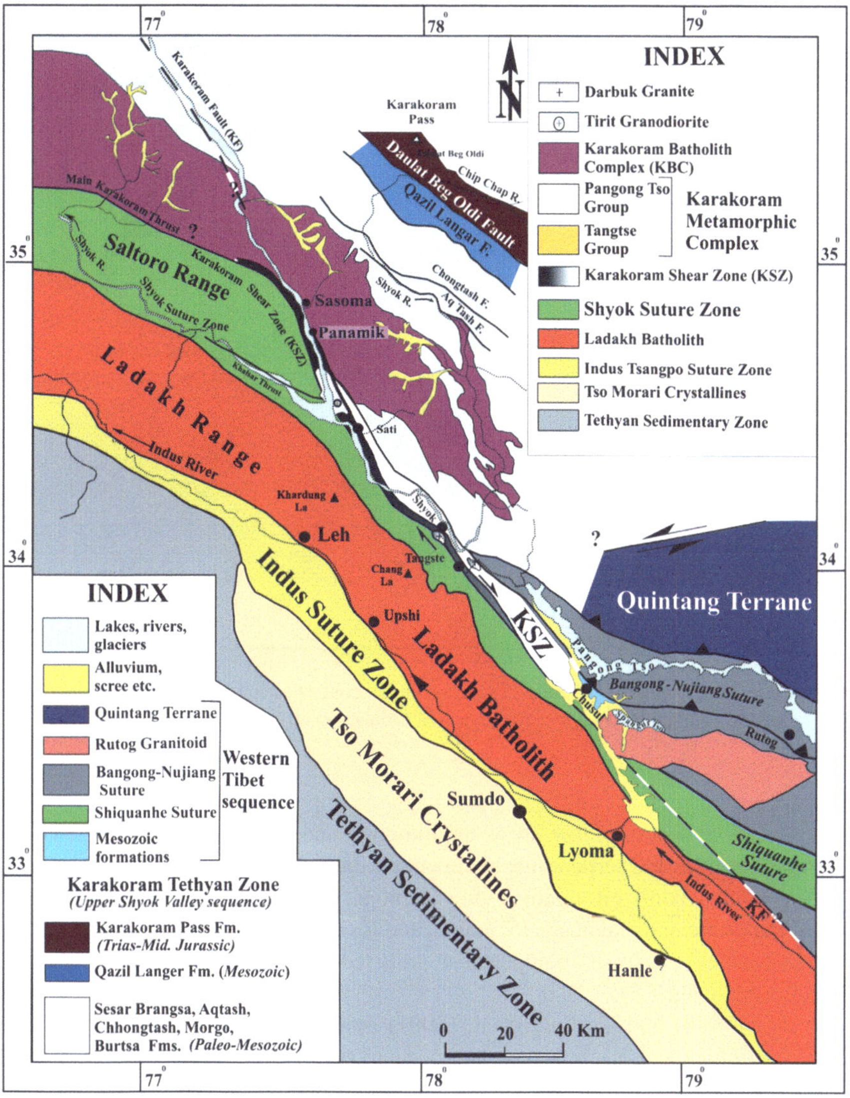

Fig. 7.2 Simplified geological map of northern margin of NW Himalaya, Trans-Himalaya and Karakoram showing extension of the Karakoram Batholith Complex into the Baltoro granite, and the Shyok Suture Zone into the Shiquanhe Suture Zone, and the location of the Karakoram Shear Zone. Approximate location of Karakoram Fault (?) is also shown for comparison with the KSZ. After Jain and Singh (2009)

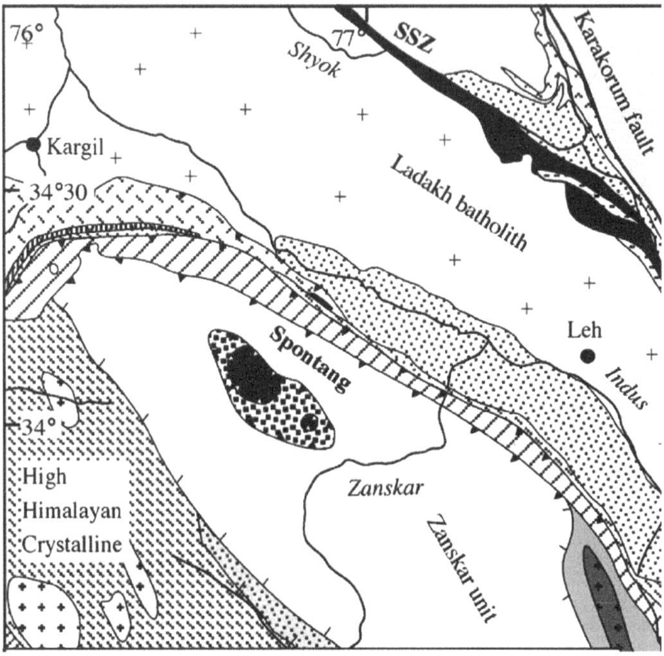

Fig. 7.3 Geological map of the Indus-Tsangpo Suture Zone showing location of the Spongtang klippe over the Tethyan Himalayan Sequence. After Maheo et al. (2004)

sub-nappe in lower parts and Spong sub-nappe in upper parts; all having thrust contacts. Basalt, andesite, pillow lavas, agglomerate and tuffaceous sediments, chert, limestone, arenite and serpentinite slivers mark the lower unit, while the upper sub-nappe contains mainly ultramafic bodies of antigorite, harzburgite, lherzolite, dunite with serpentinite and magnesite (Srikantia and Razdan 1981). These isolated eroded klippen constitute parts of an uppermost obducted ophiolite nappe in the Himalaya since Eocene (Fig. 7.3).

Reuber (1986) and Corfield et al. (2001) opined that eastern crustal unit has MORB affinities and originated at mid-oceanic ridge (also, Pedersen et al. 2001). This crust is thrust over a volcano-sedimentary sequence, which originated in an intra-oceanic supra-subduction zone (Corfield et al. 2001). U–Pb zircon dating of plagiogranite in crustal unit indicated an oceanic accretion at 177 Ma (Pedersen et al. 2001), while K–Ar hornblende ages of 170 and 125–140 Ma are interpreted as oceanic accretion and intra-oceanic thrusting (Reuber 1989). Similarly, the Spongtang diorites yielded $^{40}Ar/^{39}Ar$ hornblende age between 131 and 147, while another sample gives ages between 127 ± 3 and 138 ± 2 Ma indicating ca. 130 Ma age for post-magmatic oceanic hydrothermal alteration and minimum age for the initiation of the subduction magmatism (Maheo et al. 2004). Catlos et al. (2018) modelled origin of the Spongtang ophiolite along mid-ocean spreading center having spreading rate >2 cm/year in the Neo-Tethyan Ocean from Late Triassic to Jurassic, while ridge had increasing influence of subduction by Early Cretaceous; age of latest obduction occurred between 64.3 ± 0.8 and 42.4 ± 0.5 Ma.

7.2.1.2 Main Suture Zone

Almost continuous NW–SE trending ITSZ belt is comprised of the following litho-tectonic units (Table 7.1; see Figs. 6.2, 6.7 in Thakur 1993):

(a) **Dras Volcanics**: The Ladakh exhibits two ophiolite belts and associated mélanges: (i) a northern ophiolite belt in contact with the Ladakh Batholith and (ii) a southern Ladakh ophiolite belt. These belts are separated by an enormously wide exposure of the Dras Volcanics, containing a major association of volcanics, pyroclastics and volcaniclastic sediments with radiolarian chert for nearly 400 km. Along the northern contact, the sequence is obliquely thrust over the Kargil molasse as well as the Shergol ophiolite mélange, while the Ladakh Batholith makes the substratum for this formation with an intrusive contact. The southern contact is also tectonic with the overthrust Lamayuru Formation, the ophiolite belt and the Zanskar Supergroup sediments.

Reuber (1989) recognized two distinct lithological associations within the Dras sequence:

(i) Dras-I contains mainly basaltic, basaltic-andesite to dacite flows, volcaniclastics, black slates and *Orbitolina*-bearing limestone of Albian-Cenomanian age. Basaltic magma belongs to island arc tholeiites or calc-alkaline series (Honegger et al. 1982; Dietrich et al. 1983).
(ii) Dras-II contains tuffites, ash beds, and breccia; the sequence unconformably overlies the intruded Dras-I unit.

These rocks overlie the ocean floor sequence, which is made up of the ultramafics containing harzburgite and dunite, wehrlite and pyroxenite. Some dunites contain elongated chromite grains, indicating plastic deformation. Serpentinized peridotite and other ophiolitic facies are common as a substratum of the volcaniclastic series. Uppermost parts of ophiolite are characterized by bulbous pillow basalt (Fig. 7.4a), related to the subduction of the Indian Plate beneath the oceanic margin of the Asian Plate. It is developed as an island arc, characterized by calc-alkaline volcanic suites on oceanic crust (Dietrich et al. 1983; Reuber 1989). It is cross-cut by granodiorite plutons, dated between 103 and 70 Ma (Honegger et al. 1982). South of the Dras arc, the volcanics are replaced by the Nindam flysch, which corresponds to fore-arc area filled in with thick volcano-sedimentary deposits (Clift et al. 2000). The Nindam flysch locally overthrusts the Sapi–Shergol ophiolitic mélange and has been classically interpreted as an accretionary prism related to the Dras arc (Cannat and Mascle 1990).

(b) **Nidar Ophiolite**: Many ophiolite bodies have been observed within the ITSZ, but the most complete, undeformed and non-imbricated body in SE parts of the

Fig. 7.4 **a** Bulbous pillow lavas in the Dras Volcanics on the upper surface. **b** Typical flute marks in the Indus Group. Both indicate large-scale inversion of beds along Chiktan Nala, Ladakh

suture zone at Nidar represents the lithologies of the upper mantle to supra-ophiolitic sediments of the intra-oceanic volcanic arc affinity (Ahmad et al. 2008; Maheo et al. 2004). 3–8 km thick sequence includes chromite-bearing harzburgite, cumulate gabbro, basalt and sheeted dyke complex, pillow lava, thin shale and chert with the following lithological distribution (Satoru et al. 2001; Ahmed et al. 2008; Jain and Singh 2009; Das et al. 2015):

(i) Basal ultramafic unit contains spinel–harzburgite at the lower level, and spinel–dunite at the higher level with chromite veins and disseminations. Minor pyroxenites intrusives occur in ultramafics and the overlying gabbro.
(ii) Very coarse-grained gabbro to fine-grained microgabbro characterize the overlying sequence, with Sm–Nd WR-mineral age of 139.6 ± 32.2 Ma with an initial $^{143}Nd/^{144}Nd = 0.512835 \pm 0.000053$ or $\varepsilon_{Nd}(t) = +7.4$ (Satoru et al. 2001; Ahmad et al. 2008), like the Dras Volcanics in the west. This date lies within its biostratigraphically-determined age within error limits. For the Nidar gabbro, initial $^{87}Sr/^{86}Sr$ ratios varies from 0.703 to 0.706 and $\varepsilon Nd_t = 140$ Ma from +6.22 to +9.84, like the overlying Nidar basic and acidic rocks (+7.61 to +8.03); these values are indicative of their derivation from depleted mantle. Younger $^{40}Ar/^{39}Ar$ hornblende ages of 110 and 122 Ma may reflect either magmatic or post-metamorphic cooling (Maheo et al. 2004).
(iii) Overlying volcano-sedimentary unit comprises porphyritic basaltic–andesitic flows, acidic volcanics with preserved pillow structures; the sequence gradually passes into volcanic-derived intercalated sediments with chert, jasperite, shale, siltstone, volcanic sandstone, and conglomerate (Ahmad et al. 2008). Acid volcanics predominate at upper levels and are intercalated with pyroclastic deposits having ash flows, tuffs, lapillistone and volcanic breccia, having radiolaria of Hauterinian to Aptian (132 ± 2 to 127 ± 1.6 Ma–Early Cretaceous) age (Satoru et al. 2001; Kojima et al. 2001). High-resolution radiolarian

biostratigraphy indicated an age of an upper Barremian–upper Aptian (126–114 Ma) range for the Nidar section (Zyabrev et al. 2008), thus extrusive volcanism began by the early Aptian (ca. 124 Ma) and date the upper part of the ophiolite succession.

Geochemically, the Spongtang diorites possess limited composition range (SiO_2 between 50 and 56%), with Zr/TiO_2 ratios classifying them as sub-alkaline basalt to andesite; low TiO_2, P_2O_5 and Nb/Y ratio are characteristics of tholeiitic basalts (Maheo et al. 2004). The Nidar basalts and microgabbros have same geochemical characters as the Spongtang diorites and are, thus, tholeiitic basalt to andesites. The Karzok gabbros have low SiO_2 (42–45%), high TiO_2 (1–2%) and high CaO (11–12%) and reflect secondary enrichment during alteration. Mafic bodies from these ophiolites have common characters in discrimination diagrams where immobile elements like V/Ti, La–Nb and Th/Yb–Ta/Yb indicate tholeiite arc affinity.

REE patterns of the Spongtang diorites, Karzok gabbros, Nidar andesites and microgabbros are similar, indicating extraction of magma from the same source (Maheo et al. 2004). They are slightly depleted in LREE relative to HREE with (La/Yb) ratios between 0.4 and 0.78 and bear strong resemblance to those in N-MORBs. In some diorites and gabbros, positive Eu anomaly reflects an accumulation of plagioclase. The gabbros from Karzok and microgabbro from Nidar correspond to the least and the most differentiated magmas, respectively.

Das et al. (2017) reported in situ octahedral diamonds, their graphite pseudomorphs, hydrocarbon (C–H) and hydrogen (H_2) fluid inclusions in ultrahigh-pressure (UHP) peridotitic minerals of the Nidar ophiolite belt, indicating their sources from mantle transition zone or base of the upper mantle during mantle upwelling beneath the Neo-Tethys Ocean spreading centre.

The Nidar Ophiolite is thrust southward over the Zildat ophiolitic mélange. The available age estimation for the emplacement of the Nidar Ophiolite is Cretaceous to Eocene, based on radiolarites (Thakur and Virdi 1979).

(c) **Nindam Formation**: A volcano-sedimentary succession between the Shergol ophiolite belt was considered as lateral facies of the Dras Volcanics (Frank et al. 1977) and called as the Nindam Formation (Bassoullet et al. 1982). It consists of alternating sandstone, siltstone and shale with bedded tuffs of about 3000 m thickness (Thakur and Misra 1984). These were derived from a mafic to intermediate volcanic terrain like the Dras during Cretaceous and deposited as distal to proximal turbidites and beds of channel origin (Brookfield and Andrew-Speed 1984).

(d) **Lamayuru Unit**: Sandwiched between the Zanskar Supergroup (the Tethyan Sedimentary Zone) in the south and the Shergol ophiolite belt in the north, another 3000 m thick Lamayuru and Karamba units were deposited between Permian to Early Cretaceous on the Indian continental slope (Bassoullet et al. 1981; Robertson 2000). The units are comprised of shale, siltstone and graded sandstone, and incorporate large 'exotic' limestone block. Robertson and Degnan (1993) opined these as the remnants of the Triassic to Upper Cretaceous north-facing Indian passive margin that experienced pulsative extension and transtension since Late Permian, while ophiolitic mélanges were accreted at trench within the Tethys Ocean to the north, only during latest Cretaceous.

(e) **Indus Group**: Lying between the Ladakh Batholith in the north and the Shergol/Nidar ophiolite/Tso Morari Dome in the south, approximately 5 km-thick Indus Group trends almost NW–SE within the ITSZ. It consists mainly of distinct rhythmically alternating conglomerate, sandstone, siltstone and shale of green, grey and purple colours; sandstone develops well-preserved sole marks like flute marks on inverted beds showing northerly-flowing paleocurrents (Fig. 7.4b). A steep south-dipping thrust separates this formation from the underlying Kargil Formation, thus making it para-authochthonous in disposition (Thakur 1981). Brookfield and Andrew-Speed (1984) divided this unit as the Khalsi flysch and the Miru unit, with limestone intercalations (Khalsi limestone) yielding Aptian to lower Albian fauna; the flysch represents a deep-sea fan to basin plain deposit.

Within the ITS zone the Indus Group sediments represent the final phase of deposition in an actively subsiding Indus Basin, which represents a narrow embayment occupied by a relict shallow tidal sea with prominent energy (Singh et al. 2015). It received sediments both from the Ladakh Batholith and Karakoram block in the north as well as the Indian continental margin and Suture zone deposits in the south by different drainage systems. In some parts, large drainages met the sea, making huge braid plain extending into the sea along with some small drainage. Low-energy areas of tidal flat complex developed network of meandering creeks and channels, while sand bars and shoals developed depositing sand in the high energy areas (Fig. 7.5). During its deposition, numerous tectonic pulses contributed coarse conglomeratic alluvial fans built braided plains extending into the basin. Westwards this basin was open and connected to a diminishing Neo-Tethys Ocean. Most of the fossil records in The Indus Group contains freshwater plants, and fauna of 41–20 Ma; the latter time period represents inversion of basin and deposition over the Ladakh Batholith around 20 Ma (Singh et al. 2015).

(f) **Kargil Formation**: A long linear discontinuous sedimentary succession, called as the Kargil Formation, overlies non-conformably the Ladakh Batholith, and is also known as the Kargil molasse, the Liyan Formation, the Rumbok, Zinchon, Nimu, Hemis, the Upshi and Wakha Chu molasses (Brookfield and Andrew-Speed 1984; Garzanti and Van Haver 1988). Approximately 1700 m thick along Wakha Chu and Suru River is divided into the Kargil, Tharumsa and Pashkyum formations in the ascending order (Kumar and Virdi 1997). The oldest formation contains conglomerate, green sandstone and purple mudstone, while the overlying Tharumsa contains multi-storey grey sandstone and mudstone; youngest formation has purple and green sandstone and mudstone. The sequence was deposited during Oligocene to Pliocene by alluvial fans to braided streams between rising units of the ITSZ and the Ladakh Batholith.

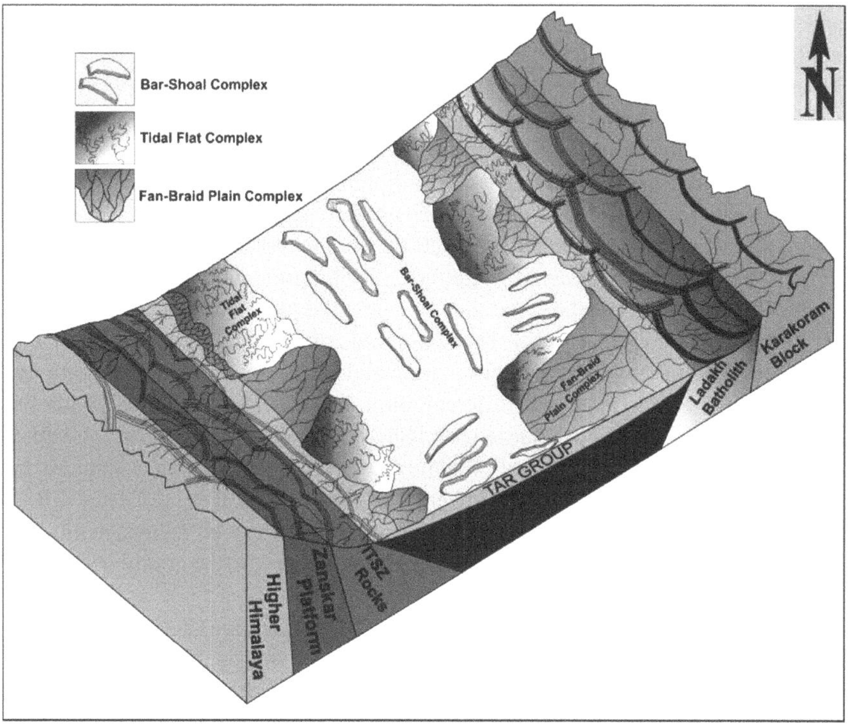

Fig. 7.5 Schematic depositional basin for the Indus Group sediments as a narrow tidal sea with depositional areas of tidal flat and bar-shoal complexes. In point source areas large braid plains were developed where deposition of conglomerate-sandstone took place. Northern margin received sediments from the Ladakh Batholith and Karakoram, while southern region had sediments derived from ITSZ rocks, Zanskar platform and High Himalaya. After Singh et al. (2015)

7.2.2 *Ladakh Batholith*

7.2.2.1 General Characters

The Trans-Himalayan batholiths, located immediately to the north of the ITSZ for almost entire length of the Himalaya, represent an Andean-type calc-alkaline magmatism due to northward subduction of the Neo-Tethyan oceanic crust beneath the Shyok–Dras Island Arc during early Cretaceous–Lower Eocene (Honegger et al. 1982; Schärer et al. 1984; Jain et al. 2002; Bouilhol et al. 2013). This belt is almost continuously developed from Astor–Deosai–Skardu region in Pakistan (Kohistan–Deosai Batholith) to Kargil–Leh–Demchuk region in India (Ladakh Batholith). It extends further southeast as the Kailash tonalite, the Gangdese pluton in southern Tibet (Scharer et al. 1984) and the Lohit Batholith in Arunachal Himalaya (Thakur and Jain 1975). Regionally, the ITSZ demarcates its southern margin , while the Main

Karakoram Thrust (MKT) and the Shyok Suture Zone (SSZ) delimits its northern boundary in Pakistan and India (Rolland et al. 2000; Jain et al. 2008; Jain and Singh 2009).

The Ladakh Batholith is a NW–SE trending 600 km long and 30–80 km wide magmatic belt with an exposed thickness of ~3 km (Honegger et al. 1982). The Ladakh Batholith is partly covered by north-dipping fore-arc and molassic sedimentary sequences, which were derived from the uplifted and eroded batholith (Garzanti and van Haver 1988; Sinclair and Jaffey 2001; Kumar et al. 2018).

The igneous complex consists of gabbro, diorite, granodiorite and granite, and intrudes the volcanics of island arc assemblages, like the Dras volcanics (Sharma and Gupta 1983; also Singh 1993). They observed zoned nature of the batholith with more mafic lithologies on its southern margin and acidic varieties further north and attributed this zoning to magmatic differentiation. The northern slope of this batholith is exposed in contact with the Khardung and Shyok Volcanics along the Shyok Valley. East of Leh, it contains about 10–20% mafic intrusives, 20–30% intermediate rocks, about 50% granodiorite and 10% biotite granite (Honegger et al. 1982). They opined that the Ladakh Batholith and similar Kailash and Lhasa Trans-Himalayan batholiths belong to the calc-alkaline series, and are clearly comparable to the granitic batholiths occurring along accretionary continental margins.

Weinberg (1996) mapped many varieties of diorite, granodiorite and granite between Leh and Khardung La, and observed magma mixing enclaves, when granitic magma intruded and disrupted the partially molten quartz-diorite, now occurring as fine-grained pillow-like enclaves (also Kumar et al. 2016). These enclaves are made up of numerous shapes with long axis parallel to the foliation, while morphology of these nearly in situ mafic inclusions was used in deciphering the relative flow directions between granite and enclaves.

A very large part of the LB has escaped deformation, but intense penetrative ductile shearing is observed north of Leh within NNW-trending dextral Thanglagso Shear Zone (Weinberg and Dunlap 2000). Along the Kharu-Chang La and Leh-Khardung La sections, it is intensely mylonitized, having 5–100 cm long, almost N90° to N120° trending ellipsoidal flattened xenoliths (Fig. 7.6a; Jain et al. 2003). Within the ductile shear zones, plagioclase and hornblende had undergone preferred orientation along NE-dipping main foliation. Localized NW-trending brittle-ductile to brittle thrusts exhibit top-to-southwest sense of imbricated shearing (Fig. 7.6b; Jain et al. 2003). Large apophyses of the batholith indicate intrusive relationships with the Shyok Suture Zone, including the Shyok Volcanics (Fig. 7.6c).

Honegger et al. (1982) noted complete differentiation series in the AFM diagram from hornblende-biotite diorite to biotite granite with negative correlation of MgO, FeO, CaO and TiO_2 with SiO_2 due to separation of plagioclase and mafic phases from the parent melt, and positive correlation with K_2O and Rb, Ba, Y and Zr from granodiorite to granite. Enrichment of Ba, Sr and Zr in late aplite and granite bodies is seen. REE pattern show negative slopes and strong LREE enrichment and HREE depletion; the LB was evolved from the partial melting of the oceanic lithosphere (Honegger et al. 1982; Kumar 2005).

Fig. 7.6 **a** Preferably oriented mafic enclaves along main foliation within the Ladakh Batholith, affected by extensional brittle-ductile shear zone. Scale-Pen 14 cm. Loc.: About 15 km from Leh on Leh-Khardung Road. **b** Brittle and listric south-verging thrusts with fault gouge in the Ladakh Batholith. Loc. About 20 km from Leh on the Leh-Khardung road. **c** Apophasis of the Ladakh Batholith within the Shyok Volcanics near Hunder in the Shyok Valley

The Ladakh Batholith is characterized by intrusive large apophyses with the Dras–Shyok Volcanic Island Arc and contains numerous mafic volcanic enclaves. Along the Indus River, north-dipping fore-arc and continental molassic sedimentary rocks cover its southern margin partially and are eroded material from this uplifted magmatic arc with subordinate components from the passive Indian margin sedimentary succession (Sinclair and Jaffey 2001; Wu et al. 2007; Henderson et al. 2011). Locally, andesitic to rhyolite volcanics along with volcaniclastic sediments nonconformably overlie the Ladakh Batholith around Khardung near its northern margin with the SSZ (Weinberg and Dunlap 2000).

7.2.2.2 Geochronology—Age of Crystallization

Pulsative crystallization and emplacement of the LB occurred in multiple stages at ~100, 72, 67, 58, 51, 41 Ma and younger times, possibly by melting of the earlier phases (Honegger et al. 1982; Scharer et al. 1984; Weinberg and Dunlap 2000; Singh et al. 2007; Upadhyay et al. 2008; St-Onge et al. 2010; White et al. 2011; Bouilhol et al. 2013; Kumar et al. 2017) (Fig. 7.7). The oldest Early Cretaceous magmatism within the LB was around Kargil in its western parts, though Kumar et al. (2017)

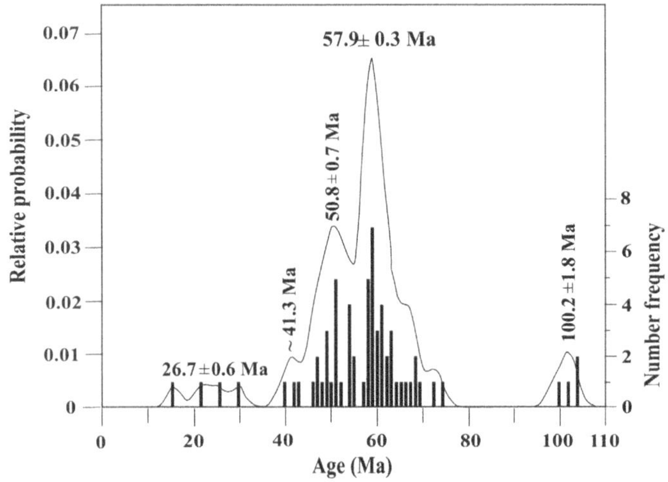

Fig. 7.7 Published geochronological data from the Ladakh batholith. U–Pb zircon ages (n = 62), U–Pb allanite and monazite (n = 1) and Rb–Sr whole-rock isochron ages (n = 2). *Data source* Jain (2014)

interpreted Early Cretaceous zircon ages around Tirit as a part of the LB rather than the Karakoram. Weinberg and Dunlap (2000) obtained SHRIMP U–Pb zircon ages of 58.1 ± 1.6 Ma from its core and relatively younger age of 49.8 ± 0.8 Ma from the rims near Leh, and interpreted these as ages of crystallization of the source igneous rocks and a later magmatic phase, respectively. Zircon crystallization ages are 58.4 ± 1.0 and 60.1 ± 0.9 Ma from the southern and northern margins of this batholith, respectively (Singh et al. 2007), while these are ~68 to 66 Ma in extreme north at Hunder along the Shyok valley (Weinberg et al. 2000; Upadhyay et al. 2008; White et al. 2011; Bouilhol et al. 2013; Kumar et al. 2017). Some younger components within the Ladakh Batholith date between 53.4 ± 1.8 and 45.27 ± 0.56 Ma along the southern margin. Relatively low initial $^{87}Sr/^{86}Sr$ (0.704 ± 0.001) and high $^{143}Nd/^{144}Nd$ ratios (0.5126) support the derivation of its magma from partial melting of subducting oceanic slab (Honegger et al. 1982).

A very detailed SHRIMP U–Pb zircon dating of the central segment of the LB revealed that the axis of the batholith has multiple zircon growths between 58 ± 0.8, 56.6 ± 0.9 and 54.8 ± 1.9 Ma between Khardung La and Chang La (Weinberg et al. 2000; Upadhyay et al. 2008; White et al. 2011; Bouilhol et al. 2013; Kumar et al. 2017), followed by multiple rim growths during 62.2 and 48.8 Ma, and another phase at around 15.6 ± 1.0 Ma. The microgranite enclaves at Chang La date almost the same as the main body. Further east, the batholith at Chumathang has zircons of 59.2 ± 0.7 Ma with rim growth at 50.4 ± 2.7, while younger intrusions date around 48.2 ± 0.8 Ma (St-Onge et al. 2010; White et al. 2011).

The southernmost parts of the Kohistan-Ladakh-Paleo-Island Arc (KLA) in immediate contact with the ITSZ between Deosai and Ladakh have yielded in situ U–Pb zircon ages between 102.1 ± 1.2 to 50.3 ± 1.2 Ma (1σ) with a homogeneous zircon age population, average 9.4 ± 0.7 $_{\varepsilon}Hf_{(i)}$, weighted mean 2.6 ± 0.7 $_{\varepsilon}Nd_{(i)}$ and 0.703744–0.704719 $^{87}Sr/^{86}Sr_{(i)}$ (Bouilhol et al. 2013). In contrast, post-50 Ma samples from the southern margin are younger to 50.4 ± 1.6 Ma (youngest being 29.6 ± 0.8 in Ladakh), and yield $_{\varepsilon}Hf_{(i)}$ as low as −15, $_{\varepsilon}Nd_{(i)}$ between −9.7 and −3.6 and $^{87}Sr/^{86}Sr_{(i)}$ ranging between 0.705862 and 0.713170 (Bouilhol et al. 2013). These ages along with a change in isotopic characters were interpreted because of the presence of a pre-50.0 Ma intra-Tethyan oceanic arc, where a mixture of components was derived from Enriched-Depleted MORB Mantle (E-DMM) and subducted oceanic lithosphere. The post-50 Ma samples exhibit strong isotopic variability and involvement of additional enriched components, e.g., isotopically evolved crust. A shift in isotopic compositions of the KLA rocks at 50.2 ± 1.5 Ma reflects the India-KLA collision, according to these authors.

Sixty-five published dates from the Trans-Himalayan Ladakh Batholith (Deosai segment included but excluding the Kohistan segment) incorporate sixty-two U–Pb multi-grain and individual robust U–Pb zircon ages, one U–Pb allanite and monazite age and two Rb–Sr whole-rock isochron ages. These zircon ages were determined by TIMS, SHRIMP and LA-MC-ICPMS, indicating a peak of 57.9 ± 0.3 Ma from the LB with minor peaks at 50.8 and 41.3 Ma (Fig. 7.7; Jain 2014). When extreme ages of some samples (~101, 15.6–30 Ma) are excluded, one obtains the oldest peak value at 72.3 ± 0.5 Ma, followed by other peaks at 67.6 ± 1.0 Ma, 50.9 ± 0.7 Ma and 40.8 ± 0.5 Ma, besides the main peak value of 57.9 ± 0.3 Ma. Following the closure of the Neo-Tethys, major pulse of uplift and erosion had occurred in Ladakh. U–Pb SHRIMP-II zircon dating from two widely-located samples along the Kharu-Chang La section provide undisputed evidences for almost simultaneous emplacement and crystallization of diorite and granodiorite, as the former gave mean $^{206}Pb/^{238}U$ age of 58.4 ± 1.0 Ma, and the granodiorite was dated as 60.1 ± 0.9 Ma (Singh et al. 2007).

There had been debate on the country rocks being intruded by the Ladakh Batholith, i.e. the continental crust or the Cretaceous Shyok–Dras Volcanic Arc (see Rolland et al. 2002, and references therein). Presence of numerous mafic enclaves within the batholiths in the upper parts containing vesicles and its intrusive relation with the Chang La ultramafics possibly indicate that the Ladakh Batholith has intruded an extensive terrain of the Dras-Shyok island arc that had an attached oceanic lithosphere to it on the northern margin (Rolland et al. 2002).

7.2.2.3 Age of Dyke Formation

Weinberg and Dunlap (2000) observed N50° to 70° trending vertical dyke swarm near Phyang around Leh and determined its $^{40}Ar/^{39}Ar$ crystallization age as 45.7 ± 0.8 Ma from hornblende phenocrysts.

7.2.2.4 Age of Deformation

Most of the Ladakh Batholith remained undeformed, though intense penetrative deformation was observed within a ductile shear zone in the Thanglagso valley on northern slopes of the Ladakh Range due to later Himalayan tectonics (Weinberg and Dunlap 2000). Nearly vertically dipping and N25° W trending shear zone is about 2 km wide and parallels the Karakoram Shear Zone with right-lateral S-C shear fabric and asymmetrical feldspar augen. K-feldspar grains from an augen gneiss yielded apparent $^{40}Ar/^{39}Ar$ ages of from 52 to 22 Ma. As the Ladakh Batholith crystallized at ~60 Ma, a rapid cooling is inferred until 52 Ma from ca. 750–300 °C at about 55 °C/m year, and then at a rate of 2 °C/m year between 52 and 22 Ma, when the cooling temperature of 260 ± 20 °C was attained. Weinberg and Dunlap (2000) concluded that the dextral shearing along this zone occurred before the closure of K-feldspar grains at ca. 22 Ma.

7.2.2.5 Exhumation Patterns

After the crystallization of the Ladakh Batholith at ~58.0 Ma, it underwent normal magmatic cooling till 45–46 Ma, as suggested by Rb–Sr and K–Ar biotite cooling ages (Kumar et al. 2017). Very fast Early–Middle Eocene exhumation of the batholith is revealed by Rb–Sr biotite and zircon fission-track ages (Fig. 7.8). Exhumation peaked at 3.5 ± 0.9 mm/a between 50–45 Ma ($^{40}Ar/^{39}Ar$ hornblende ages) and 48–45 Ma (Rb–Sr biotite ages) because of the India–Asia convergence. It was followed by deceleration at a rate of 1.2 ± 0.4 mm/a until 43–42 Ma (zircon FT ages), like the Deosai batholith in the west. Exhumation rates finally decreased

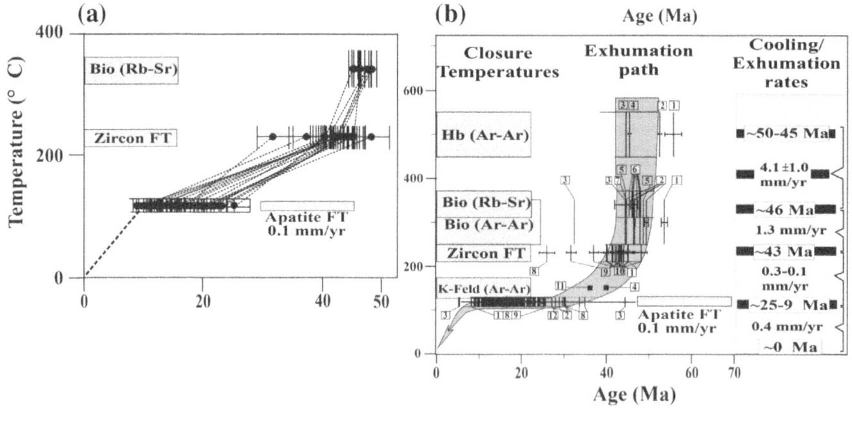

Fig. 7.8 Exhumation paths of the Ladakh Batholith. **a** New Rb–Sr biotite and fission track zircon, and apatite data. Tie lines for co-existing minerals in same samples. **b** Available published ages on hornblende K–Ar/$^{40}Ar/^{39}Ar$, biotite Rb–Sr, biotite K–Ar/$^{40}Ar/^{39}Ar$, zircon FT, K-feldspar $^{40}Ar/^{39}Ar$ and apatite FT. See Kumar et al. (2017) for data source

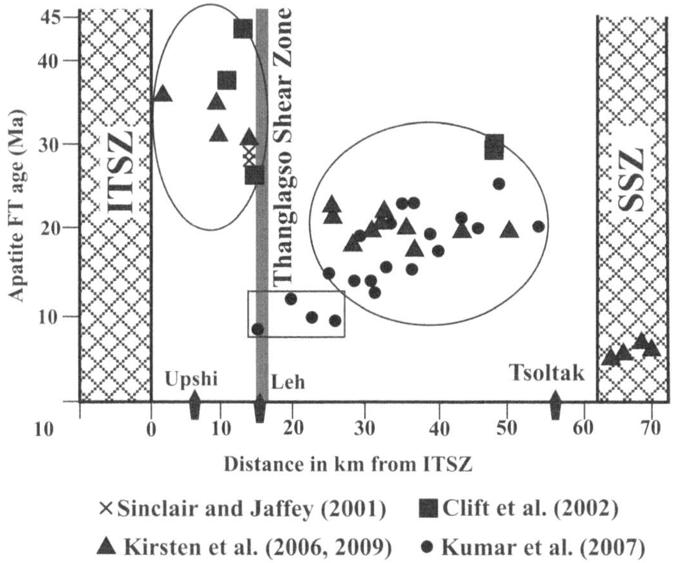

Fig. 7.9 Distance versus age plot of AFT ages from the Ladakh Batholith along Upshi-Tsoltak section paralleling the Kharu-Chang La road. All ages from the Leh-Khardung La section and adjoining regions projected along this section. Note four distinct age groups and role of the TSZ and SSZ in distribution of the AFT ages. See Kumar et al. (2017) for data source

during Oligocene to a minimum of ~0.1 mm/a before a mild late Miocene–Holocene acceleration (Fig. 7.8). A consistent pattern is observed in the zircon fission-track ages, but the LB underwent perturbations during lowering of its temperature. These apatite FT ages distribution have breaks between 44 and 5 Ma due to (i) ITSZ and Thanglasgo SZ in the southernmost parts, (ii) very young apatite FT within the Thanglasgo SZ, (iii) typical Miocene ages in the central parts, and (iv) late Miocene reactivation along the Karakoram SZ (KSZ) in the north (Fig. 7.9). Lower-Middle Eocene exhumation of the LB was tectonically controlled by slab break-off of the Neo-Tethys oceanic lithosphere and underthrusting of the Himalayan Metamorphic Belt (Jain 2014; Kumar et al. 2017).

7.2.3 Shyok Suture Zone

The Shyok Suture Zone demarcates an important subduction phase of the Indian Plate beneath the Paleozoic-Mesozoic platform sediments of the Asian Plate (Srimal 1986). It marks either the site of subduction of a wide Tethyan ocean or represents an Early Cretaceous intra-continental marginal basin along the southern margin of Asia. The Ladakh Batholith possibly intruded the interior of the Dras Island Arc in Late Cretaceous to Eocene (Weinberg et al. 2000), which is demarcated by another suture zone in the northeast—the Shyok Suture Zone (SSZ). This suture zone was first recognized by Gansser (1977), based on earlier works and satellite imageries. Though there had been controversies regarding the presence of the northern suture (Rai 1982; Srimal 1986; Thakur and Misra 1984), it appears that this suture zone was made up of dismembered bodies of ultramafics, gabbro, basalt and sediments with tectonic mélanges during Cretaceous (Matte et al. 1996). The SSZ extends to the NW beyond the Nanga Parbat spur, and has been grouped as the northern part of the Kohistan-Ladakh Arc as a reactivated steep late Main Karakoram Thrust-MKT (Rolland et al. 2000 and references therein). In the southeast, it terminates against the ITSZ in southwestern Tibet.

The Shyok Suture Zone reveals tectonic stacking of the Cretaceous volcano-sedimentary sequences from different volcanic arc and back-arc environments (Rolland et al. 2000). In the western part (Skardu area), the SSZ can be subdivided into two groups: (i) Northern Group of olistolith basaltic blocks and tuffs, where basalts are LREE depleted with a LILE enrichment and a slight Nb depletion suggesting back-arc origin, and (ii) Southern Group of predominantly of andesites with LREE enrichment, a flat HREE pattern, strong Nb–Ta depletion, and LILE enrichment, having island-arc tholeiite (IAT) to calc-alkaline affinities.

7.2.3.1 Regional Geological Setting

Unlike well-defined and better-investigated ITSZ, the SSZ is poorly known due to inadequate geological information, poor accessibility and thick alluvial cover along the Shyok and Nubra Valleys. To the north of the Ladakh Range in the Shyok sector, Dunlap and Wysoczanski (2002) have grouped the Shyok Volcanics and interstratified sedimentary sequence (conglomerate-sandstone-chert) in this belt over the Khardung Volcanics (Fig. 7.10), while other workers consider these volcanics as a part of the Ladakh Batholith (Rai 1982; Srimal 1986; Thakur 1993).

7.2.3.2 Shyok–Saltoro–Nubra Sector

On either side of the Saltoro Range, westernly-flowing Shyok River and the southeasterly-flowing Nubra River drain the region of the Shyok Suture Zone (SSZ). This region is occupied by five important sequences, namely the Khalsar Formation, the Shyok Volcanics, the Hundri Formation, the Saltoro Volcanics, molasses and ophiolites, and the Nubra Formation (Bhandari et al. 1979; Thakur et al. 1981; Rai 1982, 1983; Srimal 1986; Weinberg et al. 2000; Dunlap and Wysoczanski 2002; Juyal 2006). All these formations occur as melanges of tectonic slabs and slices without normal stratigraphic contacts.

An enormously thick sequence of acidic Khardung Volcanics and associated sediments overlies the Ladakh Batholith. SHRIMP U-Pb zircon dates from rhyolite and

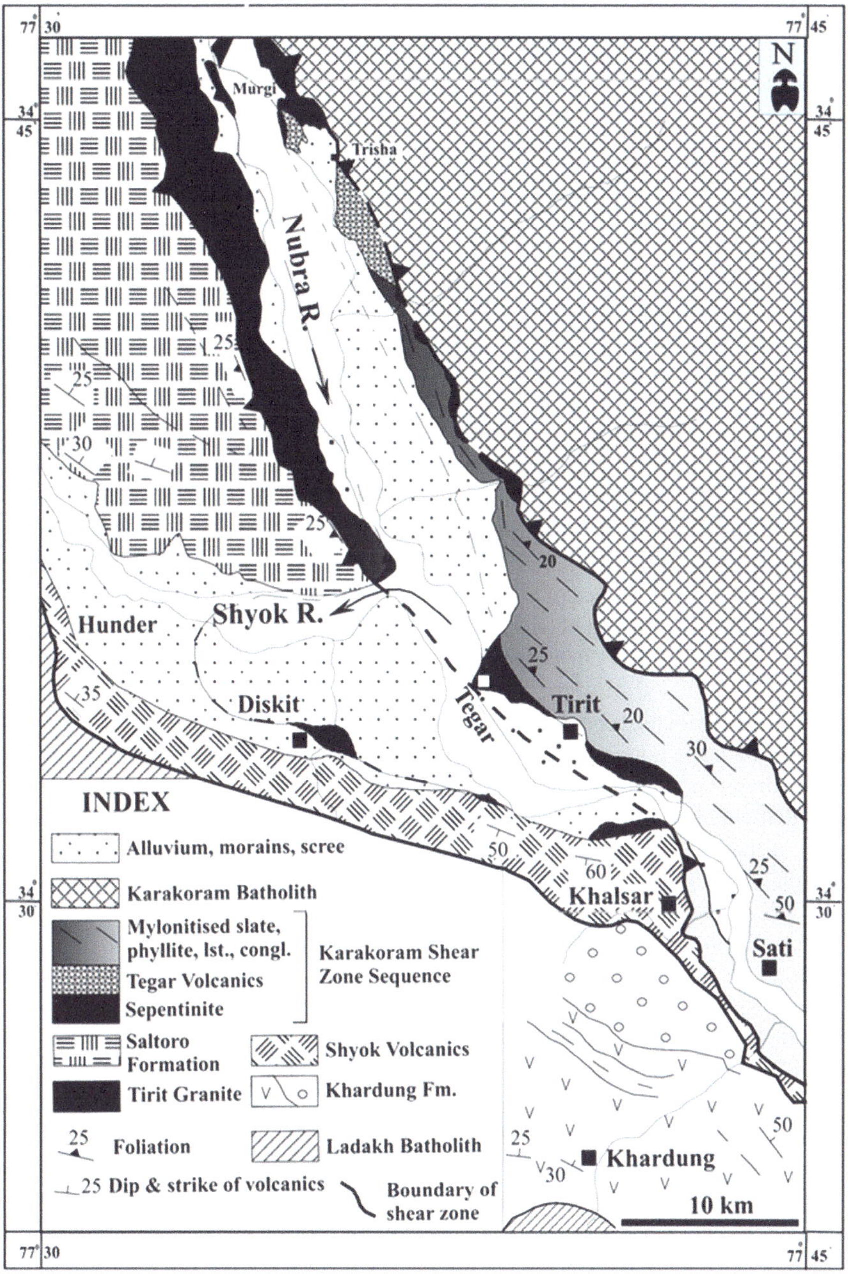

Fig. 7.10 Simplified geological map of the Nubra-Shyok Valley junction around Tirit. Modified after Jain and Singh (2009)

feeding dyke indicated flows between 67.4 ± 1.1 and 60.5 ± 1.3 Ma, respectively, thus limiting age of emplacement of batholith as Late Cretaceous–Paleogene (Dunlap and Wysoczanski 2002).

Southernmost Khalsar Formation contains calcareous phyllite, chlorite and mica schist with limestone and quartzite bands, and is tectonically separated from the underlying Khardung Volcanics by north-hading Khalsar Thrust (Thakur 1993). It is overlain by continuous enormously thick basalt and andesite of the Shyok Volcanics between Diskit, Hunder and further west along the Shyok River.

Most inaccessible and least explored parts of the SSZ of the Saltoro Mountains is exposed between Y-junction of the Shyok and Nubra Valleys. Along the southern slopes of this range and lying above the Shyok Volcanics with a thrust contact, a succession of slate, phyllite, limestone, calc schist, marls and chlorite schist is exposed that has been variously called as the Saltoro Flysch (Rai 1983), Hundri Formation (Thakur 1993), Diskit Formation (Razdan and Raina 1996), or Saltoro Formation (Upadhyay et al. 1999) of Upper Cretaceous–Lower Eocene age. Predominantly pelitic and low grade metamorphosed Hundri Formation consists of slate, phyllite, quartzite and limestone; the latter contains Upper Cretaceous–Lower Eocene foraminifera fauna (Rai 1982, 1983; Thakur 1993). This sequence is thrust over by a 3000 m-thick coarse clastics of the Saltoro molasse, which is made up of conglomerate, sandstone and variegated shale (Rai 1983). Pebbles of limestone yielded Lower Eocene *Nummulites* sp., *Orbitolina* sp. and clasts of the acid Khardung Volcanics, now dated as ~67–60 Ma (cf., Dunlap and Wysoczanski 2002), while Rai (1983) opined it of post–Oligocene age. Contemporaneous volcanism is represented by a persistent andesite horizon within the molasse. This molasse sequence has youngest detrital zircons population of approximately 92 Ma and a dike of 85 Ma (U/Pb zircon ages) that cuts basal molasse outcrops, implying that deposition of the succession began in the Late Cretaceous (Borneman et al. 2015). This minimum age for the SSZ rules out any possibility of an Eocene collision between Kohistan-Ladakh and Asia. These results support correlation of the SSZ with the Bangong suture zone in Tibet, which implies a total offset across the Karakoram Fault of approximately 130–190 km.

In the Saltoro Hills, the Shyok Volcanics occupy the core of a major synform and is thrust southward over the Saltoro Formation along a thrust. Large track of the Saltoro Range contains an ophiolite complex of low-grade metamorphosed ultramafics, cumulate to non-cumulate gabbros, diabase, pillow and massive basalt and chert as the Saltoro ophiolite of the North Saltoro belt (Srimal 1986), possibly as the repetition/correlation of more southerly Biagdang belt (also Borneman et al. 2015).

Along the Nubra Valley and to the northeast of the Saltoro Range, a succession of volcanics, shale, limestone, conglomerate, slate and imbricated thin serpentinite and peridotite bodies is exposed beneath the overthrust Karakoram Batholith Complex. This long and linear belt has been grouped into the Nubra ophiolitic mélange/Nubra Formation (cf., Thakur 1993; Weinberg et al. 2000). The Tirit Granite (75–68 Ma) intrudes the sequence. Extensive outcrops of these volcanics are seen in the upper reaches of the Nubra Valley, where these override the lithologies of the Karakoram Shear Zone mylonite along a southerly and moderately hading thrust.

7.2.3.3 Chang La Sector

In the Chang La sector of the SSZ, enormously thick imbricated ultramafics, gabbro, volcanics, conglomerate, sandstone and chert are exposed between Tsoltak and Darbuk, and represents the most accessible cross-section through this suture (Rolland et al. 2000). Clastic rocks with limestone alternate with siliceous tuff, tuffaceous mudstone, pyroclastics and basalt between Chang La and Tangtse. In the lower parts, limestone intercalations yielded early to middle Albian foraminifers like *Mesorbitolina minuta* (Douglass), *M. texana* (Roemer), *Simplorbitolina* sp. (Matsumaru et al. 2006). Ehiro et al. (2007) separated mudstone-dominated facies from the Shyok Formations as a new stratigraphic unit–the Tsoltak Formation with Callovian (late Middle Jurassic) ammonoids (*Macrocephalites*, *Jeanneticeras)* in the lower parts.

On the turn and along the left flank of the Tangtse Valley, an interesting long linear diorite-granodiorite belt is exposed and intensely sheared within the Karakoram Shear Zone (KSZ). This belt extends towards northwest intermittently along the Shyok Valley, where the Tirit granodiorite intrudes the slate-phyllite sequence of the Karakoram Shear Zone around Tirit–Diskit–Khalsar junction of Nubra–Shyok Rivers.

7.2.3.4 Chusul–Dungti Sector

In the Chusul sector, the SSZ (called as the Chusul tectonic zone in the southeast—Srikantia et al. 1982), comprises Lower to Middle Cretaceous andesite, dacite and interbedded cherts with *Orbitolina*-bearing limestone of the Luzarmu Formation, nonconformably overlying the Ladakh Batholith between Chusul–Dungti–Fukche (Thakur and Misra 1984). It is overlain by thickly bedded limestone, quartzite, sandstone, conglomerate and interbedded volcanics of the Diong Formation (= Hundri Formation). Ophiolite succession of pyroxenite, peridotite, gabbro and volcanics-chert follow west of Koyul as a part of the SSZ, which is unconformably overlain by the Kole molassic sediments (see Fig. 6.7 in Thakur 1993).

7.2.4 Karakoram Mountains: Southern Margin of the Asian Plate

The vast Tethyan Ocean occupied the Himalayan–Karakoram–Tibetan region during the Paleo-Mesozoic. It was only near the end of the Mesozoic that the Himalayan–Karakoram system evolved sequentially by southward propagating Indo–Asian convergence tectonics. The Tibetan Plateau contains remnants of older oceanic crust along major sutures zones e.g., the Bangong–Nujiang and Shiquanhe Suture Zones, which record large-scale lithospheric contraction from Permian to Cretaceous (Allegre et al. 1984; Matte et al. 1996; Murphy et al. 2002; Kapp et al. 2003). In addition, large-scale major strike-slip fault system characterizes this plateau in contrast to the thrust faults of the Himalayan convergence zone along its southern margin even though both the deformations have resulted from the India–Asia convergence (Fig. 31; Dewey et al. 1988; Peltzer and Tapponier 1988; England and Molnar 1997; Hodges 2000; Yin 2006).

Two end-member models have been proposed for the tectonic evolution of Asia from the Tethys Ocean: (i) an eastward lateral extrusion of undeformed continental segments along major strike-slip faults in Tibet, including the dextral Karakoram Fault (KF) and the sinistral Altai Tagh Fault (ATF) systems (Peltzer and Tapponier 1988; Armijo et al. 1989; Avouac and Tapponier 1993), and (ii) overthickening and distributed north–south crustal shortening due to indenting India into Asia (Coward et al. 1988; Dewey et al. 1988; England and Molnar 1997). These end-member models are significant in understanding evolution of the Himalayan–Tibet convergence from the Tethyan oceanic domain by an earlier pre-Miocene crustal shortening in Tibet and later eastward extrusion.

The sinistral ATF system along the northern edge of the Tibetan Plateau and dextral KF on its southwestern margin play significant roles in framing crustal deformational models of this domain. Though the ATF appears to cut the lithosphere to accommodate ~500 km lateral displacement of Tibet since initial collision (Matte et al. 1996), displacement along the KF system has not been well constrained along the Himalayan–Tibetan margin. It demarcates this margin for nearly 700 km and is controversial in its location, sense of shearing, large-scale geometry, offset and timing of initiation of the movement. Largely located within wide valley of the Nubra–Shyok–Indus Rivers, and forming a prominent rectilinear topographic feature, the KF has been mapped as a dextral strike–slip fault running through these 1–3 km wide alluvial valleys (Sharma and Gupta 1978; Searle 1996; Upadhyay et al. 1999; Jain and Singh 2009). High exhumation rates of ~3.0 mm/year and erosion of about 20 km thick crustal rocks are recorded between 18.0 and 11.3 Ma along this fault, which has partitioned an early transpressional strain associated with the Pangong zone from dominantly dextral strike-slip late phase motion since c. 11 Ma (Searle et al. 1998).

Estimates of contemporary and Holocene slip rates along the KF vary from ~30 mm/year on the basis of offsets of geomorphic features (Avouac and Tapponier 1993; Matte et al. 1996), 10 $\pm$ 3 mm/year since 23–24 Ma (Lacassin et al. 2004) to 3–4 mm/year (Brown et al. 2002; Murphy et al. 2002). Current rates appear to be even as low as 3.4 $\pm$ 5 mm/year from the GPS measurements (Jade et al. 2004). Along the KF, estimates of possible offsets are as follows: (i) 1000 km by tying the Ladakh and Gangdese plutons across this fault (Peltzer and Tapponinier 1988), (ii) >400 km by correlating the Bangong suture with the Rushan/Pshart suture in Pamirs (Lacassin et al. 2004), (iii) a ~300 km offset either by linking the Saltoro Range ultramafics with those of the Shiquanhe (Matte et al. 1996) or from lateral correlation of Karakoram and Lhasa blocks (Rolland et al. 2000), (iv) ~280 km by correlating the Shyok and Shiquanhe sutures (Lacassin et al. 2004), (v) ~200 km by matching offset of the Indus–Yalu suture (Ratschbacher et al. 1994) or by linking the Bangong suture with the North Saltoro ophiolite (Srimal 1986), (vi) about 150 km on the basis of displacement of Baltoro-type granite (Searle 1996), (vii) various estimates of 120, 90, and 85 km by correlating the Shyok and Bangong sutures (Searle 1996; Searle et al. 1998), (viii) 40–150 km offset from linking the Baltoro granite with Karakoram Batholith and Shyok with the Bangong suture (Phillips 2008; Phillips et al. 2004), and (viii) smaller movement of ~66 km along the southern end of the fault by pinning points on the South Kailash thrust system (Murphy et al. 2002).

7.2.4.1 Karakoram Tectonic Units

Karakoram Shear Zone (KSZ): An imbricated and intensely mylonitized granite gneiss, volcanics, conglomerate, slate-phyllite-limestone intercalations, amphibolite and serpentinite intervene the Shyok Suture Zone and the frontal Asian Plate margin along the Nubra–Shyok Valleys for nearly 200 km to form the 1–5 km wide KSZ. Previously, this sequence was either grouped as a part of the Saltoro flysch and Saltoro mélange (Sinha et al. 1999; Upadhyay et al. 1999; Upadhyay 2002) or the Nubra Formation (Weinberg et al. 2000). In the extreme northwest between Panamik and Sasoma, this narrow belt deforms the Karakoram Batholith Complex (KBC) around Kubed into augen mylonite, and extends further upstream along the Nubra Valley for another 30 km. Largely covered by alluvium and megafans of transverse tributaries, the KSZ contains massive amygdoidal basalt and andesite along road cuttings between Trisha and Tegar. Road section to the Tegar Gompa exposes sheared conglomerate at the base, followed upwards by purple and grey-green slate and limestone intercalations. It is overthrust by highly sheared serpentinized ultramafic lenses within the KSZ; all dip gently to moderately between 20° and 50° towards northeast beneath the overthrust KBC. Gently dipping dark pelites demarcate the front of this belt beneath the KBC from Sati to Shyok and further southeast along the Shyok Valley to the U-shape bend (see Van Buer et al. 2015 for more details about this region).

Within the KSZ, undeformed and fractured Tirit granodiorite intrudes the slate–phyllite–volcanic sequence around Diskit–Khalsar–Tirit junction of the Nubra–Shyok Rivers and extends northwards on western face of the hill between Trisha and Murgi. Subalkaline to calc-alkaline Tirit body is an I-type suite of trondhjemite–tonalite–granodiorite–granite of volcanic arc affinity (Upadhyay et al. 1999) and solidified at subvolcanic level between 2.5 and 3.5 km thick overburden of the Shyok volcanics (Kumar et al. 2017). Apophyses of this body are seen at the confluence of the Nubra and Shyok Rivers. Elsewhere, it is intruded by numerous dolerite dykes. U–Pb zircon ages from this granitoid have yielded two mean crystallization ages: (i) an older 109.4 ± 1.1 Ma, and younger 67.32 ± 0.66 Ma (Kumar et al. 2017); the latter is almost the same as the 68 ± 1 Ma age, determined by Upadhyay et al. (1999) and is somewhat younger to $^{40}Ar/^{39}Ar$ hornblende age of 73.6 ± 1 Ma (Weinberg et al. 2000). These data, thus, bracket an important plutonic emplacement within the KSZ between 74 and 68 Ma.

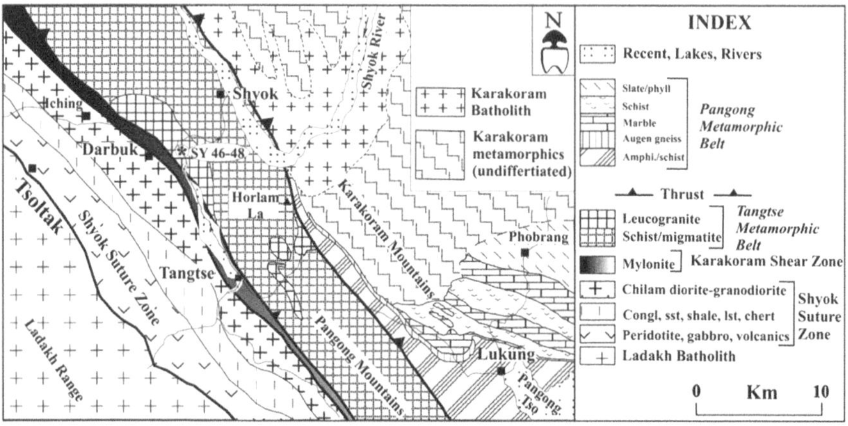

Fig. 7.11 Geological map of the Trans-Himalaya and Karakoram Mountains, Tangtse–Pangong Tso region showing main strands of the Karakoram Shear Zone (KSZ). Adopted from Roy and Jain (2010)

In the central Shyok–Tangtse–Chusul sector, the KSZ contains highly mylonitized amphibolite and granite gneiss, derived from granodiorite–diorite suite of the SSZ and granitoids of the KMB, respectively. Ultramylonite alternate with augen mylonite and are best exposed along the Tangtse Valley between Darbuk and Tangtse and marked by intense penetratively developed S-C ductile shear fabric (Figs. 7.11 and 7.12a), and steeply plunging down-dip stretching mineral lineation. Kinematic indicators include S-C foliation, synthetic C′ and C″ antithetic shear bands, Type A σ-mantled porphyroclasts, oblique quartz foliation, micro-shears with bookshelf gliding, mineral fishes including Group 2 mica fishes (Fig. 7.12c), and Type 1 and 2a pull-apart microstructures (Fig. 7.12d; Roy et al. 2010). These exhibit strong dextral sense of ductile shearing towards southeast with temperature of mylonitization ranging between 300 and 500 °C in the upper greenschist facies. Kinematic vorticity (Wk) analysis along the KSZ, using Porphyroclast Hyperbolic Distribution (PHD) and Shear Band (SB) methods yield Wk values ranging from 0.29 to 0.43 and 0.45 to 0.93, respectively, indicating distinct pure and simple shear-dominant regimes during different stages of the evolution of the KSZ (Roy et al. 2016). Pure shear strain has affected southern edge of the Asian plate when it was initially juxtaposed against the Indian plate around 70 Ma, and changed to simple shear, possibly during its reactivation during 21–13 Ma to produce the shear bands (Fig. 7.13).

It is intruded by a slightly deformed Darbuk granite stock at the confluence of Tangste and Iching Rivers with U–Pb zircon ages of 20.8 ± 0.4 Ma (Jain and Singh 2008). Thin concordant to discordant sheets of highly deformed folded and undeformed 2-mica and hornblende–bearing granitoids intrude the Tangste mylonite and metamorphics (Fig. 7.12b). U–Pb SHRIMP analysis of zircons from these sheets indicate their older crystallization ages of 106 ± 2.3 Ma, 72.8 ± 0.9 Ma, 71.4 ± 0.6 Ma and 63.0 ± 0.8 Ma (Searle et al. 1998; Reichardt et al. 2010), with a younger

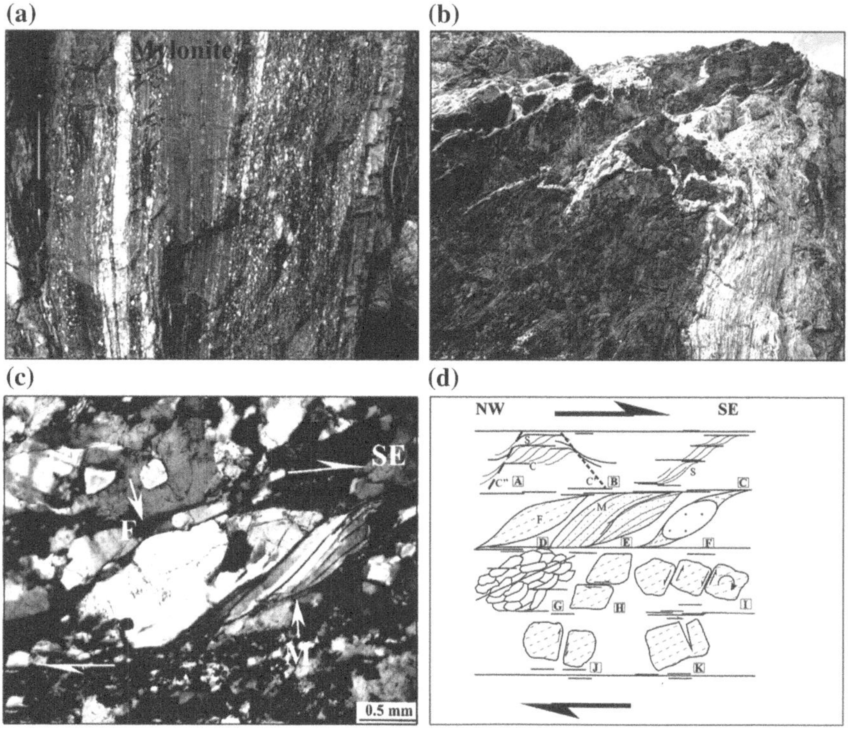

Fig. 7.12 **a** Typical mylonite from the Karakoram Shear Zone (KSZ) showing alternating ultra-mylonite and augen mylonite bands at the Tangste Gompa. **b** Apophyses of leucogranite veins intruding retrogressed amphibolite from foliated and sheared granite at Tangste Gompa. Photo length: 10 m. **c** Sigmoidal Group 2 mica fish defined by muscovite (M) with tips inclined in the direction of foliation from the protomylonitized Darbuk granite. The mica fish defines the S foliation of the shear band. Inclusion trails along opposite tips show stepping with dextral shear sense. Slide SY/49/R/133. **d** Composite diagram of the Tangste mylonite with numerous dextral shear indicators. (i) C″-foliation with respect to C- and S-foliations. (ii) C′-foliation affecting C- and S-foliations. (iii) Mutual relationships between S- and C-foliations. S-foliation defined by sigmoidal mica fish. (iv) Rhombic feldspar fish. (v) Dextral duplex in muscovite (M) fish. (vi) σ-mantled porphyroclast. (vii) Oblique foliation defined by quartz elongate grains. (viii) Cross-cutting micro-shear. (ix) Bookshelf structures. (x) Parallel pull-aparts. (xi) "V"-pull-apart. Adopted from Roy and Jain (2010)

phase between 22 and 15 Ma (Reichardt et al. 2010; Boutonnet et al. 2012). Still younger leucogranite veins yielded zircons of 15.68 ± 0.52 and 13.73 ± 0.28 Ma, even 9 Ma (Phillips et al. 2004; Horton and Leech 2013).

Karakoram Batholith Complex (KBC): Monotonous vertical cliffs of biotite–muscovite granite in upper parts of the Nubra Valley beyond Panamik constitute the main body of the KBC. Best exposures of this granite are seen along the Panamik–Sasoma section and beyond towards northwest. To the southeast, it does not extend beyond U–bend of the Shyok Valley as a monolithic body but intrudes the Karakoram Metamorphic Complex (KMC) as thin concordant bodies. The batholith can be easily

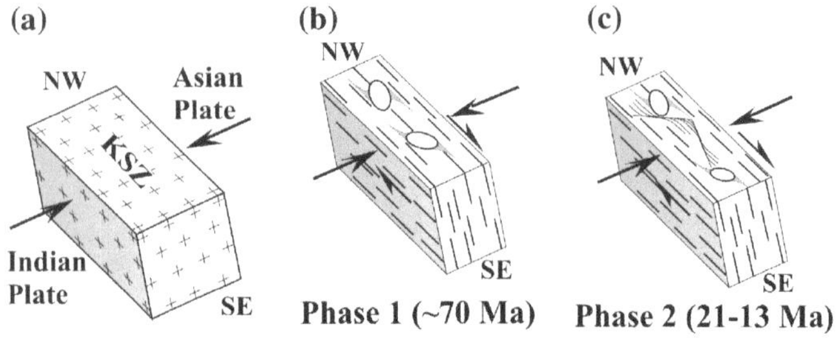

Fig. 7.13 Schematic diagrams showing variation in vorticity patterns due to the India–Asia Plate interaction. **a** Emplacement of Karakoram granitoids along the KSZ ~70 Ma during juxtaposition of the Asian Plate with the Indian Plate. **b** Initial higher pure shear component and horizontal shortening perpendicular to the shear zone boundary in initial phase ~70 Ma. **c** Superposed simple shear during dextral strike-slip movement around 21–13 Ma to produce shear band fabric. Adopted from Roy et al. (2016)

seen in IRS-1C satellite images, exposed beneath the metamorphosed Karakoram sedimentary cover along the upper reaches of this valley (Gregan and Pant 1983).

Two distinct granite suites of I- and S-type affinities are developed within the KBC (Srimal 1986; Srimal et al. 1987; Ravikant 2006). An inner belt of I-type granitoids, composed of quartz monzonite, granodiorite and tonalite, is characterized by $\delta^{18}O$ isotopic values of +6.2 to +7.7‰ (Srimal et al. 1987) and initial $^{87}Sr/^{86}Sr$ values of 0.7061 ± 0.0002 (Ravikant 2006). The younger outer peraluminous 2-mica and garnet-bearing S-type granitoids are juxtaposed linearly against the KSZ and marked by higher δ^{18} O isotopic values of +9.5 to +10.7‰ (Srimal et al. 1987) and higher initial $^{87}Sr/^{86}Sr$ ratios 0.7244 ± 0.0001 (Ravikant 2006). The batholith is regionally thrust southwest along gently to moderately dipping KSZ all along the Nubra and Shyok valleys, except in the north where the thrust zone dips vertically at Kubed and then changes its dip to the southwest. These granitoids have intruded the Permo–Carboniferous Tethyan Karakoram sequence between 130 and 50 Ma (Srimal et al. 1987; Searle 1991; Debon and Khan 1996; Weinberg and Searle 1998) and then by a younger phase between 25 and 17 Ma (Parrish and Tirrul 1989; Searle et al. 1998).

Physical continuity of the Karakoram Batholith is established with the Baltoro-type granitoids in western Karakoram along the Nubra Valley (Fig. 7.14a). No trace of the Karakoram fault could be observed on the ground for many kilometers, despite being shown on various maps of the region (Sharma and Gupta 1978; Searle 1996; Upadhyay 2002; Ravikant 2006). Instead, vertically dipping KSZ at Kubed changes to gently to moderately and southwesterly-dipping thrust due to overturning in the northwestern parts (Fig. 7.14b). This thrust appears to connect with the Main Karakoram Thrust of the Hunza–Biafond region in the west (cf., Searle et al. 1992). In the southeast along the Shyok valley, it dips gently towards northeast and continues for a very long distance until Tangste where it dips again steeply (Fig. 7.14c, d).

Karakoram Metamorphic Complex (KMC): In the Shyok–Pangong Mountain–Phobrang region, southern edge of the Asian Plate is extensively deformed and metamorphosed in two distinct metamorphic belts (Fig. 7.11).

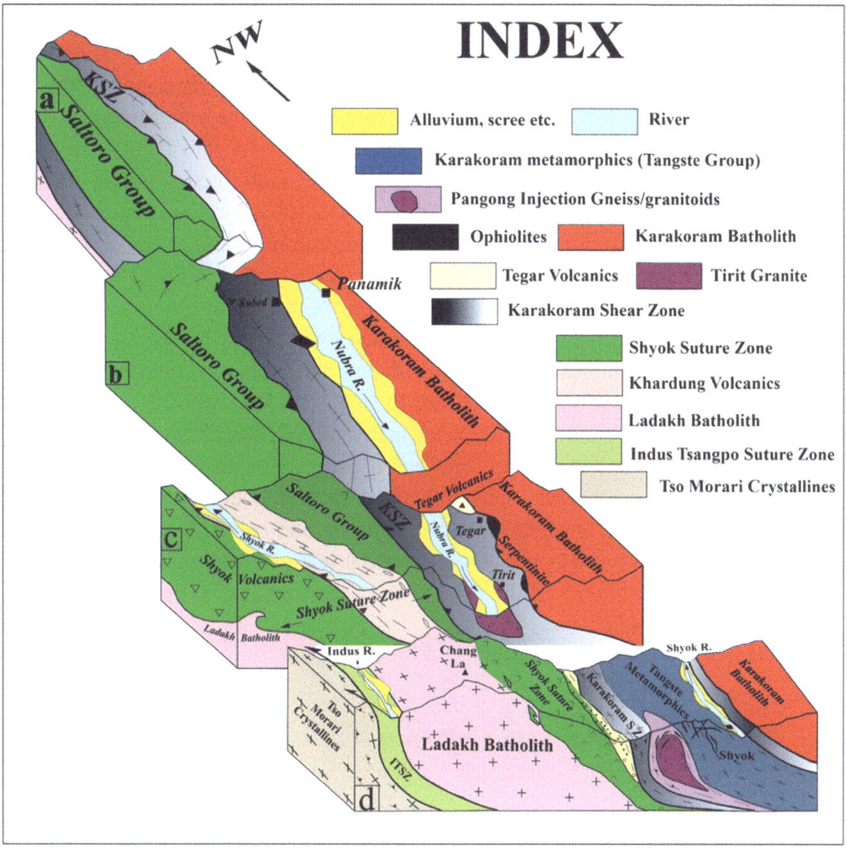

Fig. 7.14 3-dimensional cross-sections of the Karakoram Mountains depicting mutual geological and tectonic relationships along strike of the Karakoram Shear Zone (KSZ). **a** Extreme northwestern parts of the KSZ along the Nubra valley. **b** Panamik-Sasoma sector of the KSZ. **c** Tegar-Tirit area of the Nubra valley. **d** Darbuk-Shyok villages of the Tangste valley

Tangtse Group: The southern outer Tangtse Group is exposed around U-turn of the Shyok River and occupies vast expanse of the Pangong Mountains between the Tangtse gorge and Chusul in immediate vicinity of the KSZ (Fig. 7.11). It is characterized by high-grade sillimanite–K-feldspar-bearing garnetiferous schist and gneiss, amphibolite, hornblende granite gneiss and leucogranite as well as localized granulite facies metamorphics in the Pangong Range (Rolland and Pecher 2001; Jain et al. 2003). Granulite facies metamorphism was achieved at high temperature (800 °C, 5.5 kbar), and subsequently metamorphosed into amphibolite (700–750 °C, 4–5 kbar) and greenschist facies (350–400 °C, 3–4 kbar) at minimum age of 32 Ma, 20–18 and 13.6 ± 0.9 Ma, and finally at 11 Ma, respectively (Rolland et al. 2009). Presence of granulite facies rocks indicate that the KSZ might have accommodated lateral extrusion of Tibet at least at crustal or even a lithospheric-scale shear zone.

Mica schist and gneiss have undergone partial melting and prolific migmatization, which produced the Pangong Injection Complex along the Tangtse gorge (Fig. 7.11; Weinberg and Searle 1998). Numerous dome-shaped and elongated lenses of these biotite-rich granitoids reveal passive melt injection in extensively migmatized rocks. In turn, metamorphics and granitoids are cut by numerous late phase pegmatite veins, which emanate from these bodies along the least-strained southern margin. However, associated calc–silicate and amphibolite in this zone remained unaffected by migmatization and melting. Likewise, leucogranite near Sati in the Tirit sector yields a zircon SHRIMP U–Pb 15.0 ± 0.4 Ma age from well-developed zircon rims, with cores ranging between 1437 and 84 Ma (Weinberg et al. 2000).

Pangong Group: Further northeast, the inner belt of the KMC, called as the Pangong Group after the lake, is comprised of mainly slate, mica schist, greenschist/amphibolite and marble, calc–silicate and a band of mylonitized granite gneiss (Fig. 7.11). Low-grade metamorphics with many marble bands are best developed between Pangong Tso and Phobrang. This belt is extensively developed on its banks, and our reconnaissance observations confirm the earlier documentation of the presence of a dominant suite of greenschist/amphibolite and calc-silicates by Srikantia et al. (1982). Metamorphism varies from biotite grade of middle greenschist to sillimanite–muscovite subfacies of the middle amphibolite facies within this belt (Jain and Singh 2008).

7.2.5 *Timing of India-Asia Collision*

Various paleogeographic reconstructions of India and Asia since Paleozoic postulated that the vast Paleo- and Neo-Tethys Ocean spanned and separated the southern Asian Plate and northern Indian Plate margins—the region now constitutes parts of the Himalaya and Trans-Himalayan mountain ranges like the Ladakh, the Karakoram Mountains and the vast Tibet Plateau (Patriat and Achache 1984; Stampfli and Borel 2002) (Fig. 6.1a). In a relatively stationary Asian Plate reference frame, the Indian Plate converged northwards at about 180 ± 50 mm year^{-1} during 80 and 55 Ma and subsequently slowed down to 134 ± 33 mm year^{-1} at the collision not later than 55 Ma (Copley et al. 2010) (Fig. 6.1b). This movement coincided with the anticlockwise rotation of the Indian Plate, thereby impinging the Asian Plate earlier in the northwest in contrast to its hit in the east (Copley et al. 2010). Impingement of the Indian Plate even persisted during whole of the Cenozoic and is still an ongoing process. Hence, estimating the timing of the India-Asia impingement/collision is one of the most controversial topic in the Himalayan tectonics, as it is estimated anywhere between 65 and 35 Ma (Patriat and Achache 1984; Garzanti et al. 1987; Jaeger et al. 1989; Klootwijk et al. 1992; Rowley 1996, 1998; de Sigoyer et al. 2000; Guillot et al. 2003, 2008; Leech et al. 2005, 2007; Zhu et al. 2005; Aitchison et al. 2007; Green et al. 2008; Sciunnach and Garzanti 2012). Evidences for the India-Asia collision and its timing can be broadly classified into the following groups.

(i) *Paleomagnetism*: Convergence in the northward movement of the Indian Plate and its slow-down took place at around 55 ± 1 Ma from the paleomagnetic anomalies in the Indian Ocean (Copley et al. 2010; Klootwijk et al. 1992; Guillot et al. 2003).

(ii) *Paleolatitude evidences*: The India-Asia suturing of the Tethyan succession in the Himalaya with the Lhasa terrane to the north took place at 46 ± 8 Ma when these terranes started to overlap at 22.8 ± 4.2° N paleolatitude (Dupont-Nivet et al. 2010).

(iii) *Stratigraphy*: (a) 56.5–54.9 Ma as the maximum age of initiation of the India-Asia collision, deciphered from termination of continuous marine sedimentation within the Indus Tsangpo Suture Zone (ITSZ) in Ladakh, and (b) 50.5 Ma as the minimum age of closure of the Tethys along the ITSZ (Green et al. 2008) ~51 Ma (Garzanti and van Haver 1988) or 50.6±0.2 Ma (Zhu et al. 2005). In this scenario, the Indian continental lithosphere travelled to the ITSZ trench at 58 Ma (Garzanti et al. 1987). Final marine deposition in South Tibet terminated at ≤52 Ma as a consequence of initiation of collision and onset of fluvial sedimentation ca. 51 Ma (Rowley 1996, 1998).

(iv) *Sedimentology*: Closure of the Neo-Tethyan Ocean during earliest India-Asia collisional stage at ~56 Ma (Sciunnach and Garzanti 2012) is indicated from renewed clastic supply to the Tethys Himalayan margin in Zanskar, fore bulge related uplift, evaporite nodules in the Upper Paleocene and later red beds having caliche paleosols. Timing of first arrival of the Asian-derived detritus in the uppermost Tethyan sediments deposited on the Indian plate provides another indication of the collision at ~50 Ma (Najman et al. 2010) or 54 Ma using U–Pb ages and Hf signatures of zircons (Najman et al. 2017).

(v) *Magmatism*: The Trans-Himalayan Ladakh Batholith grew episodically and interruptedly with the very first small pulse at 105–100 Ma and subsequently between 70 and 50 Ma due to melting of subducting Tethyan oceanic lithosphere till 49.8 ± 0.8 Ma to indicate the timing of the India Asia continental collision (Weinberg and Dunlap 2000; Singh et al. 2007; Upadhyay et al. 2008). This age is now constrained to 50.2 ± 1.5 Ma as the initial collision age of the Kohistan–Ladakh Island Arc (KLA) with India along the ITSZ, and the final collision between the assembled India/Arc and Asia ~10 Ma later at 40.47 ± 1.3 Ma along the Shyok Suture Zone by integrating U–Pb zircon ages with their H_f, whole-rock Nd and Sr isotopic characters (Bouilhol et al. 2013).

(vi) *Metamorphism*: The Indian continental crust was eclogitized in the vicinity of the ITSZ trench, when the UHP coesite-bearing eclogite in Tso Morari were produced at 55 ± 7 Ma (de Sigoyer et al. 2000) or 53.3 ± 0.7 to be more precise (Leech et al. 2005, 2007), while these are dated at 46 ± 0.1 Ma in Pakistan at Kaghan (Tonarini et al. 1993; Kaneko et al. 2003).

(vii) *Integrated geological data*: Two-stage events have been recorded in southern Tibet and elsewhere indicating a ~55 Ma collision of an island arc system with India (Event 1) and a younger Oligocene age of the India-Asia continental collision (Event 2) (Aitchison et al. 2007). Two-stage Cenozoic collision is also postulated from estimating amount of convergence: 50 Ma collision of an extended microcontinent fragment and continental Asia, and 25 Ma hard continent-continent collision (van Hinsbergen et al. 2012).

Somewhat different approach provides valuable data for determining the initial convergence/collision timing between the Indian and the Asian Plates, which are sutured and juxtaposed across the ITSZ (Jain 2014, 2017). Timing of the first India–Asia impingement has been better constrained ~58 Ma by comparing ages and products of deep-seated and surface processes: (i) subduction and melting of the Tethyan oceanic lithosphere—the Trans-Himalayan LB; (ii) subducted continental lithospheric and the UHP metamorphosed Indian crust—the TMC, and (iii) biostratigraphy of the youngest marine sedimentation in Zanskar. Bulk intrusion of the Trans-Himalayan LB took place at 57.9 ± 0.3 Ma, while the UHP metamorphism in Tso Morari was at 53.3 ± 0.7 Ma. Both the units signify the drastic geodynamic changes within ~4–5 m year.

References

Ahmad T, Tanaka T, Sachan HK, Asahara Y, Islam R, Khanna PP (2008) Geochemical and isotopic constraints on the age and origin of the Nidar ophiolitic complex, Ladakh, India: implication for the Neo-Tethyan subduction along the Indus suture zone. Tectonophysics 451:206–224

Aitchison JC, Ali JR, Davis AM (2007) When and where did India and Asia collide? J Geophy Res 112:B05423. https://doi.org/10.1029/2006JB004706

Allegre CJ, Courtillot V, Tapponnier P et al (1984) Structure and evolution of the Himalaya-Tibet orogenic belt. Nature 307:17–22

Armijo R, Tapponnier P, Tonglin H (1989) Late Cenozoic right-lateral strike-slip faulting in southern Tibet. J Geophys Res 94:2787–2838

Avouac J-P, Tapponnier P (1993) Kinematic model for active deformation in Central Asia. Geophys Res Lett 20:895–898

Bassoullet JP, Colchen M, Marcoux J, Mascle G (1981) Field evidences for continental rifting in Triassic time in the Ladakh part of the Indus suture zone. Geol Ecol. Studies Qinghai-Xizang Plateau 1. Science Press, Beijing, pp 579–586

Bassoullet JP, Colchen M, Juteau T, Marcoux J, Mascle G, Reibel G (1982) Geological studies in the Indus suture zone of Ladakh (Himalayas). Contrib Himal Geol 2:96–124

Berthelsen A (1953) On the geology of the Rupshu district, northwest Himalaya. Medd Dan Geol Foren 12:350–414

Bhandari AK, Srimal N, Radcliff RP, Srivastava DK (1979) Geology of part of Shyok and Nubra valleys Ladakh district, Jammu and Kashmir, India: a preliminary report. Him Geol 9:158–171

Borneman NL, Hodges KV, van Soest MC, Bohon W, Wartho J-A, Cronk SS, Ahmad T (2015) Age and structure of the Shyok suture in the Ladakh region of northwestern India: implications for slip on the Karakoram fault system. Tectonics 34:2011–2033. https://doi.org/10.1002/2015TC003933

Bouilhol P, Jagoutz O, Hanchar JM, Dudas FO (2013) Dating the India-Eurasia collision through arc magmatic records. Earth Planet Sci Lett 366:163–175

Boutonnet E, Leloup PH, Arnaud N, Paquette J-L, Davis WJ, Hattori K (2012) Synkinematic magmatism, heterogeneous deformation, and progressive strain localization in a strike-slip shear zone: the case of the right-lateral Karakorum fault. Tectonics 31:TC4012. https://doi.org/10.1029/2011tc003049

Brookfield ME, Andrew-Speed CP (1984) Sedimentation in high-altitude intermontane basin—The Wakka Chu Molasse (Mid. Tertiary, Northwestern India). Bull Indian Geol Assoc 17:175–193

Brown ET, Bendick R, Bourles DL, Gaur V, Molnar P, Raisbeck GM, Yiou F (2002) Slip rates of the Karakoram fault, Ladakh, India, determined using cosmic ray exposure dating of debris flows and morains. J Geophys Res 107(B9):2192. https://doi.org/10.1029/2000JB000100

Cannat M, Mascle G (1990) Réunion extraordinaire de la société géologique de France en Himalaya du Ladakh. Bull Soc Géol France, 553–582

Catlos EJ, Pease EC, Dygert N, Brookfield M, Schwarz WH, Bhutani R, Pande K, and Schmitt A. (2018) Nature, age, and emplacement of the Spongtang Ophiolite, Ladakh, NW India. J Geol Soc. https://doi.org/10.1144/jgs2018-085

Clift PD, Degnan PJ, Hannigan R, Blusztajn J (2000) Sedimentary and geochemical evolution of the Dras forearc basin, Indus Suture, Ladakh Himalaya, India. Geol Soc Am Bull 112:450–466

Copley A, Avouac JP, Royer J-Y (2010) India-Asia collision and the Cenozoic slowdown of the Indian plate: implications for the forces driving plate motions. J Geoph Res 95:B03410. https://doi.org/10.1029/2009JB006634

Corfield RI, Searle MP, Pedersen RB (2001) Tectonic setting and obduction history of the Spontang ophiolite, Ladakh Himalaya, NW India. J Geol 109:715–736

Coward MP, Kidd WSF, Pan Y, Shackleton RM, Zhang H (1988) The structure of the 1985 Tibet Geotraverse, Lhasa to Golmud. Phil Trans Royal Soc London A 327:307–336

Das S, Basu AR, Mukherjee BK (2017) In situ peridotitic diamond in Indus ophiolite sourced from hydrocarbon fluids in the mantle transition zone. Geology. https://doi.org/10.1130/G39100.1

Das S, Mukherjee BK, Basu AR, Sen K Jr (2015) Peridotitic minerals of the Nidar Ophiolite in the NW Himalaya: sourced from the depth of the mantle transition zone and above. In: Mukherjee S, Carosi R, van der Beek PA, Mukherjee BK, Robinson DM (eds) Tectonics of the Himalaya. Geol Soc London Spec Publ 412:271–286, https://doi.org/10.1144/sp412.12y

de Sigoyer J, Chavagnac V, Blichert-Toft J, Villa IM, Luais B, Guillot S, Cosca M, Mascle G (2000) Dating the Indian continental subduction and collisional thickening in the northwest Himalaya: multichronology of the Tso Morari eclogites. Geology, 487–490

Debon F, Khan NA (1996) Alkaline orogenic plutonism in the Karakoram batholith: the Upper Cretaceous Koz Sar complex (Karamber valley, N. Pakistan). Geodyn Acta 9:145–160

Dewey JF, Bird JM (1970) Mountain belts and the new global tectonics. J Geophys Res 75:2625–2647

Dewey JF, Shackleton RM, Chang C, Sun Y (1988) The tectonic evolution of the Tibetan Plateau. Phil Trans Royal Soc London A 327:379–413

Dietrich V, Frank W, Gansser A, Honneger KH (1983) A Jurassic-Cretaceous island arc in the Ladakh-Himalayas. J Volcanol Geotherm Res 18:405–433

Dunlap WJ, Weinberg RF, Searle MP (1998) Karakoram fault zone rocks cool in two phases. J Geol Soc London 155:903–912

Dunlap WJ, Wysoczanski R (2002) Thermal evidence for early Cretaceous metamorphism in the Shyok suture zone and age of the Khardung volcanic rocks, Ladakh, India. J Asian Earth Sci 20:481–490. https://doi.org/10.1016/S1367-9120(01)00042-6

Dupont-Nivet G, Lippert PC, van Hinsbergen DJJ, Meijers MJM, Kapp P (2010) Palaeolatitude and age of the Indo-Asia collision: palaeomagnetic constraints. Geophys J Int 182:1189–1198. https://doi.org/10.1111/j.1365-246X.2010.04697.x

Ehiro M, Kojima S, Sato T, Ahmad T, Ohtani T (2007) Discovery of Jurassic ammonoids from the Shyok suture zone to the northeast of Chang La pass, Ladakh, northwest India and its tectonic significance. Island Arc 16:124–132

England P, Molnar P (1997) Active deformation of Asia: from kinematics to dynamics. Science 278:647–650

Frank W, Thoni M, Purtscheller F (1977) Geology and petrography of Kulu-south Lahul area. Colloq Int CNRS Himalaya 268:147–160

Gansser A (1964) Geology of the Himalayas. Intersci Publ. Wiley, London, 289

Gansser A (1977) The great suture zone between the Himalaya and Tibet—a preliminary account. Colloq Int CNRS Himalaya 261:181–191

Gansser A (1980) The significance of the Himalayan suture zone. Tectonophysics 62:37–52

Garzanti E, van Haver T (1988) The Indus clastics: forearc basin sedimentation in the Ladakh Himalaya (India). Sedim Geol 59:237–249

Garzanti E, Baud A, Mascle G (1987) Sedimentary record of the northward flight of India and its collision with Eurasia (Ladakh Himalaya, India). Geodyn Acta 1:297–312

Green OR, Searle MP, Corfield RI, Corfield RM (2008) Cretaceous-Tertiary carbonate platform evolution and the age of the India-Asia collision along the Ladakh Himalaya (Northwest India). J Geol 116:331–353

Gregan JT, Pant PC (1983) Geology and stratigraphy of eastern Karakoram, Ladakh. In: Thakur VC, Sharma KK (eds) Geology of Indus Suture Zone of Ladakh, Wadia Inst Himalayan Geol Dehradun, 99–106

Guillot S, Garzanti E, Baratoux,D. Marquer D, Mahéo G, de Sigoyer J (2003) Reconstructing the total shortening history of the NW Himalaya. Geochem Geophys Geosyst 4(7):1064. https://doi.org/10.1029/2002gc000484

Guillot S, Mahéo G, de Sigoyer J, Hattori KH, Pêcher A (2008) Tethyan and Indian subduction viewed from the Himalayan high-to ultrahigh-pressure metamorphic rocks. Tectonophysics 451:225–241

Heim A, Gansser A (1939) Central Himalaya, geological observations of the Swiss expeditions: Mémoire Société Helvetique Science. Naturelles 73:245

Henderson AL, Najman Y, Parrish R, Mark D, Foster GL (2011) Constraints to the timing of India-Eurasia collision; are-evaluation of evidence from the Indus Basin sedimentary rocks of the Indus-Tsangpo Suture Zone, Ladakh, India. Earth-Sci Rev 106:265–292

Hodges KV (2000) Tectonics of the Himalaya and southern Tibet from two perspectives. Geol Soc Amer Bull 92:324–350

Honegger K, Dietrich V, Frank W, Gansser A, Thoni M, Trommsdorff V (1982) Magmatism and metamorphism in the Ladakh Himalaya the Indus-Tsangpo suture zone. Earth Planet Sci Lett 60:253–292

Horton F, Leech ML (2013) Age and origin of granites in the Karakoram shear zone and Greater Himalaya Sequence, NW India. Lithosphere 5:300–320. https://doi.org/10.1130/L213

Jade S, Bhatt B, Yang Z, Bendick R, Gaur VK, Molnar P, Anand M, Kumar D (2004) GPS measurements from the Ladakh Himalaya, India: preliminary tests of plate-like or continuous deformation in Tibet. Geol Soc Amer Bull 116(11/12):1385–1391

Jaeger JJ, Courtillot V, Tapponnier P (1989) Paelontological view of the ages of the Deccan Traps, the Cretaceous/tertiary boundary, and the India/Asia collision. Geology 17:316–319

Jain AK (2014) When did India-Asia collide and make the Himalaya? Curr Sci 106(2):254–266

Jain AK (2017) Continental subduction in the NW-Himalaya and Trans-Himalaya. Ital J Geosci 136(1):89–102

Jain AK, Singh S (2008) Tectonics of the southern Asian Plate margin along the Karakoram Shear Zone: constraints from field observations and U-Pb SHRIMP ages. Tectonophysics 451:186–205

Jain AK, Singh S (2009) Geology and tectonics of the Southeastern Ladakh and Karakoram. Geological Society of India, Bangalore India, p 179

Jain AK, Singh S, Manickavasagam RM (2002) Himalayan collision tectonics. Gondwana Res Group Mem 7:114

Jain AK, Singh S, Manickavasagam RM, Joshi M, Verma PK (2003) HIMPROBE programme: integrated studies on geology, petrology, geochronology and geophysics of the Trans-Himalaya and Karakoram. Mem Geol Soc India 53:1–56

Juyal KP (2006) Foraminiferal biostratigraphy of the Early Cretaceous Hundiri Formation, lower Shyok area, eastern Karakoram, India. Curr Sci 91:1096–1101

Kaneko Y, Katayama I, Yamamoto H, Misawa K, Ishikawa M, Rehman HU, Kausar AB, Shiraishi K (2003) Timing of Himalayan ultrahigh-pressure metamorphism: sinking rate and subduction angle of the Indian continental crust beneath Asia. J Metamor Geol 21:589–599

Kapp P, Murphy P, Yin A, Harrison TM (2003) Mesozoic and Cenozoic tectonic evolution of the Shiquanhe area of western Tibet. Tectonics 22(4):1029. https://doi.org/10.1029/2001tc001332

Klootwijk CT, Gee JS, Peirce JW, Smith GM, McFadden PL (1992) An early India-Asia contact: paleomagnetic constraints from Ninetyeast Ridge, ODP Leg 121. Geology 20:395–398

Kojima S, Ahmad T, Tanaka T, Bagati TN, Mishra M, Kumar R, Islam R, Khanna PP (2001) Early Cretaceous radiolarians from the Indus suture zone, Ladakh, northern India. In: News Osaka Micropaleont (NOM) Spec vol 12:257–270

Kumar R (2005) Tectonics, geochemistry and geochronology of the Ladakh Batholith, Trans-Himalaya. Unpubl Ph.D. thesis, Indian Inst Tech Roorkee, Roorkee, India 196

Kumar R, Lal N, Singh S, Jain AK (2007) Cooling and exhumation of the Trans-Himalayan Ladakh Batholith, as constrained by fission track apatite and zircon ages. Curr Sci 92:490–496

Kumar R, Virdi NS (1997) Evolution of Inter-montane Kargil Basin (Oligo-Miocene) Ladakh Himalaya: a sedimentological approach. J Geol Soc India 49:675–686

Kumar R, Jain AK, Lal N, Singh S (2018) Early-Middle Eocene exhumation of the Trans-Himalayan Ladakh Batholith, and the India-Asia convergence. Curr Sci 113:1090–1098

Kumar S, Bora S, Sharma UK (2016) Geological appraisal of Ladakh and Tirit granitoids in the Indus-Shyok Suture Zones of Northwest Himalaya, India. J Geol Soc India 87:737–746

Kumar S, Bora S, Sharma UK, Yi K, Kim N (2017) Early Cretaceous subvolcanic calc-alkaline granitoid magmatism in the Nubra-Shyok valley of the Shyok Suture Zone, Ladakh Himalaya, India: evidence from geochemistry and U-Pb SHRIMP zircon geochronology. Lithos 277:33–50. https://doi.org/10.1016/j.lithos.2016.11.019

Lacassin R, Valli F, Arnaud N, Leloup PH, Paquette JL, Haibing L, Tapponnier P, Chevalier M-L, Guillot S, Maheo G, Zhiqin X (2004) Large-scale geometry, offset and kinematic evolution of the Karakorum fault, Tibet. Earth Planet Sci Lett 219:255–269

Leech ML, Singh S, Jain AK (2007) Continuous metamorphic zircon growth and interpretation of U-Pb SHRIMP dating: an example from the Western Himalaya. Intl Geol Rev 49:313–328

Leech ML, Singh S, Jain AK, Klemperer SL, Manickavasagam RM (2005) Early, steep subduction of India beneath Asia required by early UHP metamorphism. Earth Planet Sci Lett 234:83–97

Maheo G, Bertrand H, Guillot S, Villa IM, Keller F, Capiez P (2004) The south Ladakh ophiolites (NW Himalaya, India): an intra-oceanic tholeiitic origin with implication for the closure of the Neo-Tethys. Chem Geol 203:273–303

Matsumaru K, Ehiro M, Kojima S (2006) On *Orbitolina* (Foraminifera) from the Shyok Suture Zone, Kashmir, NW India. J Palaeont Soc India 51:43–49

Matte PH, Tapponnier P, Arnaud N, Bourjot L, Avouac J-P, Vidal PH, Liu Q, Pan Y, Wang Y (1996) Tectonics of Western Tibet, between the Tarim and Indus. Earth Planet Sci Lett 142:311–220

Murphy MA, Yin A, Kapp P, Harrison TM, Manning CE, Ryerson FJ, Ding L, Guo J (2002) Structural evolution of the Gurla Mandhata detachment system, southwest Tibet: implications for the eastward extent of the Karakoram fault system. Geol Soc Amer Bull 114:428–447

Najman Y, Appel E, Boudagher-Fadel M, Bown P, Carter A, Garzanti E, Godin L, Han JT, Liebke U, Oliver G, Parrish R, Vezzoli G (2010) Timing of India-Asia collision: geological, biostratigraphic, and palaeomagnetic constraints. J Geophys Res 115:B12416

Najmana Y, Jenksa D, Godin L, Boudagher-Fadel M, Millar I, Garzantie E, Horstwood M, Braccialia L (2017) The Tethyan Himalayan detrital record shows that India-Asia terminal collision occurred by 54 Ma in the Western Himalaya. Earth Planet Sci Lett 459:301–310

Parrish RR, Tirrul R (1989) U-Pb ages of the Baltoro granite, northwest Himalaya, and implication for zircon inheritance and monazite U-Pb systematics. Geology 17:1076–1079

Patriat P, Achache J (1984) India-Eurasia collision chronology has implications for crustal shortening and driving mechanism of plates. Nature 311:615–621. https://doi.org/10.1038/311615a0

Pedersen RB, Searle MP, Corfield RI (2001) U-Pb zircon ages from the Spontang Ophiolite, Ladakh Himalaya. J Geol Soc London 158:513–520

Peltzer G, Tapponnier P (1988) Formation and evolution of strike-slip faults, rifts, and basins during India-Asia collision: an experimental approach. J Geophys Res 93:15085–15117

Phillips RJ (2008) Geological map of the Karakoram Fault Zone Eastern Karakoram, Ladakh. NW Himalaya. J Maps 4(1):21–37. https://doi.org/10.4113/jom.2008.98

Phillips RJ, Parrish R, Searle MP (2004) Age constraints on ductile deformation and long-term slip rates along the Karakoram fault zone, Ladakh. Earth Planet Sci Lett 226:305–319

Pudsey CJ (1986) The Northern Suture, Pakistan: margin of a Cretaceous island-arc. Geol Mag 123:405–423

Rai H (1982) Geological evidence against the Shyok Palaeosuture, Ladakh Himalaya. Nature 297:142–144

Rai H (1983) Geology of the Nubra valley and its significance on the evolution of the Ladakh Himalaya. In Sharma KK (ed) Geology and geodynamic evolution of the Himalayan collision zone. Part-2, Phys Chem Earth 18:79–91

Ratschbacher L, Frisch W, Herrman U, Strecker M (1994) Distributed deformation in southern and western Tibet during and after India Asia collision: an experimental approach. J Geophys Res 99:19,917–19,945

Ravikant V (2006) Utility of Rb–Sr geochronology in constraining Miocene and Cretaceous events in the eastern Karakoram, Ladakh, India. J Asian Earth Sci 27:534–543

Razdan ML, Raina CB (1996) Geology of Khardung La-Sumur area, Shyok-Nubra tectonic belt, Ladakh (Jammu and Kashmir). Geol Surv India Spec Publ 21:99–109

Reichardt H, Weinberg RF, Andersson UB, Fanning CM (2010) Hybridization of granitic magmas in the source: the origin of the Karakoram Batholith, Ladakh, NW India. Lithos 116:249–272. https://doi.org/10.1016/j.lithos.2009.11.013

Reuber I (1986) Geometry of accretion, oceanic thrusting of the Spontang Ophiolite, Ladakh-Himalaya. Nature 321:592–596

Reuber I (1989) The Dras arc: two successive volcanic events on eroded oceanic crust. Tectonophysics 161:93–106

Reuber I, Colchen M, Mevel C (1987) The geodynamic evolution of the South-Tethyan margin in Zanskar, NW Himalaya, as revealed by the Spontang ophiolitic melange. Geodin Acta 1:283–296

Robertson AHF (2000) Formation of mélanges in the Indus Suture Zone, Ladakh Himalaya by successive subduction-related, collisional and post-collisional processes during Late Mesozoic–Late Tertiary time. In: Khan MA, Treolar PJ, Searle MP, Jan Q (eds) Tectonics of the Nanga Parbat Syntaxis and the Western Himalaya. Geol Soc London Spec Pub 170:333–374

Robertson AHF, Degnan PJ (1993) Sedimentology, tectonic implications of the Lamayuru Complex, deep-water facies of the Indian passive margin, Indus Suture Zone, Ladakh Himalaya. In: Treloar PJ, Searle MP (eds) Himalayan tectonics. Geol Soc London Spec Pub 74:299–321

Rolland Y (2002) From intra-oceanic convergence to post-collisional evolution: example of the India-Asia convergence in NW Himalaya, from Cretaceous to present. In: Rosenbaum G, Lister GS (eds) Reconstruction of the evolution of the Alpine-Himalayan Orogen. J Virtual Expl 8:185–208

Rolland Y, Mahéo G, Pêcher A, Villa IM (2009) Syn-kinematic emplacement of the Pangong metamorphic and magmatic complex along the Karakorum Fault (N Ladakh). J Asian Earth Sci 34:10–25. https://doi.org/10.1016/j.jseaes.2008.03.009

Rolland Y, Pecher A (2001) The Pangong granulites of the Karakoram Fault (Western Tibet): vertical extrusion within a lithospheric-scale fault? Comptes Rendus de l'Acad Sci Paris 332:363–370

Rolland Y, Pêcher A, Picard C (2000) Middle Cretaceous back-arc formation and arc evolution along the Asian margin: the Shyok Suture Zone in northern Ladakh (NW Himalaya). Tectonophysics 325:145–173

Rolland Y, Picard C, Pêcher A, Lapierre H, Bosch D. Keller F (2002) The cretaceous Ladakh arc of NW Himalaya—slab melting and melt-mantle interaction during fast northward drift of Indian Plate. Chem Geol 182:139–178

Rowley DB (1996) Age of initiation of collision between India and Asia: a review of stratigraphic data. Earth Planet Sci Lett 145:1–13

Rowley D (1998) Minimum age of initiation of collision between India and Asia north of Everest based on the subsidence history of the Zhepure Mountain section. J Geol 106:229–235

Roy P, Jain AK, Singh S (2010) Microstructures of mylonites along the Karakoram Shear Zone, Tangste Valley, Pangong Mountains, Karakoram. J Geol Soc India 75:679–694. https://doi.org/10.1007/s12594-010-0065-1

Roy P, Jain AK, Singh S (2016) Kinematic Vorticity Analysis along the Karakoram Shear Zone, Pangong Mountains, Karakoram: Implications for the India-Asia Tectonics. J Geol Soc India 87:249–260

Satoru K, Ahmad T, Tanaka T, Bagati TN, Mishra M, Kumar R, Islam R, Khanna PP (2001) Early Cretaceous radiolarians from the Indus suture zone, Ladakh, northern India. News Osaka Micropaleont (NOM) Spec vol 12:257–270

Scharer U, Hamet J, Allegre CJ (1984) The Transhimalaya (Gangdese) plutonism in the Ladakh region: a U-Pb and Rb–Sr study. Earth Planet Sci Lett 67:327–339

Sciunnach D, Garzanti E (2012) Subsidence history of the Tethys Himalaya. Earth-Sci Rev 111:179–198

Searle MP (1991) Geology and Tectonics of the Karakoram Mountains. Wiley, Chichester, p 358

Searle MP (1996) Geological evidence against large scale pre-Holocene offsets along the Karakoram Fault: implications for the limited extrusion of the Tibetan Plateau. Tectonics 15:171–186

Searle MP, Crawford MB, Rex AJ (1992). Field relations, geochemistry, origin and emplacement of the Baltoro granite, Central Karakoram. Trans Royal Soc Edinb: Earth Sci 83(03):519–38, http://doi.org/10.1017/s0263593300005861

Searle MP, Weinberg, RF, Dunlap WJ (1998) Transpressional tectonics along the Karakoram Fault Zone, northern Ladakh. In: Holdsworth RE, Strachan RA, Dewey JE (eds) Continental transpressional and transtensional tectonics. Geol Soc London Spec Publ 135:307–326

Searle MP, Windley BF, Coward MP, Cooper DJW, Rex AJ, Rex D, Tingdong L, Xuchang X, Jan MQ, Thakur VC, Kumar S (1987) The closing of the Tethys and the tectonics of the Himalaya. Geol Soc Amer Bull 98:678–701

Shakelton RM (1981) Structure of southern Tibet: report on a traverse from Lhasa to Kathmandu organized by Academia Sinica. J Struct Geol 3:97–105

Sharma KK, Gupta KR (1978) Some observations on the geology of the Indus and Shyok valleys between Leh and Panamik, District Ladakh, Jammu and Kashmir, India. Recent Res Geol 7:133–143

Sharma KK, Gupta KR (1983) Calc-alkaline island arc volcanism in Indus-Tsangpo suture zone. In: Thakur VC, Sharma KK (eds) Geology of Indus Suture Zone of Ladakh. Wadia Inst Him Geol, Dehradun, pp 71–78

Sinclair HD, Jaffey N (2001) Sedimentology of the Indus Group, Ladakh, northern India: implications for the timing of initiation of paleo-Indus River. J Geol So London 158(1):151–162

Singh B (1993) Geological set up of a part of the Ladakh Granitoid Complex, Ladakh Himalaya. J Him Geol 4(1):57–62

Singh IB, Sahni A, Jain AK, and 11 others (2015) Post-collision sedimentation in the Indus Basin (Ladakh, India): implications for the evolution of the northern margin of the Indian Plate. J Palaeont Soc India 60(2):97–146

Singh S, Kumar R, Barley M, Jain AK (2007) U-Pb SHRIMP ages and depth of emplacement of Ladakh Batholith, NW Himalaya. J Asian Earth Sci 30:490–503

Sinha AK, Rai H, Upadhyay R, Chandra R (1999) Contribution to the geology of the eastern Karakoram, India. In: Macfarlane A, Sorkhabi RB, Quade J (eds) Himalaya and Tibet: mountain roots to mountain tops. Geol Soc Amer Spec Paper 328:33–44

Srikantia SV, Ganesan TM, Wangdus C (1982) A note on the tectonic framework and geological set-up of the Pangong-Chusul sector, Ladakh Himalaya. J Geol Soc India 23:354–357

Srikantia SV, Razdan ML (1980) Geology of part of central Ladakh Himalaya with particular reference to Indus Tectonic Zone. J Geol Soc India 21:523–545

Srikantia SV, Razdan ML (1981) Shilakang ophiolite nappe of Zanskar mountains, Ladakh Himalaya. J Geol Soc India 22:227–234

Srimal N (1986) India-Asia collision: implications from the geology of the eastern Karakoram. Geology 14:523–527

Srimal N, Basu AR, Kyser TK (1987) Tectonic inferences from oxygen isotopes in volcano-plutonic complexes of the India-Asia collision zone, NW India. Tectonics 6:261–273
Stampfli GM, Borel GD (2002) A plate tectonic model for the Paleozoic and Mesozoic constrained by dynamic plate boundaries and restored synthetic oceanic isochrons. Earth Planet Sci Lett 196(1):17–33. https://doi.org/10.1016/S0012-821X(01)00588-X
St-Onge MR, Rayner N, Searle MP (2010) Zircon age determinations for the Ladakh batholith at Chumathang Northwest India: implications for the age of the India-Asia collision in the Ladakh Himalaya. Tectonophysics 495:171–183
Thakur VC (1981) Regional framework and geodynamic evolution of the Indus Tsangpo suture zone in the Ladakh Himalaya. Trans Roy Soc Edin (Earth Sci) 72:381–392
Thakur VC (1993) Geology of the Western Himalaya. Pergamon Press, Oxford, p 355
Thakur VC, Jain AK (1975) Some observations on deformation, metamorphism and tectonic significance of the rocks of some parts of the Mishmi Hills, Lohit district (NEFA), Arunachal Pradesh. Him Geol 5:339–363
Thakur VC, Mishra DK (1984) Tectonic framework of Indus and Shyok suture zones in eastern Ladakh, northwest Himalaya. Tectonophysics 101:207–220
Thakur VC, Virdi NS (1979) Lithostratigraphy, structural framework, deformation and metamorphism of the south-eastern region of Ladakh. Kashmir Himalaya, India. Him Geol 9:11–25
Thakur VC, Virdi NS, Rai H, Gupta KR (1981) A note on the geology of Nubra-Shyok area of Ladakh, Kashmir Himalaya. J Geol Soc India 22:46–50
Tonarini S, Villa I, Oberli M, Meier F, Spencer DA, Pognante U, Ramsay JG (1993) Eocene age of eclogite metamorphism in Pakistan Himalaya: implications for India-Eurasia collision. Terra Nova 5:13–20
Upadhyay R (2002) Startigraphy and tectonics of Ladakh, eastern Karakoram, western Tibet and western Kun Lun. J Geol Soc India 59:447–467
Upadhyay R, Frisch W, Siebel W (2008) Tectonic implications of new U-Pb zircon ages of the Ladakh batholith, Indus suture zone, northwest Himalaya, India. Terra Nova 20:309–317
Upadhyay R, Sinha AK, Chandra R, Rai H (1999) Tectonic and magmatic evolution of the eastern Karakoram, India. Geodyn Acta 12:341–358
Van Buer NJ, Jagoutz O, Upadhyay R, Guillong M (2015) Mid-crustal detachment beneath western Tibet exhumed where conjugate Karakoram and Longmu–Gozha Co faults intersect. Earth Planet Sci Lett 413:144–157
van Hinsbergen DJJ, Lippert PC, Dupont-Nivete G, McQuarrie N, Doubrovine PV, Spakmani W, Torsvika TH (2012) Greater India Basin hypothesis and a two-stage Cenozoic collision between India and Asia. Proc Nat Acad Sci 109(20):7659–7664. https://doi.org/10.1073/pnas.1117262109
Weinberg RF (1996) The ascent mechanism of felsic magmas: news and views, Third Hutton Symposium on Granites. Trans Roy Soc Edin 87:95–103
Weinberg RF, Searle MP (1998) The Pangong Injection Complex, Indian Karakoram: a case of pervasive granite flow through hot viscous crust. J Geol Soc London 155:883–891
Weinberg RF, Dunlap J (2000) Growth and deformation of the Ladakh Batholith, Northwest Himalayas: implications for timing of continental collision and origin of calcalkaline batholith. J Geol 108:303–320
Weinberg RF, Dunlap WJ, Whitehouse M (2000) New field structural and geochronological data from the Shyok and Nubra valleys, northern Ladakh: linking Kohistan to Tibet. Geol Soc London Spec Publ 170:253–275
Weinberg RF, Mark G (2008) Magma migration, folding and disaggregation of migmatites in the Karakoram shear zone, Ladakh, NW India. Geol Soc Amer Bull 120(7/8):994–1009. https://doi.org/10.1130/B26227.1
White LT, Ahmad T, Ireland TR, Lister G, Forster MA (2011) Deconvolving episodic age spectra from zircons of the Ladakh Batholith, northwest Indian Himalaya. Chem Geol 289:179–196
Windley BF (1983) Metamorphism and tectonics of the Himalaya. J Geol Soc London 140:849–865
Wu F-Y, Clift PD, Yang J-H (2007) Zircon Hf isotopic constraints on the sources of the Indus Molasse, Ladakh Himalaya, India. Tectonics 26:TC2014. https://doi.org/10.1029/2006tc002051

Yin A (2006) Cenozoic tectonic evolution of the Himalayan orogen as constrained by along-strike variation of structural geometry, exhumation history, and foreland sedimentation. Earth-Sci Rev 76:1–31

Zhu B, Kidd WSAF, Rowley DB, Currie BS, Shafique N (2005) Age of initiation of the India-Asia collision in the East-Central Himalaya. J Geol 113:265–285, http://doi.org/10.1086/428805

Zyabrev SV, Kojima S, Ahmad T (2008) Radiolarian biostratigraphic constraints on the generation of the Nidar ophiolite and the onset of Dras arc volcanism: tracing the evolution of the closing Tethys along the Indus–Yarlung-Tsangpo suture. Stratigraphy 5(1):99–112

Chapter 8
Deccan Volcanic Province

8.1 Introduction

The earliest description of the Deccan Volcanic Province (DVP) characterized by the step-like profile of the hill-slopes may be traced to Sykes (1833; cf., Pascoe 1956) who coined the name derived from (Sanskrit word) "*Dakshin*" meaning south and (Swedish word) "*Trappa*" meaning stairway. The observations by Blanford (1867a, b, 1869), Foote (1876), Bose (1884), Fermor and Fox (1916), Auden (1949a, b, 1954) and others were compiled by Pascoe (1956) in his volumes on the Geology of India. The DVP attracted attention of the world community in the 1980s when its temporal proximity to the end-Cretaceous mass extinction event was recognized. Large volumes of data on the distribution, petrological, chemical characters and age of the DVP has been published in the last four decades (Subbarao and Sukheswala 1981; Subbarao 1988, 1999; Deshmukh and Nair 1996; Sheth and Vanderklyusen 2014; Mukherjee et al. 2017). It is difficult to enumerate all details of current knowledge on the DVP in a limited space, and hence much of the petrogenetic and related information has been kept limited to that which provides clues to tectonic events. It is recommended that the literature cited in these publications be accessed by the reader to understand original observations and data.

The DVP is one of the largest continental flood basalt (CFB) provinces of the world, occupying a contiguous exposed area of around 500,000 km^2 (Fig. 8.1). Its original extent is speculated to have been more than twice this, from which large parts were eroded away or down-faulted below the Arabian Sea. Geochronological and paleomagnetic studies demonstrated the proximity of the Indian Plate over the Reunion hotspot (Courtillot et al. 1986, 1988; Duncan and Pyle 1988), and the age of the Deccan volcanism straddling the Cretaceous–Paleogene (K–Pg) Boundary. The hot-spot related volcanism, in conjunction with the rifting of the Indian Plate from Madagascar and then Seychelles during the Cretaceous and the position of this

A. K. Jain et al., *Tectonics of the Indian Subcontinent*, Society of Earth Scientists Series, https://doi.org/10.1007/978-3-030-42845-7_8

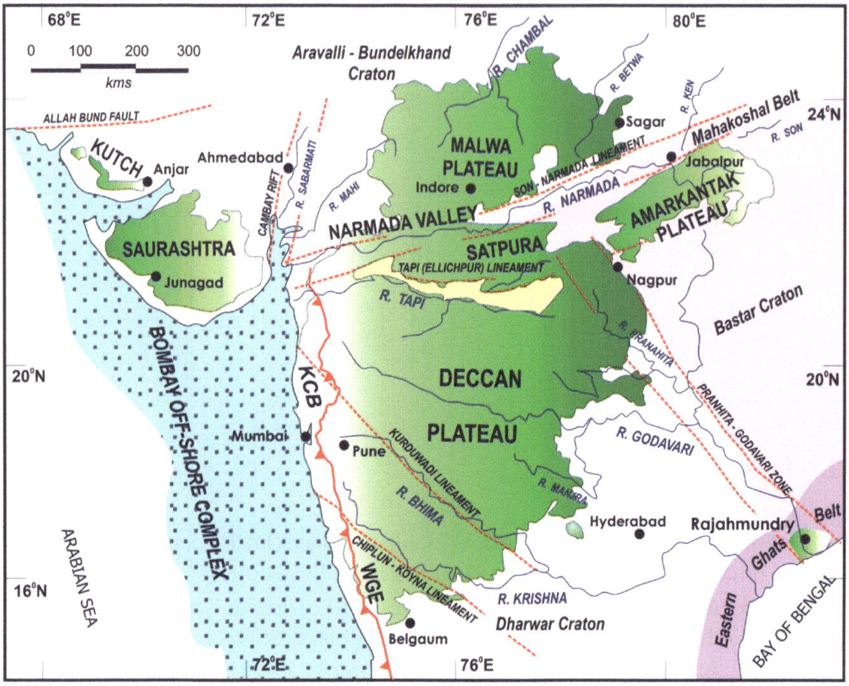

Fig. 8.1 Map depicting the geographic sectors of the Konkan Coastal Belt (KCB), Western Ghats Escarpment (WGE) and major structural zones transecting the Deccan Volcanic Province (DVP). After Kale et al. (2020). The different Precambrian blocks of Peninsular India are shown with differing shades of pink to bring out the diversity of sub-Trappean crust. The major regional lineaments (structural zones) with a Precambrian heritage along which post-Trappean deformation has been shown as red dotted lines. After Peshwa and Kale (1997) and Kale et al. (2017)

province on the western margin of India justify the recognition of the DVP as a CFB-VRM (continental flood basalts on a volcanic rifted margin) province (Bryan and Ernst 2008).

The sub-horizontal stratification of the ~2 km thick stack of basaltic flows that persists across several tens of km led to the mistaken traditional perception that it is a tectonically stable, structurally undisturbed terrain. The damaging intraplate earthquakes across this province in the last century (Bapat et al. 1983) revealed that this is not a tectonically stable continental block.

Reassessment of the stratigraphic configuration of the DVP (Kale et al. 2019) and the review of its volcanological aspects (Kale 2020) have questioned some of the established perceptions regarding it. The Deccan volcanism is recognized (Schöbel et al. 2014; Kale et al. 2019; Schoene et al. 2019; Sprain et al. 2019) to have occurred in not less than 3 phases between ~68 and ~61 Ma straddling the K–Pg Boundary (66.043 ± 0.004 Ma: Gradstein et al. 2012; Renne et al. 2013). Recent multidisciplinary studies suggest that this continental crustal block was subjected to pre-volcanic, syn-volcanic and post-volcanic tectonic events (Table 8.1). Each

Table 8.1 Chronology of events in the DVP

Age	Phase	Event
¼Holocene	Neotectonic	Differential uplift within some blocks attended by seismicity along block boundaries
Late Pliocene–Pleistocene	Post-volcanic	Scarp retreat and excavation of deep gorges & waterfalls along the WGE and SONATA. Block adjustments and local tectonic reactivations leading to antecedent drainage development, river capture and knick point development particularly along the structurally active zones transecting the DVP
Mid–Pliocene		Detrital (low-level) laterites along KCB
Miocene–Early Pliocene		Period of relative stability and amenable climate leading to wide planation surfaces, extensive (high-level) laterite formation across western India
Late Oligocene		Post-eruptive faulting (often by reactivation of ancient basement shears/zones of weakness) and differential uplift attended by development of the western coastal shelf and the passive continental margin of India
Eocene–Early Oligocene		Sedimentation in rifted zones along the Cambay rift and in Kutch
Late Paleocene (?Earliest Eocene)		Tilting of the Peninsular block towards East and establishment of the easterly drainage across the Peninsula. Volcanism in the Lakshadweep and offshore parts of Arabian Sea (southernmost Peninsular shield)
Early Paleocene (Late Danian) ~62 Ma	Syn-volcanic	Late volcanic phase of DVP (Bombay Group) with differentiates and intrusives (C28–C27)
Early Paleocene (Early Danian) ~65–64 Ma		Eruptive pulse in Mandla Subprovince & Second eruptive pulse in Western DVP (C29R–C29N transition)
Terminal Cretaceous (Late Maastrichtian) between 68 and 66 Ma		First large eruptive pulse of DVP (in Malwa, Central and Western DVP) over eroded basement (C30N–C29R)

(continued)

Table 8.1 (continued)

Age	Phase	Event
Late Cretaceous (Turonian/Santonian to? Campanian)	Pre-volcanic	Separation of India from Madagascar. Onset of the passage of the Indian continental block over the Reunion hotspot
Late Cretaceous (Late Campanian to Early Maastrichtian) ~68 ± 4 Ma		Infratrappean (Bagh Group, Lameta Formation and equivalents) sediments. Shallow rifting, local marine incursions and precursory igneous activity (intrusives and lava flows) in Gondwana basins and along Paleo-Narmada valley

Based on Gunnell et al. (2007), Hooper et al. (2010), Bonnet et al. (2016), Bhattacharya and Yatheesh (2015), Pande et al. (2017a), and Kale et al. (2019)

of these phases left hitherto unrecognized imprints on the volcanic and structural architecture of the Deccan Trap lava flows.

The pre-volcanic phase of tectonics covers events that have a connection with this large volcanic event and occurred during the Cretaceous prior to 68 Ma. This phase was unquestionably linked to the passage of the Indian Plate over the Reunion hotspot and includes continental rifting with mafic intrusions in the pre-Trappean terrains of the Indian sub-continent and seafloor spreading related to the Indo-Madagascar rifting. The syn-volcanic events cover the Maastrichtian–Danian eruption of voluminous basaltic flows associated with dramatic environmental implications and their contribution to the terminal Cretaceous mass extinction and the delayed faunal recovery in the Paleogene (Punekar et al. 2014, Font et al. 2016). The post-eruptive Cenozoic tectonism appears to include a phase of continental rifting and emergence of the western coast of India with associated seafloor spreading and a Quaternary phase of continent-interior neotectonic activity. The youngest events in this phase may be speculated to be linked to the far-field effects of the Indo-Asian collision. The available data on these three phases of tectonics that the DVP went through are discussed below.

8.2 Pre-volcanic Events

8.2.1 Antecedents

The Indian Peninsular Shield had been an emergent landmass and suffered an extensive period of erosion and denudation of the erstwhile Precambrian rocks during most of the Paleozoic Era. The breakup of Gondwanaland during the Late Paleozoic–early Mesozoic times led to the evolution of the intracontinental rifts that hosted the fluvio-lacustrine Gondwana Supergroup sediments during the Permian to Cretaceous times. The surrounding cratonic rocks continued to feed sedimentary detritus to the rifted basins during this period (Fig. 8.2). These basins gradually ceased to be active depocentres during the early Cretaceous times.

It is likely that these rifts and their extensions eventually gave way to the breakup of East Gondwana into the Indo-Madagascar-Seychelles block that drifted away from the Austro-Antarctic block. The linkage of this breakup with the Kerguelen hotspot (Coffin et al. 2002; Bredow and Steinberge 2018) is well established. In the Indian Plate, the impact of the Kerguelen hotspot is recorded in the 130–115 Ma old Rajmahal–Sylhet Traps (Kent 1991; Ray et al. 2005) and along the Ninety East Ridge in the Bay of Bengal extending further southwards into the eastern Indian Ocean.

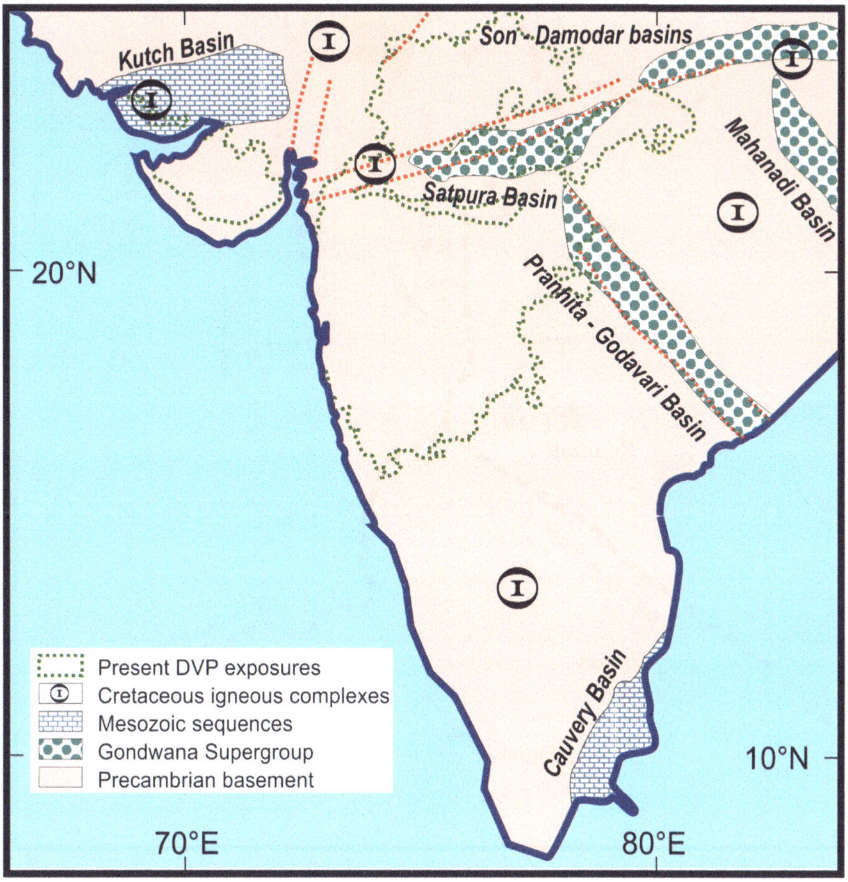

Fig. 8.2 Map depicting areas of Gondwana rifts, Mesozoic basins and rifts (with red dotted lines) in relation to the present outcrops of the DVP. The activity of sedimentation and dyke emplacement continued in some of the Gondwana basins during the Early Cretaceous times. Known occurrences of Late Cretaceous igneous complexes, discussed in the text are shown with encircled '**I**'

The separation between India and Madagascar has been established (Bhattacharya and Yatheesh 2015; Pande et al. 2017b) to have occurred during the Turonian to Santonian times between ~88 and ~83 Ma. This is followed by the northward drift of the Indian Plate that resulted in its coming under the influence of the Reunion hotspot in the southern Indian Ocean. These events led to opening of the Indian Ocean with associated oceanic volcanism (Fig. 8.3), as elaborated in Chap. 9 on the Western margin of India.

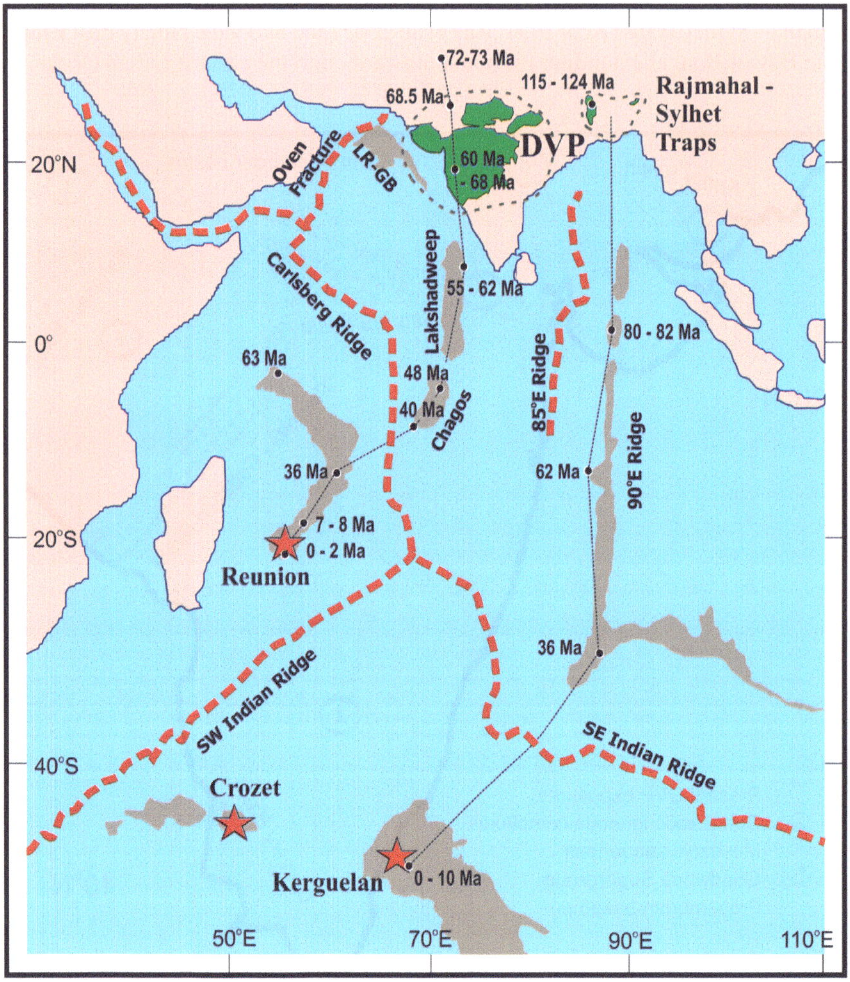

Fig. 8.3 Track of mafic magmatic activity and tectonic elements in the Indian Ocean. Location of the Laxmi Ridge–Gop Basin complex (LR-GB) that represents the Indo-Seychelles break-up (Bhattacharya and Yatheesh 2015) is depicted. Thin lines denote the inferred track of volcanics connecting the Rajmahal–Sylhet and Deccan Traps with the Kerguelan and Reunion hotspots respectively. Modified after Sheth (2000) and Kale (2020)

Isolated mafic magmatism is recorded from the different parts of Peninsular India during pre-Trappean times (encircled 'I' in Fig. 8.2). Late Cretaceous (~80 Ma) intrusives and dykes are known from the Dharwar craton (Kumar et al. 2001; Pande et al. 2001), Bastar craton (Chalapathi Rao et al. 2011, 2014), Gondwana Supergroup from the rifts in the Mahakoshal belt and Mundwara alkaline complex in Rajasthan (Pande et al. 2017a). The alkaline plugs, cones and sheet-like intrusions in the Mesozoic sequences in Kutch (Karmalkar et al. 2005, 2008, 2016) have been the early signals of a plume-lithosphere interaction and the possibility of magmatic underplating of the continental crust. Mantle xenoliths in these bodies (Karmalkar et al. 2000, 2014) have further supported this interpretation. The geographic extent and ages of such alkaline complexes (Parisio et al. 2016; Fosu et al. 2018) suggests that mafic magmatism in the north-western part of the Indian Peninsula preceded the eruptive phase of the DVP. They manifest the plume interactions, magmatic underplating of the continental crust and validate the paleogeographic position of the Indian Plate over the Reunion hotspot around 70 ± 2 Ma (Fig. 8.3).

8.2.2 Subtrappean Basement

The Deccan Trap lavas rest with an erosional nonconformity upon a variety of sequences. In Malwa Plateau, they are underlain in the north by the Archean gneissic rocks of the Aravalli–Bundelkhand Cratons as well as the Proterozoic Vindhyan Supergroup and the Mahakoshal mobile belt (Fig. 8.4a). Patchy exposures of the Late Cretaceous Bagh Group (in the western side) and the Lameta Formation (on the east) occur along the Narmada valley. The basalts in the Amarkantak Plateau are underlain by the Proterozoic mobile belts and underlying Archean gneisses of the Bastar craton in the south and the Mahakoshal mobile belt in the north; with thin patchy occurrences of the Gondwana and Lameta sediments. A complex relation induced by structural dislocations is encountered between the Deccan Traps and the Gondwana Supergroup (and its basement) along the Satpura belt between the Narmada and Tapi valleys (Fig. 8.4b). The entire southern margin of the DVP exposes a direct overlap of the Deccan basalts on the sequences from the Dharwar craton, including Archean granite greenstone belts, gneisses, and Proterozoic Kaladgi and Bhima sediments. In the Kutch and Saurashtra regions, the Deccan Traps are underlain by the Mesozoic sequences.

One of the interesting aspects of the base of the Deccan Traps is the presence of patchy exposures of Late Cretaceous sedimentary beds below them in Kutch, Saurashtra, along the Narmada valley up to Jabalpur and below the Amarkantak Plateau (Mehr 1995; Shrivastava and Ahmad 2005). In many of these exposures, except those of the Bagh Group in the central Narmada valley, the sedimentary sequences contain interbedded lava flows. This renders some of these beds to an

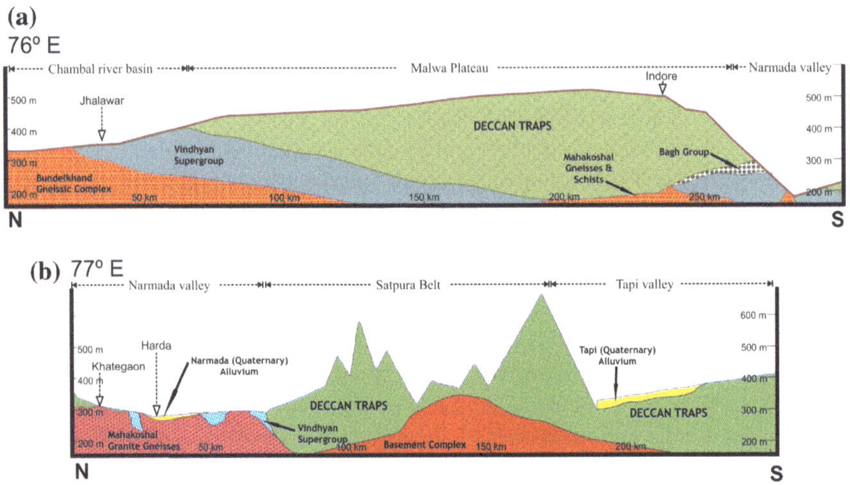

Fig. 8.4 Generalized N–S cross-sections depicting the topographic profiles across the **a** Malwa Plateau along the 76° E longitude, and **b** Satpura belt along the 77° E longitude. The thickness of the Deccan Traps has been plotted based on geophysical estimations (Kaila 1988; Murthy et al. 2014) and the geological logs based on GSI (2001) plotted by Kale et al. (2019). The dipping basalts produce a rugged topography of the Satpura belt, while they are subhorizontal in the Malwa Plateau resulting in a smooth planar topography. Complex folded, faulted and intrusive interrelationships between the older sequences that form the basement of the Deccan Traps are excluded in the sketches

'inter-trappean' rather than 'infra-trappean' stratigraphic status (Tandon et al. 1995). This shows that the early erupted Deccan lavas spread into these sedimentary basins which were coeval with the early phase of volcanism of the Deccan. Such sedimentary sequences should be better termed as Trap-related sediments to avoid this confusion. These beds host a rich assemblage of Maastrichtian flora and fauna including numerous egg-clutches of dinosaurian species (Mahobey and Udhoji 1996; Sahni and Bajpai 1988; Keller et al. 2009a, b; Samant et al. 2014; Srikarni et al. 2017; Kapur and Khosla 2018). This has been one of the cornerstones of the linkage of the Deccan volcanism with the terminal Cretaceous faunal turnover.

There is no doubt that the Precambrian rocks occurring below the Deccan Traps were subjected to a protracted period of continental weathering and erosion prior to them being covered by the lavas. Such terrains consequently display a mature peneplained topography with relict hill ranges of more resistant rocks. Several authors on the stratigraphic correlations of the Deccan lavas (e.g. Cox and Hawkesworth 1985; Jay and Widdowson 2008; Richards et al. 2015) assumed a flat peneplained surface of the basement below the Deccan lavas. They appear to have disregarded significant paleotopographic undulations (representing the relict hill ranges) with amplitudes of more than 200 m (Jain et al. 1995; Kale et al. 2014) that are evident along most of the fringes of the current exposures of the DVP and were remarked upon even by Blanford (1867a, b) and Auden (1949a). Even the elevation corrected basement–Trap contact, recorded from the 9 boreholes drilled around the Koyna Warna Seismic zone (Gupta et al. 2017; Arora et al. 2018), displays undulations of the order of +100 m within a linear distance of 50 km. Regional profiles across the DVP (using logs generated from geological maps; Kale et al. 2019) also demonstrate that the model of a smooth, gently sloping per-Trappean topography does not stand the test of factual observations. The consequent speculations on correlations of the lava flows across long distances (Self et al. 2008) are equally ambiguous.

Another aspect of the Trap–basement relation that has not been explained satisfactorily is the remarkable absence of any thermal imprint on the infra-trappean strata. These sediments have large proportions of calcareous limestones, marls, etc. that are susceptible to recrystallization and/or thermal metamorphism at temperatures of less than 200 °C. How they were insulated from any thermal imprints of the basaltic lavas, erupted at temperatures ranging between 800 and 1100 °C, remains intriguing. No baking effect or alterations have been recorded from such infra-trappean sediments that are immediately capped by the basaltic lava flows, barring the development of wollastonite at the contact between the basaltic lava and Shahabad Limestone in the Bhima basin on the southeastern edge of the DVP (Kale and Peshwa 1995).

8.2.3 Other Aspects

Whether the passage over a mantle plume below the Reunion hotspot caused the doming in the Indian Plate or not (Sheth 2007) remains ambiguous. What is certain is that during the Maastrichtian times (between ~72 and ~68 Ma), prior to the commencement of the volcanism, several small and large continental sedimentary basins dotted the central parts of the Indian Peninsular. There is evidence of a marine incursion along the (present day) alignment of the Narmada valley during this period, indicating subsidence of rift-bounded blocks of the crust.

Several off-shore marine and deep oceanic basins may also have evolved between Madagascar and the Indian Plate during this period. Calvés et al. (2011) have inferred that a large volcanic platform representing a submarine flood basalt event (estimated between 75 and 65.5 Ma) was present off the coast of India. Whether this should be treated as the early phase of Deccan volcanism or not remains to be established, but it is certainly a part of the sequence of events resulting from the passage of India over the Reunion hotspot that eventually led to the continental flood basalt eruptions of the DVP.

8.3 Deccan Volcanism

8.3.1 Aerial Extent and Volume

The piles of the basaltic flows in the 500,000 km^2 area on the Indian continental block of the DVP (Fig. 1) have a maximum exposed thickness of more than 1650 m (at Kalsubai peak = 1646 m above msl) along the Western Ghats Escarpment (WGE). Recent drilling in the Koyna-Warna seismic zone (Mishra et al. 2017) yielded an uninterrupted 1251 m thick stack of subhorizontal basaltic flows, demonstrating that the Trap–basement contact occurs up to 400 m below msl in this area. The maximum uninterrupted thickness of the Deccan basalts adds up to more than 2000 m; which has also been endorsed by geophysical methods (Kaila 1988; Patro and Sarma 2008).

This yields an estimated volume of ~2.8×10^5 km^3 for the present day exposures of these flood basalts, assuming thinning of the pile towards its fringes (Kale et al. 2020). Jay and Widdowson (2008) had estimated a volume of 13×10^5 km^3 for the Deccan volcanism, while Richards et al. (2015) suggested that the volume may be half of this estimate. It is therefore safe to assume that the volume of basalts originally erupted on the continental parts exceeded 7×10^5 km^3 of which only half is presently preserved after erosional removal. If one were to club the Maastrichtian–Paleogene marine flood basalts occurring in the Indian Ocean (Calvés et al. 2011; Pande et al. 2017b) as a part of the Deccan LIP, the estimated volume may be ~20×10^5 km^3.

Recent high precision dating has confirmed that this volcanism (including its precursors and late events) lasted between 70 and 60 Ma, with the main continental flood basalts being erupted in a shorter span of less than 5 million years (Schoene et al. 2019; Sprain et al. 2019). Whether or not there was a change in the average eruptive tempo as postulated by Richards et al. (2015), Renne et al (2015) is open to debate, given that the stratigraphic model used by them is itself questionable (Kale, et al. 2019). The eruption of ~7×10^5 km^3 in about 5 million years yields an average rate of 0.14 km^3/year; assuming a continuous streaming of lava for the entire duration.

8.4 Eruptive Style

8.4.1 Lava Flows and Flow Fields

The Deccan Traps essentially display a sheet-like geometry with the lateral spread being 50–00 times larger than the thickness. Individual lava flow ranges in thickness between 5 and 25 m, although exceptionally thick flows (~100 m) have also been recorded. Uninterrupted lateral continuity across several tens of km is best seen along the escarpment faces of hill ranges in this province. Closer examination of many of the thick, widespread 'flows' indicate that they are flow-fields of several flows and lobes, rather than singular lava flows.

The individual lava flows are often separated by 'interflow horizons' of variable thicknesses. Many are constituted of volcaniclastic tuffaceous material that may or may not be baked and oxidized; and are recognized as 'boles'. Some of them may be pedogenic altered powdery layers derived from the flow-top breccia or the chilled crusts (Sayyed 2014; Srivastava et al. 2018). These types of interflow horizons occur as thin (less than a meter thick) impersistent units within the thick piles of flows across the DVP. In some flow-fields, individual flows may not display a distinctive interflow horizon, suggesting that the lava emplacement occurred in quick succession and the capping flow spreads over while the earlier unit is cooling down. Several interflow horizons, display sedimentary stratification. Such inter-trappean sediments are more common along the fringes of the DVP and are sometimes fossiliferous. All such interflow horizons manifest a gap between the emplacements of the successive lava flows.

8.4.2 Morphological Types

The basaltic lava flows in the DVP display three internal layers, namely the crust, core and base (Kezsthelyi et al. 1999) from top to bottom, respectively. Based on their internal structure (relative proportions of the upper crust, core and base, distribution of vesicles, etc.) and geometry, Walker (1971) and Deshmukh (1988) classified the Deccan basaltic flows into (i) compound with multiple units of pāhoehoe lobes, (ii) simple and (iii) áā types.

A large variety of morphological types and their lateral transitions were recorded from the DVP by Bondre et al. (2000, 2004), Duraiswami et al. (2001, 2003a, 2004, 2008, 2014), Kashyap et al. (2010), Brown et al. (2011), Sen (2017), Sheth (2018a). Figure 8.5 provides a snapshot of the major morphological types of lavas recognized from the DVP. Most of the earlier mapped áā flows (e.g. Brown et al. 2011) can be shown to be laterally grading into slabby pāhoehoe morphologies (Duraiswami et al. 2003b, 2014), leading to doubts whether the typical Hawaiian áā flows are indeed present in this province.

Differences in volumetric rates, streaming versus pulsed emplacement, cooling and vapour loss governed by local conditions during the lateral transfer of the extruded lava result in these morphological variations as modeled in recent volcanological studies (e.g. Harris et al. 2007; Katterhorn and Schaefer 2008; Guest et al. 2012; Cashman et al. 2013; Glaze and Baloga 2013; Òskarsson and Riishuus 2014; Bernardi et al. 2015). Kale et al. (2019) suggested that the observed morphologies in the DVP represent a continuous variation series between two end-members, 'lobate' and 'sheet' flows. The latter have a magnitude smaller height–length ratio than the former.

The lobate flows are akin to the typical channel-fed pāhoehoe flows with small volumes that cool as individual lobes (Fig. 8.5). Early loss of vapour phase results in enhanced viscosity of the lava that inhibits long distance spreading, leading to a pile up close to the vent. When such lobes are emplaced in rapid succession,

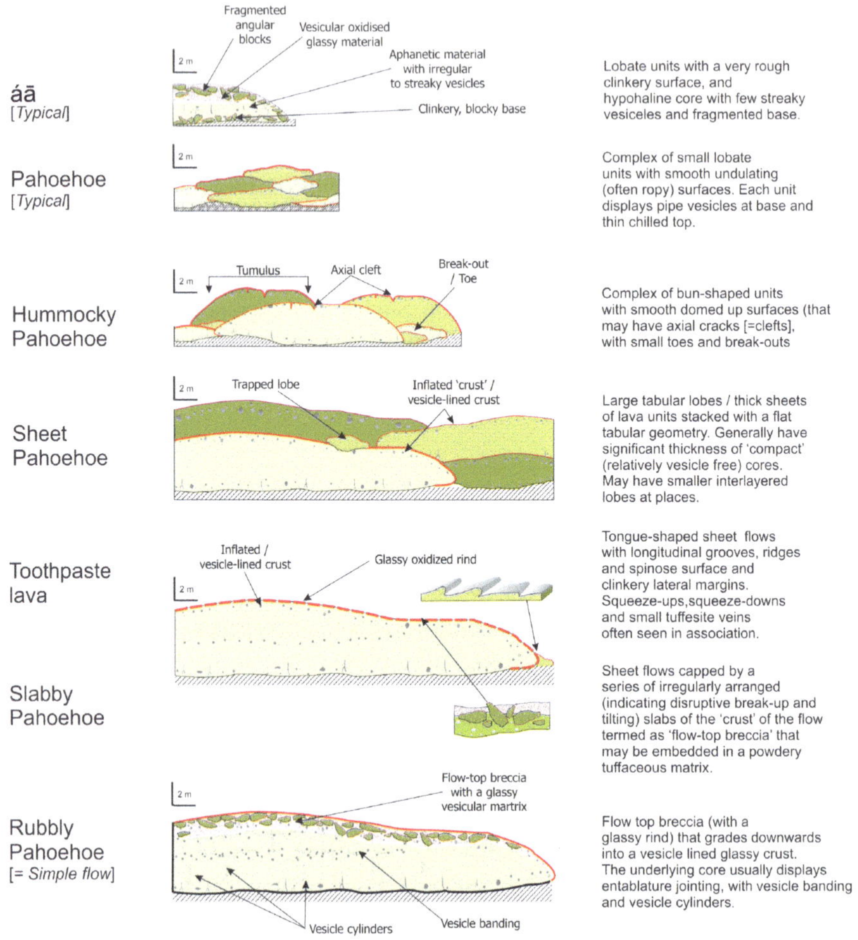

Fig. 8.5 Major morphological types recognized in basaltic flows from the DVP after Duraiswami et al. (2014). Kale et al. (2019) classified the morphologies into a continuous variation where the 'lobate flows' display the internal structure and geometry of the hummocky and typical pāhoehoe types, while the 'sheet flows' include all other morphological types.

they may be annealed together into a 'compound' flow (sensu Deshmukh 1988). The sheet flows have a much larger aerial spread and may display internal structures comparable to sheet/slabby/rubbly pāhoehoe along their length of exposures. A larger volumetric rate of semi-continuous emplacement is responsible for the development of sheet flows. The early formed rapidly chilled crust in such flows impedes vapour loss, retaining fluidity in the lava; and permits a relatively slower cooling of the core. This mechanism (demonstrated by Hon et al. 1994 and refined in subsequent years by several workers) allows a wider spread of the lava far from its eruptive edifice. Such flows are traceable continuously across long distances and were earlier mapped as 'simple' flows (GSI 2001). This emerging volcanological reassessment of the architecture of the lava flows from the DVP suggests that they are better compared with the Icelandic volcanic sequences than the traditional comparison with the Hawaiian lavas.

The distribution of these two end-member morphological types (lobate and sheet flows/flow-fields) across the province is depicted in Fig. 8.6. The morphological classification of some of the sectors of the DVP (northern Malwa and Saurashtra in particular) is unclear and hence they are shown as 'unclassified or mixed type' in this figure. The mixed type distribution indicates that both end-member types occur in subequal proportions in the sequences exposed in the Amarkantak Plateau, along the Narmada and Tapi valleys and in parts of the northern Konkan Coastal Belt (KCB) and between Mumbai and Pune.

Some of the sheet flows have been continuously mapped extending across more than 500 km^2 with thicknesses of the order of >70 m (Choubey 1973; Chatterjee and Dash 2017). Such flows and the gaps in eruptive sequence (indicated by the interflow horizons), indicate that the actual eruptive rate in the Deccan may have been much larger than the province-wide average of 0.14 km^3/year. In comparison the Grimsvötn eruptions in Iceland vented basaltic lava at the rate of about 7.25 km^3/year (Cas and Wright 1987).

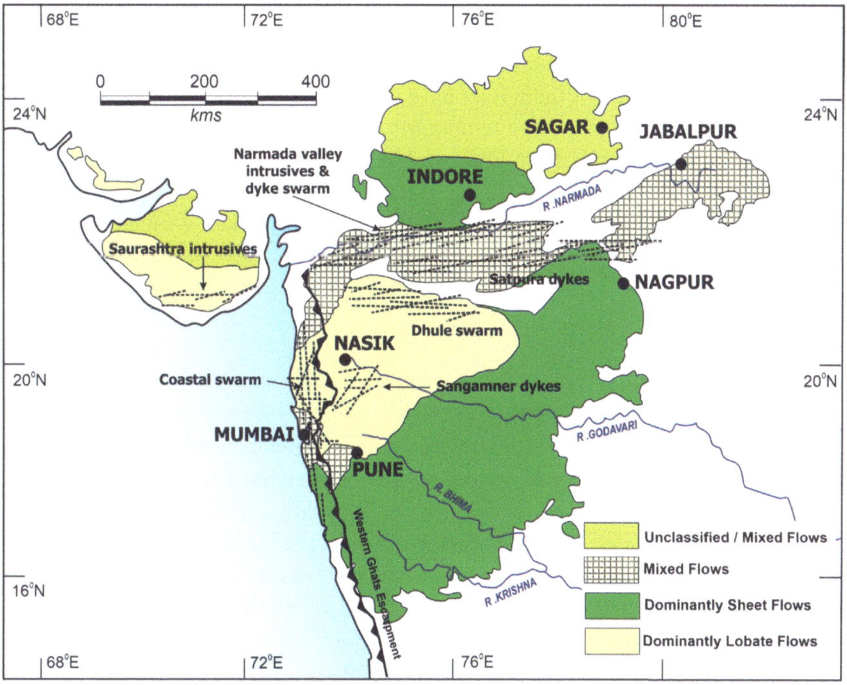

Fig. 8.6 Distribution of the flow-types (Kale et al. 2019) and the locations and trends of major dyke swarms in the DVP (Kale et al. 2020)

The chemical stratigraphy established in the western parts of the DVP in the late 1980s (Table 8.2) led to the emergence of a central shield volcanic model for this province. This appeared to be consistent with the distribution of the lavas based on the Hawaiian volcanological classification (Deshmukh 1988). Subsequent workers (e.g. Subbarao and Hooper 1988; Mitchell and Widdowson 1991; Jay and Widdowson 2008; Self et al. 2008; Shrivastava et al. 2014; Richards et al. 2015; Sprain et al. 2019) have used the assumption of a main volcanic edifice located north of Nasik and overstepping of the chemical formations radially away from it, to model the Deccan volcanism. This monocentric eruptive model did not take on board evidence of the occurrence of several eruptive vents occurring across the DVP (Srinivasan et al. 1998). With the recognition of the morphological variations and their consequent implications on the correlation of lava flows/flow-fields across long distances (Kale et al. 2019) it became evident that the architecture of the Deccan lavas has greater similarities with the Icelandic piles of lavas (e.g. Óskarsson et al. 2017; Óskarsson and Riishuus 2014; Eibl et al. 2017) than the Hawaiian volcanism. Combined with the doubts regarding the capacity of chemical typing as an index of correlation, this indicates that the DVP must have a polycentric eruptive history rather than the currently popular monocentric one. Kale et al. (2019) have defended the earlier assumption of a dyke-fed fissure eruptive model for the Deccan lavas and advocated their comparison with Icelandic volcanic systems rather than Hawaiian volcanics.

Overall, the morphology and structure of the basaltic lavas in the DVP conform to their subaerial eruption. Rare occurrences of spilitic pillow lavas from Mumbai (Sukheswala 1974) and the Mandla subprovince by GSI (Bodas–personal communication) suggest that some of them may have erupted in submarine/subaqueous conditions. Localized lakes may have temporarily developed due to volcanism-related precipitation in some parts of the DVP, yielding the lacustrine inter-trappean sediments. Such localized lakes may also have hosted some of the interflow horizons that display stratification and evidence of accumulation of reworked weathered basaltic material.

8.4.3 Dykes

The dyke-fed fissure eruption model has however suffered from the absence of a¾ny known unambiguous exposure of a feeder dyke connected to a lava flow. The importance of the dykes as potential feeders for the lava flows was discussed earlier by Blanford (1867b), Auden (1949b), Peshwa et al. (1987), Deshmukh and Sehgal (1988) and Shrivastava et al (2017). Some correlations between lava flows and intrusive dykes were attempted based on the chemical types (Sheth 2000; Sheth et al. 2009; Vanderklyusen et al. 2011; Sheth and Cañón-Tapia 2015; Shrivastava et al. 2017), but they remained inconclusive speculations.

Intrusive dykes, sills and other bodies are well known from various parts of the DVP (Blanford 1869; Auden 1949b). Figure 8.6 depicts the location and major orientations of the dykes from the DVP. In the Narmada valley, the ENE–WSW

Table 8.2 Comparison of chemostratigraphy and lithostratigraphy of the subprovinces of DVP (after Kale et al. 2019)

Chemostratigraphy			Lithostratigraphy								
			Western Subprovince			Satpura (Central) Subprovince		Malwa (Northern) Subprovince		Mandla (Eastern) Subprovince	
Group	Subgroup	Formation	Group	Subgroup	Formation	Group	Formation	Group	Formation	Group	Formation
Deccan Traps			Sahyadri	Bombay	Borivali						
					Elephanta			Malwa	Singarchori		
	Wai	Desur		Khandala	Mahabaleshwar				Bargonda		
		Panhala				Satpura	Karanja				
		Mahabaleshwar			M4 (GPB)		Buldhana		Indore	Amarkantak	Kuleru
		Ambenali			Purandargarh						Khamla/Khampla
		Poladpur			Diveghat		Chikhli		Kankaria–Pirukheri		
	Lonavala	Bushe			Karla		Ajanta				Amarwara
		Khandala			Indrayani						Multai
	Kalsubai	Giravali GPB		Kalsubai	M3 (GPB)		Upper Ratangarh		Kalisindh		Linga
		Bhimashankar			Upper Ratangarh				Mandleshwar		Pipardahi
		Manchar GPB			M2 (GPB)						Dhuma
		Thakurwadi			Lower Ratangarfh		Karanja		Gaganwara		Mandla
		Tunnel 5 GPB			M1 (GPB)		Buldhana				Unclassified
		Neral			Salher		Chikhli		Singarchori		
		Kashele GPB									
		Igatpuri									
		Thalghat GPB									
		Jawhar									

Note that no lateral correlation of lithostratigraphic units is implied between the subprovinces dyke swarm cuts through not only the basaltic flows, but also their Cretaceous and Proterozoic basement sequences (Peshwa et al. 1987). A smaller set of E–W and NW–SE dykes (Deshmukh and Sehgal 1988; Rajurkar et al. 1990) is also known from this sector.

The Satpura dykes intrude the Gondwana sediments as well as their cap of Deccan basaltic flows and are at some places transected and dislocated by faults. All the other dyke swarms (Dessai and Bertrand 1995; Peshwa and Kale 1997; Bondre et al. 2006; Ray et al. 2007; Sheth et al. 2009, 2018) are exposed cutting across only the basaltic flows, since their basement is not exposed in those sectors. The N–S to NNE–SSW trending coastal dyke swarm in the KCB and the NNE–SSW and WNW–ESE oriented Sangamner dykes display bidirectional distribution, while the Dhule (also called Nandurbar–Dhule) swarm shows a preferred E–W orientation.

If the dyke-fed fissure eruption model (Lala et al. 2011; Eibl et al. 2017; Ju et al. 2017; Kale et al. 2020) is taken into consideration, these dyke swarms represent vestiges of the fissures that fed the basaltic flows. Their orientations reflect the extensional fractures that tapped the sub-crustal magma chambers and, therefore, can be used as proxies to the syn-volcanic stress systems. The passage of the Indian continental crust over the Reunion hotspot led to fracturing and rifting with an associated re-opening of earlier formed planes of weakness in the crust. Such deep crustal fractures provided amenable channels by the upwelling lava to come to the surface.

It is, therefore, quite likely as was speculated by Auden (1949a) that the Narmada valley and its southern extension (covering parts of the Satpura ranges and the Tapi valley) were the primary eruptive axis along which the earliest Deccan Trap basalts were erupted in an (N–S oriented) extensional tectonic setting. The anomalous crustal structure along the Narmada valley, manifested as a string of gravity highs, presence of mantle and crustal xenoliths in the dykes occurring along this linear zone and its westward extension in the Saurashtra–Kutch region (Ray et al. 2008; Karmalkar et al. 2000, 2016) endorse such a model. The Coastal and Sangamner dyke swarms (Fig. 8.6) and the Panvel flexure (Dessai and Bertrand 1995; Peshwa and Kale 1997) reflect an E–W oriented extensional system. It is likely that this stress system is linked to the India–Seychelles rifting and evolution of the western continental margin of the Indian Plate (Misra et al. 2014; Misra and Mukherjee 2017), besides being the source of the western sequences of basaltic flows. Some of these dykes also have been recorded to contain lower crustal and upper mantle xenoliths (Dessai et al. 2004). Gadgil et al. (2019) concluded that the southern (Goa) dyke swarm represents fracture-filled injections in pre-existing fractures in the basement; while the northern coastal dyke swarm is likely to be fissures that fed the extrusion of the flood basalts. The presence of the dyke swarms in regions of the DVP where the flow sequences display a mixed morphology further supports the inference that they were channels from which the upwelling lavas were extruded on to the surface. The mixed morphologies would result in the vicinity of the vents depending upon the variations in the volumetric rates of eruption as sequential batches of lava were pumped up from the sub-crustal magma chambers and extruded on the surface.

8.4.4 Eruptive History

The remarkable uniformity in the petrochemical composition of these basalt flows was recognized since the time of Washington (1922) and the minor variations in them across the province were limited essentially to the minor and trace-elemental compositions. Like other continental flood basalts (CFBs) with zero to negative ε_{Nd} and $^{87}Sr/^{86}Sr > 0.704$, the primary source of the Deccan magmas from enriched mantle sources has been established over the years (Peng et al. 1994; Krishnamurthy 2008; Manu Prasanth et al. 2019). Adiabatic decompression melting of the aesthenospheric anhydrous mantle peridotites at the head of a rising mantle plume is suggested (Lai et al. 2012) to be the primary sources of melts for the CFBs. Cox and Hawkesworth (1985) proposed that the primary melt resided in a sub-crustal magma chamber and underwent cycles of progressive fractionation and variable crustal contamination, followed by release (tapping by the volcanic vents) and periodic picritic replenishment. Sheth (2016) discussed the possibility of more than one sub-crustal magma chambers in the evolution of the Deccan lavas and a polycentric eruptive history of the DVP (Kale and Pande 2017). Variable degrees of crustal contamination recorded in different parts of the DVP, reflect the heterogeneous nature of the sub-Trappean basement (see Fig. 8.1), which is further corroborated by the xenoliths in various dyke swarms.

Within this broad generalization, details regarding the nature of the primary and secondary differentiation, compositional impact of its crustal trajectory through more than 20–25 km of continental crust and signatures of the diverse crustal segments below different parts of the DVP remain poorly constrained. Non-basaltic rocks (including rhyolites, andesitic tuffs, lamprophyres, andesitic and alkaline basaltic dykes; carbonatites and related igneous complexes like Girnar, Mundhwara and Ambadongar (Chandra et al. 2018) are an integral component of the DVP although they represent less than 5% of its volume. Nonetheless, the petrogenetic relation between them and the dominant tholeiitic basaltic magma remains enigmatic.

Using the minor variations in the compositional characters of basaltic flows in the western parts of the DVP, chemical stratigraphy was established in the mid-1980s (Cox and Hawkesworth 1985; Beane et al. 1986; Bodas et al. 1988; Khadri et al. 1988; Subbarao and Hooper 1988). This stratigraphy (see Table 8.2) became the foundation of most of the subsequent models of long distance correlations of lava packets and the eruptive history of this province (Mahoney et al. 2000; Self et al. 2008b; Jay and Widdowson 2008; Richards et al. 2015; Pathak et al. 2017). Recent studies and a critical analysis of the data for this model led to its being questioned (Sheth et al. 2018) on fundamental basics of stratigraphic principles. Using geochronological, paleomagnetic data, volcanological models and critical analysis of the chemical database, Kale et al. (2019) have advocated a subprovince-wise stratigraphy of the DVP, without any implied lateral correlations between subprovinces (Table 8.2). Geochronological data compiled by them shows that volcanism did not occur across the entire province at the same time (as postulated in the monocentric models), but was limited to only some subprovinces during each of the phases. This is further elaborated by Kale et al. (2020) taking into cognizance the paleomagnetic orientations and available age data for representative sections (Fig. 8.7).

Volcanism in the continental segment of the DVP occurred in not less than three distinct phases (Schoene et al. 2019; Sprain et al. 2019) but had limited geographic extents during each phase. The oldest and largest eruptive phase (68–66 Ma) spanned the magnetic chrons 30N–29R. This widespread volcanism occurred in Kutch–Saurashtra–Malwa and the Western DVP, straddles the K–Pg Boundary and could have contributed significantly to the environmental stresses leading to mass extinctions. Some of these earliest flows occur interbedded with Maastrichtian sediments, presently exposed along the fringes of the DVP (Fig. 8.7). The next phase occurred in Mandla and Satpura subprovinces during the Danian times between 65.8 and 64.0 Ma corresponding with the magnetic chrons 29N–28R–28N (Shrivastava et al. 2014, 2015, 2017). The uppermost lava sequence in the Western subprovince (with normal polarity) were coeval with this phase of eruptions. Some of the younger flows

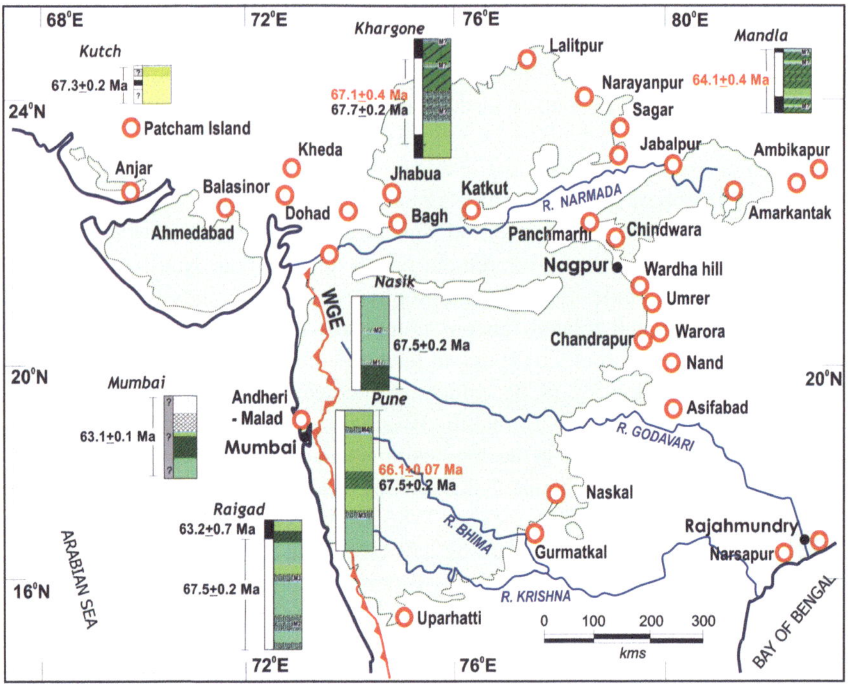

Fig. 8.7 Known localities of fossiliferous sediment associated with the Deccan Traps and the representative lithologs of the flow sequences plotted with geochronological and geomagnetic data (Kale et al. 2020). The magnetic polarity (white = Reverse; black = Normal; grey with question marks = mixed on indeterminate) is depicted alongside the lithologs. Weighted mean ages of each sector from available $^{40}Ar/^{39}Ar$ ages (normalized to FCs standard: Kale et al. 2019) are given on the side, with black lettering for whole-rock ages and red lettering for mineral ages. Note the scales of the logs of Nasik, Pune and Raigad are half that of the other columns

in the Malwa subprovince may belong to this episode as also some of the dykes occurring in the Narmada and Satpura belts that intrude the earlier flows. The youngest phase of Deccan volcanism occurred in the western coastal belt around Bombay and in the offshore region, approximately in the period of 63.5–62.0 Ma. The magnetic polarity data suggests that some of the youngest lavas from the Mandla subprovince may have been erupted during this phase, as well. This phase also appears to have yielded the non-basaltic volcanics, including the Bombay rhyolites, trachytes and some of the volcaniclastic and tuffaceous beds from parts of Saurashtra that are interbedded within the early Paleogene strata.

It can, therefore, be concluded that emerging multifaceted data on the DVP is significantly amending several earlier models and concepts regarding this continental flood basalt province. Exciting information is anticipated in the days to come, particularly with more geochronological and paleomagnetic data, which may further refine the emerging model of the Deccan volcanism.

8.5 Post-volcanic Events

8.5.1 Regional Pattern

Horizontally exposed stacks of lava flows, yielding picturesque cliffs all along the edges of peneplained plateau, are characters of the DVP that led to it being considered a structurally undeformed and stable continental block in traditional literature. Dipping flows indicating post-eruptive deformation have been known from around Mumbai; associated with the Panvel Flexure (Sukheswala and Poldervaart 1958), and in parts of the Satpura belt between the Narmada and Tapi river valleys for a long time (Auden 1949a, 1954). These two sectors were recognized as linear zones along which the Deccan lavas have been structurally deformed. Seismic events at Koyna (1967: M 6.3), Bhatsa (1983: M 4.2), Killari (1993: M 6.2), Jabalpur (1997: M 5.8) etc. within the Deccan Plateau disrupted the conventional notion of tectonic stability of the DVP, leading to a renewed interest in the structure of this province (Krishnaswamy 1981; Deshmukh and Sehgal 1988; Rastogi 1992) and its concealed basement (Brahmam and Negi 1973; Kaila1988; Ramachandran and Kesavamani 1998).

8.5.2 Regional Zones of Deformation

Some of the earlier studies (Kailasam 1979; Powar 1993; Radhakrishna 1993) concluded that these features of deformation are connected to deep crustal features that were reactivated in response to the cymatogenic uplift of the Deccan Plateau in the post-trappean times. The geomorphic studies in different parts of the Deccan Traps (e.g., Dikshit 1970, 1976; Kale and Rajaguru 1987, 1988) attributed several geomorphic anomalies in the DVP to either the step-like erosion of the stack of the basaltic flows or to climate changes. Others (Peshwa et al. 1987; Rajurkar et al. 1990; Harinarayana et al. 2007; Patro and Sarma 2008) interpreted them to be manifestations of post-trappean tectonism that was superimposed on the basaltic flows.

The average northward rate of drift of the Indian Plate was almost twice (v_{rms} > 150 mm/year) that of global peak plate motion rates (v_{rms} ~ 75 mm/year) since Triassic (Tetley et al. 2019). It is also relevant to note that geophysical data demonstrate that the sub-trappean basement is not a coherent one; rather it is fragmented into crustal blocks having differing rheological characters (Harinarayana et al. 2007; Raval and Veeraswamy 2011; Ravi Kumar et al. 2015; Rajaram et al. 2017). It is, therefore, not surprising that intra-plate stresses in this fast-moving continental plate were easily released along such pre-existing zones of weakness within the crust. Reactivation of the sub-trappean structures has concentrated deformation of the Deccan Trap basalts along them (Fig. 8.1) during the Quaternary times.

8.5.2.1 SONATA Zone

The ENE–WSE trending Son-Narmada Lineament (West 1962), also designated as the SONATA (Son–Narmada–Tapi) Zone, is a linear zone in central India, and is also referred to as the Central Indian Suture Zone. It has a prolonged geological history stretching from the Precambrian to present day (Shanker 1991; Valdiya 2016). Several regional faults of Precambrian origin have been mapped along this zone (GSI 2000; Kumar et al. 2019). A ~1000 m thick Quaternary deposit, informally called as Hoshangabad alluvium, has accumulated in the middle reaches of the Narmada valley. It is further divided into the Palikavar, Dhansi, Surajkund, Baneta and Hirdepur Formations of Pleistocene age, and is capped by the Bauras and Ramgarh Formations of Holocene age (Rahate and Solanki 2002). This is followed downstream by bedrock gorges (Kale 2003; Murthy et al. 2014; Chakraborty et al. 2019) cutting through the Precambrian rocks as well as the Deccan Traps. Several fault planes dislocating the Deccan Traps have been recorded in the Narmada valley (Fig. 8.8a).

Although known since Auden (1949b), the >250 km long Ellichpur fault, also termed as the Tapi Lineament and clubbed within the SONATA zone, is a neglected fault in the DVP (Copley et al. 2015). This southern boundary of the Satpura horst (Sheth et al. 2018) marks the boundary between the Quaternary Tapi alluvium in the south and the Deccan Traps in the north. The stratigraphy of the Deccan Traps north and south of this fault plane is different (Fig. 8.9), indicating that the sectors on either side of this fault had diverse eruptive histories during the eruptive phase of the Deccan Traps. The basaltic flows dip by more than 30° along it (Fig. 8.8b), and display regional open folds adjoining it.

The deformation and dislocation of the basaltic flows is suggestive of relative subsidence of crustal blocks in post-eruptive times along the SONATA zone. The presence of the Hoshangabad alluvium and the Tapi alluvium in the middle reaches of the Narmada and Tapi river valleys, supports this inference. These alluvial basins

Fig. 8.8 Deformations within the Deccan Traps. **a** Northward dipping and E–W trending fault plane along which the basaltic flows are dislocated >10 m. Loc. ~Sendhwa Toll booth on Mumbai Agra highway (AH4). Note displacement of interflow horizon along the contact (thin white line) between the two flows (A and B) and its down faulting along the hanging wall of the fault plane (dash line). The flows here display sub-horizontal disposition. Photo: Shilpa Patil Pillai. **b** Northward dipping flows, separated by the red interflow horizon, exposed along the Ellichpur fault north of Dhule. **c** Reverse fault in the Quaternary alluvium of the Pravara river near Chandanapuri. After Dole et al. (2000). **d** Exposure of the Sutarwadi fault in the Dhebewadi plateau reported by Bhave et al. (2017)

capping the Deccan Traps in several parts host the fossils of ancestors of several present-day animal and plant species (Salahuddin et al. 1997; Sonakia 1998; Chauhan 2008). Such a voluminous fluvial, intracontinental deposition is possible only if one accepts that the underlying block subsided progressively downwards and resulted in tectonically created intracontinental sinks into which fluvio-lacustrine sediments derived by erosion from the surrounding terrain was accumulated. Evidences of neotectonism have been recorded from the lower Naramda valley in the west (Chamyal et al. 2002; Joshi et al. 2013). The seismicity near Jabalapur at its eastern end is a recent manifestation of the activity along SONATA zone.

8.5.2.2 Kurduwadi Lineament (KL)

This lineament was first recognized as a NW–SE trending sub-Trappean rift zone (Brahmam and Negi 1973). Subsequent geological studies demonstrated that it is essentially an intracontinental shear along which dominantly strike-slip movements have dominated (Kale and Peshwa 1988; Peshwa and Kale 1997). The recent seismicity at Killari and around Bhatsa–Vaitarna valleys (Rastogi et al. 1986) in Thane

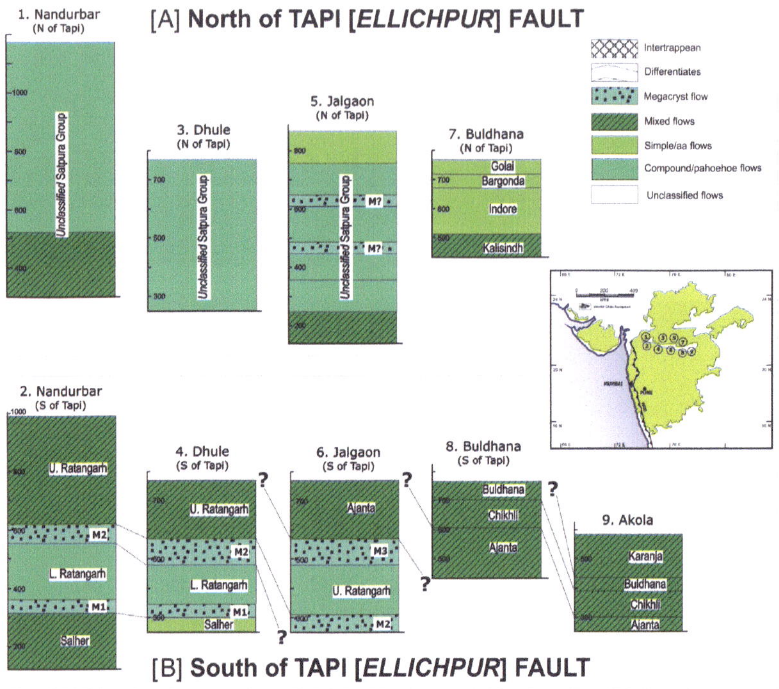

Fig. 8.9 District-wise logs based on the DRM sheets (GSI 2001) plotted against the elevations of exposure above msl. Note that the stratigraphic sequence across the Ellichpur fault is different and not comparable with each other. Derived from Kale et al. (2019)

district had been inferred by them to manifest the modern tectonic reactivation of this zone (Fig. 8.1). The tectonic atlas of this region (GSI 2000) also endorses the faults in the Thana district mapped earlier; while Misra et al. (2014) suggest that the strike-slip tectonics along this region accommodates the extensional regime associated with the India–Seychelles rifting and separation.

Evidences of neotectonic deformation of the Quaternary sediments (Fig. 8.8c) from the Pravara river valley (Dole et al. 2000, 2002) and the record of recent low intensity seismicity near Bote (in Ahmednagar district) further testify to its active nature. Based on an integrated geomorphic and structural study of the upland part of the northern Deccan Plateau, Kale et al. (2017) concluded that Quaternary movements of crustal blocks along the Kurduwadi lineament are responsible for the anomalous accumulation (and localized deformation) of sediments along the Mula and Pravara river valleys that drain this region.

8.5.2.3 Koyna-Warna Seismic Zone (KWSZ)

Following the earthquake (M = 6.3) in the vicinity of the Koyna region in December 1967, attention was focused on the possible causes of this intra-plate seismic event. This small area (Fig. 8.10) of about 40 × 20 km has been suffering from a sustained seismic activity for the last 50 years with more than 22 events of magnitude greater than 5 and nearly 200 events with magnitude greater than 4. It is believed to be one of the best documented cases of reservoir-triggered seismicity (Mallika et al. 2013; Catherine et al. 2015; Rao and Shashidhar 2016). Literature on investigations into the geological setup and the inferred causative mechanisms of this seismicity is voluminous (Yadav et al. 2016). Scientific drilling in this region was initiated in 2014 to understand this seismicity and its causes (Gupta et al. 2017). It is significant to note that notwithstanding the extensive examination of the fault-plane solutions by numerous authors over the last 50 years, there is no conclusive evidence or unambiguous conclusion on the orientation and nature of the main seismogenic fault responsible

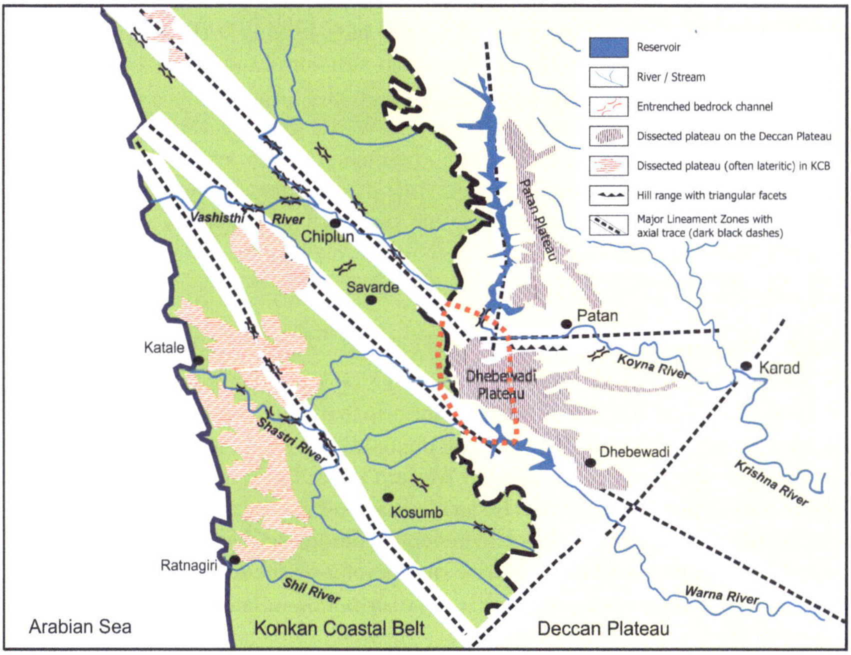

Fig. 8.10 Regional lineament zones and major geomorphic features in the Koyna-Warna seismic zone, marked by red dotted outline. Modified after Kale et al. (2014). Note that all the plateaux in this region are capped by laterites. The ones in the KCB are at elevations of <200 m above msl, while Patan and Dhebewadi plateaux occur at average elevations exceeding 700 m. The black dashed line marks the escarpment face of the WGE in this area

for these events. Absence of any surface expression of faulting in this region and the 1000 m thick pile of subhorizontal basalts (apparently structurally undisturbed) added to the uncertainties in these studies.

Recent compilations and re-examination of the geomorphic and geological setup of this region, with a focus on the structural features of this region brought out several geomorphic anomalies and fracture zones that indicate recent activity (Kale et al. 2014). It is evident that the NW–SE trending Chiplun–Warna lineament is a regional zone of shearing that has strong expressions on the Plateau region as well as the coastal tract (Fig. 8.1). Besides shear fractures recorded along this zone, it has several geomorphic anomalies, such as triangular facets of hill ranges and abrupt structurally controlled drainage diversions aligned along it that suggest that the landscape evolution has been subjected to strong structural controls along it. The presence of NW–SE trending faulting (Fig. 8.8d) with a dominant strike-slip component and low vertical displacement was recently recorded (Bhave et al. 2017) which has dislocated the Deccan Trap flows in this region. It is significant that such a shear zone trending NW–SE has been recorded from the basement of the Deccan Traps from magnetic potential mapping and its correlation with Bouguer gravity data (Rajaram et al. 2017). Arora et al. (2018) have used LIDAR mapping to infer a close correspondence between the basement structures and the mapped regional lineaments traversing the Deccan Traps in this region. These emerging studies are indicating that there is a tectonically active NW–SE trending zone underlying the KWSZ, that has roots in the Precambrian basement and has exerted control on the geomorphic development of the terrain. How this regional basement feature or its active segments contribute to focussing the sustained seismicity in a small area of the KWSZ is a question that remains open to alternative explanations as of now.

8.5.2.4 Konkan Coastal Belt

The Konkan Coastal Belt is a narrow coastal strip that is less than 100 km at its widest and runs in a N–S direction across more than 400 km within the DVP. It is flanked on its west by the Arabian Sea and the Western Ghats Escarpment (WGE) along its eastern margin that towers over it by an elevation difference of not less than 500 m. Several lines of evidence including (a) general geomorphic characters of the drainage network, (b) continuity of the lava flow stratigraphy (in the southward overstepping model of Cox and Hawkesworth 1985), (c) distribution of laterite caps, (d) physical and geomorphic continuity of the coastal strip southwards beyond the DVP into the basement terrain; etc. have been used by different authors to explain the evolution of this narrow coastal strip. The evolution of this coastal strip has been generally attributed to an erosional retreat of the hypothetical West Coast fault that developed due to the rifting between India and Seychelles (e.g. Powar 1981; Hooper et al. 2010). This model assumes that the lava flows that once covered the present exposures have been eroded away. Others believe that the coastal strip emerged during the easterly titling of the Indian Peninsula and may not be a structural feature at all.

The presence of coastal dyke swarm is significant, particularly in the northern parts of the KCB (Fig. 8.6). Their preferred N–S orientations with a minor conjugate set (NNE–SSW and NW–SE) of orientations in the Thane and Raigad districts is remarkable. Structurally, they would indicate the presence of a strong E–W extensional system leading to their development. The possibility of their being feeders to the adjoining lava flows has been discussed by Vanderklyusen et al. (2011). The presence of NW–SE trending shear faults and fractures dislocating the basaltic flows that transect across selected sectors of the KCB (adjoining or overlying the KL and KWSZ) is relevant in this context. Such features can be better explained by assuming a rejuvenation of ancient (inherited) structural features during the extensional tectonics suffered by this part of the DVP.

The basaltic flows in this belt display a distinctive westerly tilt of 3°–5°. The basaltic flows affected by the Panvel flexure have dips of up to 20° towards the west and give an appearance of being a monoclonal flexure fold. The Panvel flexure with westward monoclonal dips of the lava flows is associated with faulting along strike-slip faults and fracture zones. These are superimposed upon the basaltic flows along the KCB (Figs. 8, 12 of Misra et al. 2014) and are linked with the Indo-Seychelles rifting, seafloor spreading (Misra et al. 2014; Pande et al. 2017b; Samant et al. 2019) and northward drift of western continental margin of India (Sheth 1998).

Besides the structural manifestations recorded in the Deccan Trap basalts, the KCB has several hot-springs aligned N–S all along its length (Pitale et al. 1987; Minissale et al. 2000; Chandrasekhar et al. 2018). They suggest that the coastal belt is prone to shallow crustal tectonic activity in the present day. This aspect has been often missed out by several authors who modelled the evolution of the KCB as an artefact of ancient rifting or to climatic changes or eustatic sea-level changes.

8.6 Geomorphic Studies

The present-day geomorphology of the Deccan Volcanic Province (DVP) represents a composite result of progressive erosion of the lava pile with its structural complexities and climatic variations through time exerting their influence on landscape development. There is no doubt that the present landscape is of Holocene age, but it bears inherited signatures of the earlier denudational history of this province. A large body of archeological studies across this province (Salahuddin et al. 1997, Sonakia 1998, Chouhan 2008; Wescoat 2019) from the anomalously accumulated Quaternary alluvial deposits along different rivers flowing across the DVP has yielded a rich assemblage of fossils and associated evidences of humanoid settlements in this terrain, ranging back more than 100,000 years (Wescoat 2019). This provides the basis of examining the geomorphic features and anomalies in this province for evidences of Quaternary and Recent tectonic events. Compilations and quantifications of the geomorphology of the DVP by Dikshit (1970, 1976, 2001), Gupte and Rajaguru (1971), Vaidyanadhan (1977, 1987), and Subramanyan (1981) laid the foundations of the present knowledge of the geomorphology of the DVP that have been compiled in Gunnel and Radhakrishna (2001) and Kale and Vaidyanadhan (2014).

On a regional scale (Fig. 8.1), the DVP is characterized by three major dissected Plateaux (Malwa, Amarkantak and Deccan), besides the undulating Saurashtra Peninsula and the KCB. All the major southern peninsular rivers (except the southernmost Cauvery) have their source within the DVP (Fig. 8.1). The Malwa Plateau is drained northward by the Rivers Chambal and Betwa, which eventually join the River Ganga. Its western extremities are drained by River Mahi. The Godavari and Krishna river basins drain the Amarkantak and Deccan Plateaus southeastwards. The anomalously westward draining Narmada and Tapi basins (Auden 1949a; West 1962) occupy the inter-Plateau regions and, as enumerated above, are located along subsided linear zones of intracontinental structural deformation during the Quaternary times. The Saurashtra peninsula and the KCB drain into the Arabian Sea through several small river valleys. Steep escarpments mark the edges of the Plateaux of which the Western Ghats Escarpment (WGE) is perhaps the most spectacular stretching more than 700 km from north to south and having heights of up to 1500 m.

Recognition of geomorphic characters as indices of tectonic activity (Burbank and Anderson 2012) and the advent of various software techniques of terrain analysis coupled with the availability of digital terrain data have led to a plethora of geomorphic studies in the DVP. Geomorphic development on the subhorizontal stacks of basaltic flows led to the erstwhile perception that its present configuration is a result continental-scale isostatically driven cycles of uplift and peneplanation on a gently eastward tilted Indian Peninsula (Dikshit 1970; Vaidyanadhan 1977; Subramanyan 1981). The WGE was inferred to be a half-rift shoulder resulting from rifting and separation of India from Madagascar and Seychelles, which contributed to the Plateau uplift along with the eruption of the Deccan Traps (Kailasam 1979; Watts and Cox 1989; Powar 1993; Widdowson and Cox 1996; Harbour and Gunnel 2007; Sheth 2007). Within this framework, several geomorphic anomalies recorded from this province (Rajaguru and Kale 1985; Kale and Rajaguru 1987, 1988; Rajaguru et al. 1993; Sadakata et al. 1995; Joshi and Kale 1997; Kale 2003; Kale and Shejwalkar 2008) were attributed to climate change rather than tectonics. Reassessment of the geomorphic parameters in many cases across the DVP (e.g. Kale et al. 1986b; Kundu and Matam 2000; Misra et al. 2014; Dole et al. 2017; Misra and Mukherjee 2017) have suggested that there is sufficient geological evidence to demonstrate that these anomalous patterns reflect Quaternary tectonics and relative dislocations within adjoining blocks within the DVP. The tectonic movements resulted in creation of structural controls that altered the antecedent drainage already developed upon the DVP and resulted in such anomalous patterns.

8.7 Concluding Remarks

It is evident that newer data is that is emerging from the Deccan Volcanic Province (DVP) is opening new lines of thought on the evolution of this continental flood basalt province. Several ideas that were earlier considered resolved are being questioned by this data. Besides its large aerial spread and its association with the terminal Cretaceous environmental crisis and eventual faunal turnover (Keller et al. 2009b; Punekar et al. 2014), this province is the host to a large population of Quaternary fossils along its fluvial alluvial tracts (Chauhan 2008), including early evidence of human settlements across this province (Wescoat 2019), not to mention the human population currently supported by it. Its tectonic evolution, landscape evolution and potential natural hazards (like floods and seismicity) have therefore great significance that cannot be ignored.

It may be concluded that these aspect of the evolution of the Deccan Volcanic Province (DVP) are being fed exciting new multidisciplinary data that is challenging many established concepts of its tectonic stability. It is anticipated that future studies may bring out numerous more examples of tectonic controls on landscape development of this terrain that was once clubbed as an undeformed block of the Indian Plate. This will provide an accurate assessment of the overall Quaternary and Recent evolution of the landscape on the DVP; and enable its comparison with other tectonic events from the Indian Plate covering Peninsular India, Indo-Gangatic plains, Himalayan tectonic belt, and the marine segments from the Arabian Sea, Bay of Bengal and the Indian Ocean.

References

Arora K, Srinu Y, Gopinadh D, Chadha RK, Raza H, Mikhailov V, Ponomarev A, Kiseleva E, Smirnov V (2018) Lineaments in the Deccan Basalts: the basement connection in the Koyna-Warna RTS region. Bull Seism Soc Amer 108(5B):2919–2932. https://doi.org/10.1785/0120180011

Auden JB (1949a) A geological discussion on the Satpura hypothesis and the Garo-Rajmahal gap. Proc Indian Nat Sci Acad 15:315–340

Auden JB (1949b) Dykes in Western India–a discussion on their relationship with the Deccan Traps. Trans Nat Inst Sci 3:127–157

Auden JB (1954) Erosional patterns and fracture zones in Peninsular India. Geol Mag 91:89–101

Bapat A, Kulkarni R, Guha SK (1983) Catalogue of earthquakes in India and neighbourhood from the historical period up to 1979, p 233. Indian Soc Earthq Techn, Roorkee, pp 233

Beane JE, Turner CA, Hooper PR, Subbrao KV, Walsh JN (1986) Stratigraphy, composition and form of the Deccan basalts, Western Ghats, India. Bull 48:61–83

Bernardi MI, Bertotto GW, Jalowitzki TLR, Orihashi Y, Ponce AD (2015) Emplacement history and inflation evidence of a long basaltic lava flow located in southern Payenia volcanic province, Argentina. J Volcanol Geotherm Res 293:46–56

Bhattacharya GC, Yatheesh V (2015) Plate tectonic evolution of the deep ocean basins adjoining the western continental margin of India—a proposed model for the early opening scenario. In: Mukherjee S (ed) Petroleum geoscience: Indian contexts. Springer, Switzerland, pp 1–61

Bhave N, Siddique MI, Desai J, Patil Pillai S, Dole G, Kulkarni H, Kale VS (2017) Faulting in Deccan Traps in the vicinity of Koyna-Warna seismic Zone. Mem Geol Soc India 90:748–751
Blanford WT (1867a) On the geology of a portion of cutch. Mem Geol Surv India 6:17–38
Blanford WT (1867b) On the traps and intertrappean beds of Western and Central India. Mem Geol Surv India 6:137–162
Blanford WT (1869) On the geology of the tapti and lower Nerbuda valley and some adjoining districts. Mem Geol Surv India 6:163–384
Bodas MS, Khadri SFR, Subbarao KV (1988) Geology and stratigraphy of the Jawahar and Igatpuri Formations, Western Ghats lava pile, India. Mem Geol Soc India 10:253–280
Bondre NR, Dole G, Phadnis VM, Duraiswami RA, Kale VS (2000) Inflated pahoehoe lavas from the Sangamner area of the Western Deccan Volcanic Province. Current Sci 78:1004–1007
Bondre NR, Duraiswami RA, Dole G (2004) Morphology and emplacement of flows from the Deccan Volcanic Province, India. Bull Volcanol 66:29–45
Bondre NR, Hart WK, Sheth HC (2006) Geology and geochemistry of the Sangamner mafic dike swarm, Western Deccan Volcanic Province, India: implications for regional stratigraphy. J Geol 114:155–170
Bose PN (1884) Geology of the lower Narmada valley between Nimawar and Kawant. Mem Geol Surv India 21:1–72
Brahmam NK, Negi JG (1973) Rift valleys beneath the Deccan Traps (India). Geophys Res Bull 11:207–223
Bredow E, Steinberger B (2018) Variable melt production rate of the Kerguelen hot spot due to long-term Plume-Ridge interaction. Geophys Res Lett 45:126–136. https://doi.org/10.1002/2017/GL075822
Brown RJ, Blake S, Bondre NR, Phadnis VM, Self S (2011) A 'ā' lava flows in the Deccan Volcanic Province, India, and their significance for the nature of continental flood basalt eruptions. Bull Volcan 73:737–752
Bryan SE, Ernst RE (2008) Revised definition of large igneous provinces (LIPs). Earth Sci Rev 86:175–202
Bryan SE, Riley TR, Jerram DA, Stephens CJ, Leat PT (2002) Silicic volcanism: an undervalued component of large igneous provinces and volcanic rifted margins. Geol Soc Amer Spec Paper 362:99–120
Burbank DW, Anderson RS (2012) Tectonic Geomorphology, 2nd edn. John Wiley & Sons, UK, p 478
Calvés G, Schwab AM, Huuse M, Clift PD, Gaina C, Jolley D, Tabres AR, Inam A (2011) Seismic volcano stratigraphy of the Western Indian rifted margin: the pre-Deccan igneous province. J Geophys Res 116:28. https://doi.org/10.1029/2010JB000862
Cas RAF, Wright JV (1987) Volcanic successions modern and ancient. Allen & Unwin, London, UK, p 528
Cashman KV, Soule SA, Mackey BH, Deligne NI, Deardorff ND, Dietterich HR (2013) How lava flows: new insights from applications of lidar technologies to lava flow studies. Geosphere 9:1664–1680. https://doi.org/10.1130/GES00706.1
Catherine JK, Gahalaut VK, Kundu B, Ambikapathy A, Yadav RK, Bansal A, Narasaiah M, Naidu SM (2015) Low deformation rate in the Koyna–Warna region, a reservoir triggered earthquake site in west-central stable India. J Asian Earth Sci 97:1–9
Chakraborty PP, Tandon SK, Saha S (2019) Development of Phanerozoic basins of India. J Asian Earth Sci 184:9. https://doi.org/10.1016/j.jsaeaes.2019.103991
Chalapathi Rao NV, Burgess R, Lehmann B, Mainkar D, Pande SK, Hari KR, Bodhankar N (2011) ^{40}Ar–^{39}Ar ages of mafic dykes from the Mesoproterozoic Chattisgarh basin, Bastar craton, Central India: implication for the origin and spatial extent of the Deccan Large Igneous Province. Lithos 125:994–1005
Chalapathi Rao NV, Lehmann B, Balaram V (2014) Platinum—ground element (PGE) geochemistry of Deccan orangeites, Bastar craton, central India: implications for non-terrestrial origin of iridium enrichment at the K-Pg boundary. J Asian Earth Sci 84:24–33

Chamyal LS, Maurya DM, Bhandari S, Raj R (2002) Late Quaternary geomorphic evolution of the lower Narmada valley, Western India: implications for neotectonic activity along the Narmada–Son fault. Geomorphology 46:177–202

Chandra J, Paul D, Viladkar SG, Sensarma S (2018) Origin of the Amba Dongar carbonatite complex, India and its possible linkage with the Deccan large igneous province. In: Sensarma S, Storey BC (eds) Large igneous provinces from Gondwana and adjacent regions. Geol Soc Spec Publ 463, pp 137–169

Chandrasekhar T, Minissale A, Vaselli O, Chandrasekharam D, Singh HK (2018) Understanding the evolution of thermal fluids along the western continental margin of India using geochemical and boron isotope signatures. Geothermics 74:197–209, https://doi.org/10.1016/j.geothermics.2018.03.007

Chatterjee P, Dash S (2017) Building up of 3D volcanic facies architecture of the Diveghat formation of Deccan Traps. Geol Surv India Open-File Report, no. RP/CR/MH/2014/69, 337p

Chauhan PR (2008) Large mammal fossil occurrences and associated archeological evidence in Pleistocene contexts of peninsular India and Sri Lanka. Quatern Inter 192:20–42

Choubey VD (1973) Long-distance correlation of Deccan Basalt flows, central India. Geol Soc Amer Bull 84:2785–2790

Coffin MF, Pringle MS, Duncan RA, Gladczenko TP, Storey M, Muller RD, Gahagan LA (2002) Kerguelen hotspot magma output since 130 Ma. J Petrol 43:1121–1139

Copley A, Mitra S, Sloan RA, Gaonkar S, Reynolds K (2015) Active faulting in the apparently stable Peninsular India: rift inversion and a Holocene age great earthquake on the Tapi Fault. J Geophys Res Solid Earth 119:6650–6666

Courtillot V, Besse J, Vandamme D, Montigny R, Jaeger JJ, Cappetta H (1986) Deccan flood basalts at the Cretaceous/tertiary boundary? Earth Planet Sci Lett 80:361–374. https://doi.org/10.1016/0012-821X(86)90118-4

Courtillot V, Feraud G, Maluski H, Vandamme D, Moreau MG, Besse J (1988) Deccan flood basalts and the Cretaceous/tertiary boundary. Nature 333:843–846

Cox KG, Hawkesworth CJ (1985) Geochemical stratigraphy of the Deccan Traps at Mahabaleshwar, Western Ghats, India, with implications for open-system magmatic processes. J Petrol 26:355–377

Deshmukh SS (1988) Petrological variations in compound flows in Deccan Traps and their significance. Mem Geol Soc India 10:305–319

Deshmukh SS, Nair KKK (eds) (1996) Deccan basalts. Gondwana Geol Mag Spec Issue 2, pp 1–543

Deshmukh SS, Sehgal MN (1988) Mafic dyke swarms in Deccan Volcanic Province of Madhya Pradesh and Maharashtra. Mem Geol Soc India 10:323–340

Dessai AG, Bertrand H (1995) The "Panvel flexure" along the Western Indian continental margin: an extensional fault structure related to Deccan magmatism. Tectonophysics 241:165–178

Dessai AG, Viegas A (2010) Petrolgenesis of alkaline rocks from Murud–Janjira in the Deccan Traps, Western India. Mineral Petrol 98:297–311

Dessai AG, Markwick A, Vaselli O, Downes H (2004) Granulite and pyroxenite xenoliths from the Deccan Trap: insight into the nature and composition of the lower lithosphere beneath cratonic India. Lithos 78:263–290

Dikshit KR (1970) The Western Ghats: a geomorphic overview. In: Gunnell Y, Radhakrishna BP (eds) Sahyadri. Mem Geol Soc India 47, pp 159–184

Dikshit KR (1976) Drainage basins of Konkan. forms and characteristics. Nat Geograph J India 22:79–105

Dikshit KR (2001) The Western Ghat: a geomorphic overview. In: Gunnell Y, Radhakrishna BP (eds) Sahyadri. Geol Soc India, Mem 47, pp 159–183

Dole G, Peshwa VV, Kale VS (2000) Evidence of a paleoseismic event from the Deccan plateau uplands. J Geol Soc India 56:547–555

Dole G, Peshwa VV, Kale VS (2002) Evidence of neotectonism in Quaternary sediments from Western Deccan Upland Region, Maharashtra. Mem Geol Soc India 49:91–108

Dole G, Patil PS, Upasani D, Kale VS (2017) Triggering of the largest Deccan eruptions by the Chicxulub impact: comment. Geol Soc Amer Bull 129:253–255
Duncan RA, Pyle DG (1988) Rapid eruption of the Deccan flood basalt at the Cretaceous-Tertiary Boundary. Nature 333:841–843
Duraiswami RA, Bondre NR, Dole G, Phadnis VM, Kale VS (2001) Tumuli and associated features from the Western Deccan Volcanic Province, India. Bull Volcan 63:435–442
Duraiswami RA, Bondre NR, Dole G (2003a) Enigmatic spiracle like structures from a basaltic flow near Chekewadi, Western Deccan Volcanic Province. Curr Sci 85:101–104
Duraiswami RA, Dole G, Bondre NR (2003b) Slabby pahoehoe from the Western Deccan Volcanic Province: evidence for incipient pahoehoe–aa transitions. J Volcanol Geotherm Res 121:195–217
Duraiswami RA, Bondre NR, Dole G (2004) Possible lava tube system in a hummocky lava flows at Daund, Western Deccan Volcanic Province, India. Proc Indian Acad Sci (Earth Planet Sci) 113:818–830
Duraiswami RA, Bondre NR, Managave S (2008) Morphology of rubbly pahoehoe (simple) flows from the Deccan Volcanic Province: implications on style of emplacement. J Volcanol Geotherm Res 177:822–836
Duraiswami RA, Gadpallu P, Shaikh TN, Cardin N (2014) Pāhoehoe–a'ā transitions in the lava flow fields of the Western Deccan Traps, India—implications for emplacement dynamics, flood basalt architecture and volcanic stratigraphy. J Asian Earth Sci 84:146–166
Eibl EPS, Bean CJ, Jónsdóttir I, Höskuldsson A, Thordarson T, Coppola D, Witt T, Walter TR (2017) Multiple coincident eruptive seismic tremor sources during the 2014–2015 eruption at Holuhraun, Iceland. J Geophys Res Solid Earth 122:2972–2987. https://doi.org/10.1002/2016JB013892
Fermor LL, Fox CS (1916) The Deccan Trap flows of Linga, Chhindwara district, Central Provinces. Rec Geol Surv India 47:81–136
Font E, Adatte T, Sial AN, De Lacerda LD, Keller G, Punekar J (2016) Mercury anomaly, Deccan volcanism and the end-Cretaceous mass extinction. Geology 44:171–174
Foote RB (1876) The geological features of the south Mahratta country and adjacent districts. Mem Geol Surv India 12:70–138
Fosu B, Ghosh P, Chew DM, Viladkar SG (2018) Composition and U-Pb ages of apatite in the Amba Dongar carbonatite–alkaline complex, India. Geol J. https://doi.org/10.1002/gj.3350
Gadgil R, Viegas A, Iyer SD (2019) Structure and emplacement of the coastal Deccan tholeiitic dyke swarm in Goa, on the Western Indian rifted margin. Bull Volcanol 81:1–19. https://doi.org/10.1007/s00445-019-1297-6
Glaze LS, Baloga SM (2013) Simulation of inflated pāhoehoe lava flows. J Volcanol Geotherm Res 255:108–123
Gradstein FM, Ogg JG, Schmitz M (eds) (2012) The Geological time scale 2012. Elsevier, Boston 1 and 2, pp 1144. https://doi.org/10.1016/b978-0-44-59425-9.00004-4
GSI (2000) Seismotectonic Atlas of India and its environs. In: Narula PL, Acharyya SK, Banerjee J (eds) Geol Surv India Pub, Kolkata
GSI (2001) District resources map series. Geol Surv India Pub, Kolkata
Guest JE, Duncan AM, Stofan ER, Anderson SW (2012) Effect of slope on development of pahoehoe flow fields: evidence from Mount Etna. J Volcanol Geotherm Res 219–220:52–62
Gunnell Y, Radhakrishna BP (eds) (2001) Sahyadri: the great escarpment of the Indian subcontinent. Mem Geol Soc India 47, pp 1054
Gunnell Y, Braucher R, Bourlès D, André G (2007) Quantitative and qualitative insights into bedrock landform erosion on the South Indian craton using cosmogenic nuclides and apatite fission tracks. Geol Soc Amer Bull 119:576–585
Gupta HK, Arora K, Rao NP et al (2017) Investigations of continued reservoir triggered seismicity at Koyna, India. In: Mukherjee S, Misra A, Calves G, Nemčok M (eds) Tectonics of the Deccan large igneous province. Geol Soc London Spec Pub 445, pp 151–188
Gupte RB, Rajaguru SN (1971) Late Pleistocene geomorphologic history of rivers of Western Maharashtra. Bull Volcan 35:686–695

Harbor D, Gunnell Y (2007) Along–strike escarpment heterogeneity of the Western Ghats: a synthesis of drainage and topography using digital morphometric tools. J Geol Soc India 70:411–426

Harinarayana T, Patro BPK, Veeraswamy K, Manoj C, Naganjaneyulu K, Murthy DN, Virupakshi G (2007) Regional geoelectric structure beneath Deccan Volcanic Province of the Indian subcontinent using magnetotellurics. Tectonophysics 445:66–80

Harris AJL, Dehn J, Calvari S (2007) Lava effusion rate definition and measurement: a review. Bull Volcan 70:1–22

Hon K, Kauahikaua J, Denilinger R, Mackay K (1994) Emplacement and inflation of pahoehoe sheet flows: observations and measurements of active lava flows on Kilauea Volcano, Hawaii. Geol Soc Amer Bull 106:351–370

Hooper P, Widdowson M, Kelley S (2010) Tectonic setting and timing of the final Deccan flood basalt eruptions. Geology 38:839–842

Jain SC, Nair KKK, Yedekar DB (1995) Geology of the Son–Narmada–Tapti Lineament zone, central India. Geol Surv India Spec Publ 10:1–154

Jay AE, Widdowson M (2008) Stratigraphy, structure and volcanology of the SE Deccan continental flood basalt province: implications for eruptive extent and volumes. J Geol Soc London 165:177–188

Joshi VU, Kale VS (1997) Colluvial deposits in Northwest Deccan, India: their significance in the interpretation of late Quaternary history. J Quatern Sci 12:391–403

Joshi PN, Maurya DM, Chamyal LS (2013) Morphotectonic segmentation and spatial variability of neotectonic activity along the Narmada Son Fault, Western India: Remote sensing and GIS analysis. Geomorp 180–181:292–306. https://doi.org/10.1016/j.geomorph.2012.10.023

Ju W, Hou G, Hari KR (2017) Dyke emplacement in the Narmada rift zone and implications for the evolution of the Deccan Traps. In: Mukherjee S, Misra A, Calves G, Nemčok M (eds) Tectonics of the Deccan large igneous province. Geol Soc Spec Publ 445, pp 297–315

Kaila KL (1988) Mapping the thickness of Deccan Trap flows in India from DSS studies and inferences about a hidden Mesozoic Basin in the Narmada-Tapti Region. Mem Geol Soc India 10:91–116

Kailasam LN (1979) Plateau uplift in Peninsular India. Tectonophysics 61:243–269

Kale VS (2003) Fluvial geomorphology of Indian rivers: an overview. Prog Phys Geogr 26:400–433

Kale VS (2020) Cretaceous volcanism in Peninsular India: Rajmahal–Sylhet & Deccan Traps. In: Tandon SK, Gupta N (eds) Geodynamics of the Indian plate. Springer Geology, https://doi.org/10.1007/978-3-030-15989-4_8

Kale VS, Pande K (2017) Polycentric multi-phase deccan volcanism: is causative linkage to K-Pg Boundary valid? GSA annual convention seattle: abstracts with programs, vol 49, no 6. https://doi.org/10.1130/abs/2017am-294069

Kale VS, Peshwa VV (1987) Manifestations of the Kurduwadi lineament based on remote sensor data interpretation. Curr Sci 57:1293–1295

Kale VS, Peshwa VV (1995) Bhima Basin. Geological Survey of India, Bangalore, p 142

Kale VS, Rajaguru SN (1987) Late Quaternary alluvial history of the Northwestern Deccan upland region. Nature 325(12):612–614

Kale VS, Rajaguru SN (1988) Morphology and denudation chronology of the coastal and upland river basins of Western Deccan Trappean landscape (India): a collation. Z Geomorph NF 32(3):311–327

Kale VS, Shejwalkar N (2008) Uplift along the western margin of the Deccan Basalt Province: is there any geomorphic evidence? J Earth System Sci 117:959–971

Kale VS, Vaidyanadhan R (2014) The Indian peninsula: geomorphic landscapes. In: Kale VS (ed) Landscapes and landforms of India. Springer, Berlin, pp 65–78. https://doi.org/10.1007/978/94/017/8029/2/6

Kale VS, Ganjoo RK, Rajaguru SN, Salahuddin (1986a) A link-channel occupational site of acheulian man, upper Krishna valley, Karnataka. Current Sci 55(21):1073–1075

Kale VS, Rajaguru SN, Rajagopalan G (1986b) Late holocene evidence of neotectonics in the upper Vashishti valley (Western Maharashtra). Current Sci 55(24):1240–1241

Kale VS, Survase V, Upasani D (2014) Geological mapping in the Koyna–Warna region. ACWADAM Technical Report no. 2014–C1:163. https://doi.org/10.13140/2.1.3098.0809

Kale VS, Dole G, Upasani D, Patil Pillai S (2017) Deccan plateau uplift: insights from the Western Uplands, Maharashtra, India. In: Mukherjee S, Misra A, Calves G, Nemčok M (eds) Tectonics of the Deccan large igneous province. Geol Soc London Spec Publ 445, pp 11–46

Kale VS, Dole G, Shandilya P, Pande K (2019) Stratigraphy and correlations in Deccan Volcanic Province, India: quo vadis? Geol Soc Amer Bull. https://doi.org/10.1130/B35018.1

Kale VS, Bodas MS, Chatterjee P, Pande K (2020) Emplacement history and evolution of the Deccan Volcanic Province, India. Episodes In Press

Kapur VV, Khosla A (2018) Faunal elements from the Deccan volcano-sedimentary sequences of India: a reappraisal of biostratigraphic, palaeoecology, and palaeobiogeographic aspects. Geol J 32. https://doi.org/10.1002/gj.3379

Karmalkar NR, GriffinWL O'Reilley SY (2000) Ultramafic xenoliths from Kutch (NW India): plume related mantle samples? Int Geol Rev 42:416–444

Karmalkar NR, Rege S, Griffin WL, O'Reilley SY (2005) Alkaline magmatism from Kutch, NW India: implications for plume–lithosphere interaction. Lithos 81:101–119

Karmalkar NR, Kale MG, Duraiswami RA, Jonnalagadda M (2008) Magma underplating and storage in the crust-building process beneath the Kutch region, NW India. Curr Sci 94:1582–1588

Karmalkar NR, Kale MG, Duraiswami RA, Jonnalagadda M (2014) Mid-Cretaceous Lamproite from the Kutch region, Gujarat, India: genesis and tectonic implications. Gondwana Res 26:942–956

Karmalkar NR, Duraiswami RA, Jonnalagadda M, Griffin WL, Gregoire M, Benoit M, Delpech G (2016) Magma types and source characterization of the early Deccan magmatism, Kutch Region, NW India: insights from the geochemistry of igneous intrusion. In: Thakkar MG (ed) Recent studies on the geology of Kutch. Geol Soc India Spec Publ 6, pp 193–208

Kashyap M, Shrivastava JP, Kumar Raju (2010) Occurrence of small scale inflated pāhoehoe lava flows in the Mandla lobe of the eastern Deccan Volcanic Province. Curr Sci 98:72–76

Kattenhorn SA, Schaefer CJ (2008) Thermal–echanical modeling of cooling history and fracture development in inflationary basalt lava flows. J Volcanol Geotherm Res 170:181–197

Keller G, Adatte T, Bajpai S, Mahobey DM, Widdowson M, Khosla A, Sharma R, Khosla SC, Gertsch B, Fleitmann D, Sahni A (2009a) K–transition in Deccan Traps and intertrappean beds in central India mark major marine seaway across India. Earth Planet Sci Lett 282:10–23

Keller G, Sahni A, Bajpai S (2009b) Deccan volcanism, the KT mass extinction and dinosaurs. J Biosci: Indian Acad Sci 34:709–728

Kent R (1991) Lithospheric uplift in Eastern Gondwana: evidence for a long-lived mantle plume system? Geology 19:19–23

Keszthelyi L, Self S, Thordarson T (1999) Applications of recent studies on the emplacement of basaltic lava flows to the Deccan Traps. In: Subbarao KV (ed) Deccan Volcanic Province. Mem Geol Soc India 43, pp 485–520

Khadri SFR, Subbarao KV, Hooper PR, Walsh JN (1988) Stratigraphy of the Thakurwadi formation, Western Deccan Basalt Province, India. In: Subbarao KV (ed) Deccan flood basalts. Mem Geol Soc India 10, pp 281–304

Krishnamurthy P (2008) Cretaceous continental flood basalt magmatism in India. Mem Geol Soc India 74:261–297

Krishnaswamy VS (1981) The Deccan volcanic episode, related tectonism, and geothermal manifestations. Mem Geol Soc India 3:261–297

Kumar A, Pande K, Venkatesan TR, Bhaskar Rao YJ (2001) The Karnataka late Cretaceous dykes as products of Marion hot spot at the Madagascar-India breakup event: evidence from ^{40}Ar–^{39}Ar geochronology and geochemistry. Geophys Res Letts 28:2715–2718

Kumar P, Tewari HC, Sreenivas B (2019) Seismic structure of the central Indian crust and its implications on crustal evolution. J Geol Soc India 93:163–170

Kundu B, Matam A (2000) Identification of probable faults in the vicinity of Harnai–Ratnagiri region of the Konkan coast, Maharashtra, India. Curr Sci 78:1556–1560
Lai S, Qin J, Li Y, Li S, Santosh M (2012) Permian high Ti/Y basalts from the eastern part of the Emeishan large igneous province, Southwestern China: petrogenesis and tectonic implications. J Asian Earth Sci 47:216–230
Lala T, Choudhary AK, Patil SK, Paul DK (2011) Mafic dykes of Rewa basin, central India: implications on magma dispersal and petrolgenesis. In: Srivastava RK (ed) Dyke swarms: keys for geodynamic interpretation. Springer, Berlin, pp 144–166
Mahobey DM, Udhoji SG (1996) Fauna and flora from late Cretaceous (Maestrichtian) non-marine Lameta sediments associated with Deccan volcanic episode, Maharashtra: its relevance to the K-T boundary problem, paleoenvironment and paleogeography. Spec Publ Gondwana Geol Soc 2:349–364
Mahoney JJ, Sheth HC, Chandrasekharam D, Peng ZX (2000) Geochemistry of flood basalts of the Toranmal section, Northern Deccan Traps, India: implications for regional Deccan stratigraphy. J Petrol 41(7):1099–1120
Mallika K, Gupta H, Shashidhar D, Purnachandra Rao N, Yadav A, Rohilla S, Satyanarana HVS (2013) Temporal variation of the b value associated with M ~4 earthquakes in the reservoir-triggered seismic environment of the Koyna–Warna region, Western India. J Seismol 17:189–195
Manu Prasanth MP, Hari KR, Santosh M (2019) Tholeiitic basalts of Deccan large igneous province, India: an overview. Geol J 14. https://doi.org/10.1002/gj3497
Mehr SS (1995) Geology of Gujarat. Geological Society of India, Bangalore, p 222
Minissale A, Vaselli O, Chandrasekharam D, Magro G, Tassi F, Casiglia A (2000) Origin and evolution of 'intracratonic' thermal fluids from Central-Western Peninsular India. Earth Planet Sci Lett 181:377–394
Mishra S, Misra S, Vyas D, Nikalje D, Warhade A, Roy S (2017) A 1251 m thick Deccan flood basalt pile recovered by scientific drilling in the Koyna region, Western Maharashtra. J Geol Soc India 90:788–794
Misra AA, Mukherjee S (2017) Dyke–brittle shear relationships in the Western Deccan strike-slip zone around Mumbai (Maharashtra, India). In: Mukherjee S, Misra A, Calves G, Nemcok M (eds) Tectonics of the Deccan large igneous province. Geol Soc London Spec Publ 445, pp 269–295
Misra AA, Bhattacharya G, Mukherjee S, Bose N (2014) Near N-S paleo-extension in the Western Deccan region, India: does it link strike-slip tectonics with the India-Seychelles rifting? Int J Earth Sci 103:1645–1680. https://doi.org/10.1007/s00531.014/1021/x
Mitchell C, Widdowson M (1991) A Geological map of the Southern Deccan Traps, India and its structural implications. J Geol Soc London 148:495–505
Mukherjee S, Misra AA, Calvés G, Nemcok M (eds) (2017) Tectonics of the Deccan large igneous province. Geol Soc London Spec Publ 445, pp 345
Murthy ASN, Sarkar D, Sen MK, Sridher V, Prasad ASSSRS (2014) Mapping the sub-trappean mesozoic sediments in the Western part of the Narmada-Tapi region of Deccan Volcanic Province, India. J Asian Earth Sci 93:15–24
Óskarsson BV, Riishuus MS (2014) The mode of emplacement of neogene flood basalts in eastern Iceland: facies architecture and structure of simple aphyric basalt groups. J Volcanol Geotherm Res 289:170–192
Óskarsson BV, Anderson CB, Riishuus MS, Sørensen EV, Tegner C (2017) The mode of emplacement of neogene flood basalts in eastern Iceland: the plagioclase ultraphyric basalts in the grænavatn group. J Volcanol Geotherm Res 289:170–192
Pande K, Sheth HC, Bhutani R (2001) ^{40}Ar–^{39}Ar age of St, Mary Island volcanics, southern India: record of India-Madagascar break-up on the Indian subcontinent. Earth Planet Sci Letts 193:39–46
Pande K, Cucciniello C, Sheth H, Vijayan A, Sharma KK, Purohit R, Jagadeesan KC, Shinde S (2017a) Polychronous (early Cretaceous to palaeogene) emplacement of the Mundwara alkaline complex, Rajasthan, ^{40}Ar–^{39}Ar geochronology, petrolchemistry and geodynamics. Int J Earth Sci 106:1487–1504

Pande K, Yatheesh V, Sheth H (2017b) $^{40}Ar/^{39}Ar$ dating of the Mumbai tholeiites and Panvel flexure: intense 62.5 Ma onshore-offshore Deccan magmatism during India-Laxmi Ridge-Seychelles breakup. Geophys J Int 210:1160–1170

Parisio L, Jourdan F, Marzoli A, Melluso L, Sethna SF, Bellieni G (2016) $^{40}Ar/^{39}Ar$ ages of alkaline and tholeiitic rocks from the northern Deccan Traps: implications for magmatic processes and the K-Pg boundary. J Geol Soc London 173:679–688

Pascoe EH (1956) Manual of the geology of India and Burma. Geological Survey of India, Kolkata, pp 483–1343

Pathak V, Patil SK, Shrivastava JP (2017) Tectonomagmatic setting of lava packages in the Mandla lobe of the eastern Deccan Volcanic Province, India: paleomagnetism and magnetostratigraphic evidence. In: Mukherjee S, Misra A, Calves G, Nemčok M (eds) Tectonics of the Deccan large igneous province. Geol Soc London Spec Publ 445, pp 69–94

Patro BPK, Sarma SVS (2008) Trap thickness and the subtrappean structures related to mode of eruption in the Deccan plateau of India: results from magnetotellurics. Earth Planets Space 59:75–81

Peng ZX, Mahoney JJ, Hooper PR, Harris C, Beane JE (1994) A role for lower continental crust in flood basalt genesis? Isotopic and incompatible element study of the lower six formations of the Western Deccan Traps. Geochim Cosmochim Acta 58:267–288

Peshwa VV, Kale VS (1997) Neotectonics of the Deccan Trap province: focus on the Kurduwadi lineament. J Geophys 18(1):77–86

Peshwa VV, Mulay JG, Kale VS (1987) Fracture zones in the Deccan Traps of Western and central India: a study based on remote sensing techniques. J Indian Soc Remote Sens 15(1):9–17

Pitale UL, Dubey R, Saxena RK, Prasad JM, Muthuraman K, Thussu JL, Sharma SC (1987) Review of the geothermal studies of West Coast hot spring belt, Maharashtra, India. Geol Surv India Spec Publ 45:41–60

Powar KB (1981) Lineament fabric and dyke pattern in the western part of the Deccan volcanic province. Mem Geol Soc India 3:45–57

Powar KB (1993) Geomorphological evolution of the Konkan coastal belt and adjoining Sahyadri uplands with reference to Quaternary uplift. Curr Sci 64:793–796

Punekar J, Keller G, Khozyem H, Hamming C, Adatte T, Tantawy AAAM, Ponte J (2014) Late Maastrichtian–early Danian high-stress environments and delayed recovery linked to Deccan volcanism. Cretaceous Res 49:63–82

Radhakrishna BP (1993) Neogene uplift and geomorphic rejuvenation of the Indian Peninsula. Curr Sci 64:787–793

Rahate DN, Solanki JN (2002) District resource map: Hoshangabad District, Madhya Pradesh, DRM series, Geol Surv India Kolkata

Rajaguru SN, Kale VS (1985) Changes in the fluvial regimes of Western Maharashtra Upland rivers during late Quaternary. J Geol Soc India 26:16–27

Rajaguru SN, Kale VS, Badam GL (1993) Quaternary fluvial systems in the Upland Maharastra. Curr Sci 64:817–822

Rajaram M, Anand SP, Erram VC, Shinde BN (2017) Insight into the structure below the Deccan Trap covered region of Maharashtra, India from geopotential data. In: Mukherjee S, Misra A, Calves G, Nemčok M (eds) Tectonics of the Deccan large igneous province. Geol Soc Lond Spec Publ 445, pp 219–236

Rajurkar ST, Bhate VD, Sharma SB (1990) Lineament fabric of Madhya Pradesh and Maharashtra and its tectonic significance. Geol Surv India Spec Publ 28:241–259

Ramachandran C, Kesavamani M (1998) Fault segmentation and earthquake hazard zones in the Deccan Trap region: gravity evidence. J Geol Soc India 49(1):23–32

Rao NP, Shashidhar D (2016) Periodic variation of stress field in the Koyna-Warna reservoir triggered seismic zone inferred from focal mechanism studies. Tectonophys 679:29–40

Rastogi BK (1992) Seismotectonics inferred from earthquakes and earthquake regimes in India during the 1980s. Curr Sci 62:101–108

Rastogi BK, Chadha RK, Raju IP (1986) Seismicity near Bhatsa reservoir, Maharashtra, India. Phys Earth Planet Inter 44:177–199
Raval U, Veeraswamy K (2011) Mapping of tectonic corridors through hidden parts of the Greater Dharwar terrane, Southern India. J Asian Earth Sci 43:1210–1225
Ravi Kumar M, Singh A, Kumar N, Sarkar D (2015) Passive seismological imaging of the Narmada paleo-rift, central India. Precamb Res 270:155–164
Ray JS, Pattanayak SK, Pande K (2005) Rapid emplacement of the Kerguelen Plume related Sylhet Traps, eastern India: evidence from $^{40}Ar/^{39}Ar$ geochronology. Geophys Res Lett 32:1–4
Ray R, Sheth HC, Mallik J (2007) Structure and emplacement of the Nandurbar–Dhule mafic dyke swarm, Deccan Traps and the tectonomagmatic evolution of flood basalts. Bull Volcan 69:537–551
Ray R, Shukla AD, Sheth HC, Ray JS, Duraiswami RA, Vanderkluysen L, Rautela CS, Mallik J (2008) Highly heterogeneous Precambrian basement under the central Deccan Traps, India: direct evidence from xenoliths in dykes. Gondwana Res 13:375–385
Renne PR, Deino AL, Hilgen FJ, Kuiper KF, Mark DF, Mitchell WS, III, Morgan LE, Mundil R, Smit J (2013) Time scales of critical events around the Cretaceous–Paleogene boundary. Science 339:684–687. https://doi.org/10.1126/sci.1230492
Renne PR, Sprain CJ, Richards MA, Self S, Vanderkluysen L, Pande K (2015) State shift in Deccan Volcanism at the Cretaceous–Paleogene boundary possibly induced by impact. Science 350:76–78
Richards MA, Alvarez W, Self S, Karlstrom L, Renne PR, Manga M, Sprain CJ, Smit J, Vanderkluysen L, Gibson SA (2015) Triggering of the largest Deccan eruptions by the Chicxulub impact. Geol Soc Amer Bull 127(11–12):1507–1520. https://doi.org/10.1130/B31167.1
Sadakata N, Maemoku H, Rajaguru SN, Misra S, Fujiwara KF (1995) Late Quaternary environmental changes in the Pravara river basin, Northwestern Deccan Upland, India. Hiroshima Uni Spec Publ 24:43–56
Sahni A, Bajpai S (1988) Cretaceous-Tertiary boundary events: the fossil vertebrate, paleomagnetic and radiometric evidence from peninsular India. J Geol Soc India 32:382–396
Salahuddin, Ganjoo RK, Badam GL, Rajaguru SN (1986) On the archeological association of fossil Hominid from Hathnora, Madhya Pradesh, India. Asian Perspective 27:193–203
Samant B, Mahobey DM, Srivastava P, Thakre D (2014) Palynology and clay mineralogy of the Deccan volcanic associated sediments of Saurashtra, Gujarat: age and paleoenvironment. J Earth Syst Sci 123:219–232
Samant H, Patel V, Pande K, Sheth H, Jagadeesan KC (2019) $^{40}Ar–^{39}Ar$ dating of tholeiitic flows and dykes of Elephanta Island, Panvel flexure zone, Western Deccan Traps: a five-million-year record of magmatism preceding India-Laxmi Ridge-Seychelles breakup. J Volcanol Geotherm Res 379:12–22. https://doi.org/10.1016/j.jvolgeores.2019.05.004
Samant H, Pundalik A, D'souza J, Sheth H, Lobo KC, D'souza K, Patel V (2017) Geology of the Elephanta Island fault zone, western rifted margin, and its significance for understanding the Panvel flexure. J Earth Sys Sci 126:9. https://doi.org/10.1007/s12040-16-0793-8
Sayyed MGR (2014) Flood basalt hosted paleosols: potential paleoclimatic indicators of global climate change. Geosci Front 5:79–799. https://doi.org/10.1015/j.gsf.2013.08.005
Schöbel S, de Wall H, Ganerød M, Pandit MK, Rolf C (2014) Magnetostratigraphy and $^{40}Ar–^{39}Ar$ geochronology of the Malwa Plateau region (Northern Deccan Traps), Central-Western India: significance and correlation with the main Deccan large igneous province sequences. J Asian Earth Sci 89:28–45. https://doi.org/10.1016/j.jseaes.2014.03.022
Schoene B, Eddy MP, Samperton KM, Keller CB, Keller G, Adatte T, Khadri SFR (2019) U-Pb constraints on pulsed eruption of the Deccan Traps across the end-Cretaceous mass extinction. Science 363:862–866
Self S, Jay AE, Widdowson M, Keszthelyi LP (2008) Correlation of the Deccan and Rajahmundry trap lavas: are these the longest and largest lava flows on Earth? J Volcanol Geotherm Res 172:3–19
Sen B (2017) Lava flow transition from Pāhoehoe–dominated lower pile of the Deccan Traps from Manmad–Chandwad area, Western Maharashtra. J Geol Soc India 89:281–290

Shanker R (1991) Thermal and crustal structure of SONATA. A zone of midcontinental rifting in Indian shield. J Geol Soc India 37:211–220
Sheth HC (1998) A reappraisal of the coastal Panvel flexure, Deccan Traps, as a listric-fault-controlled reverse drag structure. Tectonophysics 294:143–149
Sheth HC (2000) The timing of crustal extension, diking, and eruption of the Deccan flood basalts. Int Geol Rev 42:1007–1016
Sheth HC (2007) Plume-related regional prevolcanic uplift in the Deccan Traps: absence of evidence, evidence of absence. Geol Soc Amer Spec Paper 430:785–813
Sheth HC (2016) Giant plagioclase basalts: continental flood basalt induced remobilization of anorthositic mushes in a deep crustal sill complex. Geol Soc Amer Bull 128:916–925
Sheth HC (2018a) A photographic atlas of flood basalt volcanism. Springer, Amsterdam, p 363. ISBN 978-3-319-67705-7
Sheth HC (2018b) Is the Satpura "horst" in fact a compressional uplift? J Geol Soc India 91:391–394
Sheth HC, Cañón-Tapia E (2015) Are flood basalt eruptions monogenetic or polygenetic. Int J Earth Sci 104:2147–2162, https://doi.org/ 10.1007/ s00531-014-1048-z
Sheth HC, Vanderklyusen L (2014) Flood basalts of Asia. J Asian Earth Sci 84:200
Sheth HC, Ray JS, Ray R, Vanderkluysen L, Mahoney JJ, Kumar A, Shukla AD, Das P, Adhikari S, Jana B (2009) Geology and geochemistry of Pachmarhi dykes and sills, Satpura Gondwana Basin, central India: problems of dyke-sill flow correlations in the Deccan Traps. Contrib Miner Petrol 158:357–380
Sheth H, Vanderklyusen L, Demonterova EI, Ivanov AV, Savatenkov VM (2018) Geochemistry and $^{40}Ar/^{39}Ar$ geochronology of the Nandurbar–Dhule mafic dyke swarm: dyke-sill-flow correlations and stratigraphic development across the Deccan flood basalt province. Geol J 21. https://doi.org/10.1002/gj.3167
Shrivastava JP, Ahmad M (2005) A review of research on late Cretaceous volcanic sedimentary sequences of the Mandla lobe: implications for Deccan volcanism and the Cretaceous/Palaeogene boundary. Cretaceous Res 26:145–156
Shrivastava JP, Mahoney JJ, Kashyap MR (2014) Trace elemental and Nd–Sr–Ph isotopic compositional variation in 37 lava flows of the Mandla lobe and their chemical relation to the Western Deccan stratigraphic succession, India. Mineral Petrol 108:801–817
Shrivastava JP, Duncan RA, Kashyap M (2015) Post–K/Pg younger ^{40}Ar–^{39}Ar ages of the Mandla lavas: implications for the duration of the Deccan volcanism. Lithos 224:214–224
Shrivastava JP, Kumar R, Rani N (2017) Feeder and post-Deccan Trap dyke activities in the northern slope of the Satpura Mountain: evidence from new ^{40}Ar–^{39}Ar ages. Geosci Front 8:483–492
Solanki JN, Bhattacharya DD, Jain AK, Mukherjee A (1996) Stratigraphy and tectonics of the Deccan Traps of Mandla. Gondwana Geol Mag Spl 2:101–114
Sonakia A (1998) Antiquity of the Narmada Homo erectus, the early man in India. Curr Sci 75:391–393
Sprain CJ, Renne PR, Vanderkluysen L, Pande K, Self S, Mittal T (2019) The eruptive tempo of Deccan volcanism in relation to the Cretaceous Paleogene boundary. Science 363:866–870
Sreenivasa Rao M, Reddy RM, Subbarao KV, Prasad CVRK, Radhakrishnamurthy C (1985) Chemical and magnetic stratigraphy of parts of Narmada region, Deccan Basalt Province. J Geol Soc India 26(9):617–639
Srikarni C, Srivastava S, Mahobey DM, Ghevariya ZG, Mathur UB, Dwivedi GN (eds) (2017) Dinosaurs of Gujarat: a compilation on the various dinosaur fossil finds and related life during the Mesozoic period in Gujarat, India. Geol Surv India Spec Publ 106, pp 158
Srinivasan R, Jaffri SH, Rao GV, Reddy GK (1998) Phreatomagmatic eruptive center from the Deccan Trap province, Jabalpur, Central India. Curr Sci 74:787–790
Srivastava P, Siva Siddaiah N, Sangode SJ, Meshram DC (2018) Mineralogy and geochemistry of various colored boles from the Deccan Volcanic Province: implications on paleoweathering and paleoenvironmental conditions. Catena 167:440–459
Subbarao KV (ed) (1988) Deccan Flood Basalts. Mem Geol Soc India 10, pp 393
Subbarao KV (ed) (1999) Deccan Volcanic Province. Mem Geol Soc India 43(2):947

Subbarao KV, Hooper PR (1988) Reconnaissance map of the Deccan Basalt group in the Western Ghats, India: scale 1:1,000,000. In: Subbarao KV (ed) Deccan Flood Basalts. Mem Geol Soc India 10, enclosure

Subbarao KV, Sukheswala RN (eds) (1981) Deccan Volcanism and related basalt provinces in other parts of the World. Mem Geol Soc India 3, pp 474

Subramanyan V (1981) Geomorphology of the Deccan Volcanic Province. In Subbarao KV (ed) Deccan Flood Basalts. Mem Geol Soc India 3, pp 101–116

Sukheswala RN (1974) Gradation of tholeiitic Deccan basalt into spilite, Bombay, India. In: Amstutz GC (ed) Spilites and spilitic rocks. Springer, Heidelberg, pp 229–250. https://doi.org/10.1007/978/3/642/88230/2/14

Sukheswala RN, Poldervaart A (1958) Deccan basalts of the Bombay area. Geol Soc Amer Bull 69:1475–1494

Tandon SK, Sood A, Andrew JE, Dennis PF (1995) Palaeoenvironments of the dinosaur-bearing Lameta Beds (Maastrichtian), Narmada valley, central India. Palaeogeogr Palaeocl Palaeoeco 117:153–184

Tetley M, Simon W, Gurnis M, Flament NE, Muller RD (2019) Constraining absolute plate motions since the Triassic. J Geophys Res Solid Earth. https://doi.org/10.1029/2019JB017442

Vaidyanadhan R (1977) Recent advances in geomorphic studies in Peninsular India—a review. Intern J Earth Sci S Ray vol:13–35

Vaidyanadhan R (1987) Quaternary planar surfaces—their mapping and dating techniques with examples from Peninsular India. Intern J Earth Sci 14:252–260

Valdiya KS (2016) The making of India: Geodynamic evolution. 2nd edn. Springer, Berlin, pp 924

Vanderkluysen L, Mahoney JJ, Hooper PR, Sheth HC, Ray R (2011) The feeder system of the Deccan Traps (India): insights from dike geochemistry. J Petrol 52(2):315–343

Walker GPL (1971) Compound and simple lava flows and flood basalts. Bull Volcan 35:579–590

Washington HS (1922) Deccan Traps and other plateau basalts. Geol Soc Amer Bull 33:765–804

Watts DR, Cox KG (1989) The Deccan Traps: an interpretation in terms of progressive lithospheric flexure in response to a migrating load. Earth Planet Sci Lett 93:85–97

Wescoat JL Jr (2019) Human use of landforms on the Deccan Volcanic Plateau: formation of a geocultural region. Geomorph 331:175–190

West WD (1962) The line of Narmada-Son Valley. Curr Sci 31:143–144

White JDL, Bryan SE, Ross PS, Self S, Thordarson T (2009) Physical volcanology of continental large igneous provinces: update and review. In: Thordarson T, Self S, Larsen G, Rowland SK, Hoskuldsson A (eds) Studies in volcanology: The legacy of George Walker. Spec Publ IAVCEI, Geol Soc, London 2, pp 291–321

Widdowson M, Cox KG (1996) Uplift and erosional history of the Deccan Traps, India: evidence from laterites and drainage patterns of the Western Ghats and Konkan Coast. Earth Planet Sci Lett 137:57–69

Yadav A, Bansal BK, Pandey AP (2016) Five decades of triggered earthquakes in Koyna- Warna region, western India – a review. Earth Sci Rev 162:433–450

Chapter 9
Tectonics of Western Margin of India

9.1 Introduction

Many texts books on the Indian geology have described the physiography and tectonics of the western margin of the Indian Peninsula (Pascoe 1956; Wadia 1978; Krishnan 1982). However, spectacular character of these *Sahyadri* mountains was highlighted by Gunnell and Radhakrishna (2001) who pointed out that they are no way less in their dimensions than the *Himadri* (=Himalayan) mountain chain, both products of Cenozoic tectonics.

North-south trending Sahyadri Ranges span more than 1600 km linearly in their geometry, parallel to the western coastline of India (Fig. 9.1); and rise to elevations above 2000 m. This range is characterized by steep west-facing escarpments up to 1200 m, leading to it being referred to as the Western Ghats or Western Ghats Escarpment (WGE) in geological literature. The eastern slopes of the Sahyadri Range are gentler and gradually merge into the peninsular plateau. Although Precambrian rocks are exposed along more than half of its length, the emergence of this range is closely associated with the Cenozoic evolution of the western margin of the Indian subcontinent.

Much of the available literature on the evolution of the WGE and the adjoining coastal belt focusses on the geological and geomorphic features exposed in continental India (Widdowson and Cox 1996; Gunnel and Radhakrishna 2001; Harbor and Gunnel 2007; Valdiya 2016; Richards et al. 2016). The paleomagnetic and gravity–seismic models of structure and evolution of western margin of the Indian continent were limited to the off-shore shelf region and the continental segments (Kailasam 1979; Radhakrishna 1993; Tiwari et al. 2006; Calvés et al. 2011). The discovery and eventual exploration of off-shore petroliferous basins (Biswas 2008; Bijesh et al. 2018) augmented this data significantly. Studies in the last few decades by scientists from the NIO, NCAOR and GSI in collaboration with the International Ocean Drilling Program (IODP) and its affiliates have mapped the marine realm in the Arabian Sea and the Indian Ocean and helped unravel the larger tectonic picture of

A. K. Jain et al., *Tectonics of the Indian Subcontinent*, Society of Earth Scientists Series, https://doi.org/10.1007/978-3-030-42845-7_9

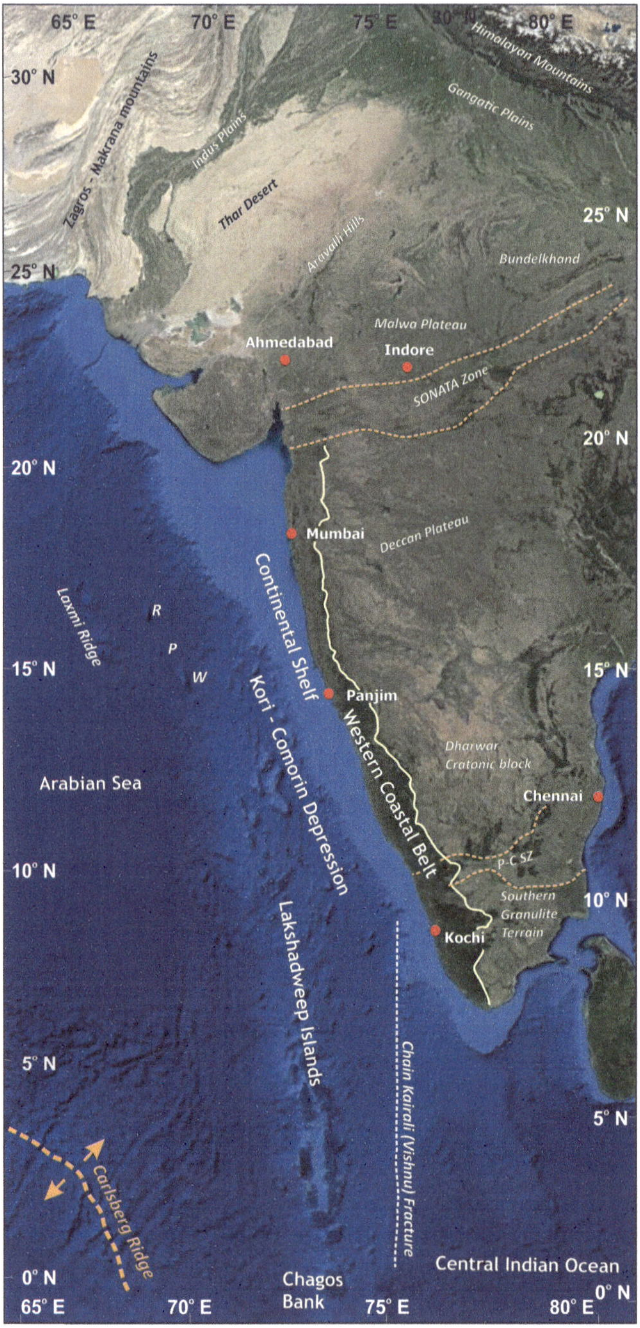

Fig. 9.1 Western margin of India. Pale yellow continuous line marks the crest of the Sahyadri ranges (=Western Ghats Escarpment (WGE)). P-C SZ=Palghat–Cauvery Shear Zone. Off-shore bathymetric depths are shaded in blue with the deepest parts displayed as darker shades. Google Earth image

the Indian continental plate and its oceanic connection (Bhattacharya and Yatheesh 2015; Pande et al. 2017b; Yatheesh et al. 2019).

9.2 Divisions

The western margin of India (Fig. 9.1) can be broadly divided into the following:

(i) Western Ghats Escarpment (WGE),
(ii) Narrow (~100 km wide) coastal strip running parallel to the WGE with average elevations of less than 200 m above msl,
(iii) Southward narrowing >250 to ~100 km continental shelf with a surface that is largely above the 200 m isobath, and
(iv) Deep marine basins and volcanic submarine plateaus; some of which occur as a string of islands.

Radhakrishna and Vaidyanadhan (2011) refer to the linear depression off the continental slope as the Kori–Comorin Depression or Ridge. Those parts of this margin that are floored by continental crust are collectively referred to as the Western Continental Margin of India (WCMI). It is evident that the geological framework and tectonics of these four components of the western margin of India are closely interlinked with each other during their Cenozoic evolution.

The WCMI displays 3 differing sectors from north to south, which appear to coincide with the nature of the rocks exposed on the continental side. The northern sector (Fig. 9.2) exposes the Deccan Trap basalts on the continental part in the Sahyadri Ranges as well as the Konkan Coastal Belt (KCB); while the adjoining wide continental shelf (Bombay off-shore) hosts several petroliferous basins of Tertiary age. The central part exposes the Archean Dharwar sequences on the edge of the Mysore Plateau, and has the lowest average altitudes of the Sahyadri Range flanking the narrow coastal strip called the Kanara Coastal Belt and a narrow continental shelf. In the southern sector the Nilgiri Hills are amongst the highest hill ranges in the Sahyadri. They are followed westward by narrow Kerala Coastal Strip displaying several lagoons and bars and has a slender continental shelf (Fig. 9.2).

Figure 9.3 gives a synoptic view of the key tectonic elements in this segment of the Indian Ocean and the Indian subcontinent, including its fringing mountain chains and major rivers. This configuration is a result of the northward flight of the Indian subcontinent and its intercontinental collision with the Asian mainland in the Cenozoic that was preceded by its rifting from Madagascar and subsequently Seychelles during the Upper Cretaceous. Each of these tectonic events has left an imprint on the evolution of the western margin of the Indian subcontinent.

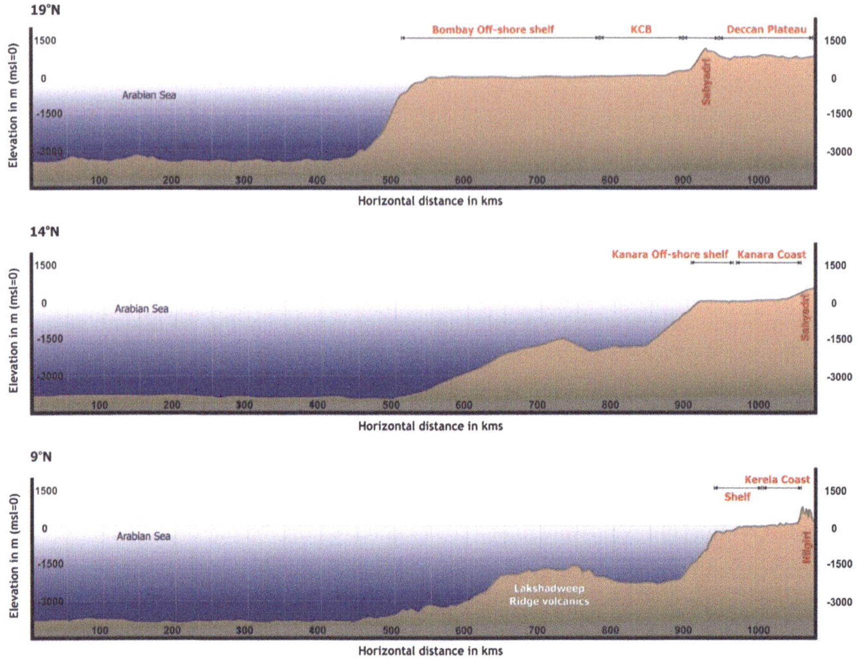

Fig. 9.2 Topographic profiles between 65° E and 75° E along the parallels of 19° N, 14° N and 9° N, depicting the western continental margin in northern, central and southern sectors, respectively. Based on SRTM IOC-IHO-BODC (2003) data and Google Earth

9.2.1 Offshore Marine Systems

Deep marine features along the western margin of India are a part of the Indian Ocean and the Arabian Sea that extends between Africa and India, and parts of the southern margin of central Asia and the Arabian Peninsula. Interest in the western continental shelf of India was driven by the discovery of petroliferous basin in the off-shore Bombay region (Biswas 1987, 2008). Wagle et al. (1994) had mapped a series of marine terraces on western continental shelf. Vora et al. (1996) described the occurrence of a barrier-reef system at the edge of this shelf extending more than 1300 km in the N–S direction (Fig. 9.3). The morphotectonic evolution of the deep marine realm, paleomagnetic characterization of the ocean floor strips and the intervening volcanic ridge complex has been elucidated in the works of Faruque and Ramachandran (2014), Bhattacharya and Yatheesh (2015), Pandey et al. (2017, 2018), Bijesh et al. (2018), Nair and Pandey (2018), Shuhail et al. (2018), Desa et al. (2019), Yatheesh et al. (2013, 2019) among others.

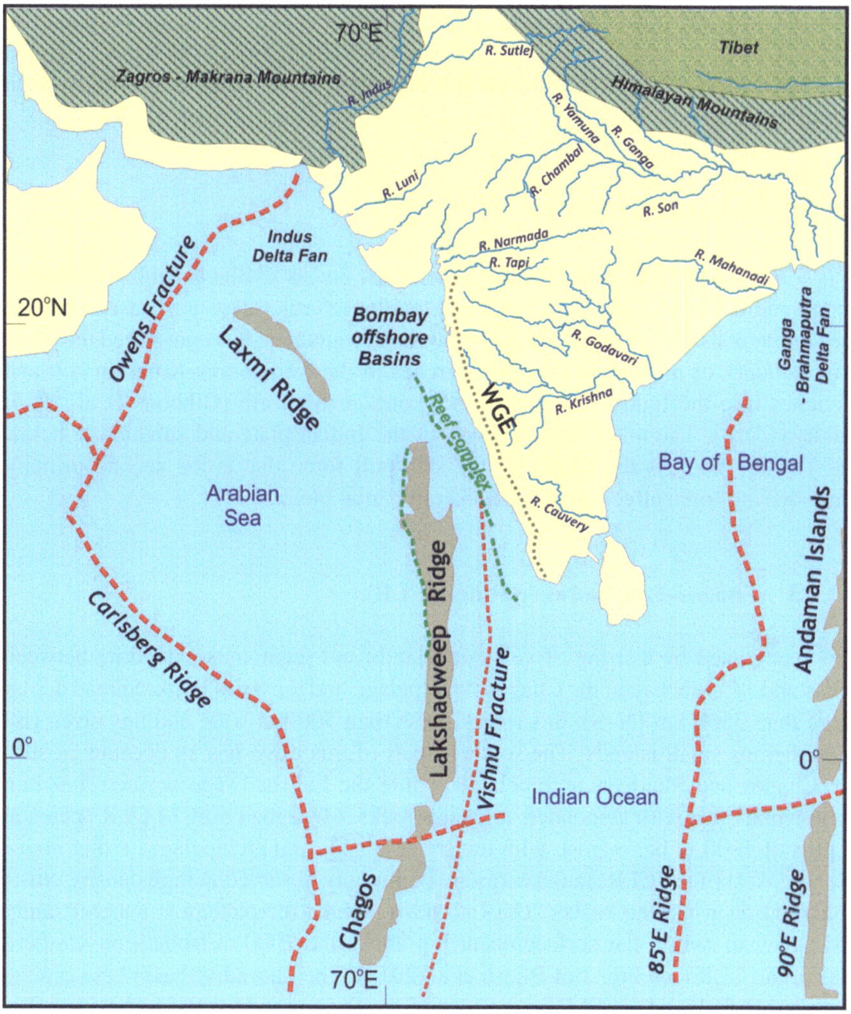

Fig. 9.3 Major tectonic elements of the Indian Ocean adjoining western margin of the Indian subcontinent. Major rivers in the Indian subcontinent (based on India-WRIS 2012) are depicted for easy reference. The Carlsberg and 85° E Ridges are active spreading mid-oceanic ridges, while other ridges are volcanic ridges. Segment of deeper shelf region between the Reef complex and the Lakshadweep Ridge is recognized as the Kori–Comorin basin or depression

9.2.1.1 Carlsberg Ridge

The Carlsberg Ridge represents the northern extension of the Central Indian Ocean Ridge) as an active spreading mid-oceanic ridge (MOR). The Indian MOR displays variable half-spreading rates between 32 and 120 mm per annum (mm/y) in the N–S direction (Desa et al. 2019), while the NNE-SSW oriented spreading along the

Carlsberg ridge is slower in comparison (Mudholkar et al. 2002; Kamesh Raju et al. 2008). This ridge push is one of the primary plate-driving forces that have affected the Indian continental block since its inception during the separation of India from West Gondwana.

9.2.1.2 Owens Fracture Zone

It is a transform fault that extends northwards, and is collinear with the Murray Ridge and Chaman Transform Fault that continue further east-northeastwards into the continental crust along the Zagros-Makrana Mountains. It is considered the western boundary of the Indian Plate; that arches in the Karakoram–Kohistan belt and stretches into the Indus–Tsangpo Suture zone in the north (Gibbons et al. 2015; Valdiya 2016). The northward collision of the Indian plate and subduction below the Tibetan block is the source of the slab-pull force that is the second primary plate-driving force affecting the Indian continental block.

9.2.1.3 Chagos–Lakshadweep Ridge (CLR)

It is recognized by a string of volcanic islands and seamounts extending between 10° S and 14° N between the Chagos Archipelago and the Adas Bank. Spread across more than 2500 km (N–S), this ridge is less than 300 km wide and has given rise to numerous small islands. The southern part of this ridge has an oceanic crust in the Chagos and Maldives archipelagos, while the Lakshadweep segment has thin continental crust with associated volcanics rocks. More than 80% of CLR occurs at depths of 1000 m below msl, with the small islands and archipelagos rising above the sea. West of the CLR, fault scarps are commonly observed in high density crusts bounding deep marine basins. The Laccadive submarine plateau is a significantly large plateau with a flat surface occurring above the 2000 m isobath on northern part of the CLR (see Fig. 1 of Bijesh et al. 2018). The Laccadive basin between the continental shelf and the CLR hosts several Late Tertiary sediments and has provided evidence of the contemporary eustatic fluctuations in conjunction with the sediments from the submarine terraces occurring on the edge of the shelf. The basaltic lavas from the Lakshadweep islands yield ages of 55–62 Ma, while those in the Chagos archipelago have been dated at around 40 Ma (Sheth 2000; Kale et al. 2019). They provide the links of the trail of sustained volcanic activity of the Reunion Hotspot with the DVP and establish the track the northward drift of the Indian Plate.

9.2.1.4 Laxmi Ridge and Basin Complex

This complex is an array of ridges and rift basins occurring in the northern parts of the Arabian Sea, southwest of the Saurashtra Peninsula (Fig. 9.4). The Laxmi Ridge

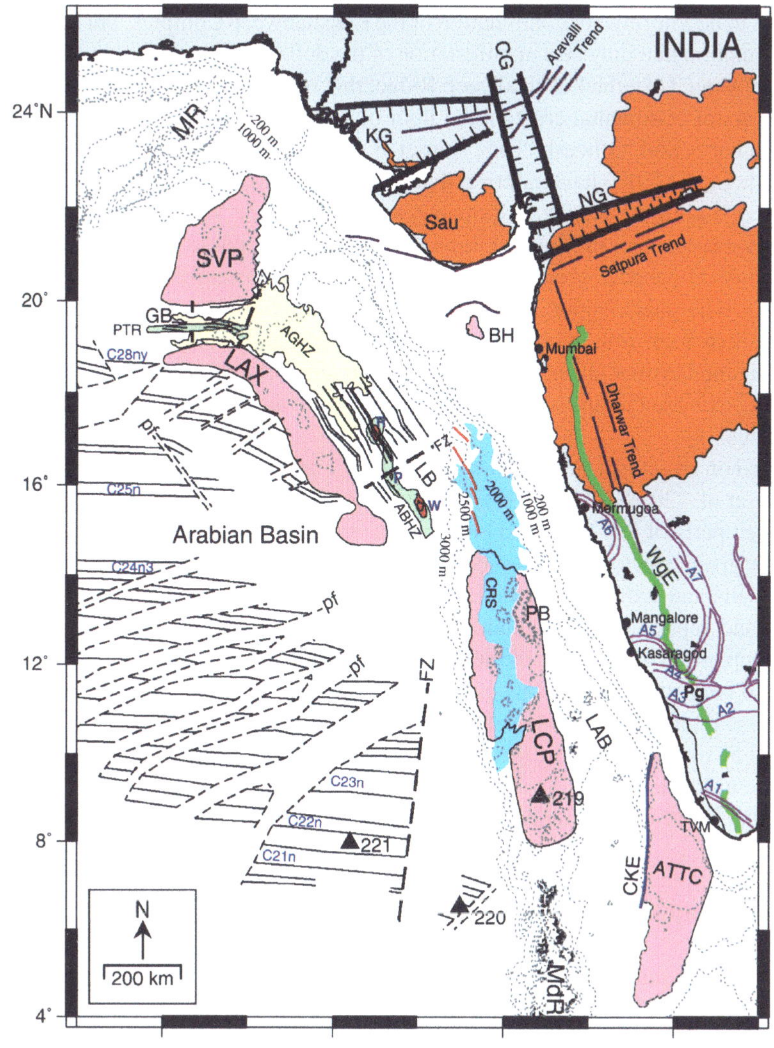

Fig. 9.4 Laxmi Ridge in the Arabian Sea and its components. ABHZ–Anomalous Basement High Zone. AGHZ–Anomalous Gravity High Zone. BH–Bombay High. GB–Gop Basin. LAX–Laxmi Ridge continental sliver. LAB–Laccadive Basin. LB–Laxmi Basin. LCP–Laccadive Plateau. MR–Murray Ridge. PTR–Palitana Ridge. SVP–Saurashtra Volcanic Platform. P–Panikkar Seamount. R–Raman Seamount. W–Wadia Guyot. Explanation of items of the legend: (a) Continental slivers. (b) Extent of ABHZ in Laxmi Basin and PTR in Gop Basin. (c) Anomalous gravity high zone (AGHZ). (d) Extents of Deccan Flood Basalts. (e) Seamounts in the Laxmi Basin. (f) Cannanore Rift System. R: Raman Seamount; P: Panikkar Seamount; W: Wadia Guyot. After Bhattacharya and Yatheesh (2015)

appears to be a northward continuation of the Lakshadweep Complex, but has a NW–SE orientation that may be a manifestation of the deflection along the Owen–Murray Fracture zone. Like the Lakshadweep Ridge, this segment of the oceanic realm also has slivers of continental crust on the floor of the sediments in the adjoining basins (Bhattacharya and Yatheesh 2015; Nair et al. 2015).

Pande et al. (2017) have inferred that the voluminous volcanism, recorded along this ridge and in the Panikkar–Raman seamounts and the Wadia Guyot along with the coeval volcanism in Mumbai, represents the syn-rift phase of volcanism related to the Indo-Seychelles separation which they conclude occurred between 62.9 and 62.1 Ma. The paleomagnetic strips adjoining the Laxmi ridge further endorse that the rift-drift transition occurred around 61–60 Ma and is supported by presence of continental crustal slivers. The Deccan continental flood basalt volcanism can be therefore extended far into the off-shore region, more than 500 km west of the present exposures.

Based on structures and sediments from the Laxmi Basin, derived from the IODP, Pandey et al. (2017, 2018, 2019) inferred that the original (? syn-rift) flexure that deepened parts of this basin and enabled accumulation of a thick pile of sediments; has now given way to incipient subduction along parts of the basin. There is no doubt that basalts and sediments in the Laxmi–Gop basins provide clear evidence for the rifting and separation of India and Seychelles in the late Paleocene–Early Eocene times, supported by mapping of the ocean floor magnetic strips (Gibbons et al. 2015). It is also interesting to note that average sedimentation rates ranging between 4 and 10 cm/ky, interpreted from the study of the cores of IODP holes, display a sudden spike of 58 cm/ky during the Pliocene times (Pandey et al. 2018). How this connects with the uplift history of the adjoining western continental margin of India and the collision history of the Himalaya remains a point of curiosity.

9.2.1.5 Continental Shelf

The Western Continental Shelf of India (WCSI) occupies an off-shore area of more than 300,000 km^2 between the western coast line of India and shelf margin (Fig. 9.5), with depths of up to the 200 m below msl. It extends from the traces of the Murray Ridge in the north till south of Kanyakumari across a N–S (to NNW–SSE) across a length of more than 2000 km. The WCSI represents a thinned and stretched continental crust covered by Cenozoic sediments (Corfield et al. 2010; Faruque and Ramachandran 2014) The western margin of this continental shelf (marked as Shelf Margin in Fig. 9.5) is the steep continental slope where the depth drops to ~2000 m below msl abruptly within a linear distance of less than 20 km (isobaths in Fig. 9.4). The WCSI becomes narrower in the southern parts and has relatively gentler continental slopes than in the north (see Fig. 9.2) .

The Ministry of Petroleum and Natural Gas of the Government of India has classified the sedimentary sequences deposited on and around the WCSI into (a) Kutch, (b) Cambay (c) Bombay High and (d) Kerala Konkan Basin, based on their potential as sources of hydrocarbon resources (NDR 2019). The Kutch Basin includes

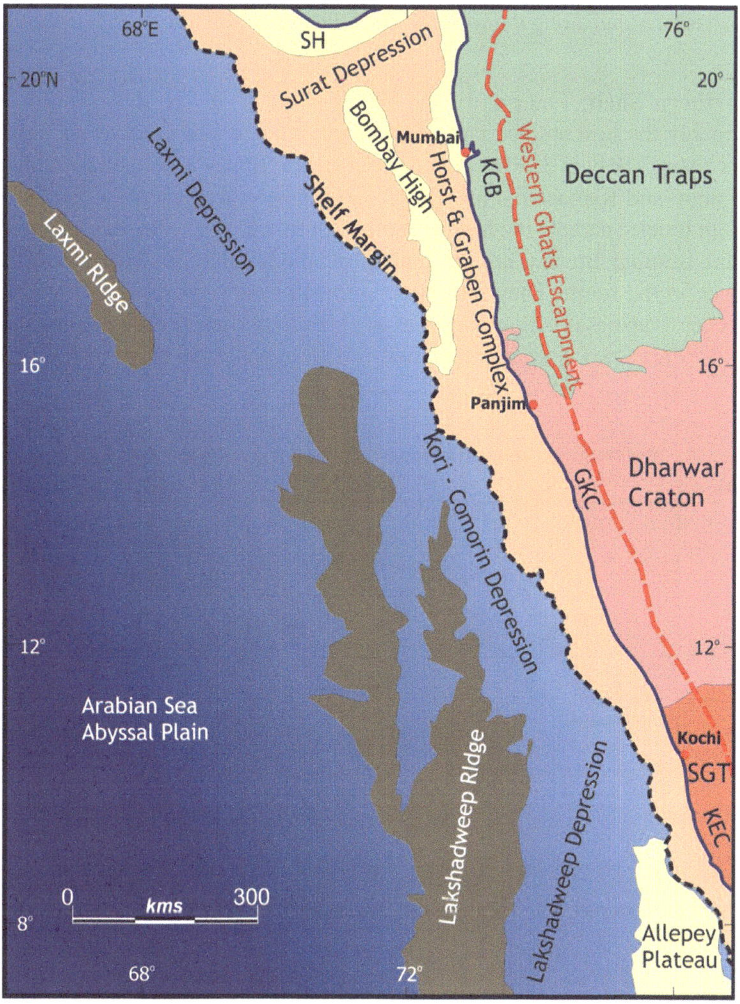

Fig. 9.5 Simplified map of morphotectonic units off the WCMI and Western Coastal Belt flanked by the Western Ghats Escarpment (red dashed line). The Western Coastal Belt includes the Konkan Coastal Belt (KCB), Goa–Karnataka Coast (GKC) and Kerala Coast (KEC). Region between the shelf margin and the coast line (blue thick line) is the Western Continental Shelf of India (WCSI). Horst and Graben complex extends all along the inner shelf. Bombay High includes the Ratnagiri Horst in southern parts. SGT = Southern Granulite Terrain. SH = Saurashtra Horst

both off-shore and on-shore sequences ranging in age from Mesozoic to Recent (Biswas 2005; Chaudhuri et al. 2020). They are essentially fault-bounded shallow marine to terrigenous sequences and include the Deccan Trap basalts and several related volcanic intrusives. The Cambay Basin is a rifted N–S trending basin that extends from the onland segment in Gujarat to the off-shore segment of the Gulf of Khambat (Bastia and Radhakrishna 2012). The Bombay High basin and the Kerala

Konkan basin extend along the western shelf and are notionally divided at the latitude 16° N.

(a) **Northern Shelf**: The part of the WCSI north of 16° N is the widest and is arguably the best studied part of the shelf, since it hosts numerous petroliferous basins (Biswas 1987; Gombos et al. 1995; Bastia and Radhakrishna 2012; Faruque and Ramachandran 2014). These basins are generally divided into 3 major blocks, namely the Kutch–Saurashtra Block (SW of the Saurashtra Peninsula), Bombay Block (Gulf of Cambay and the Bombay High) and the Ratnagiri Block in the south. They host Eocene to Miocene sediments resting either on Deccan Trap basaltic flows or occasional Precambrian crystalline gneissic basement (Fig. 9.6a). Listric faults, and roll-over anticlines, series of subparallel

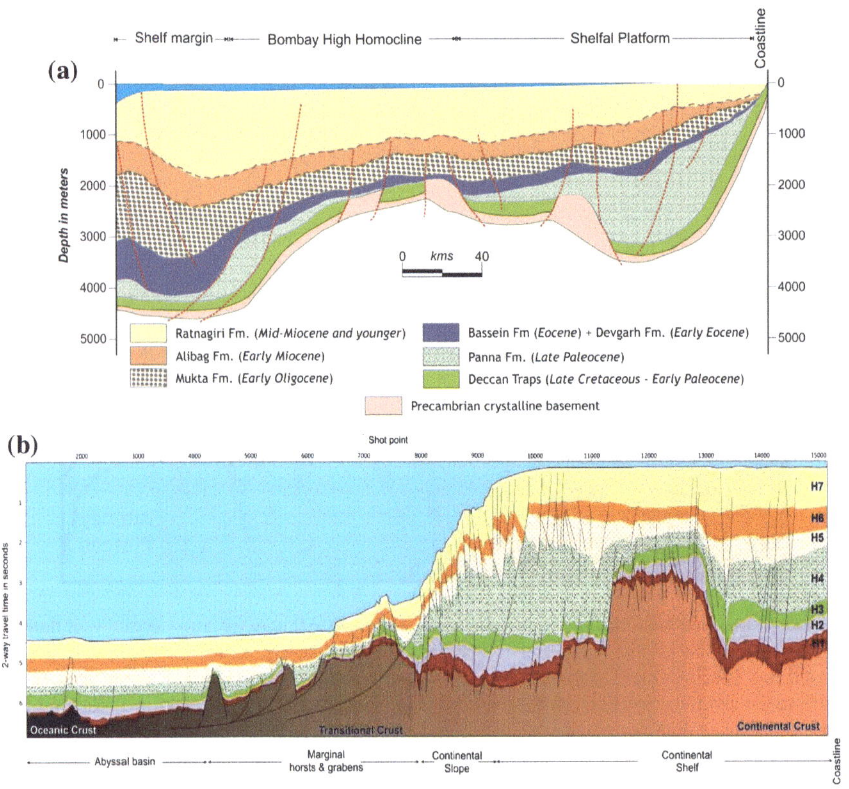

Fig. 9.6 **a** Simplified stratigraphic cross-section of the Bombay block of the WCSI. Regional erosional unconformities below and above the Alibag Formation are highlighted. Based on Mathur and Nair (1993), Bastia and Radhakrishna (2012), Nair and Pandey (2018), and NDR (2019). **b** Structural cross-section from edge of the WCSI along a NE–SW seismic profile, depicting complex faults, with listric fault complexes, rifts and horsts resulting in basement highs. Adapted from Nair and Pandey (2018); who have assigned seismic strata H1-H4 as syn-rift sequences and H5-H7 as post-rift sequences

normal and reverse faults besides typical rift-bounded sag basins with occasional mud-diapirs (Fig. 9.6b, c) are recorded from them. Recently, a regional, tectonically triggered Miocene slump has been recorded along the adjoining continental slope (Dailey et al. 2019). Except for a few cases, restricted to the basal Paleocene sediments in some of them and in the Cambay Rift, none of these sequences have any interbedded volcanics. This suggests that the Cenozoic evolution of this passive margin was a non-volcanic one. This structural setup is consistent with that encountered on a progressively subsiding shelf on trailing edge of a continental margin that suffered events of flexuring, fault-driven subsidence and load-sagging. Most of these sediments display a high degree of petrological maturity indicating their derivation from stable granite-gneiss continent interiors. Presence of mixed paralic sediments on shelf-edge in mid-shelf sectors and restricted shelf facies interbedded with carbonate ramp facies during most of the Paleogene period indicates high degree of tectonic stability punctuated by episodic off-shore faulting.

The faulting may have been sag driven or may be linked to the shelf-edge instability. Evolution of large homoclinal carbonate banks in the Saurashtra block during the Paleogene period and its on-shore representatives suggests that it was largely a stable block without significant tectonism. It is significant that the on-shore Cambay Rift has comparable petroliferous sequences and connects with this part of the shelf through the Gulf of Khambat, this may be attributed to intracontinental or continent-margin rifting with a N–S orientation. Much of the Cenozoic sediments in this sector were derived from north through continental drainage of the Mahi and Sabarmati rivers flowing into the Gulf of Khambat (Valdiya 2002) or through the Indus-Sutlej deltaic system which was then redistributed by oceanic currents.

Significant paucity of sediments derived from the Deccan Trap basalts in them indicates that the DVP did not drain westwards during much of the Cenozoic Era. Neogene and Quaternary tectonics appear to be largely responsible for the compressive structures present in these sequences along the northern parts of the shelf. Towards the shore-ward side, these sediments are capped by Holocene coastal sediments that have representatives exposed on-shore in the coastal belt as well (discussed below). This further suggests that besides eustatic and climatic adjustments the western coastal system was subjected to some tectonic events during the Quaternary period.

(ii) **Central Shelf**: This segment of the WCSI is 130–80 km wide, narrowing progressively southwards, with a fairly smooth surface. Coast-parallel reefs, submarine terraces, rocky islands and sunken structures including paleo-channels characterize this segment. Some of the submerged terraces of carbonate reefs, oolites and recycled sands occurring off the coast off Karwar at depths of 75–92 m below msl have been dated to be of Holocene age. They have been interpreted to represent products of a still-stand period during the lowered early Holocene global sea-levels (Faruque and Ramachandran 2014). These terraces and reefs (see Fig. 9.2) extend all the way to the southern tip of the WCSI.

The presence of volcanic islands on the inner shelf near the shoreline is a typical feature of this segment of the WCSI. Several of these contain silicic volcanic rocks, and the ones exposed in the St. Mary's Islands have been dated at around 89–85 Ma (Pande et al. 2001). Based on their composition and similarity with contemporary rocks in Madagascar, they provide a direct evidence of Upper Cretaceous pre-consanguinity of India and Madagascar (Melluso et al. 2009). They are considered as syn-rift magmatism related to the breakup between them.

This segment of the shelf is transected by a series of NNW–SSE trending en-echelon faults and fractures that are cross-cut by a set of WNW–ESE trending shear zones. They have controlled linear geometry of the coast-line as well as the successive coast-parallel reefs that occur at depths of ~45 m, ~75 m, 86–89 m and 96–102 m below msl. This suggests that the reef evolution was controlled by an array of step-faults that emerged parallel to the coast-line. The outer shelf hosts a carbonate bank comprising of bioherms, oolites and sandy beds.

Connection between these faults and regional structural features such as guyots, volcanic plateaus and hills (see Fig. 2 of Bijesh et al. 2018) in adjoining ocean floor that are a part of the northern Lakshadweep Ridge or the Vishnu Fracture zone (Fig. 9.1) remain open to interpretation. However, there is no doubt that this segment of the shelf was essentially depleted of detrital cratonic sediment supply through a large part of the Cenozoic. This would suggest either that it remained an emergent part of the WCMI till the Quaternary times or that a post-Miocene uplift of this part enabled removal of early deposited sediments completely. This has to be assessed in light of the Neogene sediments and laterites occurring all along the southern part of the coastal belt (Manjunatha and Shankar 1992, Nair et al. 2006).

(iii) **Southern Shelf**: The southern part of the WCSI is the narrowest and connects around the continental tip with the Indo-Sri Lankan shelf (see Fig. 9.1). It is flanked on the west by the Laccadive Basin which has the northern part of Lakshadweep Ridge on its other side. A series of off-shore marine terraces (Alleppey–Trivandrum Terrace Complex) rimmed by a stable reef complex creates a local semicircular protrusion of the shelf into the Arabian Sea. The western edge of this reef–terrace complex is a steep N–S escarpment known as the Chain–Kairali Escarpment which corresponds with the northern tip of the Vishnu Fracture Zone (see Fig. 9.3). Significant strike-slip dislocations of the ocean floor stripes has been marked along the Vishnu Fracture, that it is perhaps a transform fault.

The sedimentation on this part of the WCSI appears to have been continuous from the Oligocene onwards, only to be interrupted by either tectonism and or eustatic fluctuations in the sea-level. The carbonate reefs present along the outer shelf are a part of the larger reef complex that extends across more than 1200 km (Fig. 9.2) and also occur fringing the islands on the adjoining volcanic ridges. They testify to the rich biogenic sedimentation along the shelf and its interactions with the changes in the climate and sea-level, besides variable tectonic movements.

The carbonate reefs and older sediments of the southwestern shelf are a covered by Holocene terrigenous sediments derived from the Precambrian crystalline rocks of the Peninsular hinterland and reworked Cenozoic sediments (Rao and Wagle 1997; Faruque and Ramachandran 2014). Some of them host lignite and fossilized wood suggesting lush coastal forests in an equatorial environment (Soman 2013) that were subsequently downfaulted and eventually submerged under the oceanic waters. Most of the normal faults in this segment of the shelf strike N–S to NNW–SSE; and they are cut across by a series of WNW–ESE to E–W trending shear zones that appear to have extensions on the continental side, representing the shear zones in the Palghat Cauvery Shear Zone (Fig. 9.1). The series of coast-parallel terraces recorded on the inner shelf provide excellent clues to the interplay of tectonics and eustatic sea-level changes (Wagle et al. 1994; Nair et al. 2006).

9.2.2 *Coastal Belt*

9.2.2.1 Western Coastal Belt

Geographically, the coastal belt along the western continental margin of India (Fig. 9.5) can be divided from north to south into the (i) Konkan Coastal Belt (KCB), (ii) Kanara Coastal belt–often called as the Karwar or Goa–Karnataka coast (GKC), and (iii) the Kerala Coast (KEC). The KCB is underlain by the Deccan Trap basalts, while the GKC exposes the granite gneiss and greenstone sequences of the Dharwar cratonic block. KEC is underlain by the charnokites, granulites and gneisses of the Southern Granulitic Terrain. All the older rocks in the coastal belt are covered by laterite patches that are recognized as the coastal or Low-level laterites in literature (Widdowson and Cox 1996, and references therein) along with localized Quaternary sediment patches.

The rivers that drain this coastal strip have their origin in the WGE and flow westwards into the Arabian Sea. They display narrow linear channels with active headward erosion, cascades and waterfalls. They have relatively small catchments and have developed entrenched meanders in the middle reaches before draining into the sea through southward deflected estuaries (Mukhopadhyay and Karisiddaiah 2014). This narrow coastal strip between the WGE and the Arabian Sea has evolved due to scarp-retreat during the Late Pliocene–Recent times.

9.2.2.2 Konkan Coastal Belt (KCB)

The stretch of the coast between Tapi River and Goa, referred to as the Konkan Coastal Belt (KCB), is entirely underlain by the Deccan Trap basaltic flows, dykes and associated intrusives (Fig. 9.7a). In its southern parts, erosional removal of basaltic flows exposes underlying the Proterozoic Kaladgi Supergroup and the Archean sequences of the Dharwar craton. The KCB is characterized by rocky coast-line, punctuated with estuaries of west-flowing coastal rivers that originate in the WGE. While northern part of the KCB exposes the Deccan Traps with limited soil cover, the southern

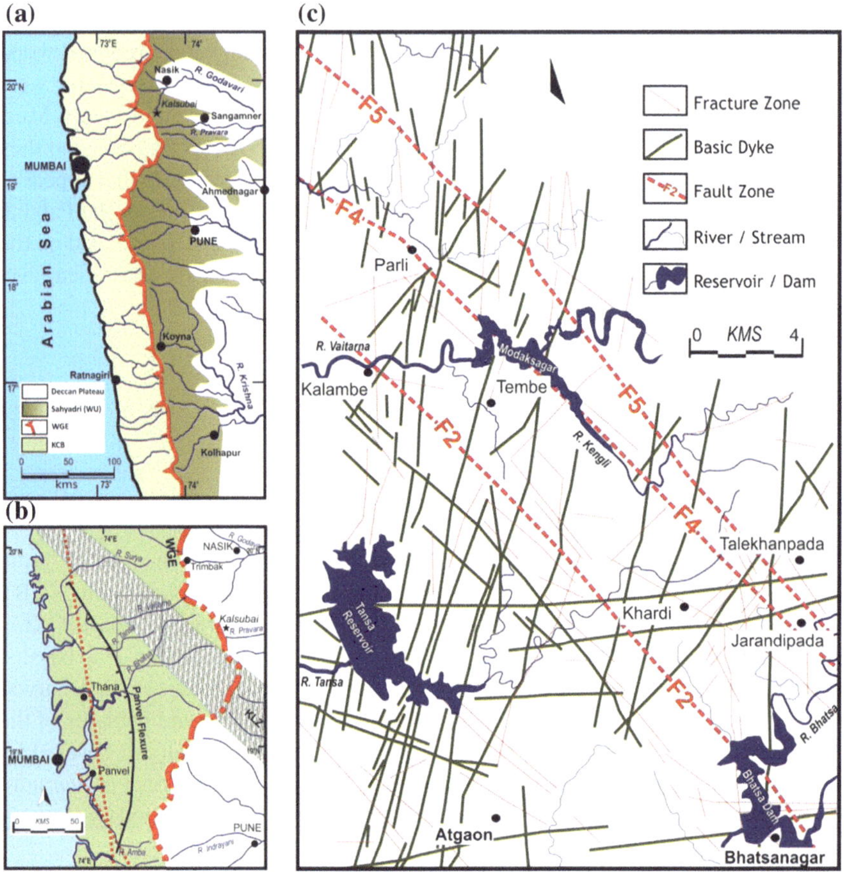

Fig. 9.7 **a** Konkan Coastal Belt (KCB) flanked by the WGE and the Arabian Sea, with major drainage lines (Kale et al. 2017). **b** Major tectonic elements of northern segment of the KCB around Mumbai. Red dotted N–S line represents the postulated trace of the "West Coast Fault" after GSI (2000) along with the Panvel Flexure. **c** Seismogenic areas around Bhatsanagar and Parli in Thana district showing NW–SE trending strike-slip faults across basaltic flows and dyke-swarms (Peshwa and Kale 1997). F4 and F5 faults are interpreted to host the foci of the Recent low-magnitude seismicity (Rastogi et al. 1986)

part displays large patches of lateritic cover over the basaltic flows. It hosts a number of hot-springs (Pitale et al. 1987; Chandrasekhar et al. 2018) with sulphurous emanations, possibly due to frictional heating of connate waters indicating active movements along the fractures. The overall topography of the northern KCB is rugged and displays sharp structural controls on drainage development.

The northern KCB exposes the oldest and the youngest Deccan Trap lava flows that steepen westwards up to 20° from the axis of monoclinal Panvel Flexure (Dessai and Bertrand 1995; Sheth 1998). A hypothetical (buried) fault trending N–S has been mapped as the West Coast Fault (Fig. 9.7b) in this part of the KCB (GSI 2000). The northern KCB hosts a dyke-swarm of N–S, E–W and NNW–SSE trending mafic dykes (Fig. 9.7c; Deshmukh and Sehgal 1988), some of which are known to host mantle and lower crustal xenoliths (Dessai et al. 2004).

Hooper et al. (2010) opined that the dyke swarm and Panvel flexure represent rift-related fractures that opened up during the Indo-Madagascar rifting and were eventually used as pathways for the eruption of the lava. Peshwa and Kale (1997) demonstrated the presence of several subparallel NW–SE trending faults with a dominant strike-slip component that have dislocated these dykes (Fig. 9.7c) indicating that structural imprint on the local geology is a later phenomenon and not a synvolcanic feature. Srinivasan (2002) demonstrated the westward downthrow of a series of N–S trending faults in this region and inferred that faulting in these parts is of post-Deccan Trap age and not related to its eruptive history. This transtensional deformation was endorsed by Misra et al. (2014) who suggested that these stresses are linked to the India–Seychelles rifting.

The southern KCB exposes subhorizontal Deccan Trap basaltic flows with an extensive lateritic cappings. Recent structural studies in this sector have demonstrated the presence of regional fracture zones that have modified and controlled drainage development in this coastal tract (Kundu and Matam 2000). Studies in the Koyna Warna Seismic Zone (Kale et al. 2014) indicated the presence of a major regional shear zone from the Precambrian basement, namely the Chiplun–Warna Lineament (Fig. 9.8), which extends further inland as key structural elements in this seismic zone (Rajaram et al. 2017; have et al. 2017; Arora et al. 2018). All these features suggest that buried structures in the Precambrian basement beneath the Deccan Traps have been reactivated during the Quaternary and have controlled the local landscape evolution in this rocky coastal belt.

9.2.2.3 Kanara Coast

This segment of the western coastal belt, also often recognized as the Goa-Karnataka Coast (GKC), exposes deeply weathered sequences of the Western Dharwar Craton which is extensively capped by laterite, having a number of narrow linear beaches. In this segment rivers flow westwards down the Sahyadri ranges and then across the coastal strip into the Arabian Sea. Headward active erosion and river piracy near their source regions where the tributaries flow eastward; are indicated by presence of waterfalls and cascades after turning west across the Sahyadri ranges (Valdiya 1998, 2016). Valdiya and Samwal (2017) indicated that ponding along the Gangavali and Netravati Rivers has resulted from downstream neotectonic uplift along NNW–SSE trending faults that transect the Kanara coast.

Local rocky headlands are observed at places along this coastal belt, where deeply weathered basement rocks are exposed. St Mary's and adjoining rocky islands near Udupi expose a variety of acidic and basic volcanics (Bhushan et al. 2010), which erupted between 89 and 85 Ma as a part of continental magmatism related to the Indo-Madagascar separation (Pande et al. 2001).

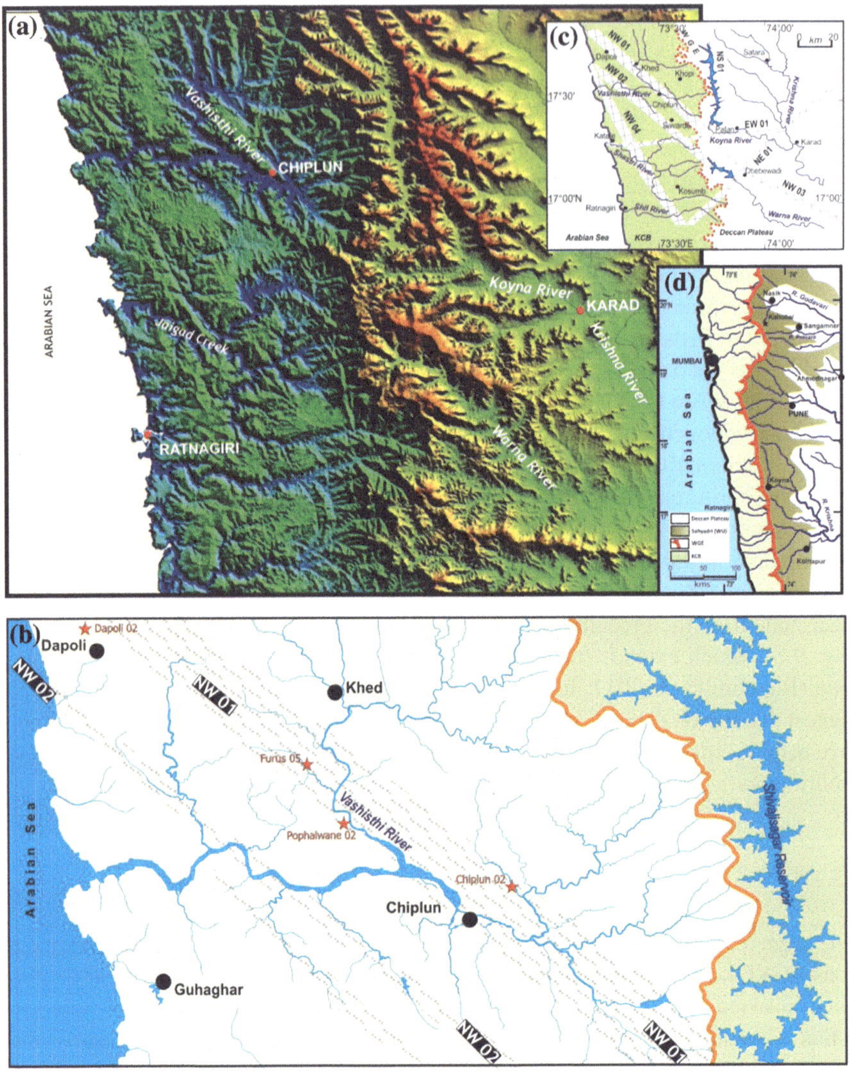

Fig. 9.8 **a** DEM of southern KCB and adjoining region of the Koyna–Warna Seismic Zone. The Chiplun–Warna Lineament is manifested locally as Lineaments NW01, NW02 and NW03; the latter controls the channel of the Warna River on the Deccan Plateau. **b** Lineament NW01 controls drainage system of the Vasishti River channel. NW–SE trending fracture zones control knee-bend of this river near Chiplun, some of which also display intense dextral shearing, and evidences of neotectonic movements (Kale et al. 1986). **c** Regional lineaments NW01 etc. (Kale et al. 2014). **d** Inset depicting regional drainage and features along the KCB and WGE

9.2.2.4 Kerala Coast

The Kerala coastal strip (KEC - see Fig. 9.5) is underlain by the Archaean–Proterozoic hornblende biotite gneiss, Dharwarian (Sargur) schist belt and charnokite–khondalite massif, which are separated by the E–W trending deep crustal Moyar–Attur, Bhavani and Palghat–Cauvery Shear Zones. On it's east, this narrow coastal strip is fringed along the foothills of the Sahyadri Ranges by lateritic plateaux that occasionally extend all the way up to the coast. The laterites are of Paleogene–Neogene age (Soman 2013), although several younger events of lateralization have also been noted in the Neogene sediments, exposed in Kerala.

Several of the off-shore Neogene–Quaternary sequences, occurring in the off-shore shelf region, have small patchy exposures in this coastal strip (Manjunatha and Shankar 1992; Nair et al. 2006; Campanile et al. 2008). The Middle Miocene to Pliocene Quilon and Warkal (=Warkalli) formations have exposed thickness of ~70 and ~80 m, respectively (Reauter et al. 2010). These sediments are commonly flanked by a series of subparallel N–S to NNW–SSE trending normal faults on landward side with progressive westwards downthrow. These sediments display presence of intercalated lateritic development, and bear testimony to the interplay of eustatic sea level changes, climatic changes and tectonic uplifts during the Neogene period. They also suggest that this segment of the coast was an active depocentre with supply of the sediments coming from the adjoining Sahyadri (Nilgiri) Hills from Paleocene (=Cochin and Kasargod formations: see Fig. 9.8) till Holocene times. This contrasts with the northern part of the west coast and shelf, which did not apparently receive any significant sediments from the Sahyadri ranges till the Late Miocene times.

9.2.3 Sahyadri Ranges

As depicted in Fig. 9.1, the 1600 km long Sahyadri Ranges run parallel to the western coastline of India and rises to a peak elevation of more than 2600 m above msl. It is characterized by steep escarpment face towering over the western coastal belt, while its eastern slopes are gentler with rugged hilly terrain gradually merging into the adjoining Deccan and Karnataka plateaus across most of its length. Only in its southern extremity in the SGT does one encounter a very rough terrain of dissected structural hill ranges with equally steep slopes on either side.

This range of mountains is unique in several ways, and underlain by three different geological domains (Fig. 9.9), the Deccan basaltic lava flows of Maastrichtian to Danian age in the north, the Dharwar cratonic terrain of Archean age in the middle and the Late Proterozoic South Indian Granulitic Terrain in the south.

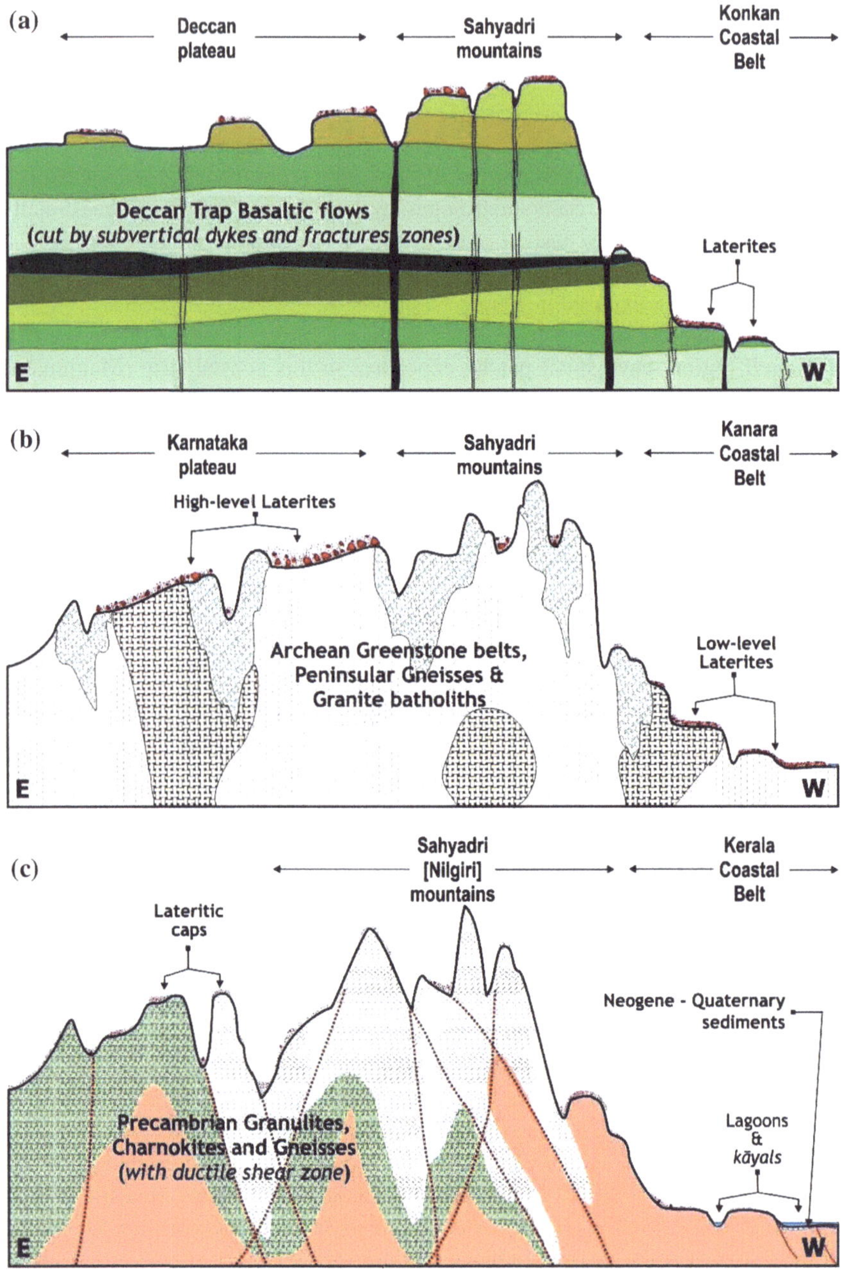

Fig. 9.9 Schematic East–West sections depicting the profiles of the WGE and the bedrock geology for the **a** northern, **b** central and **c** southern sectors of the Sahyadri mountains. Vertical scale is exaggerated to highlight its topographic contrast with adjoining western coastal plains

9.2.3.1 Topographic Characters

The overall topographic characters of the Sahyadri Range, WGE and adjoining Deccan plateau have been recently reviewed using geomorphic characters and digital elevation models (Radhakrishna 1993; Vaidyanadhan 1977, 1987; Widdowson 1997; Gunnell and Radhakrishna 2001; Dikshit 1976, 2001a; Gunnell and Michom 2001; Harbor and Gunnell 2007; Kale and Shejwalkar 2008; Ramakrishnan and Vaidyandhan 2008; Vaidyanadhan and Ramakrishnan 2010; Kale and Vaidyanadhan 2014; Richards et al. 2016). While its western sides display active erosional incision by streams, including well documented examples of river capture (e.g. Sheravati and Vaitarna rivers), the eastern sides display mature meandering river channels very close to the crest. The apparent paucity of significant geomorphic heterogeneity across its length is significant barring local features, induced by differential resistance of bedrocks to weathering. The drainage patterns in all three domains appear to have responded to identical changes in base-levels through the Quaternary times (Kale and Rajaguru 1987; Gunnell 2001b; Richards et al. 2016). This geomorphic consistency across bedrock domains with differing ages and structural patterns is indicative of a unified Quaternary denudational history for this mountain range.

Dikshit (2001b) discussed the three possible origins of this razor-sharp escarpment, namely (i) Fault escarpment, (ii) Erosional escarpment, and (iii) Dead cliff (ancient marine cliff). Kailasam (1979) had earlier used gravity data to suggest that the peninsular plateaux were uplifted during the Cenozoic times. Epeirogenic uplift was used to explain the uplift history of the Sahyadri (Powar 1993; Radhakrishna 1993). In an attempt the explain a structural model of the Deccan Traps, Widdowson and Cox (1996) speculated that the structure may be a combined effect of the original rifting of the western margin of India, crustal doming over a plume-head, coastal monocline and flexure associated with isostatic rebound. The possibility of crustal doming was discounted by Sheth (2007), who suggested that there has been a major Neogene uplift of the Indian Peninsula, that is unrelated with the Deccan Traps. This observation is supported by subsequent studies by Gunnell (2001b) and Richards et al. (2016) that delink the Sahyadri uplift from the Deccan Traps volcanism.

The lateritic caps on the Sahyadri Ranges and the adjoining coastal plains on the west (Fig. 9.9) have been recognized as High-level and Low-level Laterites respectively (Sahasrabudhe and Deshmukh 1981; Widdowson and Cox 1996; Gunnell 2001a); with the latter being possibly reworked material derived by erosion of the former (Ollier and Sheth 2008; Liu et al. 2019). The widespread development of in-situ laterite (irrespective of the bedrock lithology) is controlled by a relatively flat topography, mature river valleys and hot and wet climate (Wimpenny et al. 2007). Recent dating of the duricrust and laterites across the Indian Peninsula (Beeauvais et al. 2016; Bonnet et al. 2016; Jean et al. 2019) has shown that intense weathering occurred in two phases namely the Early Eocene (53–45 Ma) and then during Late Eocene - Oligocene (37–23 Ma) when most of the High-level Laterites may have developed. They assign the dates of ~9 Ma (Late Miocene) and ~ 2.5 Ma (Late Pliocene) for the Low-level Laterites indicating a second pulse of intense weathering. This shows that the entire peninsular block was subjected to a continent-scale denudational stand-still in a hot and wet equatorial climate enabling the development of the duricrusts and laterites (and also locally bauxite).

The fluvial systems that drain the Sahyadri ranges on either side have only Quaternary sediments deposited along them. The present-day landscape of the Indian Peninsula has inherited significant features of its Precambrian basement, Mesozoic rifting and the Cenozoic tectonics, but the present-day drainage systems and landforms have been largely controlled by Quaternary eustacy and climatic changes (Kale 2014). Emerging evidence from these terrains attribute some of the anomalous geomorphic features to neotectonic activity (Dole et al. 2000; Kale et al. 2017).

9.2.3.2 Structural Evidences

As pointed out above, three sectors of the Sahyadri ranges have three differing bedrock structures. Main objection to their 'faulted' origin is the complete absence of any N–S trending dislocation plane(s) along these ranges or their western foothills in the coastal plains. Geophysical data has demonstrated that the continental crust thins out west of the WGE (Kailasam 1979; Singh et al. 2007; Nemcok and Rybar 2017). Various positions of the postulated N–S trending West Coast Fault between the western margin of the Peninsula and the WGE have been plotted (e.g. GSI 2000; Mukherjee et al. 2017; Fig. 9.1). However, there is no exposed evidence of such a fault surface nor a dislocation recorded along its alignment, except for the westward tilt of basaltic flows along the Panvel Flexure. This has been pointed out by several authors as the primary lacuna in the faulted origin of the Sahyadri ranges. Widdowson and Cox (1996) and Sheth (2007) discussed the possibility of N–S doming axis resulting from the passage of the Indian Plate over the Reunion hotspot and consequent magmatic underplating. In this the WGE becomes an erosional escarpment with a history that ranges back to the early Cenozoic times.

It has been argued that the major continent-scale faulting is perhaps located offshore within the shelf. Consequently, only weak sympathetic N–S fault planes and flexure axes (Figs. 9.7 and 9.8), evidenced by intense fracturing and shearing of basaltic flows and gneisses (that promote rapid erosion and anomalous drainage on the landward side) are available as indirect evidences of faulting. The off-shore structural patterns along the WCSI (Fig. 9.6), transverse strike-slip faults (e.g. Kundu and Matam 2000; Misra et al. 2014; Rajaram et al. 2017) and the subparallel faults controlling the Palaeogene–Neogene sediments along the Kerala coastal belt appear to support this assumption.

Ajaykumar et al. (2017) have inferred that a group of the deep-seated fractures, identified using geophysical data, were reactivated during episodic break-up of Gondwanaland between ~90 and 65 Ma and led to distension surface faulting and associated dyke emplacement. They describe a second group of younger geophysical lineaments devoid of dykes that were reactivated from weak transformation planes in basement gneiss and charnockite in response to intraplate high intensity stress fields due to back-thrust of the Himalayan collision, epeirogenic forces associated with uplift of the Western Ghats and flexural isostatic uplifts during the period 20 Ma to the present day. The latter event could have sculpted high-relief, rugged terrain of the Southern Sahyadri. Geophysical data across the DVP have demonstrated that ancient zones of shearing and faulting have been reactivated during and after the eruption of the Deccan Traps leading to the present-day structural configuration of the northern Sahyadri and adjoining regions (Peshwa et al. 1987; Veeraswamy and Raval 2005; Ravi Kumar et al. 2015; Kale et al. 2017).

If these supporting evidences and presence of hidden active N–S shear planes, as indicated by line of thermal springs along the Western coast, are considered, conclusion that the current location of the WGE represents a receded fault-scarp appears to be plausible. In this context, 'step-faulting' model of the KCB along the Panvel flexure (Dessai and Bertrand 1995; Srinivasan 2002) and structural patterns recorded in the off-shore region are consistent with the 'rift margin' origin for the western coast of India, though it provides only limited clues to age of uplift of the Sahyadri.

9.2.3.3 Controls on Evolution of the Sahyadri Ranges

It is therefore evident that the Sahyadri may have originated earlier; but its present physiography represents a surface, drained by an antecedent drainage reflecting inherited patterns, which are modified by the Neogene and Quaternary tectonics and climatic changes. Earlier studies in the Indian Peninsula tended to assume its tectonic stability and ascribed the modifications to climatic and eustatic changes. Emerging evidence of neotectonic activity (Valdiya and Samwal 2017) and recurring seismicity pose doubts about the tectonic stability of this continental block.

The compilation by Gunnell and Radhakrishna (2001) encapsulates the then available knowledge and models on this range of mountains. The linkage of the easterly tilt of the Indian Peninsular, resulting in drainage network in the same direction with the rise of the Sahyadri, has been continuously debated upon based on geomorphic evidences (Subramanyan 1981; Radhakrishna 1993; Widdowson 1997; Valdiya 2001; Kale 2003, 2010; Richards et al. 2016). Association of the Sahyadri uplift linked to the crustal doming related to the Deccan Traps volcanism (Chandrasekharam 1985; Gombos et al. 1995; Widdowson and Cox 1996; Sheth 2007; Singh et al. 2007; Hooper et al. 2010) fails to explain satisfactorily the uplift history beyond the limits of this volcanism. Geophysical data (Kailasm 1979; Tiwari et al. 2006) supports the rifted western margin, but is inconclusive on the uplift history.

The 'razor-sharp' WGE does give an impression of being a fault-scarp; but fails to provide any geological evidence of faulting and dislocation along it or in its immediate vicinity. The elevations achieved by the Sahyadri ranges are the result of uplift with complementary subsidence of the coastal and continental margin regions which may have its roots in the rifting of the Indian subcontinent from Madagascar and subsequently Seychelles continental blocks. The key question that remains unrequited is whether the elevations were achieved at the time of rifting or did uplift continue subsequently during the drift of the Indian plate. Kale and Vaidyanadhan

(2014: p. 65) aptly state that: "*Major geological and tectonic events such as separation from Gondwanaland and migration of the Indian landmass through different latitudinal (climatic) zones, rifting along the eastern and western margins, Deccan volcanism, and superimposed effects of climatic and tectonic vicissitudes have all left an indelible imprint*" on the Indian subcontinent. The emergence and evolution of the Sahyadri Ranges must be therefore understood within this framework and cannot be delinked either from the geology, structure and tectonics of the off-shore segments, Indian Ocean (including the segment of the Arabian Sea); or from the far-field effects of the Indo-Asian collision that yielded the Himadri (Himalayan) ranges in the north.

9.3 Synthesis

Emerging multidisciplinary studies have now provided a far clearer picture of the tectonic evolution of the Western margin of the Indian subcontinent than was available a few decades back. Integrated evidences from off-shore regions with that from the continental shelf, coastal belt and flanking mountain ranges enables an obvious model within the framework of global tectonics (Muller et al. 2019). There cannot be any doubt that the Western Margin of India is a rift margin that evolved because of the break-up of the Gondwana Supercontinent, particularly the fragmentation of the East Gondwana.

The chronology of events that have impacted the Western Margin of India is as follows:

(i) The Australia–Antarctica block rifted away from the Indo–Madagascar–Seychelles block earlier during the Aptian–Albian times (~113–118 Ma) in response to the Kerguelan Hotspot. This led to the opening of the Eastern Indian Ocean, eruption of the Rajmahal–Sylhet–Bengal Traps and the evolution of Cretaceous sedimentary basins along the eastern flank of the Indian subcontinent (Coffin et al. 2002; Chatterjee and Bajpai 2016; Bredow and Steinberger 2018; Kale 2020).

(ii) India and Madagascar rifted and eventually drifted apart during the Turonian–Campanian times (~90–84 Ma). Volcanism associated with this rifting has imprints in parts of the Indian Peninsula and Madagascar (Fig. 9.10a). This led to the early opening of the Indian Ocean and northward drift of Greater India, including Seychelles and other smaller segments (Storey et al. 1995; Yatheesh et al. 2013). As depicted in Fig. 9.10a, the western continental margin of Greater India had started emerging during this period and had its limits further west of the present bounds.

(iii) The northward trajectory of Greater India led to its interaction with the Reunion Hotspot during the Maastrichtian times (~70 Ma). Early imprints of this interaction are found in the early igneous activity in Saurashtra and the Narmada valley (Karmalkar et al. 2000, 2014), which appear to be the

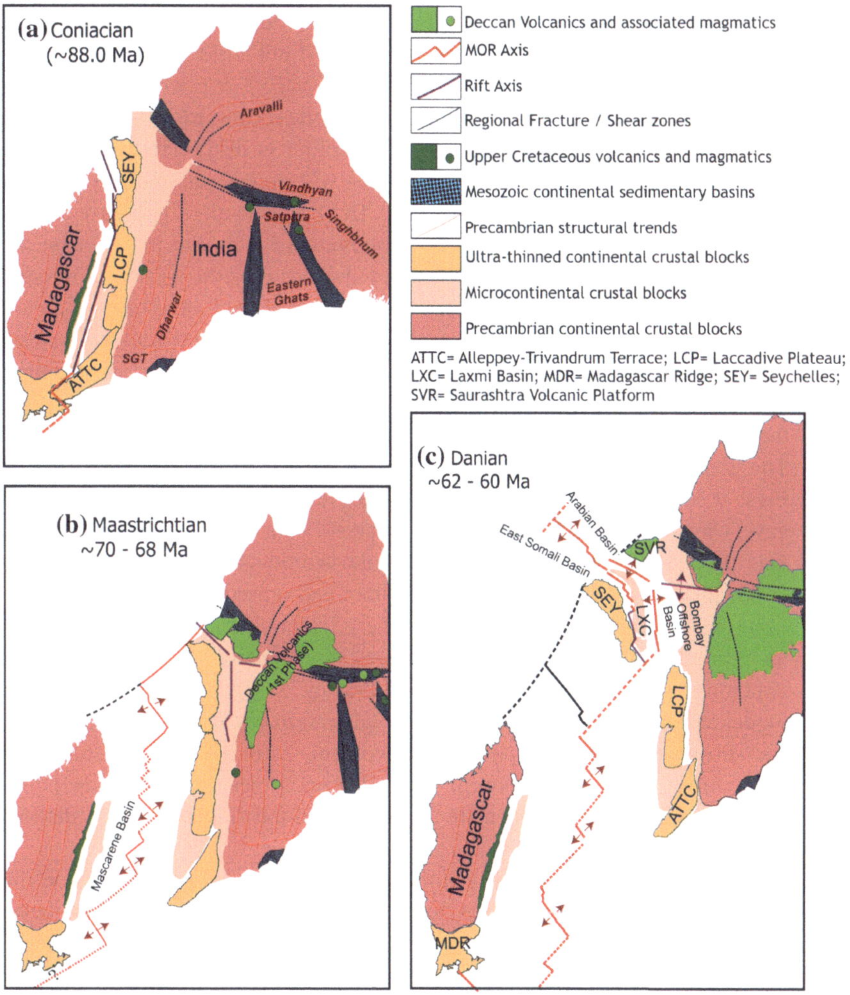

Fig. 9.10 Simplified sequence of event during the Late Cretaceous–Paleocene depicting **a** rifting of the Indian Plate from Madagascar, **b** onset of the first phase of Deccan volcanism and development of the southern Indian Ocean, followed by **c** second (?largest) phase of Deccan volcanism, rifting and separation of India from the Seychelles microcontinent. Sequence adapted from Pande et al. (2017) and modified to depict additional features of significance

precursors to the Deccan Volcanism. Sea-floor spreading and further separation of the India–Africa blocks, as recorded in the North Arabian and Somali basins (Eagles and Hoang 2013), show slow plate divergence between 71 and 69 Ma.

(iv) The terminal Createaceous (Late Maastrichtian) Deccan flood basalts on the continental block between 69–65 Ma are associated with extensive flood basalt volcanism in the Saurashtra High and Somnath Ridge (Fig. 9.10b). This volcanic pulse was as large as the ensuing pulse, if not larger, going by the area occupied by it.

(v) Early Paleocene separation of India and Seychelles occurred during the second phase of Deccan volcanism (Fig. 9.10c; Collier et al. 2008). This pulse is suggested to represent the largest and most widespread phase of flood basalt volcanism in this province (Renne et al. 2015). Kale et al. (2019) have argued that this is more widespread on the continental block, but may not be the largest pulse. Late non-basaltic volcanics also occur on the westernmost edge of the DVP between 62 and 60 Ma. The opening of the Laxmi basin (Samant et al. 2019), and the subsidence of the WCSI that hosted sediments were derived from the Indus delta as well as the Narmada–Mahi rivers draining into the Gulf of Khambat. This indirectly provides an indicative timing for the emergence of the coastal highlands fringing the continental shelf.

(v) The Paleogene syn-rift sediments on various blocks of the WCSI (see Figs. 9.7b and 9.8) are dominated by detrital siliciclastics that eventually gave way to calcareous sediments. Presence of a large Eocene submarine mass-transport complex at the base of the continental slope indicates sustained instability of the off-shore basin-floor and synsedimentary deformation of the sediments (Dailey et al. 2019). These features and structural characters including listric faults, roll-over anticlines, mud-intrusions, and horst-graben structures (Pandey et al. 2017; Nair and Pandey 2018) are consistent with the model of evolution of a passive continental margin passing through the rifting and subsequently drifting phases. The volcanic activity in the Lakshadweep ridge during the Late Palaeocene–Early Eocene times (55–62 Ma) is a contemporary ocean floor volcanism that permitted the development of shallow islands, guyots and related features along the ridge.

(vi) The absence of Trap-derived sediments in the Paleogene sequences on the northern shelf demonstrate that the DVP had not emerged as a provenance for them, indicating that there was very little west-flowing drainage from the continental flood basalt province. This appears to be consistent with easterly tilting of the Indian Peninsula and appears to have occurred during the Early Eocene times (Radhakrishna 1993). The recent age of the first phase of lateritization (45–53 Ma; Bonnet et al. 2016; Jean et al. 2019) further supports this concept.

(vii) With the establishment of the passive margin continental shelf along the western margin of India, further drifting of the Indian plate is manifested in the ocean-floor (Bhattacharya and Yatheesh 2015; Bijesh et al. 2018) with the widening of the Indian Ocean, although at a much slower rate than that encountered at the K-Pg Boundary concurrent with the Deccan volcanism (Cande and Patriat 2015). It is significant that peak spreading rate (estimated at around 200 mm/year at 66 Ma) drops to less than 40 mm/year around 45 Ma by the end of the Eocene. It is curious that this corresponds very well with the ages of early laterite formation on the Peninsular Shield (Jean et al. 2019). Significantly, major hiatal breaks in sedimentation on the WCSI (NDR 2019) in Early Eocene (Ypresian), Late Eocene (Priabonian) and Middle Oligocene correspond with the ages of lateritization on the Peninsular block.

(viii) It is significant to note that there appears to be a close temporal relation between the events of Himalayan collision along northern margin of the Indian Plate and the sea-floor spreading rates in the Indian Ocean (see Gibbons et al. 2015; Tetley et al. 2019). One of the results of the Himalayan orogeny was the establishment of monsoon climatic domain across the Indian Peninsula with alternations of wet and dry phases. The other major consequence is the onset of slab-pull as additional tectonic force acting on the Indian plate, besides the ridge push from the Carlsberg and Central Indian Ocean ridges.

(ix) The Late Oligocene emerges as a time when sea-floor spreading rates along the ridges appear to again display an increase to around 60–65 mm/year. This period also corresponds to tectonic movements along the sequences on the WCSI including some of their onshore exposures along the Kerala coast.

(x) The Miocene to Early Pliocene times (~22–3.6 Ma) represents a period of relative stability of the depositional and erosional system across the western margin. Wide planation surfaces with laterite evolution in the Sahyadri as well as the coastal belts, peat deposits in contemporary sediments along the coast, besides wide-spread development of reefs along the shelf-edge and in the adjoining volcanic islands testify to this period of quiescence and drainage maturation.

(xi) Major structural upheavals, both in the off-shore and on-shore segments, are recorded during the Late Pliocene and Holocene times. They include differential block movements along the WGE and the SONATA as well. The relative subsidence of blocks bounded by reactivated basement structural planes enabled thick accumulation of intracontinental fluvial sediments along middle reaches of rivers Narmada, Tapi, Pravara. Such differential block movements during the Quaternary not only rejuvenated the fluvial domains, but also appear to have aggravated the scarp retreat along the Sahyadri ranges (Joshi et al. 2013; Kale et al. 2017; Valdiya and Samwal 2017).

The present-day configuration of the Western Margin of India is a consequence of this long tectonic history, modified by the influences of eustatic sea-level changes, climatic variability over time and finally subjected to various phases of uplift and subsidence. There is no doubt that it is a typical rifted passive margin of a continental block which evolved during the Cenozoic times, but had roots of its origin in the late Mesozoic times. Part of this rifted margin is volcanic but part of it is devoid of any magmatic activity. One of the questions that have no satisfactory answers is regarding the relative uplift of the Sahyadri ranges. Did it originate as a rift-edge escarpment and sustain over time; or did it rise far beyond its original configuration. The model that links it to plume-related doming has limited applicability, because the DVP occupies only a third of its length; and the fact that the highest peaks of this range are not in any way floored by volcanic or magmatic rocks. Associating it with purely isostatic uplift has several issues, besides the absence of any significant variation in the crustal thickness across its axis. The westerly thinning of the continental crust has more to do with the rift-drift evolution than isostatic rebound.

While we now have a fairly well-constrained understanding of the evolution of the Western margin of India in context of the rifting from Gondwana, sea-floor spreading and associated pulses of magmatic activity. much more needs to be done in context of the uplift history of the Sahyadri and the neotectonics evolution of the adjoining continental regime.

References

Ajaykumar P, Rajendran S, Mahadevan TM (2017) Geophysical lineaments of Western Ghats and adjoining coastal areas of central Kerala and their temporal development. Geosci Front 8:1089–1104

Arora K, Srinu Y, Gopinadh D, Chadha RK, Raza H, Mikhailov V, Ponomarev A, Kiseleva E, Smirnov V (2018) Lineaments in the Deccan Basalts: the basement connection in the Koyna–Warna RTS region. Bull Seism Soc Amer 108(5B):2919–2932. https://doi.org/10.1785/0120180011

Bastia R, Radhakrishna M (2012) Basin evolution and petroleum prospectivity of the continental margins of India. Dev Pet Sci Ser 59:432

Beeauvais A, Bonnet NJ, Chardon D, Arnaud N, Jayananda M (2016) Very long term stability of passive margin escarpment constrained by $^{40}Ar/^{39}Ar$ dating of K-Mn oxides. Geology 44:299–302

Bhattacharya GC, Yatheesh V (2015) Plate-tectonic evolution of the deep ocean basins adjoining the western continental margin of India—a proposed model for the early opening scenario. In: Mukherjee S (ed) Petroleum geoscience: Indian contexts. Springer, Berlin, Switzerland, pp 1–61

Bhave N, Siddique MI, Desai J, Patil Pillai S, Dole G, Kulkarni H, Kale VS (2017) Faulting in Deccan Traps in the vicinity of Koyna-Warna Seismic Zone. Mem Geol Soc India 90:748–751

Bhushan SK, Rao KN, Vidyadharan KT (2010) Petrography and geochemistry of St. Mary Islands near Malpe, Dakshina Kananada district. Karnataka J Geol Soc India 76:155–163

Bijesh CM, John Kurian P, Yatheesh V, Tyagi A, Twinkle D (2018) Morphotectonic characteristics, distribution and probable genesis of bathymetric highs off southwest coast of India. Geomorp 315:33–44

Biswas SK (1987) Regional tectonic framework, structure and evolution of the western marginal basins of India. Tectonophysics 135:307–327

Biswas SK (2005) A review of structure and tectonics of Kutch basin, Western India, with special reference to earthquakes. Curr Sci 88:1592–1600

Biswas SK (2008) Petroliferous basins of India—fifty years' history and perspective. Mem Geol Soc India 66:159–201

Bonnet NJ, Beauvais A, Arnaud N, Chardon D, Jayananda M (2016) Cenozoic lateritic weathering and erosion history of Peninsular India from 40Ar/39Ar dating of supergene K–Mn oxides. Chem Geol 446:33–53, https://doi.org/10.1016/j.chemgeo.2016.04.018

Bredow E, Steinberger B (2018) Variable melt production rate of the Kerguelen hot spot due to long-term Plume-Ridge interaction. Geophy Res Lett 45:126–136. https://doi.org/10.1002/2017/GL075822

Calvés G, Schwab AM, Huuse M, Clift PD, Gaina C, Jolley D, Tabres AR, Inam A (2011) Seismic volcanostratigraphy of the western Indian rifted margin: the pre-Deccan igneous province. J Geophy Res 116(B01101):28. https://doi.org/10.1029/2010JB000862

Campanile D, Nambiar GC, Bishop P, Widdowson M, Brown R (2008) Sedimentation record in the Konkan–Kerala Basin: implications for the evolution of the Western Ghats and the Western Indian passive margin. Basin Res 20:3–22
Cande SC, Patriat P (2015) The anticorrelated velocities of Africa and India in the late Cretaceous and early Cenozoic. Geophy J Int 200:227–243. https://doi.org/10.1093/gji/ggu392
Chandrasekhar T, Minissale A, Vaselli O, Chandrasekharam D, Singh HK (2018) Understanding the evolution of thermal fluids along the western continental margin of India using geochemical and boron isotope signatures. Geotherm 74:197–209. https://doi.org/10.1016/j.geothermics.2018.03.007
Chandrasekharam D (1985) Structure and evolution of the western continental margin of India deduced from gravity, seismic, geomagnetic and geochronological studies. Phys Earth Planet Int 81:186–198
Chatterjee S, Bajpai S (2016) India's northward drift from Gondwana to Asia during the Late Cretaceous–Eocene. Proc Indian Nat Sci Acad 82:479–487
Chaudhuri A, Banerjee S, Chauhan G (2020) Compositional evolution of siliciclastic sediments recording tectonic stability of a pericratonic rift: Mesozoic Kutch Basin, Western India. Marine Petrol Geol 111:476–495
Coffin MF, Pringle MS, Duncan RA, Gladczenko TP, Storey M, Muller RD, Gahagan LA (2002) Kerguelen hotspot magma output since 130 Ma. J Petro 43:1121–1139
Collier JS, Sansom V, Ishizuka O, Taylor RN, Minshull TA, Whitmarsh RB (2008) Age of Seychelles–India break-up. Earth Planet Sci Lett 272:264–277
Corfield RI, Carmichael S, Bennett J, Akhter S, Fatimi M, Craig T (2010) Variability in the crustal structure of the West Indian Continental Margin in the northern Arabian Sea. Petrol Geosci 16:257–265. https://doi.org/10.1144/1354-079309-902
Dailey SK, Clift PD et al (2019) Large-scale mass wasting on the Miocene continental margin of Western India. Geol Soc Amer Bull 28. https://doi.org/10.1130/b35158
Desa MA, Ramprasad T, Kamesh Raju KA (2019) An integrated geophysical study east of the southern Chagos Laccadive Ridge complex, Central Indian Ocean Basin: implications for the Rodriguez Triple Junction dynamics in the Late Cretaceous. Marine Geol 414:47–63
Deshmukh SS, Sehgal MN (1988) Mafic dyke swarms in Deccan Volcanic Province of Madhya Pradesh and Maharashtra. Mem Geol Soc India 10:323–340
Dessai AG, Bertrand H (1995) The "Panvel flexure" along the western Indian continental margin: an extensional fault structure related to Deccan magmatism. Tectonophysics 241:165–178
Dessai AG, Markwick A, Vaselli O, Downes H (2004) Granulite and pyroxenite xenoliths from the Deccan Trap: insight into the nature and composition of the lower lithosphere beneath cratonic India. Lithos 78:263–290
Dikshit KR (1976) Drainage basins of Konkan: forms and characteristic. Nat Geograph J India 22:79–105
Dikshit KR (2001a) Polycyclic landscape and the surface of erosion in the Deccan country with special reference to upland Maharashtra. Nat Geographic J India 16:236–252
Dikshit KR (2001b) The Western Ghats: a geomorphic overview. In: Gunnell Y, Radhakrishna BP (eds) Sahyadri. Geol Soc of India, Memoir 47(1), pp 159–183
Dole G, Peshwa VV, Kale VS (2000) Evidence of a palaeoseismic event from the Deccan plateau uplands. J Geol Soc India 56:547–555
Eagles G, Hoang HH (2013) Cretaceous to present kinematics of the Indian, African and Seychelles plates. Geophy J Int 196(1):1–14. https://doi.org/10.1093/gji/ggt372
Faruque BM, Ramachandran KV (2014) The continental shelf of western India. In: Chiocci FL, Chivas AR (eds) Continental shelves of the World: their evolution during the last glacio-eustatic cycle. Mem Geol Soc London 41, pp 213–220
GSI (2000) Seismotectonic Atlas of India and its environs. In: Narula PL, Acharyya SK, Banerjee J (eds) Geol Surv of India Publ, Kolkata

Gibbons AD, Zahirovic S, Muller RD, Wittaker JM, Yatheesh V (2015) A tectonic model reconciling evidence for the collisions between India, Eurasia and intra-oceanic arcs of the central-eastern Tethys. Gondwana Res 28:451–482
Gombos AM, Powell WG, Norton IO (1995) The tectonic evolution of western India and its impact on hydrocarbon occurrences: an overview. Sedim Geol 96:119–129
Gunnell Y (2001a) Laterites of the Konkan and Kanara coastal plateaux as keys to understanding the denudational chronology of the Western Ghats. Mem Geol Soc India 47:719–750
Gunnell Y (2001b) Morphology and scarp recession rates of the Western Ghats Escarpment. Mem Geol Soc India 47:89–117
Gunnell Y, Michom R (2001) Exploring the relief of Peninsular India using digital elevation models. Mem Geol Soc India 47:11–37
Gunnell Y, Radhakrishna BP (eds) (2001) Sahyadri: the great escarpment of the Indian subcontinent. Mem Geol Soc of India 47(1–2), pp 1054
Harbor D, Gunnell Y (2007) Along-strike Escarpment heterogeneity of the Western Ghats: a synthesis of drainage and topography using digital morphometric tools. J Geol Soc of India 70(3):411–426
Hooper P, Widdowson M, Kelley S (2010) Tectonic setting and timing of the final Deccan flood basalt eruptions. Geology 38:839–842
IOC-IHO-BODC (2003) General Bathymetric Chart of Oceans Digital Atlas (Centenary Edition: on CD-ROM). Intergovernmental Oceanographic Commission and International Hydrographic Organization with British Oceanographic Data Centre, Liverpool, UK
India–WRIS (2012) River Basin Atlas of India. [Central Water Commission, Ministry of Water Resources, Government of India & Regional Remote Sensing Centre-West, National Remote Sensing Centre, Indian Space Research Organization, Department of Space, Government of India] Jodhpur, 144
Jean A, Beauvais A, Chardon D, Arnaud N, Jayananda M, Mathe P (2019) Weathering history and landscape evolution of Western Ghats (India) from 40Ar/39Ar dating of supergene K-Mn oxides. J Geol Soc London, doi:10.1144/jgs2019-048
Joshi PN, Maurya DM, Chamyal LS (2013) Morphotectonic segmentation and spatial variability of neotectonic activity along the Narmada Son fault, western India: remote sensing and GIS analysis. Geomorph 180–181:292–306. https://doi.org/10.1016/j.geomorph.2012.10.023
Kailasam LN (1979) Plateau uplift in peninsular India. Tectonophysics 61:243–269
Kale VS (2003) Fluvial geomorphology of Indian rivers: an overview. Prog Phys Geogr 26:400–433
Kale VS (2010) The Western Ghats: the great escarpment of India. In: Migon P (ed) Geomorphological landscapes of the world. Springer, Berlin, pp 257–264
Kale VS (ed) (2014) Landscapes and Landforms of India. Springer, Dordrecht, pp 1–271
Kale VS (2020) Cretaceous volcanism in Peninsular India: Rajmahal–Sylhet and Deccan Traps. In: Tandon SK, Gupta N (eds) Geodynamics of the Indian plate. Springer, Berlin. https://doi.org/10.1007/978-3-030-15989-4_8
Kale VS, Dole G, Shandilya P, Pande K (2019) Stratigraphy and correlations in Deccan Volcanic Province, India: Quo vadis? Geol Soc Amer Bull 132:588–607, https://doi.org/10.1130/b35018.1
Kale VS, Dole G, Upasani D, Patil PS (2017) Deccan Plateau uplift: insights from the Western Uplands, Maharashtra, India. In: Mukherjee S, Misra A, Calves G, Nemcok M (eds) Tectonics of the Deccan Large Igneous Province. Geol Soc London Spec Publ 445:11–46
Kale VS, Rajaguru SN (1987) Late Quaternary alluvial history of the northwestern Deccan upland region. Nature 325(12):612–614
Kale VS, Rajaguru SN, Rajagopalan G (1986) Late Holocene evidence of neotectonics in the upper Vashishti valley (Western Maharashtra). Curr Sci 55(24):1240–1241
Kale VS, Shejwalkar N (2008) Uplift along the western margin of the Deccan Basalt Province: is there any geomorphic evidence? J Earth Sys Sci 117:959–971
Kale VS, Survase V, Upasani D (2014) Geological mapping in the Koyna–Warna region. ACWADAM Technical Report no. 2014-C1, 163. https://doi.org/10.13140/2.1.3098.0809

Kale VS, Vaidyanadhan R (2014) The Indian Peninsula: geomorphic landscapes. In: Kale VS (ed) Landscapes and Landforms of India. Springer, Berlin, pp 65–78. https://doi.org/10.1007/978-94-017-8029-2_6

Kamesh Raju KA, Chaubey AK, Amarnath D, Mudholkar A (2008) Morphotectonics of the Carlsberg Ridge between 62°20′ and 66°20′ E, Northwest Indian Ocean. Marine Geol 252:120–128

Karmalkar NR, Griffin WL, O'Reilley SY (2000) Ultramafic xenoliths from Kutch (NW India): Plume related mantle samples? Int Geol Rev 42:416–444

Karmalkar NR, MG, Duraiswami RA, Jonnalagadda M (2014) Mid-Cretaceous lamproite from the Kutch region, Gujarat, India: genesis and tectonic implications. Gondwana Res 26:942–956

Krishnan MS (1982) Geology of India and Burma, 6th edn. CBS Publishers and Distributors, Delhi, p 53

Kundu B, Matam A (2000) Identification of probable faults in the vicinity of Harnai–Ratnagiri region of the Konkan coast, Maharashtra, India. Curr Sci 78:1556–1560

Liu X, Mao X, Yuan Y, Ma M (2019) Aeolian accumulation: an alternative origin of laterite on the Deccan Plateau, India. Palaeogeog Palaeoclim Palaeoeco 518: 34–44

Manjunatha BR, Shankar R (1992) A note on the factors controlling the sedimentation rate along the western continental shelf of India. Marine Geol 104:219–224

Melluso L, Sheth HC, Mahoney JJ, Morra V, Petrone CM, Sotery M (2009) Correlation between silicic volcanic rocks of the St Mary's Islands (southwestern India) and eastern Madagascar: implication for Late Cretaceous India–Madagascar Reconstructions. J Geol Soc London 166:283–294

Misra AA, Bhattacharya G, Mukherjee S, Bose N (2014) Near N–S paleo-extension in the western Deccan region, India: does it link strike-slip tectonics with the India-Seychelles rifting? Int J Earth Sci 103:1645–1680. https://doi.org/10.1007/s00531.014-1021-x

Mudholkar AV, Kodagali VN, Kamesh Raju KA, Valsangkar AB, Ranade GH, Ambre NV (2002) Geological and geophysical observations along a segment of slow-spreading Carlsberg Ridge. Curr Sci 82:2982–2989

Mukherjee S, Misra A, Calves G and Nemcok M (eds) (2017) Tectonics of the Deccan Large Igneous Province: An Introduction. Geol Soc of London Spec Publ 445, pp 1–9

Mukhopadhyay R, Karisiddaiah SM (2014) The Indian Coastline: processes and landforms. In: Kale VS (ed) Landscapes and Landforms of India. Springer, Dordrecht, pp 91–101

Muller RD, Zahirovic S, Williams SE, Cannon J, Seton M, Bower DJ, Tetley MG, Heine C, Le Breton E, Liu S, Russel SHJ, Yang T, Leonard J, Gurnis M (2019) A global plate model including lithospheric deformation along major rifts and Orgens since the Triassic. Tectonics 38–24. https://doi.org/10.1029/2018tc005462

Nair KM, Padmalal D, Kurmaran KPN (2006) Quaternary geology of south Kerala sedimentary basin—an outline. J Geol Soc India 67:165–179

Nair N, Anand SP, Rajaram M, Rama Rao P (2015) A relook into the crustal architecture of Laxmi Ridge, northeastern Arabian Sea from geopotential data. J Earth Sys Sci 124:613–630

Nair N, Pandey DK (2018) Cenozoic sedimentation in the Mumbai offshore basin: implications for tectonic evolution of the western continental margin of India. J Asian Earth Sci 152:132–144

National Data Repository (2019) Directorate General of Hydrocarbons, Ministry of Petroleum & Natural Gas, Government of India: www.ndrdgh.gov.in. Accessed Oct 2019

Nemcok M, Rybar S (2017) Rift-drift transition in a magma-rich system: the Gop rift-Laxmi basin case study: western India. In Mukherjee S, Misra A, Calves G, Nemcok M (eds) Tectonics of the Deccan Large Igneous Province. Geol Soc London Spec Publ 445, pp 95–117

Pande K, Sheth HC, Bhutani R (2001) ^{40}Ar–^{39}Ar age of the St. Mary's Islands volcanics, southern India: record of India-Madagascar break-up on the Indian subcontinent. Earth Planet Sci Lett 193:39–46

Pande K, Yatheesh V, Sheth H (2017) ^{40}Ar/^{39}Ar dating of the Mumbai tholeiites and Panvel flexure: intense 62.5 Ma onshore-offshore Deccan magmatism during India–Laxmi Ridge-Seychelles breakup. Geophy J Int 210:1160–1170

Pandey DK, Nair N, Pandey A, Sriram G (2017) Basement tectonics and flexure subsidence along western continental margin of India. Geosci Front 8:1009–1024. https://doi.org/10.1016/j.gsf.2016.10.006
Pandey DK, Pandey A, Clift PD, Nair N, Ramesh P Kulhanek DK, Yadav R (2018) Flexure subsidence analysis of the Laxmi basin, Arabian Sea and its tectonic implications. Geol Mag 14. https://doi.org/10.1017/s0016756818000833
Pandey DK, Pandey A, Whattam S (2019) Relict subduction initiation along a passive margin in the northwest Indian Ocean. Nat Commun 10(2248). https://doi.org/10.1038/s41476-019-10227-8
Pascoe EH (1956) Manual of the Geology of India and Burma. Geol Surv India 483–1343
Peshwa VV, Kale VS (1997) Neotectonics of the Deccan Trap Province: Focus on the Kurduwadi Lineament. J Geophy 18(1):7786
Peshwa VV, Mulay JG, Kale VS (1987) Fracture zones in the Deccan Traps of western and central India: a study based on remote sensing techniques. J Ind Soc Remote Sensing 15(1):9–17
Pitale UL, Dubey R, Saxena RK, Prasad JM, Muthuraman K, Thussu JL, Sharma SC (1987) Review of the geothermal studies of West Coast hot spring belt, Maharashtra, India. Geol Surv India Spec Publ 45:41–60
Powar KB (1993) Geomorphological evolution of the Konkan coastal belt and adjoining Sahyadri uplands with reference to Quaternary uplift. Curr Sci 64:793–796
Radhakrishna BP (1993) Neogene uplift and geomorphic rejuvenation of the Indian Peninsula. Curr Sci 64:787–793
Radhakrishna BP, Vaidyanadhan R (2011) Geology of Karnataka, 2nd edn. Geol Soc India, Bangalore
Rajaram M, Anand SP, Erram VC, Shinde BN (2017) Insight into the structure below the Deccan Trap covered region of Maharashtra, India from geopotential data. In: Mukherjee S, Misra A, Calves G, Nemcok M Eds. Tectonics of the Deccan Large Igneous Province. Geol Soc London Spec Publ 445, pp 219–236
Ramakrishnan M, Vaidyanadhan R (2008) Geology of India 1. Geol Soc India
Rao VP, Wagle BG (1997) Geomorphology and surficial geology of the western continental slope of India: a review. Curr Sci 73:330–350
Rastogi BK, Chadha RK, Raju IP (1986) Seismicity near Bhatsa reservoir, Maharashtra, India. Phys Earth Planet Inter 44:177–199
Ravi Kumar M, Singh A, Kumar N, Sarkar D (2015) Passive seismological imaging of the Narmada paleo-rift, Central India. Precamb Res 270:155–164
Renne PR, Sprain CJ, Richards MA, Self S, Vanderkluysen L, Pande K (2015) State shift in Deccan Volcanism at the Cretaceous-Paleogene boundary possibly induced by impact. Science 350:76–78
Reuter M, Piller WE, Harzhauser M, Kroh A, Rögl F and Ćorić S (2010) The Quilon Limestone, Kerala basin, India: an archive for Miocene Indo-Pacific sea-grass beds. Lethaia 44(1)76–86. https://doi.org/10.1111/j.1502-3931.2010.00226.x
Richards FD, Hoggard MJ, White NJ (2016) Cenozoic epeirogeny of the Indian peninsula. Geochem Geophy Geosystems 17(12):4920–4954. https://doi.org/10.1002/2016gc006545
Samant H, Patel V, Pande K, Sheth H, Jagadeesan KC (2019) ^{40}Ar–^{39}Ar dating of tholeiitic flows and dykes of Elephanta Island, Panvel flexure zone, western Deccan Traps: a five-million-year record of magmatism preceding India–Laxmi Ridge–Seychelles breakup. Ournal Volcanol Geotherm Res 379:12–22. https://doi.org/10.1016/j.jvolgeores.2019.05.004
Sahasrabudhe YS, Deshmukh SS (1981) Bauxite deposits of Gujarat, Maharashtra and parts of Karanataka. Bull Geol Surv India Econo Geol Ser 39:129
Sheth HC (1998) A reappraisal of the coastal Panvel flexure, Deccan Traps, as a listric-fault-controlled reverse drag structure. Tectonophysics 294:143–149
Sheth HC (2000) The timing of crustal extension, diking and eruption of the Deccan Flood basalts. Int Geol Rev 42:1007–1016
Sheth HC (2007) Plume-related regional prevolcanic uplift in the Deccan Traps: Absence of evidence, evidence of absence. Geol Soc Am Spec Pap 430:785–813

Singh K, Radhakrishna M, Pant AP (2007) Geophysical structure of the western off-shore basins of India and its implications on the Western Ghats. J Geol Soc India 70:445–458
Shuhail M, Yatheesh V, Bhattacharya GC, Dietmar Muller R, Kamesh Raju KA, Mahender K (2018) Formation and evolution of the Chain-Kailarli Escarpment and the Vishnu Fracture Zone in the western Indian Ocean. J Asian Earth Sci 164:307–321
Soman K (2013) Geology of Kerala, 2nd edn. Geological Society of India, Bangalore
Srinivasan V (2002) Post-Deccan Trap faulting in Raigad and Thana districts of Maharashtra. J Geol Soc India 59:23–31
Storey M, Mahoney JJ, Saunders AD, Duncan RA, Kelley SP, Coffin MF (1995) Timing of hotspot related volcanism and the breakup of Madagascar and India. Science 267:852–855
Subramanyan V (1981) Geomorphology of the Deccan Volcanic Province: In: Subbarao KV (ed) Deccan flood basalts. Mem Geol Soc India 3, pp 101–116
Tetley M, Simon W, Gurnis M, Flament NE, Muller RD (2019) Constraining absolute plate motions since the Triassic. J Geophy Res: Solid Earth. https://doi.org/10.1029/2019JB017442
Tiwari PK, Surve G, Mohan G (2006) Crustal constraints on the uplift mechanism of the Western Ghats of India. Int Geophy J 167:1306–1316. https://doi.org/10.1111/j.1365-246x2006.03093.x
Vaidyanadhan R (1977) Recent advances in geomorphic studies in Peninsular India–a review. Indian J Earth Sci Ray volume:13–35
Vaidyanadhan R (1987) Quaternary planar surfaces: their mapping and dating techniques with examples from Peninsular India. Indian J Earth Sci 14:252–260
Vaidyanadhan R, Ramakrishnan M (2010) Geology of India 2. Geol Soc India, Bangalore, pp 557–994
Valdiya KS (1998) Late Quaternary movements and landscape rejuvenation in southeastern Karnataka and adjoining Tamil Nadu in Southern Indian Shield. J Geol Soc India 51(2):139–166
Valdiya KS (2001) River response to continuing movements and the scarp development in the central Sahyadri and adjoining coastal belt. J Geol Soc India 57:13–30
Valdiya KS (2002) Saraswati—the river that disappeared. University Press, Hyderabad, pp 1–116
Valdiya KS (2016) The making of India: geodynamic evolution, 2nd edn. Springer, Berlin pp 924
Valdiya KS, Samwal J (2017) Neotectonism in the Indian Subcontinent: landscape evolution. Elsevier, Amsterdam, pp 436
Veeraswamy K, Rawal U (2005) Remobilization of the palaeoconvergent corridors hidden under the Deccan trap cover and some major stable continental region earthquakes. Curr Sci 89(3):522–530
Vora KH, Wagle BG, Veerayya M, Almeida F, Karisidaiah SM (1996) 1300 km long late Pleistocene–Holocene shelf edge barrier reef system along the western continental shelf of India: occurrence and significance. Marine Geol 134:145–162
Wadia DN (1978) Geology of India, ELBS edn. Macmillan, London, p 460
Wagle BG, Vora KH, Karisiddaiah SM, Veerayya M, Almeida F (1994) Holocene submarine terraces on the western continental shelf of India: implications for sea-level changes. Marine Geol 117:207–225
Widdowson M (1997) Tertiary paleosurfaces of the SW Deccan, Western India: Implications for passive margin uplift. Geol Soc London Spec Publ 120:221–248
Widdowson M, Cox KG (1996) Uplift and erosional history of the Deccan Traps, India: evidence from laterites and drainage patterns of the Western Ghats and Konkan Coast. Earth Planet Sci Lett 137:57–69
Wimpenny J, Gannoun A, Burton K W, Widdowson M, James RH, Gislason SR (2007) Rhenium and osmium isotope and elemental behaviour accompanying laterite formation in the Deccan region of India. Earth Planet Sci Lett 261:239–258
Yatheesh V, Dyment J, Bthattacharya GC, Royer JY, Kamesh Raju KA, Ramprasad T, Chaubey AK, Patriat P, Srinivas K, Choi Y (2019) Detailed structure and plate reconstructions of the Central Indian Ocean between 83.0 and 42.5 Ma (Chrons 34 and 20). J Geophy Res Solid Earth 124:4305–4322. https://doi.org/10.1029/2018JB016812
Yatheesh V, John Kuran P, Bhattacharya GC, Rajan S (2013) Morphotectonic architecture of an India–Madagascar breakup related anomalous submarine terrace complex on the southwest continental margin of India. Marine Petrol Geol 46:304–318

Chapter 10
Geology and Tectonics of Bangladesh

10.1 Location

Bangladesh is in the northeastern parts of the Indian Subcontinent between the Indian Shield to the west and the Indo-Myanmar Ranges to the east and shares the geology of the Bengal Basin. In the north, the basin is bounded by the Himalayan Foredeep, the Shillong Plateau and the Assam Basin (Roy and Chatterjee 2015; Najman et al. 2016). The entire country, barring a few small tract in the east and NE, has been confined within the Great Bengal Delta Plain (Umitsu 1993; Goodbred and Kuehl 2000), also known as the Ganga–Brahmaputra–Meghna Delta (GBD), formed by the alluvial sedimentary deposits of the Ganga, the Brahmaputra and Meghna rivers. These delta plain deposits are spread over parts of the Indian state of West Bengal in the west and Tripura in the east. The western and southwestern parts of the Bengal Basin consist of an easterly inclined shelf, separated from the Singhbhum Craton of the northeastern part of the Indian Shield (Hossain et al. 2017). The southern boundary of the basin is marked by the northern edge of the Bengal Fan (Fig. 10.1; Curray et al. 2003; Stecklet et al. 2008), while southeastern boundary of this basin coincides with the eastern margin of the Chittagong Tripura Fold Belt (CTFB) along the western fringe of the Indo-Burman Ranges (IBR) (Acharyya 2007; Wang et al. 2014).

Configuration of the Bengal Basin and its fluvio-deltaic sedimentary fill is intimately linked to the world's largest Himalayan orogenic system, covering approximately 200,000 km^2 of area. World's largest submarine fan system, the Bengal Deep Sea Fan, extends far to the south in the Bay of Bengal. The geology is characterized by the rapid subsidence and filling of this basin in which huge thickness of deltaic sediments were deposited as a mega-delta out built which progressed southwards. The delta building is continuing into the present Bay of Bengal (Islam and Tooley 1999). A broad fluvial front of the Ganges–Brahmaputra–Meghna river system gradually follows it from behind. Only the eastern part of Bangladesh has been uplifted into hilly landform incorporating itself into frontal belt of the Indo-Myanmar Range

A. K. Jain et al., *Tectonics of the Indian Subcontinent*, Society of Earth Scientists Series, https://doi.org/10.1007/978-3-030-42845-7_10

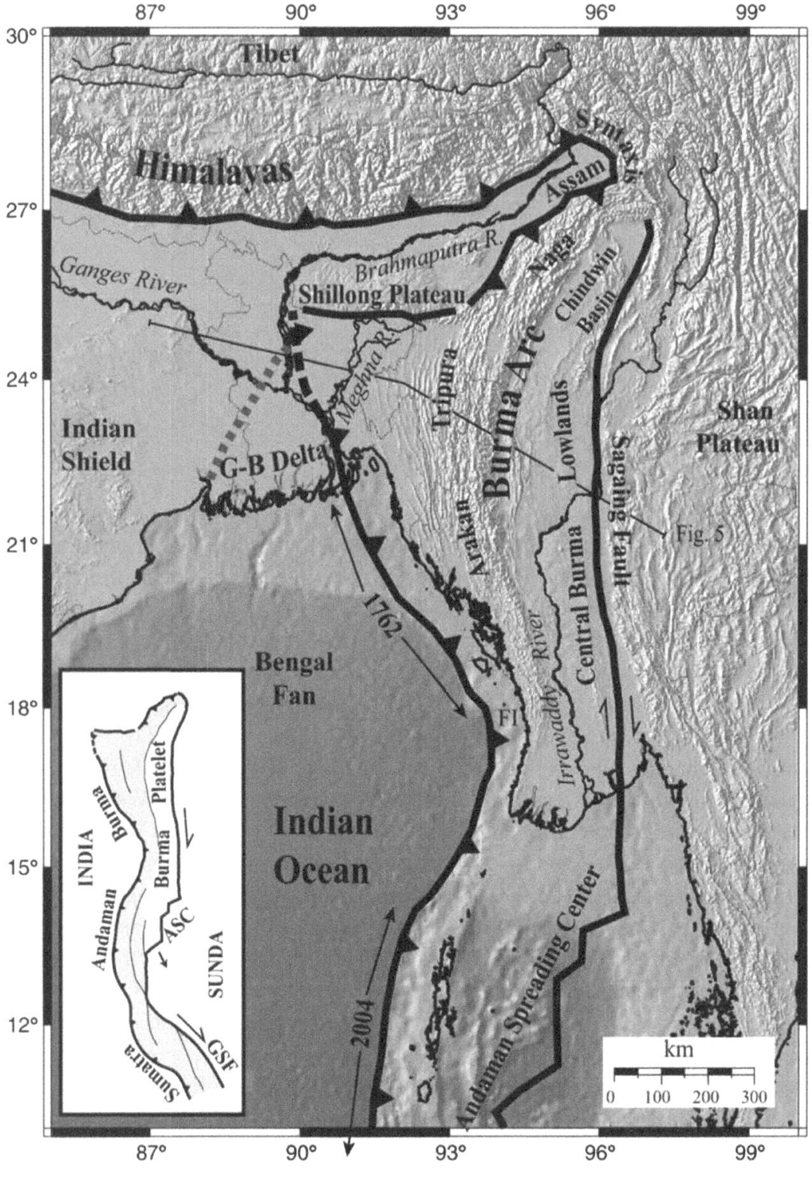

Fig. 10.1 Plate tectonic setting of Bangladesh between Ganges–Brahmaputra Delta and the Indo-Myanmar/Burma Arc. Heavy black lines are major faults and tectonic boundaries with thrusts barbed on the upper plate. Source areas 1762 and 2004 earthquakes are shown by arrows. FI = Foul Island. Inset map shows the outline of the Burma Platelet between the Sunda and Indian Plates where shaded region is the deforming forearc region. ASC = Andaman Spreading. After Steckler et al. (2008)

to the east. The Bengal Basin is well known for the development of a thick Early Cretaceous–Holocene sedimentary succession (cf. Alam et al. 2003). The regional stratigraphic and tectonic scenario unraveled by earlier workers (e.g. Evans 1964; Sengupta 1966; Raju 1968) helped in building up the sub-surface stratigraphy of Bangladesh. Credit goes to Bakhtine (1966) for initiating the identification of the tectonic elements within the Bangladesh part (then East Pakistan) of the Bengal Basin. Investigations by Moore et al. (1974), Graham et al. (1975), Paul and Lian (1975), Curray et al. (1982) Reimann (1993), Shamsuddin and Abdullah (1997), Uddin and Lundberg (1999) contributed immensely to our basic understanding of the evolution of this basin. Alam (1989, 1997) described the stratigraphic and tectonic history of the basin. Johnson and Alam (1991), Alam (1995), Gani and Alam (1999) carried out studies related to the sedimentation and basinal tectonics. It is important to point out that the geology of Bangladesh and its basin-fill history relied to a large extent on seismic records instead of outcrop evidence (Fig. 10. 2).

10.2 Plate Boundaries

The Ganga–Brahmaputra Delta (GBD) is the world's largest delta and is gradually building up by sediments eroded from the Himalaya. These sediments prograded the continental margin of the Indian subcontinent by ~400 km, forming a huge sediment pile that is now entering the Indo-Myanmar Arc subduction zone (Streckler et al. 2008). Thick sediment cover in the Bengal Basin conceals the basement configuration and makes the reconstruction and precise location of plate boundaries and sutures difficult. Plate movement patterns and evolution of the Bengal Basin and the Bay of Bengal are carried out mostly with data and interpretation from the Indian Ocean, following early work by McKenzie and Sclater (1971), Sclater and Fisher (1974) and others.

One of the problems of plate reconstruction for the Indian subcontinent is determining the eastern limit of the Indian continental crust. Curray and Moore (1974), Graham et al. (1975), Curray et al. (1982) considered the eastern limit of the Indian continental crust to be approximately along the Hinge Zone, which lies above the Calcutta–Mymensingh Gravity High, with the oceanic part of the Indian Plate subducting beneath the Indo-Myanmar Ranges west of the Burma Block. The Burma Block was inferred to be of continental origin from Gondwana which show the area between the Hinge Zone and Barisal–Chandpur Gravity High to be attenuated or thinned continental crust, so that the continent–ocean crust boundary lies along the Barisal–Chandpur Gravity High. Alam et al. (2003) conclude that the continent ocean boundary (COB) beneath the Bengal Basin is located between the Hinge Zone and Barisal–Chandpur Gravity High, so that it lies along the northwest side of the gravity high and passes offshore approximately down the axis of the upper part of the Swatch-of-no-Ground submarine canyon

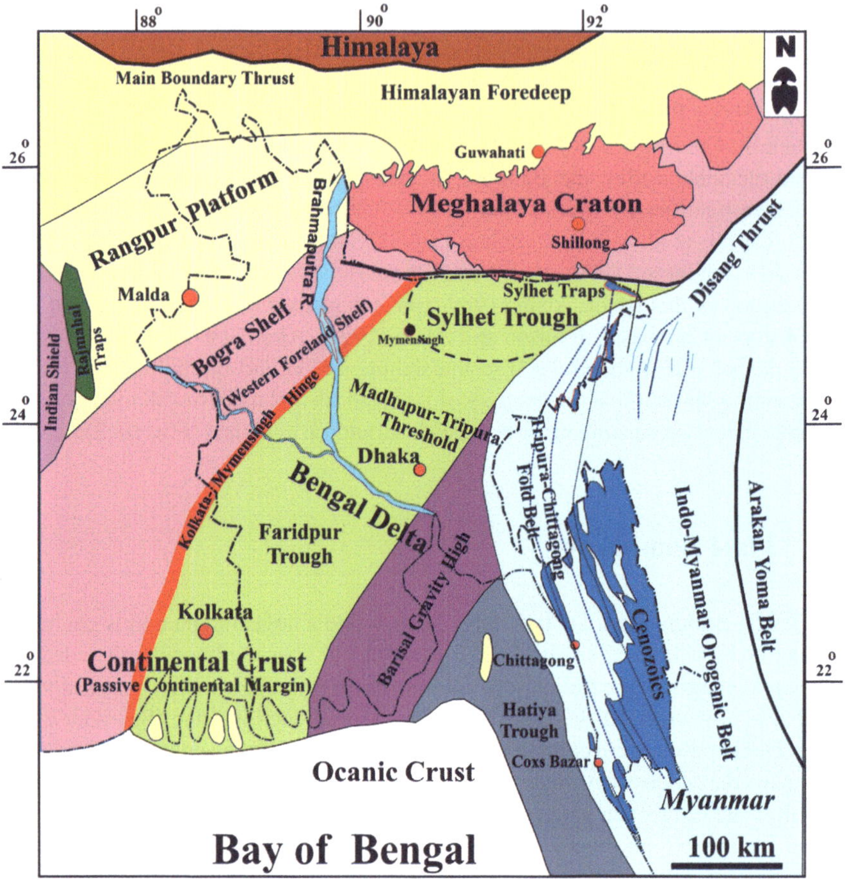

Fig. 10.2 Geological and tectonic framework of Bangladesh and adjoining areas, showing SE sloping basement, its thick cover and deformation near the Indo-Myanmar Ranges. Redrawn and simplified after Geological map of Bangladesh (2011) and http://en.banglapedia.org/index.php?title=Tectonic_Framework

10.3 Stratigraphy

The Hatia Trough in the Bengal Basin is the main sediment depocenter (±22 km) where it is floored by a large subsurface Precambrian Platform, which is the eastward extension of the Archean/Proterozoic suites of rock belonging to the Singhbhum Craton, Chotanagpur belt and CITZ on the Indian side. The stratigraphy of Bengal Basin in the Bangladesh can be studied under three stratigraphic domains ((Alam 1972; Alam et al. 2003) corresponding to the geo-tectonic provinces (i) Stable Shelf or Geotectonic Province 1 (=Precambrian Platform of Qazi 1986), (ii) Central Deep Basin (including the Sylhet and Hatia Troughs) or Geotectonic Province 2; and (iii)

Chittagong–Tripura Trough or Geotectonic Province 3. The Bengal Deep Sea Fan may also be considered an extension of the present sedimentation regime.

Lithostratigraphy of the Bengal Basin in Bangladesh was established on sporadic outcrops in fold belt in eastern and northeastern parts of the country, based on their correlatability with the sections exposed in Assam, NE India (Evans 1932). Extensive subsurface explorations in the basin, mostly by the petroleum companies have brought new data for the entire basin. Palynological and micropaleontological studies (e.g. Chowdhury 1982; Uddin and Ahmed 1989; Reimann 1993; Ahmed 1968; Ismail 1978) and seismo-stratigraphic studies (Lietz and Kabir 1982; Salt et al. 1986; Lindsay et al. 1991) helped in establishing more precise stratigraphy of the region. The basic premise of this revision was the realization that the shelf facies may not show the same order of superposition as those in the deeper basin due to differences in rates of sedimentation and configuration of sedimentation floor. Stratigraphic names used for the Sylhet Trough are extended to the stable shelf and to the fold belt regime (Table 10.1). Alam et al. (2003) identified five distinct phases of sediment accumulation: (i) Late Paleozoic to early Cretaceous, (ii) Late Cretaceous–Mid-Eocene, (iii) Mid-Eocene–Early Miocene, (iv) Early Miocene–Mid-Pliocene, and (v) Mid-Pliocene–Quaternary.

10.3.1 Late Palaeozoic

The Palaeozoic cover sediments starts to show up only during the Permian when the oldest sedimentary succession was deposited in the Bengal Basin of Bangladesh. They belong to the Gondwana Group and, rests unconformably on the Precambrian crystalline basement. The Gondwana Group (Raniganj?) is composed of hard sandstone with coal and shale layers. The Permian is about 1000 m thick and found in fault bounded graben basins. The Kuchma Formation (490 m thick) of sandstone, carbonaceous shale and white sandstone represents the early Permian, while the late Permian Paharpur Formation (465 m) containing fine to coarse arkosic sandstone with kaolin pockets and thick coal beds is recognized in the Paharpur coal field in isolated grabens in the basement (Zaher and Rahman 1980; Khan 1991). At the base tillite indicates major glaciation in the region indicating cold climatic conditions (Wardell 1999) while coal deposits suggest interglacial sedimentation. These Permian sediments are interpreted to have been deposited in low-sinuosity braided fluvial systems flanked by vegetated overbank and swampy flood plains (Uddin and Islam 1992; Uddin 1994).

10.3.2 Jurassic–Cretaceous

Above the Gondwana Group (Raniganj?) lies a sequence of volcanic basalts of Late Jurassic age called the Rajmahal Trap formation. The unit is about 610 m thick and

Table 10.1 Stratigraphy of the Bengal Basin, Bangladesh

Age	Group/Formation (Stable Precambrian Shelf) (Alam et al. 2003)		Group/Formation (Sylhet Trough/Northeastern Deep Basin) (Qazi 1986; Hiller and Elahi 1988; Alam et al. 2003)		Lithology	Depositional environment	Tectonic events
Holocene (Including Meghalayan)	Alluvium		Alluvium		Silt, clay, gravel	Fluvial-alluvial Rapidly prograding Delta	Folding in Eastern Bangladesh
Pleistocene-Late Pliocene	Barind	Barind Clay (50 m)	Madhupur Clays		Yellowish brown sand, clay, Silicified wood		
		Dihing (150 m)	Dihing				
Unconformity						Fluvial and prograding Delta-shelf	
Early Pliocene-Late Miocene	Dupi Tilla Fm (280 m)		Dupi Tilla Fm	Upper Dupi Tilla Lower Dupi Tilla	Claystone, siltstone, sandstone, gravel		
Unconformity						Delta front to shelf and slope	Folding in Indo-Burma Ranges
Early to Middle Miocence	on rateJamalganj (415 m)		Tipam Fm	Girujan Clay Tipam Sst	Alternating sandstone, siltstone, shale		
			Surma Group	Bokabil Bhuban			
Middle to Upper Oligocene	Bogra (165 m)		Barail Fm	Renji Jenam Lesong	Siltstone, carbonaceous shale, fine sandstone	Increasing sedimentation rate	

(continued)

Table 10.1 (continued)

Age	Group/Formation (Stable Precambrian Shelf) (Alam et al. 2003)		Group/Formation (Sylhet Trough/Northeastern Deep Basin) (Qazi 1986; Hiller and Elahi 1988; Alam et al. 2003)		Lithology	Depositional environment	Tectonic events
Unconformity							
Upper Eocene to Paleocene	Jaintia (735 m)	Kopili Shale (240)	Jaintia	Kopili	Sandstone, locally glauconitic Highly fossiliferous, shale	Deltaic to slope	Hard collision Soft collision (Closure sutures)
		Sylhet Limestone (250 m)		Sylhet Lst Jaintia	Nummulitic limestone,sandstone interbeds	Carbonate platform	
		Tura Sandstone (245 m)		Tura Sst	Sandstone, coal and shale	Deltaic to outer shelf	
Unconformity							
Middle to Late Cretaceous	Rajmahal (840 m)	Sibganj Trapwash (230 m)	Cretaceous Series		Coarse brown sandstone, volcanic material and clays	Coastal to fluvial-alluvial	
Early Cretaceous Late Jurassic		Rajmahal Traps (610 m)	Rajmahal Traps		Amygdoidal basalt, andesite, Sepentinized shale, agglomerate	Subaerial lava flows Fluvial-deltaic to shallow marine	
Unconformity							Main break

(continued)

Table 10.1 (continued)

Age	Group/Formation (Stable Precambrian Shelf) (Alam et al. 2003)		Group/Formation (Sylhet Trough/Northeastern Deep Basin) (Qazi 1986; Hiller and Elahi 1988; Alam et al. 2003)	Lithology	Depositional environment	Tectonic events
Permian Upper Carboniferous	Gondwana (955 m)	Paharpur Fm	Raniganj	Feldspathic sandstone, thick coal seams	Fluvial to delta plain, coal swamps	(Unconformity) Rapid extension Graben formation
		Kuchma Fm		Coarse sandstone, conglomerate Thin coal seams		
Unconformity						
Precambrian Platform	Precambrian metamorphic and magmatic rocks		Archean		Stable Gondwana Continent	Gondwana Continent

encountered in drill holes in the Rajshahi-Bogra area. The Rajmahal Trap formation differs from other primarily sedimentary sequences due to its association with hornblende basalt, olivine basalt, andesite, ash beds, tuff and agglomerate (Khan 1991). It also harbors a meter-thick sedimentary bed with plant fossils. Kent (1991) and Curray and Munasinghe (1991) described Rajmahal Traps and its equivalent Sylhet Traps on the eastern side of the Permian province and found them to be more extensively developed than that is revealed by limited outcrops and drill hole data. The Rajmahal trap is overlain by the early Cretaceous Shibganj Trapwash Formation, a relatively thin cover of the weathered product of volcanic rocks consisting of red ferruginous sandstone and mudstone. The formation includes poorly sorted coarse sandstone, white kaolinitic sandstone, iron-rich trapwash sandstone, red shale, claystone, tuff and sedimentary beds with fragmentary Gondwana plant fossils. These rocks were deposited in fluvial and coastal settings, particularly tidal flats, deltaic and lagoonal environments.

10.3.3 Early Cenozoic

The marine Paleocene-Eocene sedimentary sequence in all the geotectonic basin of the Bengal basin is represented by the Jaintia Group which is subdivided into three stratigraphic units:

(i) the Tura Sandstone (named after the Tura range in the Tura district) at the base consists of sandstones, siltstone, carbonaceous mudstone and thin coal seams. The sandstones often contain foraminifera, shell debris and glauconite. The Tura sandstone is a very thick unit (245 m) formed during the early and middle Cenozoic. Very little of the sandstone is visible at the surface. It forms three small hills at Takerghat, although these are blanketed in alluvium. A small pillar in Lalghat and an exposure by a stream are the only places where the Tura sandstone is visible.

(ii) the Midle Eocene Sylhet Limestone (250 m thick) characterizes both the maximum marine transgression on the stable shelf of the Bengal Basin as well as defines the southern limit of the shelf. The carbonate rocks are dolo-biomicritic with foraminifera and algal fragments (Banerji 1981). Crinoids, corals and bryozoans are also reported from the Sylhet Limestone in the Lower Assam Basin. The Sylhet limestone preserves additional rock units such as shale and sandstone, indicative of a shallow marine environment of deposition. The seismic data suggest that the lower part of Sylhet Limestone forms is a marker horizon in the seismic section and the broad time-transgressive facies extends from the shelf edge to the upper shelf area.

(iii) The Upper Eocene Kopili formation (240 m) overlies the Sylhet Limestone and is composed of dark grey to black fossiliferous shale with a few limestone beds and marks the end of open marine conditions of deposition. These rocks are interpreted as deposits of distal deltaic to shelf and/or slope environments.

10.3.4 Late Cenozoic

The Oligocene Barail Group unconformably overlies the Jaintia Group in the Geotectoniic Province 2 or the Folded Trough of the eastern flank. It is made up of 150–200 m thick sandstone, shale and siltstone with occasional carbonaceous layer indicating deltaic conditions of deposition. These rocks on the Indian side are represented by the Burdwaran Formation (200 m thick) and Memari formation (150 m thick) and show signs of ocean transgression and estuarine environment. The Burdwan Formation represents the sandy proximal facies, while the Memari Formation is the more distal and muddy deltaic facies (Lindsay et al. 1991).

The Barail group is formally subdivided into the Lisong, Jenam and Renji formations in the ascending order in the Sylhet Trough. These rocks have been interpreted as deposits of predominantly tide-dominated shelf environments (Alam 1991). On the basis of facies analysis, Dasgupta et al. (1991) identified the Barail Group (3500 m thick) in the basinal facies of the Lower Assam Basin, India and interpreted them to be the deposits of a basin-ward migrating progradational submarine fan complex, and thereby suggested that sedimentation continued into the overlying Surma Group. The contact of the Surma Group with the underlying Barail Group appears to be a transgressive onlap, approximately at the Oligocene–Miocene boundary (Banerji 1984; Salt et al. 1986). This marine transgression on the shelf may be the result of a major upthrust movement along the Dauki Fault in the Early Miocene.

Over the Precambrian Stable Platform (Geotectonic Province 1) similar lithology with little enlarged thickness is given a name of Bogra Group (Khan and Muminullah 1980; Alam et al. 2003), which appears to be a facies variant of the Barail Group of the east. Depositional environment in both the units are almost identical. With an unconformity, a 415 m thick deltaic sandstone, siltstone and shale of the Lower Middle Miocene Jamalgunj Formation (Alam et al. 2003) overlies the Bogra Group in the Geotectonic Province 1. The Surma Group of Province 2 occupies the same stratigraphic position as that of the Jamalgunj Formation, which is deposited in a large delta complex. The Surma Group has a thickness of about 3500–4500 m, and is composed of alternating sandstone, shale, siltstone and conglomerate. The Surma Group is divided into a lower sandy Bhuban Formation and an upper argillaceous Bokabil Formation. These are largely deposited in deltaic to shallow marine environments, although recent studies indicated deep marine turbidite origin of the basal parts of the Upper Bhuban formation as evident from the sediment cores in the Chittagong Hill Tracts. Johnson and Alam (1991) have interpreted the Bhuban Formation as prodelta and delta-front deposits of a mud-rich delta system like the modern Bengal delta. The sediments of the Bokabil Formation represent deposits of subaerial to brackish environments, based on mudrocks and pollen types (Hystrichosphroedis; Holtrop and Keizer 1970). Alderson (1991) noted marine influence within the Bokabil Formation in eastern Sylhet Trough. The Surma Group is probably the most important stratigraphic unit in Bangladesh because it is represented by great thickness in all the wells drilled in the deeper parts of the Basin and forms the backbone of the eastern hilly areas of Bangladesh, including those of Sylhet and Chittagong Hills where it is extensively exposed. These are overlain by the Middle Miocene Tipam Group and is subdivided into a lower Tipam Sandstone and upper Girujan Clay. The 1200–2500 m thick Tipam Sandstone formation comprises coarse-grained, cross-bedded and pebbly sand, with common carbonized wood fragments and coal interbeds and interpreted as deposits of bedload dominated braided-fluvial systems (Johnson and Alam 1991). The overlying Girujan Clay formation (100–1000 m thick) is composed mainly of mottled clay, accumulated in subaerial conditions as lacustrine and fluvial overbank deposits (Reimann 1993). The Surma and Tipam groups are not identified sensu stricto in the Stable Precambrian Platform succession where these are collectively represented by the Jamalgunj Group. After a short hiatus, 270 m thick Dupi Tila Formation of the Pliocene-Pleistocene age is deposited over the Jamalgunj Group/Formation and is composed of coarse and pebbly sandstone. These are riverine sediments spread over the deltaic plain. In sectors where Surma Group of sediments are identified, 2–3 thousand meters thick Dupi Tila Sandstone formation is divisible into the Lower and Upper units and are made up of medium to coarse loosely compacted cross-bedded sandstone, which also show the fluvial conditions of deposition. Lying unconformably over the Upper Dupi Tilla Formation, the Dihing Formation is also fluvial deposit of pebbly to gravelly sands. The widespread Barind Clay (Madhupur Clay) predominantly consists of yellowish to reddish brown clay, silty-clay and silty-sand with minor pebbly beds. Thickness of the Barind Group in Stable Platform Province 1 is about 200 m while thickness of Madhupur Clay in eastern province remains uncertain. The Bengal Alluvium of Meghalayan age engulfs all older lithologies over the entire Bengal Delta Plain.

Alam et al. (2003) contended that stratigraphy and sedimentation pattern in the southern part of the Central Deep Basin of the Hatia and Faridpur Troughs is likely to have different stratigraphic order as these troughs are centrally located with respect to the other geotectonic provinces and therefore received sediments from all the directions at different times and rates.

Stratigraphy of the sedimentary fill in the deeper basins reflect the presence of ~20 km of Cenozoic sedimentary succession. Rapid subsidence and concomitant sedimentation allowed storage of such great thickness of sediments. The country around Sylhet evolved from a Quaternary lagoon filled up by the debris from the Rajmahal Hills located in the west. The much recent change of course of Brahmaputra led to the creation of an independent river Jamuna while the Tista displayed some changes in its course during Holocene tectonics (Quazi 1986). The eastern sea-face of the delta is changing at a rapid rate by the formation of new ground and new islands, while the western portion of the deltaic coastline has remained practically unchanged (Islam and Tooley 1999).

10.4 Tectonics

The tectonics involved in the development of the Bengal Basin include (Alam 1972): (i) Syn-rift stage, (ii) Drifting stage, (iii) Early collision stage, and (iv) Late collision stage. The initial Permo-Carboniferous to early Cretaceous sedimentation started within the graben basins on the Precambrian platform (Province 1). The breakup of Gondwanaland, further advanced sedimentation in the Cretaceous-Mid Eocene times (123–132 Ma). This was the time of subsidence and marine transgression into area of Provinces 1 and 2 of the Bengal Basin (Banerji 1981) where rate of sedimentation increased in the Santonian (about 84–88 Ma) with sediment accumulation on northern continental margin of India. Deeper parts of the basin were dominated by hemipelagic deposits. The mild Paleocene collision between the northern Indian continental crust and its subduction zone in the Indo-Myanmar Ranges did not affect this early Bengal basin. In the Mid-Eocene, a marine transgression is recognized in carbonate-dominated Precambrian platform. The marine regression accumulated shelfal to deltaic deposits on Provinces 1 and 2, whereas submarine fan turbidite sedimentation dominated southwestern portion of Province 2 and in Province 3.

In Early Miocene, the Bengal Basin assumed the configuration of a remnant ocean basin, so that Provinces 2 and 3 started to act as active sediment depocenters. Major collision of the Indian plate with South Tibet and Burma plates took place in the Early Miocene accompanied by uplift of sedimentary pile to form the Himalaya. In the eastern sector, a major upthrust movement produced the Dauki Fault resulting in separation of the Sylhet Trough from the stable Precambrian shelf, and the trough formed an important sediment depocenter. Repetitive marine transgressions and regressions dominated the depositional processes over the entire Bengal Basin during Late Miocene. At the beginning of Mid-Pliocene sedimentation, the final marine regression from most of the Bengal Basin concomitant with the tectonic upheaval of the eastern Bengal Basin, established a fluvio-deltaic environment of deposition. With continued Plio-Pleistocene collision of India with Tibet and Burma, and rapid rise of the Himalaya, the sediment depositional center was shifted further to the south and the present Faridpur and Hatiam Troughs became the major sediment depositional centers.

10.4.1 Precambrian Platform–Geotectonic Province: A Stable Shelf

Our understanding of geology of the Bangladesh is largely based on data recovered during subsurface exploration in search of hydrocarbon and hydrological resources (Khan 1991). The Precambrian rocks are identified on the submerged **Dinajpur Slope**, which connects with the Indian Deccan Shield to the west and the Shillong Massif in the north. A part of the Shillong Massif continues into the Bangladesh. It is a large tectonically exhumed block of the basement, that got detached from the main Indian peninsular block.

A thick pile of Cenozoic sediments is identified as the Eocene Hinge Belt which separates the Precambrian platform from the Deeper Basin. This belt passes under the city of Kolkata, just south of Bogra and Mymensingh in the north. More than 6000 m of sediments rest on the crystalline basement occupying the space in the ridges and depressions. They show gradual thickening to the southeast. Based on the characteristics of the crystalline basement and sedimentary packages, the Precambrian Platform is further subdivided (Quazi 1986) into the following:

(i) The **Platform Slope** of the Himalayan Foredeep,
(ii) The **Garo-Rangpur Saddle** or the Gap, and
(iii) The **Stable Shelf Zone** or the Bogra Flank.

10.4.1.1 Himalayan Foredeep

The northwestern part of the Bengal Basin lies south of the Main Frontal Thrust (MFT) and is covered by the Recent to sub-Recent piedmont plain deposits. This element is marked by the high negative Bouguer gravity anomaly from −110 to −150 mGal in extreme northwest of Bangladesh, with thickening of basinal strata northwards into the Siwalik foreland basin (Uddin and Lundberg 2004). The −110 mGal contour near Panchagarh represents the approximate southern boundary of the Himalayan Foredeep with the Dinajpur Slope.

It is comprised of ~3–4.5 km Neogene sediments of sandstone, subordinate shale and gravel beds and attain a thickness of (Guha et al. 2010). The only well drilled in this tectonic element is located on the north-western most tip of Bangladesh at Salbanhat in Tetulia by Shell Oil Co. in 1988. The well touched the basement at 2518 m depth penetrating the Mio-Pliocene sequence but did not encounter the Eocene Limestone (Rabbani et al. 2000). Two major lineaments transect the Himalayan Foredeep in the Bengal Basin along NW–SE direction, namely the Tista and Gangtok lineaments (Baruah et al. 2016).

10.4.1.2 Dinajpur Shelf/Dinajpur Slope

The Dinajpur Slope is characterized by almost E–W trending linear Bouguer gravity contours, with values ranging between −50 and −110 mGal as as a separate tectonic element in between Himalayan Foredeep to the north and Rangpur Saddle to the south (Guha et al. 2010). Some fault bounded graben basins in the Precambrian basement acted as repositories for the Permian coal.

10.4.1.3 Platform Slope of the Himalayan Foredeep–Garo-Rangpur Saddle

The Rangpur Saddle is characterized by irregular shaped Bouguer gravity contours with several closed gravity highs and lows typical of a shallow basement in and around Rangpur region. The northern Rangpur Saddle rests on a very shallow Precambrian basement (130–1000 m), while the southern Bogra Flank occurs at moderate depth of 1–6 km. Sedimentary layers in the Bogra shelf dip very gently towards the southeast until it reaches the hinge zone when the dip suddenly increases to 15–20° and the sedimentary units plunge to a great depth into the deep geosynclinal basin to the southeast. In the eastern part of this tectonic element, the SE end of the Tista and Gangtok lineaments is terminated by the N–S oriented Jamuna Fault.

The Precambrian Platform is recognized subsurface at Rajshahi, Bogra, Rangpur and Dinajpur areas and is characterized by limited to moderate thickness of sedimentary rocks above the Precambrian igneous and metamorphic basement. This unit is geologically stable and has not been affected by structural deformation. The basement is considered as an eastward subsurface extension of the Indian Shield, though Ameen et al. (2007, 2016) and Tapu et al. (2016) considered it as separate microcontinental blocks. In the Rangpur Saddle, the basement is made up of diorite, tonalite and granodiorite, dissected by pegmatite and mafic/ultramafic dykes (Ameen et al. 1998; Kabir et al. 2001; Hossain et al. 2007). SHRIMP zircon U–Pb age dating of these rocks including the pegmatite yielded ages of both reported identical Paleoproterozoic ages of 1.72–1.73 Ga (Ameen et al. 2007; Hossain et al. 2007, 2017). It is however debated whether the basement constituted a Paleoproterozoic micro-continental Dinajpur block (Ameen et al. 2007) and Tapu et al. (2016), or continuation of the Central Indian Tectonic Zone (CITZ) (Chatterjee 2017; Hossain et al. 2017). Alternatively, it may even be an extension of the Meghalaya Craton.

10.4.1.4 Eocene Hinge Zone

This zone is also known as shelf break or paleo-continental slope or trace of the Eocene shelf edge (Reimann 1993; Uddin and Lundberg 2004; Singh et al. 2016) and lies between the Stable Shelf/Foreland Shelf to the west and the Foredeep Basin to the east. This tectonic element is characterized by almost ENE-WSW trending linear Bouguer gravity contours, with values ranging approximately between −30 and −15 mGal and sloping towards southeast (Khan and Rahman 1992). The Precambrian basement dips southeast abruptly from 2–3° to 6–12° at the contact of the stable platform and the Eocene Hinge Zone to the west, and then dips more gently 1–2° again in southeast (Uddin and Lundberg 2004). At its upper northwestern edge, recorded seismic depth on top of the Eocene Sylhet Limestone is 3500 m, whereas it deepens to 5000 m at its lower southeastern edge. Sengupta (1966) and Khandoker (1989) named this NNE–SSW running narrow 25–100 km Hinge zone as the ‘Calcutta–Mymensingh gravity high’ (Figs. 10.2 and 10.3). Although it truncates against the Dauki Fault in the northeast, many consider that this zone progressively convexes basinward to the northeast and possibly continues towards the Haflong Thrust at northeastern corner of the Bengal Basin (Alam et al. 2003).

The Eocene Hinge Zone marks the structural as well as depositional transition between central foredeep basin to the southeast and stable shelf to the northwest. This tectonic element possibly marks the transition from the thick continental crust (west) to extended thinned crust of the continental margin (east) (Singh et al. 2016). Most

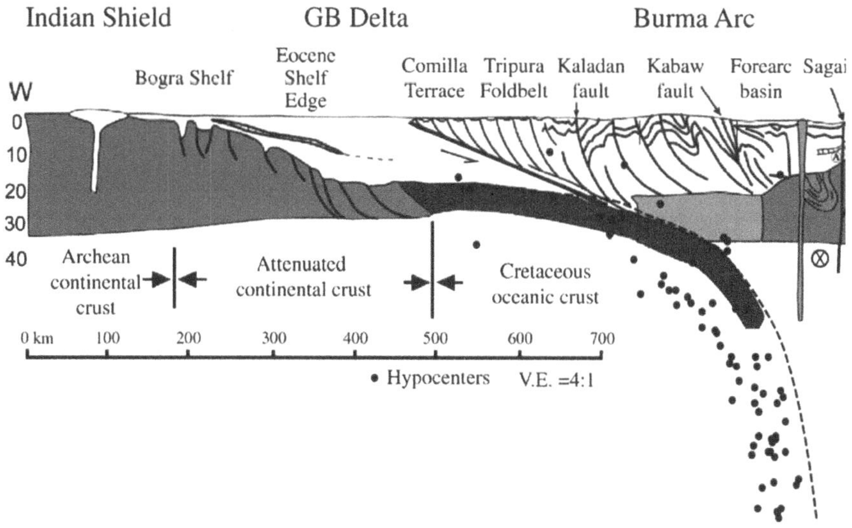

Fig. 10.3 Schematic cross-section of subduction zone from the Indian Craton to the Ganga–Bhramaputra Delta (GBD) and Indo-Myanmar Arc to the Sunda plate. Also shown are the hypocenters (1960–2000 USGS) to interpret the top of basement in the downgoing slab and project the megathrust to the near surface. After Steckler et al. (2008)

prominent seismic reflector of the Eocene Sylhet Limestone in the Bengal Basin implies this shelf break (Fig. 10.3; Reimann 1993).

10.4.1.5 Central Deep Basin or Geotectonic Province 2

This structural unit, includes huge area between the the Eocene Hinge Zone to the west, the Shillong Plateau to the north and the Chittagong-Tripura Fold Belt (CTFB) to the east. It occupies areas of greater Dhaka, Faridpur, Noakhali and Chittagong and the Bay of Bengal. It is approximately 200 km wide to the north, narrowed at the middle, and then gradually widens to about 500 km to the south with an overall NE trend (Reimann 1993). In this area, the basement is at a much greater depth and is covered by the sandy Tertiary molasse deposits. Deep drilling in many parts of this country helped in dividing this basin into: (i) Surma Basin/Sylhet Trough, (ii) Faridpur Trough, (iii) Barisal uplift, (iv) Tripura uplift, and (v) Hatia Trough, (vi) **Chittagong Trough**, and (vi) Bay of Bengal (Uddin and Lundberg 1999).

The Sylhet and Faridpur troughs are elliptical depressions almost adjacent to the Eocene Hinge Belt and are identified on gravity minima. The hinge zone is a 25-km wide northeast-southwest zone that separates the Precambrian platform in the northwest from the deeper basin to the southeast. There is no surface expression of this unit but it is marked by the sudden increase of dip in subsurface sedimentary layers as shown strongly by the seismic marker at the top of the Sylhet Limestone unit of

Eocene age. The Cenozoic sediments have filled up these troughs. The uplifted zones of Barisal and Tripura divide the Bengal Foredeep into two parts: the northwestern and the southeastern. The southeastern portion is occupied by the **Chittagong Trough** and the eastern part of the Chittagong Trough is called the **Chittagong Hill tracts**.

The **Hatia Trough** and the adjacent fold belt of the Chittagong Hill Tracts forms the outermost part of the west-propagating Indo-Burmese wedge. Basin-wide seismic stratigraphic framework for the Neogene rocks, calibrated by biostratigraphy, divides the succession into three seismically distinct and regionally correlatable megasequences. Seismic mapping, thermochronological analyses of detrital mineral grains, isotopic analyses of bulk rock, heavy mineral and petrographic data, show that the Neogene rocks of the Hatia Trough and Chittagong Hill Tracts are predominantly Himalayan-derived, with a subordinate arc-derived input possibly from the Paleogene Indo-Burman Ranges as well as the Trans-Himalaya (Najman et al. 2012)

The huge thickness of sediments in this basin is the result of tectonic mobility or instability of the areas causing rapid subsidence and sedimentation in a relatively short span of geologic time. The foredeep unit is characterized by mild or no folding., and the sedimentary layers are horizontal to sub-horizontal, free from major tectonic deformation in the central part of the basin. This is expressed as river to delta plain topography of the land (Umitsu 1993).

10.4.1.6 Folded Flank of the Chittagong-Tripura Trough (Geotectonic Province 3)

This fold belt is characterized by folding of the Cenozoic sedimentary sequences into a series of long curvilinear anticlines and synclines, forming hills and valleys, respectively in the eastern Chittagong-Comilla-Sylhet regions. Intensity of folding gradually increases eastwards, causing higher topographic elevation in the eastern Chittagong Hill Track. As the folding intensity decreases westwards, the folded unit merges among themselves.

The Chittagong Trough is subdivided into three subzones:

(i) **Western Subzone**: This zone lies with box like structure along the eastern coast of the Bay of Bengal and includes Dakhin Nila, Inani, Jaldi, Sitakund, Semutang and Patiya anticlines and their corresponding synclines. They are characterized by gentle dips between 3°–5° and 10°–12° in the crestal parts being rather wide, 1–2 miles. The dip on their flanks ranges from 45° to 55°. The cores of the anticlines expose sediments from upper Bhuban to Tipam, while the synclinal troughs are composed of Dupi Tilla sediments having very gentle dips (Quazi 1986).

(ii) **Middle Subzone**: It is tectonically affected by complex folding styles. In the south, this subzone is made up of structures with large amplitudes and asymmetry such as Matamuhri, Bandarban, Gilasari while northern part includes structures of smaller sizes, namely, the Sitapahar, Belassari, Changhotaung, Kassalong and Shishak. Matamuhri, Bandarban and Gilasari anticlines. In the northern portion of the subzone, anticlines are shorter, having box-like form and their eastern limbs are thrust. On the crests, mainly upper Bhuban rocks are exposed.

(iii) **Eastern Subzone**: This subzone comprises sharp faulted anticlines like Mowdak, Barkal, Dumba etc. with limbs dipping at 70°–90°. The cores of the anticlines are formed by the middle Bhuban rocks, while the synclines are filled up by the Tipam and Bokabil rocks.

It should be noted that the Precambrian basement is not exposed anywhere in Bangladesh but is identified only in the drill cores while large outcrops of these rocks are seen on the Indian side around Shillong Plateau in the north and Birbhum and the Chotanagpur granite massif in the west. These are either Archaean metasediments with granite, dolerite and lamprophyric intrusions or an agglomeration of several gneissic and metasedimentary bodies namely the Garo block, the Khasi-Jainta block, the Kopili and the Mikir blocks. The basement rocks in the buried Indian Shield consist of crystalline gneiss, granitoids, pegmatites and high grade metamorphics. These rocks underlie the northern parts of Rajshahi and Bogia districts and the whole of Dinajpur district at depths ranging from ~200 to about 350 m. In all the other districts these basement is buried much deeper under thousands of meters of much younger Cenozoic sediments period (Rashid 1991).

References

Acharyya SK (2007) Collisional emplacement history of the Naga-Andaman ophiolites and the position of the eastern Indian suture. J Asian Earth Sci 29:229–242

Ahmed ST (1968) Cenozoic fauna of the cox's bazar coastal cliff. Unpublished M.Sc. thesis, University of Dhaka, Dhaka

Alam M (1972) Tectonic classification of the Bengal Basin. Geol Soc Amer Bull 83:519–522

Alam M (1989) Geology and depositional history of Cenozoic sediments of the Bengal Basin of Bangladesh. Palaeogeogr Palaeoclim Palaeoecol 69:125–139

Alam MM (1991) Paleoenvironmental study of the Barail succession exposed in northeastern Sylhet, Bangladesh. Bang J Sci Res 9:25–32

Alam MM (1995) Tide-dominated sedimentation in the upper tertiary succession of the sitapahar anticline, Bangladesh. In: Flemming BW, Bortholma A (eds), Tidal signatures in modern and ancient sediments, vol 24. International Association of Sedimentologist, pp 329–341 (Special Publication)

Alam M (1997) Bangladesh. In: Moores EM, Fairbridge RW (eds) Encyclopedia of European and Asian regional geology. Chapman Hall, London, pp 64–72

Alam M, Mustafa A, Joseph R, Curray JR, Chowdhury MLR, Gani MR (2003) An overview of the sedimentary geology of the Bengal Basin in relation to the regional tectonic framework and basin-fill history. Sedim Geol 155:179–208

Alderson A (1991) The Surma Basin, Bangladesh: a palynostratigraphical manual, geological synthesis and comparison with the adjacent hydrocarbon provinces of NE India and Myanmar

Ameen SMM, Khan MSH, Akon E, Kazi AI (1998) Petrography and major oxide chemistry of some Precambrian crystalline rocks from Maddhapara, Dinajpur. Bang Geosci J 4:1–19

Ameen SMM, Wilde SA, Kabir MZ, Akon E, Chowdhury KR, Khan MSH (2007) Paleoproterozoic granitoids in the basement of Bangladesh: a piece of the Indian shield or an exotic fragment of the Gondwana jigsaw? Gondwana Res 12:280–387

Ameen SMM, Hossain MS, Hossain MS, Abdullah R, Bari Z, Uddin MN, Das SC, Tapu AT, Jahan H and Shahriar MS (2016) Deciphering the Precambrian crust between Shillong Massif and northeast India. In: International Association of Gondwana Reserach (IAGR) Convention. Trivandum, India, pp 75–79

Bakhtine MI (1966) Major tectonic features of Pakistan: part II. Eastern Prov Sci Ind 4:89–100

Banerji RK (1981) Cretaceous-Eocene sedimentation, tectonism and biofacies in the Bengal Basin, India. Palaeogeogr Palaeoclim Palaeoecol 34:57–85

Banerji RK (1984) Post-Eocene biofacies, palaeoenvironments and palaeogeography of the Bengal basin, India. Palaeogeogr, Palaeoclim, Palaeoecol 45:49–73

Baruah S, Saikia S, Baruah S, Bora PK, Tatevossian R, Kayal JR (2016) The September 2011 Sikkim Himalaya earthquake Mw 6.9: is it a plane of detachment earthquake? Geomatics Natural Hazards Risk 7(1) 248–263, https://doi.org/10.1080/19475705.2014.895963

Chatterjee N (2017) Constraints from monazite and xenotime growth modelling in the Mn CKFMASH-PYC e system on the P–T path of a metapelite from Shillong-Meghalaya Plateau: implications for the Indian shield assembly. J Metamor Geol 35:393–412

Chowdhury SQ (1982) Palynostratigraphy of the Neogene sediments of the Sitapahar anticline (western flank), Chittagong Hill Tracts, Bangladesh. Bang J Geol 1:35–49

Curray JR, Moore DG (1974) Sedimentary and Tectonic Processes in the Bengal Deep-Sea Fan and Geosyncline. In: Burk CA, Drake CL (eds) The Geology of Continental Margins:617–627, Springer, Berlin, Heidelberg, https://doi.org/10.1007/978-3-662-01141-6_45

Curray JR, Munasinghe T (1991) Origin of the Rajmahal traps and the 85°E ridge: preliminary reconstructions of the trace of the Crozet hotspot. Geology 19:1237–1240

Curray JR, Emmel FJ, Moore DG, Raitt RW (1982) Structure, tectonics and geological history of the northeastern Indian ocean. In: Nairn AEM, Stehli FG (eds) The ocean basins and margins, vol 6. The Indian Ocean, Plenum, New York, pp 399–450

Curray JR, Emmel FJ, Moore DG (2003) The Bengal fan: morphology, geometry, stratigraphy, history and processes. Mar Petrol Geol 19:1191–1223

Dasgupta PK, Chakrabarti PK, Datta D (1991) Basinward migrating submarine fan environments from the Barail group in the north Cachar Hills, Assam–Arakan Orogen, India. In: Bouma AH, Carter RM (eds) Facies models, pp 195–217

Evans P (1932) Tertiary succession in Assam. Trans Min Geol Inst India 27:155–260

Evans P (1964) Tectonic framework of Assam. J Geophys Soc India 5:80–96

Gani MR, Alam MM (1999) Trench-slope controlled deep-sea clastics in the exposed lower Surma group in the southeastern fold belt of the Bengal Basin, Bangladesh. Sedim Geol 127:221–236

Geological map of Bangladesh (2011), https://commons.wikimedia.org/wiki/File:Geological_map_of_Bangladesh.PNG

Goodbred SL Jr, Kuehl SA (1999) Holocene and modern sediment budgets for the Ganges Brahmaputra river system: evidence for highstand dispersal to floodplain, shelf and deep-sea depocenters. Geology 27:559–562

Goodbred SL Jr, Kuehl SA (2000) The significance of large sediment supply, active tectonism and eustasy on marginal sequence development: late quaternary stratigraphy and evolution of Ganga–Brahmaputra delta. Sedim Geol 133:227–248

Graham SA, Dickinson WR, Ingersoll RV (1975) Himalayan–Bengal model for flysch dispersal in the Appalachian–Ouachita system. Geol Soc Amer Bull 94:273–286

Guha DK, Henkel H, Imam B (2010) Geothermal potential in Bangladesh—results from investigations of abandoned deep wells. In: Proc World Geothermal Congr 2010 Bali, Indonesia:1–8

Hiller K, Elahi M (1988) Structural growth and hydrocarbon entrapment in the Surma basin, Bangladesh. In: Wagner HC, Wagner LC, Wang FFH, Wong FL (eds), Petroleum resources of China and related subjects, vol 10. Circum-Pacific council for energy and mineral resources Earth science series, Houston, Texas, pp 657–669

Holtrop JF, Keizer J (1970) Some aspects of the stratigraphy and correlation of the Surma basin wells, East Pakistan. ESCAFE Miner Resour Dev Ser 36:143–154

Hossain I, Tsunogae T, Rajesh HM, Chen B, Arakawa Y (2007) Palaeoproterozoic U–Pb SHRIMP zircon age from basement rocks in Bangladesh: a possible remnant of the Columbia supercontinent. Comptes Rendus Geosci 339:979–986

Hossain I, Tsunogae T, Tsutsumi Y, Takahashi K (2017) Petrology, geochemistry and LA-ICP-MS U–Pb geochronology of Paleoproterozoic basement rocks in Bangladesh: an evaluation of calc-alkaline magmatism and implication for Columbia supercontinent amalgamation. J Asian Earth Sci 157:22–39

http://en.banglapedia.org/index.php?title=Tectonic_Framework

Islam MS, Tooley MJ (1999) Coastal and sealevel changes during the holocene in Bangladesh. Quatr Intern 55:61–75

Ismail M (1978) Stratigraphical position of Bogra limestone of the platform area of Bangladesh. In: Proceedings of 4th annual conference Bangladesh geological society, pp 19–26

Johnson SY, Alam AMN (1991) Sedimentation and tectonics of the Sylhet Trough, Bangladesh. Geol Soc Amer Bull 103:1513–1527

Kabir MZ, Chowdhury KR, Akon E, Kazi AI, Ameen SMM (2001) Petrogenetic study of Precambrian basement rocks from Maddhapara, Dinajpur, Bangladesh. Bang Geosci J 7:1–18

Kent R (1991) Lithospheric uplift in eastern Gondwana: evidence for a long-lived mantle plume system? Geology 19:19–23

Khan FH (1991) Geology of Bangladesh. The University Press Ltd, Dhaka

Khan MR, Muminullah M (1980) Stratigraphy of Bangladesh. In: Petroleum and mineral resources of Bangladesh, seminar and exhibition. Dhaka, pp 35–40

Khan AA, Rahman T (1992) An analysis of gravity field and tectonic evaluation of the northwestern part of Bangladesh. Tectonophysics 206:351–364

Khandoker RA (1989) Development of major tectonic elements of the Bengal Basin: a plate tectonic appraisal. Bang J Sci Res 7:221–232

Lietz JK, Kabir J (1982) Prospects and constraints of oil exploration in Bangladesh. In: Proceedings of 4th offshore southeast Asia conference. Singapore, pp 1–6

Lindsay JF, Holliday DW, Hulbert AG (1991) Sequence stratigraphy and the evolution of the Ganges–Brahmaputra Delta complex. Amer Assoc Petrol Geol Bull 75:1233–1254

Mckenzie D, Sclater JG (1971) The evolution of the Indian Ocean since the late cretaceous. R Astron Soc Geophys J 24:437–528

Moore DG, Curray JR, Raitt RW, Emmel FJ (1974) Stratigraphic-seismic section correlations and implications to Bengal Fan history. In: von der Borch CC, Sclater JG, et al. Initial reports of the Deep Sea Drilling Project, v. XXII: Washington, D.C., U.S. Govt. Printing Office, 403–412

Najman YR, Allen R, Willett EAF, Carter A, Barfod D, Garzanti E, Wijbrans J, Bickle MJ, Vezzoli G, Ando S, Oliver G, Uddin MJ (2012) The record of Himalayan erosion preserved in the sedimentary rocks of the Hatia Trough of the Bengal Basin and the Chittagong Hill Tracts, Bangladesh. Basin Res 24(5):499–519. https://doi.org/10.1111/j.1365-2117.2011.00540.x

Najman Y, Bracciali L, Parrish RR, Chisty E, Copley A (2016) Evolving strain partitioning in the eastern Himalaya: the growth of the Shillong Plateau. Earth Plan Sci Lett 433:1–9

Paul DD, Lian HM (1975) Offshore basins of southwest Asia—Bay of Bengal to south sea. In: Proceedings of 9th world petrol congress, vol 3. Tokyo, pp 1107–1121

Quazi AH (1986) Geological framework of Bangladesh, GEOSEA V Proceedings vol. II. Geol Soc Malaysia Bull 20:73–80

Rabbani MG, Chowdhury KR, Huq MM (2000) Stratigraphic analysis by interpretation of seismic and drill hole data of Rangpur Dinajpur area, Bangladesh. Bangladesh Geosci J 6:1–16

Raju ATR (1968) Geological evolution of Assam and Cambay Basins of India. Am Assoc Pet Geol Bull 52:2422–2437

Rashid H (1991) Geography of Bangladesh. The University Press Ltd, Dhaka

Reimann KU (1993) Geology of Bangladesh. Gebrueder Borntraeger, Berlin, p 160

Roy AB, Chatterjee A (2015) Tectonic framework and evolutionary history of the Bengal Basin in the Indian subcontinent. Curr Sci 109:271–279

Salt CA, Alam MM, Hossain MM (1986) Bengal Basin: current exploration of the hinge zone area of southwestern Bangladesh. In: Proceedings of 6th offshore southeast Asia conference. Singapore, pp 55–57

Sclater JG, Fisher RL (1974) Evolution of the east central Indian Ocean, with emphasis on the tectonic setting of the ninety east ridge. Geol Soc Amer Bull 85:683–702

Sengupta S (1966) Geological and geophysical studies in the western part of Bengal Basin, India. Amer Assoc Petrol Geol Bull 50:1001–1017

Shamsuddin AHM, Abdullah SKM (1997) Geological evolution of the Bengal Basin and its implication in hydrocarbon exploration in Bangladesh. Indian J Geol 69:93–121

Singh A, Bhushan K, Singh C, Steckler MS, Akhter SH, Seeber L, Kim W-Y, Tiwari AK, Biswas R (2016) Crustal structure and tectonics of Bangladesh: new constraints from inversion of receiver functions. Tectonophysics 680:99–112

Steckler MS, Akhter SH, Seeber L (2008) Collision of the Ganges–Brahmaputra delta with the Burma Arc. Earth Planet Sci Lett 273:367–378

Tapu AT, Ameen SMM, Abdullah R, Zaman MN (2016) Geochemical evaluation of the diorite basement in Barapaharpur, Rangpur, northwest Bangladesh. Bang Geosci J 22:17–36

Uddin MN (1994) Structure and sedimentation in the Gondwana basins of Bangladesh. In: Gondwana nine, 9th international Gondwana symposium, vol 2. Hyderabad, India, pp 805–819

Uddin MN, Ahmed Z (1989) Palynology of the Kopili formation at GDH-31, Gaibandha District, Bangladesh. Bang J Geol 8:31–42

Uddin MN, Islam MSU (1992) Depositional environments of the Gondwana rocks in the Khalashpir basin, Rangpur District, Bangladesh. Bang J Geol 11:31–40

Uddin A, Lundberg N (1999) A paleo-Brahmaputra? Subsurface lithofacies analysis of miocene deltaic sediments in the Himalayan–Bengal system, Bangladesh. Sedim Geol 123:239–254

Umitsu M (1993) Late quaternary environment and landforms in Ganges Delta. Sedim Geol 83:177–186

Uddin A, Lundberg N (2004) Miocene sedimentation and subsidence during continent–continent collision, Bengal basin, Bangladesh. Sedimen Geol 164:131–146

Wang Y, Sieh K, Tun ST, Lai K-Y, Myint T (2014) Active tectonics and earthquake potential of the Myanmar region. J Geophy Res 119:3767–3822

Wardell A (1999) Barapukuria Coal Deposits, Stage 2, Feasibility Study, Geological Survey Bangladesh–UK Coal Project, No. 4298/7B. Petrobangla, Dhaka

Zaher MA, Rahman A (1980) Prospects and investigations for minerals in the northwestern part of Bangladesh. Petrol Mineral Resources Bangladesh. Sem Exhibition, Dhaka, 9–18

The manufacturer's authorised representative in the EU is Springer Nature Customer Service Centre GmbH, Europaplatz 3, 69115 Heidelberg, Germany. If you have any concerns regarding our products, please contact ProductSafety@springernature.com

Printed and bound by CPI Group (UK) Ltd, Croydon, CR0 4YY
30/09/2026
02225647-0017